# CELL CYCLE CLOCKS

# CELL CYCLE CLOCKS

edited by

*Leland N. Edmunds, Jr.*

Department of Anatomical Sciences
School of Medicine
Health Sciences Center
State University of New York
Stony Brook, New York

MARCEL DEKKER, INC. New York and Basel

**Library of Congress Cataloging in Publication Data**
Main entry under title:

Cell cycle clocks.

Includes indexes.
1. Cell cycle. 2. Biological rhythms. I. Edmunds, Leland N.
QH605.C425 1984 574.87'6 83-26343
ISBN 0-8247-7060-7



Marcel Dekker, Inc.
270 Madison Avenue, New York, New York 10016

Current printing (last digit):
10 9 8 7 6 5 4 3 2 1

Printed in the United States of America

Let a hundred flowers blossom,

Let a hundred schools of thought contend.

Mao Tse-tung, Chairman
Peoples' Republic of China
From a proposal to the Supreme State Conference,
May 2, 1956

# Preface

Our understanding of the cell division cycle has come a long way since Howard and Pelc more than 30 years ago divided it into four consecutive intervals—$G_1$, S, $G_2$, and M. Progress has not been confined, however, to a mere filling in of these gaps and subdividing them into smaller steps. Rather, more recent theoretical and experimental work has focused on mechanisms that regulate the cell cycle.

The cell division cycle may be loosely considered to be a clock in the sense that it "measures" time, albeit imperfectly, under a given set of environmental conditions. In turn, discussions of mechanisms that control microbial cell cycles include with increasing frequency the notion of timers and clocks that would regulate cell cycle sequences that culminate in mitosis and cell division. Both the title of this book and the selection of chapters contributed by an international cast of authors reflect this emphasis on biochemical and macromolecular oscillators and other timing devices that have been hypothesized to underlie the various events, or "landmarks," that comprise the cell division cycle.

The volume is divided into six main sections linked by introductory bridges. Part I considers the more general features of cellular temporal organization, its origin and its breakdown. The next part addresses the problem of cell cycle variability and its implications and challenges for modeling the cell cycle. Part III introduces specific events, sequences, pathways, and timers that play a role in the temporal coordination of both prokaryotic and eukaryotic cell cycles and leads logically into a consideration (Part IV) of the timing of the cell division cycle by autonomous oscillators and their interaction with environmental periodicities. Finally, the last two sections deal with the implications of cell cycle clocks for pathology and chronotherapy (Part V) and for development and aging (Part VI). Each contributor has thoroughly but concisely reviewed his or her field and has included new and previously unpublished material to give a critical and up-to-date summary of the subject.

This book should appeal to cellular, developmental, and molecular biologists as well as to many others in the life and biomedical sciences. It is aimed not only at the specialist in cell cycles and in chronobiology but also should be of value to both researchers and students who have a general interest in cellular control mechanisms. We hope that it will provide a background for a new synthesis of the vast amount of data in the fields of cell cycle regulation and biological timekeeping and will inspire others to study and contribute to this exciting but still largely unexplored interface.

I wish to thank all of the authors, who enthusiastically agreed to contribute to this book without knowing what the final product would resemble; the fruit of their labors is now the reader's feast. I am particularly indebted to Dr. D. L. Laval-Martin, who re-read the manuscripts and page proof and was instrumental in ruthlessly rooting out the bugs that escaped earlier extermination.

**Leland N. Edmunds, Jr.**

# Contributors

**Kenneth J. Adams** Department of Biology, North East London Polytechnic, Stratford, London, England

**B. Wolfanger Belisle*** Department of Biology, Georgetown University, Washington, D.C.

**William Byrd** Department of Obstetrics and Gynecology, The University of Texas Medical School, Houston, Texas

**Sallie W. Chisholm** Ralph M. Parsons Laboratory, Massachusetts Institute of Technology, Cambridge, Massachusetts

**Ole Petter Fraas Clausen** Department of Pathology, Rikshospitalet, Oslo, Norway

**Stephen Cooper** Department of Microbiology and Immunology, University of Michigan Medical School, Ann Arbor, Michigan

**Leland N. Edmunds, Jr.** Department of Anatomical Sciences, School of Medicine, Health Sciences Center, State University of New York, Stony Brook, New York

**Steven W. Edwards†** Department of Microbiology, University College, Cardiff, Wales

---

*Current affiliations:*

*Department of Microbiology, Uniformed Services University of the Health Sciences, Bethesda, Maryland

†Department of Medical Biochemistry, University of Wales Hospital, Cardiff, Wales

**Peter A. Fantes** Department of Zoology, University of Edinburgh, Edinburgh, Scotland

**Donald A. Gilbert*** Virus Cancer Research Unit, National Institute for Virology, Sandringham, South Africa

**Marguerite Guiguet** Unité de Recherches Biomathématiques et Biostatistiques, Université Paris 7 and INSERM, Paris, France

**Niels Rudolph Hartmann** The Finsen Laboratory, The Finsen Institute, Copenhagen, Denmark

**Stuart A. Kauffman** Department of Biochemistry and Biophysics and Department of Biology, University of Pennsylvania School of Medicine, Philadelphia, Pennsylvania

**Niels Keiding** Statistical Research Unit, Danish Medical Research Council and Danish Social Science Research Council, Copenhagen, Denmark

**Robert R. Klevecz** Department of Cell Biology, Division of Biology, The Beckman Research Institute of the City of Hope National Medical Center, Duarte, California

**Jean-Jacques Kupiec** Unité de Recherches Biomathématiques et Biostatistiques, Université Paris 7 and INSERM, Paris, France

**Ole Didrik Laerum** Department of Pathology, The Gade Institute, University of Bergen, Bergen, Norway

**Donald J. R. Laurence** Ludwig Institute for Cancer Research (London Branch), Royal Marsden Hospital, Sutton, Surrey, England

**Danielle L. Laval-Martin** Department of Anatomical Sciences, School of Medicine, Health Sciences Center, State University of New York, Stony Brook, New York, and Laboratoire des Membranes Biologiques, Université Paris 7, Paris, France

**David Lloyd** Department of Microbiology, University College, Cardiff, Wales

**Neil H. Mendelson** Department of Cellular and Developmental Biology and Graduate Committee on Genetics, University of Arizona, Tucson, Arizona

**J. M. Mitchison** Department of Zoology, University of Edinburgh, Edinburgh, Scotland

**Ulla Møller** The Finsen Laboratory, The Finsen Institute, Copenhagen, Denmark

**Robert J. Olson** Ralph M. Parsons Laboratory, Massachusetts Institute of Technology, Cambridge, Massachusetts

**Claudine L. Oudet** Unité 213 de l'INSERM and Institut de Physiologie, Faculté de Médecine, Strasbourg, France

**Page R. Painter** Department of Bacteriology, University of California, Davis, California

**Alexandre G. Petrovic** Unité 213 de l'INSERM and Institut de Physiologie, Faculté de Médecine, Strasbourg, France

---

*Current affiliation:*

*Department of Biochemistry, University of the Witwaterstrand, Braamfontein, Johannesburg, South Africa

**Robert K. Poole** Department of Microbiology, Queen Elizabeth College, University of London, London, England

**Simone Puiseux-Dao** Laboratoire de Biologie Cellulaire Végétale, Université Paris 7 and ERA C.N.R.S. 325, Paris, France

**Wilhelm Sachsenmaier** Institute for Experimental Cancer Research, University of Innsbruck, Innsbruck, Austria

**Noriyuki Satoh** Department of Zoology, Faculty of Science, Kyoto University, Kyoto, Japan

**Lawrence E. Scheving** Department of Anatomy, University of Arkansas College of Medicine, Little Rock, Arkansas

**Robert E. Scott** Section of Experimental Pathology, Departments of Pathology and Cell Biology, Mayo Clinic/Foundation, Rochester, Minnesota

**Ronald M. Shymko** Department of Radiation Research, City of Hope National Medical Center, Duarte, California

**Bonnie Smith** John Muir Memorial Hospital, Walnut Creek, California

**John A. Smith** Ludwig Institute for Cancer Research (London Branch), Royal Marsden Hospital, Sutton, Surrey, England

**Jeanne J. Stutzmann** Unité 213 de l'INSERM and Institut de Physiologie, Faculté de Médecine, Strasbourg, France

**Erik Thorud** General Department, The Norwegian Radium Hospital, Montebello, Oslo, Norway

**John J. Tyson** Department of Biology, Virginia Polytechnic Institute and State University, Blacksburg, Virginia

**Alain-Jacques Valleron** Unité de Recherches Biomathématiques et Biostatistiques, Université Paris 7 and INSERM, Paris, France

**Daniel Vaulot*** MIT/WHOI Joint Program in Oceanography, Massachusetts Institute of Technology, Cambridge, Massachusetts

**Caroline S. Weiler** Department of Biology, Whitman College, Walla Walla, Washington

**John J. Wille, Jr.** Section of Experimental Pathology, Department of Cell Biology, Mayo Clinic/Foundation, Rochester, Minnesota

**Arthur T. Winfree** Department of Biological Sciences, Purdue University, West Lafayette, Indiana

**Glenn A. Zorn** John Muir Memorial Hospital, Walnut Creek, California

*Current affiliation:*
*On leave from Laboratoire d'Hydrobiologie, ERA 467 CNRS, U.S.T.L., Montpellier, France

# Contents

# PART I
# TEMPORAL ORGANIZATION IN CELLS
## GENERAL FEATURES

The biologist is confronted with a continuously reproducing and evolving set of highly organized living systems. An organism that has thrived by differential reproductive success is said to be adapted, and its adaptation is reflected in its total organization. This organization is strongly history dependent, having arisen through the twin processes of natural selection and adaptation. Biological problems, therefore, pivot on the complexities of biological organization (Pittendrigh, 1961).

It is almost self-evident that the spatial organization and the functioning of living forms are inextricably intertwined. Of equal importance, however, is the temporal dimension: at the physiological level, for example, not only must the "right" amount of the "right" substance must be at the "right" place, but this must also occur at the "right" time (Halberg, 1960). This also holds true for the organism itself, which often must be positioned in time with respect to favorable biotic or physical conditions. Inasmuch as the environment is highly periodic with respect to many of its variables, it would not be surprising (and, indeed, would be essential) for the organism to adapt to these periodicities.

That organisms can and do measure time in some manner is explicitly demonstrated by four categories of phenomena occurring throughout the animal and plant kingdoms: (1) persistent rhythms, having daily (circadian), tidal, lunar monthly, and yearly (circannual) periods; (2) the *Zeitgedächtnis,* or time sense of bees and humans; (3) seasonal photoperiodism, wherein many organisms perform a certain function at a quite specific time of the year by what may well be essentially a daily measurement of the length of the day (or night); and (4) celestial orientation and navigation, in which the sun, moon, or stars are used as direction givers, implicating a timing system to compensate for their continuously, but predictably, shifting positions (Bünning, 1973; Palmer et al., 1976; Brady,

1982). All four types of timekeeping, or functional biochronometry, have external correlates (generated by the movements of the earth, moon, and sun) to which the organism has adapted. Although the last three kinds are commonly found only in higher organisms and are probably relatively recent, more sophisticated variations on a more ancient evolutionary theme, the first category of persistent rhythms are commonly displayed in many eukaryotic (but not prokaryotic) unicells.

The underlying biological clocks that generate the foregoing types of rhythmicity all possess considerably longer periods than those that give rise to ultrafast and rapid chemical reactions, to biochemical rhythmicities of intermediary metabolism (e.g., glycolytic oscillations), and to those rhythms (such as respiration) commonly observed in the epigenetic time domain. The duration of the cell division cycle in the vast majority of microbes and cultured cells ranges from about 10-30 min in certain prokaryotes, 2-3 hr in yeasts and many ciliates, to 15-25 hr in cultured mammalian cells under ideal conditions, or even several days in algal flagellates. The time domain of the cell cycle, therefore, falls somewhere in between the metabolic and circadian (and longer) time domains in the temporal hierarchy of living systems; it is this type of periodicity which is the primary subject matter of this book.

In Part I, the general features of temporal organization (and, conversely, disorganization) are considered by Gilbert (Chapter 1). Many of the themes introduced here will be found interwoven throughout the 26 other chapters comprising the book. In Chapter 2, Lloyd treats the thorny problem of the relation of epigenetic oscillations to the cell division cycle of lower eukaryotes, such as *Saccharomyces* or *Acanthamoeba*) and their coupling to a clock, a question reechoed later for prokaryotes by Poole (Chapter 11). Chapter 3, by Klevecz, provocatively discusses the evolution of oscillators in a changing physical environment (as the earth's rotational period slowed considerably through the eons), speculating that present-day cellular oscillators may represent the vestiges of a primitive circadian clock. Finally, Winfree provides (Chapter 4) a theoretical and mathematical background for the experimental testing of various oscillator models of the cell cycle (e.g., limit cycle and relaxation oscillators) that will be a prevalent theme throughout Part III and Part IV, with particular emphasis on phase-resetting curves, singularity points, and phase-compromise experiments.

Perhaps editorial license will permit an appropriate poetic closing (for which I am indebted to D. A. Gilbert, who had used up his space allotment but still felt unfulfilled):

The living cell is most surely dynamic,
Its behaviour so very erratic;
Why is it so true
That all but a few
Insist on treating it static?

There is an awfully big schism
Between those proposing a rhythm
For the cell as it grows
And all of their foes
Who just won't agree with 'em.

## REFERENCES

Brady, J. (ed.) (1982). *Biological Timekeeping,* Soc. Exp. Biol. Seminar Series no. 14, Cambridge Univ. Press, Cambridge, England.

Bünning, E. (1973). *The Physiological Clock,* 3rd ed. Springer-Verlag, New York.
Halberg, F. (1960). Temporal coordination of physiologic function. *Cold Spring Harbor Symp. Quant. Biol. 25*:289-310.
Palmer, J. D., Brown, F. A., Jr., and Edmunds, L. N., Jr. (1976). *An Introduction to Biological Rhythms.* Academic Press, New York.
Pittendrigh, C. S. (1961). On temporal organization in living systems. *Harvey Lect. 56*: 93-125.

# 1

# Temporal Organization, Reorganization, and Disorganization in Cells

**Donald A. Gilbert*** / National Institute for Virology, Sandringham, South Africa

## INTRODUCTION

It cannot be fortuitous that the major outstanding biological enigmas are those that involve time-dependent processes. Indeed, all living entities are dynamic and the fundamental form of life, the individual cell, is no exception. Dynamic biological attributes imply that the underlying biochemistry is also dynamic. On this basis, several crucial questions come to mind, in particular: (1) How does the dynamic state arise? (what is its nature?) (2) How is it structured and maintained? (3) Does the dynamic state play any significant role in determining the properties, behavior, characteristics, and functions of cells, or is it merely coincidental?

It is inconceivable that fluctuations in the levels of various constitutents can occur without them affecting cellular properties, if only to some degree, for the many reactions of which the cell is composed are highly integrated as a result of coupling through common coenzymes, cofactors, substrates, and products. In addition, there is no logical reason to believe that the process responsible form some completely independent systems that have no role other than to be dynamic. It also seems quite clear that the overall dynamic state must (1) be essentially autonomic, deterministic, and sustained, (2) be heritable from cell to cell and hence endogenous (though modifiable by external factors), (3) involve a number of dynamic subsystems, and (4) be accounted for in kinetic terms with the reactions obeying the same biochemical and physical laws as any inanimate system.

The replicating cell is obviously more dynamic than its quiescent counterpart and clearly exhibits temporal organization. The relevant reactions differ from those presumably associated with viability in that proliferation is one major dynamic attribute that can

*Current affiliation: University of the Witwaterstrand, Johannesburg, South Africa.

be readily regulated, and even reversibly inhibited. Replication is related to each of the other kinetic enigmas (differentiation, aging, and cancer) but classic views fail to indicate the nature of the dependencies. Although a detailed understanding of the various phenomena requires the reactants to be identified, it is the kinetics of the pertinent reactions which determine their temporal aspects. The latter can thus be discussed in general terms without reference to the identity of the reactants. In the following, these thoughts and ideas are amplified and then experimental data are presented in support of some of the conclusions. Associated technical and analytical problems are also outlined.

## BASIC THEORETICAL ASPECTS

### Nature of the Dynamic State

From these considerations it seems evident that the cell must exhibit temporal organization as well as spatial organization and that the former not only confers the attribute of life but also play a role in influencing, if not totally determining, the properties and behavior of a cell. As a result I suggested (Gilbert 1966, 1968) that each cell species has its own basic pattern of temporal organization that reflects its own unique pattern of metabolism. It followed that changes in the pattern of temporal organization must either accompany differentiation and the malignant transformation or, a far more important possibility, be at least partly responsible for these phenomena (Gilbert, 1968) (see Fig. 1). In coming to these conclusions it was obviously necessary to (1) take into account the existence and role of cellular control systems (normally considered to maintain constancy), and (2) note that the only true sustained dynamic steady state is one where the behavior is periodic, otherwise the state would vary with time over a prolonged interval and could not be regarded as steady. These and other considerations suggested that oscillatory biochemical reactions are responsible for the dynamic behavior. In which case, what significance do we attach to the dynamic characteristics of the rhythms, that is, to their mean levels, frequencies, amplitudes, and phasings? What importance must we place on the interactions between distinct periodicities and the effect on them, individually and collectively, of external agents?

For the most part we merely assume that cellular properties reflect levels of constituents, whether static or dynamic, and that therefore any factor influencing the metabolic state will have some action on those properties. All disturbances will provoke transient responses, but only a few will produce permanent changes. What factors turn a transient effect into a permanent one? The various aspects can be treated quite generally in terms of the actions of disturbances (signals!) on biochemical systems, since even the effect of one rhythm on another can be viewed in this light. It is stressed that disturbances cannot arise de novo: they must have a cause. In the following the terms *disturbance* and *system* are used in a general sense and the latter covers, according to context, any group of reactions within the cell or the cell as a whole.

### Source and Significance of Cellular Rhythms

Much has already been written about possible sources and existence of cellular oscillations (Brodsky, 1975; Rapp, 1978). It is believed that those rhythms of major significance to this discussion arise within control systems and especially those determining the levels of cofactors, such as metal ions, thiols-disulfides, and nicotonamide dinucleotides. In the following we make no further reference to the origins of various oscillations, this being immaterial insofar as present considerations are concerned, but membrane reactions are sources of potential significance (Mustafin and Volkov, 1982; Chay, 1981). The dynamic characteristics may be regulated through modulation of the rates and kinetics of the pertinent reactions by agents of internal or external origin.

The roles of cellular oscillations can be considered from several points of view. Tornheim (1979) and also Termona and Ross (1981) have concluded that oscillating glycolysis can yield a higher adenylate charge than aperiodic metabolism, and Richter and Ross (1981) and Termona and Ross (1981) have suggested that such behavior may have evolved in order to reduce the dissipation of free energy. It seems reasonable to expect similar considerations to apply to oscillatory processes in general. Thus, the total amount of free energy dissipated may be dependent on the set of biochemical rhythms and hence, on the present basis, some cells are probably more efficient than others in this respect. For example, a factor of possible significance: if both glycolysis and respiration oscillate, it does not seem unreasonable to expect the overall state of energy metabolism and the loss of free energy to depend on the dynamic characteristics of both rhythms and, particularly, on their coordination.

Information can be contained in any oscillatory process. The necessary information for differentiation may thus not lie entirely in the genome, e.g., which acts more like a memory bank (read-only memory; or ROM) (Apter and Wolpert, 1965), but partly in the pattern of cellular rhythms. The former sets limits to the possible steady states rather than determines which one exists. It is proposed that no program exists, only a potential to follow particular time courses, as is the case with any chemical reactions. Which courses are actually followed and which state is eventually reached is considered to depend on disturbances to which each cell is subject, their nature, magnitude, and timing in relation to the dynamic state of the system at the time. No gene can turn itself on or off, and specific inducers require their own specific inducers. Which genetic information is expressed is considered to be directly determined by a limited number of constituents with a broad spectrum of actions. Their levels are believed to oscillate continuously, with specificity of action being conferred through factors which include the previous history (e.g., which genes are already expressed), the requisite dynamic behavior, and the relevant kinetics, as discussed later. The dynamic state is likened to a running computer program awaiting inputs in order to alter its output. These come, initially, from without the system, but the total response also depends on what information is thereby called from long-term memory as a result of the original disturbance just as decisions are made in a computer program according to the information called from the ROM. In essence, therefore, such a program as exists is gradually modified by the circumstances to which each individual cell is exposed. The information content is thus also forced to vary with time, and cellular characteristics are correspondingly altered through the changing dynamic state of one or more of the oscillations. These effects will be transient unless relevant stability limits are thereby exceeded, when the system will move to another steady state with distinct information content (Kauffman, 1971).

### Temporal Organization

We thus equate the overall dynamic state with a set of metabolic oscillations that is characteristic of the cell species concerned and the conditions under which it then exists. It is suggested that at least some of the components whose levels oscillate are capable of producing permanent changes at both metabolic and genetic levels. It is convenient to discuss some issues as if the individual rhythms are independent, but because of coupling

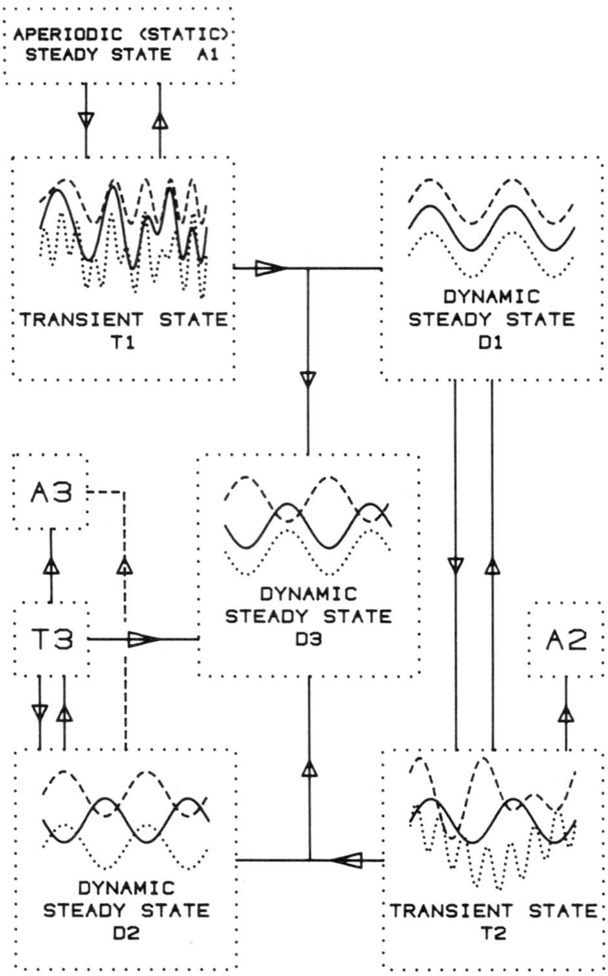


**FIGURE 1** Temporal structure in cells. The processes believed to underlie fertilization, differentiation (or the establishment of a cell line), the malignant transformation, and cell death are shown diagrammatically in this figure, where A1, A2, and A3 represent distinct aperiodic (static) steady states; T1, T2, and T3 are transient dynamic states, wherein the frequencies, amplitudes, mean levels, and phasings of intracellular metabolic rhythms are varying with time because of interactions between the relevant control processes; and D1, D2, and D3 are different dynamic steady states attained wherein the overall pattern of cellular rhythms is itself periodic. Cells in an effective steady state can only change their state as a result of a disturbance of external origin that, in particular circumstances, may be regarded as a "signal." Following such a perturbation, the cell will either relax back into its original state, or move to another stable state. Which outcome is achieved depends on (1) whether the amplitudes, frequencies, and phasings of the oscillations during the transient state are such that relevant stability thresholds are exceeded, and (2) the amplitude, frequency, and phase discriminatory properties of the systems being affected by them. It is suggested that the new stable dynamic steady states reflect distinct patterns of temporal organization corresponding to new patterns of metabolism. Where the disturbances are normal, the resultant dynamic steady states are considered to be differentiated states (e.g., D1, D2). Where the perturbation is abnormal, the outcome is a malfunctional

reactions this is not likely to be the general case. Indeed, their interaction is of major significance to present arguments but discussion is simplified if this aspect is considered separately.

It is doubtful if any cell is ever in a true steady state. It will be continually subject to both random and deterministic disturbances of low magnitude. When referring to steady-state situations we are really concerned with pseudo-steady states wherein these perturbations are minor and have no irreversible effects. On this basis a fully differentiated cell must be in a stable steady state, otherwise its composition and hence its properties would vary continuously with time. This implies that such cells must have an *overall* dynamic state that is also periodic (Gilbert, 1968), i.e., where all the rhythmic processes are fully coordinated, having either the same frequency or frequencies that are an integral multiple of some base frequency (Fig. 1). This latter conclusion is again drawn on the argument that if this is not so, then the rhythms and properties of the cells will be changing with time. Coordination would be achieved through interactions among the various rhythms such that there is mutual adaptation of their dynamic characteristics. Because the overall pattern of rhythms is also periodic, gene expression stays the same and no metabolic switches are irreversibly triggered through some systems may be switched on and off repeatedly (Sanglier and Nicolis, 1976). It is well known from observations in totally different fields that under suitable conditions, interacting oscillations of similar frequencies can become entrained with one another; such effects must surely occur in cells too. It is suggested that (1) certain phase relationships between particular rhythms are more favored than others, and (2) as the steady state is approached, both frequency and phase locking occur between these periodicities, thereby stabilizing the pattern (Guevera et al., 1981; Ermentrout, 1981). The different stable phase relationships may correspond to dissipative minima.

### Temporal Reorganization

To induce differentiation or transformation, it is necessary to disturb the preexisting dynamic steady state (Fig. 1). Eventually this can result in another dynamic steady state

---

cell. If the replicative processes of the latter are suitably modified by the changes in the state produced (see text), the cell also becomes tumorigenic (e.g., D3). The totally aperiodic state is considered to be nonviable. The sequence A1-T1 might thus represent fertilization, T1-D1 an initial stage of differentiation, and D1-T2-D2 a subsequent stage. Aging is considered to be the result of natural parametric dampening of all rhythms due to the cell not being in a true steady state (D2-A3), thus leading, eventually, to total aperiodicity. Malignant cells in vivo, and tissue cells recently established in culture, are likely to be in transient states due to the perturbations received as a result of the oncogenic stimulus or change in conditions. Delays in the establishment of the differentiated state, the establishment of a cell line or of a stable tumor line, are considered to be due to the time taken for the mutual adaptation of interacting biochemical control processes to the altered circumstances, e.g., a gene or metabolic switch effect. A cell already in a disturbed state might be more susceptible to permanent changes in state as the result of a secondary or teriary disturbance. This might account for interactions between "signals" or disturbances, e.g., oncogenic and promoter actions. If the cell is in the process of differentiation, it might thus be more easily directed away from the original target state, or the switching process may be inhibited, preventing the normal course of events from taking place (T2-D3 instead of T2-D2).

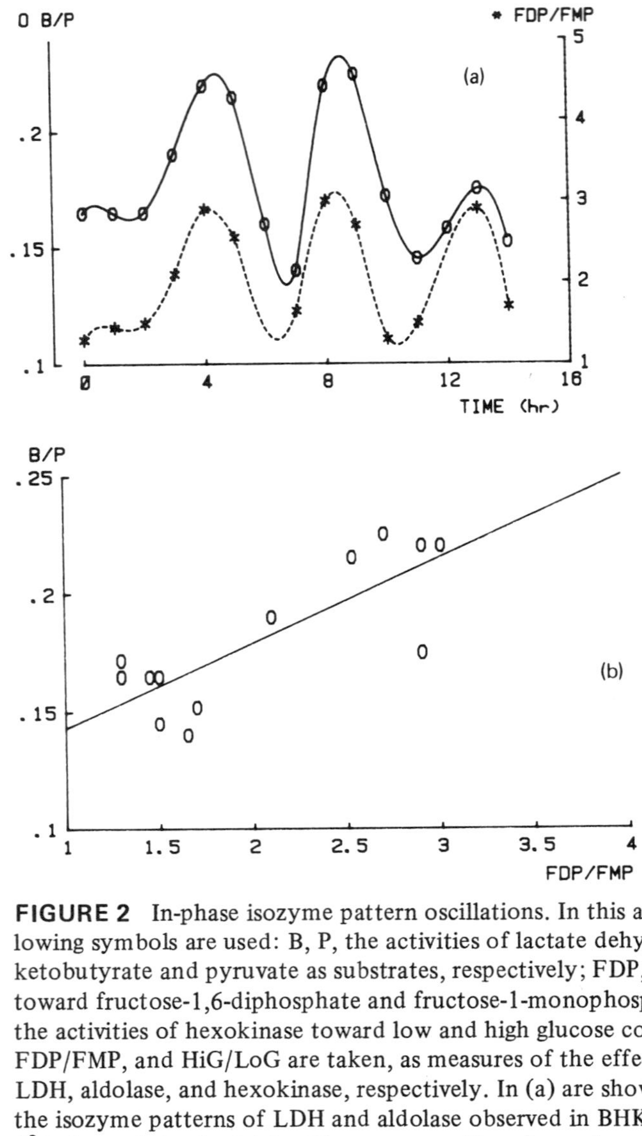


**FIGURE 2** In-phase isozyme pattern oscillations. In this and subsequent figures the following symbols are used: B, P, the activities of lactate dehydrogenase (LDH) toward $\alpha$-ketobutyrate and pyruvate as substrates, respectively; FDP, FMP, the activities of aldolase toward fructose-1,6-diphosphate and fructose-1-monophosphate as substrates; LoG, HiG, the activities of hexokinase toward low and high glucose concentrations. The ratios B/P, FDP/FMP, and HiG/LoG are taken, as measures of the effective isozyme patterns of LDH, aldolase, and hexokinase, respectively. In (a) are shown the temporal variations in the isozyme patterns of LDH and aldolase observed in BHK-C13 cells after release from a 4°C thermal shock of 1 hr. The corresponding phase plane representation of the data (relating the different enzyme values determined on the same sample) is given in (b). The in-phase relationship between the two isozyme oscillations is evident from the positions of the peaks and troughs in the time curves and from the fact that the points in the phase plane graph are positively correlated ($r = 0.81$). See also Figs. 3 and 4. (Exp. 604, 1 hr sampling interval). All time curves presented here have been drawn using a running, fourth degree polynomial fit to five data points. [Curves in (a) adapted from Gilbert, 1968.]

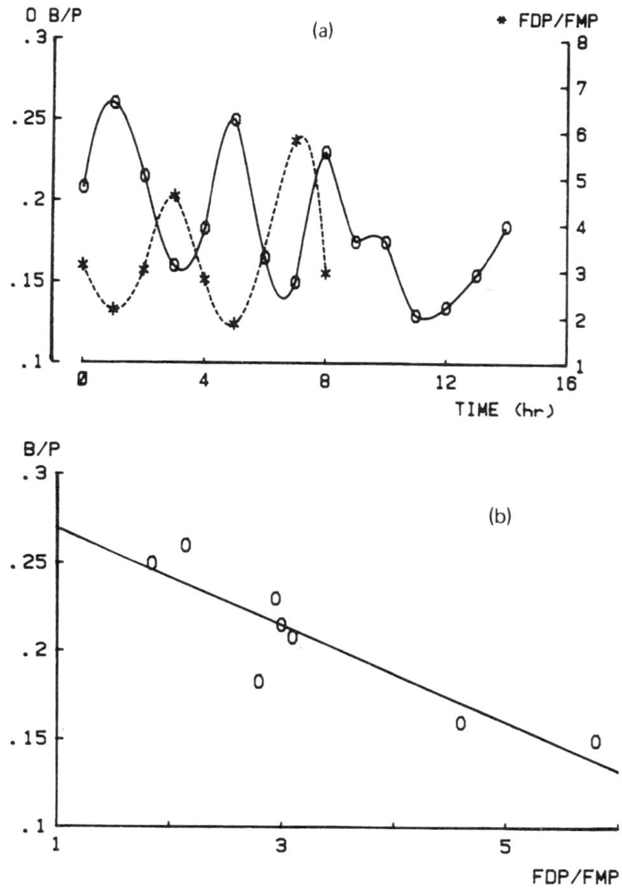


**FIGURE 3** Out-of-phase isozyme pattern oscillations. In contrast to the results presented in Fig. 2, obtained with untransformed cells, the isozyme pattern oscillations seen with polyoma virus-transformed BHK cells (BHK-PYY), after thermal shock, proved to be out of phase, as shown by the relative positions of the peaks and troughs in the two time curves (a) and by the negative correlation (r = 0.90) in the phase plane graph (b). Other examples are given in Fig. 4. (Exp. 604.) [Curves in (a) adapted from Gilbert, 1968.]

with the new pattern of temporal organization reflecting the exact circumstances. Differentiation is regarded not simply in terms of switching but as that period during which the metabolic systems and the individual periodicities are adapting to one another. I proposed (Gilbert, 1968) that oncogens also produce permanent modification of the pattern of temporal organization, but in these instances the outcome is abnormal *for the cells concerned* (Figs. 2 and 3). The new functional (e.g., replicative) characteristics of the cell thus reflect the actual changes in temporal organization. It is also implied that malfunctioning cells can be produced in the same way, but only when the replicative characteristics are suitably altered does the cell give rise to a tumor (see A Limit Cycle Oscillator Concept). Extending these views led to the additional proposals (Gilbert, 1968) that fertilization is merely a means of initiating the dynamic (i.e., oscillatory) state (Webb and Nuccitelli, 1980). A system may be more susceptible to reorganization when it is already in a disturbed state (e.g., when differentiating) or when in, or approaching, a static steady state. This could explain the action of tumor promoters and also the prevalence of particular tumors in certain extreme age groups. Hormones (Gilbert, 1968, 1974b) and also mitogens may act through disturbances of temporal organization, but their action is believed to be generally reversible.

Interaction between different periodicities will (1) determine the eventual overall steady-state pattern attained, and (2) be responsible for the complexity of the rhythms when the system is disturbed (in a transient state). The question arises as to whether differentiation necessarily involves the appearance of a completely new constituent or the total disappearance of one that preexisted in the cell. It is considered that heritable changes can occur in the various parameters either discontinuously (through metabolic or genetic switching) or continuously (because of the integrated nature of the subsystems and the overall complexity of the cell). For example, characteristic metabolic differences between the glycolytic behavior of different cell species are not due to all-or-none changes in the levels of any enzymes. The distinctions cannot be due to a simple on-off effect but must involve (1) mutual quantitative adaptation of the different metabolic systems, or, possibly, (2) a change in the frequency at which the relevant genes are repeatedly switched on and off, and/or (3) changes in the coordination of rhythmic processes (Fig. 1).

Our thesis is thus that (1) a changing pattern of rhythms implies varying properties and functions, and (2) different temporal organization implies nonidentical metabolic states and therefore distinct characteristics. Whether the latter differences are qualitative or merely quantitative depends on the particular circumstances. Marked disturbances might be more likely to produce the former effects, but action on coupling processes may be required to permit discrete shifts in pattern. Controlled reorganization might be achieved by subjecting the cells to temporal stress, e.g., by applying two external periodic pulsed disturbances (e.g., light and temperature) and gradually altering the phasing between them or the frequency of one.

### Temporal Disorganization and Stasis

Oscillatory behavior of control systems can be regarded as a compromise between no control and maximum control and, as such, allows changes to occur but only within limits. Stasis can result when the control exerted is effectively maximal (wherein the regulatory processes act in the manner usually associated with them) or when an equilibrium state exists with no through flux (the system is no longer open).

Death is believed to be due to eventual stasis (the lack of any temporal organization). It can arise gradually from parametric dampening of rhythmic behavior, e.g., the aging process (Gilbert, 1968, 1980; Mosko et al., 1980), or, initially, to the total disorganization of temporal processes by some agent (Gilbert, 1974b). In the latter case the system fails to attain a new dynamic steady state. As with physiological aging (believed to involve the same process), cellular death may be effective before absolute stasis is reached. Although aging may be viewed as a smooth transition toward a static state, it is possible that at some point the pattern of rhythms could become unstable as their amplitudes decrease and the phase and frequency locking actions become weaker. According to circumstance, this could alter the rate at which stasis is approached and might lead to a "permanently" viable but malfunctional or malignant cell. One can thus explain "spontaneous" transformation in both in vivo and in vitro situations.

### The Overall Dynamic State

The overall dynamic state will be determined by the interaction between (1) individual dynamic systems within the cell, (2) the systems and their genome, (3) the systems of different cells in the population, and (4) the systems and their micro- and macroenvironment. Although attention is usually centered on the identity of specific reactions and whether they are totally inhibited or activated, the overall kinetics are generally of major importance regarding stability and the net response of a system to the action of an agent. Here we examine some of the factors involved.

Of particular interest is the effect of one rhythm on another. Unless the overall pattern of such periodicities are fully coordinated, one rhythm can be expected to disturb another until one or other, or both, adapt with regard to frequencies, amplitudes, phasing, and mean levels, or until such disorganization ensues that they are totally inhibited. The results of such interactions depend on the relative characteristics of the disturbance and the system on which it acts.

The effects of the initial reactions will be propagated through the system by changes in the levels of substrates, etc., which will constitute secondary disturbances. The pattern of variations (the waveform) will, in general, differ from those involved in the initial reactions and, in addition, will normally be delayed to some extent, depending on the nature and kinetics of intervening reactions. If the number of the latter is large, or a switch reaction is involved (see Biochemical Switching), then the delay will be effectively finite. Intermediate reactions thus introduce waveform distortion and delays. Both of these effects may be significant in determining the overall response of the system to the disturbance.

### Biochemical Switching

Of special importance to an understanding of the behavior of biochemical systems in relation to differentiation and the malignant transformation is the biochemical switch reaction involving not a smooth transition between distinct stable states but a discontinuous response. The existence of such phenomena is usually implied rather than experimentally proven. We can distinguish two types of switch reaction, those involving hysteresis and those depending on bifurcations. The first, which does not require the continued presence of an agent to maintain the change in state, is now well established in *theoretical* biochemistry: transitions between alternate states are governed by distinct threshold conditions, one for forward and one for reverse transitions in state. I have previously considered the malignant transformation and its reversal in terms of switching in systems exhibiting hysteresis (Gilbert, 1973a). The thresholds give the stability limits of such a system to one kind of perturbation, but it must be remembered that some disturbances are likely to alter the values of the thresholds and even completely eliminate hysteretic behavior. As indicated, it is believed that sets of rhythmic processes can effectively exhibit distinct alternate stable phase relationships and that switching between such states is possible (Figures 2 and 3). Reversibility, induced or spontaneous (Gilbert, 1973a), is feasible providing no secondary effects are produced that alter the system in question.

Shifts in the mean levels and amplitudes of rhythms can thus trigger switching, but it should be noted that even small random disturbances can be critical if the mean state of the system lies close to a threshold or bifurcation (Gilbert, 1973a). I have suggested that such processes are responsible for cell cycle and tumor variability and for the stochastic element in differentiation (Gilbert, 1978c, 1980, 1982a). With regard to the second of these it could perhaps be that the higher degree of variability of cells in a tumor result from enhanced energy utilization, causing more marked thermal dissipation and hence larger amplitude random fluctuations.

In its simplest form, at least, the bifurcation represents a simple threshold, a critical set of conditions that separate not discrete states but different behavior patterns, for example, a static and oscillatory mode of operation of a control system (see The Cell Cycle). The forward and reverse thresholds are usually the same.

## Amplitude, Frequency, and Phase Discrimination

The response of a system (e.g., a switch process) to a disturbance is generally considered only in terms of (1) its magnitude (Rosen, 1971), and (2) the effect of one such perturbation acting alone. Triggering only occurs if the peak (or trough) attained is sufficient: the system exhibits amplitude discrimination. In addition, though, systems will normally also exhibit frequency and phase discrimination (Gilbert, 1973a,b). If the response time of a reaction is high, then a sudden change in, for example, the level of a component, can be absorbed without triggering some switch action, whereas if it is low then a pertinent threshold can be exceeded. If the disturbance is of sufficient magnitude, the switch could be triggered by a single pulse, but otherwise triggering will depend on the pulse frequency and the number of cycles of some rhythm (a biochemical counting process). If the frequency of the latter is high, compared with the relaxation time of the system, then the mean level of the compound will gradually increase until triggering is permitted (Gilbert, 1973a). A slow response time can thus convey stability, and with moderate disturbances, the system may require to be perturbed several times in quick succession before it will respond significantly. Thus, triggering may depend on the existence of an underlying periodicity with suitable dynamic characteristics. The requisite behavior may depend on the system being disturbed correctly. A possible advantage of control by frequency modulation is that it may be less sensitive to noise (Rapp et al., 1981). This may also be true for phase modulation.

What if the system is affected by two distinct disturbances (perhaps arising from the same primary perturbation through two pathways, each introducing distortions and delays) which act at different points in the system? Elsewhere (Gilbert, 1973a,b) we have examined this kind of situation in relation to differentiation and the malignant transformation by considering the modulation of a metabolic switch reaction by two rhythms. In addition to confirming that the system exhibits both amplitude and frequency discrimination, it was shown that the relative timing (phasing) and the sequence of action of the two periodicities can each determine whether triggering takes place. On the other hand, such factors can also govern spontaneous reversal, a subject of concern with regard to the malignant transformation (Gilbert, 1973a).

In the steady state, the amplitudes, frequencies, and phases of the rhythms will be such that no new triggering occurs. Once the system is disturbed these various parameters, and also the relative timings of the periodicities, may be transiently sufficient to trip new switch processes, one perhaps involving the phasings of the rhythms themselves. Through the involvement of such factors, a few components with generalized metabolic or genetic action (e.g., metal ions) might exhibit differential and specificity of action on the patterns of gene expression and metabolic behavior. It is known that the site of chromosome puffing is very dependent on not only the nature of the ionic environment but also the absolute and relative levels of particular ions (Lezzi, 1970). If the local concentrations oscillate, then the set of active genes could reflect the pattern of temporal organization, and vice versa.

## THE CELL CYCLE

### General Considerations

Having suggested that the malignant transformation involves a derangement of temporal organization, it became necessary to explain how such a lesion could produce the observed biological effects. Since cancer primarily involves a change in the ability of the cell to replicate under conditions where the corresponding normal cells do not, this meant providing an account of the way a change in the pattern of cellular rhythms could modify the proliferative behavior. In order to understand the aberrant nature of the transformed cell, it is necessary to know what makes any cell replicate. Models of the cycle that are based on sequences of dependent reactions require (Gilbert, 1981) (1) delineation of closure reactions by which the sequence is completed, (2) an account of the dissociability of different phases, and (3) provision for refractory behavior that prevents the sequence (or supposed random event) from being reinitiated while the cell is at another stage of the cycle. The lack of any adequate concept of the phenomenon of replication led me to revive the idea that a rhythmic process underlies the cell cycle (Gilbert, 1968). Following this, and in view of a certain amount of evidence indicating the involvement of thiols (SH) and disulfides (SS) in the latter, Sel'kov (1970) presented a mathematical analysis of the sulfur metabolism of cells. The result was a pair of differential equations that yield limit cycle oscillatory solutions in the levels of SH and SS for some sets of the parameters (the rates and kinetics of the relevant reactions). I have since developed this model and shown that it mimics the cell with respect to most, perhaps all, of the major facets of replication (Gilbert, 1974a, 1977, 1978a-c, 1980, 1982a,b).

### A Limit Cycle Oscillator Concept

The limit cycle concept does not have to meet the above requirements, being based on a continuous rather than a discontinuous process. Quiescence is considered to reflect the existence of a static steady state and replication the occurrence of a cycle of the oscillation that initiates and coordinates the overt events. The behavior of the system is determined by the set of reaction rates and kinetics (the parameter values) of the reactions forming the control system. The switchlike characteristics of the initiation and suppression of replication is accounted for in terms of a bifurcation in the values of the parameters that separate static and autodynamic behavior. This concept thus requires the existence of thresholds in the levels of regulators (both mitogenic and inhibitory), which may be negligible in particular situations. On this basis it is possible to account for (1) the phenomenon of density-dependent limitation of proliferation and its facets and (2) the nature of the effect of the malignant transformation on replication characteristics in terms of discrete or dynamic changes in the relevant threshold (Gilbert, 1977, 1978a, 1980). In addition, the concept also accounts for the commitment to replication, the mechanism of action and interaction of regulators, the control of rate or frequency of replication, all methods of synchronization, the distinct shapes of phase response curves, variability in cell cycle phases, and yet other aspects. Here we briefly consider the interdependence of replication and other phenomena in terms of interacting rhythms.

### Replication, Differentiation, Aging, and Cancer

There are three basic situations: (1) the cell cycle rhythm affects the levels of other constituents (some also dynamic); (2) the cell cycle control system is modulated by other periodicities; and (3) both effects occur, so that the reactions outside the cell cycle system form a self-regulating loop (Gilbert, 1978d, 1980). Thus, component levels (and hence cell mass) can depend on those of SH and SS and therefore vary in a replication-dependent manner, with the pattern reflecting all the relevant rates and kinetics. On this basis it is possible to account for peak, linear, and step changes in the levels or activities of components in response to the SH-SS rhythm. This concept does not require (nor, in general, expect) these changes to necessarily yield duplication of components but suggests that such an outcome is possible as a result of the mutual adaptation of the dynamic systems. By modulating the other dynamic systems, the cell cycle oscillation can thereby alter their behavior and thus contribute to differentiation, tumor progression, and the development of a cell line. Once new triggering has occurred, however, the cycle rhythm may no longer exert such an effect.

As the component levels vary in response to the changing levels of SH and SS (or similar factors) some, in turn, can modify the values of the cell cycle control system parameters (Gilbert, 1978d). The frequency of the latter rhythm can thus become self-modulated. Depending on the kinetics involved, these secondary control loops can cause a progressive change in replication characteristics and, eventually, lead to inhibition of proliferation in a cell cycle-dependent manner, such as in cell aging (Gilbert, 1980).

If the rates and kinetics of the control system reactions are modulated by other rhythmic processes, then the replicative characteristics will vary periodically with time and can thus depend on cellular temporal organization. In this way we may account for (1) the view that the cell cycle is a multiple of some high-frequency rhythm (Klevecz, 1976; Gilbert, 1982a), (2) contact inhibition of growth (Burton and Canham, 1973), and (3) the dependence of replicative ability on, for example, the phasing of two such periodicities and hence for the way in which transformation can produce the effects recognized as cancer (Gilbert, 1977, 1978a, 1980).

## EXPERIMENTAL ASPECTS

### Basic Issues

The concepts of differentiation and oncogenesis outlined here require the existence in all cells of several distinct biochemical rhythms which (1) continue to exist under adverse conditions (i.e., while the cell is viable); (2) are essentially characteristic of the cell concerned (under defined conditions); and (3) respond to some disturbances by temporary changes in frequencies, mean levels, amplitudes, and phasings, and to other stimuli, by permanent modification in these features. In Figs. 2 through 9, results are presented here that conform to these requirements, emphasis being placed on switching of oscillatory states and irregular dynamics (Gilbert, 1968, 1969, 1974b; Gilbert and Tsilimigras, 1981; Tsilimigras and Gilbert, 1977).

### Practical Problems

Due to the nature of the processes being invoked and the relevant factors concerned, direct confirmation of the concepts is difficult, if not impossible, to obtain. This is especially true of differentiation in the higher organisms and oncogenesis in vivo. Initially there

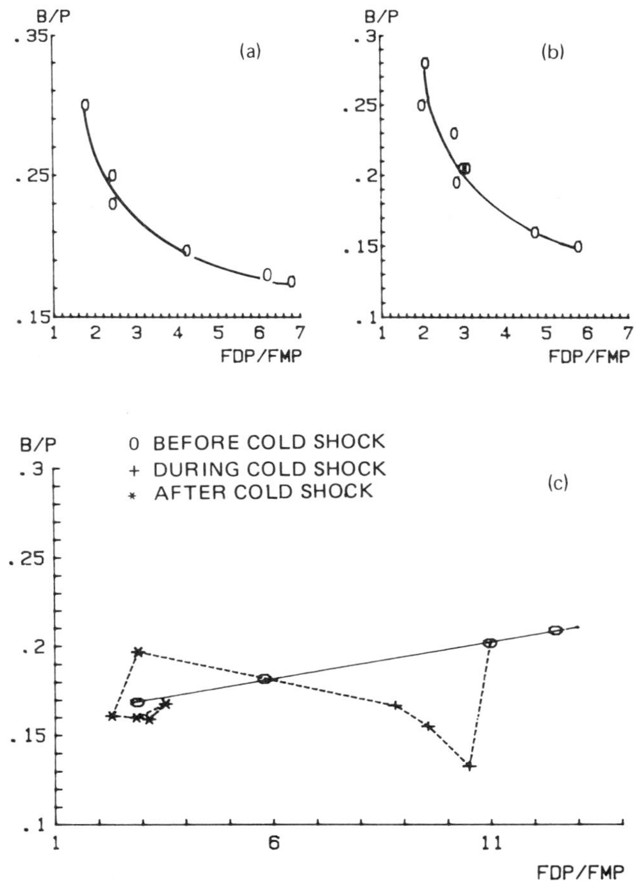


**FIGURE 4** In-phase and out-of-phase isozyme oscillations. In (a) and (b) are shown phase plane plots obtained for LDH and aldolase isozyme pattern rhythms in BHK-PYY cells not subjected to thermal shock. The correlations are still negative, but the relationship is nonlinear. Data obtained with BHK-C13 cells before, during, and after thermal shock (4°C, 1 hr) are presented in (c). (Exps. 604, 611, 1 hr sampling period.) [Curve (c) adapted from Gilbert, 1969.]

are the questions: (1) How many oscillations exist in each cell species? (2) Which are primary, and which occur in response to others? (3) Which are relevant in any particular situation? Irrespective of the answers, there are various practical issues. The number of cells required for determinations and the prolonged period involved in these processes precludes suitable studies, as does the need for observations that do not disturb the reactions taking place. The heterogeneity of the cell population, asynchrony in events within the population, and lack of information regarding the identity of target cells are yet other complicating factors. Culture systems, though artificial, would seem to offer advantages, but even then many problems exist. When compared with the time to trip a switch or with the period of a rhythm, differentiation and transformation are rarely, if ever, rapidly achieved. Heterogeneity is still a problem, and disturbances often enhance the degree of variability. Yields of transformed cells are frequently low, and the use of

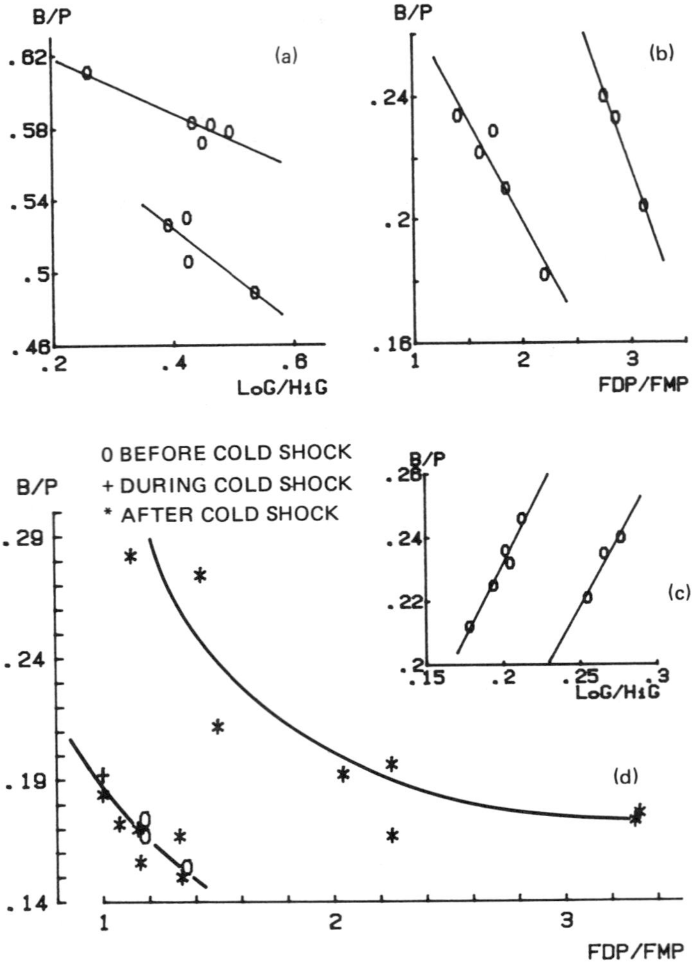


**FIGURE 5** Metabolic switching in oscillating systems. In a number of instances, especially when the cells concerned had previously been disturbed, the relationship between two rhythms appears to shift in a discrete manner with respect to the mean level of one or other or both of the oscillations. Phase plane examples of such behavior are given in (a), (b), and (c). Particularly where prolonged observations have been made, the jumps in behavior have occurred repeatedly, as in (d), where the cells had been subject to thermal shock (4°C, 2 hr). These effects are tentatively attributed to the existence of hysteresis in the kinetics of the metabolic systems. Note the different phase relationships, as evident from the slopes of the plots, between (1) the B/P and LoG/HiG rhythms in (a) and (c), and (2) the B/P and FDP/FMP periodicities in (b) and the B/P and Log/HiG variations in (c). The latter observation indicates that whereas the LDH and hexokinase rhythms were always in phase, the aldolase and LDH oscillations were always out of phase, as were the aldolase and hexokinase periodicities. The linear correlation coefficients were (a) 0.87 and 0.94; (b) 0.96 and 0.98; (c) 0.98 and 0.96. Interestingly, where, as in (d), the phase plots are nonlinear, highly linear correlations (r = 0.90-0.98) are obtained if the inverse rate ratio (FMP/FDP) is used. [(a), human erythrocytes, Exp. 810. (b) and (c), HaK cells, Exp. 872. (d), BHK-PYY cells, Exp. 608.] [(a) adapted from Gilbert, 1974b. (d) adapted from Gilbert, 1969.]

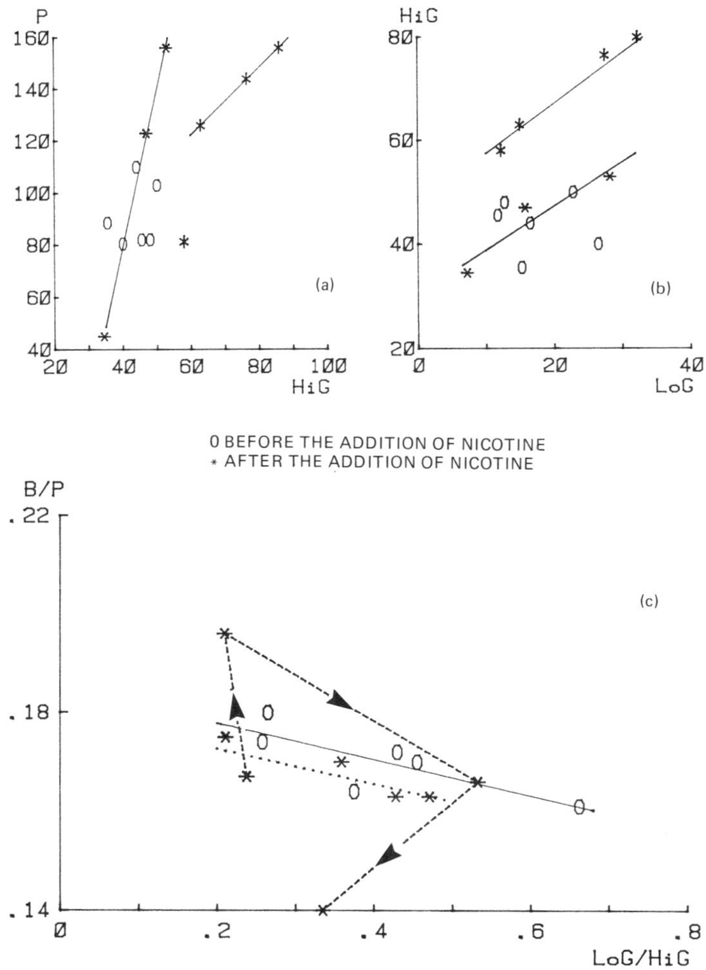


**FIGURE 6** Differential effects of agents. In general, agents have distinct effects on different rhythms, even those involving a given enzyme, as illustrated here, where nicotine has been added to HaK cells. The two activities of both LDH and hexokinase were immediately affected by the agent (but not identically) and, subsequently, switching apparently occurred in both systems [r values: (a) 1.0, 1.0; (b) 0.99, 0.95]. Apart from a slight shift in the mean level, the relationship between the two isozyme pattern changes (c) remained the same for about 5 hr. The linear correlation coefficients for the full line and dotted line (covering the first five points after the addition of agent) are identical ($r = 0.8$). In (c) the last three data point have been joined by broken lines in order to illustrate the change in behavior: they are not intended to indicate the true time course. (Exp. 824.) [(a) and (c) adapted from Gilbert, 1974b.]

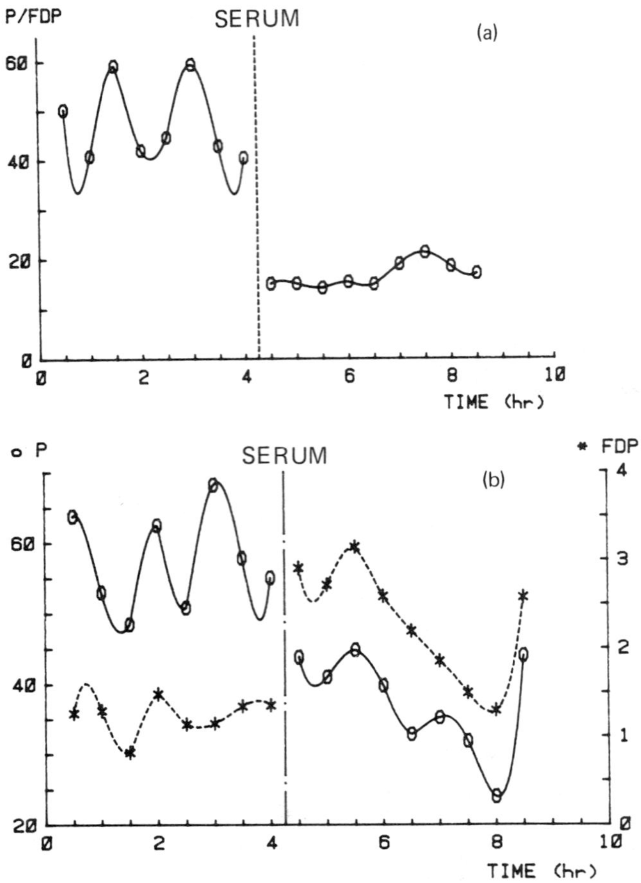


**FIGURE 7** The effect of mitogens on temporal organization. In the last three diagrams we present results of preliminary studies on the effect of serum on quiescent cells, partly in support of the proposal that hormones and mitogens act by disturbing the pattern of temporal organization (Gilbert, 1974b) but also to illustrate the complexities of transient states and difficulties in their analysis and intrepretation. In (a) is shown, as a function of time, the variation in the ratio of the activities of two distinct enzymes in HaK cells, maintained overnight on a serum-free medium, and the initial effect of serum added to a level of 10%. The corresponding time curves for the individual activities are given in (b). See also Fig. 8 (Exp. 872). [Curves in (a) adapted from Gilbert and Tsilimigras, 1981.]

prolonged culture to increase their number can result in selection of particular and perhaps irrelevant species, thereby confusing the issues. With these and other factors in mind, we have adopted a broad experimental approach, merely attempting to obtain various pieces of supporting evidence while gleaning information regarding the mode of action of various agents (e.g., mitogens, viruses, tumor promoters), particularly those known to act at the membrane level. With both in vivo and in vitro systems, a number of analytical and intepretative problems limit the amount and extent of the information that can be obtained.

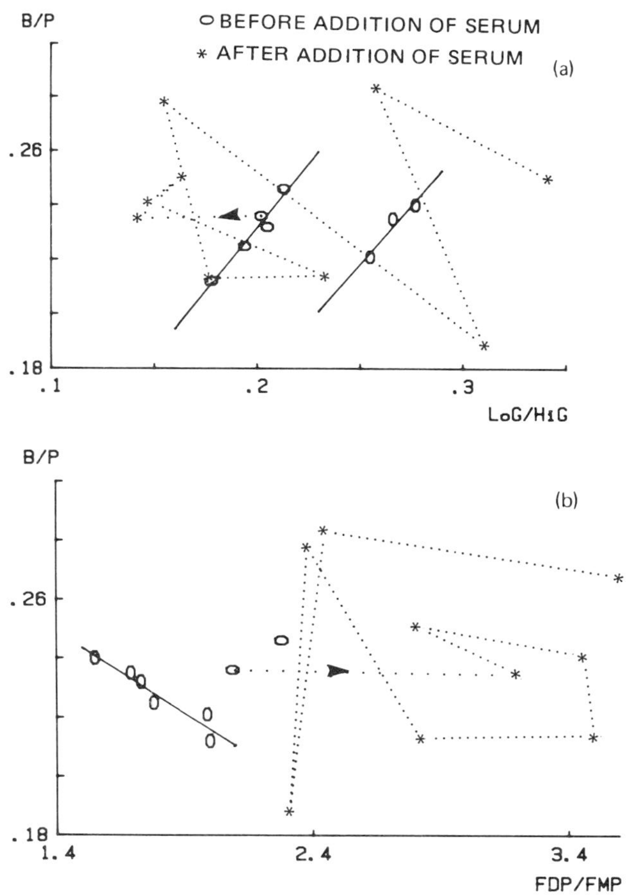


**FIGURE 8** The effect of mitogens on temporal organization. In (a) we replot the data of Fig. 5c (obtained with HaK cells, maintained overnight without serum and showing the switching discussed earlier) and the effect of adding serum. It can be seen from the dotted lines (not intended to indicate the actual time course) that the agent had a rapid (<30 min) effect on the mean level of the hexokinase rhythm, although initially, the phasing between the two oscillations remained the same (r = 0.95). Thereafter, marked variations occurred in the amplitudes and phase relationships due to differential actions on the two systems. In (b) is shown the phase plot for adolase and LDH isozyme in the same experiment. In this case all but two of the values for the serum-starved cells were negatively correlated (r = 0.95; see Figs. 3 and 5), and the serum rapidly altered both the mean level of the aldolase oscillation and also the phase relationship between the two isozyme pattern rhythms, probably as a result of frequency changes. In this experiment the "filter" method was used to eliminate the contribution of high-frequency rhythms (Gilbert, 1974b): three samples were taken at 5 min intervals and pooled, this procedure being repeated every 30 min to obtain the data points. (Exp. 872.)

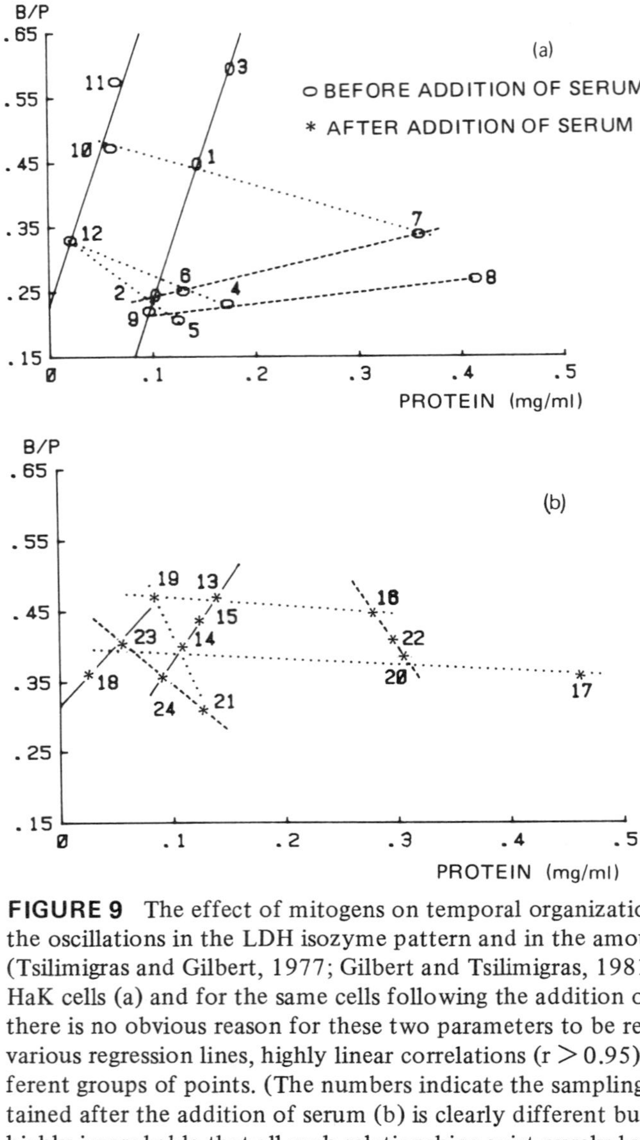


**FIGURE 9** The effect of mitogens on temporal organization. The phase plane plots for the oscillations in the LDH isozyme pattern and in the amount of extractable protein (Tsilimigras and Gilbert, 1977; Gilbert and Tsilimigras, 1981) are shown for serum-starved HaK cells (a) and for the same cells following the addition of serum to 10% (b). Although there is no obvious reason for these two parameters to be related, as can be seen from the various regression lines, highly linear correlations ($r > 0.95$) can be detected between different groups of points. (The numbers indicate the sampling sequence.) The pattern obtained after the addition of serum (b) is clearly different but similarities exist. It seems highly improbable that all such relationships exist purely by chance (especially with certain lines being essentially parallel), but the pattern of kinetic restraints leading to these complex results is by no means evident. They may arise through multiple switch reactions. (Exp. 872.)

Ideally, one wishes to know what rhythms exist in the cell in question, what their frequencies, amplitudes, and phasings are under the particular conditions, and the temporary and permanent effects thereon of specific disturbances. Even with cells existing in a steady state, the determination of these parameters is not always simple and straightforward (Gilbert, 1974b). When the system is disturbed it becomes even more difficult to obtain definitive information due to (1) the appearance of additional frequencies, (2) the resultant change of dynamic characteristics with time, and (3) the short duration of most experiments. Curve fitting and various other techniques (e.g., Fourier, autocorrelation, and periodogram analyses) are often necessary to estimate frequencies of rhythms present in the data, but the values obtained may not be the true ones. They assume steady-state situations: when the system is disturbed it is necessary to apply the methods sequentially to sections of the data (Gilbert and Tsilimigras, 1981).

Frequently a periodicity may exist and not be recognized or it may even be ignored: this can give rise to a degree of variability that is greater than otherwise expected but is wrongly attributed to "biological error." The poor reproducibility may stem from the fact that observations are not being made at the same time relative to the cycle concerned and/or to the fact that, unwittingly, the agent is not being added at the same phase of the rhythm in the different experiments. Even where the existence of a rhythm is known or suspected, it can still prove difficult to be sufficiently systematic, or to obtain reproducible results, simply because the state of the cells cannot be ascertained quickly enough. Without some rapidly ascertained reference point, one is unable to sample or perturb the cells at a specific and predetermined phase of the rhythm. If the agent causes a discrete phase shift in a rhythm, then the immediate effects on a parameter may be a rapid increase in one experiment, a decrease in another, or there may be no immediate effect, purely because of the phase-dependent nature of the response. Only if the observations before and after the addition of a reagent cover several cycles of the rhythm could this be obvious. A good maxim is to suspect an oscillatory process if the reproducibility is poor.

When observations are made discontinuously, particularly if they are also irregular and the number of observations per cycle is small (less than three), erroneous conclusions may be drawn regarding the true time course followed and hence to the frequency, amplitude, and phasing of the rhythm (the problem of aliasing) (Gilbert, 1974b). What initially may be a suitable sampling interval need not be so after the system has been disturbed. It seems highly likely that much published information is of doubtful value simply because rhythmic behavior was not suspected and/or because these various factors have not been appreciated.

## ACKNOWLEDGMENTS

Thanks are due to the cooperation of Carol Tsilimigras in some of the experimental studies and Dr. I. M. Spence for reading and improving the manuscript. This work was supported in part by grants from the National Cancer Association of South Africa.

## REFERENCES

Apter, M. J., and Wolpert, L. (1965). Cybernetics and development. I. Information theory. *J. Theor. Biol.* 8:244-257.

Brodsky, W. Y. (1975). Protein synthesis rhythm. *J. Theor. Biol.* 55:167-200.

Burton, A. C., and Canham, P. B. (1973). The behaviour of coupled biochemical oscillators as a model of contact inhibition of cellular division. *J. Theor. Biol.* 39:555-580.

Chay, T. R. (1981). A model for biological oscillations. *Proc. Natl. Acad. Sci. USA* 78: 2204-2207.

Ermentrout, G. B. (1981). n:m Phase locking of weakly coupled oscillators. *J. Math. Biol.* 12:327-342.

Gilbert, D. A. (1966). Isozymes and cell regulation. *Discovery* 27:23-26.
Gilbert, D. A. (1968). Differentiation, oncogenesis and cellular periodicities. *J. Theor. Biol.* 21:113-122.
Gilbert, D. A. (1969). Phase plane analysis of periodic isozyme pattern changes in cultured cells. *Biochem. Biophys. Res. Commun.* 37:860-866.
Gilbert, D. A. (1973a). The malignant transformation as a metabolic steady state transition: The possible significance of the phasing of enzyme synthesis and related aspects. *Biosystems* 5:128-139.
Gilbert, D. A. (1973b). Biochemical phase discrimination in relation to differentiation and development. *S. Afr. J. Sci.* 69:348-349.
Gilbert, D. A. (1974a). The nature of the cell cycle and the control of cell proliferation. *Biosystems* 5:197-206.
Gilbert, D. A. (1974b). The temporal response of the dynamic cell to disturbances and its possible relationship to differentiation and cancer. *S. Afr. J. Sci.* 70:234-244.
Gilbert, D. A. (1977). Density dependent limitation of growth and the regulation of cell replication by changes in the triggering level of the cell cycle switch. *Biosystems* 9: 215-228.
Gilbert, D. A. (1978a). The malignant transformation: The nature of its effects on cell replication characteristics. *S. Afr. J. Sci.* 74:48-49.
Gilbert, D. A. (1978b). The mechanism of action and interaction of regulators of cell replication. *Biosystems* 10:227-233.
Gilbert, D. A. (1978c). The relationship between the transition probability and oscillator concepts of the cell cycle and the nature of the commitment to replication. *Biosystems* 10:234-250.
Gilbert, D. A. (1978d). Feedback quenching as a means of effectively increasing the period of biochemical and biological oscillations. *Biosystems* 10:241-244.
Gilbert, D. A. (1980). Mathematics and cancer. In *Mathematical Modelling in Biology and Ecology, Lecture Notes in Biomathematics,* vol. 33, W. Getz (ed.). Springer-Verlag, Berlin, pp. 97-115.
Gilbert, D. A. (1981). The cell cycle 1981: One or more limit cycle oscillations? *S. Afr. J. Sci.* 77:541-546.
Gilbert, D. A. (1982a). Cell cycle variability: The oscillator model of the cell cycle yields transition probability alpha and beta type curves. *Biosystems* 15:317-320.
Gilbert, D. A. (1982b). An oscillator cell cycle model needs no first or second chance event. *Biosystems* 15:331-339.
Gilbert, D. A., and Tsilimigras, C. W. A. (1981). Cellular oscillations: Relative independence of enzyme activity rhythms and periodic variations in the amount of extractable protein. *S. Afr. J. Sci.* 77:66-73.
Guevera, M. R., Glass, L., and Shrier, A. (1981). Phase locking, period doubling bifurcations and irregular dynamics in periodically stimulated cardiac cells. *Science* 214: 1350-1353.
Kauffman, S. (1971). Differentiation of malignant to benign cells, *J. Theor. Biol.* 31: 429-451.
Klevecz, R. R. (1976). Quantized generation time in mammalian cells as an expression of the cellular clock. *Proc. Natl. Acad. Sci. USA* 73:4012-4016.
Lezzi, M. (1970). Differential gene activation in isolated chromosomes. *Int. Rev. Cytol.* 29:127-168.
Mosko, S. S., Erickson, G. F., and Moore, R. Y. (1980). Dampened circadian rhythms in reproductively senescent femal rats. *Behav. Neural Biol.* 28:1-14.
Mustafin, A. T., and Volkov, E. (1982). On the distribution of cell generation times. *Biosystems* 15:111-126.

Rapp, P. E. (1978). Bifurcation theory, control theory and metabolic regulation. A mathematical investigation of biological and biochemical oscillators. In *Biological systems, modelling and control,* D. A. Linkins (ed.). Peter Peregrinus, London, pp. 1-83.

Rapp, P. E., Mees, A. I., and Sparrow, C. T. (1981). Frequency encoded biochemical regulation is more accurate than amplitude dependent control. *J. Theor. Biol.* 90:531-544.

Richter, P. H., and Ross, J. (1981). Concentration oscillations and efficiency: Glycolysis. *Science* 211:715-716.

Rosen, R. (1971). *Dynamical System Theory in Biology,* vol. 1. Wiley-Interscience, New York.

Sanglier, M., and Nicolis, G. (1976). Sustained oscillations and threshold phenomena in an operon control circuit. *Biophys. Chem.* 4:113-121.

Sel'kov, E. E. (1970). Two alternative, self oscillating stationary states in thiol metabolism—two alternative types of cell division—normal and malignant ones. *Biophysika* 15:1065-1073.

Termona, Y., and Ross, J. (1981). Oscillations and control features in glycolysis: Numerical analysis of a comprehensive model. *Proc. Natl. Acad. Sci. USA* 78:2952-2956.

Tornheim, K. (1979). Oscillation of the glycolytic pathway and the purine nucleotide cycle. *J. Theor. Biol.* 79:491-541.

Tsilimigras, C. W. A., and Gilbert, D. A. (1977). High frequency, high amplitude oscillations in the amount of protein extractable from cultured cells, *S. Afr. J. Sci.* 73: 123-125.

Webb, D. J., and Nuccitelli, R. (1981). Direct measurements of intracellular pH changes in xenopus eggs at fertilisation and cleavage. *J. Cell Biol.* 91:562-567.

# 2

# Epigenetic Oscillations during the Cell Cycles of Lower Eukaryotes Are Coupled to a Clock

## Life's Slow Dance to the Music of Time

**David Lloyd and Steven W. Edwards*** / University College, Cardiff, Wales

> Philosophers and physicists vie with poets like Lucretius and T. S. Eliot, and historians and psychologists, to drain time of its mysteries.
>
> Landsberg (1981)

### TIME DOMAINS OF LIVING SYSTEMS

The ever increasingly detailed picture of cell ultrastructure, gleaned from molecular studies, represents only the static shell of the living system. It is but one of two hierarchies that cannot be dissociated. Yet the temporal hierarchy is, at best, only sketchily charted and little understood, despite a current burgeoning interest in many differing time scales.

The central reactions of metabolism providing material and energy for cellular processes have a characteristic time scale that varies with metabolic activity but is usually within the range of seconds or minutes (the "metabolic domain"; Fig. 1). Some metabolites, as well as participating in these interconverions, are also involved in more rapid reactions (e.g., binding to proteins, formation of ionic complexes, protonation) that approach equilibrium within milliseconds or less. Membrane-associated reactions that are not diffusion limited (e.g., electron transport reactions) and processes that involve large-scale conformational changes in proteins or membranes also proceed on this time scale. The internal mobility of proteins or membranes can now be examined in times of nanoseconds or picoseconds (Hayward et al., 1981).

More relevant to the present discussion, however, are those reactions proceeding more slowly than those of the metabolic domain. The total content of enzyme, protein, and cofactors, and of stored macromolecules, changes only slowly, and measurable altera-

*Current affiliation: University of Wales Hospital, Cardiff, Wales

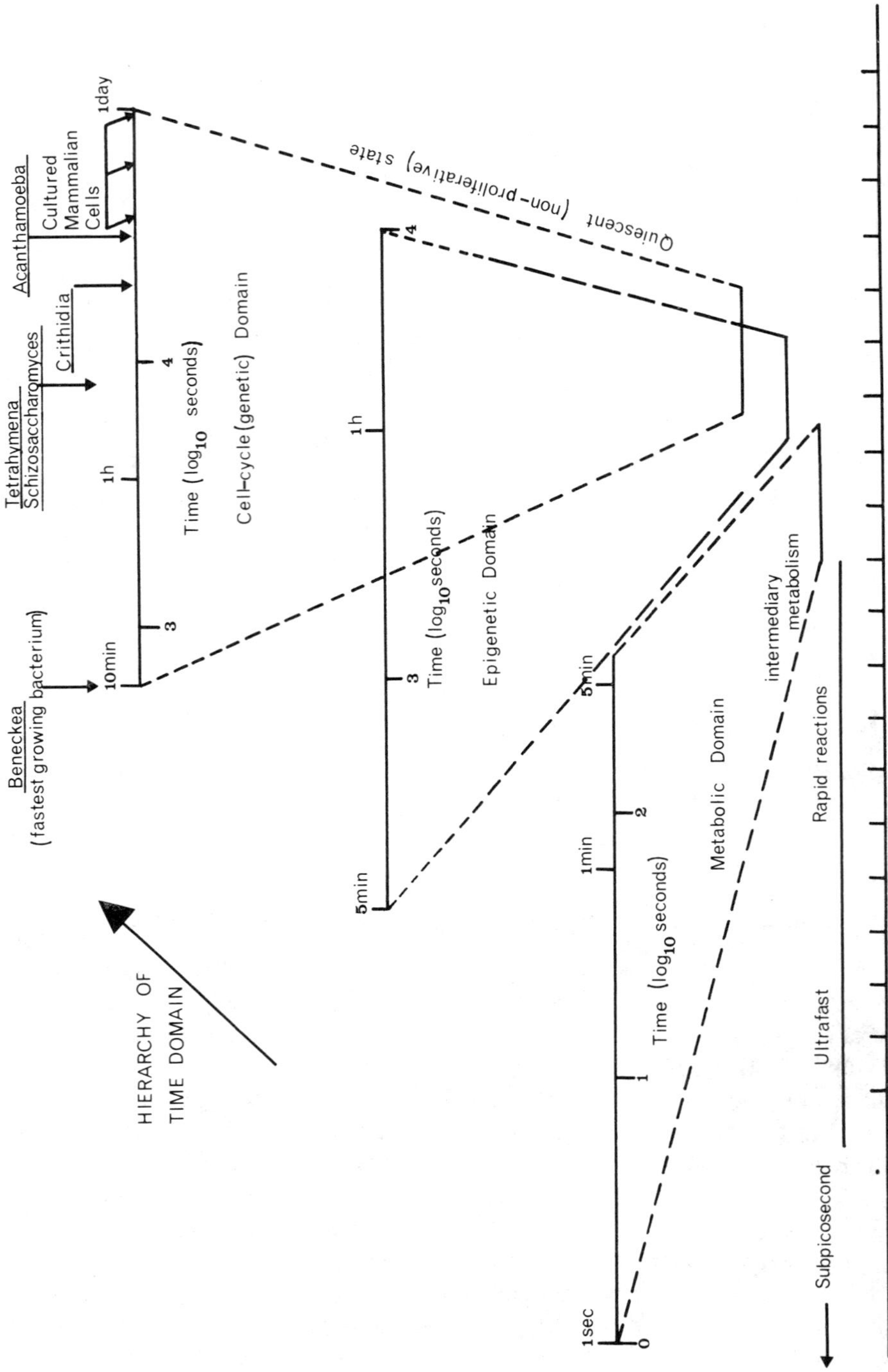


**FIGURE 1** The time domains of living systems. The cell cycle times of some organisms used in these studies are included.

tions in these quantities may be disregarded if our prime focus is on the events of intermediary metabolism. It is, however, these slowly drifting parameters of the "epigenetic domain" which become of major concern when we study cellular growth or slow adaptive processes. An even longer time scale must be used to encompass the events of cell multiplication: the "cell cycle (or genetic) domain," which can last just 10 min for some prokaryotes, or which, at the other end of the scale, is often dominated in nature by circadian timekeeping (see Chap. 17). Viewed in this way, the dynamic perspective of the living organism is revealed to remind us of van Bertalanffy's concise appraisal: "What are called structures are slow processes of long duration, functions are quick processes of short duration."

## OSCILLATIONS ON THE METABOLIC TIME SCALE

The metabolic network, composed of the reaction sequences and cycles of intermediary metabolism, is kinetically controlled for the most part by activation and inhibition by low molecular weight effectors. The $10^4$-$10^5$ kinds of interaction possible in the network comprise the integrated whole, and modeling of its separate subsystems aids understanding (Reich and Sel'kov, 1982). The turnover rate of substrate molecules ($10^2$-$10^5$ molecules per enzyme molecule) determines the time course of establishment of steady states in the metabolic system. Using an average intracellular enzyme concentration of $10^{-8}$ M, and an average turnover number of $10^3$ $sec^{-1}$, it is evident that measurable changes in substrate concentration can occur in a period of a second, typical relaxation times (the time taken for the effect of perturbation to have decreased to $e^{-1}$ of its initial value) are of the order of 2 min (Goodwin, 1976).

Feedback inhibition, allosteric and cooperative enzyme control, and nucleotide coupling of pathways provide many opportunities for oscillations in systems with nonlinear kinetics maintained far from equilibrium (Higgins, 1967; Sel'kov, 1968; Chance et al., 1973).

The vast majority of the cellular oscillations studied [over four hundred listed by Rapp (1979)], including the most thoroughly explored system, glycolysis, are *metabolic* oscillations with periods most often lying in the range 1 sec to 5 min. Many of these oscillations may represent inevitable consequences of control within the network. Three possible functions have also been suggested: (1) It may be necessary to separate incompatible reactions catalyzed by enzymes located in a single subcellular compartment: this can only be achieved by having the reactions occur at different times (Sel'kov, 1972). Switching of antagonist enzymes on and off reciprocally leads to self-oscillatory changes in enzyme activity so as to minimize futile cycling (Sel'kov, 1980). (2) Metabolic oscillations can lead to spatial dissipative structures which may possibly play a role in the generation of biological form (Boiteux and Hess, 1980). (3) It has been repeatedly suggested that metabolic oscillations may have a time-keeping function, i.e., represent high-frequency ticks of the circadian clock (Chance et al., 1964; Boiteux et al., 1975). The most recent model (Sel'kov, 1980) highlights precisely a possible frequency reduction mechanism dependent on reversible interaction between a metabolic intermediate (e.g., hexose monophosphate) and a huge amount of a reserve energy store (e.g., glycogen).

Although impressive in experimental ingenuity and theoretical elegance, work on metabolic oscillations presents no convincing evidence for their definite involvement in

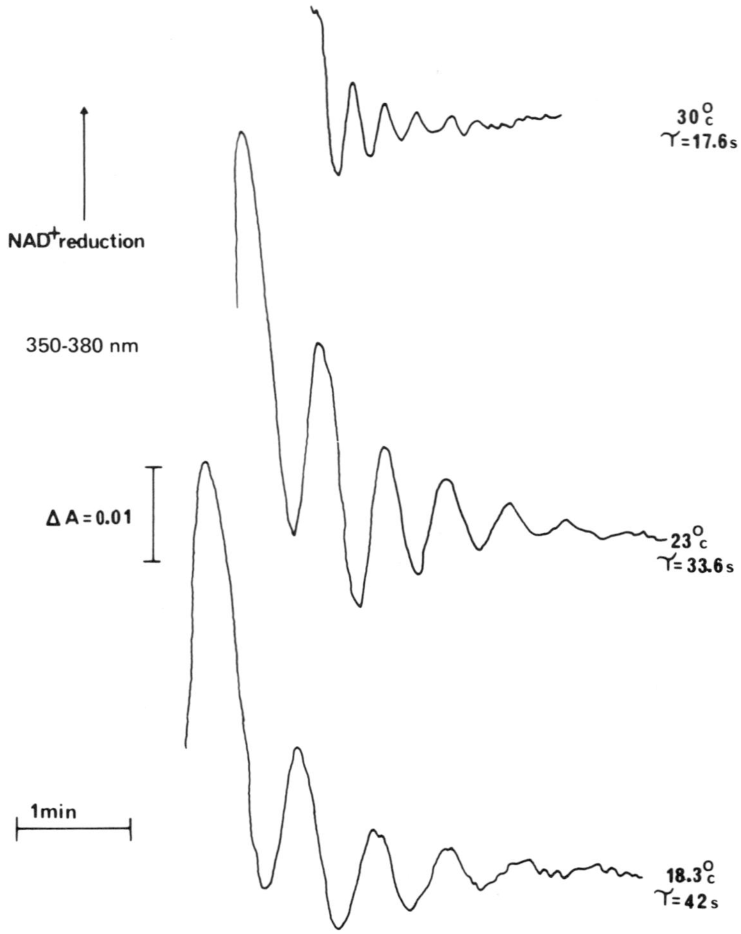


**FIGURE 2** Temperature dependence of the period of glycolytic oscillations in *Saccharomyces uvarum.* Starved cells (about 1 × $10^9$ cells per milliliter) in 100 mM K phosphate buffer (pH 6.5) were kept anaerobic for 10 min, exposed to $O_2$ for 2 min, and 100 mM glucose was added before switching the gas phase back to $N_2$.

biological timekeeping. Unless special conditions are met and maintained, e.g., cell concentration or rate of nutrient supply (von Klitzing and Betz, 1970), these oscillations are typically represented in vivo (Fig. 2) as highly damped, and therefore short-lived oscillations, with periods of the order of minutes; most importantly, this period is highly temperature sensitive.

## OSCILLATIONS ON THE EPIGENETIC TIME SCALE

The synthesis of macromolecules, their modification (processing), transport, and interaction, and their integration into membranes are all epigenetic processes. Broadly defined, these are the reactions involved in gene expression. The time domain occupied by these processes lies between the metabolic and the cell cycle domains. Thus, the cellular content of macromolecules changes slowly by comparison with the speed of metabolic interconversions that make up the metabolic map. Even so, the pace of this slow "adaptive" change is itself rapid by comparison with the time taken for a cell to double in volume and divide, the cell cycle time. Experimental studies on the kinetics of epigenetic reactions are less well advanced than those with the reactions of intermediary metabolism. In eukaryotes, the epigenetic system is composed of reactions that have relaxation times lying within the range $10^2$-$10^4$ sec (1.5 min to 3 hr), depending on the type of cell studied. Generally it might be expected that some correlation would be observed between the values of these relaxation times and cell cycle times of different organisms or cell lines, but this seems not to be always the case.

In order to study the temporal organization of the expression of gene activity it would be best to use a single cell. This would relieve us of the problem that data obtained from a population are necessarily time averaged. However, the limited analytical means as yet at our disposal to deal with the single cell makes compromise unavoidable. The fall-back position, that of using synchronous cultures, has provided a wealth of information on the time course of epigenetic change during the cell cycle (Edwards, 1981; Lloyd et al., 1982a). Unfortunately, the reliability of many of these results is questioned; many of the observed changes may have resulted from perturbative disturbance during procedures employed for the establishment of synchrony. At least two methods based on centrifugal velocity sedimentation, differential sedimentation in the culture medium (Chagla and Griffiths, 1978) and continuous flow size selection (Lloyd et al., 1975), have been shown to produce little perturbation as determined by control experiments with asynchronous populations subjected to similar centrifugal procedures. These methods for synchronization of cell populations under conditions of minimal perturbation have been employed throughout the work described here. They are rapid procedures that avoid removal of organisms from the growth medium; anaerobiosis, nutrient deprivation, and temperature changes are avoided, and the physical forces of viscous shear, centrifugation, and hydrodynamic pressure are kept to a minimum by using the lowest possible rotor speeds (Lloyd and Ball, 1979; Lloyd et al., 1982a).

During a study of the development of respiration during the growth of synchronous cultures of protozoa and yeasts, we observed that respiration does not increase continuously so as to double in rate over the period of the cell cycle, but rather shows oscillatory behavior (Lloyd, 1974). Early work with the fission yeast, *Schizosaccharomyces pombe,* and the kinetoplastid protozoan, *Crithidia fasciculata,* showed oscillatory energy metabolism in cultures synchronized after size selection by velocity sedimentation through density gradients.

Oscillations in respiration and in several enzyme activities, including cytochrome *c* oxidase and the amounts of cytochromes $a + a_3$ and $b_{563}$, were observed in *S. pombe* growing with glucose (Poole and Lloyd, 1973, 1974; Poole et al., 1973, 1974). In *C. fasciculata,* hourly maxima in $O_2$ uptake rates were correlated with changes in pools of adenine nucleotides (Edwards et al., 1975). Application of continuous flow size selection to *S. pombe* indicated that synchronous populations of glucose-grown cells showed oscillatory changes in mitochondrial ATPase activity that correspond temporally with altered sensitivity to oligomycin inhibition and the effects of 10 other inhibitors of the ATP-synthetase complex (Edwards and Lloyd, 1977; Lloyd and Edwards, 1977). Oscillations in $O_2$ consumption and adenine nucleotide levels were observed in synchronous cultures of the ciliate *Tetrahymena pyriformis* (Phillips and Lloyd, 1978; Lloyd et al., 1978), and respiratory oscillations also occurred in cultures of the budding yeast *Candida utilis,* irrespective of whether glucose, acetate, or glycerol was used as sole carbon source (Kader and Lloyd, 1979).

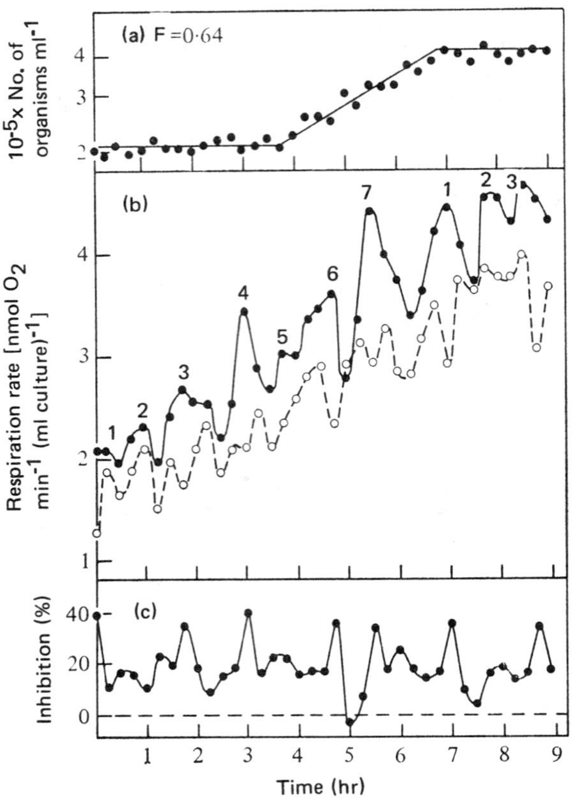


**FIGURE 3** Effect of cyanide on respiratory oscillations in a synchronous culture of *A. castelanii.* (a) Cell numbers and synchrony index, F (Blumenthal and Zahler, 1962); (b) respiration rate in the presence (●) or absence (o) of 0.5 mM cyanide; (c) inhibition of respiration (%) resulting from the addition of cyanide. (From Edwards and Lloyd, 1978.)

Further insights into the mechanisms that underlie these oscillating reactions of energy metabolism come from work with the soil amoeba *Acanthamoeba castellanii* (Edwards and Lloyd, 1978). This organism has a cell cycle time of 7.8 ± 0.5 hr (14 experiments) at 30°C, and size selection by low-speed centrifugation of exponentially growing cultures that show a complete doubling in cell numbers (Fig. 3). The mean respiration rate of the synchronous culture doubled over the cell cycle time and showed seven maxima; peak to trough amplitude of the oscillation accounted for between 18 and 31% of the minimum respiration.

Addition of 0.5 mM KCN to samples taken from the culture gave differing effects at different times. The greatest respiratory inhibition occurred when cyanide was added to organisms showing maximum respiratory activities and was least at troughs of respiration. Addition of 2.4 μM FCCP, an uncoupler of mitochondrial energy conservation, to organisims withdrawn from a synchronous culture gave a preferential stimulatory effect on respiration rates when organisms showed minimal respiratory activity. These observations suggest that the cycles of increased and decreased cellular respiratory activity correspond with cycles of energization and deenergization of mitochondria in vivo. Thus, the

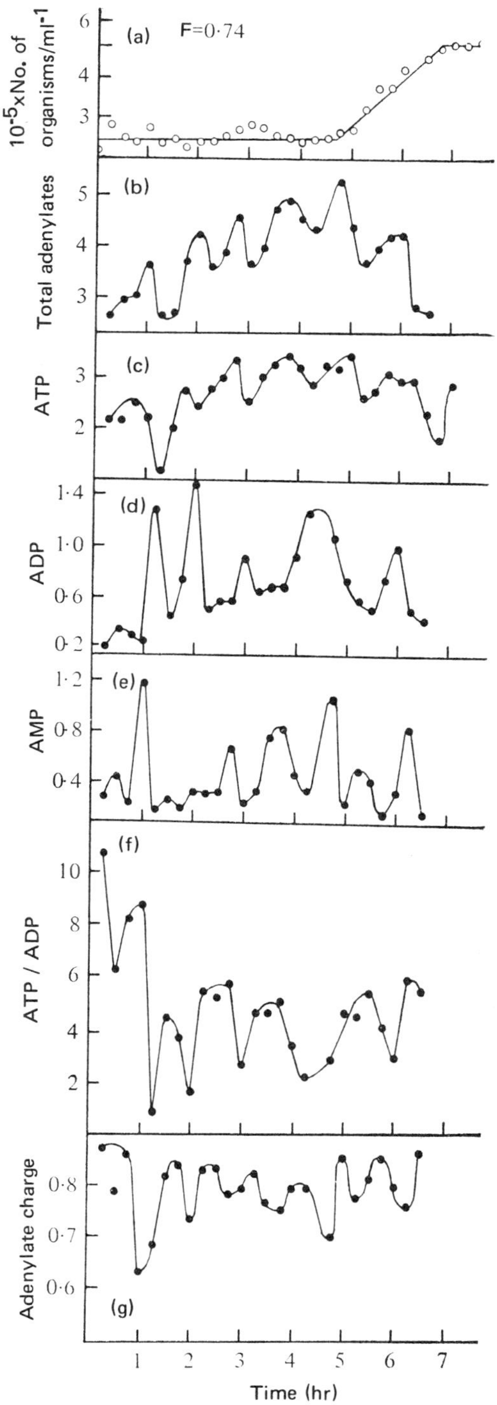


**FIGURE 4** Changes in adenine nucleotide pool levels in a synchronous culture of *A. castellanii.* (a) Cell numbers and synchrony index, F; (b) total adenylates; (c) ATP; (d) ADP; (e) AMP; (f) ATP/ADP ratio; (g) adenylate charge. Adenylate concentrations are expressed as nanomoles per milliliter culture. (From Edwards and Lloyd, 1978.)

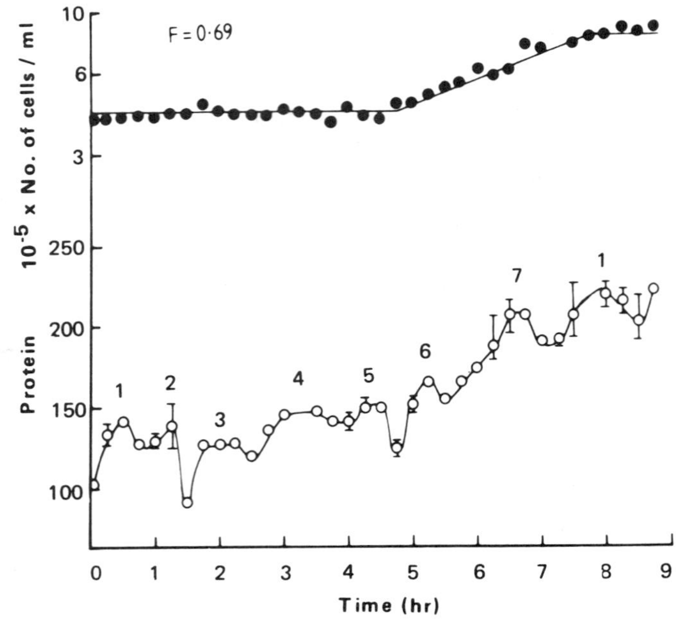


**FIGURE 5** Changes in total protein in a synchronous culture of *A. castellanii*: (●) cell numbers; (o) micrograms protein per milliliter after washing cells twice with 10 vol 50 mM $MgCl_2$ (pH 7.4). Each point is the mean of three separate protein estimations on each sample, and the range of values are presented where these are greater than the diameter of the symbol. (From Edwards and Lloyd, 1980.)

respiration of undamaged mitochondria isolated from *A. castellanii,* in common with those from many other sources, can be stimulated by addition of ADP in the presence of $P_i$ provided that energy conservation is tightly coupled to electron flow. In this energized state (state 3) cyanide sensitivity is greatest, whereas the acceleration of $O_2$ consumption due to uncoupler is most in the deenergized state (state 4), where adenine nucleotide acceptor control (respiratory control) limits respiration.

That the respiration of intact organisms shows two distinct energy states analogous to those of the isolated respiratory organelles was confirmed by measurement of adenine nucleotide pool sizes (Fig. 4). The periodicity of the oscillations was similar to that observed for respiration. The maximum observed amplitudes (peak-trough) for ATP, ADP, and AMP, expressed as percentages of the minimum values, were 108, 194, and 520, respectively. Adenylate change values varied between 0.63 and 0.88, and the ratio ATP/ADP varied between values of 0.9 and 10.9. Cell cycle maps showed a clear phase relationship: maxima of respiratory activity correspond to maximum ADP levels and are out of phase with ATP/ADP ratios. Therefore, at those times in the cell cycle when respiration is at a maximum, growing organisms are in state 3, whereas minimal respiratory activity corresponds to state 4. The redox state of mitochondria in thin films of cultures that were only one cell deep was assessed with a flying-spot fluorimeter using excitation with a scanning argon-ion laser beam. Cyclic changes in flavoprotein fluorescence also confirmed that synchronous cultures of *A. castellanii* oscillate between the ADP-limited (state 4)

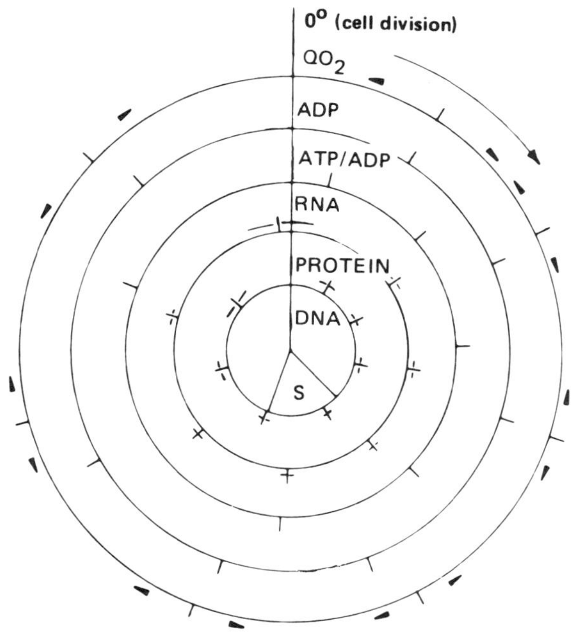


**FIGURE 6** Timings of cell cycle events in *A. castellanii.* 0° represents time zero for a synchronous culture and 360° the time when the cells have completed division: maxima of cell cycle events (and ranges) shown as phase angles. (From Edwards and Lloyd, 1980.)

and ADP-sufficient (state 3) respiratory states, i.e., exhibit in vivo respiratory control (Bashford et al., 1980).

At this point we asked why the periodicities of the oscillations are between one and two orders of magnitude longer than typical metabolic oscillations; switching between energy states in isolated mitochondria is fast, with half-times of the order of seconds. The answer to this paradox must be that the pacemaker in the integrated system of energy supply and energy demand in the growing organism does not reside in energy metabolism per se, but rather in the load reactions of biosynthesis. Cellular protein synthesis is by far the largest energetic sink, as more than 50% of the dry weight of cells is protein; by comparison, RNA synthesis accounts for only a small (and DNA replication a trivial) proportion of the total energetic demand.

Changes in levels of accumulated protein during the growth of a synchronous culture (Fig. 5) indicated not just that this major source of energy demand was discontinuous (stepped) but, more unexpectedly, that it *too* showed oscillations with maximal amplitudes of 54% (Edwards and Lloyd, 1980). Even more pronounced variations in cellular (i.e., mostly ribosomal) RNA levels were evident. Phase relationships between maxima of $O_2$ uptake rates, ADP, and ATP/ADP ratios, together with those for accumulated protein and RNA, were mapped from many experiments and were also determined during part of the cell cycle of a single synchronous culture with frequent sampling (every 5 min). Maximum levels of total protein corresponded to those points in the cell cycle where respiration rates and ADP levels are highest; at 30°C the cell cycle is therefore comprised of a sequence of seven subcycles (Fig. 6).

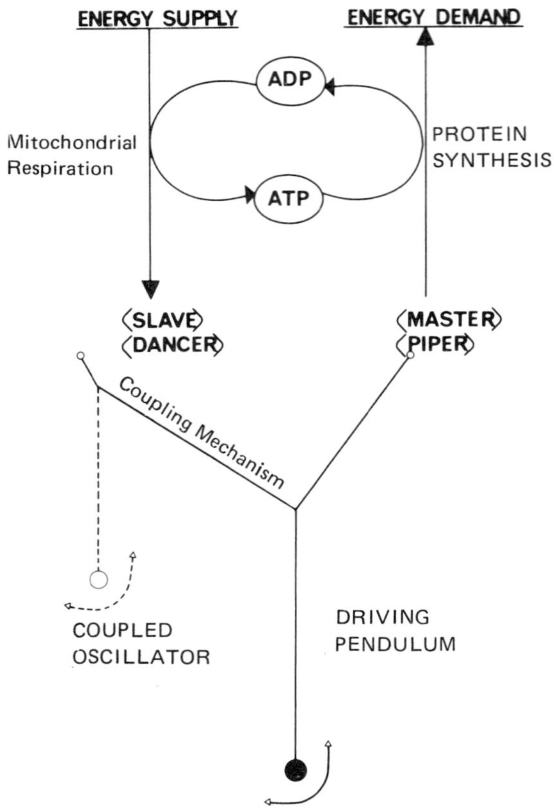


**FIGURE 7** Energy demand and its supply as coupled oscillators: the period of the master oscillator determines that of the slaved (mitochondrial) system (or the piper plays the tune).

Time resolution of the process of cellular growth thus reveals a system of coupled oscillators. The master oscillator (analogous to the more massive pendulum bob in the model, Fig. 7) that determines the overall frequency of the coupled system, is based in the epigenetic time domain. These epigenetic oscillations originate in feedback loops resulting from temporal changes in the control of gene expression (i.e., they involve transcription, translation, and perhaps modification of proteins). The energy-yielding reactions of respiration are enslaved by the master oscillator, and thus do not exhibit their own characteristically rapid dynamics; they must instead also run through hourly cycles. So "life's slow dance to the music of time" has the mitochondrial activities entrained to an epigenetic piper who plays a slow tune. The dynamics of demand predominate, and the *energetic capacity is never limiting.* Close connection between energy demand and supply is provided by an ADP pool (the string linking the two pendulums).

The oscillation in total protein indicates that a major proportion of individual protein species are extensively and rapidly turned over, even under optimal conditions of growth. Phases of synthesis alternate with phases of degradation in temporally discrete compartments. Catalase, a predominantly peroxisomal enzyme, accumulates in this

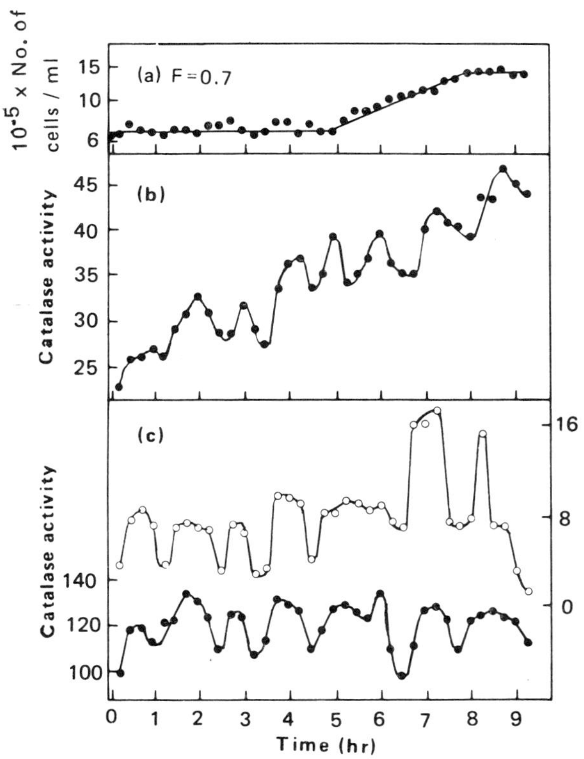


**FIGURE 8** Changes in cell numbers, catalase activity, and catalase protein in synchronous cultures of *A. castellanii*: (a) cell numbers and synchrony index, F; (b) catalase activity (milliunits per milliliter culture); (c) catalase activity (milliunits per milligram, ●), and immunologically determined catalase protein (o). (From Edwards et al., 1981.)

manner: during the 8 hr cell cycle seven maxima in activity and in immunologically assayed determinants were observed (Edwards et al., 1981) (Fig. 8). Experiments in which extracts containing different levels of enzyme activities were mixed showed that no catalase inhibitor was present. Similar oscillations were observed in the specific activities of mitochondrial ATPase and in its naturally occurring protein inhibitor (Edwards et al., 1982) (Fig. 9). Again, variations in enzyme activity reflected changes in amount of enzyme protein as assayed by an immunological technique. Large variations in $I_{50}$ values (micrograms inhibitor per milligram protein) for inhibition of ATPase activity by seven different inhibitors of energy conservation were observed. Activity was most sensitive to inhibition by oligomycin, efrapeptin, citreoviridin, and quercetin when values were highest; these changes in inhibitor sensitivity may be partly due to alterations in the amount of bound naturally occurring inhibitor protein, but this explanation cannot account for the observed parallel changes in enzyme activity and enzyme protein. Total spectrophotometrically detectable cellular content of *a, b,* and *c* cytochromes (Fig. 10), cytochrome *c* oxidase activity, and protein reacting with antibody raised to purified cytochrome *c* oxidase also oscillated with an hourly period, and a cell cycle map indicates close phase correlation with the oscillation in total cellular protein. Amplitudes (peak to trough as a

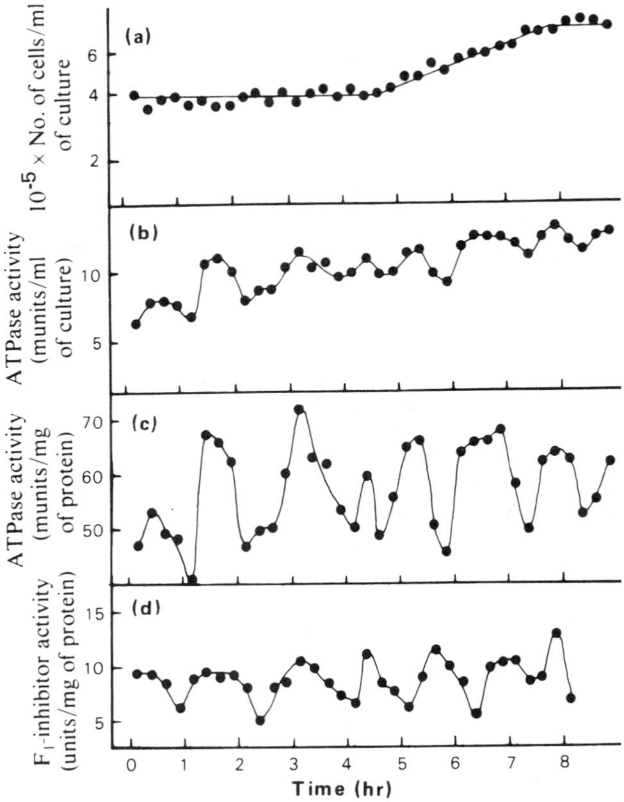


**FIGURE 9** Changes in cell numbers, ATPase activity, and $F_1$ inhibitor protein activity during synchronous growth of *A. castellanii*: (a) cell numbers; (b) ATPase activity per milliliter culture; (c) ATPase activity per milligram protein, and (d) $F_1$ inhibitor activity per milligram protein. The synchrony index was 0.68. (From Edwards et al., 1982.)

percentage of minimal values) were as follows [coefficients of variation v for data points in control experiments with asynchronous cultures (v = 100 × standard deviation/mean) are in parentheses] : catalase 21 (4.2), ATPase 37 (3.8), $F_1$-ATPase inhibitor protein 74 (7.2), cytochrome *c*-549 52 (7.6), cytochrome *b*-562 48 (7.1), cytochrome $a + a_3$ 37 (5.1).

Similar patterns of enzyme accumulation were observed in synchronous cultures established by different procedures. Both the timings of maxima and the amplitude for catalase and cytochrome c oxidase activities and for total cytochromes (as determined in low-temperature redox difference spectra) were similar, irrespective of whether continuous flow selection was employed (flow rate 0.38 liters/min, rotor speed 1000 rev $min^{-1}$, or whether differential centrifugation in growth medium employed 300 rev $min^{-1}$ (10 *g*, $r_{av}$ 10 cm) for 2 min in 50 ml tubes or 800 rev $min^{-1}$ (150 *g*, $r_{av}$ 21 cm) for 2 min in 1 liter centrifuge pots. No evidence for perturbative disturbance of enzyme accumulation was ever observed in corresponding control asynchronous cultures exposed to similar centrifugation conditions: we always observed smoothly increasing enzyme activities and amounts. In another type of control experiment, disruption of

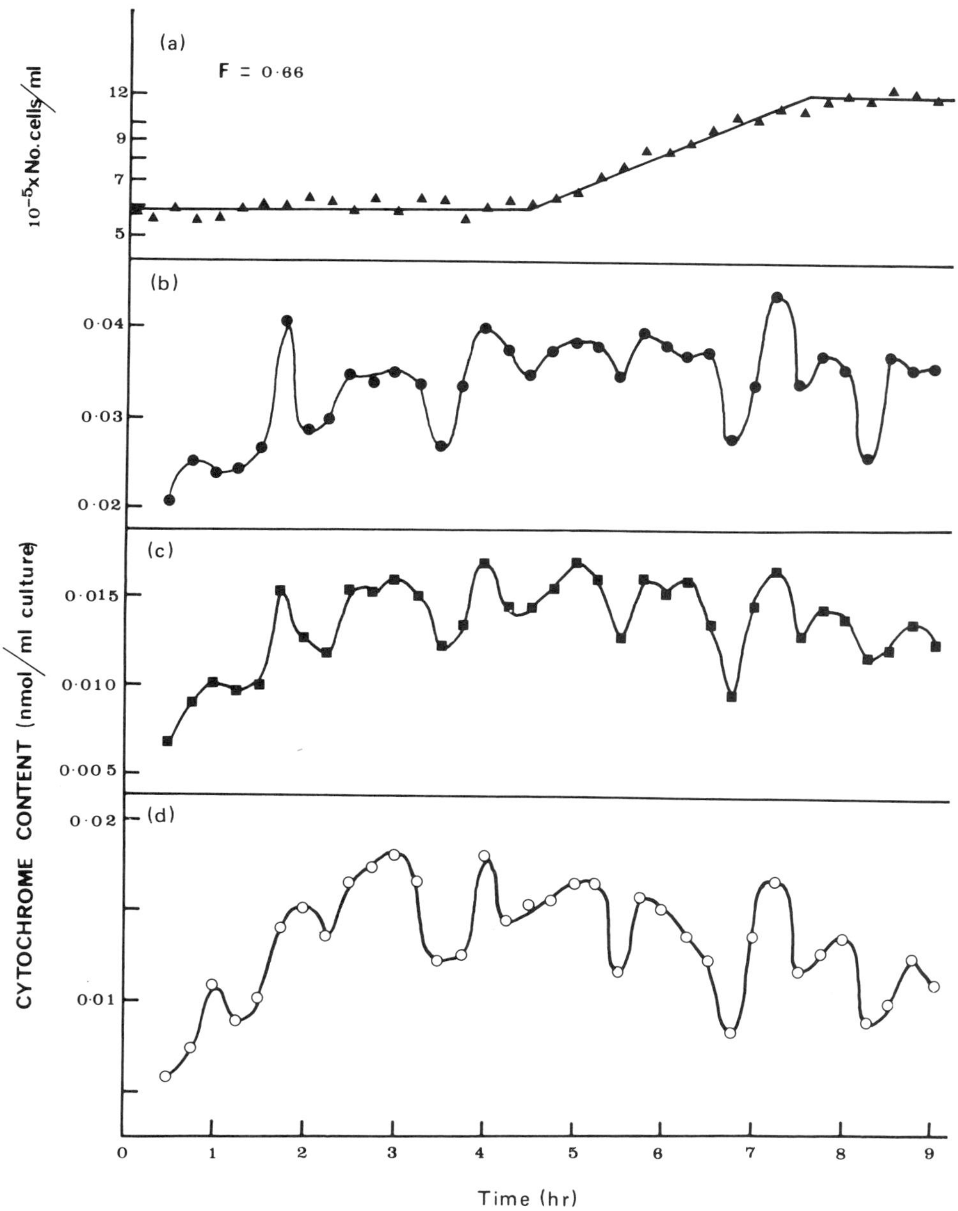


**FIGURE 10** Changes in total cytochromes in *A. castellanii* during growth of a synchronous culture: (a) cell numbers and synchrony index F; (b) cytochrome *aa*3; (c) cytochrome *b*-562; (d) cytochrome *c*-549. (From Lloyd et al., 1983.)

organisms from synchronous cultures was carried out in the presence of a mixture of protease inhibitors (2 mM phenylmethylsulfonyl fluoride, 5 mM p-aminobenzamidine HCl, and 5 mM ϵ-aminocaproic acid); oscillatory accumulation of catalase, ATPase, and cytochrome *c* oxidase was still observed, indicating that differential proteolytic degradation in different extracts between the time of cell disruption and assay was unlikely.

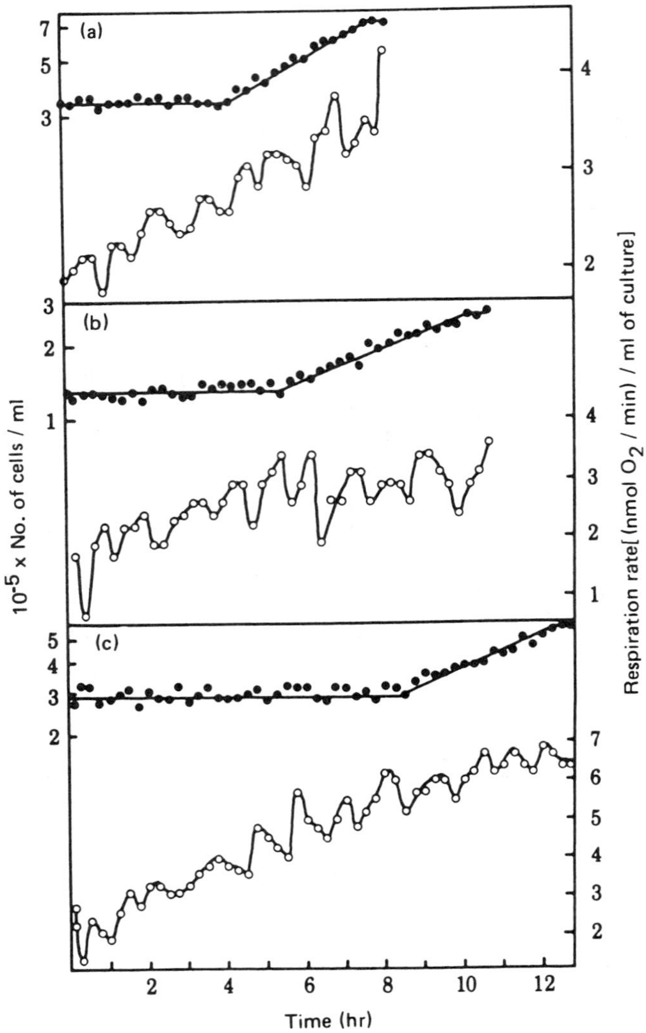


**FIGURE 11** Growth and respiration of synchronous cultures of *A. castellanii* at 30°C (a), 27°C (b) and 25°C (c): ●, growth; ○, respiration rate. (From Lloyd et al., 1982b.)

The cell cycle time for *A. castellanii* is 7.8 hr at 30°C, 10 hr at 27°C, and 12 hr at 25°C (Fig. 11); at 20°C it increased to about 16 hr (Lloyd et al., 1982b). Respiration doubled overall during the cell cycle in each culture, and a period of about 1 hr is evident at all three temperatures from visual inspection of the oscillations. After making appropriate parabolic or linear corrections for trend effects, spectral analysis showed up to about four peaks in the power spectrum. Only one frequency showed a peak at all three temperatures; this was clearly the major peak at 25 and 30°C and the second largest peak, by only 15% at 27°C. These peaks were consistent with a cycle having a period of 65 ± 3 min but were not significant ($P > 0.05$). Significant ($P < 0.05$) cycles were, however, de tected in time-series analysis of data from experiments in which sampling was more frequent during the first 2-5 hr of growth of synchronous cultures (Fig. 12); all but one data

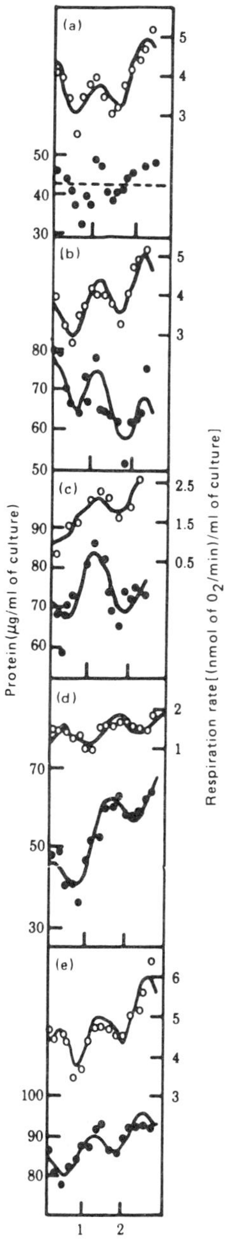


**FIGURE 12** Protein accumulation and respiration rate in synchronous cultures of *A. castellanii* at 30°C (a), 27°C (b), 25°C (c), 22.5°C (d), and 20°C (e); ●, protein accumulation; ○, respiration rate. (From Lloyd et al., 1982b.)

set showed significant cycles (Table 1). There was little difference between the periods of the major cycles (72 ± 10 min for respiration and 80 ± 15 min for cellular protein), and the mean values were not significantly different ($P > 0.1$). Phase correspondence occurred between respiratory rate and protein accumulation, and tight coupling between the processes was evident at each of the five temperatures; temperature compensation of the period was observed over the ten degrees centigrade range. In these short experiments with frequent

**TABLE 1** Periods of Significant Cycles[a] in Respiration and Cellular Protein over 2.5 hr in Synchronous Cultures of *A. castellanii*

| Period (min) at | 20°C | 22.5°C | 25°C | 27°C | 30°C |
|---|---|---|---|---|---|
| Cellular protein | 73(74)[b] | 90(93) | 86(70)[c] | 71(61) | NSC[d] |
| Respiration | 69(81) | 80(73) | 81(85) | 66(89) | 65[e]; 23(88) |

[a]Cycles were deemed significant if they removed all the significant autocorrelation from the residuals or, in data sets having < 15 values, when the Durbin-Watson statistic was >1.8.
[b]Values in parentheses are percentages of the variation in the data explained by the time series model.
[c]Data set having <15 values.
[d]NSC, no significant cycles present in this sample.
[e]Two periods are given because two cyclic components were necessary to remove all significant serial autocorrelation from the residuals.
*Source*: Lloyd et al. (1982b).

sampling, the cultures were always allowed to grow through a complete cell cycle, and in all cases doubling in total cell protein over the cell cycle time was confirmed. Asynchronous cultures treated by identical procedures and grown at different temperatures showed no visually obvious oscillations and did not show either a peak in its power spectrum close to 65 min or any other significant peaks.

Thus although the duration of the cell cycle in *A. castellanii* is highly temperature dependent ($Q_{10} \simeq 2$), the period of the oscillations remained approximately constant. The effect of lowering temperature is thus to increase the number of subcycles per cell cycle time.

## POSSIBLE INTERACTIONS BETWEEN THE DOMAINS

In the experiments described, the synchronous population is *selected* from the temporally heterogeneous, exponentially growing culture at zero time. Mapping of events in the cell cycle uses this reference point. That mapping of maxima of respiratory oscillations in many separate experiments gives consistent phase angles on the cell cycle map (Fig. 6) indicates that size selection provides organisms at the start, not only of a cell cycle but also of a subcycle. That this is so at different temperatures also suggests that there must be an integral number of subcycles in a cell cycle. The temperature-compensated epigenetic oscillations may be acting as a timer for the cell cycle. This commensurate relationship, integral winding number in the topological sense (Winfree, 1980), would predict quantized cell cycle times; a similar model has been suggested for cultured mammalian cells (Klevecz, 1976). The cell cycle map plotted in real time increases by quantal units (each equal to one subcycle time) as the temperature is decreased, until eventually circadian controls may determine the length of the cell cycle at very low growth rates (Fig. 13).

The epigenetic oscillations are not hidden by the apparent dispersion in cell cycle times. A number of possible alternative explanations must be considered: (1) the degree of synchrony is underestimated by cell counts, (2) dispersion arises late in the cell cycle,

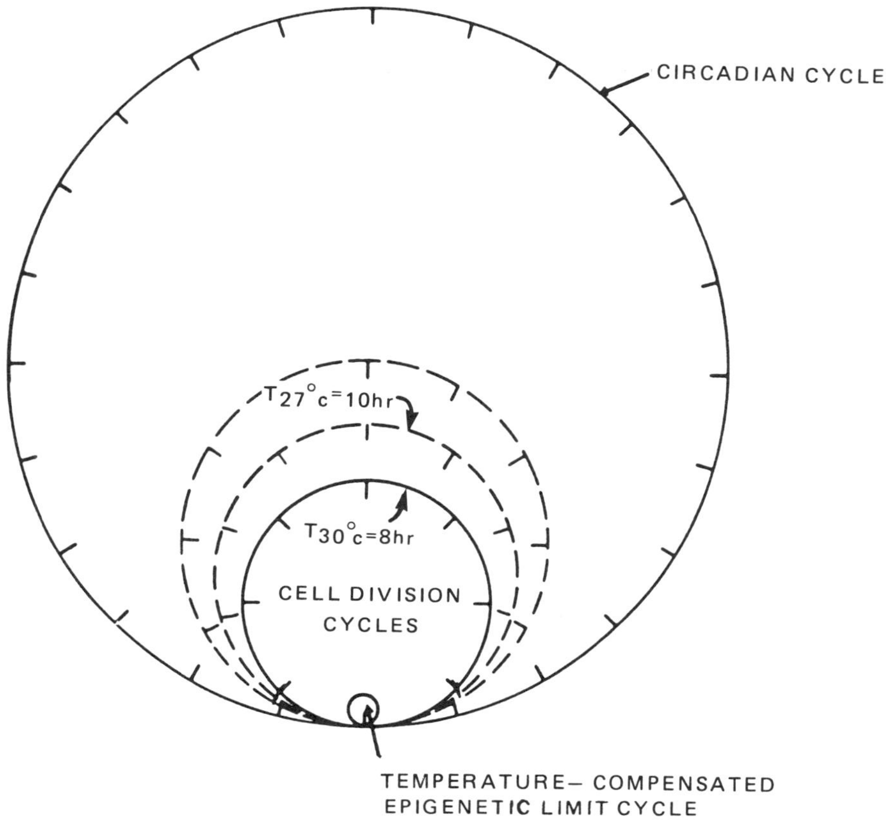


**FIGURE 13** Model for epigenetic oscillations that serve the dual role of cell cycle and circadian timekeeping. The outer circle represents the circadian cycle (period ≃ 24 hr), the small inner circle represents the temperature-compensated epigenetic limit cycle (period ≃ 1 hr), and the intermediate circles represent cell division cycles at different temperatures. (From Lloyd et al., 1982b.)

or (3) mutual entrainment of individuals of the population occurs by cell-cell interactions. Evidence for (1) comes from DNA measurements that indicate an S phase that lasts for only 1 hr in synchronous cultures at 30°C (Edwards and Lloyd, 1978, 1980). Although we have no evidence for (3), mutual interaction of individual oscillators is a well-established physical (Wiener, 1961; Yamaguchi et al., 1981) and biological phenomenon (Aldrich, 1976).

Other reports of oscillatory accumulation of cellular protein with periods in the range 0.3-4 hr have come from a diversity of systems, including *Euglena gracilis,* mammalian cells, and sea urchin embryos (Edmunds and Funch, 1969; Klevecz and Ruddle, 1968; Brodsky, 1975; Gilbert and Tsilimigras, 1981; Mano, 1970, 1975); further examples undoubtedly await more detailed examination in experimental systems designed to minimize time averaging within a cell population. Oscillatory accumulation of total cellular protein in synchronous cultures of *C. utilis* (Lloyd et al., 1981) confirms that the phenomenon is not restricted to *A. castellanii.*

## CONCLUDING REMARKS

The epigenetic oscillations revealed in synchronous cultures of *A. castellanii* differ from metabolic oscillations in the following respects:

1. They show longer relaxation times (of the order of an hour as opposed to the usual relaxation time of minutes for metabolic oscillations).
2. They show no signs of attenuation during a 9 hr experiment, whereas metabolic oscillations induced in intact organisms are usually highly damped, unless conditions are arranged so as to sustain them (Chance et al., 1964; Boiteux et al., 1975).
3. They have a period that is temperature compensated in the range 20-30°C, whereas no temperature compensation of metabolic oscillations in vivo has been reported.

It is widely believed that the relatively long period of circadian time must be derived from a higher frequency rhythm (Hastings and Schweiger, 1976), although it has also been suggested that this need not be so (Winfree, 1980). The clocklike property of the temperature-compensated period of the epigenetic oscillations makes them possible candidates in the role of time-unit measurement for circadian events. That phase correspondence between these oscillations and the cell cycle has been established also may indicate a role in cell cycle timing. This hypothesis fits the suggestion that there is just one cellular clock (Wille, 1979), and that a mitotic oscillator and the circadian oscillator are related (or identical) limit cycles operating in two different modes.

## REFERENCES

Aldrich, J. (1976). Short-range intercellular communication, biochemical oscillations, and circadian rhythms. In *Handbook of Engineering in Medicine and Biology*, D. G. Fleming and B. N. Feinberg (eds.). CRC Press, Cleveland, Ohio, pp. 57-147.

Bashford, C. L., Chance, B., Lloyd, D., and Poole, R. K. (1980). Oscillations of redox states in synchronously dividing cultures of *Acanthamoeba castellanii* and *Schizosaccharomyces pombe. Biophys. J.* 29:1-12.

Blumenthal, L. K., and Zahler, S. A. (1962). Index for measurements of synchronization of cell populations. *Science* 135:724.

Boiteux, A., Goldbetter, A., and Hess, B. (1975). Control of oscillating glycolysis of yeast by stochastic, periodic and steady source of substrates: A model and experimental study. *Proc. Natl. Acad. Sci. USA* 72:3829-3833.

Boiteux, A., Goldbeter, A., and Hess, B. (1975). Control of oscillating glycolysis of yeast *Bunsenges. Phys. Chem.* 84:392-398.

Brodsky, W. Y. (1975). Protein synthesis rhythm. *J. Theor. Biol.* 55:167-200.

Chagla, A. H., and Griffiths, A. J. (1978). Synchronous cultures of *Acanthamoeba castellanii* and their use in the study of encystation. *J. Gen. Microbiol.* 108:39-43.

Chance, B., Estabrook, R. W., and Ghosh, A. (1964). Damped sinusoidal oscillations of cytoplasmic reduced pyridine nucleotides in yeast cells. *Proc. Natl. Acad. Sci. USA* 51:1244-1251.

Chance, B., Pye, E. K., Ghosh, A. K., and Hess, B. (eds.) (1973). *Biological and Biochemical Oscillators.* Academic Press, New York.

Edmunds, L. N., Jr., and Funch, R. R. (1969). Circadian rhythm of cell division in *Euglena*: Effects of a random illumination regime. *Science* 165:500-503.

Edwards, C. (1981). *The Microbial Cell Cycle.* Nelson, Surrey, England.

Edwards, C., Statham, M., and Lloyd, D. (1975). The preparation of large-scale synchronous cultures of the trypanosomatid, *Crithidia fasciculata,* by cell-size selection:

Changes in respiration and adenylate charge through the cell-cycle. *J. Gen. Microbiol.* 88:141-152.
Edwards, S. W., Evans, J. B., and Lloyd, D. (1981). Oscillatory accumulation of catalase during the cell cycle of *Acanthamoeba castellanii. J. Gen. Microbiol.* 125:459-462.
Edwards, S. W., Evans, J. B., Williams, J. L., and Lloyd, D. (1982). The mitochondrial adenosine triphosphatase of *Acanthamoeba castellanii*: Oscillatory accumulation of enzyme activity, enzyme protein and $F_1$-inhibitor during the cell cycle. *Biochem. J.* 202:453-458.
Edwards, S. W., and Lloyd, D. (1977). Mitochondrial adenosine triphosphatase of the fission yeast, *Schizosaccharomyces pombe* 972 $h^-$. Changes in activity and oligomycin-sensitivity during the cell cycle of catabolite-repressed and -derepressed cells. *Biochem. J.* 162:39-46.
Edwards, S. W., and Lloyd, D. (1978). Oscillations of respiration and adenine nucleotides in synchronous cultures of *Acanthamoeba castellanii*: Mitochondrial respiratory control in vivo. *J. Gen. Microbiol.* 108:197-204.
Edwards, S. W., and Lloyd, D. (1980). Oscillations in protein and RNA content during synchronous growth of *Acanthamoeba castellanii*: Evidence for periodic turnover of macromolecules during the cell cycle. *FEBS Lett.* 109:21-26.
Gilbert, D. A., and Tsilimigras, C. W. A. (1981). Cellular oscillations: Relative independence of enzyme activity rhythms and periodic variations in the amounts of extractable protein. *S. Afr. J. Sci.* 77:66-73.
Goodwin, B. C. (1976). *Analytical Physiology of Cells and Developing Organisms.* Academic Press, London.
Hastings, J. W., and Schweiger, H.-G. (1976). *The Molecular Basis of Circadian Rhythms.* Dahlem Konferenzen, Berlin.
Hayward, G., Carlsen, W., Siegman, A., and Stryer, L. (1981). Retinal chromophore of rhodopsin photoisomerized within picoseconds. *Science* 211:942-944.
Higgins, J. (1967). The theory of oscillating reactions. *Ind. Engin. Chem.* 59:19-62.
Kader, J., and Lloyd, D. (1979). Respiratory oscillations and heat evolution in synchronous cultures of *Candida utilis* NCYC 193. *J. Gen. Microbiol.* 114:455-461.
Klevecz, R. R. (1976). Quantized generation time in mammalian cells as an expression of the cellular clock. *Proc. Natl. Acad. Sci. USA* 73:4012-4016.
Klevecz, R. R., and Ruddle, F. H. (1968). Cyclic changes in enzyme activity in synchronized mammalian cell cultures. *Science* 159:633-634.
Landsberg, P. T. (1981). All time is unredeemable. *Times Higher Educational Supplement 17 April,* pp. 12-13.
Lloyd, D. (1974). *The Mitochondria of Micro-organisms.* Academic Press, London.
Lloyd, D., and Ball, J. (1979). Perturbation of respiration in *Candida utilis*: Induction of metabolic oscillations. *J. Gen. Microbiol.* 114:463-466.
Lloyd, D., and Edwards, S. W. (1977). Mitochondrial adenosine triphosphatase of the fission yeast *Schizosaccharomyces pombe* 972 $h^-$. Changes in inhibitor sensitivities during the cell cycle indicate similarities and differences in binding sites. *Biochem. J.* 162:581-590.
Lloyd, D., Edwards, S. W., and Fry, J. C. (1982b). Temperature-compensated oscillations in respiration and cellular protein content in synchronous cultures of *Acanthamoeba castellanii. Proc. Natl. Acad. Sci. USA* 79:3785-3788.
Lloyd, D., Edwards, S. W., and Williams, J. L. (1981). Oscillatory accumulation of total cell protein in synchronous cultures of *Candida utilis. FEMS Microbiol. Lett.* 12: 295-298.
Lloyd, D., Edwards, S. W., Williams, J. L., and Evans, J. B. (1983). Mitochondrial cytochromes of *Acanthamoeba castellanii*: oscillatory accumulation of haemoproteins,

immunological determinants and activity during the cell cycle. *FEMS Microbiol. Lett.* 16:307-312.

Lloyd, D., John, L., Edwards, C., and Chagla, A. H. (1975). Synchronous cultures of micro-organisms: Large-scale preparation by continuous-flow size selection. *J. Gen. Microbiol.* 88:153-158.

Lloyd, D., Phillips, C. A., and Statham, M. (1978). Oscillations of respiration, adenine nucleotide levels and heat evolution in synchronous cultures of *Tetrahymena pyriformis* ST prepared by continuous-flow selection. *J. Gen. Microbiol.* 106:19-26.

Lloyd, D., Poole, R. K., and Edwards, S. W. (1982a). *The Cell Division Cycle. Temporal Organization and Control of Cellular Growth and Reproduction.* Academic Press, London.

Mano, Y. (1970). Cytoplasmic regulation and cyclic variation in protein synthesis in the early cleavage stage of the sea urchin embryo. *Dev. Biol. 22*:433-460.

Mano, Y. (1975). Systems constituting the metabolic sequence in the cell cycle. *Biosystems* 7:51-65.

Phillips, C. A., and Lloyd, D. (1978). Continuous-flow size selection of *Tetrahymena pyriformis* ST: Changes in volume, DNA, RNA and protein during synchronous growth. *J. Gen. Microbiol.* 105:95-103.

Poole, R. K., and Lloyd, D. (1973). Oscillations of enzyme activities during the cell cycle of a glucose-repressed fission-yeast *Schizosaccharomyces pombe* 972 h⁻. *Biochem. J.* 136:195-207.

Poole, R. K., and Lloyd, D. (1974). Changes in respiratory activities during the cell cycle of the fission yeast *Schizosaccharomyces pombe* 972 h⁻ growing in the presence of glycerol. *Biochem. J.* 144:141-148.

Poole, R. K., Lloyd, D., and Chance, B. (1974). The development of cytochromes during the cell cycle of a glucose-repressed fission yeast *Schizosaccharomyces pombe* 972 h⁻. *Biochem. J.* 138:201-210.

Poole, R. K., Lloyd, D., and Kemp, R. B. (1973). Respiratory oscillations and heat evolution in synchronously dividing cultures of the fission yeast *Schizosaccharomyces pombe* 972 h⁻. *J. Gen. Microbiol.* 77:209-220.

Rapp, P. E. (1979). An atlas of cellular oscillations. *J. Exp. Biol.* 81:281-306.

Reich, J. G., and Sel'kov, E. E. (1982). *Energy Metabolism of the Cell.* Academic Press, London.

Sel'kov, E. E. (1968). Self-oscillations in glycolysis. 1. A simple kinetic model. *Eur. J. Biochem.* 4:79-86.

Sel'kov, E. E. (1972). Non-linearity of multienzyme systems. In *Analysis and Simulation of Biochemical Systems,* H. C. Hemker and B. Hess (eds.). North Holland, Amsterdam, pp. 145-161.

Sel'kov, E. E. (1980). Instability and self-oscillations in the cell energy metabolism. *Ber. Bunsenges. Gesell. Phys. Chem.* 84:399-402.

von Klitzing, L., and Betz, A. (1970). Metabolic control in flow systems. I. Sustained glycolytic oscillations in yeast suspension under continual substrate infusion. *Arch. Microbiol.* 71:220-225.

Wiener, N. (1961). *Cybernetics.* MIT Press, Cambridge, Massachusetts.

Wille, J. J., Jr. (1979). Biological rhythms in protozoa. In *Biochemistry and Physiology of Protozoa,* 2nd ed., vol. 2, M. Levandowsky and S. H. Hutner (eds.). Academic Press, New York, pp. 67-146.

Winfree, A. T. (1980). *The Geometry of Biological Time.* Springer-Verlag, Berlin.

Yamaguchi, Y., Komentani, K., and Shimizu, H. (1981). Self-synchronization of nonlinear oscillations in the presence of fluctuations. *J. Statistical Physics* 26:719-743.

# 3
# Cellular Oscillators as Vestiges of a Primitive Circadian Clock

**Robert R. Klevecz** / The Beckman Research Institute of the City of Hope National Medical Center, Duarte, California

## INTRODUCTION

The macroscopic properties of circadian rhythms have been characterized by Pittendrigh, Bünning, Aschoff, and their many students in the course of the past 30 years. While it is generally agreed that underlying the overt circadian rhythms there exists an endogenous, cellularly based oscillator or ensemble of oscillators with stable trajectories that specify the phase and period of the clock, identifying these oscillators with any particular element of cellular or organismic chemistry has proven elusive (Robertson, 1975). Even in the case where arrhythmic mutants of the *Drosophila melanogaster* clock have been isolated (Konopka and Benzer, 1971), it is not clear whether these are mutants of "the" clock or merely the result of the uncoupling of the endogenous oscillator, or the disintegration of a set of oscillators, from one of the systems it drives. In other words, the cellular clock may be intact but not expressed in an organized way in the whole animal. In other cases clock mutants are exclusively period ($\tau$) mutants and the periods of these mutants appear to cluster around values that vary in quantal increments from the wild type. This is particularly true of *D. melanogaster* (Konopka and Benzer, 1971), excluding the arrhythmic mutant, and *Chlamydomonas reinhardi* (Bruce, 1972) but less perfectly so of *Neurospora crassa* where periods cluster at both 2.5 and 5 hr intervals (Feldman and Hoyle, 1976; Feldman et al., 1978). It is not yet universally agreed whether the stable period and phase relationships of circadian rhythms arise by differential strengths of coupling between endogenous rhythms that are themselves of circadian duration (Pittendrigh, 1976) or whether inhibitory coupling of higher frequency oscillations is involved (Pavlidis, 1973).

In order to proceed it is necessary for the reader to know what is meant by a "clock." A biological clock engenders time sense; it does not merely record time's passage. In preparing this study it was interesting to note that in the paleontological literature manifestations of rhythmicities were sometimes discussed in ways that are not meaningful in the context of modern biological rhythm literature (Rosenberg, 1974; Rosenberg and Runcorn, 1974). Although biological rhythms may have arisen as passive responses to the environment, the daily rhythms in mineral metabolism of corals, bivalves, and stromatolites so useful to paleontologists are not now passively driven, and it seems most unlikely that these rhythms were other than endogenous as late as the Devonian period.

According to Pittendrigh's generalizations (1960), circadian rhythms are ubiquitous, endogenous, self-sustaining, innate, and occur at all levels of organization (integral); they are species specific (heritable), precise, temperature compensated, phase responsive, and entrainable. In addition, they show afteraffects (inertial), light intensity dependence (obey Aschoff's rule), phase lability but period stability (robust), and the one generalization which is clearly no longer correct, intractability to chemical perturbation. Historically, circadian rhythms are defined as those rhythms whose $\tau_f$ (free running period) is an approximation to the period of the earth's rotation (Brady, 1979). Indeed, the evolutionary argument runs vigorously through the whole of the circadian rhythm literature. There has never been a question about the function of circadian and other biological rhythms (Pittendrigh, 1981):

> In all cases—daily, tidal, lunar and annual—the temporal program is based on undamped biological oscillators whose period (frequency) is an evolved approximation ("circa-") to the environmental cycle it copes with and exploits.

Circadian rhythms appear not only heritable but, judging from the difficulty in obtaining other than quantal variants of $\tau$, multi- or totigenic expressions of cellular metabolism.

Given the stability of circadian rhythms and the difference between day length now and that of the earth in Precambrian times, a question occurs as to the ability of the biological clock to have adapted to changes in the rotation rate of the earth through the history of evolution. If the temporal environment of the earth was constant, or nearly so, through all of chemical and biological evolution, then the emergence of biological time sense could have occurred at any of a number of points and perhaps independently in several systems. Alternatively, if the fundamental periodicities of the physical world themselves changed through the course of evolution, then we are forced to consider that circadian rhythms have not always been approximations to 24 hr, and beyond that to consider more specifically what properties of oscillatory reaction schemes would have led to their successful adaptation to an environment that may have changed relatively rapidly on the geological time scale. In the popular view, rhythms expressed now as endogenous and heritable properties of each species may have begun as passive rhythms driven initially by the major environmental periodicities, until through time they became fixed in the chemistry of the organism.

The earth's rotational period and the accompanying variations in tides, temperature, and insolation are among the most universal environmental periodicities to which living systems must adapt. Most theories of planetary evolution place the earth and moon in close proximity and give the earth a much shortened day length at some distant time. By extrapolation using present-day values for tidal friction, the earth and moon can be reckoned to have been in close proximity at some time less than three eons ago. Under those circumstances the solar day would have been less than 10 hr, possibly as short as 4-5 hr, and punctuated by boiling and condensing tides of great amplitude. The geological record of such a violent event in the earth's history should be clear and serves as a boundary condition for the time of occurrence. The rapid rotation of the primitive earth and the subsequent slowing of rotation rate due to tidal friction in the earth-moon system present a problem in the evolution of an oscillator generating the biological clock. If a short period clock became fixed in the chemistry of early life forms prior to significant retardation of the rotation rate of the earth, then any subsequent rapid slowing of the earth's rotation would have made continuous adaptation of the clock difficult. If the clock evolved via saltatory increases in period, then this primitive oscillator must underlie the modern circadian clock and vestiges of the primitive clock period may be expressed in ultradian rhythms and $\tau$ mutants of the modern clock. Alternatively, if slowing in the earth's rotation commenced soon after accretion of the planets some 4-5 billion years before the present (BYBP), then somewhat less serious problems in tuning of the oscillator period to the variation in day length from 16-18 hr in the Precambrian to the present 24 hr period need to be explained. In either case the limits of entrainment as they are presently understood would have been exceeded.

In the section that follows, considerable attention has been given to geophysical considerations regarding the state of the earth's environment and rotational rate during the important moments in biological evolution. This background, although it is somewhat tangential to the major point, seemed essential in order to establish a sound basis for argumentation. Of the two views that come out of this effort, one requires a profound reconsideration of the nature and origin of the modern clock. The critical unanswered questions in this exploration have to do with the time scale of change, the value for tidal friction, and the heritability and mutability of clock period.

## PRIMITIVE ENVIRONMENT

Evidence for the appearance of photoautotrophs just after 3.4 BYBP (Oehler et al., 1972) fixes the latest time during which chemical evolution might have occurred and suggests that the rotational rate of the earth during the era immediately preceding and following this point requires the closest security. Three independent lines of evidence can be used to estimate the rotational period of the earth in the Precambrian and Cambrian eras. They include (1) empirical generalizations regarding the evolution of rotational period in planetary and stellar bodies, (2) geophysical calculations extrapolated from present-day rates of earth-moon recession using best estimates of tidal friction, and (3) paleontological studies of annual and daily growth rings in corals, bivalves, and stromatolites.

### Paleorotation of the Earth

#### *Mass-Independent Constancy of Period*

As a first step to understanding the geophysical environment in the Archeozoic period it would be useful to fix initial angular velocity of the earth soon after its accretion. We are aided in this estimate by the fact that the present rate of rotation of six planets in this solar system are similar in spite of differences in orbital characteristics and by the very remarkable fact that angular momentum density is a simple function of mass ($A = kM^{2/3}$) for bodies of mass varying by 14 orders of magnitude (MacDonald, 1964; Alfven, 1964;

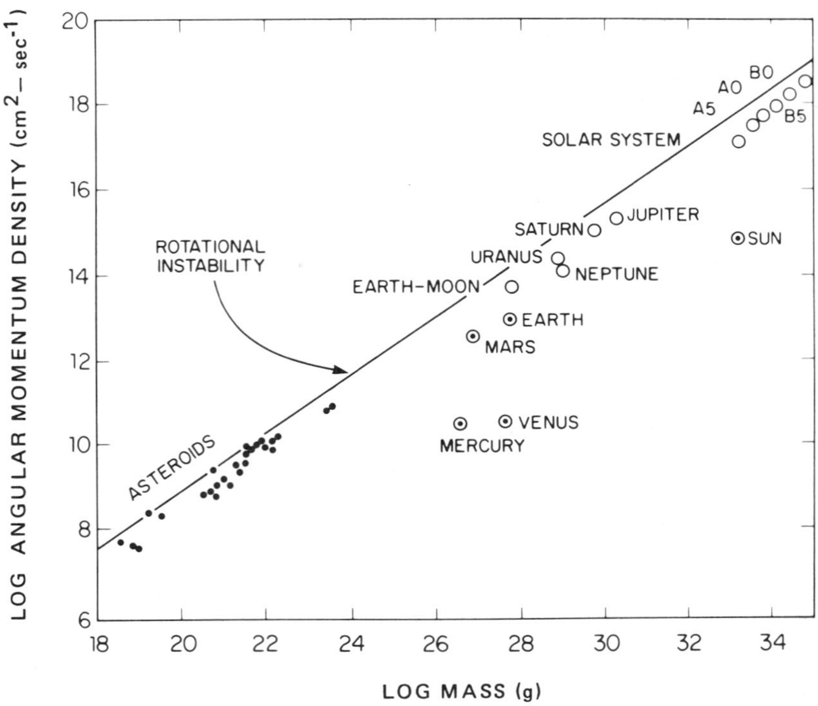


**FIGURE 1** Angular momentum density versus mass for rapidly rotating stars, the solar system, the sun, planets, and asteroids: o and • follow a line (not shown) given by $A = k'M^{2/3}$, where $k' = 1.6 \times 10^{-5}$, and one displaced downward from the line representing rotational instability $A = k''M^{2/3}$, when $k'' = 3k'$. (From Ringwood, 1979.)

Fish, 1967; Ringwood, 1972). For bodies as massive as Type O and B stars or as small as asteroids, rotational period is limited by rotational instability and independent of mass. In other words, in the absence of gravitational interaction, heavenly bodies form and rotate at a mass-independent constant period of between 2.5 and 4.5 hr (Fig. 1).

The relationship suggests the operation of some fundamental physical process. Alfven (1964), Fish (1967), and Hartmann and Larson (1967) pointed out that rotational instability led to a mass-independent constant period, and Ringwood (1979) further suggested that all the planets of our solar system were formed near their limiting rotational stability. Few planetary bodies today display such rapid rotation, and the simple relationship between angular momentum density and mass is displaced downward from the line of rotational instability. Most bodies in this solar system as well as nearby stars presently have periods of 8-10 hr. For all planets in this solar system but the earth, the orbital angular momentum of the satellites is a small fraction of the rotational momentum of the planet. The large deviation from the simple $A = kM^{2/3}$ relationship observed for Mercury and Venus can be explained by solar tidal friction (Goldreich and Soter, 1966), while the smaller uniform deviation of all bodies between mass $10^8$ *g* and $10^{34}$ *g* may be due to hydromagnetic coupling to ejected material and the transfer of angular momentum to the light gases as they were lost (Ringwood, 1979).

*Tidal Friction and Tidal Torque*

The rapid rotation of the protoearth as it grew by accretion, the formation of the moon as a separate body in close earth orbit, and the subsequent recession of the earth and moon due to tidal friction are processes predicted by most post-Apollo Mission theories of planetary formation (Ringwood, 1979; Taylor, 1979). The greatest uncertainty remains in obtaining values for the initial earth-moon distance once accretion was substantially completed and in choosing the time of appearance of a significant earth ocean and thence a set of values for the coefficient of tidal friction (Sundermann and Brosche, 1977).

From the time of origin of the primitive ocean, tides raised by the moon have played a major role in the evolution of the earth-moon system. The flow of tides across the ocean floor in response to the gravitational pull of the moon creates a turbulent boundary layer in which the frictional loss in mechanical energy of the waters' movement is dissipated as heat. The presence of friction causes tidal distortion to lag with respect to that achieved in a perfectly elastic planet. Consequently, if the rotational period of the planet is shorter than the satellite's orbital period (as is the case in the earth-moon system), the delay will result in displacement of the tidal maximum in the direction of the earth's rotation. This asymmetry in position of the tidal bulge with respect to the line of centers of the two bodies results in a net torque on the satellite which, because it leads the satellite in phase, acts, due to the conservation of angular momentum, to increase the satellite's orbital angular momentum at the expense of that in the rotating planet. As a result of tides raised by the moon on the earth, angular momentum is transferred from the earth's rotation into the moon's orbital motion with the effect that the moon recedes from the earth and the earth's rotational rate is slowed (MacDonald, 1965; Goldreich, 1966). This process continues today as it did at the beginning, though perhaps at differing rates. Direct extrapolation using appropriate values for tidal friction place the earth and moon in close proximity some 2-3 BYBP and yields a rotational period for the earth of $\simeq 5$ hr.

*Paleontological Evidence*

Precambrian stromatolites and Devonian and Ordovician corals and bivalves have recorded the recession of the moon and the slowing of the earth's rotation in daily periodicities in mineral deposits and growth patterns. Wells (1963) was the first to realize that the history of the earth's rotational period and the evolution of the earth-moon system could be charted by counting the number of daily increments per annual series added to skeletons of living and fossil corals. He found that whereas living corals add 365 daily increments per year, Devonian fossils added 400, yielding a day length at that time of 21.9 hr (Fig. 2). Subsequently, Pannella (1972) extended this record back by examining the daily increments in Bulawayan algal stromatolites dated at $2.8 \pm 0.25$ BYBP (Fig. 3) whereas Mohr (1975) found that day length 3 BYBP could have been as short as 8-9 hr. Growth ring analysis also suggests a variable rate of slowing earth rotation that correlates with the extent of shallow seas (Pannella et al., 1968). Turcotte et al. (1977) and Clark et al. (1975) have used the present value for tidal friction to show that the predicted curve for earth-moon separation with time is in close agreement with the value of 8-9 hr for day length estimated using the Bulawayan stomatolite growth ring data (Pannella, 1972) (Fig. 3). Interestingly, they point out that the modern stromatolitic forms are flagellated photosynthetic bacteria or algal colonies found only in hot mineral pools (Walter et al.,

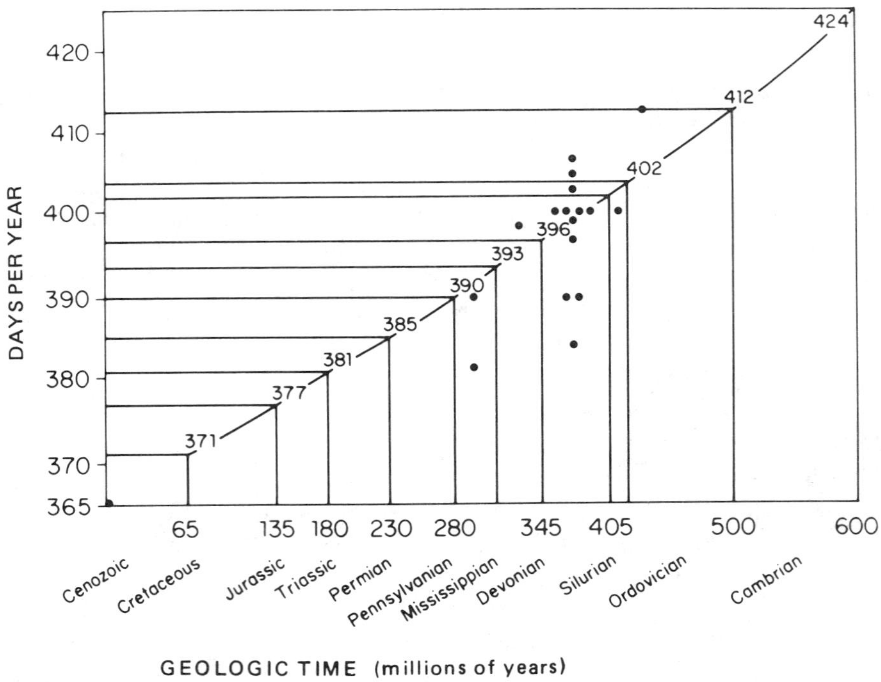


**FIGURE 2** Relationship between days per year and geological time as determined from coral specimens. (Modified from Wells, 1963.)

1972), an environment which they consider to mimic that of the earth at the time of close lunar proximity. It should be noted, though, that the agreement is critically dependent on the periodicity in Bulawayan stromatolites dating from 2.85 ± 0.25 BYBP, and that recently paleontologists have become more conservative in the assessment of growth ring data (Pannella, 1976). Taken at face value it puts the earth and moon in close proximity just 3 BYBP and yields a day length of 8-9 hr.

#### Lunar Origins, Shallow Seas, and the Time Scale Problem

Earlier hypotheses for the origin of the moon and the evolution of the earth-moon system include fission from the earth (Darwin, 1898; O'Keefe, 1969), capture (Gerstenkorn, 1955, 1967; Goldreich, 1966), or accretional capture (Opik, 1972; Mitler, 1975; Clark et al., 1975). Until recently the recession of the earth and moon was the dominant geophysical reality and theories of lunar origin were designed largely to reconcile the fact that back-extrapolation from their present positions put the earth and moon in close proximity 1-3 BYBP (MacDonald, 1964). Such calculations were also in agreement with some interpretations of growth ring data (Mohr, 1975; Pannella, 1976; Turcotte et al., 1977), though the close approach at any time more recently than 3.3 BYBP finds little support in the geological record (Kaula, 1971).

Evidence gained from the Apollo Missions has placed certain constraints on theories of lunar origin (Taylor, 1979). Primary among these constraints are the compositional similarities and the lack of evidence for mare volcanism and basaltic flooding after 3.3 BYBP as would be predicted by theories involving lunar capture (Taylor, 1979; Ringwood,

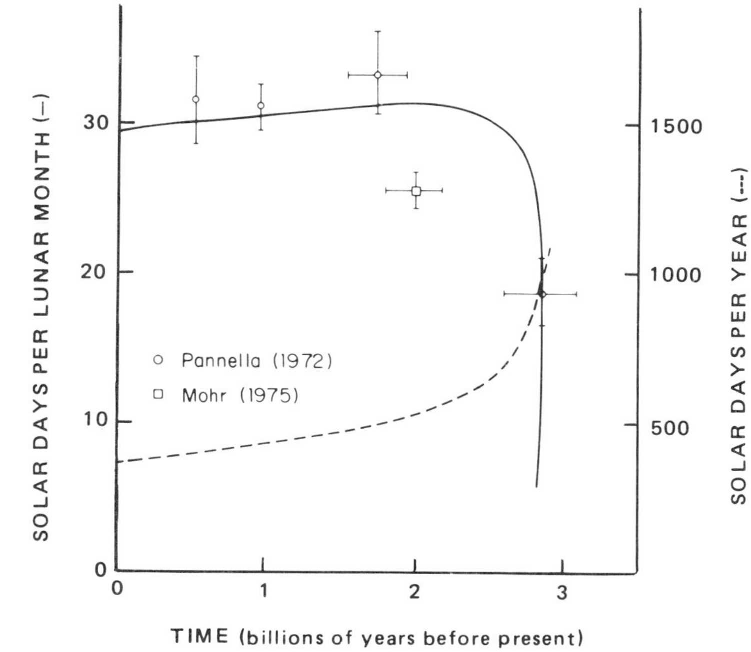


**FIGURE 3** Data points shown are for the estimated number of solar days per lunar month as determined from depositional periodicities in stromatolites by Pannella (1972) and Mohr (1975). The solid and dotted lines are the calculated change in solar days per lunar month and per year, caused by tidal dissipation of kinetic and potential energy and the changing distribution of angular momentum in the earth-moon system as calculated by Turcotte et al. (1977), respectively.

1978). The currently favored hypothesis proposes homogeneous accretion of the moon in close earth orbit soon after 4.4 BYBP followed by extensive melting. In the accretion hypotheses the moon is thought to have begun formation from smaller differentiated debris and satellites not later than $10^8$ years after the earth's formation. Once begun the process was largely completed in a time span of $10^2$-$10^3$ years, with subsequent additions of material within $10^7$ years. If earth-moon recession began soon after formation with a value for tidal friction similar to the present value, then they would not occupy their present position. We can only speculate that rapid separation of the earth and moon probably would not have begun until the earth's oceans appeared and were of significant extent.

Continuous slow recession of the moon from the earth is in contradiction with geophysical calculations of tidal torque. Deceleration due to tidal friction, though dependent upon the initial conditions, and particularly on place and time of lunar origin, places the earth and moon much closer together at some point in time than they are at present and more problematically places them in proximity relatively late in the history of the earth. For example, although most recent theories of lunar origin make it a contemporary of the earth, or 4.5 billion years old, extrapolation of current rates of tidal evolution places the moon close to the earth as recently as 1.5-3 BYBP. In consequence there is an unexplained period of 1.5-3 billion years between formation of the moon and calculated separation of the earth-moon system. Under those conditions tidal friction should have produced marked heating and volcanism, but little geological evidence exists in support of such a recent occurrence.

To account for the disparity between the compositional evidence suggesting that the moon formed as a contemporary of the earth and that tidal evolution places the two bodies in proximity rather recently, Opik (1972) suggested that the moon existed transiently as a circumterrestrial ring similar to but more massive than the rings of Saturn and coagulated first into a set of small moons and then through several fragmentations coagulated into its present form. He suggests that the original orbit of the moon was at 5 earth radii and that in this synchronous or nearly synchronous orbit tidal friction was minimal. A simpler hypothesis to account for the time scale problem holds that the values for tidal friction were considerably less in the past than they are at present due to changes in the extent and location of shallow seas. The effect of this would be the gradual recession of the moon at some time after its formation, followed, upon the appearance of significant oceans, by a more rapid separation and slowing of the earth's rotation rate at a later time in the history of the two bodies (3.8-2.8 BYBP). Arguments regarding the evolution of the clock are based on this later hypothesis.

## PRIMITIVE CLOCKS AND OSCILLATORS

Having created a problem where there was none, I feel obliged with some trepidation, to suggest how clock evolution might have proceeded given the consequences of tidal evolution for the earth's rotation. At first glance it would seem that geophysical calculations and the paleontological record provide more substantial clues to the evolutionary pathway of biological oscillators than do the oscillators themselves. Still, it may be premature to conclude that no vital evidence of their early history exists today. If a primitive short-period oscillator with the capacity to entrain to sunlight or temperature emerged early, and if the ensemble of reactions specifying phase and period were fixed and integrated into the chemistry of the organism, as "frozen accidents" are fixed in primordial DNA sequences (Ohno, 1973), it is just conceivable that some vestige of this primitive clock might still be expressed as an ultradian rhythm in unicellular eukaryotes (Chisholm and Costello, 1980), simple metazoans (Jauker and Cleffman, 1970), or mammalian cells (Klevecz, 1976). One might also consider that any process causing the disintegration of circadian rhythms, such as occurs in mutants of *Drosophila, Chlamydomonas,* and *Neurospora* (Feldman and Hoyle, 1976; Feldman et al., 1978; Bruce, 1972; Konopka and Benzer, 1971), or perhaps in suprachiasmatic nucleus (SCN)-lesioned mammals (Rusak, 1977) might lead to expression of a vestigial period.

### Vestiges of the Primitive Clock

Biological rhythmicities that qualify as clocks by virtue of being temperature compensated have been detected over a range of periodicities from a few seconds, as in the courtship of the male *Drosophila* (Kyriacou and Hall, 1980), through the cell cycle (Klevecz, 1976; Klevecz et al., 1982) to period mutants (Bruce, 1972) and natural quantization of circadian rhythms (Chisholm and Costello, 1980). Hence the argument that a particular subset of these rhythms represents the vestiges of the primoridal oscillator is inclined to be specious. Nevertheless, they are intuitively the place to start such a search.

In *Drosophila,* Pittendrigh (1967) found by continuous selection for early eclosion applied for 50 generations that emergence of a mutant "early" strain of flies occurred at a phase 4 hr earlier than the wild type; however $\tau$ was unchanged in these animals. As mentioned previously, by picking phototaxic variants of *Chlamydomonas* exposed to nitrosoguanidine, Bruce (1972) was able to isolate mutants with periods of 16, 20, 23.5, and 28 hr (wild type = 24 hr). Similarly, Konopka and Benzer (1971) isolated long (28 hr) and short (19 hr) period mutants of *D. melanogaster.* In the most complete study to date, Feldman and coworkers (Feldman and Hoyle, 1973, 1976; Feldman et al., 1978) have isolated and mapped a number of clock period mutants in *Neurospora* which appear to cluster both temporally—at periods differing from wild type by 2.5 and 5 hr, with $\tau$ values of 16.5, 19, 21.5, 24, and 29 hr—and genetically, at two loci.

When the suprachiasmatic nucleus in hamsters is ablated and the integrating capacity of the animal is destroyed, no 24 hr rhythm endures. Rather, higher frequency oscillations with 8 and 12 hr periods are expressed in running wheel and drinking activity (Rusak, 1977). Apropos of Pavlidis (1973), analysis of the behavior of populations of oscillators and diffusive coupling (Shymko, personal communication), Jacklet and Geronimo (1971) observed that as the size of the *Aplysia* retinal tissue was surgically diminished, so was periodic firing frequency. With only a slight reduction in the tissue size, free running period was reduced from 28 to 24 hr and remained unchanged until 80% of the tissue was removed, at which time period declined first to 12 hr, then to 6 hr, and then to 3 hr as additional tissue was removed. In mammalian cells the distribution of cell cycle times of a number of cell types and the interval between modal generation times of differing cell types, have been observed to show quantization at nominal 4 hr intervals (Klevecz, 1976). When the cells are intentionally perturbed their pattern of phase reset has a periodicity that again approximates 4 hr (Klevecz et al., 1980). Cell cycle time appears to be controlled by a timekeeping oscillator of 4 hr period which is bounded in some way by cell size or tissue extent (Shymko and Klevecz, 1981). Similarly, in the circadian growth mode, the cell division rhythm of the autotrophic phytoplankton *Thalassiosira fluviatilis* displays a naturally occurring quantization of generation time with intermodal period of 4-6 hr (Chisholm and Costello, 1980).

## Oscillator Dynamics

In attempts to deal with the molecular biology of cellular oscillators it may be equally productive to approach the problem via analysis and computer simulation of the response of individual oscillators and populations of oscillators to changing coupling strength, or to changing values for diffusion or tissue extent (oscillator domain), or via how period might be specified stably in an oscillation involving a large number of variables. The behavior of populations of interacting oscillators has been considered analytically by a number of workers (Goodwin, 1963; Winfree, 1967; Pavlidis, 1969). Stimulated by the observations of Chance et al.(1964) on high-frequency oscillations in glycolytic intermediates from yeast extracts, early analyses were directed toward producing frequency reduction in the observable product oscillations through frequency beats (Higgins, 1967). That such a mechanism for generating circadian rhythms is fraught with difficulties has been pointed out on a number of occasions (Winfree, 1967). If frequencies of the individual oscillator are high and coupling is strong (to produce rhythms of circadian frequency), then mutual entrainment at a high frequency occurs. On the other hand, if they are coupled weakly to avoid mutual entrainment, phase is determined by initial conditions and will be very subject to perturbation.

Goodwin (1963) calculated that a macromolecular oscillator whose constituents had half-lives on the order of a few minutes to an hour or two and which were controlled by negative feedback would have a period length of several hours, but the notion that sustained oscillations could occur in this controlled oscillator was refuted by Morales and McKay (1967), who claimed that unless significant delays, due perhaps to diffusion, were included in the scheme such an oscillator would not be self-sustaining. Most recent analyses of this problem have focused on nonlinear oscillations of the general form described by Prigogine and his coworkers (Prigogine and Lefever, 1968; Nicolis and Prigogine, 1977). The properties that make them appropriate as biological timekeepers have been discussed elsewhere (Winfree, 1970; Pavlidis, 1973; Klevecz et al., 1982) (see also Chap. 16). Pavlidis (1969) has considered in some detail the consequences of weak or strong inhibitory coupling in populations of oscillators, though analytically his study dealt only with linear harmonic oscillators. He found that when coupling is strong individual oscillators synchronize to a new lower frequency. For weaker coupling some of the high-frequency oscillations are expressed along with the lower frequency output.

## Hypercycles: The Origin of Epigenetic Information

There is every reason to question whether the reaction schemes of primitive prebiotic systems had the type of feedback that would lead to expression of a stable period. Certainly, though, the extremes in environment that existed then would have favored any system with the capacity for temporal coordination. Surprisingly, some work that may bear on the origins and molecular basis of primitive oscillators was begun a number of years ago by Eigen (Eigen, 1971; Eigen and Schuster, 1977). The motivation for exploring systems that displayed "self-organization" was to devise a means of resolving the cause-effect or, more accurately, the informational relationship between proteins and nucleic acids and hence to discern the origins of genetic information. The system that Eigen (1971) devised is remarkable in a number of respects, most notably a certain Lamarckian element. Though it is clearly beyond the scope of this short chapter to adequately describe hypercycle behavior, the scheme bears mentioning since in addition to providing a specific model for the origin of genetic information from an epigenetic process, many membered hypercycles (>4) also display stable trajectories and thereby take on the properties of stable attractors (oscillators) (Eigen and Schuster, 1978). The essence of the problem lies in the fact that even with a small number of codons accurate replication requires a specific replicase enzyme, but the lengthy nucleic acid sequence necessary to specify an accurate replicase is too long to be accurately copied in the absence of the enzyme it codes for. It is thought that replication of very short nucleic acid sequences (10-100 bases) could well have occurred without specific enzymes whereas sequences of sufficient length to code for and maintain the information in an accurate replication complex (5000 bases) might be too long to be replicated accurately. The problem of bridging this sequence gap is resolved by initially reducing the full sequence size and hence information content and tendency to errors to a set of shorter messages whose length is such as to ensure an acceptable error rate (on the average less than one per molecule). However, a problem arises in this system. Given a vast excess of precursor nucleotides and adequately accurate replication, competition will occur within the unique sets of molecules for nucleotides, with the result that there is selection for more and more, shorter and shorter sequences. That such an event occurs in nature was shown by Mills et al. (1967) studying in vitro Q$\beta$ replication. The essential trick in Eigen's scheme is to arrange the set of sequences in a "hypercycle" and require that the replicase coded by sequence 1 function in the replication of the next sequence, and so on, with the replicase specified by the nth sequence functioning in the replication of sequence 1.

### Entrainment to a Changing Zeitgeber

Assuming an oscillator had been selected which appropriately coordinated the chemistry of the system with daily variations in temperature and sunlight and that this primitive limit cycle clock had some of the same properties as modern clocks, then as zeitgeber period lengthened so did clock period, for it is a property of limit cycles that their period can be tuned over a considerable range. Eventually one of two things happened: limits of entrainment were reached or day length reached twice the previous value, at which time subharmonic entrainment took place. In contrast, modern circadian oscillators appear to be bounded in their ability to tune or entrain to an environmental periodicity. Empirically, the limit of entrainment in circadian rhythms is ±4-5 hr, or 20% of the free-running period. When the limits of entrainment are exceeded the system alternately breaks away and entrains or displays subharmonic entrainment (Pittendrigh and Daan, 1976). If primitive clocks showed similar behavior, then as day length increased the period of the endogenous clock would be similarly lengthened without maladaptive effect but also without selective pressure, until the limits of entrainment were reached; the timekeeping system would then break away from entrainment and alternatively free run and lock on to environmental time cues as the two periods moved in and out of phase. Due to the property of tunability in the limits of entrainment, this maladaptive condition would not have arisen until the free-running period of the oscillator was very different from the day length of the earth, so that continuous selective pressure for a slightly increased period might never have occurred. If the earth-moon separation took place rapidly, with doubling of period in the course of $10^4$ years as some have estimated (MacDonald, 1965), a continuous genetic change in period would have been difficult. The rapid change in rotational period would have forced the system to find new adaptive, nongenetic ways to accomodate the change. It is most important to note that the problem of entrainment and adaptation to a changing day length would have occurred whether the day length was very short or as long as 16 hr, a value for which there is solid paleontological evidence. A solution relies on two lines of evidence, one gained from simulations of the behavior of populations of interacting oscillators and the other from a consideration of the role of tissue domain or cell size in determining oscillator period. As discussed above, when Pavlidis investigated the behavior of a set of linear oscillators of slightly dissimilar period, mutually coupled in an inhibitory fashion, he found that the period length would be equal to the average period times the number of individual oscillators considered. Jacklet and Geronimo (1971) found by experimentation that as the size of a tissue showing circadian oscillations was diminished, period length was likewise shortened. I suggest that primitive short period oscillators resolved the problem of changing day length physiologically either by the simple expedient of increasing cell size or more efficiently by fusion of organisms with similar but nonidentical circadian periods.

## SUMMARY

In eukaryotes, endogenous oscillators capable of anticipating and entraining to daily variations are universally observed and are exploited by the organism to anticipate and coordinate essentially all physiologically important functions. The evolution of the earth-moon system and the consequent changes in day length through geological time and, in particular, the possibility that during prebiotic evolution day length was quite short, form the matrix in which evolution of the timekeeping oscillator must be considered. By most reckoning, the earth and moon were in close proximity at some time more than three eons ago. At that time the solar day was probably less than 10 hr, possibly as short as 4-5 hr, and was punctuated by boiling tides and intense ultraviolet radiation. If a primitive oscillator with the capacity to entrain to sunlight or temperature changes had evolved at that time it would have had a relatively short free-running period. The selection of an autocatalytic chemical system displaying stable temporal trajectories and permitting coordination of the reaction sequence with the extreme environment of late prebiotic evolution would certainly have been favored. Rapid changes in rotational period combined with the ability of oscillators to adapt and tune to period changes may have resulted in the lack of selective pressure for continuous incremental increases in period. The limits of entrainment once reached would force upon the system a requirement for sudden increases in period. The property of populations of oscillators coupled in an inhibitory manner to yield saltatory increases in period may have led to the fusion of two or more oscillatory systems with similar but nonidentical period into a single integrated system and may have allowed these "dynamic symbionts" to reachieve appropriate coordination with the environment. Based on paleontological observations of growth ring differences, the existence of ultradian, often quantal, rhythmic expression in unicellular eukaryotes, simple metazoans, and mammalian cells in culture, and the clustering of $\tau$ mutants in *Drosophila, Chlamydomonas,* and *Neurospora,* I would like to suggest that modern circadian rhythms have evolved by organismic or organellar fusion and coupling of systems with short period oscillators that first emerged at a time when the earth's rotation was rapid and the moon was in close proximity.

It goes without saying that the ideas expressed here differ from the view of Pittendrigh and others (1981), who suggest that the clock, or more properly clocks, arose by a variety of means and perhaps exploited a number of molecular mechanisms to accomplish their aim of coordinating the internal and external worlds. There is no problem with their view if the environmental periodicities have always been as they are today. It is less apparent how successful such systems would have been at adapting to a changing day length once they became an integral element of organismic chemistry. The view presented here fails to deal with the fact that prokaryotic systems have rarely, if ever, been reproducibly shown to express circadian rhythmicity. Prokaryotic systems with their very short cell cycle times and generally more rapid macromolecular catabolism may well have exploited rapid relaxation time as a means of escaping from an overbearing environment. One would hope that a closer examination of rhythmicity in stromatolitic algae and the most primitive eukaryotes may help to elevate the question of clock origins to the level of enlightened speculation.

## REFERENCES

Alfven, H. (1964). On the origin of the asteroids. *Icarus* 3:52-56.

Brady, J. (1979). *Biological Clocks.* University Park Press, Baltimore.

Bruce, V. G. (1972). Mutants of the biological clock in *Chlamydomonas reinhardi. Genetics* 70:537-548.

Chance, B., Estabrook, R. W., and Ghosh, A. (1964). Damped sinusoidal oscillations of cytoplasmic reduced pyridine nucleotide in yeast cells. *Proc. Natl. Acad. Sci. USA* 51:1244-1251.

Chisholm, S. W., and Costello, J. C. (1980). Influence of environmental factors and population composition on the timing of cell division in *Thalassiosira fluviatilis* (Bacillariophyceae) grown on light/dark cycles. *J. Phycol.* 16:375-383.

Clark, S. P., Turcotte, D. L., and Nordmann, J. C. (1975). Accretional capture of the moon. *Nature* 258:219-220.

Darwin, G. H. (1898, reprinted 1962). *The Tides.* W. H. Freeman and Cc., San Francisco.

Eigen, M. (1971). Self-organization of matter and the evolution of biological macromolecules. *Naturwissenschaften* 58:465-522.

Eigen, M., and Schuster, P. (1977). The hypercycle. A principle of natural self-organization. Part A: Emergence of the hypercycle. *Naturwissenschaften* 64:541-565.

Eigen. M., and Schuster, P. (1978). The hypercycle. A principle of natural self-organization. Part B: The abstract hypercycle. *Naturwissenschaften* 65:7-41.

Feldman, J. F., Gardner, G., and Denison, R. (1978). Genetic analysis of the circadian clock of *Neurospora.* In *Biorhythm and Its Central Mechanism.* Naito Int. Symposium on Biorhythm and Its Central Mechanism, Tokyo, Japan.

Feldman, J. F., and Hoyle, M. N. (1973). Isolation of circadian clock mutants of *Neurospora crassa. Genetics* 75:605-613.

Feldman, J. F., and Hoyle, M. N. (1976). Complementation analysis of linked circadian clock mutants of *Neurospora crassa. Genetics* 82:9-17.

Fish, F. F. (1967). Angular momenta of the planets. *Icarus* 7:251-256.

Gerstenkorn, H. (1955). Uber Geizeitenreibung beim Zweikorper problem. *Z. Astrophys.* 36:245-274.

Gerstenkorn, H. (1967). On the controversy over the effect of tidal friction upon the history of the earth-moon system. *Icarus* 7:160-167.

Goldreich, P. (1966). History of the lunar orbit. *Rev. Geophys.* 4:411-439.

Goldreich, P., and Soter, S. (1966). Q in the solar system. *Icarus* 5:275-289.

Goodwin, B. C. (1963). *Temporal Organization in Cells.* Academic Press, New York, p. 21.

Hartmann, W. K., and Larson, S. M. (1967). Angular momenta of planetary bodies. *Icarus* 7:257-260.

Higgins, J. (1967). The theory of oscillating reactions. *Ind. Engin. Chem.* 59:19-62.

Jacklet, J. W., and Geronimo, J. (1971). Circadian rhythm: Population of interacting neurons. *Science* 174:299-302.

Jauker, F., and Cleffmann, G. (1970). Oscillation of individual generation times in cell lines of *Tetrahymena pyriformis. Exp. Cell Res.* 62:477-480.

Kaula, W. M. (1971). Dynamical aspects of lunar origin. *Rev. Geophys. Space Phys.* 9: 217-238.

Klevecz, R. R. (1976). Quantized generation time in mammalian cells as an expression of the cellular clock. *Proc. Natl. Acad. Sci. USA* 73:4012-4016.

Klevecz, R. R., King, G. A., and Shymko, R. M. (1980). Mapping the mitotic clock by phase perturbation. *J. Supramol. Struct.* 14:329-342.

Klevecz, R. R., Shymko, R. M., and Kauffman, S. A. (1982). Cellular clocks and Oscillators. *Int. Rev. Cytol.* in press.

Konopka, R. S., and Benzer, S. (1971). Clock mutants of *Drosophila melanogaster. Proc. Natl. Acad. Sci. USA* 68:2112-2116.

Kyriacou, C. P., and Hall, J. C. (1980). Circadian rhythm mutations in *Drosophila melanogaster* affect shorter term fluctuations in the male's courtship song. *Proc. Natl. Acad. Sci. USA* 77:6729-6733.

MacDonald, G. J. F. (1964). Earth and moon: Past and future. *Science* 145:881-890.

MacDonald, G. J. F. (1965). Origin of the moon: Dynamical considerations. *Ann. N.Y. Acad. Sci.* 118:739-782.

Mills, D. R., Peterson, R. L., and Spiegelman, S. (1967). An extracellular Darwinian experiment with a self-duplicating nucleic acid molecule. *Proc. Natl. Acad. Sci. USA* 58:217-224.

Mitler, H. E. (1975). Formation of iron-poor moon by partial capture, or: yet another exotic theory of lunar origin. *Icarus* 24:256-268.

Mohr, R. E. (1975). In *Growth Rhythms and the History of the Earth's Rotation,* G. D. Rosenberg and S. K. Runcorn (eds.). Wiley, London.

Morales, M., and McKay, D. (1967). Biochemical oscillations in "controlled" systems. *Biophys. J.* 7:621-625.

Nicolis, G., and Prigogine, I. (1977). *Self-Organization in Nonequilibrium Systems.* John Wiley and Sons, New York.

Oehler, D. Z., Schopf, J. W., and Kuenvolden, K. A. (1972). Carbon isotopic studies of organic matter in precambrian rocks. *Science* 179:1246-1248.

Ohno, S. (1973). Ancient linkage groups and frozen accidents. *Nature* 244:259-262.

O'Keefe, J. A. (1969). Origin of the moon. *J. Geophys. Res.* 74:2758-2767.

Opik, E. J. (1972). Comments on lunar origin. *Irish Astron. J.* 10:190-238.

Pannella, G. (1972). Paleontological evidence on the earth's rotational history since early precambrian. *Astrophys. Space Sci.* 16:212-237.

Pannella, G. (1976). Palaeontological clocks and the history of the earth's rotation. In *Growth Rhythms and the History of the Earth's Rotation.* John Wiley and Sons, New York, p. 253.

Pannella, G., MacClintock, C., and Thompson, M. N. (1968). Paleontological evidence of variations in length of synodic month since late cambrian. *Science* 162:792-796.

Pavlidis, T. (1969). Populations of interacting oscillators and circadian rhythms. *J. Theor. Biol.* 22:418-436.

Pavlidis, T. (1973). *Biological Oscillators: Their Mathematical Analysis.* Academic Press, New York.

Pittendrigh, C. S. (1960). Circadian rhythms and the circadian organization of living systems. *Cold Spring Harbor Symp. Quant. Biol.* 25:159-184.

Pittendrigh, C. S. (1967). Circadian systems. I. The driving oscillation and its assay in *Drosophila pseudoobscura. Proc. Natl. Acad. Sci. USA* 58:1762-1767.

Pittendrigh, C. S. (1976). In *Dahlem Workshop on the Molecular Basis of Biological Clocks,* J. W. Hastings and H.-G. Schweiger (eds.). Dahlem Press, Berlin, p. 11.

Pittendrigh, C. S. (1981). Circadian organization and the photoperiodic phenomena. In *Biological Clocks in Seasonal Reproductive Cycles,* B. K. Follett and D. E. Follett (eds.). Wright, Bristol, England, p. 37.

Pittendrigh, C. S., and Daan, S. (1976). A functional analysis of circadian pacemakers in nocturnal rodents. *J. Comp. Physiol.* 106:223-252.

Prigogine, I., and LeFever, R. (1968). Symmetry breaking instabilities in dissipative systems. *J. Chem. Phys.* 48:1695-1700.

Ringwood, A. E. (1972). Some comparative aspects of lunar origin. *Phys. Earth Planet. Int.* 6:366-376.

Ringwood, A. E. (1978). Origin of the moon. *Lunar Planet. Sci.* 9:961-963. (Lunar and Planetary Institute, Houston.)

Ringwood, A. E. (1979). *Origin of the Earth and Moon.* Springer-Verlag, New York.

Robertson, M. (1975). Circadian rhythms. *Nature* 258:291.

Rosenberg, G. D. (1974). In *Growth Rhythms and the History of the Earth's Rotation,* G. D. Rosenberg and S. K. Runcorn, (eds.). John Wiley and Sons, New York, p. 1-9.

Rosenberg, G. D., and Runcorn, S. K. (eds.) (1974). In *Growth Rhythms and The History of The Earth's Rotation.* John Wiley and Sons, New York, p. 536.

Rusak, B. (1977). The role of the suprachiasmatic nuclei in the generation of circadian rhythms in the golden hamster, *Mesocricetus auratus. J. Comp. Physiol.* 118:145-164.

Shymko, R. M., and Klevecz, R. R. (1981). In *Biomathematics and Cell Kinetics,* M. Rotenberg (ed.). Elsevier/North-Holland Biomedical Press, Amsterdam, p. 329.

Sundermann, J., and Brosche, P. (1977). *Tidal Friction and the Earth's Rotation.* Springer-Verlag, New York.

Taylor, S. R. (1979). Structure and evolution of the moon. *Nature* 281:105-110.
Turcotte, D. L., Cisne, J. L., and Nordmann, J. C. (1977). On the evolution of the lunar orbit. *Icarus* 30:254-266.
Walter, M. R., Bauld, J., and Brock, T. D. (1972). Siliceous algal and bacterial stromatolites in hot spring and geyser effluents of Yellowstone National Park. *Science* 178: 402-405.
Wells, J. W. (1963). Coral growth and geochronometry. *Nature* 197:948-950.
Winfree, A. T. (1967). Biological rhythms and the behavior of populations of coupled oscillators. *J. Theor. Biol.* 16:15-42.
Winfree, A. T. (1970). Integrated view of resetting a circadian clock. *J. Theor. Biol.* 28: 327-374.

# 4

# Discontinuities and Singularities in the Timing of Nuclear Division

**Arthur T. Winfree** / Purdue University, West Lafayette, Indiana

Rules of Reasoning in Philosophy. Rule II: . . . We are certainly not to relinquish the evidence of experiments for the sake of dreams and vain fictions of our own devising; nor are we to recede from the analogy of Nature, which is wont to be simple and always consonant to itself.

Isaac Newton, *The System of the World* (1730)

## INTRODUCTION

In no case is the process well understood whereby the growing cell "decides" to replicate its genome, segregate its chromosomes into two nuclei, and wall them off from each other by cell fission. Such a fundamental biological process presumably has some universal aspects. Its appeal as an object of investigation is further enhanced by the seductive belief that the mechanism of replication is constrained to some kind of simplicity by these facts:

1. The end state is close to the initial state (namely, a fresh divided cell).
2. The whole process repeats at fixed intervals of time, at least in certain kinds of cells in optimal growth conditions.

A brief foray into this intriguing puzzle is included here for three reasons:

At least three influential notions about clocklike dynamics arose first in this context and later found application in studies of circadian rhythms.

---

Adapted from A. T. Winfree, The cell mitotic cycle, Chap. 22 In *The Geometry of Biological Time,* Springer-Verlag, New York, 1980, pp. 435-449.

There is a close but poorly understood causal link between circadian rhythms and the mitotic cycle.

Techniques of analysis evolved in circadian experiments later stimulated parallel enquiries into the cell cycle (mostly by Kauffman et al.: see below), with results that bear upon the exploration of phase singularities (Winfree, 1980).

I devote a few pages to each of these three topics. The last section goes into a little more detail about one specific experimental system, the true slime mold *Physarum*. Like the three other members of kingdom fungi (yeasts, cellular slime molds, and colonial ascomycetes) the true slime mold is conspicuously periodic in its biochemical habits. In this case a clocklike regularity of mitosis is most prominently featured. With this periodism comes a crisis of logic about periodically timed biochemistry. The crisis is resolved not by a phase singularity but by a discontinuity that appears to reveal an honest physiological cataclysm.

## THREE BASIC CONCEPTS AND SOME MODELS

### A Sequence of States

The earliest inquiries into the cell cycle led to a paradigm that later pervaded much of the circadian literature, though only in recent years has it become explicit enough to suggest experimental tests capable of excluding diverse qualitatively distinct alternatives. This paradigm is best summarized graphically (Figs. 1 and 2). As generally envisioned, the cell cycle is a series of discrete steps or states that a cell must execute in sequence if it is to

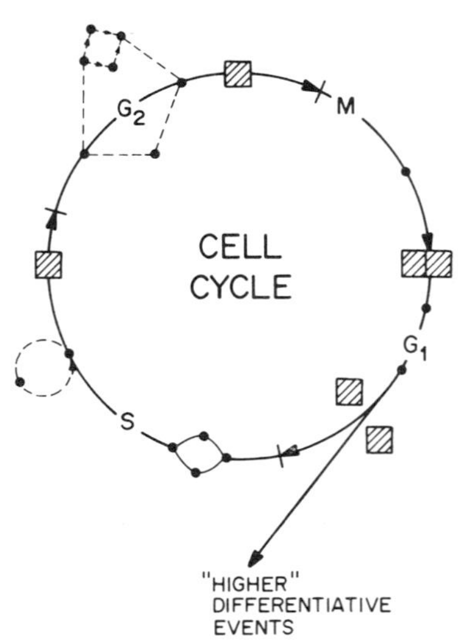


**FIGURE 1** Diagram of a generalized cell cycle. The four classic phases are shown: $G_1$ (gap 1), S (synthetic), $G_2$ (gap 2), and M (mitotic). In addition, a number of arbitrarily chosen points are designated by dots in order to illustrate possible alternative pathways, branched network, loops, and blockage points. Any precise resemblance to a particular cell is purely coincidental. (From Edmunds, 1975.)

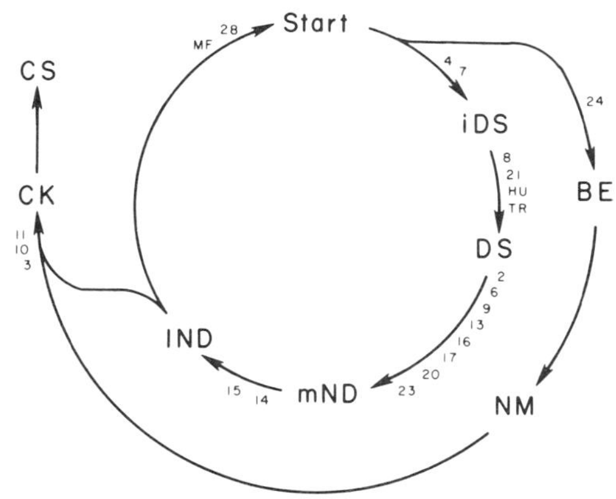


**FIGURE 2** The circuitry of the yeast cell cycle. Events connected by an arrow are proposed to be related such that the distal event is dependent for its occurrence upon the prior completion of the proximal event. Numbers refer to *cdc* genes that are required for progress from one event to the next; HU and TR refer to the DNA synthesis inhibitors hydroxyurea and trenimon, respectively; MF refers to the mating factor, $\alpha$ factor. (From Hartwell et al., 1974. Copyright 1974 by the American Association for the Advancement of Science.)

replicate normally. Pictured abstractly, these steps constitute a one-way cycle though possibly adorned with baroque appendages. There may be loops in the cycle, or alternate pathways here and there. There may be stopping places at which the cell is unable to go on to the next step until some permissive signal is given. But the central point for my purpose here is the intrinsically one-dimensional character of the process. The dynamics more resemble driving a car on roadways than piloting a sailboat or airplane in continua with more degrees of freedom.

It is this vision of the cell cycle that gave rise to the concept of a "simple clock" (Campbell, 1964), and, by analogy, to much thinking about circadian clocks, and to diverse experimental efforts to ask whether circadian clocks are in fact simple clocks.

### Phase-Specific Arrest Within the Cycle

If the cell cycle consists of a sequence of processes, some of which are necessary and sufficient causes of others, then the first job in preparing a "troubleshooter's manual of the cell cycle" is to construct a flow chart of these cause-and-effect connections, similar to a flow chart of metabolic pathways (Figs. 1 and 2). With the normal cycle thus verbally divided into stages, it would be illuminating to seek mutants incapable of each process named, just as biochemical geneticists unraveled the intricate contingencies that constitute the manufacture of small molecules in intermediary metabolism by finding out which subsequent events are also arrested in such a mutant.

Hartwell et al. (1974) discovered in this way that the cell cycle in yeasts consists of at least two parallel sequences of events, one of which can proceed even repetitiously, while the other is blocked. One involves DNA synthesis and nuclear division, and the other involves budding and cytoplasmic division.

### Rephasing as a Quantitative Test

Systematic phase-resetting experiments to probe the mechanism underlying a physiological rhythm made their first appearance in the cell cycle literature. The protocol was typically as follows:

1. Synchronize mitoses by selecting cells that are a certain size or sticky in a certain range of phases, or by synchronous release of a population from arrest by an inhibitor, or by fusion of randomly phased plasmodia.
2. At some initial phase $\phi$ (measured as hours elapsed since mitosis), interfere with normal operation of the cell cycle by changing the temperature, introducing a poison, or whatever.
3. Continue the stimulus for a time M (for magnitude of disturbance) and then return the cells to standard culture conditions.
4. Observe the setback or excess delay of mitosis time beyond what would be expected if the cell cycle were simply paralyzed during treatment time $M$. Excess delay $D$ is defined as the actual period $P$ (measured in units of one normal cycle) from the last mitosis before perturbation to the first mitosis after perturbation, minus the usual period between mitoses (which we take for our unit of time), minus stimulus duration $M$:

$$D = P - 1 - M = P - M \text{ (modulo 1 period)} \tag{1}$$

Converting to terms used elsewhere: Phase shift, with advances taken as positive, is

$$\Delta\phi = 1 - (D + M) = 1 - P = -P \text{ (modulo 1 period)} \tag{2}$$

Final phase at the end of $M$,

$$\phi' = \phi - D \tag{3}$$

where $\phi$ is the initial phase at which the stimulus begins. Cophase at the end of the stimulus is

$$\theta = 1 - \phi' = D - \phi \tag{4}$$

5. Plot $D$ against the initial $\phi$ (when the treatment began). Many such plots are available in the literature since Thormar published the first in 1959. For example, see Fig. 3. Their main uses are twofold:

To indicate whether a given treatment affects the mechanism of the mitotic timer, and if so during what phases.

To detect any physiological discontinuity in the cells' sensitivity. There are typically two discontinuities: a point of no return in the cycle, beyond which the very next mitosis occurs on schedule, though phase shifts may be evident in subsequent mitoses; and, near mitosis time, an abrupt change in the new phase to which the whole subsequent rhythm is reset if the stimulus is given near that time.

### Testing Models

Before passing on to the next section, I would like to elaborate a bit on rephasing as a quantitative test of models. Though resetting curves are available, it is not usually possible to make rigorous use of the "circular logic" presented in Winfree (1980) because

1. These mitotic resetting curves usually leave a substantial gap surrounding mitosis, so it is impossible to determine the topological type of resetting.

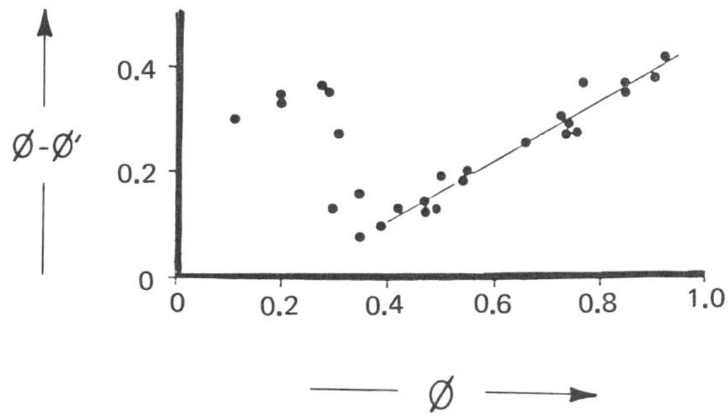


**FIGURE 3** Excess mitotic delays in *Physarum* induced by 30 min pulses of 10 μg/ml cycloheximide (abscissa): fraction of cell cycle completed when pulses begin. Mitosis at 0 and 1.0. DNA synthesis ends around 0.3. (ordinate): excess delay (delay in excess of pulse duration) plotted as fraction of a control cell cycle. (From Scheffey and Wille, 1978.)

2. Measurements usually neglect mitoses after the one that immediately follows the stimulus, leaving us to assume that this first mitosis reflects a rephasing of the mitotic timer for all cycles to follow. In circadian rhythms, in glycolysis in yeast, and in neural pacemakers, this is not always a sound assumption. Nor is it in the case of mitosis in *Physarum* (Brewer and Rusch, 1968; Devi et al., 1968; Sachsenmaier et al., 1970; Kauffman and Wille, 1975). It is sound and requires no empirical check if and only if (as seems widely assumed) the mitotic cycle is a simple clock and mitosis is part of its mechanism.
3. At present, there is no complete set of measurements in which both initial phase $\phi$ and stimulus duration $M$ are varied.
4. It is not always clear that the population of cells observed were perfectly synchronous to start with, or that their mitoses were unimodally distributed when the reset phase was later determined. This is essential for interpreting the topological features of resetting curves.

### The Simple Clock

Setting aside these caveats for a moment, it would appear that excess delay data might nevertheless be capable of excluding the notion that the cell cycle is a simple clock, i.e., a continuum or finely spaced sequence of successive states along which the cell moves more quickly or slowly according to environmental conditions. If the cell cycle *were* a simple clock, then the slope of a final phase versus initial phase plot could not be negative anywhere (Winfree, 1980). In other words, using the conventions of this field together with Eq. (3), the excess delay curve's slope could not exceed 45° anywhere. In fact, it usually doesn't; for example, see Mitchison (1971, pp. 204-240). One gets the impression from looking at a lot of these curves that there is a distinct limit at 45° against which many excess delay curves press, without transgressing it. To this extent, the cell cycle in several organisms does give an appearance of restriction to a fixed cycle, varying only its rate of progress or of retrogression in response to external conditions.

But there do exist counterexamples in which the slope runs as high as 2 or 3: specifically, (1) sea urchin eggs irradiated with x-rays (Rustad, 1970) and (2) *Tetrahymena* shocked with chemicals (Rasmussen and Zeuthen, 1962, Figs. 9 and 11; Hamburger, 1962; Frankel, 1969), and heat (Frankel, 1962, Figs. 4 and 5).

Unless these can be somehow explained away (e.g., on the basis of population schism, or interpreting the apparent phase shift as only a transient of the first cycle after treatment), then they constitute clear violations of the simple clock model. Thus, for example, we cannot think of mitosis in these cells as being governed by any single "division protein" or "mitogen" accumulating or decomposing at a rate which is merely altered, directly or indirectly, by these stimuli, or as governed by any complex of chemical factors that always change in strict mutual coordination, at varying rates, along the closed ring of states.

The simple clock paradigm has served so well that subversive observations such as these should not be allowed to pass unchallenged.

### Feedback Oscillation

Alternatives have been prepared for consideration. Goodwin (1963) finds periodic dynamics implicit in the regulation of transcription and translation, which he suggests may have a period long enough to account for the cell cycle or for the circadian cycle. Goodwin's principal innovation is emphasis on feedback regulation as a source of instability in a biochemical steady state, quite a different notion from progress around a ring. In this respect, this notion resembles (1) the cytoplasmic oscillator proposed later by Goodwin (1966) to account for the serial ordering of enzyme activities in *Escherichia coli* during its repeated fissions;* (2) the circadian oscillator which commonly gates successive rounds of mitosis; (3) the growth cycle distinguished from the DNA division sequence in Mitchison (1971, pp. 244-249); (4) the cell clock biochemical attracting cycle proposed by Selkov (1973); (5) the "budding" oscillator proposed by Hartwell et al. (1974) from his observations of mutant yeast; (6) the cytoplasmic oscillator observed by Mano (1975) and by Yoneda et al. (1978) in association with early divisions of the urchin egg and by Hara et al. (1980) and Kirschner and Gerhardt (1981) in *Xenopus* eggs; (7) the cytoplasmic attracting cycle proposed for mitosis in the acellular slime mold by Kauffman (1974), Kauffman and Wille (1975), and Tyson and Kauffman (1975); and (8) the attracting cycle of 4 hr period supposed by Klevecz et al. (1978) to underlie mitosis in mammalian cells.

From these papers, it would appear that if there is anything about the cell cycle that resembles an oscillator, it plays a role accessory to the one-dimensional sequence of processes that constitute nuclear replication: Perhaps the oscillator governs cytoplasmic replication and division, and initiates the next start of nuclear replication.

It should be noted that, apart from Goodwin's first heresy of 1963, all the investigators cited above built their mitotic timers outside the actual mechanisms of nuclear replication. For the best update, see Chap. 10 of this volume and Tyson (1983a,b).

These models, and less specifically detailed suggestions toward the same ends by Engelberg (1968), Burton (1971), Burton and Canham (1973), and Gilbert (1974), all emphasize continuous aspects of biochemical dynamics.

---

*Note that this chapter concerns mitosis in eukaryotes exclusively. The corresponding process, called "cell division" or "nuclear division," in prokaryotes, need not have similar dynamics.

### Biochemical Switching Networks

At the opposite extreme, one could emphasize *switchlike* aspects of reaction kinetics involving, say, allosteric enzymes and/or mechanical events (like polymerase attachment) on the DNA itself. This interpretation is encouraged by the observation of stepwise doubling of membrane transport in mammalian cells (Jung and Rothstein, 1967; Sander and Pardee, 1972) and of enzyme activity in *E. coli* and in yeast (Halvorson et al., 1971; Elliott and McLaughlin, 1978; Tyson, 1979), each enzyme showing its increase at a characteristic phase.

This idea was pioneered in Kauffman's Boolean logic network model of the cell (1969). According to the logical net model, everything important about the state of the cell is known when one knows which genes are active and which are inactive. There are no intermediate levels of activity. Each activity state results in biochemical processes which directly or indirectly shut off some genes and induce or depress others. As in a switching network, the cell's state keeps changing until some prior state recurs, completing one cycle of gene activity and of gene replication, and thus permitting one cell division. *There are many alternative cycles,* each corresponding to a different state of differentiation (liver cell, muscle cell, and so on). According to this model, cycling persists in nondividing cells, but each gene is transcribed without being replicated. Further exploitation of this stimulating suggestion awaits evidence of such regularly periodic genetic activity in nondividing cells.

A related model by Ehret and Trucco (1967) supposes a unique order of replication of the genome, its stages separated by delays while substances diffuse across the nuclear membrane. Each stage consists of processing one "chronon," tailored by natural selection to the right length for the required timing.

## REGULATION OF MITOSIS BY THE CIRCADIAN CLOCK

The average rate at which cells run through their cycle of division is governed by many factors, including hormones, some of which (e.g., epinephrine) are secreted in a daily rhythm. Epinephrine has been shown to affect the average rate of mitosis in the epithelium of a mouse's ear (Bullough and Laurence, 1964). The result is that while mitosis continues at all hours of the day, the mitotic index of this tissue is very much greater during some parts of day than during others (Moller et al., 1974; Taguchi and Tabachnick, 1974). Circadian regulation of mitotic rates can even go to the extreme of completely suppressing division during part of the circadian cycle. Mitoses in *Euglena* (Edmunds, 1974a, 1975) and in *Tetrahymena* (Edmunds, 1974b) and in *Gonyaulax* (Sweeney and Hastings, 1958), for example, occur only during certain phases of their circadian cycle. In these cases it is inferred that only a small fraction of the population divides during any one daily opportunity. The same inference follows from the extreme daily variation of mitotic index in rat cornea (and other tissues), in which a cell divides only every month or two but finally does so at a particular time of day (Scheving and Pauly, 1973).

Complete inhibition of mitosis during certain hours of the day can be important for therapeutic applications (Llanos and Nash, 1970). Timed administration helps cytotoxic medication to discriminate between two issues, selectively killing a proliferating tumor tissue while leaving the normal healthy tissue unpoisoned. See Halberg (1977) for applications.

The cell clock models of Sel'kov and Sozinov (1970), Sel'kov (1970, 1973), and Wille (1979) look on the circadian cycle and the mitotic cycle as essentially the same biochemical process. Similarly, the chronon model, also mentioned above, envisions a close relationship between the mechanisms of the eukaryotic cell's mitotic cycle and of its circadian cycle. Admittedly not very much is known about the mechanisms of either cycle. Still, one might take the fact that the circadian clock can go while the cell cycle is inhibited, and vice versa (Bruce, 1965; Barnett, 1966; Edmunds, 1975), to suggest two distinct coexisting mechanisms, each capable of affecting the other under suitable conditions. Mitosis is sometimes gated at circadian intervals as we saw above, and eukaryotic cells are incapable of circadian output while mitosing at intervals shorter than 24 hr (Ehret and Wille, 1970; Ehret, 1974; Wille, 1979). For an excellent review of the relationships between the mitotic and circadian cycles, see Edmunds (1977).

## FURTHER DEVELOPMENTS IN THE AREA OF CIRCADIAN RHYTHMS, APPLIED BACK TO THE CELL CYCLE

### A Phase Singularity?

It was in context of a circadian rhythms that type 0 or "strong" resetting (Winfree, 1980) was first noticed as something qualitatively distinctive and impossible to account for in terms of movement along a fixed cycle. This led to development of experimental criteria for distinguishing between qualitatively different kinds of clock mechanisms in terms of topological features of their resetting behavior. Such methods are capable of revealing whether the cell cycle really ought to be thought of strictly as a cycle (as in Figs. 1 and 2). An alternative would be to endow it with "insides" necessarily including a state of indeterminate or ambiguous phase. Such a state would have no clear interpretation in terms of the sequence-of-stages interpretation of the cell cycle. During the past decade at least seven models of this sort have been advanced in the scholarly literature. Though there are some experimental grounds for taking this possibility seriously (see the last section), the results seem more compatible with an alternative advanced in several forms by Rasmussen and Zeuthen (1962), Rusch et al. (1966), Zeuthen and Williams (1969), Pritchard et al. (1969), Sachsenmaier et al. (1972), Cohen (1972), Fantes et al. (1975), and Sudbery and Grant (1976). According to this class of models, some single structure (division protein) or substance (mitogen) accumulates or is diluted steadily during the cell's growth. When enough has accumulated or when cell volume has sufficiently increased, nuclear division is triggered and the substance is made anew or accumulation is restarted, initiating the next cycle. An appealing version due to Sachsenmaier is that the accumulating initiator substance combines with nuclear receptor sites, which double stepwise at each nuclear division. Nuclear division is triggered as often as all available sites are saturated. Thus the control mechanism of nuclear division acts like an hourglass or relaxation oscillator. In terms of the categories of model mentioned above, it acts like a simple clock with a single switch. The abrupt resetting characteristic of such models is inferred from the abrupt change typically observed in the plasmodium's response to fusion with another plasmodium at a particular phase in the cycle.

This is not a necessary inference from the observations, since the same data *could* indicate a phase singularity associated with a smooth cycle of a correctly chosen shape (Winfree, 1974; Kauffman, 1974). But experiments have not yet produced compelling reasons to doubt the earlier interpretation (see the last section).

### Chaotic Dynamics Outside the Nucleus?

Advocates of oscillator mechanisms for mitotic timing usually place the oscillator outside the nucleus, leaving the nucleus to initiate its chain of necessary events upon signal from the triggering oscillator (citations under Feedback Oscillation). It seems to me likely that such a parochial "gatekeeping" aspect of the cell cycle might vary in behavior and in mechanism from one cell type to the next. In the extreme case of very complicated or irregular (chaotic) dynamics, such a gatekeeper dynamic might account for the great variability of intervals between mitoses typical of mammalian cells. Conjectures of this sort began with Engelberg (1968), who suggested that mitosis might be triggered when the sum of many separate harmonic oscillations in cell metabolism momentarily exceeds a threshold. Burton (1971) and Burton and Canham (1973) elaborated on this notion. Leigh (1965), studying superimposed predator-prey oscillations in ecological context, worked out the statistics of such threshold crossings. But comparisons with data on mitotic variability have not yet been carried out. The first explicit efforts to distinguish between models on the basis of their statistical behavior are due to Burns and Tannock (1970) and Smith and Martin (1973). They suggest that the variability of mitotic intervals is almost all due to a variable delay of unknown cause between the completion of one cycle and the initiation of the next. They found that the delay can be described by a Poisson process. Subsequent experiments bear this out (Minor and Smith, 1974; Brooks, 1975, 1976; Shields, 1976, 1977; Shilo et al., 1977).

Models have thus far attempted most combinations of continuity versus discreteness of state; of involvement versus independence of the stages of nuclear replication; of one state variable versus two; of determinism versus probablism. The next innovation will probably have to exploit the fashionable behavior of "strange attractors" and "cusp catastrophes," by accounting for both mitotic repetitiousness and for variability in terms of three state variables, two of which cycle smoothly while the third undergoes a jump at mitosis time. The raw materials for such a model can be found in the papers of Zeeman (1977) and of Rössler (1976a-d, 1978). The stochastic-start models originated by Burns and Tannock (1970) and Smith and Martin (1973) sound like natural candidates for formalization in terms of a chaotic cycle or an attracting cycle tangent to a homoclinic point.

It should be noted in connection with such a phylogenetically eclectic literature that experiments on bacteria, slime molds, yeasts, and mammalian cells might not all require the same interpretation.

## PHYSARUM POLYCEPHARUM

### Synchronous Nuclear Division

Howard discovered in 1932 that the true slime mold, *Physarum polycephalum,* behaves in nuclear division much like a single gigantic cell. All its nuclei coexist in one blob of jelly, undivided by plasma membranes or cell walls. A similar situation obtains in other members of the kingdom fungi, in multinucleate tumor cells, in multinucleate frog eggs, in the malaria plasmodium, in artificially fused cells, and in early insect eggs. In all these cases, nuclear division among the several nuclei occurs synchronously. In *Physarum,* the effect is more dramatic and biochemically more convenient for its sheer massiveness, as millions of nuclei cleave within 15 min during a 9-12 hr nuclear division cycle. Such synchrony is difficult to obtain and is ephemeral in suspensions of independent cells.

The fact of nuclear synchrony encourages belief that division is triggered by soluble factors equally distributed throughout the cytoplasm. The causes of the observed synchrony are still not understood, but have been subjected to much experimental study, especially since Guttes et al. (1969) discovered the technique for fusing two plasmodia. Though each of the two parent plasmodia is synchronous within itself, their phases may be quite different. Fusion results in thorough mixing of the two parents. As they now constitute one plasmodium, nuclear fission is again synchronous. Rao and Johnson (1970) and Rao et al. (1974) carried out fusion in the same format using mammalian cells, with results much like those obtained in the more extensive experiments with *Physarum.*

## Phase Compromise Experiments

One question asked in such experiments is, "What will be the compromise phase?" Guttes et al. observe that the compromise phase is simply the average of the parent phases, or a weighted average in case of unequal weights of the two parent plasmodia. In taking this average, they consider phase as a number from zero just after nuclear division increasing to one just before nuclear division. A unit discontinuity occurs at nuclear division. In a critical series of experiments, Sachsenmaier et al. (1972) took two plasmodia one-half cycle apart, and fused samples of each from time to time throughout the 10 hr cycle. For the first 5 hr of fusion time, the fused plasmodia eventually reached nuclear division at a certain fixed compromise time (Fig. 4). But fusion after the leading parent reaches nuclear division resulted in eventual phase compromise 5 hr shifted relative to that observed in the first half of the experiment. For the remaining 5 hr of fusion time, the result stayed the same, then it shifted back after the lagging parent reached nuclear division. This all suggests a startlingly simple interpretation. In the words of Tyson and Sachsenmaier (1978):

> An unstable activator, which is synthesized at a rate proportional to cell mass, binds reversibly to certain nuclear sites. When enough activator has accumulated to titrate all these sites, mitosis and another round of DNA synthesis are initiated. . . . (or). . . an unstable cytoplasmic inhibitor is in equilibrium with a nuclear component so that its total amount is proportional to the number of genome equivalents in the plasmodium. During growth, inhibitor concentration decreases until it reaches a critical value below which nuclear divisions and DNA synthesis are no longer suppressed.

The question of phase compromise is made even more tantalizing by the observation (Winfree, 1980) that it is impossible to construct even a hypothetical table of compromise phase as a function of parent phases without including either a large discontinuity or a phase singularity.

I recall the excitement of Sachsenmaier's year at Princeton in 1969 and our conversations about the prospects of someday completing the phase compromise measurements. The experiments we outlined would fill out the ($\phi_A$, $\phi_B$, weight ratio) cube, plainly exposing the anatomy of the unavoidable discontinuity. When these laborious experiments are done in such a way that fusion is prompt and synchrony is complete, then the critical questions will be as follows:

1. Does the dependence of compromise phase on parent phases and bulk ratios have any of the following characteristics?

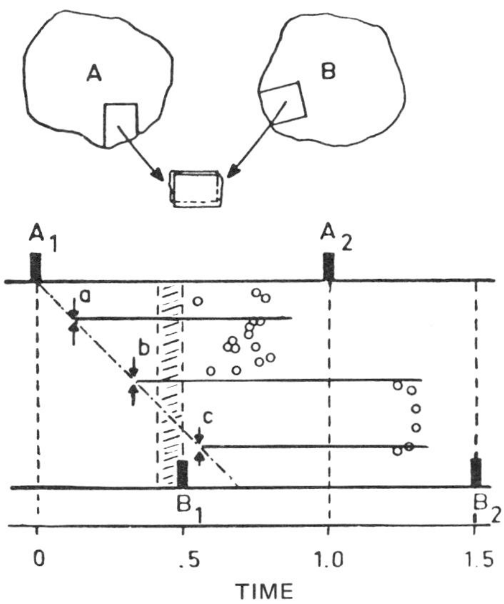


**FIGURE 4** Fusion of plasmodial pieces derived from two differently phased macroplasmodia A and B at various time intervals. Small plasmodial pieces ($\simeq$ 1.5 $cm^2$) were excised from A and B and placed on top of each other. The horizontal projection of each point onto the oblique line indicates the time when fusion was started. Three examples are marked with double arrows (a, b, c). *Fusion is complete within approximately 1 hr.* Time unit: duration of one division cycle ($\simeq$ 10 hr). ▮, Time of mitoses in parent cultures (telophase); o, mitoses in mixed plasmodia. Shaded area, entire period of mitosis ($B_1$) from prophase to telophase. (From Sachsenmaier, 1976.)

a. A diagonal discontinuity plane (two-dimensional), indicating a perfectly circular attracting cycle with a perfectly centered stationary state (Winfree, 1974).
b. Two perpendicular discontinuity planes (two-dimensional) near mitosis time, indicating a critical phase, as in the hourglass or sawtooth relaxation oscillator model (Winfree, 1974). (Note that both this case and (a) may be viewed as extreme limiting cases of an attracting cycle.)
c. A filamentlike (one-dimensional) phase singularity indicating an asymmetric attracting cycle (Winfree, 1974).
d. Some other presently unforeseen shape, possibly because some state variables equalize much more slowly than others after plasmodia are fused.

2. Is there any locus of fixed $\phi_A$ or $\phi_B$ crossing the ($\phi_A$, $\phi_B$) unit cell at equal parent weights along which no discontinuity is encountered (Fig. 5)? This would be awkward to account for by any model involving a physiological jump at a critical phase in the nuclear division cycle.

3. If not, does the compromise phase vary along that locus with winding number $W = 0$, implicating a phase singularity?

This measurement would be unbelievably laborious if carried through the second or third nuclear division following fusion, as a purist would require. So it is unlikely ever to be done. However, at equal weight enough data have been available since 1972 to strongly suggest (1b) and practically rule out (1c) (Remy, 1968; Sachsenmaier et al., 1972; Chin et al., 1972; Tyson and Sachsenmaier, 1978). Contour lines through these measurements

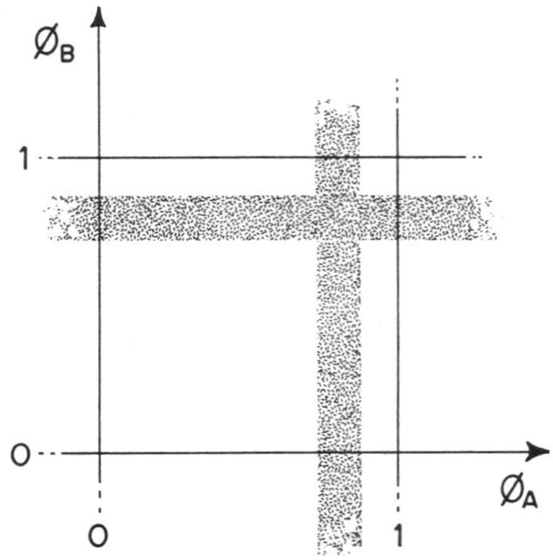


**FIGURE 5** The ($\phi_A$, $\phi_B$) torus, skinned and flattened as a unit cell in a square lattice. Does the compromise phase jump discontinuously somewhere in every band crossing the square, e.g., the shaded zones? Compare Figure 6.

clearly show perpendicular one-half cycle discontinuities. The data are sometimes irregular, but there is no clear hint of a phase singularity (1c) penetrating the equal weight plane.

More recently, Kauffman et al. (see below) reexamined the question with particular alertness to evidence of attractor cycle kinetics in the nuclear division cycle. They pursued the two main classes of experiment that have become traditional in this field:

1. Two plasmodia at different phases are fused, resulting in a phase compromise as their two nuclear division rhythms synchronize to one. The results are examined for evidence of critical phases, at which either cell's physiology abruptly changes, or for a combination of phases such that the compromise phase is unpredictable in the ways anticipated from the attractor cycle conjecture (Winfree, 1974; Kauffman, 1974; Tyson and Kauffman, 1975; Kauffman and Wille, 1975).
2. A physiological shock, such as incubation at 37°C, retards or advances the next nuclear division. Clues are sought in the amount and direction of phase shift as a function of the initial phase when the shock is administered (Winfree, 1974; Wille et al., 1977).

Both experiments are potentially capable of revealing a phase singularity if one exists, and both are capable of demonstrating type 0 resetting.

However, at this writing neither phenomenon has emerged from *Physarum*. In fact, those of the fusion data from Kauffman and Wille (1975) in which thorough mixing and synchrony are reported do fill out the 1972 picture more thoroughly and quite consistently. Kauffman and Wille (1975) conclude that mitosis in *Physarum* behaves more like a relaxation oscillator than does their model. So do Loidl and Sachsenmaier (1982).

#### Phase Compromise Among *Physarum* Nuclei

The dependence of hybrid phase on parent phases cannot be continuous (Winfree, 1974). In those arguments we assumed that the hybrid's phase typically varies smoothly as one parent's phase is varied through a full cycle. This was assumed because the alternative is

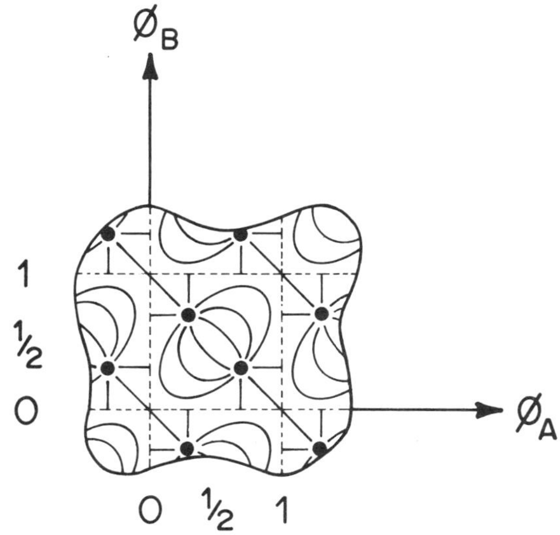


**FIGURE 6** Each contour line (solid) runs through those combinations of parent phase ($\phi_A$ and $\phi_B$) that result in identical phase compromise ($\phi_3$). $\phi_3$ increases smoothly along the diagonal, $\phi_A = \phi_B$, being equal to the parent phases. The pattern is symmetrical about the diagonal, since parent A is equivalent to parent B. It also repeats in consecutive cycles of either parent phase, so that it resembles wallpaper; here we show only one unit cell and portions of adjacent unit cells. No continuous pattern of phase compromise can fit those constraints. Here we show a possible discontinuous pattern that has a singularity point (dot) as well as a special phase (dashed lines at 0 = 1) where the phase compromise changes abruptly. Plasmodia A and B are here taken to have equal masses.

to assume a discontinuity at the outset, leaving nothing to prove. But that may be exactly what *Physarum* offers us. It appears that there is a phase somewhat before nuclear division when the *Physarum* nuclear division clock changes its reactions to perturbation so rapidly that no experiment has yet succeeded in resolving the changes as a process in time. For practical purposes of measurement, it is a discontinuity.

Does this mean that attractor cycle interpretations are excluded in favor of single-variable relaxation oscillators? Nothing of the sort. Nor does a rapid jump in the course of an attracting cycle necessarily imply anything about dynamics elsewhere in state space, or about isochron geometry and singularities (Winfree, 1980). It only prevents our using purely topological arguments.

Winfree (1974) took an alternative approach by supposing a particular dynamic, interpreting the state variables as chemical concentrations, and supposing that oscillators combine by pooling their corresponding chemicals. That particular model could be modified to give the attracting cycle an abrupt jump of some sort at phase 0 (Tyson and Kauffman, 1975). Then the contour map of hybrid phase $\phi_3$ would resemble Fig. 6: $\phi_3$ has a phase singularity in each triangle *and* a discontinuity along the perpendiculars $\phi_1 = 1$ and $\phi_2 = 0$. These features are by no means mutually exclusive. Finding one does not eliminate the other.

But *Physarum* evades this test, too, for whatever may be the nuclear state variables governing the mitotic clock, it seems unlikely that they all mix freely when plasmodia are fused. That means that the hybrid's state is given unambiguously only in the *product* space of two oscillators: a space of at least four dimensions in which the geometry of isochrons and their singularities is harder to visualize. Hence category (1d): If there is another discontinuity besides the one associated with the jump in the normal mitotic cycle, it may take unforeseen forms.

For the present, no such objects have been found in any shape or form. But the data are imprecise and much remains for eager hands to undertake.

In this circumstance I feel compelled for the present to abide by the first of Newton's two exhortations at the opening of this chapter, despite the attractiveness of theoretical schemes involving phase singularities. At the same time it must be remembered that *Physarum* fusion experiments can never resolve events more closely than the 45 min required for plasmodia to fuse. Any number of singularities might lurk unresolved in the hour of ostensible discontinuity preceding mitosis.

### Who Cares?

Knowledge of cell dynamics has obvious practical importance and obviously involves considerations of circular logic (Winfree, 1980). I remember with pleasure recurrent discussions on this fact with Graeme Mitchison in 1971, continuing hours after tables had been cleared in the Cambridge MRC Lab cafeteria. We wondered whether phase singularities could reasonably be pursued in this matter, and decided against it. But Kauffman decided on a more aggressive approach to the question, taking it to the master, *Physarum.* If Kauffman and friends have accurately interpreted *Physarum's* answer, then cell division is regulated by a continuous attractor cycle oscillation which operates independently of nuclear division itself. Such a discovery, if substantiated, might even be the main fruit of the notion of "phase singularities." I have chosen not to put it that much in the spotlight because I am apprehensive that *Physarum's* behavior may be very much more complicated than *any* of the simplistic models here entertained. But more data may come, and I do recommend the matter to your attention, as the outcome may be important.

## REFERENCES


Barnett, A. (1966). A circadian rhythm of mating type reversals in *Paramecium multimicronucleatum* Syngen 2, and its genetic control. *J. Cell Physiol.* 67:239-270.

Brewer, E. N., and Rusch, H. P. (1968). Effects of elevated temperature shocks on mitosis and on the initiation of DNA replication in *Physarum polycephalum. Exp. Cell Res.* 49:79-86.

Brooks, R. F. (1975). The kinetics of serum induced initiation of DNA synthesis in BHK-21-C-13 cells and the influence of exogenous adenosine. *J. Cell. Physiol.* 86:369-377.

Brooks, R. F. (1976). Regulation of the fibroblast cell cycle by serum. *Nature* 260: 248-250.

Bruce, V. G. (1965). Cell division rhythm and the circadian clock. In *Circadian Clocks,* J. Aschoff (ed.). North-Holland, Amsterdam, pp. 125-138.

Bullough, W., and Laurence, E. (1964). Mitotic control by internal secretion: The role of the chalone-adrenalin complex. *Exp. Cell Res.* 33:176-194.

Burns, F. J., and Tannock, I. F. (1970). On the existence of a G phase in the cell cycle. *Cell Tissue Kinet.* 3:321-334.

Burton, A. C. (1971). Cellular communication, contact inhibition, cell clocks, and cancer: The impact of the work and ideas of W. R. Loewenstein. *Perspect. Biol. Med.* 14: 301-318.

Burton, A. C., and Canham, P. B. (1973). The behavior of coupled biochemical oscillators as a model of contact inhibition of cellular division. *J. Theor. Biol.* 39:555-580.

Campbell, A. (1964). The theoretical basis of synchronization by shifts in environmental conditions. In *Synchrony in Cell Division and Growth,* E. Zeuthen (ed.). Wiley, New York, pp. 469-484.

Chin, B., Friedrich, P. D., and Bernstein, I. A. (1972). Stimulation of mitosis following fusion of plasmodia in the myxomycete *Physarum polycephalum. J. Gen. Microbiol.* 71:93-101.

Cohen, M. H. (1972). Models of clocks and maps in developing organisms. In *Some Mathematical Questions in Biology,* Jack Cowan (ed.). American Mathematics Society, Prividence, Rhode Island, pp. 3-32.

Devi, V. R., Guttes, E., and Guttes, S. (1968). Effects of ultraviolet light on mitosis in *Physarum polycephalum. Exp. Cell Res.* 50:589-598.

Edmunds, L. N. (1974a). Circadian clock control of the cell developmental cycle in synchronized cultures of *Euglena.* In *Mechanisms of Regulation of Plant Growth,* R. L. Bieleski, A. R. Ferguson, and M. M. Cresswell (eds.). Bulletin 12, The Royal Society of New Zealand, Wellington, pp. 287-297.

Edmunds, L. N. (1974b). Phasing effects of light on cell division in exponentially increasing cultures of *Tetrahymena* grown at low temperatures. *Exp. Cell Res.* 83:367-379.

Edmunds, L. N. (1975). Temporal differentiation in *Euglena*: Circadian phenomena in non-dividing populations and in synchronously dividing cells. *Colloques Internationaux du CNRS les Cycles Cellulaires et Leur Blochee Chez Pouesieurs Protistes* 240:53-67.

Edmunds, L. N. (1977). Clocked cell cycle clocks. *Waking Sleeping* 1:227-252.

Ehret, C. F. (1974). The sense of time: Evidence for its molecular basis in the eukaryotic gene-action system. In *Advances in Biology and Medical Physics,* vol. 15, J. H. Lawrence and J. W. Gofman (eds.). Academic, New York, pp. 47-77.

Ehret, C. F., and Trucco, E. (1967). Molecular models for the circadian clock. I. The chronon concept. *J. Theor. Biol.* 15:240-262.

Ehret, C. F., and Wille, J. J. (1970). The photobiology of circadian rhythms in protozoa and other eucaryotic micro-organisms. In *Photobiology of Micro-organisms,* (Chap. 13), Per Halldal (ed.). Wiley, New York, pp. 370-416.

Elliott, S. G., and McLaughlin, C. S. (1978). Rate of macromolecular synthesis through the cell cycle of the yeast *Saccharomyces cerevisiae. Proc. Natl. Acad. Sci. USA* 75:4384-4388.

Engelberg, J. (1968). On deterministic origins of mitotic variability. *J. Theor. Biol.* 20: 249-259.

Fantes, P. A., Grant, W. D., Pritchard, R. H., Sudbery, P. E., and Wheales, A. E. (1975). The regulation of cell size and the control of mitosis. *J. Theor. Biol.* 50:213-244.

Frankel, J. (1962). The effect of heat, cold and p-fluorophenylalanine on morphogenesis in synchronized *Tetrahymena pyriformis. CR Trav. Lab. Carlsberg* 33:1-52.

Frankel, J. (1969). The relationship of protein synthesis to cell division and oral development in synchronized *Tetrahymena pyriformis* GL-C: An analysis employing cycloheximide. *J. Cell. Physiol.* 74:135-148.

Gilbert, D. A. (1974). Nature of the cell cycle and the control of cell proliferation. *Biosystems* 5:197-206.

Goodwin, B. C. (1963). *Temporal Organization in Cells: A Dynamic Theory of Cellular Control Processes.* Academic, New York.

Goodwin, B. C. (1966). An entertainment model for timed enzyme synthesis in bacteria. *Nature* 209:477-481.

Guttes, E., Devi, V. R., and Guttes, S. (1969). Synchronization of mitosis in *Physarum polycephalum* by coalescence of postmitotic and premitotic plasmodial fragments. *Experientia* 25:615-616.

Halberg, F. (1977). Implications of biologic rhythms for clinical practice. *Hosp. Pract.* 12:139-149.

Halvorson, H., Carter, B., and Tauro, P. (1971). Synthesis of enzymes during the cell cycle. In *Adv. Microb. Physiol.* 6:47-106.

Hamburger, K. (1962). Division delays induced by metabolic inhibitors in synchronized cells of *Tetrahymena pyriformis. CR Trav. Lab. Carlsberg* 32:359-370.

Hara, K., Tydeman, P., and Kirschner, M. (1980). A cytoplasmic clock with the same period as the division cycle in *Xenopus* eggs. *Proc. Natl. Acad. Sci. USA* 77:462-466.

Hartwell, L., Culotti, J., Pringle, J., and Reid, B. (1974). Genetic control of the cell division cycle in yeast. *Science* 183:46-51.

Howard, F. (1932). Nuclear division in plasmodia of *Physarum. Ann. Bot.* 46:451-458.

Jung, C., and Rothstein, A. (1967). Cation metabolism in relation to cell size in synchronously grown tissue culture cells. *J. Gen. Physiol.* 50:917-932.

Kauffman, S. (1969). Metabolic stability and epigenesis in randomly constructed genetic nets. *J. Theor. Biol.* 22:437-467.

Kauffman, S. (1974). Measuring a mitotic oscillator: The arc discontinuity. *Bull. Math. Biol.* 36:171-182.

Kauffman, S., and Wille, J. J. (1975). The mitotic oscillator in *Physarum polycephalum. J. Theor. Biol.* 55:47-93.

Kirschner, M. W., and Gerhardt, J. C. (1981). Spatial and temporal changes in the amphibian egg. *Bioscience* 31:381-387.

Klevecz, R. R., Kros, J., and Gross, S. D. (1978). Phase response versus positive and negative division delay in animal cells. *Exp. Cell Res.* 116:285-290.

Leigh, E. G. (1965). On the relation between the productivity, diversity, and stability of a community. *Proc. Natl. Acad. Sci. USA* 53:777-781.

Llanos, J. M. E., and Nash, R. E. (1970). Mitotic circadian rhythm in a fast-growing and a slow-growing hepatoma: Mitotic rhythm in hepatomas. *J. Natl. Canc. Inst.* 44: 581-585.

Loidl, P., and Sachsenmaier, W. (1982). Control of mitotic synchrony in *Physarum polycephalum.* Phase shifting by fusion of heterophasic plasmodia contradicts a limit cycle oscillator model. *Eur. J. Cell Biol.* 28:175-179.

Mano, Y. (1975). Systems constituting the metabolic sequence in the cell cycle. *Biosystems* 7:51-65.

Minor, P. D., and Smith, J. A. (1974). Explanation of degree of correlation of sibling generation times in animal cells. *Nature* 248:241-243.

Mitchison, J. M. (1971). *The Biology of the Cell Cycle,* Cambridge University Press, Cambridge, England.

Møller, U., Larsen, J. K., and Faber, M. (1974). The influence of injected tritiated thymidine on the mitotic circadian rhythm in the epithelium of the hamster cheek pouch. *Cell Tissue Kinet.* 7:231-239.

Newton, Isaac (1730). *Opticks,* 4th ed., reprinted by Dover, New York (1952), p. 401.

Pritchard, R. H., Barth, P. T., and Collins, J. (1969). Control of DNA synthesis in bacteria. *Symp. Soc. Gen. Microbiol.* 19:263-297.

Rao, P. N., Hittelman, W. N., and Wilson, B. A. (1974). Mammalian cell fusion. Part 6. Regulation of mitosis in binucleate HeLa cells. *Exp. Cell Res.* 90:40-46.

Rao, P. N., and Johnson, R. (1970). Mammalian cell fusion: Studies on the regulation of DNA synthesis and mitosis. *Nature* 255:159-164.

Rasmussen, L., and Zeuthen, E. (1962). Cell division and protein synthesis in *Tetrahymena,* as studied with p-fluorophenylalanine. *CR Trav. Lab. Carlsberg* 32:333-358.

Remy, U. (1968). Ph.D. Thesis. Mediz. Fak., Ruprecht-Karl-Universität, Heidelberg.

Rössler, O. E. (1976a). An equation for continuous chaos. *Phys. Lett.* 57A:397-398.

Rössler, O. E. (1976b). Chaos in abstract kinetics: Two prototypes. *Bull. Math. Biol.* 39: 275-289.

Rössler, O. E. (1976c). Chaotic behavior in simple reaction systems. *Z. Naturforsch.* 31a: 259-264.

Rössler, O. E. (1976d). Chemical turbulence: Chaos in a simple reaction-diffusion system. *Z. Naturforsch.* 31a:1168-1172.

Rössler, O. E. (1978). Chemical turbulence—a synopsis. In *Synergetics,* H. Haken (ed.). Lecture Notes in Physics, Springer-Verlag, Berlin.

Rusch, H. P., Sachsenmaier, W., Beherns, K., and Gruter, V. (1966). Synchronization of mitosis by the fusion of the plasmodia of *Physarum polycephalum. J. Cell. Biol.* 31:204-209.

Rustad, R. C. (1970). Variations in the sensitivity to X-ray induced mitotic delay during the cell division cycle of the sea urchin egg. *Radiat. Res.* 42:498-512.

Sachsenmaier, W. (1976). Control of synchronous nuclear mitosis in *Physarum polycephalum.* In *The Molecular Basis of Circadian Rhythms,* J. W. Hastings and H. G. Schweiger (eds.). Dahlem Konferenzen, Berlin, pp. 409-420.

Sachsenmaier, W., Donges, K. H., Rupff, H., and Czihak, G. (1970). Advanced initiation of synchronized mitosis in *Physarum polycephalum* following UV-irradiation. *Z. Naturforsch.* 256:866-871.

Sachsenmaier, W., Remy, U., and Plattner-Schobel, R. (1972). Initiation of synchronous mitosis in *Physarum polycephalum. Exp. Cell Res.* 73:41-48.

Sander, G., and Pardee, A. (1972). Transport changes in synchronously growing CHO and L cells. *J. Cell. Physiol.* 80:267-272.

Scheffey, C., and Wille, J. J. (1978). Cycloheximide-induced mitotic delay in *Physarum polycephalum. Exp. Cell Res.* 113:259-262.

Scheving, J. E., and Pauly, J. E. (1973). Cellular mechanisms involving biorhythms with emphasis on those rhythms associated with the S and M stages of the cell cycle. *Int. J. Chronobiol.* 1:269-286.

Sel'kov, E. E. (1970). Two alternative auto-oscillatory steady states in the metabolism of the thiols; Two alternative types of cell multiplication: Normal and malignant. *Biofizika* 15:1065-1073.

Sel'kov, E. E. (1973). Analysis of temporal organization of multienzyme systems. In *Proceedings of the IV International Biophysics Congress, Puschino.* Academy of Sciences USSR, Moscow, pp. 437-475 (in Russian).

Sel'kov, E. E., and Sozinov, L. A. (1970). Stable circadian rhythms as a property of cell populations. *Life Sci. Space Res.* 8:157-167.

Shields, R. (1976). New view of the cell cycle. *Nature* 260:193-194.

Shields, R. (1977). Transition probability and the origin of variability in the cell cycle. *Nature* 267:704-707.

Shilo, B., Shilo, V., and Simchen, G. (1976/77). Cell cycle initiation in yeast follows first-order kinetics. *Nature* 264:767-770. Matters arising. *Nature* 267:647-649.

Smith, J. A., and Martin, L. (1973). Do cells cycle? *Proc. Natl. Acad. Sci. USA* 70: 1263-1267.

Sudbery, P. E., and Grant, W. D. (1976). The control of mitosis in *Physarum polycephalum. J. Cell Sci.* 22:59-65.

Sweeney, B. M., and Hastings, J. (1958). Rhythmic cell division in populations of *Gonyaulax polyedra. J. Protozool.* 5:217-224.

Tagauchi, Y. H., and Tabachnick, J. (1974). The effect of clipping guinea-pig hair and chronic radio dermatitis on diurnal circadian rhythms in epidermal labeling and mitotic indices. *Arch. Dermatol. Forsch.* 249:167-177.

Thormar, H. (1959). Delayed division in *Tetrahymena pyriformis* induced by temperature changes. *CR Trav. Lab. Carlsberg* 31:207-225.
Tyson, J. J. (1979). Periodic enzyme synthesis: Reconsideration of the theory of oscillatory repression. *J. Theor. Biol.* 80:27-38.
Tyson, J. J. (1983a). Unstable activator models for size control of the cell cycle. *J. Theor. Biol.* submitted.
Tyson, J. J. (1983b). Periodic enzyme synthesis and oscillatory repression. *J. Theor. Biol.* submitted.
Tyson, J. J. (1979). Periodic enzyme synthesis: Reconsideration of the theory of oscillatory repression. *J. Theor. Biol.* 80:27-38.
Tyson, J. J., and Kauffman, S. (1975). Control of mitosis by a continuous biochemical oscillation: synchronization; spatially inhomogeneous oscillations. *J. Math. Biol.* 1:289-310.
Tyson, J. J., and Sachsenmaier, W. (1978). Is nuclear division in *Physarum* controlled by a continuous limit cycle oscillator? *J. Theor. Biol.* 73:723-737.
Wille, J. J. (1979). Biological rhythms in protozoa. In *Biochemistry and Physiology of Protozoa,* vol. 2, M. Levandowsky (ed.). Academic, New York, pp. 67-144.
Wille, J. J., Scheffey, C., and Kauffman, S. A. (1977). Novel behavior of mitotic clock in *Physarum. J. Cell Sci.* 27:91-104.
Winfree, A. T. (1972). Oscillatory glycolysis in yeast: The pattern of phase resetting by oxygen. *Arch. Biochem. Biophys.* 149:388-401.
Winfree, A. T. (1974). Patterns of phase compromise in biological cycles. *J. Math. Biol.* 1:73-95.
Winfree, A. T. (1980). *The Geometry of Biological Time.* Springer-Verlag, New York.
Yoneda, M., Ikeda, M., and Washitani, S. (1978). Periodic change in the tension at the surface of activated non-nucleate fragments of sea-urchin egg. *Dev. Growth. Diff.* 20:329-336.
Zeeman, E. C. (1977). *Catastrophe Theory,* Addison Wesley, Reading, Massachusetts.
Zeuthen, E., and Williams, N. E. (1969). Division-limiting morphogenetic processes in *Tetrahymena.* In *Nucleic Acid Metabolism, Cell Differentiation and Cancer Growth,* E. V. Cowdry and S. Seno (eds.). Pergamon Press, Oxford, pp. 203-216.

# PART II

# MODELING OF THE CELL DIVISION CYCLE

## CELL CYCLE KINETICS AND VARIABILITY

One theme that occurs throughout many chapters in this book is that the cell division cycle (CDC) is a time-measuring process of sorts that under certain circumstances may itself be modulated or controlled by an oscillator or clock that appears to lie outside the CDC (see Part IV). Implicit in such considerations is the notion of some degree of precision and accuracy on the part of the basic oscillator or timing sequences (Edmunds and Adams, 1981), yet we are faced paradoxically with a disturbing amount of variability in the length of the CDC of most cells (typically the standard deviation of cell cycle times is 10-20% of the mean). This variability in traverse time of the complete CDC has been observed with bacteria, yeasts, algae, and cultured mammalian cells, and has been extended to other CDC stages, such as the entry of mammalian $G_1$ cells into the S phase. The sources of this variability are unknown, but the undeniable fact of its existence has important implications from both a theoretical standpoint (e.g., in attempts to model the CDC) and a practical point of view (e.g., for cancer chemotherapy utilizing sequential treatments on partially synchronized cell populations).

In the lead-off chapter of this second part (Chap. 5), Smith* and Laurence review the various types of deterministic and probabilistic models which have been posited to explain CDC variability and discuss the experimental and statistical limitations inherent

---

**Not* the Jim A. Smith of "transition probability" fame (Smith, J. A., and Martin, L., 1973) but the John A. Smith of the Ludwig Institute for Cancer Research and known for his concern about the limitations of cell kinetics in distinguishing among models for the CDC (Smith, J. A. et al., 1981)!

in the interpretation of data on the kinetics of cell division. This is followed (Chap. 6) by a comprehensive literature survey and statistical analysis by Marguerite Guiguet et al., demonstrating that variabilities in the durations of not only the $G_1$ phase, but also of the S and $G_2$ phases, are quite significant and must be incorporated into models for the control of the CDC. The last two chapters, by Erik Thorud et al.* (Chap. 7) and by Niels Keiding et al. (Chap. 8), give detailed experimental evidence of pronounced variations in cell population kinetics for a number of mammalian tissues.

Indeed, it is clear from these papers and the work of others that this variation may be periodic: statistically significant diurnal (and circadian) rhythms are often observed in the rate of cell flux, as measured by the rate of mitotic (or S-phase) influx and efflux, and in the distribution and duration of the phases of the CDC in a number of different types of mammalian tissues. These findings perhaps help resolve earlier conflicting reports (Gelfant et al., 1982) and are quite important not only because of their implications for the regulation of the CDC by autonomously oscillatory clocks (Part IV), but also because of their chronobiological consequences for cancer chemotherapy (the theme of Part V).

## REFERENCES

Edmunds, L. N., Jr., and Adams, K. A. (1981). Clocked cell cycle clocks. *Science* 211: 1002-1013.

Gelfant, S., Ozawa, A., Chalker, D. K., and Smith, G. (1982). Circadian rhythms and differences in epidermal and dermal cell proliferation in uninvolved and involved psoriatic skin in vitro. *J. Invest. Dermatol.* 78:58-62.

Smith, J. A., Laurence, D. J. R., and Rudland, P. S. (1981). Limitations of cell kinetics in distinguishing cell cycle models. *Nature* 293:648-650.

Smith, J. A., and Martin, L. (1973). Do cells cycle? *Proc. Natl. Acad. Sci. USA* 70: 1263-1267.

*Included in the "et al." is the very same Ole Didrik Laerum who, perhaps, should be better recognized for his startling observation in 1978 that chalones reverse the cell cycle. He has considered the far-ranging biological implications of this discovery for aging, the mysterious disappearance of persons, and even the abandonment of newly fertilized eggs by sperm ("leaving only the ectasy behind") in a recent archival paper (*J. Irreproducible Results* 26:21-23, 1980).

# 5

# Models of the Cell Cycle

## The Role of Kinetic Studies

**John A. Smith and Donald J. R. Laurence** / Ludwig Institute for Cancer Research, Royal Marsden Hospital, Sutton, Surrey, England

### INTRODUCTION

It is now 50 years since the first attempts were made with bacterial populations to construct a model of the cell cycle from observations on the kinetics of cell division (Rahn, 1932). After an initial period in which very few studies were made, there has recently been a steady increase in the number of attempts to define the cell cycle in terms of measurements of cell cycle time distribution and of times of transit through the phases of the cycle.

Until the 1960s, modeling was directed toward an interpretation of the percentage of labeled mitosis (PLM) curves resulting from pulse-labeling studies in vivo in order to define the kinetics of cell division in normal tissues and tumors subjected to various therapeutic measures. However, as demonstrated by Hartmann et al. (1975), adequate representation of the PLM curves could be obtained from a variety of different models. The PLM curve was therefore an insensitive method for defining cell models or determining their adequacy at a biological level.

Recently, therefore, efforts have continued with in vitro systems in which the behavior of individual cells can be observed. It is the purpose of this chapter to discuss the implications of kinetics as a possible guide to cell cycle mechanisms.

### MODELS OF THE CELL CYCLE

Experimental studies on a wide range of bacterial and animal cells have emphasized the variability of cell cycle times within an apparently uniform population of cells. Typically, the standard deviation of cell cycle times is 10-20% of the mean. The source of this variation is unknown. According to the *probabilistic* models, this variation is the result of the existence within the cycle of a number of critical events or transitions that must occur if the cell is to progress through the cycle. A contrasting group of *deterministic* models attributes the variation to variation in those parameters, environmental and within the cell, that determine the cycle overall.

### Probabilistic Models

The probabilistic models divide the cell cycle into a number of phases associated with events that determine whether the cell will remain within the phase or will progress to a subsequent phase. In the simple theory the probability of an event is independent of the time that the cell has been resident within the phase. Each event then gives rise to an exponential distribution of residence times within the phase.

In the earlier models the cell cycle was assumed to be filled with such critical events that were completed either in an overlapping manner (Rahn, 1932) or sequentially (Kendall, 1948). On these assumptions, with the additional proviso that the probabilities of the events were equal, it was possible to estimate the number of such events from the coefficient of variation of the distribution of cell cycle lines. For the overlapping model 260-130,000 events were required, while for the sequential model 23-93 events were indicated.

The later discovery that DNA synthesis (S phase) occupied only part of the total time between divisions led to analyses of the distribution of residence times in the S phase and in the gaps $G_1$ and $G_2$ before and after S phase. This led to the consensus that S and $G_2$ phases together with subsequent mitosis M and cytokinesis D had relatively limited variability. The sequence $S + G_2 + M + D$ was then considered to be a "determined" sequence (Sisken and Kinoshita, 1961), with the variability of the cell cycle resident mainly in $G_1$.

Under these circumstances it was no longer necessary or possible to relate the number of events to the coefficient of variation. In 1970, Burns and Tannock proposed that a single transition together with a parameter expressing cell loss could account for the PLM curves that they examined. Smith and Martin (1973) were led to a similar model, which they called the transition probability (TP) model from a study of data from in vitro observations on a variety of cell systems. The overall distribution of cell cycle times was clearly not defined by a single exponential distribution and some additional, at that time undefined, variation was postulated in the determined phase. However, if attention was given to differences between daughter cells (siblings) from a single division (Minor and Smith, 1974) a clear correspondence with an exponential generator was obtained.

Later Brooks et al. (1980) modified this model by suggesting that this additional variation was due to a second exponential generator located in the mother cell but expressed in the siblings. The original generator defined an A phase, and the additional generator was associated with a Q phase.

### Deterministic Models

The large variation of cell cycle times would appear to be a difficulty in the application of deterministic models to cell cycle data. The behavior of a determined part of the cell cycle provides an example of the way a deterministic model should perhaps be expected to behave. However, attention can be focused on the complexity of cellular processes determining the division cycle and the resultant variation attributed to a large number of small perturbations operating within a determined sequence.

Addition of a large number of small errors is associated with a normal distribution of errors even if the individual errors are not themselves normally distributed. There is still some evidence that cell cycle times or entry into the determined sequence may be distributed normally (Hersh and Kitos, 1980). Kubitschek (1966, 1971) has taken the rate of passage through the cell cycle to be normally distributed, in which case the distribution of cell cycle times is distributed inverse normally. If the composition of errors were by multiplication rather than addition a log normal distribution might be expected (Koch, 1966a).

These *transformed* normal distributions accord with the probabilistic models in predicting a positive skew to the distribution of cell cycle times. Unlike the probabilistic models, these alternative models do not require the enumeration of the critical events that determine the variation. In effect the variation is not an essential part of the deterministic model but is rather an observable feature. Correlates of this variation can be useful indications of the nature of the determination in the model.

A deterministic model can be pictured as a tunnel into which the cell enters at birth and from which it emerges at division. A probabilistic phase is more like a waiting room from which the cell may emerge with constant probability at any time during residence therein. This sort of analogy may explain the insight gained by Smith and Martin (1973) in developing the transition probability model. It had been observed that cells with a long cell cycle time reacted much more rapidly than would be expected to changes in culture conditions. This would not perhaps be expected from a tunnel model unless the tunnel was everywhere permeable and infinitely long.

## LIMITATIONS IN INTERPRETATION

Observations and interpretations of kinetics of cell division present difficult problems. There is a complex interaction between limitations of technique and statistical analysis. As most of the statistical limitations are a consequence of technique we shall discuss the practical difficulties first.

### Experimental Limitations

The failure of PLM curves to reflect a change in cell cycle model effectively rules out the possibility of choosing a model on the basis of in vivo studies. With time-lapse cinematography it is possible in vitro to obtain detailed information on the variation in cycling behavior of cell populations. The criteria necessary for making such studies were discussed by Powell (1955) and more recently by Skehan and Friedman (1979).

Homogeneity of the cellular populations is extremely desirable. It is sometimes tacitly assumed that well-established cell lines can be considered as homogeneous with respect to heritable differences in cell cycle behavior. If any such difference exists it may be expected that the rapidly growing strains would soon predominate, thereby removing the heterogeneity. Variability of cell cycle time exists even when the population is derived from a single ancestral cell (Shields, 1977). However, inherent differences may occur in the observation chamber if the conditions there differ significantly from those obtaining in the cultures in which the stock is maintained (Hughes, 1955). Pardee et al. (1979) and Koch (1980) have illustrated the difficulties in model making that can arise if the population is heterogeneous.

A related problem is to discriminate dead or moribund cells from those cells capable of generating a large progeny (McQuilkin and Earle, 1962). Different fibroblast cell lines can differ in the number of nondividing cells they contain (Shields, 1977; Kapp and Klevecz, 1976).

Statistical analysis requires an adequate unbiased sample of cell cycle times. The adequacy of the sample is limited by the low densities at which some cells must be grown, so that increases in density during the experiment do not inhibit cell division (Castor, 1968). It is usually necessary to observe cells over several generations in order to obtain enough observations. This can lead to interaction among the progeny that do not always separate to give a uniform density (Froese, 1966). In old cultures there is evidence of changes in the correlations between successive generations in their cell cycle times (Collyn-d'Hooghe et al. 1977).

Powell (1955) recommended that if a generation were included in the analysis it was important to ensure that all branches of the cell pedigree down to that generation were complete. This is a counsel of perfection that is rarely attainable with mammalian cell cultures. If the experiment is terminated before this is achieved there is danger of cutoff bias in the sample. With an exponential distribution of cell cycle times and 200 cells, all but 10 will have divided at three times the average duration of the exponential phase (equal to the standard deviation for an exponential distribution). To pronounce the last cell "dead" (at the 5% level of significance) it would be necessary to wait a further five times the average duration. Some cell systems show a progressive decrease in rate of cycling with time in culture, and the conflict between a requirement for a constant rate of division and the need to avoid cutoff bias is sometimes irreconcilable (Skehan and Friedman, 1979). A complete determination of the profile of cell division is impossible unless the average cycle time is about 24 hr or less, and little can be deduced about the mechanism of slowly cycling cells.

In order to avoid some of these problems the alternative of using synchronized cell populations seems attractive. Recent work suggests that it is difficult to synchronize a cellular population without perturbing the cells' metabolism to some extent (Mitchison, 1981).

### Statistical Limitations

The main limitations to any statistical analysis are shortage of material for analysis and an unrepresentative sample. The problem of bias in the samples was discussed in the previous section. The small sample size restricts discrimination between models to fairly simple alternatives. If the number of parameters in alternative models is large, it is possible to adapt them to match a small amount of data but we may not expect the cell to conform with a simple model as a compensation for our technical limitations.

The statistics used to analyze the data need to reflect the shape of the cell cycle distribution rather than the absolute dimensions of it, though it is useful to discuss magnitudes as well as shape in considering location in time (see Location of Variation) and control (see Control Mechanisms). Statistics relating to shape are dimensionless quantities, such as skewness $\gamma_1$, as given by $\mu_3{}^{2/3}/\mu_2$, and kurtosis $\gamma_2$, which is defined as $\mu_4/\mu_2{}^2-3$. Under some circumstances the coefficient of variation $\gamma_0$, or $\mu_2{}^{1/2}/\mu_1$ can also be helpful. Here, $\mu_i$ is the ith moment (about the mean for $i > 1$). Castor (1980). used a measure of curvature defined as the ratio of slopes in two selected regions of the cumulative probability curve, but we (Smith et al., 1981) have used differences between sample and theoretical curves on the cumulative probability axis—the Kolmogorov method. For grouped data the chi-squared test can be useful (Nelson and Green, 1981).

An additional complication of the cell cycle time distributions is the correlations that exist between siblings and, in some cell systems, between mother and daughter cells. The mother-daughter correlations appear to be weak and variable, but the intrasibling pair correlations are important. Correlation can interfere with tests of statistical significance. In the extreme case of complete correlation, if one sibling time were measured accurately, the other would supply no additional information and should be discarded from the sample. A further problem is that the overall cell cycle time distribution, that we shall term the *α distribution* for historical reasons (see Graphic Representations) could be considered as complex, containing two sources of variation (inter- and intrasibling pair variation) that may be independent. Analysis of the α distribution would then be difficult as only very simple models could be tested without introducing a large number of parameters.

In order to isolate a single source of variation, Minor and Smith (1974) analyzed intrasibling pair differences. An additional advantage of this approach is that estimation of a determined phase common to the sibling is avoided. This estimation is not difficult if the residual variation is symmetrical, but many models predict a skewed α distribution. The distribution of intrasibling pair differences we shall call a *β distribution,* once again for historical reasons. If the α distribution is exponential, then so is the β distribution. A disadvantage of the analysis of β distributions is that they are by their nature positively skewed and somewhat leptokurtic like the exponential curve, and this feature can be confusing if it is required to use the exponential distribution as a null hypothesis. The coefficient of variation $\gamma_0$ of a β distribution is a more useful statistical parameter (see Convergence of Models) as it involves a moment of the distribution less subject to cutoff bias.

The assumption of independence in the inter- and intrasibling variation can be tested by partitioning the data according to the minimum intrasibling pair times before generating β distributions (Shields, 1978).

## MODELING WITH A COMPUTER

Cell cycle studies are very labor intensive, and it is desirable before starting such studies to establish whether the proposed experiments are capable of making the desired discriminations. Such feasibility studies are easily made using a computer where the data from a few hundred "model" experiments can be transferred into disk-based files and subsequently analyzed by computer using a variety of statistical procedures with an alternative model as reference set. Generally such comparisons between models will be somewhat optimistic, so that if a discrimination fails for the computer-derived data it is likely to fail experimentally in the terms of the original intention of the experiment.

### Graphic Representations

The histogram is a relatively unbiased method of representing cell cycle time data and has been widely used for comparing grouped α distributions with model-generated envelopes (Nelson and Green, 1981). With near-normal distributions a probit plot can be helpful (Kubitschek, 1966).

Recently, the α and β distributions have been plotted on semilogarithmic cumulative frequency plots, known as α and β curves (Shields, 1978). This type of plot tends to emphasize exponential features in the tail of the distribution that is presented on an expanded scale. However, the upper 50% of the distribution is consequently contracted and deviations of as much as 5% on the cumulative probability axis can be no larger than the dimensions of a sample point in some published data.

Graphs can be helpful in orientating further research, but a linear graph can arise from an essentially "nonlinear" model, or vice versa if the sample size is small. The task of scoring such errors by eye on a regular statistical basis is formidable and requires at least a double-blind trial with a panel of judges using nonparametric statistics.

### The Classic Models

Features such as curvature and nonparallelism that are appreciated by visual observations can be converted to numerical values and examined statistically. A computer-generated sample of 1000 "experiments" with the proposed number of cells in each experiment can be used to establish confidence limits. For the reference model (equivalent to the null hypothesis of statistical theory) the limit can be set by choosing the frequency of type I errors, namely, rejection of the chosen reference model if it is true. Then comparison can be made with an alternative model to find out how often this different model will be accepted as within limits for the reference model (type II errors).

Shields (1978) showed that on large samples it was possible to distinguish readily between a normal distribution and an exponential distribution. A discrimination based on skewness $\gamma_1$ for $\alpha$ distributions on relatively small samples is shown in Fig. 1. Included in this diagram are the confidence limits for the one (E1) and two (E2) transition exponential models with a two-sided frequency of type I errors of 5% in each case. The model of Brooks et al. (1980) would lie between the E1 and E2 limits, depending on the relative value of the time constants of the two transitions.

A clear separation between normal and exponential distributions is possible on samples of more than 100 cells. The data quoted by Hersh and Kitos (1980) based on time-lapse methods fall into the region between the lower limit for the E2 model and the upper limit for the normal distribution. However, it is difficult to estimate what influence cutoff bias has had on these data.

Discrimination between the inverse and log normal models was attempted by Schmid (1967) and by Kubitschek (1971) using the chi-squared test. Subsequent developments, outlined in Convergence of Models, have tended to undermine the assumptions on which these comparisons were made. Nelson and Green (1981) have found that a number of different models can be made to fit experimental $\alpha$ distributions by suitable choice of parameters using the chi-squared test as an estimate of goodness of fit.

### Location of the Variation

As noted in Deterministic Models, the introduction of a determined sequence ($S + G_2 + M + D$) into the mitotic cycle resulted in concentration of the transition postulated by the TP model to within the remaining $G_1$ phase of the cycle. In a sequence of transitions the kinetic data cannot by themselves help to decide the order or location of the transitions. The variation ascribed to the A phase was firmly located in $G_1$ on the assumption that sibling pairs are at the same stage of their preparation for cell division at their formation by division of the mother cell. In other cell systems it has been found that sibling pair differences are associated with the $G_2$ rather than the $G_1$ phase (van Wijk et al., 1977; Nurse and Fantes, 1981). With the inverse and log normal distributions, the

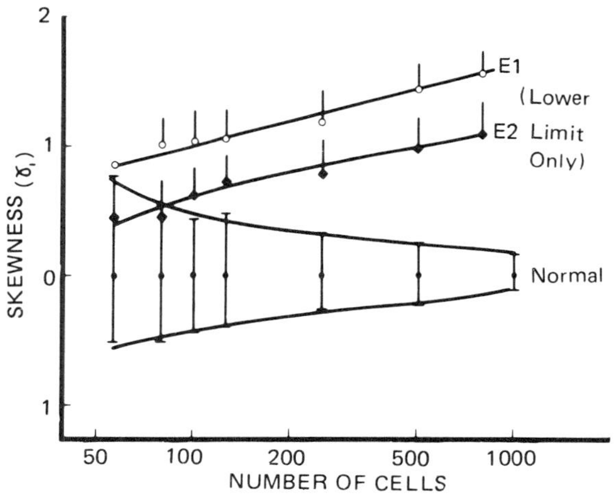


**FIGURE 1** Computer-derived confidence limits for the skewness $\gamma_1$ as a function of sample size for the normal distribution and for exponential generators with one generator (E1) or two with equal time constants in sequence (E2). The one-sided probability of a sample laying outside a limit is 2.5% in each case.

divisions of the cell cycle into a variable phase (usually $G_1$) and a determined phase produces a different result. The transformation of the normal curve requires the location of an "origin" for the transformation, known as the threshold, that locates the point on the transformed curve that is converted to a point at infinity when the reverse transformation (log or inverse) is applied to convert the curve to the normal form. The distance between this point and the mean determines the shape of the transformed normal curve.

If the variation is transferred from the cycle as a whole to a part of the cycle, such as $G_1$, a paradox is created, since the resulting contraction of the interval appears to result in a change of shape of the transformed normal distributions, whereas we know that subtraction of a constant phase cannot change the overall variation. The paradox can only be resolved if the assumption made in the classic inverse and log normal models is incorrect at least for animal cells, namely, that the threshold coincides with the beginning of the cycle and the mean with the end of the cycle.

Unlike the TP model, the deterministic models tend to assume that intrasibling pair variation is predetermined at birth by unequal division of the mother cell resulting in a difference of size between the siblings or some inequality in their internal composition. When differences in size are not apparent, as with *Tetrahymena* species (Prescott, 1959), it is possible that the variation may be determined prior to cell division by the unequal assembly of structural components that are duplicated and passed on to each daughter cell. Environmental factors operating on the siblings may also contribute to the variability (Froese, 1966) in some systems.

### Convergence of Models

The apparently exponential nature of intrasibling pair variation (as revealed by the $\beta$ curves) has been considered as unequivocal evidence of a transition point in the cell cycle. Figure 2 shows that the appearance of these $\beta$ curves cannot be considered as a certain criterion of a single transition, as a two-transition model (E2) for the intrasibling variation

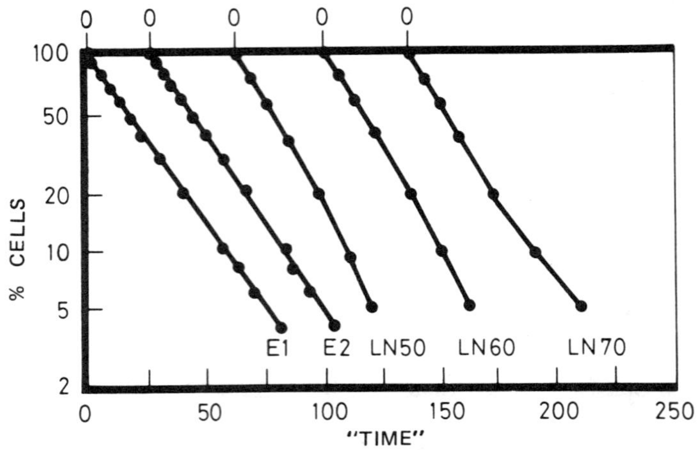


**FIGURE 2** Computer-derived $\beta$ curves based on 500 sibling pairs for the exponential generators described in Fig. 1 and for log normal generators with thresholds 50, 60, and 70 on a scale described by Smith et al. (1981). The origin for each curve is given on the top axis.

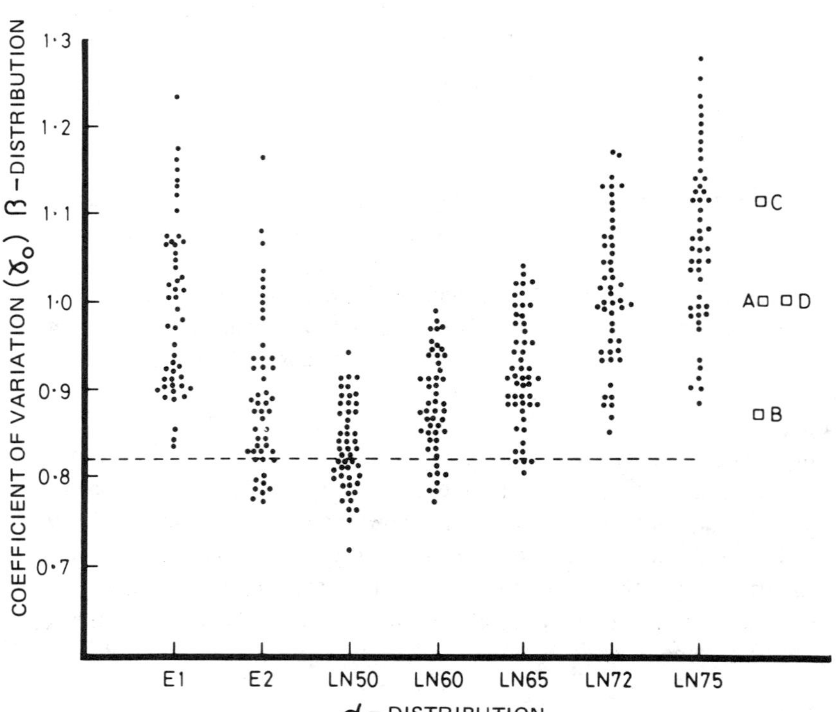


**FIGURE 3** Computer-derived coefficients of variation $\gamma_0$ for exponential and log normal generators on the convention indicated in Fig. 2. The dotted line is the lower confidence for the E1 model. Points A, B, C, and D are values for experiments quoted by Brooks et al. (1980, Table 1). The computer-generated points were calculated on samples of 200 "cells" each.

also gives a fairly linear $\beta$ curve. We have seen that the deterministic model does not define the location of variation within the cycle with any certainty. If the threshold or the mean of the log or inverse normal transformation is treated as a variable, it is also possible to produce quasilinear $\beta$ curves or else plots curved in either sense depending on the value of the threshold (Castor, 1980; Smith et al., 1981) (see also Fig. 2). In Fig. 3, a computer simulation of the coefficient of variation $\gamma_0$ of the $\beta$ distribution is plotted for the E1 and E2 models together with selected log normal models. At the edge of this figure are the values of $\gamma_0$ calculated from Table 1 of Brooks et al. (1980). Good agreement is found with the E1 model that they proposed for the $\beta$ distribution and moderate agreement with an E2 model. The log normal models can also overlap the experimental findings for selected values of the thresholds. An alternative way of producing convergence between the probabilistic and deterministic models was suggested by Svetina (1977). It was proposed that the transition probability is not constant as assumed in the TP model but is a function of the time of residence in the phase. On the basis of the tunnel and waiting room analogy, we can think of this model as a tunnel (like the deterministic model) but with walls of graded permeability without a defined exit point at the end of the tunnel.

As the models become similar in their statistical profile a larger number of observations of individual cells is required in order to distinguish between them (Smith et al., 1981). Unfortunately, real cell systems rarely follow simple models exactly, so extremely fine discriminations necessary for distinguishing between the different models cannot be made.

## THE CELL CYCLE IN REALITY

A cell cycle model should do more than determine the probable locations of successive cell divisions. If some additional landmarks were available within the period defined by the divisions it would be possible to obtain a more detailed account of cell kinetics and to approach the definition of cell age based on phase-specific markers. Much of the interest in cell cycle model kinetics is in establishing control mechanisms with the possibility of exerting control of proliferation on an experimental or therapeutic basis. Finally, we need to consider whether there is evidence for longer term mechanisms that continue to exert controls of the cell division cycle over many cell generations.

### Landmarks in the Cell Cycle

The concept of cell cycle landmarks was developed by Hartwell (1974) in connection with the yeast cell cycle. In practical terms it is important to select those aspects of the cycle that can be timed with greatest accuracy and least effort. With multiple landmarks the internal structure of the cycle can be explored and expressed in kinetic terms.

Mitchison (1981) has reviewed the use of enzyme levels as landmarks in animal cells. At one time it was considered that enzyme levels varied in a regular sequence during the cycle, but as already mentioned many of these findings were artefacts due to the methods for synchronizing the cell population. The $G_1$ phase has usually been regarded as a necessary preparation for S phase, but Mariani et al. (1981) found that an enzyme needed for DNA synthesis can be made in adequate amounts while synthesis is occurring. Much effort has been given to defining physiological or genetic factors that will block the cell cycle, thereby establishing landmarks at the site of the block. Unfortunately not all such blocks are stable (Das, 1981), and contradictions can arise if different criteria are used to interpret the data. Shilo et al. (1976), using yeast, obtained evidence of a probabilistic transition by studying variability of budding after release from a block. Naha et al. (1975), using mammalian cells and delay in entering S phase, excluded a probabilistic step by filling $G_1$ with a series of deterministic phases.

It is tacitly assumed in interpretation of block-release experiments that cycle phases tend to synchronize at a block. However, it is possible that cells become similar in some features of their cycle-related behavior but increasingly different in other features. The variation and lag experienced after release could then reflect the need of a cell to reco-ordinate its cycle-related functions. Block-release experiments are useful in demonstrating for a variety of blocks, differing quantitatively or qualitatively, that the variability after release was expressed prior to S phase with a constant progression through the rest of the cell cycle (Rudland et al., 1974).

### Control Mechanisms

According to the transition probability models, one or both of the transitions are also important control points. It was observed that for a number of mammalian cell populations an increase in cell cycle time is not accompanied by a corresponding increase in minimum cycle time. The increase is associated with an increase in variability, and this association of variability with control gives further interest to the study of origin of the variability. The deterministic models would place control in a system monitoring cell size (Nurse and Fantes, 1981), RNA content (Darzynkiewicz et al., 1979), or some related parameter capable of coupling growth with initiation of DNA synthesis.

Both models tend to place control in $G_1$, but the kinetics of entry into DNA synthesis appear to differ in some systems from the kinetics of the cell division cycle (Sisken and Morasca, 1965; van Wijk et al., 1977; Ronning et al., 1980).

### Long-Term Correlation and Clocks

Koch, in 1966, made the surprising observation that in bacteria the timing of successive generations did not show increasing variation if time was measured from the start of the first division. Bremer and Chuang (1981) showed that if bacteria were grown on a minimal glucose medium in order to separate the waves of DNA synthesis for successive generations, the synthesis of DNA was much more regular than the timing of cell division. This accords with the current view that the end of DNA synthesis gives a signal for subsequent cell division in bacteria. These observations appear to be in striking contrast with the probabilistic approach initiated 50 years ago by Rahn (1932). Schafer and Cleffmann (1982) studied a *Tetrahymena* mutant with consistently unequal division into siblings of different size. Regulation toward the normal size in this system was achieved over several generations by slowing the rate of division of the small cell, all phases of the cycle being prolonged. With mammalian cells it is rare to find systems in which there is a consistent pattern of cell division over several generations. However, van Wijk and van de Poll (1979) and van Zoelen et al. (1981) have described hepatoma and neuroblastoma lines with significant correlation between the cycle times of cousins, suggesting a "grandmaternal" contribution to the control of the cell cycle.

Whether a long term periodicity exists in the cell cycle and a clock mechanism is implicated cannot be decided by this type of observation alone. It would not be difficult to specify a clock mechanism that accords with the existing observations on the distribution of cycle times over a single generation, but this would just add further candidates to the existing rush of models that can explain the limited data now available. Perhaps the process of cell division is in any case too long established and too important to all of us to be left to chance.

## ACKNOWLEDGMENTS

We would like to thank Professor A. M. Neville and Dr. P. S. Rudland for their interest and encouragement.

## REFERENCES

Bremer, H., and Chuang, L. (1981). The cell cycle in *Escherichia coli*. B/r. *J. Mol. Biol.* 88:47-81.

Brooks, R. F., Bennett, D. C., and Smith, J. A. (1980). Mammalian cell cycles need two random transitions. *Cell* 19:493-504.

Burns, F. J., and Tannock, I. F. (1970). On the existence of a $G_0$-phase in the cell cycle. *Cell Tissue Kinet.* 3:321-334.

Castor, L. N. (1968). Contact regulation of cell division in an epithelial cell line. *J. Cell. Physiol.* 72:161-172.

Castor, L. N. (1980). A $G_1$ rate model accounts for cell-cycle kinetics attributed to "transitional probability'. *Nature* 287:857-859.

Collyn-d'hooghe, M., Valleron, A., and Malaise, E. P. (1977). Time lapse cinematography studies of cell cycle and mitosis duration. *Exp. Cell Res.* 106:405-407.

Darzynkiewicz, Z., Evenson, D. P., Staiano-Coico, L., Sharpless, T. K., and Melamed, M. L. (1979). Correlation between cell cycle duration and RNA content. *J. Cell. Physiol.* 100:425-438.

Das, M. (1981). Initiation of nuclear DNA replication: Evidence for formation of committed prereplicative cellular state. *Proc. Natl. Acad. Sci. USA* 78:5677-5681.

Froese, G. (1966). Division delay in HeLa cells and Chinese hamster cells. A time-lapse study. *Int. J. Radiat. Biol.* 10:353-367.

Hartmann, N. R., Gilbert, C. W., Jansson, B., MacDonald, P. D. M., Steel, G. G., and Valleron, A. (1975). A comparison of computer methods for the analysis of fraction labelled mitosis curves. *Cell Tissue Kinet.* 8:119-124.

Hartwell, L. H. (1974). *Saccharomyces cerevisiae* cell cycle. *Bacteriol. Rev.* 38:164-198.

Hersh, R. T., and Kitos, P. A. (1980). Is $G_1$ normally distributed? *J. Theoret. Biol.* 86: 117-122.

Hughes, W. H. (1955). The inheritance of differences in growth rate in *Escherichia coli*. *J. Gen. Microbiol.* 12:265-268.

Kapp, L. N., and Klevecz, R. R. (1976). The cell cycle of low passage and high passage human diploid fibroblasts. *Exp. Cell Res.* 101:154-158.

Kendall, D. G. (1948). On the role of variable generation time in the development of a stochastic birth process. *Biometrika* 35:316-330.

Koch, A. L. (1966a). The logarithm in biology. I. Mechanisms generating the log-normal exactly. *J. Theor. Biol.* 12:276-290.

Koch, A. L. (1966b). On evidence supporting a deterministic process of bacterial growth. *J. Gen. Microbiol.* 43:1-5.

Koch, A. L. (1980). Does the variability of the cell cycle result from one or many events? *Nature* 286:80-82.

Kubitschek, H. E. (1966). Normal distribution of cell cycle generation rates. *Nature* 209: 1039-1040.

Kubitschek, H. E. (1971). The distribution of cell generation times. *Cell Tissue Kinet.* 4: 113-122.

Mariani, B. D., Slate, D. L., and Schimke, R. T. (1981). S-phase specific synthesis of dihydrofolate reductase in Chinese hamster ovary cells. *Proc. Natl. Acad. Sci. USA* 78:4985-4989.

McQuilkin, W. T., and Earle, W. R. (1962). Cinemicrographic analysis of cell populations in vitro. *J. Natl. Cancer Inst.* 28:763-799.

Minor, P. D., and Smith, J. A. (1974). Explanation of degree of correlation of sibling generation times in animal cells. *Nature* 248:241-243.

Mitchison, J. M. (1981). Changing perspectives in the cell cycle. In *The Cell Cycle,* P. L. C. John (ed.). Cambridge University Press, Cambridge, pp. 1-10.

Naha, P. M., Meyer, A. L., and Hewitt, K. (1975). Mapping of the G1 phase of a mammalian cell cycle. *Nature* 258:49-53.

Nelson, S., and Green, P. J. (1981). The random transitional model of the cell cycle, a critical review. *Cancer Chemother. Pharmacol.* 6:11-18.

Nurse, P., and Fantes, P. A. (1981). Cell cycle controls in fission yeast: A genetic analysis. In *The Cell Cycle,* P. L. C. John (ed.). Cambridge University Press, Cambridge, pp. 85-98.

Pardee, A. B., Shilo, B.-Z., and Koch, A. L. (1979). Variability of the cell cycle. *Cold Spring Harb. Conf. Cell Proliferation* 6:373-392.

Powell, E. O. (1955). Some features of the generation times of individual bacteria. *Biometrika* 42:16-44.

Prescott, D. M. (1959). Variations in the individual generation times of *Tetrahymena gelii* HS. *Exp. Cell Res.* 16:279-284.

Rahn, O. (1932). A chemical explanation of the variability of the growth rate. *J. Gen. Physiol.* 15:257-276.

Ronning, O. W., Lindmo, T., Petterson, E. O., and Seglen, P. O. (1980). Kinetics of entry into S-phase and into the G1 phase of the subsequent cell cycle for synchronized NHK cells. In *Flow Cytometry IV.* D. D. Laerum, T. Lindmo, and E. Thorud (eds.). Universitetsforlaget, Bergen, Norway, pp. 350-353.

Rudland, P. S., Seifert, W., and Gospodarowicz, D. (1974). Growth control in cultured mouse fibroblasts: Induction of the pleiotropic and mitogenic responses by a purified growth factor. *Proc. Natl. Acad. Sci. USA* 71:2600-2604.

Schafer, E., and Cleffmann, G. (1982). Division and growth kinetics of the division mutant *conical* of *Tetrahymena. Exp. Cell Res.* 137:277-284.

Schmid, P. (1967). Temperature adaption of the growth and division process of *Tetrahymena pyriformis. Exp. Cell Res.* 45:471-486.

Shields, R. (1977). Transition probability and the origin of variation in the cell cycle. *Nature* 267:704-707.

Shields, R. (1978). Further evidence for a random transition in the cell cycle. *Nature* 273:755-758.

Shilo, B., Shilo, V., and Simchen, G. (1976). Cell cycle initiation in yeast follows first order kinetics. *Nature* 264:767-770.

Sisken, J. E. and Kinoshita, R. (1961). Variations in the mitotic cycle *in vitro. Exp. Cell Res.* 22:521-525.

Sisken, J. E., and Morasca, L. (1965). Intrapopulation kinetics of the mitotic cycle. *J. Cell. Biol.* 25 (part 2):179-189.

Skehan, P., and Friedman, S. J. (1979). Cellular senescence and the analysis of generation time distributions. *Cell Biol. Int. Rep.* 3:535-542.

Smith, J. A., Laurence, D. J. R., and Rudland, P. S. (1981). Limitations of cell kinetics in distinguishing cell cycle models. *Nature* 293:648-650.

Smith, J. A., and Martin, L. (1973). Do cells cycle? *Proc. Natl. Acad. Sci. USA* 70:1263-1267.

Svetina, S. (1977). An extended transition probability model of the variability of cell cycle generation times. *Cell Tissue Kinet.* 10:575-581.

van Wijk, R., and van de Poll, K. W. (1979). Variability of cell generation times in a hepatoma cell pedigree. *Cell Tissue Kinet.* 12:659-663.

van Wijk, R., van de Poll, K. W., Amecz, W. J. C., and Geilenkirchen, W. L. M. (1977). Studies of the variations in generation times in rat hepatoma cells in culture. *Exp. Cell Res.* 109:371-379.

van Zoelen, E. J. J., van der Saag, P. T., and de Laat, S. W. (1981). Family tree analysis of a transformed cell line and the transition probability model for the cell cycle. *Exp. Cell Res.* 131:395-406.

# 6

# A Systematic Study of the Variability of Cell Cycle Phase Durations in Experimental Mammalian Systems

**Marguerite Guiguet, Jean-Jacques Kupiec, and Alain-Jacques Valleron** / Université Paris 7 and INSERM, Paris, France

## INTRODUCTION

This chapter is aimed at presenting quantitative data concerning the variability in cell cycle and cell phase durations within a given cell system (intrapopulation variability) and the variations in cell kinetic parameters from one cell system to another (interpopulation variability). The study of these two sources of variability is important from both a practical and a theoretical point of view.

From a practical point of view, these two sources of variability explain to a great extent the lack of success of anticancer sequential therapies based on the idea of synchronization. Indeed, the differences between cell phase durations within cell populations, either normal or tumoral, lead to a rapid desynchronization after the use of an age-specific drug, and the gain in choosing an optimal time for giving a second therapy is therefore small. The tumor-to-tumor variability makes it difficult to choose standardized treatment protocols for patients, because this optimal time between two successive therapies is very dependent on the patient under treatment.

From a theoretical point of view, many modelers of the cell cycle, on the assumption that the most variable phase in the cell cycle was $G_1$, have claimed that the major control of the cell cycle was located in $G_1$ (Smith and Martin, 1973; Shields et al., 1978). Still more recently, Cooper (1979), although denying any role to the $G_1$ phase in the control of the cell cycle, wrote: "To summarize a large amount of data, it seems that the S and $G_2$ periods are fairly constant time periods." But this assertion needs to be qualified, and "in spite of all this, it is clear that, the duration of B phase which includes the S, $G_2$ and M phases is not fixed and its variability may have a more important part in proliferative control than has been realized" (Brooks et al., 1980).

In spite of this reservation, we are still faced with a stereotype according to which the $G_1$ phase is the most variable phase in the cell cycle and the S and $G_2$ durations are relatively constant. As our practical experiences of analyzing cell kinetics of various systems did not fit with this stereotype, we have undertaken a systematic review of the literature presenting data on the mean durations and on the variabilities of the durations of the phases of the cell cycle. Due to the enormous number of papers related to the subject, we restricted this review to our current field of research: mammalian tumor and normal cells in experimental systems.

## MATERIAL AND METHODS

### Material

The technique that is widely used to provide information on the mean and variability of the phase durations is the fraction-labeled mitosis technique (FLM) using computer-analyzed results. Our data have been obtained through an automatic search in literature bases (MEDLINE-ELHILL, BIOSIS-ORBIT, SCI-DIALOG). All papers concerning experimental studies on mammalian cells quoting one of the six computer methods of FLM analysis have been sorted. The six computer methods are those of Hartmann and Pedersen (1970), Gilbert (1972), Jansson (1974), MacDonald (1970), Steel and Hanes (1971), and Valleron et al. (1973). They have been widely used in the analysis of experimental data. In addition they have been intercompared in a cooperative work (Hartmann et al., 1975).

In addition, we have manually reviewed three journals (Cell Tissue Kinetics, since 1970, and *Cancer* and *Cancer Research,* since 1975), and the book by Steel (1977). We obtained a total of 186 data, which are summarized in Table 1.

### Expression of Variabilities

We deal in this chapter with variability from two different sources: the intrapopulation and interpopulation. Intrapopulation variability concerns the individual differences in phase durations between cells of the same system and is estimated from the automatic analysis of the FLM experiments. It can be expressed as the standard deviation (s) of the statistical distribution of the durations. We have chosen to express it by using the coefficient of variation (CV), which is the ratio of the standard deviation (s) to the mean m of distribution. This indeed allows an easier comparison of the variabilities of variables of different orders of magnitude. For example a phase with a mean duration of 20 hr and a standard deviation of 10 hr is obviously as variable as a phase with a mean duration of 2 hr and a standard deviation of 1 hr. This is well expressed by considering their coefficients of variation, both of which are 0.50.

Interpopulation variability concerns the differences in the kinetic parameters of the various systems studied. The kinetic parameters which we choose to characterize each cell population are the mean durations of the phases of the cell cycle and the coefficients of variation of these durations. It is worthwhile noting that it would have been possible—although troublesome—to examine the variability (between systems) of the variability (within the cell populations). It would indeed be very interesting, for example, to find that in different systems with different mean $G_1$ durations, the cell-to-cell variability is the same or—conversely—that in different systems with the same mean $G_1$ duration, the cell-to-cell variability is very different. More simply, we have presented the interpopulation variability by using conventional cumulative histograms in appropriate scales.

**TABLE 1** Mean and Variability of Phase Durations[a]

| $G_1$ | $CVG_1$ | S | CVS | $G_2$ | $CVG_2$ | Cell system | Reference[b] |
|---|---|---|---|---|---|---|---|
| | In vivo tumor systems | | | | | | |
| | Human (growing in mice) | | | | | | |
| 18.7 | 1.7 | 8.1 | 0.1 | 4.2 | 0.4 | Adenocarcinoma | Houghton, J. A., et al., *BJC,* 37:213, 1978 |
| 16.7 | 1.7 | 9.4 | 0.3 | 4.0 | 0.4 | Adenocarcinoma | Houghton, J. A., et al., *BJC,* 37:213, 1978 |
| 46.0 | 0.7 | 17.1 | 0.4 | 2.4 | 0.5 | Leukemia | Hariss et al., *CTK,* 4:433, 1971 |
| 7.4 | 1.7 | 14.4 | 0.3 | 5.6 | 0.5 | Adenocarcinoma | Pickard et al., *BJC,* 31:36, 1975 |
| 12.9 | 1.2 | 12.6 | 0.2 | 7.4 | 0.5 | Adenocarcinoma | Pickard et a.., *BJC,* 31:36, 1975 |
| 10.7 | 0.4 | 16.8 | 0.6 | 5.3 | 0.3 | Adenocarcinoma | Pickard et al., *BJC*, 31:36, 1975 |
| 20.1 | 0.1 | 24.1 | 0.7 | 4 | 0.4 | Carcinoma | Pickard et al., *BJC*, 31:36, 1975 |
| 20.9 | 0.6 | 9.8 | 0.2 | 5.1 | 0.5 | Carcinoma | Pickard et al., *BJC*, 31:36, 1975 |
| 16.8 | 0.6 | 11.5 | 0.5 | 7.8 | 0.4 | Carcinoma | Pickard et al., *BJC*, 31:36, 1975 |
| 9.7 | 1 | 16.4 | 0.5 | 4.1 | 0.5 | Carcinoma | Pickard et al., *BJC*, 31:36, 1975 |
| 12.4 | 1.3 | 12.6 | 0.4 | 5.2 | 0.5 | Carcinoma | Pickard et al., *BJC*, 31:36, 1975 |
| 19.1 | 0.8 | 12.8 | 0.4 | 5.5 | 0.6 | Carcinoma | Pickard et al., *BJC*, 31:36, 1975 |
| 21.4 | 0.5 | 15.1 | 0.5 | 5.7 | 0.5 | Melanoma | Rofstad et al., *ARORPB,* 19:261, 1980 |
| 15.5 | 0.5 | 17 | 0.9 | 5 | 0.6 | Melanoma | Rofstad et al., *CTK,* 13:163, 1980 |
| 42 | 0.7 | 23 | 0.8 | 10 | 0.8 | Melanoma | Rofstad et al., *Vir Ar B,* 24:219, 1977 |
| 21 | 0.7 | 20 | 0.7 | 9 | 0.33 | Fibrosarcoma | Rofstad et al., *Vir Ar B,* 24:219, 1977 |
| | | Mouse | | | | | |
| 15.9 | 1.2 | 8.4 | 0.1 | 3 | 0.3 | Carcinoma | Féaux de Lacroix et al., *CTK,* 14:135, 1981 |
| 7.9 | 1 | 7.8 | 0.2 | 1.2 | 1.1 | Carcinoma | Féaux de Lacroix et al., *CTK,* 14:135, 1981 |
| 6.8 | 0.8 | 6.9 | 0.2 | 1.9 | 1.2 | Carcinoma | Féaux de Lacroix et al., *CTK,* 14:135, 1981 |
| 12.5 | 0.5 | 4.6 | 0.2 | 1.9 | 0.6 | Carcinoma | Féaux de Lacroix et al., *ZKKO,* 87:181, 1976 |
| 11.9 | 1.9 | 11.9 | 0.0 | 2.3 | 2.4 | Carcinoma | Hazegawa et al., *CR,* 36:359, 1976 |
| 11.2 | 0.6 | 9.9 | 0.4 | 1.9 | 1.3 | Anaplastic | Hazegawa et al., *CR,* 36:359, 1976 |
| 2.3 | 2.4 | 11.5 | 0.3 | 2.3 | 0.7 | Lewis lung carcinoma | Simpson-Herren et al., *CTR,* 60, p1749, 1976 |
| 0.7 | 4.1 | 9.9 | 0.1 | 2 | 0.9 | Lewis lung carcinoma | Hill RP et al., *CTR,* 61:29, 1977 |
| 0.6 | 7.5 | 10.2 | 0.1 | 1.8 | 1.3 | Lewis lung carcinoma | Hill RP et al., *CTR,* 61:29, 1977 |
| 1.6 | 3.2 | 10.1 | 0.1 | 2 | 1.1 | Lewis lung carcinoma | Hill RP et al., *CTR,* 61:29, 1977 |

**TABLE 1** (Continued)

| $G_1$ | $CVG_1$ | S | SVS | $G_2$ | $CVG_2$ | Cell system | Reference[b] |
|---|---|---|---|---|---|---|---|
| 0.4 | 12.8 | 9.2 | 0.2 | 1.8 | 0.4 | Lewis lung carcinoma | Hill RP et al., *CTR,* 61:29, 1977 |
| 0.5 | 4.6 | 9.4 | 0.2 | 1.7 | 1.1 | Lewis lung carcinoma | Hill RP et al., *CTR,* 61:29, 1977 |
| 9.3 | 2.2 | 9.5 | 0.3 | 1.5 | 0.4 | Adenocarcinoma | Simpson-Herren, L., et al., *CTR,* 62:519, 1978 |
| 17 | 0.3 | 32.8 | 0.4 | 9.3 | 0.5 | Ascitic tumor nh/ly | Kopper, L., et al., *EJC,* 14:59, 1978 |
| 14.9 | 0.1 | 30.2 | 0.3 | 6.1 | 0.8 | Ascitic tumor nh/ly | Kopper, L., et al., *EJC,* 14:59, 1978 |
| 8.3 | 0.1 | 19.7 | 0.3 | 3.4 | 0.6 | Ascitic tumor nh/ly | Kopper, L., et al., *EJC,* 14:59, 1978 |
| 4 | 0.9 | 12.8 | 0.2 | 2 | 0.5 | Ascitic tumor nh/ly | Kopper, L., et al., *EJC,* 14:59, 1978 |
| 9.4 | 1.1 | 11.6 | 0.6 | 2.6 | 0.7 | EMT6 | Rockwell, S., *LAS,* 27:838, 1977 |
| 7.3 | 1.5 | 11.9 | 0.3 | 1.9 | 0.4 | EMT6 | Rockwell, S., *LAS,* 27:838, 1977 |
| 6.5 | 0.7 | 7.7 | 0.5 | 2.4 | 0.6 | Mammary tumor | Dethlefsen, L. A., et al., *CTK,* 6:173, 1973 |
| 8.3 | 0.7 | 7.2 | 0.2 | 1.3 | 0.2 | Hind limb tumor | Twentyman et al., *JNCI,* 64:595, 1980 |
| 9.3 | 0.5 | 8.5 | 0.5 | 1.1 | 0.7 | Mammary carcinoma | Wakisaka et al., *Gann,* 71:1, 1980 |
| 10.6 | 0.3 | 9.1 | 0.4 | 1.1 | 0.7 | Mammary carcinoma | Wakisaka et al., *Gann,* 71:1, 1980 |
| 51.8 | 0.6 | 14.9 | 0.7 | 1 | 0.7 | Mammary carcinoma | Wakisaka et al., *Gann,* 71:1, 1980 |
| 6.4 | 0.5 | 9.6 | 0.2 | 2.2 | 0.6 | Lung tumor | Gunduz, N. *CTK,* 14:343, 1981 |
| 7.1 | 0.4 | 8.9 | 0.2 | 2.5 | 0.5 | Lung tumor | Gunduz, N. *CTK,* 14:343, 1981 |
| 6.8 | 0.7 | 8.7 | 0.2 | 2.6 | 0.5 | Lung tumor | Gunduz, N. *CTK,* 14:343, 1981 |
| 6 | 0.4 | 8.8 | 0.2 | 2.7 | 0.6 | Lung tumor | Gunduz, N. *CTK,* 14:343, 1981 |
| 6.3 | 0.4 | 9.1 | 0.2 | 2.7 | 0.7 | Lung tumor | Gunduz, N., *CTK,* 14:343, 1981 |
| 7.7 | 0.7 | 9 | 0.2 | 2.5 | 0.5 | Lung tumor | Gunduz, N., *CTK,* 14:343, 1981 |
| 15 | 1 | 14 | 1 | 2 | 1.2 | MPSV-induced tumor | Smajda-Joffe, F., et al., *EH,* 9:137, 1981 |
| 1.4 | 1.2 | 7 | 0.5 | 0.8 | 1 | MPSV-induced tumor | Smajda-Joffe, F., et al., *EH,* 9:137, 1981 |
| 1 | 0 | 7.5 | 0.5 | 1.3 | 1 | MPSV-induced tumor | Smajda-Joffe, et al., *CTK,* 15:169, 1982 |
| 4.4 | 0.5 | 6.3 | 0.6 | 4.7 | 0.7 | MPSV-induced tumor | Smajda-Joffe et al., *CTK,* 15:169, 1982 |
| 5.6 | 0.4 | 25.1 | 0.3 | 5.9 | 0.5 | Erlich ascitic cells | |
| 14.5 | 0.8 | 6.1 | 0.3 | 1.3 | 0.5 | 1,2DMH-induced colonic tumor | Chang et al., *Vir Ar B,* 30:349, 1981 |
| 2 | 0 | 8.6 | 0 | 3.7 | 1 | Sarcoma SA180 | Simpson-Herren et al., *CCR,* 54:143, 1970 |
| 4.6 | 0.7 | 9.4 | 0.2 | 2.2 | 0.5 | Carcinoma KHJJ | Rockwell et al., *JNCI,* 49:735, 1972 |
| 9.4 | 1 | 11.6 | 0.6 | 2.6 | 0.7 | EMT6 sarcoma | Rockwell et al., *JNCI,* 49:735, 1972 |

| | | | | | | | |
|---|---|---|---|---|---|---|---|
| 5.6 | 0.5 | 7.3 | 0.2 | 1.5 | 0.6 | EMT6 sarcoma | Watson et al., *CTK*, 9:147, 1976 |
| 6 | 0.8 | 7.8 | 0.2 | 1.7 | 0.8 | EMT6 sarcoma | Watson et al., *CTK*, 9:147, 1976 |
| 7.4 | 1.2 | 8.4 | 0.3 | 1.9 | 0.3 | EMT6 sarcoma | Watson et al., *CTK*, 9:147, 1976 |
| 7.5 | 0.4 | 9.9 | 0.3 | 1.8 | 0.5 | EMT6 sarcoma | Watson et al., *CTK*, 9:147, 1976 |
| | | Mouse | | | | | |
| 4 | 0.8 | 10.5 | 0.3 | 3 | 0.7 | Carcinoma EAT solid | Tannock, *CR*, 29:1527, 1969 |
| 15 | 1.6 | 6 | 0.2 | 2 | 0.5 | Carcinoma CH3 | Denekamp, J., *CR*, 30:393, 1970 |
| 8.8 | 1 | 9.5 | 0.5 | 2.2 | 0.4 | Carcinoma NT1 | Begg, A. C., *CTK*, 4:401, 1971 |
| 9 | 1.4 | 20 | 0.5 | 4 | 0.4 | Sarcoma NT1 | Begg, A. C., *CTK*, 4:401, 1971 |
| 3 | 0.7 | 7.2 | 0.2 | 3 | 0.3 | Carcinoma C2H/HE | Tannock, I. F., *CR*, 30:2470, 1970 |
| | | Rat | | | | | |
| 8 | 0.5 | 8 | 0.2 | 3 | 0.5 | Sarcoma BICR/M1 | Steel et al., *BJC*, 20:784, 1966 |
| 5.6 | 0.1 | 10.2 | 0.5 | 3 | 0.3 | Sarcoma BICR/M1R | Rajewski, M. F., *ECR* 60:269, 1970 |
| 4 | 0.5 | 8 | 0.1 | 2 | 1 | Fibrosarcoma RUB5 | Denekamp, J., *CR*, 30:393, 1970 |
| 10 | 0.5 | 8 | 0.2 | 3 | 0.5 | Sarcoma SSO | Denekamp, J., *CR*, 30:393, 1970 |
| 26 | 0.3 | 11 | 0.4 | 3 | 0.5 | Sarcoma SSB1 | Denekamp, J., *CR*., 30:393, 1970 |
| 30.5 | 0.9 | 6 | 0.3 | 2.1 | 0.6 | Hepatoma | Evans et al., *CTK*, 10:233, 1977 |
| 15.2 | 0.8 | 6.5 | 0.2 | 4.3 | 0.3 | Hepatoma | Evans et al., *CTK*, 10:233, 1977 |
| 34.4 | 0.9 | 7.4 | 0.4 | 6.1 | 0.4 | Hepatoma | Evans et al., *CTK*, 10:233, 1977 |
| 3.5 | 1 | 5.4 | 0.1 | 1.8 | 0.6 | Adenocarcinoma | Braunschweiger et al., *CTR*, 64:293, 1980 |
| 10.8 | 1.2 | 7.8 | 0.3 | 2.5 | 0.4 | Sarcoma | Lang, W., et al., *EJC*, 14:431, 1978 |
| 16.3 | 0.9 | 10.5 | 0.5 | 3 | 0.8 | Sarcoma | Lang, W., et al., *EJC*, 14:431, 1978 |
| 10.8 | 1.2 | 8.3 | 0.2 | 1.8 | 0.8 | Adenocarcinoma | Moore, et al., *CTK*, 10:583, 1977 |
| 8.7 | 0.4 | 6.5 | 0.5 | 2.1 | 0.9 | Adenocarcinoma | Moore, et al., *CTK*, 10:583, 1977 |
| 1.5 | 1.3 | 10.2 | 0.3 | 1.1 | 0.6 | Leukemia | Hariss et al., *CTK*, 4:71, 1971 |
| 3.3 | 1.1 | 7.7 | 0.3 | 2.1 | 0.3 | Leukemia | Vilpo et al., *CTK*, 6:489 |
| 6.4 | 1.1 | 9.8 | 0.4 | 1.5 | 0.5 | Mammary tumor | Fu KF et al., *CTK*, 12:493, 1979 |
| 15.6 | 0.6 | 7.2 | 0.3 | 2.4 | 0.4 | Mammary tumor | Fu KF et al., *CTK*, 12:493, 1979 |
| 16.3 | 0.7 | 8.1 | 0.5 | 1.9 | 0.4 | Mammary tumor | Fu KF et al., *CTK*, 12:493, 1979 |
| 10.5 | 0.6 | 9.5 | 0.2 | 4 | 0.3 | Chondrosarcoma CSDC11 | Dawson et al., *CTK*, 9:547, 1976 |
| 50 | 0.8 | 10 | 0.3 | 3 | 0.3 | Fibrosarcoma | Steel et al., *BJC*, 20:784, 1966 |
| 26 | 0.9 | 19 | 0.4 | 2.6 | 0.3 | Fibrosarcoma BICR/A9 | Steel et al., *BJC*, 25:802, 1971 |
| 15 | 0.5 | 15 | 0.3 | 2.6 | 0.2 | Fibrosarcoma BICR/A9 | Steel et al., *BJC*, 25:802, 1971 |
| 9.3 | 2.4 | 9.2 | 0.3 | 2.7 | 0.4 | Fibrosarcoma BICR/A9 | Steel et al., *BJC*, 25:802, 1971 |

**TABLE 1** (Continued)

| $G_1$ | $CVG_1$ | S | SVS | $C_2$ | $CVG_2$ | Cell system | Reference[b] |
|---|---|---|---|---|---|---|---|
| 8 | 1.5 | 6.2 | 0 | 3.5 | 0.3 | Nitrosamine-induced hepatoma | Rabes, H., *ZKKO*, 72:176, 1969 |
| 5.6 | 0.2 | 12 | 0.4 | 1.4 | 1.2 | Mammary carcinoma | Simpson-Herren et al., *CR*, 33:2415, 1973 |
| 6.6 | 0.6 | 9.3 | 0.4 | 2.4 | 0.5 | Mammary carcinoma | Simpson-Herren et al., *CR*, 33:2415, 1973 |
| 9.5 | 0.9 | 8.1 | 0.3 | 2 | 0.4 | Mammary carcinoma | Simpson-Herren et al., *CR*, 33:2415, 1973 |
| 10.8 | 1.2 | 7.8 | 0.3 | 2.5 | 0.4 | Induced kidney carcinoma | Zobl et al., *EJC*, 11:159, 1975 |
| 16.5 | 1.2 | 6.5 | 0.3 | 3.3 | 0.4 | Induced kidney carcinoma | Zobl et al., *EJC*, 11:159, 1975 |
| 16.3 | 0.9 | 10.5 | 0.5 | 3 | 0.8 | Induced kidney carcinoma | Zobl et al., *EJC*, 11:159, 1975 |
| | | Hamster | | | | | |
| 6.5 | 0.7 | 7.6 | 0.4 | 2.6 | 0.3 | Cheek-pouch tumor | Reiskin et al., *CR*, 28:898, 1968 |
| 4.5 | 0.5 | 7.4 | 0.1 | 2.6 | 0.3 | Cheek-pouch tumor | Reiskin et al., *CR*, 28:898, 1968 |
| 3 | 0.5 | 5.7 | 0.2 | 2.2 | 0.7 | Cheek-pouch tumor | Reiskin et al., *CR*, 28:898, 1968 |
| | | Dog | | | | | |
| 61 | 0.8 | 5.3 | 0.3 | 5.4 | 0.3 | Mammary carcinoma | Owen et al., *BJC*, 23:493, 1969 |
| | | Cultured neoplastic cells | | | | | |
| | | Human | | | | | |
| 10.6 | 0.3 | 9.9 | 0.3 | 9.9 | 0.3 | Lymphoma | Drewinko, B., et al., *CTK*, 11:177, 1978 |
| 45.2 | 0.6 | 35.9 | 0.5 | 8.3 | 0.6 | Leukemia | Fried, J., et al., *CBR*, 7:333, 1974 |
| 5 | 0.3 | 12.1 | 0.3 | 2.1 | 0.3 | Leukemia | Fried, J., et al., *CBR*, 7:333, 1974 |
| 8.1 | 0.7 | 11.4 | 0.2 | 3.1 | 0.5 | Leukemia | Fried, J., et al., *CBR*, 7:333, 1974 |
| 30.2 | 0.7 | 16.2 | 0.2 | 3.8 | 0.5 | Leukemia | Fried, J., et al., *CBR*, 7:333, 1974 |
| 47.7 | 0.6 | 31 | 0.6 | 7.3 | 0.7 | Leukemia | Fried, J., et al., *CBR*, 7:333, 1974 |
| 61.3 | 1 | 42.3 | 0.4 | 3.9 | 0.6 | Leukemia | Fried, J., et al., *CBR*, 7:333, 1974 |
| 3.7 | 1 | 16.5 | 0.4 | 4.6 | 0.2 | Leukemia | Yen, A., et al., *ECR*, 95:295, 1975 |
| 3.8 | 0.6 | 15.1 | 0.2 | 5.4 | 0.2 | Leukemia | Foa et al., *CTK*, 15:399, 1982 |
| 2.4 | 1.1 | 9.8 | 0.1 | 4 | 0.5 | Hepatoma AH-601 | Okumura, Y., et al., *Growth*, 39:331, 1975 |
| 0.2 | 10.5 | 9.6 | 0 | 4.2 | 0.3 | Hepatoma AH-130 | Okumura, Y., et al., *Growth*, 39:331, 1975 |
| 14.4 | 0.2 | 10.5 | 0.1 | 3.2 | 0.3 | Sarcoma LY-742 | Okumura, Y., et al., *Growth*, 39:331, 1975 |

| | | | | | | | |
|---|---|---|---|---|---|---|---|
| 14 | 0.2 | 12.9 | 0.1 | 6 | 0.2 | Sarcoma LY-336 | Okumura, Y., et al., *Growth,* 39:331, 1975 |
| 11.8 | 0.3 | 10 | 0.1 | 3 | 0.3 | Sarcoma LY-7 | Okumura, Y., et al., *Growth,* 39:331, 1975 |
| 13.8 | 0.3 | 11 | 0.1 | 4.2 | 0.3 | Sarcoma LY-5 | Okumura, Y., et al., *Growth,* 39:331, 1975 |
| 9.8 | 0.6 | 10.8 | 0.1 | 3.1 | 0.3 | Sarcoma LY-1 | Okumura, Y., et al., *Growth,* 39:331, 1975 |
| 1.4 | 1.2 | 11.1 | 0.1 | 3.1 | 0.3 | Sarcoma YS | Okumura, Y., et al., *Growth,* 39:331, 1975 |
| 9 | 0.7 | 10. | 0.2 | 2.4 | 0.5 | Glioma C-6 | Liao, C., et al., *Lab. Invest.,* 38:122, 1978 |
| 114 | 1.8 | 7.3 | 0.1 | 6.1 | 0.4 | Hepatoma H-4-II-E | Kovacs et al., *CTK,* 10:245, 1977 |
| 7.6 | 0.6 | 6.8 | 0.1 | 2.3 | 0.4 | Hepatoma H-4-II-E | Kovacs et al., *CTK,* 10:245, 1977 |
| 2.4 | 1.4 | 9.3 | 0.2 | 2.3 | 0.2 | Shay leukemia | Toivonen, H., et al., *CTK,* 14:9, 1981 |
| | | Hamster | | | | | |
| 7.4 | 0.3 | 6 | 0.2 | 5.1 | 0.4 | Carcinoma | Tetsuno, O., *Gann,* 71:381, 1980 |
| In vivo normal mammalian tissues | | | | | | | |
| | | Mouse | | | | | |
| 0.3 | 1.7 | 4 | 0.2 | 1 | 0.2 | Erythroblasts | Tarbutt et al., *CTK,* 5:491, 1972 |
| 0.5 | 0.2 | 5.5 | 0.2 | 1.3 | 0.2 | Erythroblasts | Tarbutt et al., *CTK,* 5:491, 1972 |
| 4.4 | 1 | 5.2 | 0.2 | 0.8 | 0.6 | Duodenal crypt | Dethlefsen, L. A., et al., *CTK,* 6:3, 1973 |
| 5.9 | 1 | 6.9 | 0.2 | 0.7 | 0.4 | Duodenal crypt | Dethlefsen, L. A., et al., *CTK,* 6:3, 1973 |
| 26.1 | 0.5 | 12.6 | 0.3 | 3.7 | 0.6 | Periodontal fibroblasts | Perea, et al., *J. Anat.* 133:281, 1981 |
| 20.8 | 0.3 | 10.1 | 0.3 | 3 | 0.5 | Periodontal fibroglasts | Perea, et al., *J. Anat.* 133:281, 1981 |
| 2.9 | 0.9 | 4.8 | 0.3 | 1.9 | 0.5 | Thymus | Zaitoun, A. M., et al., *CTK,* 12:191, 1979 |
| 3 | 0.8 | 4.3 | 0.3 | 2.1 | 0.4 | Thymus | Zaitoun, A. M., et al., *CTK,* 12:191, 1979 |
| 2.9 | 0.8 | 4.6 | 0.3 | 2 | 0.4 | Thymus | Zaitoun, A. M., et al., *CTK,* 12:191, 1979 |
| 9.5 | 0.2 | 8.7 | 0.2 | 4.6 | 1.3 | Uterine cervix | Hazegawa, K., *CR,* 36:359, 1976 |
| 10.1 | 0.4 | 10 | 0.3 | 1.7 | 2.3 | Colon | Chang, W. L., et al., *Vir Ar B,* 30:349, 1979 |
| 10.4 | 0.3 | 9.8 | 0.4 | 1 | 0.8 | Colon | Chang, W. L., et al., *Vir Ar B,* 30:349, 1979 |
| 10.1 | 0.3 | 9.8 | 0.4 | 1 | 0.8 | Colon | Chang, W. L., et al., *Vir Ar B,* 30:349, 1979 |
| | | Rat | | | | | |
| 30.5 | 0.6 | 6.5 | 0.2 | 1.1 | 0.3 | Bone | Kimmel, D. B., et al., *CTI,* 32:123, 1980 |
| 6.8 | 0.4 | 11.6 | 0.2 | 4.6 | 0.4 | Cartilage cells | Dixon, *CTK,* 4:21, 1971 |
| 44.5 | 0.9 | 6.5 | 0 | 3.2 | 1.1 | Tibia cartilage | Walker et al., *CTK, 5:401, 1972* |
| 9.1 | 0.1 | 21.6 | 0.6 | 3.9 | 1 | Leukocytes | Spry, C. J. F., *CTK,* 4:351, 1971 |
| 0.9 | 0.6 | 7.4 | 0.1 | 2.1 | 0.5 | Leukocytes | Spry, C. J. F., *CTK,* 4:351, 1971 |

**TABLE 1** (Continued)

| $G_1$ | $CVG_1$ | S | CVS | $G_2$ | $CVG_2$ | Cell system | Reference[b] |
|---|---|---|---|---|---|---|---|
| 0.6 | 0.5 | 6.9 | 0.1 | 0.1 | 1 | Erythroblasts | Porcellini, C., et al., *PSEBM*, 153:125, 1976 |
| 1 | 0.4 | 6.6 | 0.1 | 0.1 | 3 | Erythroblasts | Porcellini, C., et al., *PSEBM*, 153:125, 1976 |
| 0.4 | 1.5 | 6.8 | 0.1 | 0.2 | 0.5 | Erythroblasts | Porcellini, C., et al., *PSEBM*, 153:125, 1976 |
| 0.8 | 0.8 | 6.7 | 0.1 | 0.1 | 1 | Erythroblasts | Porcellini, C., et al., *PSEBM*, 153:125, 1976 |
| 0.5 | 0.4 | 7.1 | 0.2 | 0.3 | 0.3 | Erythroblasts | Porcellini, C., et al., *PSEBM*, 153:125, 1976 |
| 0.5 | 0.4 | 7.1 | 0.1 | 0.3 | 0.7 | Erythroblasts | Porcellini, C., et al., *PSEBM*, 153:125, 1976 |
| 1.9 | 0.7 | 7 | 0.3 | 1.2 | 1.3 | Erythroblasts | Porcellini, C., et al., *PSEBM*, 153:125, 1976 |
| 1.4 | 1.1 | 7.9 | 0.3 | 1.8 | 0.6 | Erythroblasts | Porcellini, C., et al., *PSEBM*, 153:125, 1976 |
| 107 | 1.35 | 8.7 | 0.4 | 5.2 | 0.3 | Thyroid follicle | Christov-Tzelkov et al., *CTK*, 15:275, 1982 |
| 5.7 | 0.8 | 8.6 | 0.1 | 1.2 | 0.2 | Jejunal crypt | Hutaifa, S. A. D., et al., *CTK*, 7:587, 1974 |
| 4 | 1.1 | 7.1 | 1 | 1.2 | 0.3 | Jejunal crypt | Hutaifa, S. A. D., et al., *CTK*, 7:587, 1974 |
| 3.5 | 0.4 | 6.4 | 0.2 | 1.3 | 0.3 | Jejunal crypt | Hutaifa, S. A. D., et al., *CTK*, 7:587, 1974 |
| 3.6 | 0.4 | 5.9 | 0.2 | 1.4 | 0.3 | Jejunal crypt | Hutaifa, S. A. D., et al., *CTK*, 7:587, 1974 |
| 3.9 | 0.4 | 5.9 | 0.2 | 1.2 | 0.3 | Jejunal crypt | Hutaifa, S. A. D., et al., *CTK*, 7:587, 1974 |
| 3.2 | 0.4 | 6 | 0.2 | 1.5 | 0.4 | Jejuanl crypt | Hutaifa, S. A. D., et al., *CTK*, 7:587, 1974 |
| 3.5 | 0.6 | 6.5 | 0.2 | 1.3 | 0.3 | Jejunal crypt | Hutaifa, S. A. D., et al., *CTK*, 7:587, 1974 |
| 2.8 | 0.8 | 6.4 | 0.1 | 1.3 | 0.4 | Jejunal crypt | Hutaifa, S. A. D., et al., *CTK*, 7:587, 1974 |
| 3.1 | 0.6 | 6.3 | 0.1 | 1.3 | 0.5 | Jejunal crypt | Hutaifa, S. A. D., et al., *CTK*, 7:587, 1974 |
| 2.2 | 0.3 | 6.4 | 0.2 | 1.4 | 0.4 | Jejunal crypt | Hutaifa, S. A. D., et al., *CTK*, 7:587, 1974 |
| 1.9 | 0.3 | 6.7 | 0.2 | 1.4 | 0.5 | Jejunal crypt | Hutaifa, S. A. D., et al., *CTK*, 7:587, 1974 |
| 2.7 | 0.6 | 6.5 | 0.2 | 1.2 | 0.5 | Jejunal crypt | Hutaifa, S. A. D., et al., *CTK*, 7:587, 1974 |
| 2.8 | 0.6 | 6.5 | 0.1 | 1.3 | 0.5 | Jejunal crypt | Hutaifa, S. A. D., et al., *CTK*, 7:587, 1974 |
| 3.6 | 0.7 | 6.8 | 0.2 | 1.6 | 0.4 | Jejunal crypt | Hutaifa, S. A. D., et al., *CTK*, 7:587, 1974 |
| 4.5 | 0.6 | 8.1 | 0.2 | 1.2 | 0.3 | Jejunal crypt | Hutaifa, S. A. D., et al., *CTK*, 7:587, 1974 |
| 5.3 | 0.8 | 9.9 | 0.3 | 1.8 | 0.3 | Brain | Lewis, P. D., et al., *BR*, 104:49,1976 |
| 6.3 | 0.8 | 9.9 | 0.1 | 1.6 | 0.4 | Brain | Lewis, P. D., et al., *BR*, 104:49, 1976 |
| 5.6 | 1.1 | 10.7 | 0.2 | 1.6 | 0.6 | Brain | Lewis, P. D., et al., *BR*, 104:49, 1976 |
| 4 | 1.5 | 11.9 | 0.1 | 1.7 | 0.5 | Brain | Lewis, P. D., et al., *BR*, 104:49, 1976 |
| 5 | 1 | 11.3 | 0.1 | 2.2 | 0.4 | Brain | Lewis, P. D., et al., *BR*, 104:49, 1976 |
| 2.2 | 0.9 | 12.9 | 0.2 | 2.7 | 0.4 | Brain | Lewis, P. D., et al., *BR*, 104:49, 1976 |
| 4.9 | 1.7 | 11.4 | 0.3 | 2.6 | 0.4 | Brain | Lewis, P. D., et al., *BR*, 104:49, 1976 |
| 4.4 | 1.3 | 11.1 | 0.2 | 1.8 | 0.7 | Brain | Lewis, P. D., et al., *BR*, 104:49, 1976 |

| | | | | | | | |
|---|---|---|---|---|---|---|---|
| 3.6 | 1.6 | 11.1 | 0.2 | 1.8 | 0.7 | Brain | Lewis, P. D., et al., *BR*, 104:49, 1976 |
| 2.6 | 2 | 12 | 0.2 | 2.3 | 0.2 | Brain | Lewis, P. D., et al., *BR*, 104:49, 1976 |
| 7.1 | 2.1 | 13.9 | 0.3 | 2.3 | 0.4 | Brain | Lewis, P. D., et al., *BR*, 104:49, 1976 |
| 0.8 | 2.4 | 12.3 | 0.2 | 3.6 | 0.5 | Brain | Lewis, P. D., et al., *BR*, 104:49, 1976 |
| 5.5 | 0.4 | 14.4 | 0.2 | 3 | 1.4 | Brain | Lewis, P. D., et al., *BR*, 104:49, 1976 |
| 6 | 1.6 | 10.4 | 0.3 | 4 | 0.4 | Brain | Lewis, P. D., et al., *BR*, 104:49, 1976 |
| 5.3 | 0.5 | 10 | 0.3 | 2.2 | 0.5 | Brain | Lewis, P. D., et al., *BR*, 104:49, 1976 |
| 8.5 | 0.7 | 9.3 | 0.2 | 3.1 | 0.4 | Brain | Lewis, P. D., et al., *BR*, 104:49, 1976 |
| 5 | 0.6 | 8.7 | 0.2 | 2.4 | 0.2 | Brain | Patel et al., *Br*, 172:57, 1979 |
| 1.1 | 2.7 | 12 | 0.2 | 2.2 | 0.1 | Brain | Patel et al., *BR*, 172:57, 1979 |
| 5.5 | 1.1 | 11.3 | 0.2 | 1.5 | 0.4 | Brain | Lewis, P. D., et al., *BR*, 202:415, 1980 |
| 7 | 0.6 | 10 | 0.1 | 1.9 | 0.6 | Brain | Lewis, P. D., et al., *BR*, 202:415, 1980 |
| 3.2 | 1.4 | 11.4 | 0.2 | 2.3 | 0.5 | Brain | Lewis, P. D., et al., *BR*, 202:415, 1980 |
| 3.6 | 0.8 | 12.6 | 0.2 | 3.1 | 0.6 | Brain | Lewis, P. D., et al., *BR*, 202:415, 1980 |
| 2.8 | 2.4 | 10.2 | 0.2 | 3.4 | 0.2 | Dentate gyrus | Lewis, P. D., et al, *NAN*, 4:191, 1978 |
| 3.8 | 1.2 | 12.1 | 0.3 | 2.8 | 0.2 | Dentate gyrus | Lewis, P. D., et al., *NAN*, 4:191, 1978 |
| 1.6 | 0.9 | 11.7 | 0.3 | 2.5 | 0.2 | Dentaty gyrus | Lewis, P. D., et al., *NAN*, 4:191, 1978 |
| | In vitro cultured Cells | | | | | | |
| 2.9 | 0.6 | 5.8 | 0.2 | 3 | 0.4 | Chinese Hamster cells | Mauro, F., et al., *JNCI*, 52:705, 1974 |
| 10.3 | 1 | 7.4 | 0.4 | 3.4 | 0.7 | Chinese Hamster cells | Mauro, F.,et al., *JNCI*, 52:705, 1974 |
| 0.6 | 5 | 6.9 | 0.2 | 2.5 | 0.3 | Mouse cells | Marschall et al., *JCP*, 102:209, 1980 |
| 2.6 | 0.5 | 9 | 0.2 | 2.5 | 0.3 | Mouse-Chinese hamster hybrid | Marschall et al., *JCP*, 102:209, 1980 |
| 17.1 | 0.7 | 10.8 | 0.4 | 5.5 | 0.2 | Mouse-Chinese hamster hybrid | Marschall et al., *JCP*, 102:209, 1980 |
| 6.4 | 0.7 | 9.1 | 0.2 | 4.2 | 0.3 | Mouse-Chinese hamster hybrid | Marschall et al., *JCP*, 102:209, 1980 |
| 5.4 | 1 | 9.3 | 0.3 | 3.2 | 0.4 | Mouse-Chinese hamster hybrid | Marschall et al., *JCP*, 102:209, 1980 |

[a]Data collected are presented as explained in text: each line gives parameters of the cell cycle estimated in the sytem studied. The following parameters are given: in the first, third, and fifth columns, the mean durations of $G_1$, S, and $G_2$, respectively; in the second, fourth, and sixth columns are the variabilities of these durations inside the cell population. Variabilities are expressed by coefficients of variation (CV = s/m).

[b]For convenience only the first author is indicated in the reference of the article. Journals are indicated according to the following abbreviated form: *ARORPB* (*Acta Radiol. Oncol. Radiat. Phys. Biol.*); *BJC* (*Br. J. Cancer*); *BR* (*Brain Res.*); *CBR* (*Comput. Biomed. Res.*); *CR* (*Cancer Res.*); *CTI* (*Calcif. Tissue Int.*); *CTR* (*Cancer Treat. Rep.*); *EH* (*Exp. Hematol.*); *ECR* (*Exp. Cell Res.*); *EJC* (*Eur. J. Cancer*); *J. Anat.* *JNCI* (*J. Nat. Cancer Inst.*); *Lab. Invest.*; *PSEBM* (*Proc. Soc. Exp. Biol. Med.*); *Vir. Ar. B* (Virchows Arch. [Cell Pathol]); ZKKO (*Z. Krebsforsch. Klin. Onkol.*).

**TABLE 2** Mean Durations of Phases in Tumoral and Normal Systems

| | Tumor cell systems (n = 114) | | | Normal cell systems (n = 72) | | |
|---|---|---|---|---|---|---|
| Phase | Median (hr) | 95% confidence interval (hr) | Variability | Median (hr) | 95% confidence interval (hr) | Variability |
| $G_1$ | 9.0 | 1.8-45.0 | 0.90 | 3.5 | 0.7-30.0 | 1.26 |
| S | 10.5 | 4.4-25.0 | 0.45 | 7.8 | 5.0-15.0 | 0.32 |
| $G_2$ | 3.2 | 1.2-8.0 | 0.50 | 1.9 | 0.7-4.8 | 0.55 |

## RESULTS

Table 1 gives the individual results obtained for the 186 systems. The results have been arbitrarily arranged according to the tumor versus normal nature of the system, then to the in vivo versus in vitro nature of the experiment.

### System-to-System Variability

The overall impressions are, first, that the figures are extremely different from one line of the table to another (interpopulation variability) and that no obvious grouping of data could make any homogeneity appear, either in the mean durations of the phases, or in the coefficients of variation of these phases. These impressions are confirmed by the statistical analysis we made (not shown) which reveals few differences between any groups of systems, and—if any—these of unclear meaning.

The range containing 95% of the values of the mean durations of $G_1$, S, and $G_2$ are represented in Table 2, and the detailed distribution of values is shown in Fig. 1. There is a striking similarity between Figs. 1a and b, which represent the distributions of the mean values of the durations of $G_1$, S, and $G_2$ in tumor and normal systems, respectively. In both cases, the cumulative histogram of $G_2$ is on the left of the figure, which confirms that, as a mean, $G_2$ is shorter than the other phases in most of the systems. The cumulative histogram of $G_1$ is more horizontal than the other two, which indicates that there is greater variation in the mean duration of $G_1$ from one system to another than in S and $G_2$.

The corresponding analysis concerning the system-to-system variation in coefficients of variation were made. The 95% ranges are shown in Table 3. The range in S coefficients of variation is noticeably smaller than concerning G1 and G2. This is well demonstrated in Fig. IIa and IIb where the cumulative histogram of S is noticeably more vertical than the other two.

### Variability Within a Given Cell System

Whatever line we look at in Table 1, it is clear that the durations of the phases are very variable from one cell to another in the system we consider. Fewer than 20% of the systems have a coefficient of variation of $G_1$, or S, or $G_2$ durations less than 0.2 (standard deviation = 0.2 mean). Moreover, there is no system in which all three phases $G_1$, S, and $G_2$ have together a CV $< 0.2$, and there is not even any system in which both S and $G_2$ fit this criterion. This is worthwhile noting because the constancy of S and $G_2$ is very often assumed by modelers of the cell cycle.

To visualize the importance of the cell-to-cell variability in phase durations, we have computed the means of the parameters given in Table 1. We obtain an average "experi-

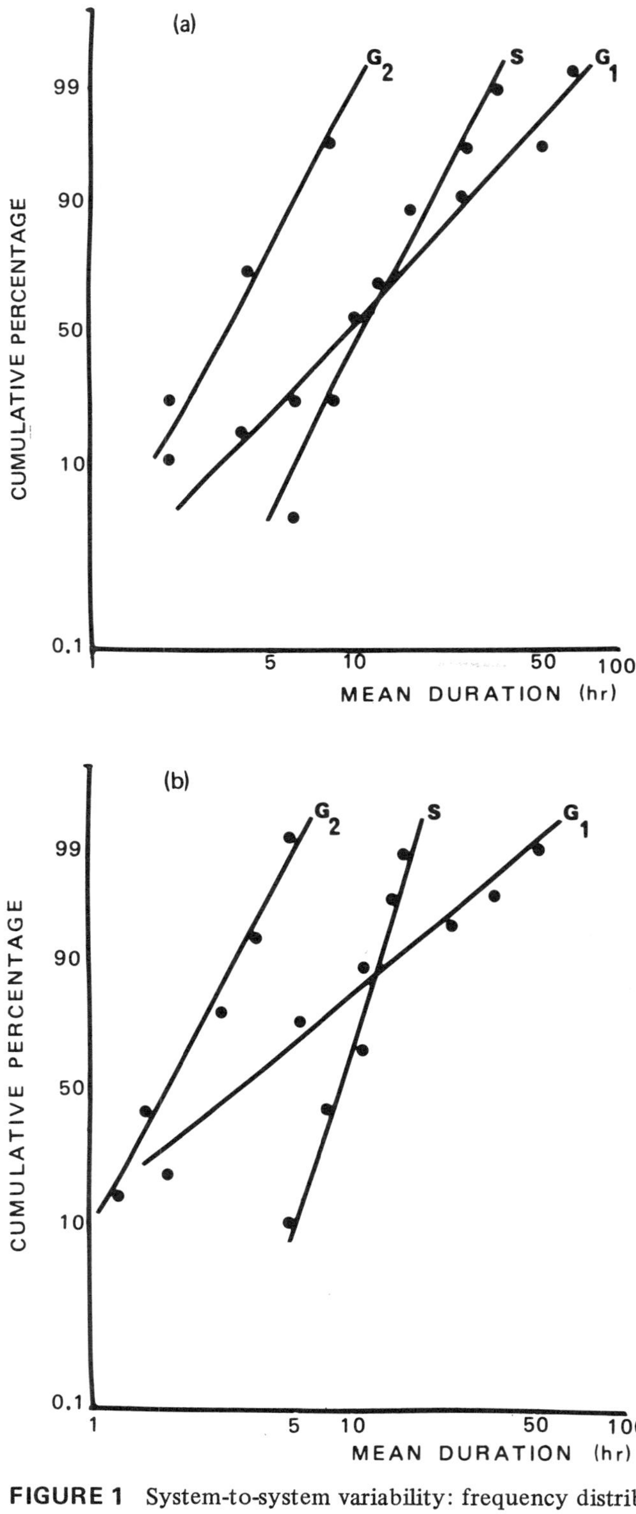


**FIGURE 1** System-to-system variability: frequency distribution of the cumulative percentage (probit scale) of (a) tumor cell systems and (b) normal cell systems as a function of the logarithm of mean phase durations.

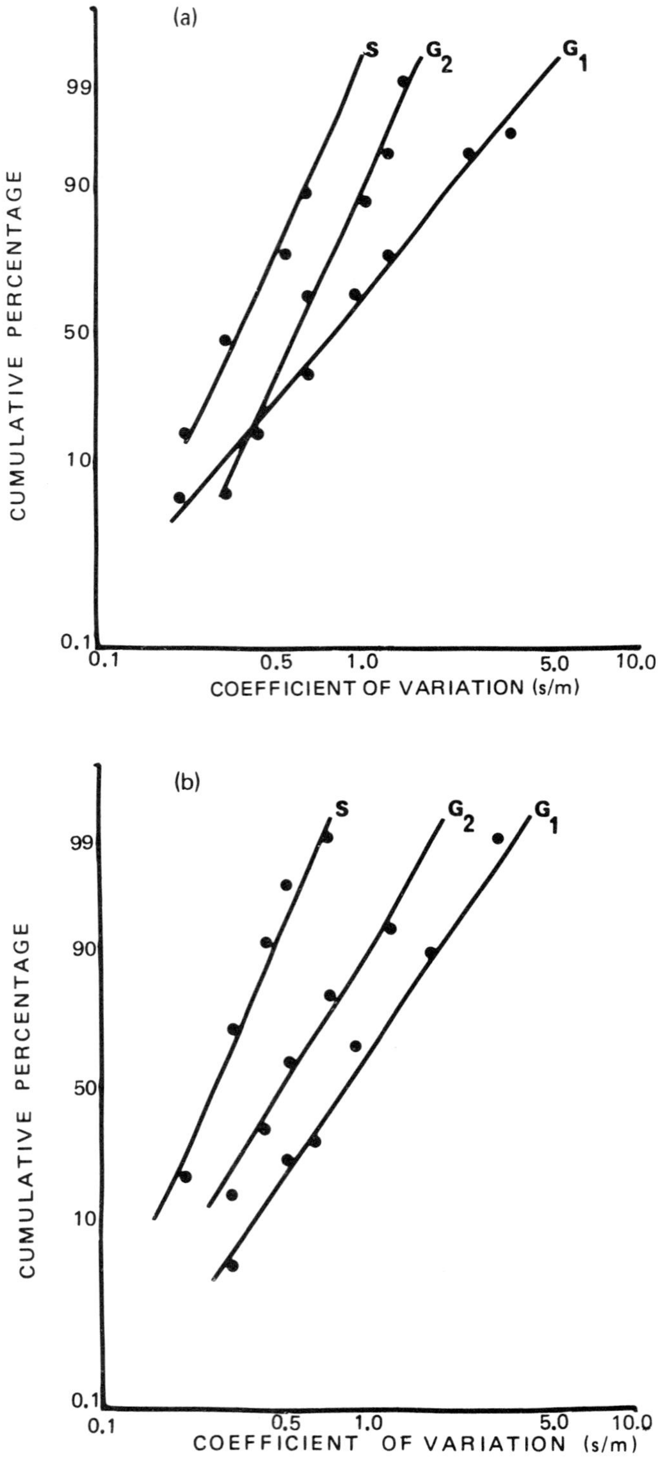


**FIGURE 2** System-to-system variability: frequency distribution of the cumulative percentage (probit scale) of (a) tumor cell systems (b) normal cell systems as a function of the logarithm phase coefficients of phase coefficients of variation.

**TABLE 3** Variabilities (CV) of Phase Durations in Tumoral and Normal Systems

| | Tumor cell systems (n = 114) | | | Normal cell systems (n = 72) | | |
|---|---|---|---|---|---|---|
| Phase | Median (%) | 95% confidence interval (%) | Variability | Median (%) | 95% confidence interval (%) | Variability |
| $G_1$ | 78 | 18-330 | 0.80 | 80 | 24-280 | 0.65 |
| S | 33 | 15-76 | 0.42 | 26 | 23-56 | 0.42 |
| $G_2$ | 56 | 25-130 | 0.44 | 48 | 18-150 | 0.60 |

mental mammalian system," which of course has no biological reality at all but which will provide the best possible stereotype of the cell cycle in a mammalian cell population. The corresponding distribution, assumed to be log normal, of $G_1$, S, and $G_2$, are shown in Fig. 3. The $G_1$ phase has the largest mean duration with the largest spread, but it is worthwhile noting, first, that the mean duration of the S phase is much longer than commonly assumed, and second, that the cell-to-cell variations in S and $G_2$ durations are very important and cannot be neglected (see the ranges in individual S and $G_2$ durations, legend of Fig. 3).

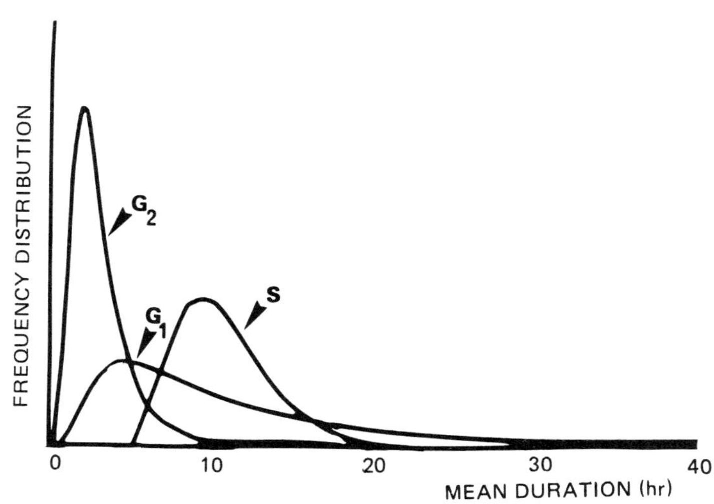


**FIGURE 3** The intrapopulation variability: by pooling together data obtained in all systems (Table 1) a stereotype of an experimental mammalian system is represented. Density probability distributions (assumed to be log normal) of $G_1$, S, and $G_2$ durations are calculated. The mean durations of $G_1$, S, and $G_2$ are equal to 11.4, 10.6, and 3.0 hr, respectively. The coefficients of variation are equal to 0.90, 0.30, and 0.55. In this stereotyped system, the phase duration could be seen in the range of 1.8-39.0 hr for $G_1$, 6.0-17.5 hr for S, and 1.0-7.0 hr for $G_2$ durations.

## DISCUSSION AND CONCLUSIONS

This chapter is based on bibliographical research concerning experimental mammalian systems, which may be not quite exhaustive. However, our data are very unlikely biased toward any given overrepresentation of systems with particular kinetic parameters.

The data concerning the variabilities of the cell phase durations have been obtained through different automatic methods of FLM analysis. These methods have been compared in the same set of experimental data (Hartmann et al., 1975), and it has been found that their final results agree closely, despite the fact that some of their assumptions differ importantly. All these methods assume that S and $G_2$ durations of individual cells are independent, and it is difficult to test this hypothesis, both experimentally and mathematically. However, it is obvious that one cannot assume that S and $G_2$ are negatively correlated to such an extent that, being variable, their sum could be constant. Indeed, in such a case the first drops of the FLM curves should be vertical and they never are.

From this review of existing data in experimental mammalian systems we can conclude that the individual differences in G1, S, and $G_2$ durations are, quite generally, very important. It is true, as often claimed, that the most variable phase is $G_1$, but the variabilities of S and $G_2$ are very important, too. Again, one can keep in mind that the coefficients of variation of the durations of $G_1$, S, and $G_2$ are 0.90, 0.30, and 0.55, respectively, in this stereotyped average experimental mammalian system. This variability of S and $G_2$ has to be taken into account to predict the desynchronization of cell populations after synchronization procedures. Moreover, cell cycle models should not assume that the control of the cell cycle is predominantly located in G1 phase on the unique basis of its variability. The important variabilities of the S and $G_2$ phases also have to be incorporated in models of cell cycle control, and the late part of the cell cycle should also on this basis have a major role in the determinism of mitosis, in the light of recent work (Brooks et al., 1980; Valleron et al., 1981).

## ACKNOWLEDGMENT

This work was supported by Association pour le Developpement de la Recherche sur le Cancer (Villejuif).

## REFERENCES

Brooks, R. F., Bennett, D. C., and Smith, J. A. (1980). Mammalian cell cycles need two random transitions. *Cell* 19:493-504.

Cooper, S. (1979). A unifying model for the G1 period in prokaryotes and eukaryotes. *Nature* 280:17-19.

Gilbert, C. W. (1972). The labelled mitoses curve and the estimation of the parameters of the cell cycle. *Cell Tissue Kinet.* 5:53-63.

Hartmann, N. R., Gilbert, C. W., Jansson, B., MacDonald, P. D. M., Steel, G. G., and Valleron, A. J. (1975). A comparison of computer methods for the analysis of fraction labelled mitoses curves. *Cell Tissue Kinet.* 8:119-124.

Hartmann, N.R., and Pedersen, T. (1970). Analysis of the kinetics of granulosa cell populations in the mouse ovary. *Cell Tissue Kinet.* 3:1-12.

Jansson, B. (1974). Mathematical models in cell cycle kinetics. In *Mathematical Models in Biology and Medicine,* North Holland, Amsterdam, pp. 21-39.

MacDonald, P. D. M. (1970). Statistical inference from the fraction labelled mitoses curve. *Biometrika* 57:489-503.

Shields, R., Brooks, R. F., Riddle, P. N., Cappellaro, D. F., and Delia, D. (1978). Cell size, cell cycle and transition probability in mouse fibroblasts. *Cell* 15:469-474.

Smith, J. A., and Martin, L. (1973). Do cells cycle? *Proc. Natl. Acad. Sci. USA* 70:1263-1297.

Steel, G. G. (1977). *Growth Kinetics of Tumours.* Clarendon Press, Oxford.

Steel, G. G., and Hanes, S. (1971). The technique of labelled mitoses: Analysis by automatic curve-fitting. *Cell Tissue Kinet.* 4:93-105.

Valleron, A. J., Guiguet, M., and Puiseux-Dao, S. (1981). Scheduling of cell cycle events by two proteins: A mathematical model. In *Biomathematics and Cell Kinetics.* Elsevier, North Holland, Amsterdam, pp. 195-210.

Valleron, A. J., Mary, J. Y., and Frindel, E. (1973). Méthode d'analyse sur ordinateur des courbes de mitoses marquées. *Biomedicine* 18:118-123.

# 7

# Circadian Rhythms in Cell Population Kinetics of Self-Renewing Mammalian Tissues

**Erik Thorud** / The Norwegian Radium Hospital, Oslo, Norway

**Ole Petter Fraas Clausen** / Rikshospitalet, Oslo, Norway

**Ole Didrik Laerum** / The Gade Institute, University of Bergen, Bergen, Norway

## INTRODUCTION

Continuously dividing, or labile cell populations are generally characterized by rapid cell proliferative activity and represent the main cell populations in the epidermis and in hair follicles, in the mucosa of the gastrointestinal and respiratory tracts, in the bone marrow, and in the corneal epithelium. The so-called stable cell populations, mainly represented by the cells in the parenchymatous organs, have a much slower proliferative activity under normal conditions, but may be triggered into rapid cell proliferation, e.g., after cell damage. In this chapter only the liver tissue will be dealt with in this group.

Circadian rhythms in cell cycle progression have been identified in almost all types of proliferating mammalian tissues. Such rhythms have also been observed in neoplastic tissues (Badran and Echave Llanos, 1965). This will not be dealt with in this chapter.

Since the early works of Fortuyn-van Leiden (1971) on circadian rhythms in the mitotic index of the corneal epithelium of newborn cats, rhythms in proliferation-related parameters have been extensively studied in many tissues. Storey and Leblond (1951) used Colcemid to determine circadian variations in the mitotic rate in murine epidermis. The existence of circadian fluctuations in DNA synthesis was first shown by Pilgrim et al. (1963) in various mouse tissues. Farsund (1975) used flow cytometry in cell kinetic studies of the mouse urinary bladder epithelium, and demonstrated circadian variations in DNA ploidy classes.

## EPIDERMIS

The proliferating cell compartment of epidermis, the basal cell layer, is attached to a basement membrane that separates dermis from epidermis. The maturing compartment covers the basal cell layer and consists of several layers of differentiating cells having reached various stages of differentiation. The surface layer (horny layer) is made up of the ultimate differentiation product, keratin, with its barrier function clearly reflected in its highly resistant structure. Keratin formation starts in the basal cell layer, and continues as the cells move toward the surface. Under normal conditions there is an equilibrium between the loss of squames from the surface and the formation of new keratinocytes in the basal cell layer.

The general structure of stratified epithelium, partly with exception of a horny layer, is found in a number of tissues in the mammalian body, e.g., cornea, oral mucosa, hamster cheek pouch, upper part of the esophagus, anus, vagina, and cervix uteri. Epidermis has been the most extensively studied organ with regard to circadian rhythms in cell population kinetics, and circadian changes have been found in epidermis situated at different parts of the mammalian body, e.g., ear epidermis (Bullough, 1948), plantar epidermis (Storey and Leblond, 1951), dorsal epidermis (Evensen, 1963), penile foreskin epidermis (Cooper and Schiff, 1938; Scheving, 1959), abdominal epidermis (Thorud and Volden, 1980), and arm epidermis (Fisher, 1968).

### Circadian Rhythms

#### *Cell Cycle Phase Distributions*

Circadian changes in the mitotic index (MI) were reviewed by Scheving (1959). The early works on mouse epidermis showed circadian variations that differed somewhat in phasing (Ortiz-Picón, 1934; Carleton, 1934; Cooper and Franklin, 1940). Later studies on nocturnal animals (Blumenfeld, 1939; Bullough, 1948; Clausen et al., 1979) showed the acrophase in MI (Fig. 1) in the early morning at the transition between the dark and the light period. Some conflicting reports with regard to the mitotic acrophase, however, have been published (Evensen, 1963; Tvermyr, 1969). Different laboratory conditions, feeding times, and animal strains may explain some of these differences. Bullough (1948) observed that the changes varied with the degree of physical activity of the mice. A high mitotic index was seen during the resting period, and a low mitotic index during the period of increased physical activity.

In the human epidermis the acrophase of the MI was generally found during the early night in both infants and in adults (Cooper and Schiff, 1938; Broders and Bublin, 1939; Cooper, 1939; Scheving, 1959; Fisher, 1968). This corresponds to a phase difference in the mitotic activity between human (diurnal) and rodent (nocturnal) epidermis of about 12 hr.

The fraction of cells in DNA synthesis phase (S phase) is by most investigators studied by means of tritiated thymidine ([$^3$H]-TdR) incorporation into nuclear DNA. Several authors (Pilgrim et al., 1963; Grube et al., 1970; Potten et al., 1971; Tvermyr, 1972; Hegazy and Fowler, 1973; Thorud et al., 1978; Clausen et al., 1979) all found a high number of cells incorporating [$^3$H] TdR, ([$^3$H] TdR labeling index, LI), around midnight, i.e., during the waking period of nocturnal animals (Fig. 1). Microspectrophotometric (Olsson, 1976) and flow cytometric (Thorud et al., 1978; Clausen et al., 1978) determinations of the proportion of mouse epidermal basal cells in S phase revealed significant circadian variations that were parallel to the variations in the [$^3$H] TdR

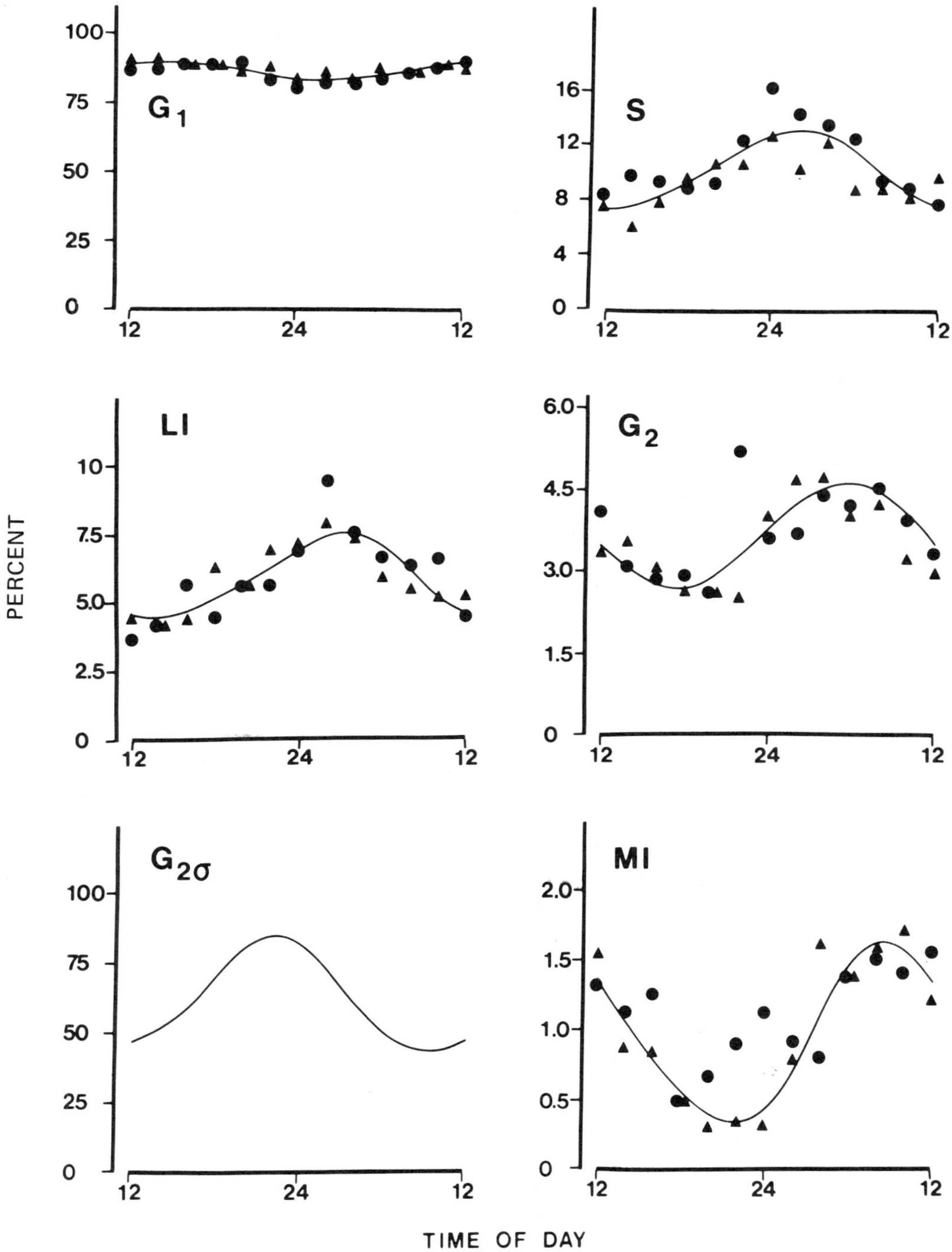


**FIGURE 1** Proportions of cells in $G_1$, S, $G_2$, and mitotic phase (MI) in the epidermal basal cell layer of hairless mice. LI represents the [$^3$H] TdR labeling index. The $G_{2\sigma}$ curve represents the number of cells in a resting $G_2$ state expressed as a proportion of cells in $G_2$ phase. Triangles and dots show the mean values of four mice in separate studies on two successive days. The solid curves represent the fitting of a mathematical model to the data.

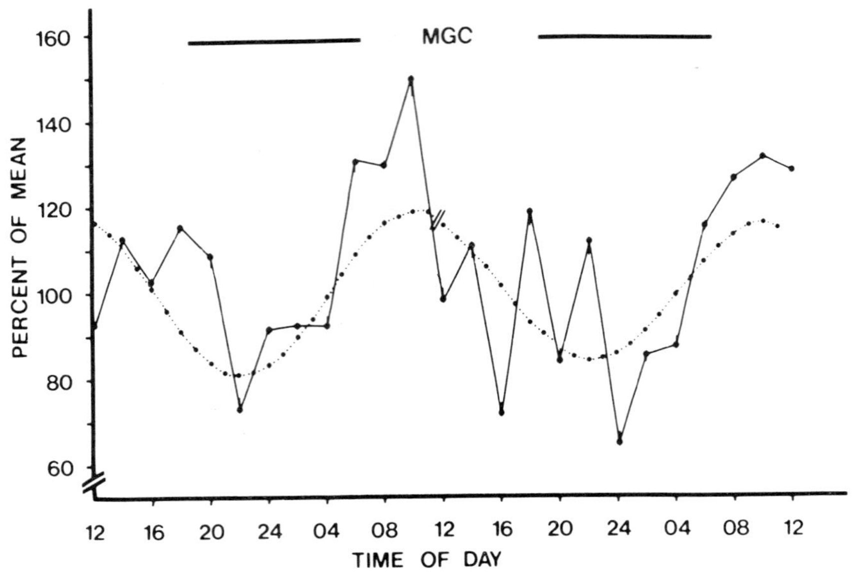


**FIGURE 2** Variations in the mean grain count over [$^3$H] TdR-labeled nuclei as obtained by autoradiography during two successive 2 hr periods of isolated epidermal basal cells of hairless mice. The dotted line indicates the first harmonic sinus fit to the data. The solid horizontal lines indicate periods of darkness.

LI, with high values at night (Fig. 1). In human epidermis, reports on circadian variations in [$^3$H] TdR incorporation data are conflicting (Kahn et al., 1968; Pullmann et al., 1974; Schell et al., 1977; Gelfant et al., 1982). Thorud and Volden (1980) found a peak value of cells with S-phase DNA content around noon by using flow cytometry. A discrepancy between the proportion of cells in S phase, as determined by DNA cytometry and the [$^3$H] TdR LI, was observed in mouse epidermis (Clausen et al., 1976, 1979, 1980). A higher proportion of cells with S-phase DNA content than the corresponding [$^3$H] TdR LI was also shown in the squamous epithelium of the hamster cheek pouch (Møller and Larsen, 1978). This indicates the existence of a subpopulation of cells blocked in S phase or with a very slow rate of DNA synthesis (Aarnaes et al., 1981; Møller et al., 1982). It was further shown that the mean grain count (Fig. 2) over [$^3$H] TdR-labeled nuclei, indicating the rate of DNA synthesis in individual cells, was lower in the dark period than during daytime in mouse epidermis (Clausen et al., 1979).

By the use of DNA flow cytometry, circadian changes in the proportion of cells with $G_2$-phase DNA content (4c) (Fig. 1) were shown for the first time in the basal cells of mouse epidermis (Thorud et al., 1978; Clausen et al., 1979). In the human epidermis no clear pattern was seen with regard to variations in cells with $G_2$ DNA content (Thorud and Volden, 1980).

The vast majority of the cells in the basal cell layer of mouse epidermis were in $G_1$ phase, which showed only minor changes during a 24 hr period (Aarnaes et al., 1981; Clausen et al., 1979) (Fig. 1).

The total cell number and the cell density of mouse epidermis showed circadian variations. The changes in total cell numbers were reflected mainly by changes in the proportion of cells in the differentiating cell layers (Clausen et al., 1982).

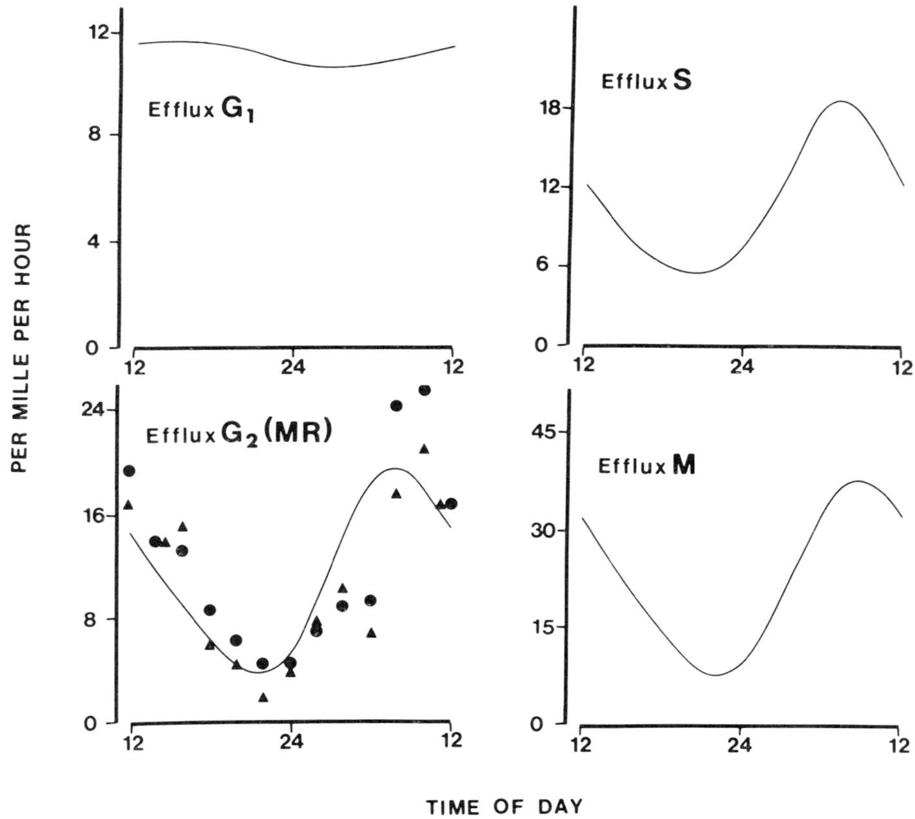


**FIGURE 3** Variations in cellular flux through the different cell cycle phases of epidermal basal cells of hairless mice. The solid lines and triangles and dots are obtained as indicated in Fig. 1.

*Cell Flux*

A direct measure of the rate of cell proliferation (mitotic rate) can be obtained by the use of the stathmokinetic method. Storey and Leblond (1951) showed pronounced circadian variations in the mitotic rate in the plantar epithelium of the rat, with a low proliferative activity during the dark period. Under standardized laboratory conditions, Tvermyr (1972) and Clausen et al. (1979) observed similar variations in the dorsal epidermis of the mouse, with high values in the early morning (Fig. 3). Some differences with regard to phasing of the variations in the mitotic rate have been observed (Evensen, 1963; Tvermyr, 1969). This is not surprising, considering that the laboratory conditions in these studies were somewhat different, particularly with regard to the lighting regimens.

In mouse epidermis, the calculated efflux values from mitosis showed circadian variations that were parallel to those of the influx. The acrophase of the efflux occurred shortly after that of the influx (Clausen et al., 1979; Aarnaes et al., 1981) (Fig. 3).

Tvermyr (1972), using a double [$^3$H] TdR-labeling technique showed circadian variations in the influx to and efflux from the S phase. The amplitude of the influx was less pronounced and the peak occurred somewhat earlier than that of the efflux, which peaked in the early morning. A similar tendency was found by Clausen et al. (1979) and

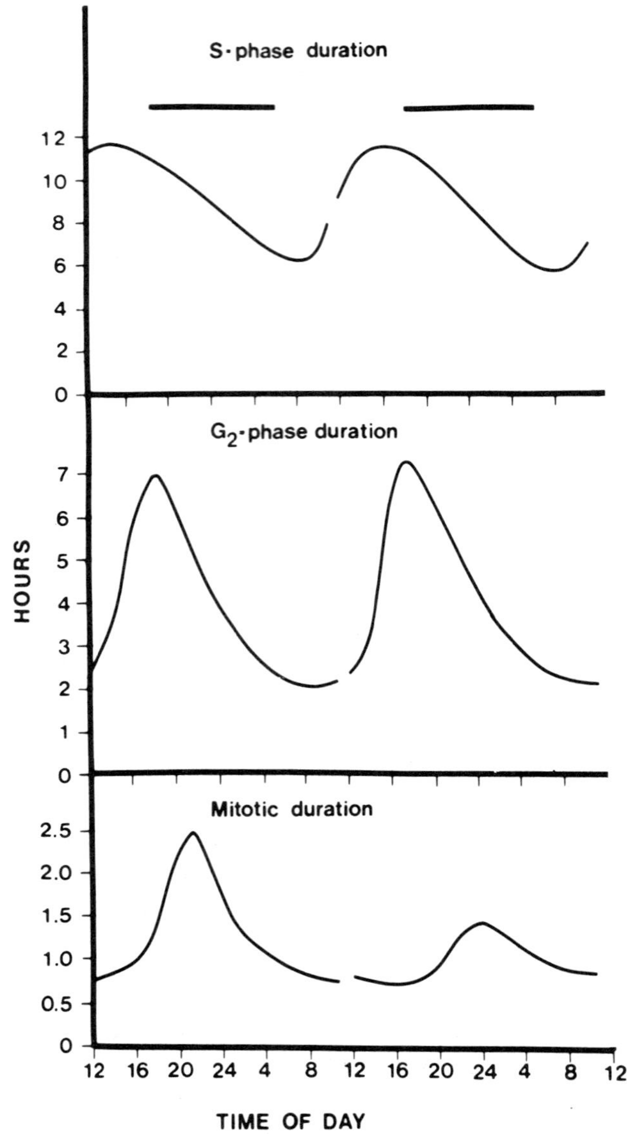


**FIGURE 4** Variations in cell cycle phase durations in the basal cell layer in the hairless mouse epidermis. The data are obtained by the use of a method proposed by Refsum (Tvermyr, 1972). Horizontal bars indicate periods of darkness.

Aarnaes et al. (1981), who based their results on calculations from changes in the magnitude of the S and $G_2$ phase compartments and the mitotic rate. The influx to S phase showed only minor changes during a 24 hr period, whereas the efflux showed changes of the same amplitude as that observed in the mitotic rate (Fig. 3).

*Cell Cycle Phase Durations*

The existence of circadian changes in the duration of the cell cycle phases in squamous epithelia was disputed for a long time (Pilgrim et al., 1963; Izquierdo and Gibbs, 1974).

This may partly be due to the use of inadequate methods. The percentage labeled mitosis (PLM) method (Quastler and Sherman, 1959), which is widely used to determine the duration of the different cell cycle phases, is by definition inadequate when the system shows circadian variations in cell proliferation. It has also been shown that slowly proliferating subpopulations will hardly influence the estimates of cell cycle phase durations in a PLM study, even if they are represented in high numbers (Hudson and Hahn, 1977; Clausen et al., 1981).

Circadian variations in the duration of the DNA synthesis phase have been demonstrated by several authors (Tvermyr, 1972; Clausen et al., 1979; Aarnaes et al., 1981). The mean duration of the S phase in the dorsal murine epidermis is about 8-9 hr varying from 14 hr around midnight to 6 hr at the beginning of the light period (Fig. 4). These values represent the estimates of the duration that is to be expected from the rate of cell progression through S phase. Some methods used tend to reduce the amplitude of the variations. This is due to calculations that imply averaging the S-phase duration over a certain period of time (Tvermyr, 1972; Clausen et al., 1979; Aarnaes et al., 1981; Thorud et al., 1983). This is especially true for cell cycle phases of long duration, e.g., the S phase.

Circadian changes in the duration of the $G_2$ phase in mouse epidermis were first indicated by Tvermyr (1972). It was shown to vary from about 7-8 hr in the dark period to 2 hr at the beginning of the light period (Clausen et al., 1979). These variations show approximately the same phasing as those observed in the duration of the S phase (Fig. 4).

Several authors have shown data indicating the existence of circadian variations in the mitotic duration (Evensen, 1963; Bullough and Laurence, 1966; Clausen et al., 1979). These studies have been based on very simple calculation methods that are sensitive to methodological and experimental errors (Thorud et al., 1983). By using several different methods it is indicated that the duration of mitosis varies from 0.7 hr in the early morning to about 2.5 hr during the dark period (Fig. 4). Methodological uncertainties, however, are also connected with these sets of data.

Estimated mean values of the $G_2$ duration from PLM measurements were much shorter than expected from a [$^3$H] TdR double-labeling experiment and from a multiparameter cell kinetic study in hairless mouse epidermis. These differences were best explained by postulating two distinct subpopulations of $G_2$ cells, one with a high and another with a low rate of cell cycle progression (Clausen et al., 1981).

### Regulatory Aspects

Evensen (1963) proposed that the circadian variations in the mitotic index were the results of variations mainly in the mitotic duration, and not of variations in the mitotic rate. Bullough and Laurence (1966) and Møller et al. (1979) have suggested that variations in the mean transit time through mitosis may be involved in the homeostasis of growth in squamous epithelia. The assumption that the mitotic process itself represents the key point for regulation of circadian variations in cell proliferation, however, may be questioned. Circadian variations in premitotic cell kinetic parameters, including transit times through S phase and $G_2$ phase, argue against the hypothesis that the transit through mitosis plays a major role in the induction of circadian rhythmicity in the rate of cell division in mouse epidermis.

During the last two decades, several reports on circadian variations in cell kinetics of squamous epithelia conclude that the initiation, or the point of origin for such changes, is in the $G_1$ phase or at the $G_1$-S transition (Brown and Berry, 1968; Izquierdo and Gibbs, 1974; Klein and Valleron, 1977; Hopper and Brockwell, 1978; Møller and Lar-

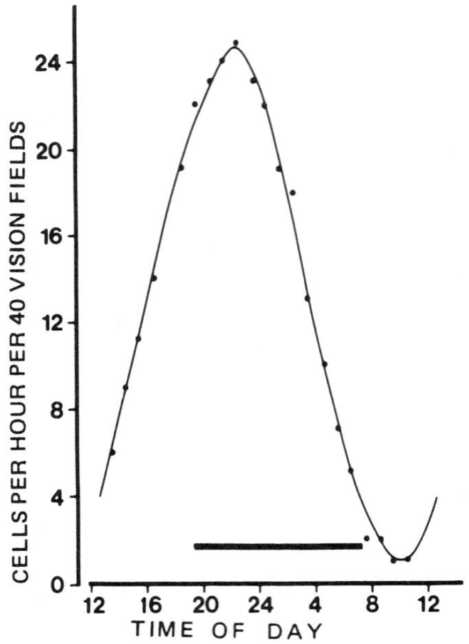


**FIGURE 5** Variations in cell loss (loss of nucleated cells) in the hairless mouse epidermis. Horizontal bar indicates period of darkness.

sen, 1978). It has been suggested that the circadian variations in corticosteroid levels in serum could modify the cellular flux at the $G_1$-S transition (Schell et al., 1980). A high concentration might block the cells at this point in the cell cycle (Hennings and Elgjo, 1970), followed by a synchronous release of cells from the $G_1$-S transition with decreasing corticosteroid levels. This synchronized cohort of cells might later give rise to the observed wave of cell division.

When mice were stressed by intermittent loud noise, the amplitude of the circadian rhythm in mitotic rate was reduced, indicating that stress hormones might modulate the rhythmicity. The 24 hr mean value did not differ significantly compared to unperturbed conditions (Kreyberg et al., 1965).

In a multivariable study in hairless mouse epidermis (Clausen et al., 1979; Aarnaes et al., 1981), the influx to S phase showed minor circadian variations, whereas efflux from S phase, showed more distinct circadian variations. Based on the analysis of these findings we concluded that the origin of the circadian variations in cell division could not exclusively be in the $G_1$ phase or at the $G_1$-S transition in the mouse. The results were best explained by a regulation of the cellular flux through the S phase or at the S-$G_2$ transition and thus were not consistent with being induced by changes in cortisol levels. The data were also consistent with a mechanism simultaneously affecting the cell cycle progression through the S, $G_2$, and M phases.

Locally acting growth inhibitory substances (chalones) are endogenous candidates for the regulation of circadian rhythms in epidermal cell population kinetics. A relatively high number of differentiating cells with an assumed high amount of growth inhibitory substances are present in the evening, and may induce reduction in the mitotic rate and cellular flux through S and $G_2$ phase during the night (Clausen et al., 1982; Elgjo et al., 1981). Low numbers of highly mature keratinocytes may thus allow reestablishment of the high mitotic rate in the morning. However, whether locally acting growth inhibitors are involved in the maintenance of circadian rhythms in cell division remains to be established.

The circadian variations in loss of nucleated cells in the differentiating cell layer of the epidermis (Fig. 5) may result from the circadian variations in cell division activity occurring 6 days in advance, since the turnover time of the murine dorsal epidermis is about 6 days (Iversen et al., 1968). This requires a high degree of synchrony in maturation of keratinocytes during this period.

Our concept of the organization of the epidermal cell cycle shows an increasing complexity. Several authors have demonstrated or postulated the existence of resting or slowly cycling cells in S and $G_2$ phase (Aarnaes et al., 1981; Clausen et al., 1981; Møller et al., 1982). It is at the present time unclear in what way these subcompartments influence the circadian changes in cell population kinetics. A simulation made by Aarnaes et al. (1981) indicated circadian changes in the subpopulation of slowly cycling cells with $G_2$ DNA content (Fig. 1). The input and output of such subcompartments may play a role in the mechanisms of circadian cell kinetic rhythms in mouse epidermis.

## CORNEAL EPITHELIUM

The circadian rhythmicity observed in the mitotic index in the corneal epithelium of the rat is similar to that observed in epidermis, with high values in the early morning and low values in the evening (Vasama and Vasama, 1956; Gololbova, 1958; Alov, 1959). Chumak (1963) observed circadian rhythms in the proportion of DNA-synthesizing cells after injection of tritiated thymidine. The peak value occurred in the early morning and coincided with that of the mitotic index. Scheving and Pauly (1967) performed a very detailed study, in which the $[^3H]$TdR labeling index (LI), the $[^3H]$TdR uptake (scintillation counting), and the mean grain count (MGC) over $[^3H]$TdR-labeled nuclei were measured. The peaks and troughs of all three DNA parameters occurred at about the same time of the day. High values were found around noon, whereas low values were found near midnight. This indicates that the cells are synthesizing DNA at a higher rate than normal when the peak in the $[^3H]$TdR LI occurs. The appearance of the peaks in the proportion of cells in S phase and cells in mitosis almost coincides. A certain lag between the peaks should, however, be expected if they were the result of cells moving in synchrony through S phase, $G_2$ phase, and mitosis. Scheving and Pauly (1973) therefore suggested that the mechanism responsible for the conversion of $G_1$ to S and $G_2$ to M must be exerting its action on both compartments at the same time.

Circadian influence on the waveform of the frequency of labeled mitoses in mouse corneal epithelium was clearly demonstrated by Burns and Scheving (1975). It could not be validated from this study, however, to what extent differences between PLM curves initiated at different hours were reflections of circadian changes in cell cycle phase durations, or a result of synchrony in the cell cycle traverse.

When the animals were maintained in darkness during the day and in light during the night (dark/light, DL 12:12), the mitotic rhythm of the corneal epithelium showed a phase shift of 180° compared with the LD 12:12 regimen. In animals maintained in constant darkness (DD) only minor changes in the amplitude and phasing of the rhythm compared with the LD 12:12 regimen were seen. In those maintained in constant light (LL), a considerable flattening of the curve occurred, which still, however, showed indications of a rhythm. This flattening is assumed mainly to be due to desynchronization of rhythms between animals, which still maintain their individual circadian rhythms (Scheving et al., 1974c). After adrenalectomy, the circadian changes in the number of corneal mitoses were preserved with a somewhat reduced amplitude (Alov, 1959). Only small changes in the phasing of the rhythm resulted from restricted feeding (Scheving et al., 1974b).

Epidermal growth factor (EGF) affected DNA synthesis (scintillation counting) and the mitotic index in mouse corneal epithelium in a circadian phase dependent manner (Yeh et al., 1981).

## HEMOPOIESIS

The hemopoietic tissue is a labile organ system with a rapid cell turnover. Parallel to the high rate of cell proliferation, there is a rapid flux of newly formed cells into the bloodstream. Although erythrocytes remain a long time in the circulation, the half-life of polymorphonuclear granulocytes and monocytes is very short, after which they enter the tissues to exert their functions. Therefore, the cell renewal in the bone marrow as well as the cell numbers in peripheral blood are strongly variable. Due to pronounced circadian rhythmicity in all compartments, quantitative measurements of these activities will strongly depend on the time of day the sample is taken.

### Stem Cells

By the use of newer assay methods for the different classes of stem cells, detailed information is now available on stem cell physiology of humans as well as animals, although the data on circadian rhythmicity in this compartment have so far been sparse.

The relative numbers of multipotent stem cells ($CFU_S$) in the marrow of mice are known to undergo circadian variations with maximum values at 06.00 (Stoney et al., 1975). Recently, circadian variations in the numbers of stem cells committed for myelopoiesis ($CFU_C$) have been reported (Bartlett et al., 1982; Aardal and Laerum, 1983). In one study the maximum value was found in the afternoon (Aardal and Laerum, 1983). Erythroid stem cells showed similar variations, but with acrophase in the early morning. However, estimation of a circadian rhythmicity in the committed stem cell compartment is complicated by the fact that the rhythms also undergo circannual variations (Laerum and Aardal, 1981). When one looks only at values from the first part of the day, maxima are found in the spring and autumn (Aardal and Laerum, 1983). Possibly, this is mainly due to the drifting of the acrophase of circadian rhythmicity throughout the year.

In humans, most of the studies have been carried out on circulating stem cells in the peripheral blood, and due to practical limitations, not in the bone marrow. Meytes et al. (1980) found pronounced individual variations in numbers of erythroid burst-forming units ($BFU_E$)–which is an early precursor of erythropoiesis–mainly due to individual and not to circadian changes. Similarly, the normal range of $CFU_C$ in peripheral blood was strongly variable (Ponassi et al., 1979), but a conclusive circadian pattern could not be found. In contrast, another investigation showed significant circadian variations in the numbers of $CFU_C$ in peripheral blood with acrophase at 09.00. Strongest changes were found in the number of eosinophil colonies (Ross et al., 1980).

### The Maturing Compartments of Hemopoiesis

When one studies the progeny of the different classes of stem cells, the investigations are hampered by the complexity of the hemopoietic cell population. Not only are there several different maturing lines to take into consideration (erythropoiesis, myelopoiesis, thrombo- and lymphopoiesis), but in addition each compartment has definite maturation stages from the blasts to the nonproliferative end cells.

Furthermore, the different phases of the cell cycle have to be recognized. Depending on the method which has been used very different results may come out. Thus, one study showed that the $[^3H]$ TdR labeling index, as an estimate of the numbers of DNA-synthesizing cells in mice, had acrophase at 22.00, and the mitotic index at 02.00 (Pizzarello and Witcofski, 1970). By the use of liquid scintillation counting, Burns (1981) found maximum uptake of $[^3H]$ TdR at 02.00, but Scheving et al. (1978) found acrophase in the middle of the day. Using the same method, Scheving and Pauly (1974) reported maximum $[^3H]$ TdR uptake at 18.00 and minimum at midnight in rat bone marrow, with an amplitude of more than twice the mean value (Scheving et al., 1974a).

One of us (ODL) has recently investigated the proliferative activity in bone marrow of female mice. This was because circadian variations in hemopoietic tissues have mainly been observed in male and only occasionally in female mice (Sharkis et al., 1971, 1974). The fractions of cells in $G_1$, DNA synthetic, and $G_2$ phase of the cell cycle were determined by measurements of relative cellular DNA content, using flow cytometry (Laerum and Aardal, 1981). In May and July the maximal proliferative activity was found about 10.00, and the differences between maximum and minimum varied between 50 and 200%. Similar to the stem cell activity, the maximum proliferative activity in mouse bone marrow on a standardized 12 hr light/darkness regimen had the maximum in the afternoon during the winter (unpublished results).

In addition, the total cell numbers per femur underwent circadian variations, usually with one peak in the early part of the day and one during the night (Laerum and Aardal, 1981). It was therefore of critical importance to express the proliferative activity in the marrow as total numbers in each cell cycle phase per femur. This phenomenon introduces a potential source of error when measurements are only expressed as relative values.

In the human bone marrow, the $[^3H]$ TdR labeling index of myelopoiesis has been found higher in the morning and first part of the day, and lower at midnight (Mauer, 1965; Killman et al., 1962). The mitotic index was highest at midnight and lower in the morning. Corresponding variations in lymphocytes and erythropoietic cells have been found in mice (Sharkis et al., 1971, 1974), although significant variations were only present in males. Circadian rhythms in mitotic activity as well as duration of mitosis and DNA synthetic phase have been reported by others (Moskalik, 1976).

The $[^3H]$ TdR labeling in erythropoiesis also undergoes circadian variations in mice, with acrophase at 03.00 (Dörmer et al., 1970). Corresponding variations in mitotic activity of erythropoietic cells have also been reported (Sharkis et al.,1974). In this connection it is interesting that males and females had different time of acrophase, 12.00-16.00 and 20.00-24.00, respectively (Sharkis et al., 1974). In contrast to this, another study did not find any circadian variations of $[^3H]$ TdR labeling index in erythropoiesis of humans, although the labeling index of myelopoietic cells as well as mitotic index in the whole marrow showed circadian rhythmicity (Mauer, 1965).

In conclusion, definite circadian variations in proliferative activity are found in the different compartments of hematopoiesis, both in laboratory animals and in humans. However, the time of the acrophase varies from investigation to investigation. In light of the seasonal variations in circadian rhythmicity reported from our laboratory, the widely differing results may easily be explained. In addition, it shows that a rapidly renewing organ system like hemopoiesis does not need to have a constant type of rhythmicity over a longer time period. Superimposed are variations in the total numbers of hemopoietic cells in the body, for example, expressed as cell numbers per femur, that may interfere with the relative estimate of proliferative activity. It shows the necessity of taking the system as a whole into consideration. Erythropoiesis seems to undergo less pronounced circadian variations than myelopoiesis, and in addition there seem to be sex differences in circadian rhythmicity.

### Regulatory Aspects

With the complex cellular composition of hematopoietic tissues, it is no wonder that a multitude of regulatory mechanisms operate. They must be able to rapidly switch over to another rate of cell proliferation according to actual demands. It is known that a series of different factors may not only have effects on hemopoiesis, but also be of regulatory importance, including hormones as well as other types of regulators; for surveys, see Gordon (1970), Cline and Gold (1979), Baum et al. (1981), and Iversen (1981).

In recent years there has been an increasing interest in endogenous regulators acting by a negative feedback mechanism. One of these, granulocyte chalone, is a small peptide which appears to be secreted by mature granulocytes (Rytömaa and Kiviniemi, 1968; Paukovits and Hinterberger, 1978). Investigations with semipurified chalone material indicate that not only the maturing compartment of myelopoiesis rapidly reacts by lowered cell proliferation, but also the renewal of the committed stem cell (Laerum and Maurer, 1973; Aardal et al., 1977). This may indicate a dual type of mechanism, one leading to a rapid reduction of proliferative rate in maturing cells, as well as a more long-term effect where several days will elapse before the progeny of a lowered stem cell number prevails.

On the other hand, production of myelopoietic cells seems to be stimulated by the glycoprotein colony stimulating factor, although the evidence for this being a physiological regulator is mainly obtained in cell culture (Burgess and Metcalf, 1980). In this connection it might be an analogy to erythropoiesis, where a specific stimulator, erythropoietin, seemed to be a critical regulator, although specific inhibitors have also been reported (Lindemann, 1971; Kivilaakso and Rytömaa, 1971). To what extent the proliferative activity in hemopoiesis really is due to variations in the different types of physiological regulators remains to be proven, although there is evidence that this may be the case for erythropoietin (McKee et al., 1974).

## GASTROINTESTINAL TRACT

The rapid cell turnover in the epithelial layer of the gastrointestinal tract is the combined result of a high rate of cell production, followed by cell migration and eventual shedding from the cell surface (Appelton et al., 1981). The proliferating cells are confined to localized areas, such as the basal cell layer in the squamous epithelium of the esophagus and the lower parts of the crypts of the gastric and intestinal mucosa. In the esophagus, keratinization ends with pyknosis and death of the cells as they migrate toward the surface, and eventual shedding of keratinized material. In the gastric mucosa newly formed cells from the bottom of the crypts migrate upward and are lost from the surface. In the small intestine, cells proliferate mainly in the lower third of the crypts and thereafter either migrate downward to form Paneth cells, or upward toward the surface. Cells from several crypts surrounding one villus move together to form the surface epithelium of the villi, where they are finally lost at the tip (Bjerknes and Cheng, 1981a-d). In the large intestine, cells are formed in a corresponding way in the lower third of the crypts and thereafter migrate upward to become lost on the cell surface (Appleton et al., 1981).

Hence, even a homogeneous suspension of epithelial cells from a limited area in the gastrointestinal tract forms a complex cell population, consisting of proliferating and nonproliferating cells, as well as young and old cells that will be shed after different time periods. With this background it is no wonder that the studies on circadian rhythms in the gut tend to show heterogeneous results.

### Circadian Variations

It has long been known that the proliferative activity of cells in the gastrointestinal tract undergo strong circadian variations (Scheving et al., 1972, 1978; Burns et al., 1972; Tutton and Barkla, 1980). One main factor contributing to this phenomenon seems to be circadian fluctuations of the transit times of the $G_1$ and $G_2$ phases of the cell cycle (Al-Dewachi et al., 1976). These rhythms seem to be of importance not only for the reaction to different types of perturbation, such as wounding (Reeve, 1975) and cytostatic therapy (Pallavicini et al., 1979), but can also be subject to alterations and phase shift due to changes in the lighting regimen (Sigdestad and Lesher, 1971). Circadian rhythms have been reported not only in the tongue and esophageal epithelium, but also in the stomach, duodenum, ileum, jejunum, colon, and rectum (Scheving et al., 1972, 1978, 1979; Burns et al., 1972; Tutton and Barkla, 1980; Al-Dewachi et al., 1976; Reeve, 1975). In mice, the acrophase has been found at the end of the dark period and in the early morning (Scheving et al., 1978). By doing comparative studies on the different parts of the gut, Scheving et al. (1978) found that the [$^3$H] TdR uptake had the same phasing in the small intestine, but different in stomach and rectum of mice. They also described low- and high-amplitude rhythms, where their lowest amplitude was found in duodenum as contrasted to a high-amplitude rhythmicity in the other parts of the intestine (Scheving et al., 1978).

Pilgrim et al. (1963) were unable to find rhythms of cell proliferation in jejunum in mice, although both DNA synthesis and mitosis underwent strong circadian variations in the esophagus and forestomach.

Chang (1971) and Chang and Nadler (1975) have done extensive studies on cell population kinetics of the crypt epithelium in colon of mice. They found that the peak of mitotic activity occurred during the first part of the day, and similar findings were reported in another study by Hamilton (1979). In line with this, the latter author could show that the peak of [$^3$H] TdR labeling index occurred at 08.00, and the maximal cell number per crypt was reached at 12.00. In addition, she could show that the cell loss from the crypts was highest immediately before maximal feces production was reached at the beginning of the dark period.

### Regulatory Aspects

The circadian rhythms in intestinal cell proliferation can be influenced by several factors, both of endogenous and exogenous origin. Thus, the rhythmicity can be reversed by altering the lighting regimen (Sigdestad and Lesher, 1971). In addition, there are several regulators of hormone type which seem to have a strongly stimulating effect in the small as well as large intestinal epithelium, including urogastrone (EGF) and glucagon (Yeh et al., 1981; Scheving et al., 1979, 1980). The importance of chalones in the intestine is not yet assessed, although several reports have indicated such a mechanism to be operating (Thompson, 1974; Brugal and Pelmont, 1975; Sassier and Bergeron, 1977, 1980; Iversen, 1981). Feeding habits may also influence the proliferative pattern. In sheep a peak is observed in the mitotic activity of the ruminal epithelium 4-8 hr after feeding (Sakata and Tamata, 1978). These authors found a maximum mitotic activity in the afternoon and early night. Sheep are diurnal animals, and the difference in phasing compared to rodents (nocturnal animals) may be explained by their reversed pattern of body activity and eating hours. Thus, the circadian variations in intestinal cell proliferation probably is a combined result of changes induced by exogenous influences, including feeding habits, as well as internal regulators.

## LIVER

As a consequence of a low cellular turnover rate, cell proliferation is slow in hepatocytes of the normal adult liver. This makes it difficult to quantitate proliferation-related parameters in the mature organ. In the immature and regenerating liver, however, the rate of hepatocyte proliferation is considerably enhanced compared with that of the adult liver. High numbers of mitoses were observed in the immature mouse liver between 05.00 and 09.00, and low numbers between 17.00 and 21.00 (Wilson, 1948), and after partial hepatectomy high rates of mitotic activity in rat liver parenchymal cells were seen in the morning, and low rates in the evening (Jaffe, 1954). Both these reports indicate the existence of a circadian rhythm in hepatocyte cell division. Similar changes in mitotic indices during liver regeneration were reported by Barnum et al. (1957), who also extended their studies to livers of normal immature mice. By means of $^{32}P$ incorporation into DNA, they found significant changes in the DNA synthesis rate over the 24 hr period, with high vaues in the early morning and low values at night. More extensive studies on circadian changes in DNA synthesis and mitotic activity of hepatocytes were performed by Echave Llanos et al. (1971) in young immature male mice. The mitotic rate was estimated by counting the proportion of hepatocytes arrested in metaphase after Colcemid injection, and DNA synthesis by scoring the proportion of labeled hepatocytes in autoradiographs after incorporation of [$^3$H] TdR. The proportion of labeled cells showed high values in the early morning and low values in the evening, whereas the mitotic rate peaked at noon. These data were consistent with earlier reports from growing liver. Circadian variations in hepatocyte numbers were also observed, showing high values in the evening. Barbason and Cantfort (1976) confirmed that high mitotic numbers were seen during daytime and low ones at night during liver regeneration. They showed that differentiated function, as indicated by synthesis of a liver enzyme, and cell division activity were mutually exclusive. By analyzing a very high number of cells, Mustafin (1976) was able to show circadian variations in the mitotic index and [$^3$H] TdR labeling index in mature mouse hepatocytes.

In a recent work from the Institute of Pathology, Rikshospitalet, Oslo, DNA flow cytometry of hepatocytes from 10-week-old mice showed clearly circadian changes in the proportion of cells with S-phase DNA content, with high values at night. Only minor circadian changes in diploid, tetraploid, and octoploid cell classes were seen (Digernes, personal communication).

Circadian changes in the mitotic rate and in the proportion of [$^3$H] TdR-labeled hepatocytes were observed in transplanted, slowly growing mouse hepatomas, and were similar to corresponding parameter values in hepatocytes of immature mouse livers (Echave Llanos and Nash, 1970; Nash and Echave Llanos, 1971). The mitotic rate and proportion of [$^3$H] TdR labeled hepatocytes in rapidly growing transplanted hepatomas showed circadian variations that were less distinct than those of slowly growing hepatomas and immature liver.

An interval of 8-12 hr was observed between the circadian peak in the proportion of DNA-synthesizing hepatocytes and that in the mitotic rate in both immature mouse liver and in slowly growing hepatomas. This interval was considered a good estimate of the lag period between DNA synthesis and mitosis. This is only true if no significant changes in cell cycle phase durations occur during the 24 hr span. Data from mouse epidermis (Clausen et al., 1979) and the hamster cheek pouch epithelium (Møller and Keiding, 1982), however, show that such variations exist in these cell populations. Estimates of cell cycle phase durations from differences in peak positions of proliferative parameters measured over a 24 h period is therefore of questionable value. Further studies are needed before any conclusions can be reached about the origin and nature of circadian changes in hepatocyte proliferation.

## REFERENCES


Aardal, N. P., and Laerum, O. D. (1983a). Circadian variations in mouse bone marrow. *Exp. Hematol.* (In press).

Aardal, N. P. and Laerum, O. D. (1983b). Circannual variations of circadian rhythmicity in murine colony forming cells ($CFU_C$). *Exp. Hematol.* (Submitted for publication).

Aardal, N. P., Laerum, O. D., and Paukovits, W. R. (1982). Biological properties of partially purified granulocyte extract (chalone) assayed in soft agar culture, *Virchows Arch [Cell Pathol.]* 38:253-261.

Aardal, N. P., Laerum, O. D., Paukovits, W. R., and Maurer, H. R. (1977). Inhibition of agar colony formation by partially purified granulocyte extracts (chalone). *Virchows Arch. [Cell Pathol.]* 24:27-39.

Aarnaes, E., Thorud, E., and Clausen, O. P. F. (1981). Model analysis of circadian rhythms in mouse epidermal basal cell proliferation. *J. Theor. Biol.* 88:355-370.

Al-Dewachi, H. S., Wright, N. A., Appleton, D. R., and Watson, A. J. (1976). Studies on the mechanism of diurnal variation of proliferative indices in the small bowel mucosa of the rat. *Cell Tissue Kinet.* 9:459-467.

Alov, I. A. (1959). The mechanism of the diurnal periodicity of mitosis. *Bull. Exp. Biol. Med.* 48:1418-1423.

Appleton, D. R., Sunter, J. P., and Watson, A. J. (1981). *Cell Proliferation in the Gastrointestinal Tract.* London, Pitman Medical Ltd.

Badran, A. F., and Echave Llanos, J. M. (1965). Persistence of mitotic circadian rhythm of a transplantable mammary carcinoma after 35 generations: Its bearing on the success of treatment with Endoxan. *J. Natl. Cancer Inst.* 35:285-290.

Barbason, H., and Cantfort, J. V. (1976). Nyctohemeral rhythms in the liver. *Prog. Liver Dis.* 5:136-148.

Barnum, C. P., Jardetzky, C. D., and Halberg, F. (1957). Nucleic acid synthesis in regenerating liver. *Texas Rep. Biol. Med.* 15:134-147.

Bartlett, P., Haus, E., Tuason, T., Sackett-Lundeen, L., and Lakatua, D. (1982). Circadian rhythm in number of erythroid and granulocytic colony forming units in culture (ECFU-C and GCFU-C) in bone marrow of $BDF_1$ male mice. In *Proceedings of the XV International Conference on Chronobiology*, E. Haus and H. F. Kabat (eds.). S. Karger, Basel, in press.

Baum, S. J., Lednye, G. D., and Khan, A. (eds.) (1981). *Experimental Haematology Today.* S. Karger, Basel.

Bjerknes, M., and Cheng, H. (1981a). The stem-cell zone of the small intestinal epithelium. I. Evidence from Paneth cells in the adult mouse. *Am. J. Anat.* 160:51-63.
Bjerknes, M., and Cheng, H. (1981b). The stem-cell zone of the small intestinal epithelium. II. Evidence from Paneth cells in the newborn mouse. *Am. J. Anat.* 160: 65-75.
Bjerknes, M., and Cheng, H. (1981c). The stem-cell zone of the small intestinal epithelium. III. Evidence from columnar, enteroendocrine, and mucous cells in the adult mouse. *Am. J. Anat.* 160:77-91.
Bjerknes, M., and Cheng, H. (1981d). The stem-cell zone of the small intestinal epithelium. V. Evidence for controls over orientation of boundaries between the stem-cell zone, proliferative zone and maturation zone. *Am. J. Anat.* 160:105-112.
Blumenfeld, C. M. (1939). Periodic mitotic activity in the epidermis of the albino rat. *Science* 90:446-447.
Broders, A. C., and Bublin, W. B. (1939). Rhymicity of mitosis in the epidermis of human beings. *Proc. Staff Meetings Mayo Clin.* 14:423-425.
Brown, J. M., and Berry, R. J. (1968). The relationship between diurnal variations of the number of cells in mitosis and of the number of cells synthesizing DNA in the epithelium of the hamster cheek pouch. *Cell Tissue Kinet.* 1:23-33.
Brugal, G., and Pelmont, J. (1975). Existence of the two chalone-like substances in intestinal extract from the adult newt, inhibiting embryonic intestinal cell proliferation. *Cell Tissue Kinet.* 8:171-187.
Bullough, W. S. (1948). Mitotic activity in the adult male mouse *Mus musculus* L. The diurnal cycles and their relation to waking and sleeping. *Proc. R. Soc. Lond. [Biol]* 135:212-232.
Bullough, W. S., and Laurence, E. B. (1966). The diurnal cycle in epidermal mitotic duration and its relation to chalone and adrenaline. *Exp. Cell Res.* 43:343-350.
Burgess, A. W., and Metcalf, D. (1980). The nature and action of granulocyte-macrophage colony stimulating factors. *Blood* 56:947-958.
Burns, E. R. (1981). Circadian rhythmicity in DNA synthesis in untreated and saline-treated mice as a basis for improved chronochemotherapy. *Cancer Res.* 41:2795-2802.
Burns, E. R., and Scheving, L. E. (1975). Circadian influence on the waveform of the frequency of labelled mitosis in mouse epithelium. *Cell Tissue Kinet.* 8:61-66.
Burns, E. R., Scheving, L. E., and Tsai, T.-H. (1972). Circadian rhythm in uptake of tritiated thymidine by kidney, parotid, and duodenum of isoproterenol-treated mice. *Science* 175:71-73.
Carleton, A. (1934). A rhythmical periodicity in the mitotic division of animal cells. *J. Anat.* 68:251-263.
Chang, W. W. L. (1971). Renewal of the epithelium in the descending colon of the mouse. III. Diurnal variation in the proliferative activity of epithelial cells. *Am. J. Anat.* 131:111-120.
Chang, W. W. L., and Nadler, N. J. (1975). Renewal of the epithelium in the descending colon in the mouse. IV. Cell population kinetics of vacuolated–columnar and mucous cells. *Am. J. Anat.* 144:39-56.
Chumak, N. G. (1963). A study of the mitotic cycle in the corneal cells of mice with the aid of tritium-labeled thymidine. *Dokl. Akad. Nauk SSSR* 149:960-962.
Clausen, O. P. F., Lindmo, T., Sandnes, K., and Thorud, E. (1976). Separation of mouse epidermal cells for microflow fluorometric measurements. *Virchows Arch. [Cell Pathol.]* 20:261-275.
Clausen, O. P. F., Thorud, E., and Aarnaes, E. (1981). Evidence of rapid and slow progression of cells through $G_2$ phase in mouse epidermis: A comparison between phase durations measured by different methods. *Cell Tissue Kinet.* 14:227-240.

Clausen, O. P. F., Thorud, E., Bjerknes, R., and Elgjo, K. (1979). Circadian rhythms in mouse epidermal basal cell proliferation. Variations in compartment size, flux and phase duration. *Cell Tissue Kinet.* 12:319-337.

Clausen, O. P. F., Thorud, E., and Bolund, L. (1980). DNA synthesis in mouse epidermis. Labelled and unlabelled basal cells in S phase after administration of tritiated thymidine. *Virchows Arch. [Cell Pathol.]* 34:1-11.

Clausen, O. P. F., Thorud, E., and Kaufman, S. L. (1982). Circadian rhythms in mouse epidermis: Variations in the numbers of basal and differentiating cells, cell density and cell loss. *Int. J. Chronobiol.* 8:83-93.

Cline, M. J., and Gold, D. W. (1979). Controlling the production of blood cells. *Blood* 53:157-165.

Cooper, Z. K. (1939). Mitotic rhythm in human epidermis. *J. Invest. Dermatol.* 2:289-300.

Cooper, Z. K., and Franklin, H. C. (1940). Mitotic rhythms in the epidermis of the mouse. *Anat. Rec.* 78:1-8.

Cooper, Z. K., and Schiff, A. (1938). Mitotic rhythm in human epidermis. *Proc. Soc. Exp. Biol.* 39:323-324.

Dörmer, P., Schmolke, W., Muschalik, P., and Brinkmann, W. (1970). Die DNS-Synthesegeschwindigkeit im Verlauf der DNS Synthesephase von Erythroblasten der Maus in vivo. *Beitr. Pathol.* 141:174-186.

Echave Llanos, J. M., Aloisso, M. D., Souto, M., Balduzzi, R., and Surur, J. M. (1971). Circadian variations of DNA synthesis, mitotic activity, and cell size of hepatocyte population in young immature male mouse growing liver. *Virchows Arch. [Cell Pathol.]* 8:309-317.

Echave Llanos, J. M., and Nash, R. E. (1970). Mitotic circadian rhythm in a fast-growing and slow-growing hepatoma: Mitotic rhythm in hepatomas. *J. Natl. Cancer Inst.* 44:581-585.

Elgjo, K., Clausen, O. P. F., and Thorud, E. (1981). Epidermis extracts (chalone) inhibit cell flux at the $G_1$-S, S-$G_2$ and $G_2$-M transitions in mouse epidermis. *Cell Tissue Kinet.* 14:21-29.

Evensen, A. (1963). Kinetics of epidermal cell proliferation in experimental skin carcinog enesis. *Bull. WHO* 28:513-515.

Farsund, T. (1975). Cell kinetics in mouse urinary bladder epithelium. 1. Circadian and age variations in cell proliferation and the nuclear DNA content. *Virchows Arch. [Cell Pathol.]* 18:35-49.

Fisher, L. B. (1968). The diurnal mitotic rhythm in the human epidermis. *Br. J. Dermatol.* 80:75-80.

Fortuyn-van Leyden, C. E. D. (1917). Some observations on periodic nuclear division in the cat. *Proc. Kon. Akad. van Wetenschappen Amsterdam* 19:38-44.

Gelfant, S., Ozawa, A., Chalker, D. K., and Smith, G. (1982). Circadian rhythms and differences in epidermal and dermal cell proliferation in uninvolved and involved psoriatic skin in vitro. *J. Invest. Dermatol.* 78:58-62.

Gololbova, M. H. (1958). Changes in mitotic activity in rats in relation to the time of day or night. *Bull. Exp. Biol. Med.* 46:143-146.

Gordon, A. S. (ed.) (1970). *Regulation of Hematopoiesis.* Appleton-Century-Crofts, New York.

Grube, D. D., Auerbach, H., and Brues, A. M. (1970). Diurnal variations in the labelling index of mouse epidermis: A double isotope autoradiographic demonstration of changing flow rates. *Cell Tissue Kinet.* 3:363-373.

Hamilton, E. (1979). Diurnal variation in proliferative compartments and their relation to cryptogenic cells in the mouse colon. *Cell Tissue Kinet.* 12:91-100.

Hegazy, M. A. H., and Fowler, J. F. (1973). Cell population kinetics of plucked and unplucked mouse skin. I. Unirradiated skin. *Cell Tissue Kinet.* 6:17-33.

Hennings, H., and Elgjo, K. (1970). Hydrocortisone: Inhibition of DNA-synthesis and mitotic rate after local application to mouse dorsal epidermis. *Virchows Arch. [Cell Pathol.]* 8:42-49.

Hopper, J. L., and Brockwell, P. J. (1978). A stochastic model for cell populations with circadian rhythms. *Cell Tissue Kinet.* 11:205-255.

Hudson, H. M., and Hahn, G. M. (1977). The labelled mitosis curve for a population consisting of fast and slowly cycling cells. *J. Theor. Biol.* 66:63-79.

Iversen, O. H. (1981). The chalones. In *Tissue Growth Factors,* vol. 57, R. Baserga (ed.). Springer Verlag, Berlin, pp. 491-550.

Iversen, O. H., Bjerknes, R., and Devik, F. (1968). Kinetics of cell renewal, cell migration and cell loss in the hairless mouse dorsal epidermis. *Cell Tissue Kinet.* 9:351-367.

Izquierdo, J. N., and Gibbs, S. J. (1974). Turnover of cell-renewing populations undergoing circadian rhythms in cell proliferation. *Cell Tissue Kinet.* 7:99-111.

Jaffe, J. J. (1954). Diurnal mitotic periodicity in regenerating rat liver. *Anat. Rec.* 120: 935-953.

Kahn, G., Weinstein, G. D., and Frost, P. (1968). Kinetics of human epidermal cell proliferation. diurnal variations. *J. Invest. Dermatol.* 50:459-462.

Killmann, S. A., Cronkite, E. P., Fliedner, T. M., and Bond, V. P. (1962). Mitotic indices of human bone marrow cells. I. Number and cytologic distribution of mitoses. *Blood* 19:743-750.

Kivilaakso, E., and Rytömaa, A. J. (1971). Erythrocyte chalone, a tissue-specific inhibitor of cell proliferation in the erythron. *Cell Tissue Kinet.* 4:1-9.

Klein, B., and Valleron, A. J. (1977). A compartment model for the study of diurnal rhythms in cell proliferation. *J. Theor. Biol.* 64:27-42.

Kreyberg, L., Evensen, A., and Iversen, O. H. (1965). Influence of stress on the diurnal rhythm in mitotic activity in the epidermis of hairless mice. *Acta Pathol. Microbiol. Scand.* 64:176-184.

Laerum, O. D., and Aardal, N. P. (1981). Chronobiological aspects of bone marrow and blood cells. In *Biological Rhythms in Structure and Function,* Eleventh International Congress of Anatomy, H. von Mayersbach, L. E. Scheving, and J. E. Pauly (eds.). Alan R. Liss, Inc., New York, pp. 87-97.

Laerum, O. D., and Maurer, H. R. (1973). Proliferation kinetics of myelopoietic cells and macrophages in diffusion chambers after treatment with granulocyte extracts (chalone). *Virchows Arch. [Cell Pathol]* 15:293-305.

Lindemann, R. (1971). Erythropoiesis inhibiting factor (EIF). I. Fractionation and demonstration of urinary EIF. *Br. J. Haematol.* 21:623-631.

Mauer, A. M. (1965). Diurnal variation of proliferative activity in the human bone marrow. *Blood* 26:1-7.

McKee, L. C., Ensign Johnson, L., and Lange, R. D. (1974). Circadian variation in the reticulocyte counts and immunodetectable erythropoietin titers (37997). *Proc. Soc. Exp. Biol. Med.* 145:1284-1287.

Meytes, D. M. A., Powell, W. B., Ortega, J. A., Shore, N. A., and Dukes, P. P. (1980). Constancy of erythroid burst forming Unit ($BFU_E$) levels in the blood of hematologically normal individuals. *Exp. Hematol.* 8:641-644.

Møller, U., Hartmann, N. R., and Faber, M. (1979). Mitotic index, influx and mean transit time in hamster cheek pouch epitelium, a partially synchronized cell system. *Cell Tissue Kinet.* 12:581-596.

Møller, U., and Keiding, N. (1982). Circadian variations in influx and efflux of the S phase in a partially synchronized cell system. *Cell Tissue Kinet.* 15:341-350.

Møller, U., Keiding, N., and Engel, F. (1982). Subclassification of cells in S phase in a partially synchronized cell system. *Cell Tissue Kinet.* 15:157-168.
Møller, U., and Larsen, J. K. (1978). The circadian variations in the epithelial growth of the hamster cheek pouch. Quantitative analysis of DNA distributions. *Cell Tissue Kinet.* 11:405-413.
Moskalik, K. G. (1976). Diurnal rhythm of mitotic activity, DNA synthesis, and duration of mitoses in mouse bone marrow cells. *Bull. Exp. Biol. Med.* 81:594-596.
Mustafin, A. G. (1976). Diurnal rhythms of cell proliferation in the early stages of liver carcinogenesis induced in mice by orthoaminoazotoluene. *Bull. Exp. Biol. Med.* 82: 1066-1068.
Nash, R. E., and Echave Llanos, J. M. (1971). Twenty-four-hour variations in DNA synthesis of a fast-growing and slow-growing hematoma: DNA synthesis rhythm in hepatoma. *J. Natl. Cancer Inst.* 47:1007-1012.
Olsson, L. (1976). Effects of tritium-labeled pyrimidine nucleosides on epithelial cell proliferation in the mouse. I. Cytodynamic perturbations in normal circadian rhythms after a single injection of $^3$H-thymidine. *Radiat. Res.* 68:258-274.
Ortiz-Picón, J. M. (1934). Über Zellteilungsfrequenz und Zellteilungsrhytmus in der Epidermus der Maus. *Z. Zellforsch. Mikroskop. Anat.* 19:488-509.
Pallavicini, M. G., Cohen, A. M., Dethlefsen, L. A., and Gray, J. W. (1979). In vivo effects of 5-fluorouracil and ftorafur [1. (tetrahydrofuran-2-71)-5-fluorouracil] on murine mammary tumors and small intestine. *Cell Tissue Kinet.* 12:177-189.
Paukovits, W. R., and Hinterberger, W. (1978). Molecular weight and some chemical properties of the granulocytic chalone. *Blut.* 37:7-18.
Pilgrim, C., Erb, W., and Maurer, W. (1963). Diurnal fluctuations in the numbers of DNA synthesizing nuclei in various mouse tissues. *Nature (Lond.)* 199:863-865.
Pizzarello, D. J., and Witcofski, R. L. (1970). A possible link between diurnal variations in radiation sensitivity and cell division in bone marrow of male mice. *Radiology* 97:165-167.
Ponassi, A., Morra, L., Bonanni, F., Molinari, A., Gigli, G., Mercelli, M., and Sacchetti, C. (1979). Normal range of blood colony-forming cells (CFU-C) in humans: Influence of experimental conditions, age, sex, and diurnal variations. *Blut* 39:257-263.
Potten, C. S., Jessup, B. A., and Croxton, M. B. (1971). Incorporation of tritiated thymidine into the skin and hair follicles. *Cell Tissue Kinet.* 4:413-421.
Pullman, H., Lennartz, K. J., and Steigleder, G. K. (1974). In vitro examination of cell proliferation in normal and psoriatic epidermis, with special regard to diurnal variations. *Arch. Derm. Forsch.* 250:177-184.
Quastler, H., and Sherman, F. G. (1959). Cell population kinetics in the intestinal epithelium of the mouse. *Exp. Cell Res.* 17:420-438.
Reeve, D. R. E. (1975). A study of mitotic activity and the diurnal variation of the epithelial cells in wounded rectal mucous membrane. *J. Anat.* 119:333-345.
Ross, D. D., Pollack, A., Akman, S. A., and Bachur, N. R. (1980). Dirunal variation of circulating human myeloid progenitor cells. *Exp. Hematol.* 8:954-960.
Rytömaa, T., and Kiviniemi, K. (1968). Control of granulocyte production. I. Chalone and antichalone, two specific humoral regulators. *Cell Tissue Kinet.* 1:329-340.
Sakata, T., and Tamata, H. (1978). Presence of circadian rhythm in the mitotic index of the ruminal epithelium in sheep. *Res. Vet. Sci.* 24:1-3.
Sassier, P., and Bergeron, M. (1977). Specific inhibition of cell proliferation in the mouse intestine by an aqueous extract of rabbit small intestine. *Cell Tissue Kinet.* 10:223-231.
Sassier, P., and Bergeron, M. (1980). Existence of an endogenous inhibitor of DNA synthesis in rabbit small intestine specifically effective on cell proliferation in adult mouse intestine. *Cell Tissue Kinet.* 13:251-261.

Schell, H., Hornsetin, O. P., and Schwarz, W. (1980). Human epidermal cell proliferation with regard to circadian variation of plasma cortisol. *Dermatologica* 161:12-21.

Schell, H., Rosenberger, H., Hornstein, O. P., and Wawra, E. (1977). Autoradiographic in vitro studies on diurnal variations in human epidermal cell proliferation. *Arch. Dermatol. Res.* 257:265-272.

Scheving, L. (1959). Mitotic activity in the human epidermis. *Anat. Rec.* 135:7-19.

Scheving, L. E., Burns, R. E., and Pauly, J. E. (1972). Circadian rhythms in mitotic activity and $^3$H-thymidine uptake in the duodenum: Effect of isoproterenol on the mitotic rhythm. *Am. J. Anat.* 135:311-318.

Scheving, L. A., Burns, E. R., Pauly, J. E., and Tsai, T. H. (1978). Circadian variation in cell division of the mouse alimentary tract, bone marrow and corneal epithelium. *Anat. Rec.* 191:479-486.

Scheving, L. E., von Mayersbach, H., and Pauly, J. E. (1974a). An overview of chronopharmacology. *J. Eur. Toxicol.* 7:203-227.

Scheving, L. E., and Pauly, J. E. (1967). Circadian phase relationships of thymidine-$^3$H uptake, labeled nuclei grain counts, and cell division rate in rat corneal epithelium. *J. Cell. Biol.* 32:677-683.

Scheving, L. E., and Pauly, J. E. (1973). Cellular mechanisms involving biorhythms with emphasis on those rhythms accociated with the S and M stages of the cell cycle. *Int. J. Chronobiol.* 1:269-286.

Scheving, L. E., and Pauly, J. E. (1974). Cellular mechanisms involving biorhythms with emphasis on those rhythms associated with the S and M stages of the cell cycle. *Int. J. Chronobiol.* 1:269-283.

Scheving, L. E., Pauly, J. E., Burns, E. R., Halberg, F., Tsai, S., and Betterton, H. O. (1974b). Lighting regimen dominates interacting meal schedules and synchronizes mitotic rhythm in mouse corneal epithelium. *Anat. Rec.* 180:47-52.

Scheving, L. E., Pauly, J. E., von Mayersbach, H., and Dunn, J. D. (1974c). The effect of continuous light or darkness on the rhythm of the mitotic index in the corneal epithelium of the rat. *Acta Pathol.* 88:411-423.

Scheving, L. A., Yeh, Y. C., Tsai, T.-H., and Scheving, L. E. (1979). Circadian-phase dependent stimulatory effects of epidermal growth factor on deoxyribonucleic acid synthesis in the tongue, esophagus and stomach of the adult mouse. *Endocrinology* 105:1475-1480.

Scheving, L. A., Yeh, Y. C., Tsau, T.-H., and Scheving, L. E. (1980). Circadian-phase dependent stimulatory effects of epidermal growth factor on deoxyribonucleic acid in the duodenum, jejunum, ileum, caecum, colon, and rectum of the adult male mouse. *Endocrinology* 106:1498-1505.

Sharkis, S. J., LoBue, J., Alexander, P., Jr., Rakowitz, F., Weitz-Hamburger, A., and Gordon, A. S. (1971). Circadian variations in mouse hemetopoiesis. II. Sex differences in mitotic indices of femoral diaphyseal marrow cells (35936). *Proc. Soc. Exp. Biol. Med.* 138:494-496.

Sharkis, S. J., Palmer, J. D., Goodenough, J., LoBue, J., and Gordon, A. S. (1974). Daily variations of marrow and splenic erythropoiesis, pinna epidermal cell mitosis and physical activity in C57B1+6J mice. *Cell Tissue Kinet.* 7:381-387.

Sigdestad, C. P., and Lesher, S. (1971). Photo-reversal of the circadian rhythm in the proliferative activity of the mouse small intestine. *J. Cell. Physiol.* 78:121-126.

Stoney, P. J., Halberg, F., and Simpson, H. W. (1975). Circadian variation in colony-forming ability of presumable intact murine bone marrow cells. *Chronobiologia* 2: 319-324.

Storey, W. F., and Leblond, C. P. (1951). Measurement of the rate of proliferation of epidermis and associated structures. *Ann. N.Y. Acad. Sci.* 53:537-545.

Thompson, J. C. (1974). Antral chalone. *Prog. Gastroenterol.* 67:752-754.

Thorud, E., Aarnaes, E., Clausen, O. P. F., Bjerknes, R., and Anker, P. (1983). Estimation of circadian variations in cell cycle phase durations in murine epidermal basal cells. Submitted for publication.

Thorud, E., Clausen, O. P. F., Bjerknes, R., Elgjo, K., and Lindmo, T. (1978). Circadian rhythms in the proliferative compartment of epidermis in the hairless mouse demonstrated by pulse cytophotometry, autoradiography and a stathmokinetic method (Colcemid). In *Pulse-Cytophotometry, Part III,* D. Lutz (ed.) European Press, Ghent, Belgium, pp, 359-364.

Thorud, E., and Volden, G. (1980). Flow cytometry (FCM) of human epidermal cells. A preparation method for epidermal cells and demonstration of circadian variations in the proportion of S-phase cells. *Arch. Dermatol. Res.* 269:137-145.

Tutton, P. J., and Barkla, D. H. (1980). The influence of dibutyryl adenosine cyclic monophosphate on cell proliferation in the epithelium of the jejunal crypts and in colonic carcinomata of rat. *Clin. Exp. Pharmacol. Physiol.* 7:275-280.

Tvermyr, E. M. F. (1969). Circadian rhythms in epidermal mitotic activity. Diurnal variations in the mitotic index, the mitotic rate and the mitotic duration. *Virchows Arch. [Cell Pathol.]* 2:318-325.

Tvermyr, E. M. F. (1972). Circadian rhythms in hairless mouse epidermal DNA synthesis as measured by double labelling with $^3$H-thymidine (H3-TdR). *Virchows Arch. [Cell Pathol.]* 11:43-54.

Vasama, R., and Vasama, R. (1957). Diurnal mitotic activity in the corneal epithelium of a mouse. *Ann. Med. Exp. Biol. Fenn.* 35:363-370.

Wilson, J. W. (1948). Diurnal rhythm of mitotic acitvity in the liver of the mouse. *Anat. Rec.* 101:672-673.

Yeh, S. C., Scheving, L. A., Tsai, T.-H., and Scheving, L. E. (1981). Circadian stage dependent effects of epidermal growth factor on deoxyribonucleic acid synthesis in ten different organs of the adult male mouse. *Endocrinology* 109:644-651.

# 8

# Diurnal Variation in Influx and Transition Intensities in the S Phase of Hamster Cheek Pouch Epithelium Cells

**Niels Keiding** / Danish Medical Research Council, and Danish Social Science Research Council, Copenhagen, Denmark

**Niels Rudolph Hartmann and Ulla Møller** / The Finsen Institute, Copenhagen, Denmark.

## INTRODUCTION

A general survey of cell kinetic modeling of circadian rhythms was recently provided by Macdonald (1981), who also discussed various experimental techniques capable of identifying the parameters in such models, among these the use of the labeling index to determine the fraction of synthesizing cells—the S fraction. By administering two injections of the same isotope at different times and taking two separate biopsies with a certain interval of time, the labeling technique may be generalized to also allow for estimation of influx (Tvermyr, 1972). Although Tvermyr was concerned with the epidermis of hairless mice, we performed a similar experiment to study diurnal variations in the cell kinetics of the cheek pouch epithelium of strongly synchronized Syrian hamsters.

The aim of this chapter is to assess the cell kinetic characteristics of the S phase from a rather detailed statistical analysis of this double-labeling experiment. We use the experience from this case study as a starting point for some critical comments on related literature.

We have earlier provided a description of the biological background and the immediate conclusions from the experiments (Møller and Keiding, 1982), including references to earlier work, and a preliminary modeling approach based on the same data has been presented (Hartmann and Møller, 1978). Data on mitotic influx and mitotic rates will be analyzed in forthcoming reports.

The model, which may be considered a variant of that proposed by Klein and Valleron (1977), is specified by the temporal variation of the total number of cells, the influx into S phase, and the transition intensitites over each of a number of hypothetical subphases. These are all parameterized by an expansion of sine waves, the two former combined with a linear trend. Nonlinear least-squares analysis is used for estimating the coefficients and identifying the minimal model that fits satisfactorily. For reasons of space, many details of the statistical analysis are omitted; a fuller version is available in research report form (Keiding et al., 1982).

## THE MODEL AND THE EXPERIMENT

### The Cell Kinetic Model

In this chapter we pass over the possibility of formulating a complete stochastic model for the cell cycle such as indicated by, e.g., Jagers (1975, Chap. 9), being concerned only with providing a rather detailed description of the influx of cells into S phase and the transition behavior of cells through this phase. By treating influx and transition as separate entities we disregard the biological feedback from mother to daughter cells, and we assume no cell loss (due to death or differentiation) in S phase. Further comments on these assumptions and of the choice of framework for the cell kinetic model are given in Cell Kinetic Models Including Circadian Variation.

However, we do want to indicate that one may well embed the classic deterministic theory into a stochastic model.

Assume that cells enter S phase according to an inhomogeneous Poisson process $Z_i(t)$ with intensity function $\tau(t)$. [This process is regarded as being given on the whole line $(-\infty, \infty)$ (Brown, 1969)]. If the distribution of the transit time T for a cell entering S phase at time t is given by the survivorship function $R(t, u) = P_t \{T > u\}$, it follows that the expected number of cells in S phase at time t is given by

$$\nu(t) = \int_0^\infty \zeta(t-u)\, R(t-u, u)\, du \qquad (1)$$

The *exit process* $Z_e(t)$ will then be an inhomogeneous Poisson process on $(-\infty, \infty)$ with intensity (called the *efflux intensity*)

$$\eta(t) = \int_0^\infty \zeta(t-u)\, f(t-u, u)\, du$$

where f(t, u) is the probability density corresponding to R (t,u). This was essentially shown by Karlin (1966, p. 497); see also Brown (1969).

The total number of cells at time t is a stochastic process which in a full cell kinetic model would be considered governed by the reproduction mechanism derived from the description of the cell cycle. In this chapter, however, we shall parameterize the expected total number of cells independently of the $G_1$ to S flux.

The survival function R (t, u) of the transit time distribution is modeled by postulating that the cells have to move through a fixed (but unknown) number M of (fictitious) subphases. In each subphase j there is a time-dependent intensity $\lambda_j(t)$ of completing the sojourn in that state; we shall be content with the simplification that $\lambda_1(t) = \ldots = \lambda_M(t)$, denoted $\lambda(t)$ in the following. It is now seen that the transition time T is specified by the waiting time until the Mth jump in a Poisson process Y(t) with time-dependent intensity $\lambda(t)$, so that the distribution of T for a cell entering S phase at time $t_0$ is given by

$$\begin{aligned} R(t_0, t) &= P_{t_0}\{T > t\} \\ &= P\{Y(t + t_0) - Y(t_0) < M\} \end{aligned}$$

$$= \exp\left[-\int_{t_0}^{t_0+t} \lambda(x)\,dx\right] \sum_{m=0}^{M-1} \frac{\left[\int_{t_0}^{t_0+t} \lambda(x)\,dx\right]^m}{m!}$$

that is, density function

$$f(t_0, t) = -\frac{d}{dt} R(t_0, t)$$

$$= \lambda(t_0 + t) \exp\left[-\int_{t_0}^{t_0+t} \lambda(x)\,dx\right] \frac{\left[\int_{t_0}^{t_0+t} \lambda(x)\,dx\right]^{M-1}}{(M-1)!}$$

Using different arguments, this formula was given earlier by Klein and Guiguet (1978).

The biological interpretation is that in each subphase a certain amount of DNA is assumed to be synthesized (not necessarily the same amount in different subphases). As a consequence, it is assumed that the DNA synthesis rate varies diurnally in the same general way for cells in different parts of S phase, and that the *DNA synthesis rate for cells of a given age in S phase is* always *proportional to* $\lambda(t)$.

### The Double Labeling Experiment

As mentioned in the introduction the biological conditions and immediate results of the experiment have been described earlier (Møller and Keiding, 1982) and we shall therefore be content with a brief summary here.

The basic experimental unit consists of the following steps. At a certain time t a hamster was injected with the radioactively labeled DNA precursor tritiated thymidine ([$^3$H] TdR), as well as with Colcemid. After 3 hr the right cheek pouch was removed under ether anesthesia, a second dose of [$^3$H] Tdr was given, and the animal was replaced in the cage. The animal was sacrificed 50 min later (that is, at time t + 3.8 hr) and the left cheek pouch removed. The purpose of the Colcemid injection is to block cells in mitosis from dividing into daughter cells.

The procedure was performed for each of seven animals (occasionally only six) starting every 1.5 hr during a 48 hr period. The design is illustrated in Fig. 1.

The data recorded for each cheek pouch were U = number of labeled cells [defined as cells with at least three grains (traces of the tritium isotope) over the nucleus] per millimeter sectioned epidermis, and C = total number of cells per millimeter. The *labeling index* (LI) is U/C, representing the proportion of labeled cells among all cells, and in a sense to be further specified below measuring the proportion of cells in S phase, the *S fraction.*

### Cell Kinetic Modeling of the Double-Labeling Experiment

In the following we use the convention of measuring t in hours, letting t = 18 correspond to the start of the experiment; the last observation thus corresponds to t = 64.5. Let $U_h(t)$ [$C_h(t)$] denote the number of marked cells per millimeter (cells/mm) in the hth biopsy (h = 1, biopsy at time t + 3; h = 2; biopsy at time t + 3.8) for a hamster injected at time t.

Since cell divisions are inhibited by the Colcemid "block" and since the growth of the epithelial tissue is assumed to proceed by new cells "pushing" their way to the surface,

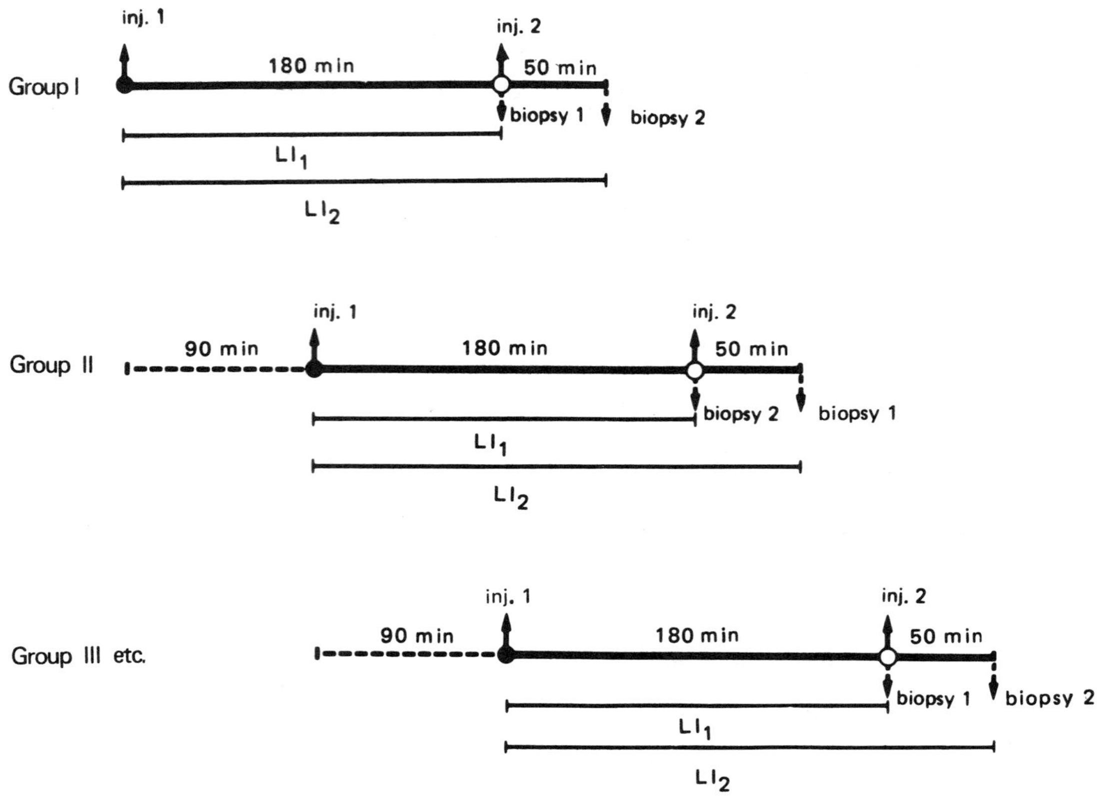


**FIGURE 1.** The design of the double-labeling experiment. Every 1.5 hr the following procedure was followed: seven animals were injected with [$^3$H] TdR and colcemid. In the figure this is symbolized by ↑ 1 injection. This injection forms the basis of $LI_1$, determined by the first removed cheek pouch. After 3 hr the first cheek pouch ($LI_1$) was removed; in the figure this is symbolized by ↓. At the same time a second dose of [$^3$H] TdR was given, labeling cells which in the 3 hr had entered the S phase. In the figure this injection is symbolized by ↑ 2 injection. The second cheek pouch was removed 50 min later, which is symbolized by ↓. $LI_2$ was determined by the second cheek pouch. (From Møller and Keiding, 1982.)

the total number of cells as reflected in the cells/mm count will in this report be referred to the time of injection. These general considerations are supported by least-squares fits of simple harmonic functions with trends to $C_1$ (t) and $C_2$ (t). There was no significant difference between these, and the common expected value, denoted by $\gamma$ (t), is estimated as

$$\hat{\gamma}(t) = 138.4 + 0.54 \cos \frac{2\pi}{24} t + 7.82 \sin \frac{2\pi}{24} t + 0.268 (t - 42).$$

The fit is shown in Fig. 2. Both the diurnal oscillation and the trend are clearly significant here.

To obtain a model for the labeling index U/C we recall that according to the traditional interpretation of pulse-labeling experiments the labeled cells are those and only those that actively synthesize DNA at the time t of injection. Interpreting the (unspecified)

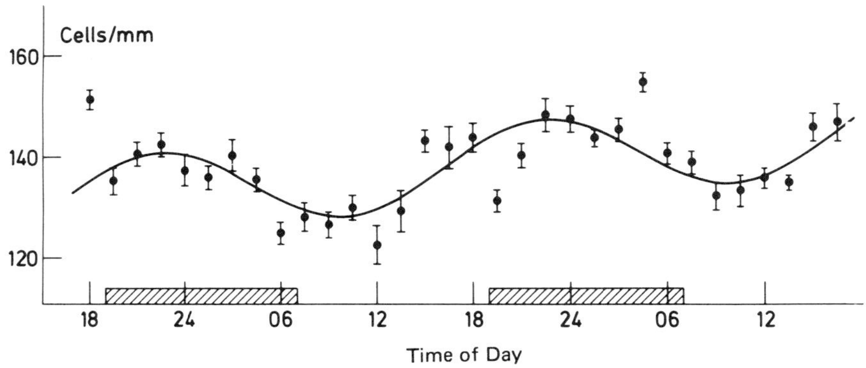


**FIGURE 2** Cells per millimeter in the epithelium of the removed cheek pouches. The values for the right and left cheek pouches are pooled and plotted at time of injection (mean ±2 SE). A simple harmonic function superimposed by a linear trend is fitted.

population of cells referred to as the cells per millimeter epidermis, the expectation of the number U of labeled epithelial cells per millimeter at the time of the first biopsy is exactly $\nu$ (t) as given by Eq. (1). The cells that are labeled at the time of the second biopsy were either in S phase at time t or in $G_1$ at t and in S at time t + 3. The expected number of such cells is accordingly given by $\nu(t) + \mu(t, t+3)$, where

$$\mu(t, t+3) = \int_0^3 \zeta(t+3-u)\, R(t+3-u, u)\, du \tag{2}$$

If U/C and C are independent (see below), then

$$E(U_1(t)) = E\,\frac{U_1(t)}{C_1(t)}\;E(C_1(t))$$

$$= E(LI_1(t))\;E(C_1(t))$$

so that we define

$$\alpha'(t) = E(LI_1(t))$$

$$= \frac{\nu(t)}{\gamma(t)} \tag{3}$$

and similarly we define

$$\beta(t) = E(LI_2(t))$$

$$= E\,\frac{U_2(t)}{C_2(t)}$$

$$= \frac{\nu(t) + \mu(t, t+3)}{\gamma(t)} \tag{4}$$

In fact, the correlations between U/C and C were 0.1 (right cheek pouch) and -0.1 (left cheek pouch), indicating no significant departure from independence.

Introducing the *relative influx* $\iota(t) = \zeta(t)/\gamma(t)$, we have according to Eqs. (3) and (1)

$$\alpha'(t) = \gamma(t)^{-1} \int_0^\infty \gamma(t-u)\,\iota(t-u)\,R(t-u,u)\,du \qquad (5)$$

and similarly by Eqs. (1), (2), and (4),

$$\beta(t) = \alpha'(t) + \gamma(t)^{-1} \int_0^3 \gamma(t+3-u)\,\iota(t+3-u)\,R(t+3-u,u)\,du \qquad (6)$$

Note that if R (t, u) is approximated by 1 for $0 \leq u \leq 3$ (that is, it is assumed that no cell completes S phase in less than 3 hr), then

$$\beta(t) - \alpha'(t) = \frac{1}{\gamma(t)} \int_t^{t+3} \zeta(u)\,du$$

expresses an *average relative influx* into S phase in the period (t, t + 3), which may be directly estimated from $LI_2(t) - LI_1(t)$.

As mentioned in the introduction, this possibility of "directly measuring influx" from the double labeling is an important feature of the design as first proposed by Tvermyr (1972).

Similarly, recalling that the *efflux* at time t is given by

$$\eta(t) = \int_0^\infty \gamma(t-u)\,\iota(t-u)\,f(t-u,u)\,du$$

and again assuming $R(x, u) = 1, 0 \leq u \leq 3$, it may be shown that

$$\int_t^{t+3} \eta(v)\,dv = \gamma(t)\,\beta(t) - \gamma(t+3)\,\alpha'(t+3)$$

Tvermyr (1972), assuming constant population size, remarked that $LI_2(t) - LI_1(t+3)$ estimates an average relative efflux in the interval (t, t + 3). In fact, we see that if $\gamma(t) = \gamma$, then

$$LI_2(t) - LI_1(t+3) \simeq \beta(t) - \alpha'(t+3)$$
$$= \gamma^{-1} \int_t^{t+3} \eta(v)\,dv$$

this fact is paraphrased by the statement that "efflux is directly measurable from the double-labeling experiment.

At this point it is necessary to mention that the preliminary analysis showed that the rationale for the double-labeling experiment as described here requires some modification (Møller and Keiding, 1982). A supplementary experiment (Møller et al., 1982b) indicated the existence of what was termed $S_0$ cells, operationally defined as cells that do not incorporate [$^3$H] TdR for the first hour after injection but which do so in the interval between 1 and 3 hr after injection. Based on this supplementary experiment we postulated a constant $S_0$ fraction of 1.9% at injections in the dark period (19.00–07.00) and no $S_0$ cells at injections in the light period (07.00–19.00).

The definition of the $S_0$ cells is tailored to restrict their appearance to the $LI_1$ index: the corrected expected $LI_1$ (t), to be denoted $\alpha(t)$, should be increased by the correction for t in the dark periods:

$$E(LI_1(t)) = \alpha(t)$$
$$= \alpha'(t) + 0.019 \qquad t = 19.5, \ldots, 30, 43.5, \ldots, 54 \tag{7a}$$

but not in the light periods:

$$\alpha(t) = \alpha'(t) \qquad t = 18, 31.5, \ldots, 42, 55.5, \ldots, 64.5 \tag{7b}$$

The expected $LI_2$ should not be changed from $\beta(t)$ as defined above. First, no $S_0$ cells at time t + 3 can be recorded in the biopsy at time t + 3.8 (because less than an hour has elapsed), and second, $S_0$ cells at time t are, by definition, S cells at time t + 3, so that they are being marked "in their own right" at the second injection anyway.

### Statistical Analysis of the Labeling Indices

We choose to parameterize the *transition intensity* $\lambda(t)$ by an expansion of harmonic functions

$$\lambda(t) = \tau_0 + \sum_{i=1}^{4} \left( \tau_i \cos \frac{2\pi i}{24} t + \theta_i \sin \frac{2\pi i}{24} t \right)$$

and the relative influx $\iota(t)$ is similarly parameterized, adding a linear trend:

$$\iota(t) = \mu_0 + \sum_{i=1}^{4} \left( \mu_i \cos \frac{2\pi i}{24} t + \kappa_i \sin \frac{2\pi i}{24} t \right) + \mu_5 (t - 42)$$

where the trend term contains t - 42, 42 being the midpoint of the times where measurements were obtained.

The expected cells/mm $\gamma(t)$ was parameterized and fitted to the observed cells/mm above. In the following we choose to regard the ratios $\gamma(t - u)/\gamma(t)$ and $\gamma(t + 3 - u)/\gamma(t)$ [see Eqs. (5) and (6)], *known* as estimated, and the correction of 0.019 linking $\alpha'(t)$ and $\alpha(t)$ [see Eq. (7)] is also supposed *known* without error. (See the end of this Section for estimation of this correction term from the present experiment.) Finally, a series of preliminary fits using different values of the integer parameter M (the number of subphases) indicated little dependence of the fitted values on M. We take M = 9 as fixed in the following.

The model thus specified was fitted by nonlinear least squares to the observed ($LI_1$, $LI_2$), taking the correlation between $LI_1$ and $LI_2$ into consideration; for details, see Keiding et al. (1982).

An approximative test for goodness of fit may be obtained by comparing the minimal value of the relevant quadratic form Q to a $\chi^2$ distribution with $2 \times 32 - (6 + 4 + 5 + 4) = 45$ degrees of freedom (d.f.). We found $Q_{min} = 97$, $P < 0.0001$, and thus the overall fit is not statistically acceptable. However, inspection of the observed and fitted values (Keiding et al., 1982, Figs. 3 and 4) indicates that biological variation beyond the reach of systematic modeling is responsible for this. We therefore proceed by assuming the existence of intraclass correlation. As discussed by Keiding et al. (1982), it is safe to use the same estimates, but the test statistics and standard errors have to be modified in a conservative direction, a crude approximate way of remedying the estimates being to inflate all estimated standard errors by $\sqrt{Q_{min}/d.f.} = \sqrt{2.2} = 1.5$ and similarly deflate the $\chi^2$ statistics by 2.2.

In searching for models with fewer parameters that still adequately fit the data, we reduce the number of superharmonic terms in the parameterizations of influx and transit time intensity as described in Table 1. Fortunately, even in view of the above remarks on

**TABLE 1** Tests for Reduction of Model

| | Influx parameters | | | Transition intensity parameters | | | |
|---|---|---|---|---|---|---|---|
| Model | Level | Oscillations | Trend | Level | Oscillations | $Q_{min}$ | d.f. |
| I | $\mu_0$ | $\mu_1, \kappa_1, \ldots, \mu_4, \kappa_4$ | $\mu_5$ | $\tau_0$ | $\tau_1, \theta_1, \ldots, \tau_4, \theta_4$ | 97 | 45 |
| II | $\mu_0$ | $\mu_1, \kappa_1, \mu_2, \kappa_2$ | $\mu_5$ | $\tau_0$ | $\tau_1, \theta_1, \tau_2, \theta_2,$ | 110 | 53 |
| III | $\mu_0$ | $\mu_1, \kappa_1$ | $\mu_5$ | $\tau_0$ | $\tau_1, \theta_1, \tau_2, \theta_2$ | 118 | 57 |
| IV | $\mu_0$ | $\mu_1, \kappa_1$ | $\mu_5$ | $\tau_0$ | $\tau_1, \theta_1$ | 153 | 59 |

anticonservatism, the statistical analysis rather unambiguously leads to choosing model III (harmonic influx with trend and harmonically varying transition intensity) among the models listed there. The estimates under this model are given in Table 2 and the fit is illustrated in Figs. 3 and 4.

The standard errors and correlation coefficients quoted in Table 2 were estimated from the ordinary inverse information matrix approximation, and being derived from the within-group variance, the standard errors are too small and the Wald test too sharp. Also, the variabilities in cells/mm, in the $S_0$-cell correction, and in the integer parameter M are excluded.

However, only the statistical significance of the trend $\mu_5$ depends on the amount of correction for intraclass correlation. But since Figs. 5 and 6 show that the deviations from the consistent negative difference in labeling index from day 1 to day 2 are rather haphazard, it seems preferable to keep the trend parameter in the model.

Throughout the statistical analysis, the $S_0$ correction was assumed fixed and equal to 1.9%. This estimate was obtained by Møller et al. (1982b) from a supplementary experiment, with a standard error of 0.42%. In order to obtain an independent assessment of the correction from the present data, model III was fitted for varying values of the correction term.

Key values of $Q_{min}$ with associated values of $\mu_0$ (estimated mean relative influx) are shown in Table 3. It is seen that the correction of 1.9% drastically improves the fit

**TABLE 2** Estimates from Final Fit (Model III)[a]

| Parameters | | | Estimate (SE) | Correlation | Wald test $\chi^2$ | d.f. | P |
|---|---|---|---|---|---|---|---|
| Influx | Constant | $\mu_0$ | 0.554 (0.033) | | | | |
| | Harmonic | $\mu_1$ | 0.469 (0.051) | 0.06 | 99.6 | 2 | $10^{-10}$ |
| | | $\kappa_1$ | 0.228 (0.051) | | | | |
| | Trend | $\mu_5$ | -0.00162 (0.0063) | | 6.61 | 1 | 0.01 |
| Transition intensity | Constant | $\tau_0$ | 1.059 (0.098) | | | | |
| | Harmonic | $\tau_1$ | 0.220 (0.110) | 0.07 | 27.8 | 2 | $10^{-6}$ |
| | | $\theta_1$ | 0.668 (0.133) | | | | |

[a]M = 9 (fixed a priori).

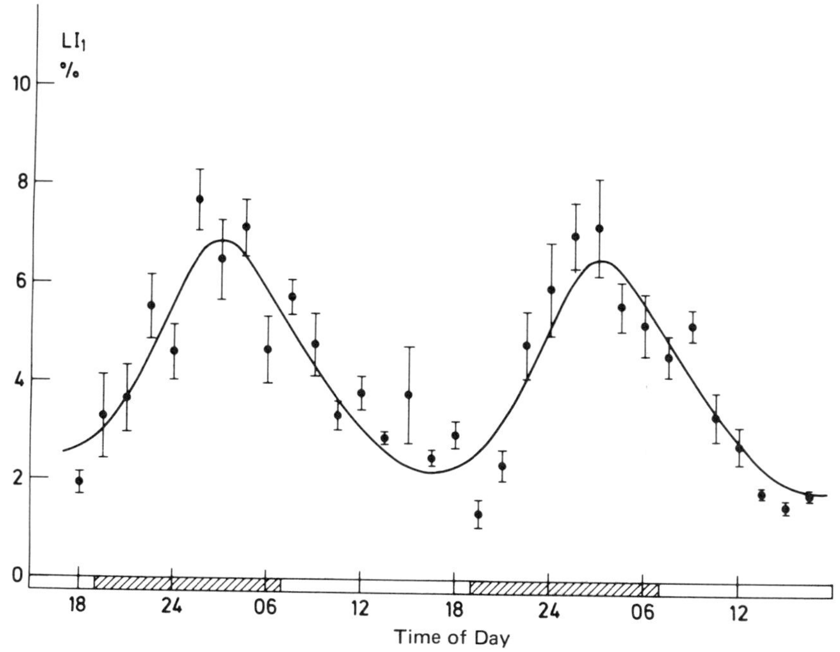


**FIGURE 3** Corrected $LI_1$ and definitive fit (model III) based on harmonic relative influx with trend and harmonically varying transition intensity.

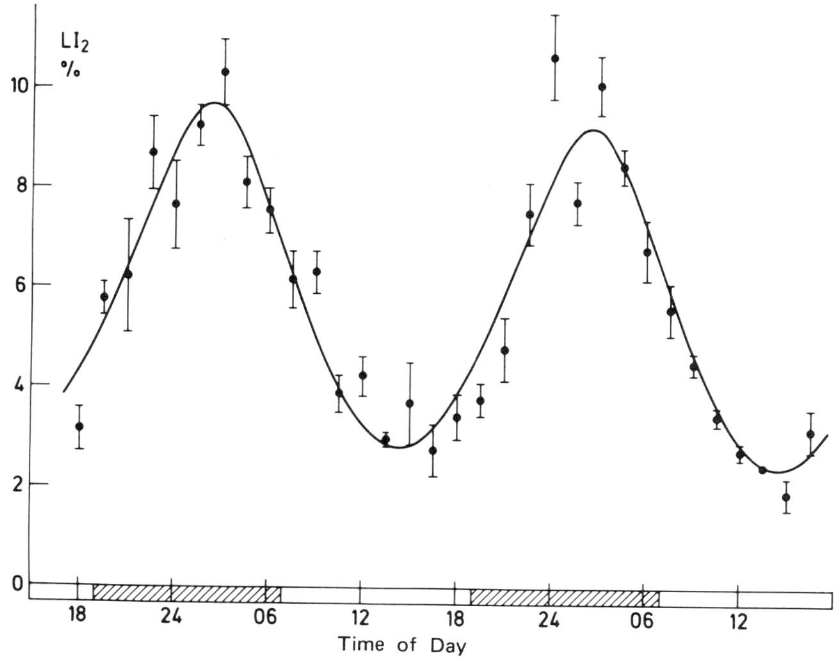


**FIGURE 4** $LI_2$ and definitive fit (model III).

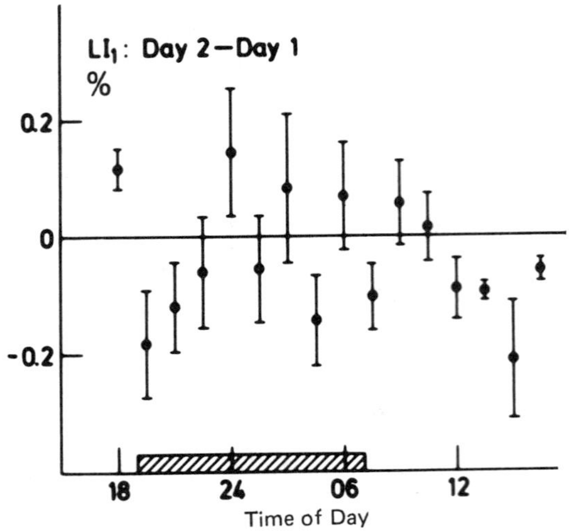


FIGURE 5 The difference between corresponding values of $LI_1$ on day 2 and day 1. Mean ± 2 SE.

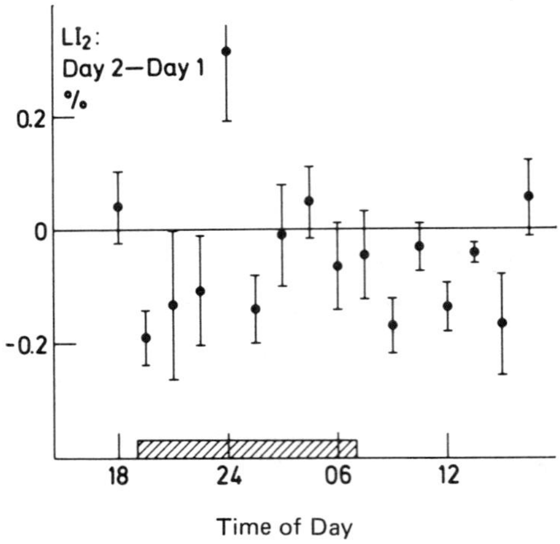


FIGURE 6 The difference between corresponding values of $LI_2$ on day 2 and day 1. Mean ± 2 SE.

TABLE 3 Internal Estimation of $S_0$ Correction Using Model III

| | | | | | |
|---|---|---|---|---|---|
| $S_0$ correction (%) | 0.0 | 1.8 | 1.9 | 2.0 | 2.1 |
| $Q_{min}$ | 203.0 | 118.6 | 118.1 | 117.9 | 118.1 |
| $\mu_0$ (%/hr) | 0.41 | 0.55 | 0.55 | 0.57 | 0.58 |

of the data from the double-labeling experiment compared to no correction, and that in fact, this externally estimated correction almost coincides with the internally optimal value of 2.0%.

## IMPLICATIONS OF THE MODEL

In the following we quote estimates derived under model III, unless otherwise stated.

### Influx into S Phase

One result of the analysis is the estimation of the relative influx density

$$\hat{\iota}(t) = 0.554 + 0.469 \cos \frac{2\pi}{24} t + 0.228 \sin \frac{2\pi}{24} t - 0.00162\,(t - 42)$$

This is plotted in Figure 7. In the original rationale of the double-labeling experiment it was emphasized that $LI_2(t) - LI_1(t)$ "directly measures influx." As discussed, this is still approximately true, after correction for $S_0$ cells, in our model, as also illustrated in Figure 7. It is seen that the influx increases from very low values at noon to a maximum just after midnight, to drop off again during the early morning hours. Møller and Keiding (1982, Fig. 6) displayed the estimated S-phase influx from Model I and commented on the further superharmonic terms which have been left out here as statistically insignificant. Keiding et al. (1982) commented on the trend terms in cells/mm and relative influx.

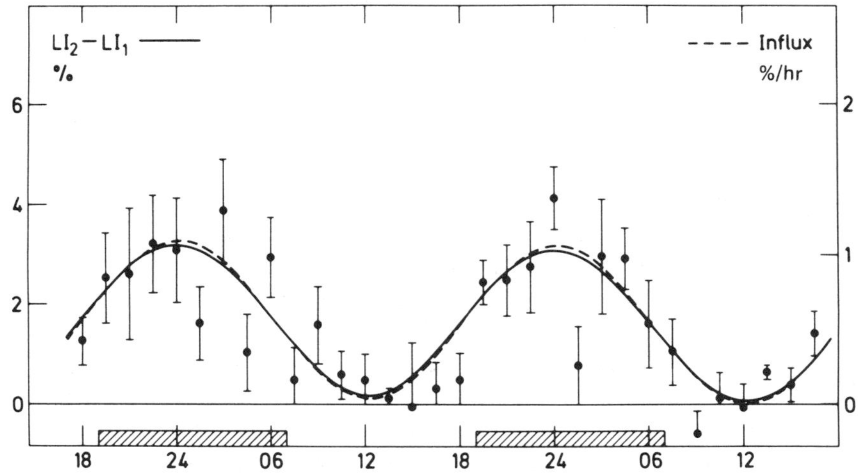


**FIGURE 7** The estimate of relative influx used by Tvermyr (1972) is $[LI_2(t)$ - corrected $LI_1(t)]/3$. Empirical values (mean ± 2 SE) and fitted values from model III (solid curve) of this quantity are compared to the direct estimate from the present model of relative influx (broken line). It is seen that influx is high in the dark period and low in the light period.

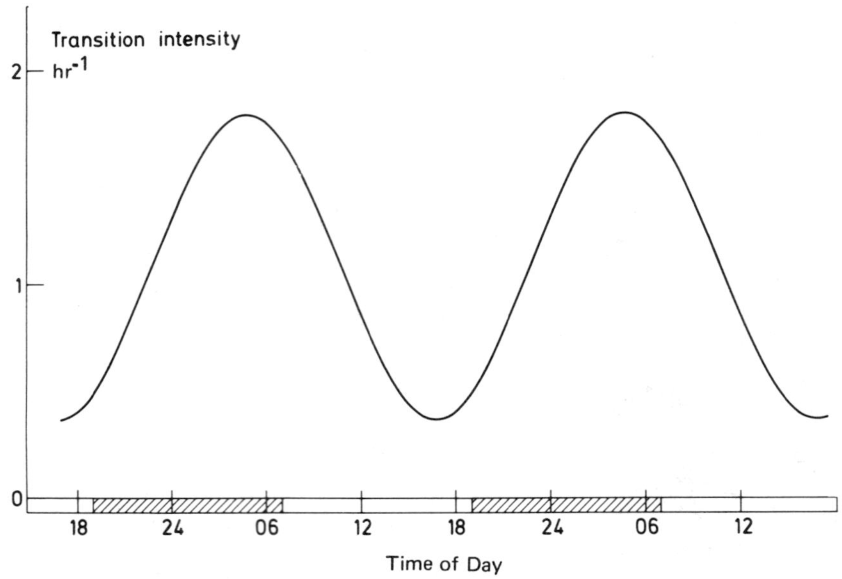


**FIGURE 8** The estimated harmonically varying transition intensity function under model III. The transition intensity begins to increase around the start of the dark period and reaches its maximum shortly before the end of this period.

### The Transit Time Distribution

Figure 8 illustrates the estimated transition intensity function $\lambda(t)$, and in Fig. 9 we plot, for selected values of t, the estimated density function f(t, u) for the transit time distribution of a cell entering S phase at time t. A three-dimensional plot (Fig. 10) illustrates f(t, u) for all t. Finally, Fig. 11 illustrates the diurnal variation of mean and coefficient of variation of the transit time. (The diurnal variation of the coefficient of variation is a consequence of the particular parameterization. Though this variation seems quite plausible, we do not claim that the experiment contains independent evidence of it.)

The diurnal variation is most directly understood from the *transition intensity* $\lambda(t)$, which in the sense specified above is proportional to *the DNA synthesis rate.* This is seen to vary considerably, with a maximum at about 5 hr and a minimum at 17 hr. The *mean transit time* represents, for each time t, an integrated summary of the development of the intensity function in the hours after t. Thus it has its minimum shortly before the maximum of $\lambda(t)$ (cells are pushed fastest through S phase in late night and early morning), and the mean transit time is highest for cells entering S phase around noon. Notice that the harmonic variation of the transition intensity *does not* imply harmonic variation of mean transit time; on the contrary, the latter rises steeply and falls off more gradually, a tendency already apparent in the preliminary step-function approach to the fit (Hartmann and Møller, 1978). Keiding et al. (1982) illustrated the qualitative effects of including the (insignificant) second harmonic terms (model II). In this case, $\lambda(t)$ rises steeply in the very beginning of the dark period, with a rather gradual decline during the day. However, the mean transit time changes little from model III to model II.

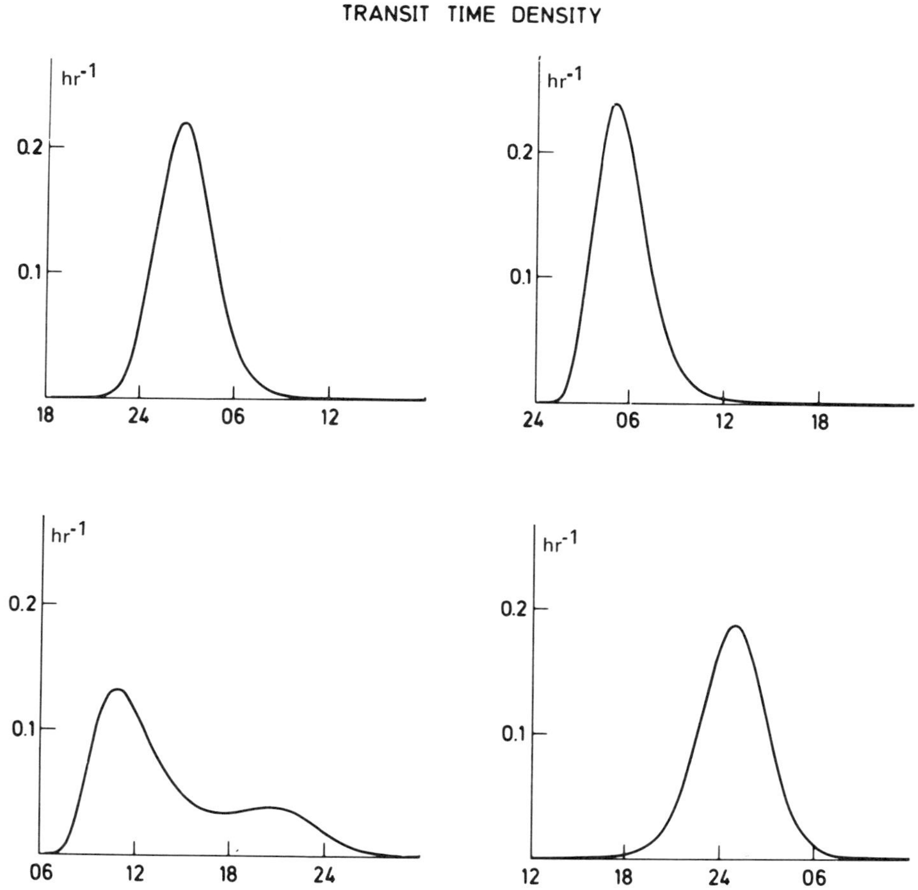


**FIGURE 9** Examples of the transit time distribution estimated under model III. Transition is fast (mode about 6 hr) for a cell entering at midnight and slower (mode about 12 hr) for a cell entering at noon. The stack plot in Fig. 10 provides more complete information on the transit time distribution.

### S-$G_2$ Flux (Efflux from S Phase)

The *relative efflux intensity* $\epsilon(t)$ from S phase may be defined in a similar way as the relative influx intensity $\iota(t)$ was defined above:

$$\epsilon(t) = \frac{\eta(t)}{\gamma(t)}$$

$$= \frac{1}{\gamma(t)} \int_0^\infty \gamma(t-u)\, \iota(t-u)\, R(t-u,u)\, du$$

an estimate of which may thus be directly derived from the estimated parameters. In Fig. 12 we have plotted the estimated $\epsilon(t)$ together with the "direct estimate" $[LI_2(t)$ - corrected $LI_1(t+3)]/3$ and its fit $[\hat{\beta}(t) - \hat{\alpha}'(t+3)]/3$. It is seen that the efflux varies even more dramatically than the influx. This is a consequence of influx and transition intensity

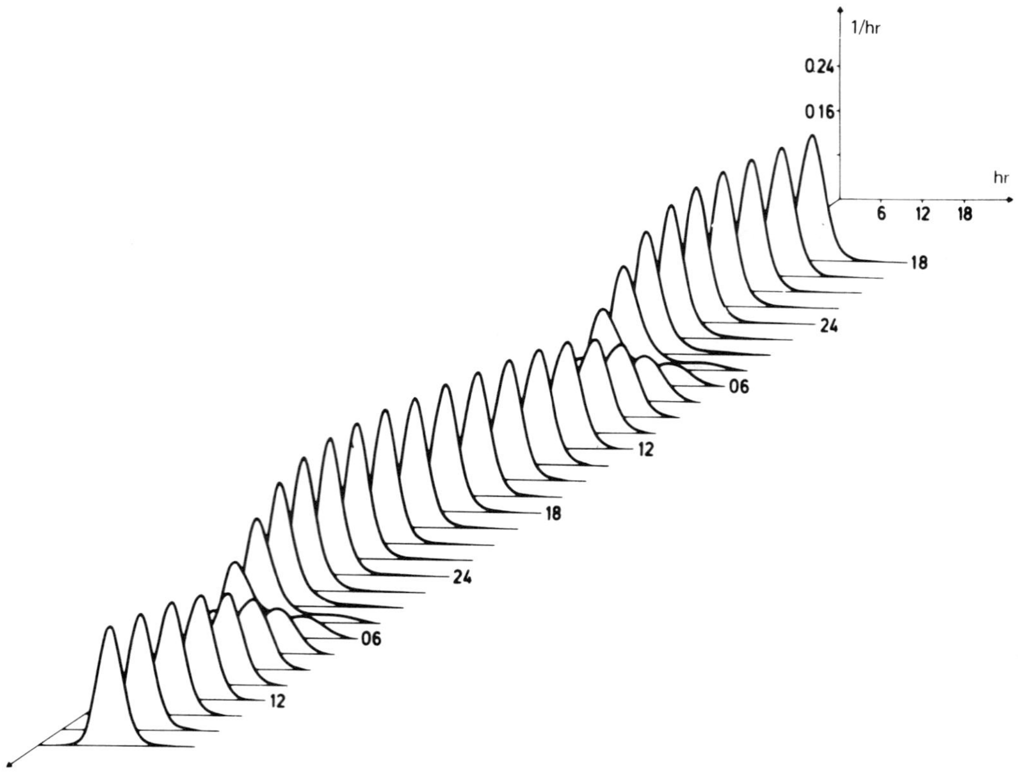


**FIGURE 10** Three-dimensional plot of transit time density estimated under model III.

being nearly in phase; compare Figs. 7 and 8. In fact, if the "filter" relating influx and efflux (and specified by the kernel $\omega(t)\, f(t, u)$) was time independent (that is, independent of t), then the oscillations in influx would be *damped* such that the efflux fluctuated *less* than the influx (Keiding et al., 1982, Fig. 17). We may take the rather large oscillations in the difference between $LI_2(t)$ and the corrected $LI_1(t + 3)$ as rather conspicuous evidence of the fluctuations in transition intensity. Møller and Keiding (1982, Fig. 7) illustrated the efflux that would have been estimated from model I.

### Age Distribution in S Phase

We discussed the representation of the transit time distribution in terms of the intensity, survivorship, and density functions, as well as mean transit time and coefficient of variation. It is, however, also of interest to combine this variation with the variation in influx to obtain the *age distribution in S phase* at different times t of the day. This is given by

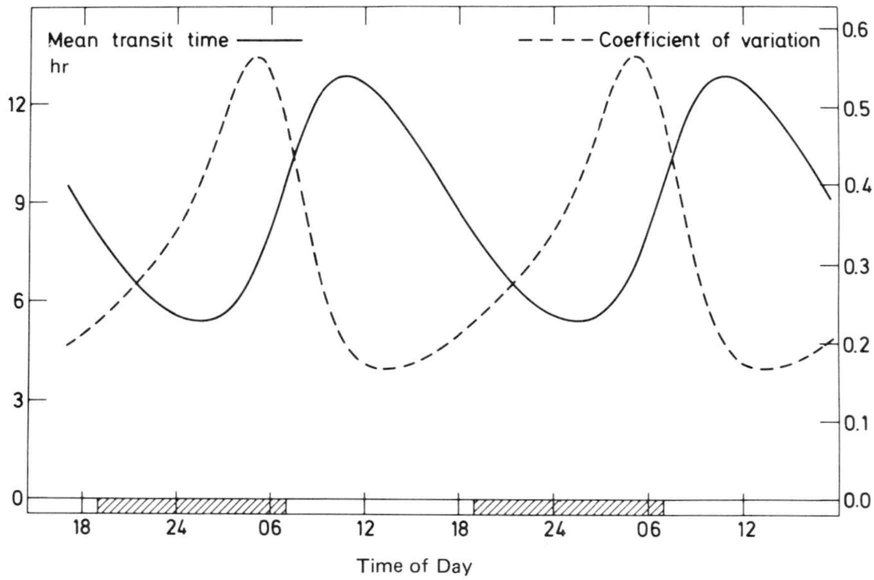


**FIGURE 11** Estimated mean transit time and coefficient of variation under model III.

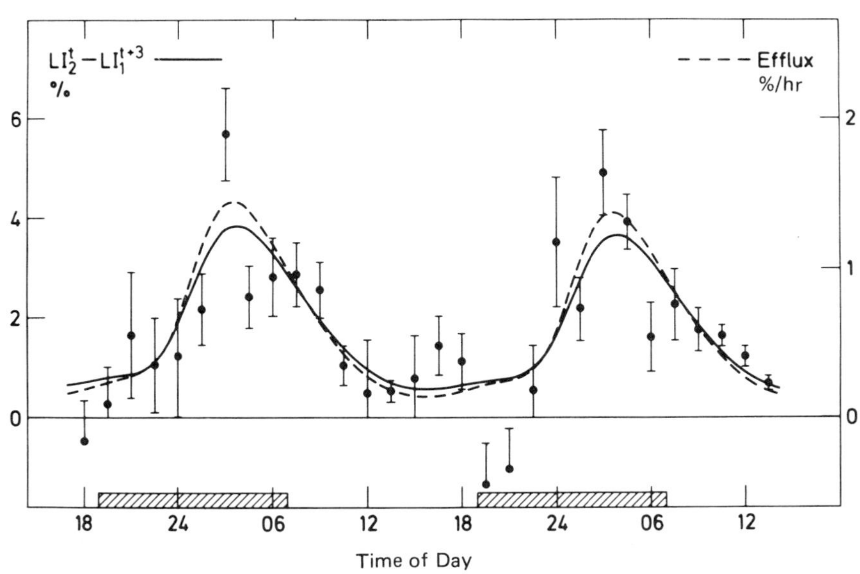


**FIGURE 12** The estimate of relative efflux used by Tvermyr (1972) is [$LI_2(t)$ - corrected $LI_1(t + 3)$]/3. Empirical values (mean ± 2 SE) and fitted values from model III (solid curve) of this quantity are compared to the direct estimate from the present model of relative efflux (broken line). It is seen that efflux varies more than influx (Fig. 7), as expected with some delay in the peaks and troughs compared to the latter.

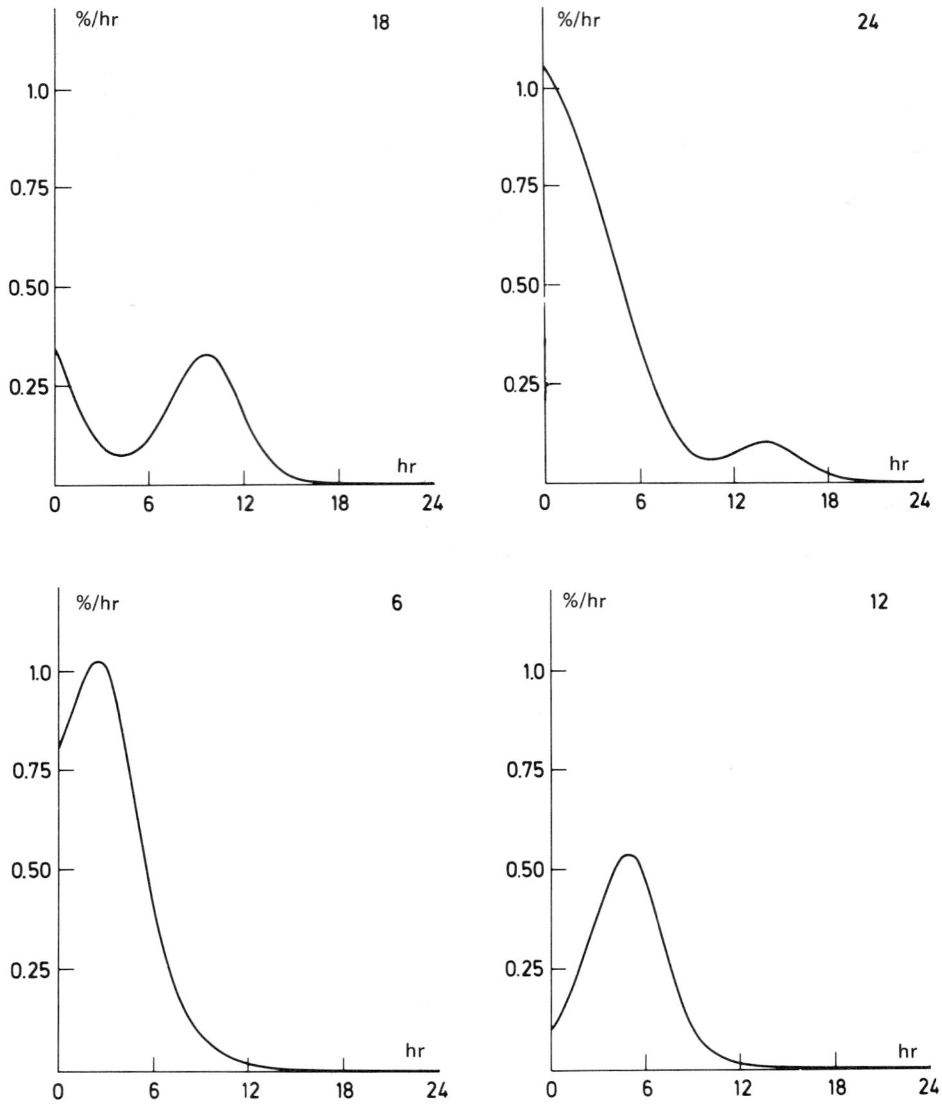


**FIGURE 13** Examples of the (absolute) age distribution estimated under model III. At 18 hr, there are a few but old (around 10 hr) cells in S phase, and at midnight this cohort is about to disappear, but a considerable influx of young cells has appeared. At 6 hr there are many cells in S phase, all of which are young, and at noon most have left S, a cohort of age around 5 hr remaining. The picture in the second 24 hr period is almost but not exactly the same, the difference being due to the trend in relative influx. The stack plot in Fig. 14 provides more complete information.

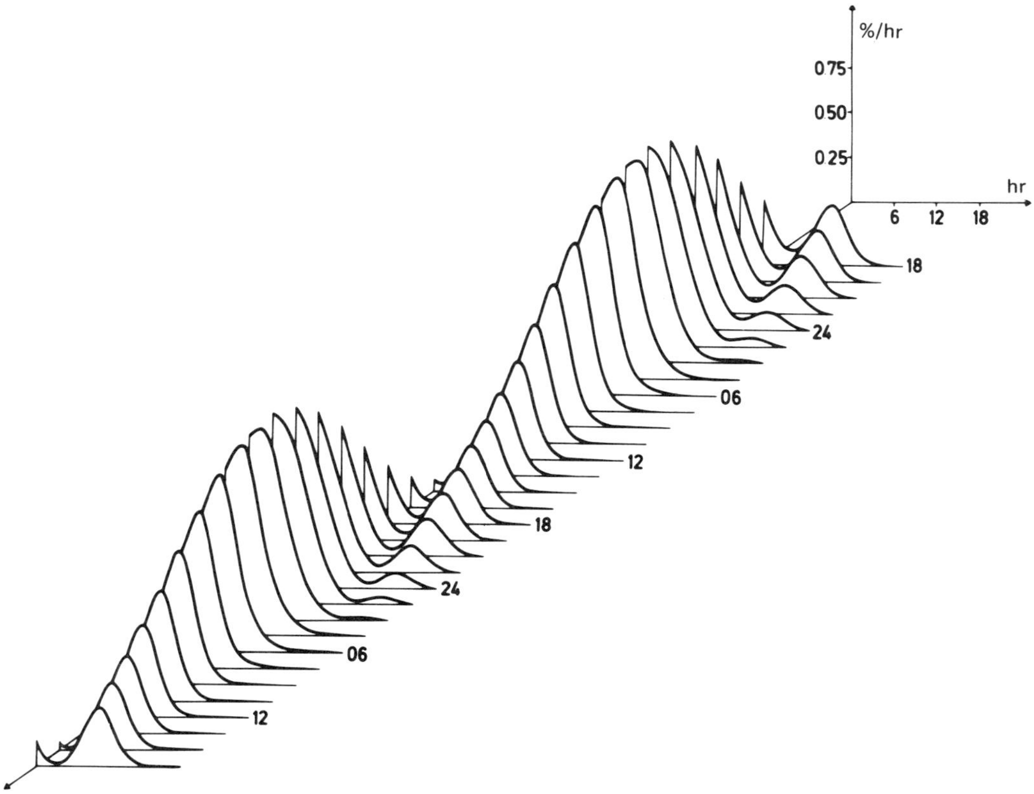


**FIGURE 14** Three-dimensional plot of the age distribution in S phase estimated under model III.

the density function $\zeta(t - u)\, R\,(t - u, u)$, $0 < u < \infty$, where $\zeta(t)$ is the absolute influx intensity.

Notice that this distribution has total mass

$$\int_0^\infty \zeta(t - u)\, R\,(t - u, u)\, du = \upsilon(t)$$

the expected number of S phase cells per millimeter at time t.

In practice we plot the estimate of $\gamma(t)^{-1}\gamma\,(t - u)\,\iota\,(t - u)\, R\,(t - u, u)$, which has total mass equal to $\alpha'\,(t) = E\,(\text{corrected } LI_1\,(t))$ (see Fig. 13). In Fig. 14 a three dimensional plot illustrates the diurnal variation of the age distribution.

### DNA Distribution

The S phase is biologically defined as the part of the cell cycle where the cell synthesizes DNA in order to double the DNA content. Under specific assumptions on the DNA synthesis rate in different parts of the S phase the model provides possibilities of deriving conclusions about the DNA distribution in S phase at different times of the day. We shall briefly illustrate one approach and refer to Hartmann and Christensen (1981) for an alternative approach based on partial differential equations for the relevant probability density functions.

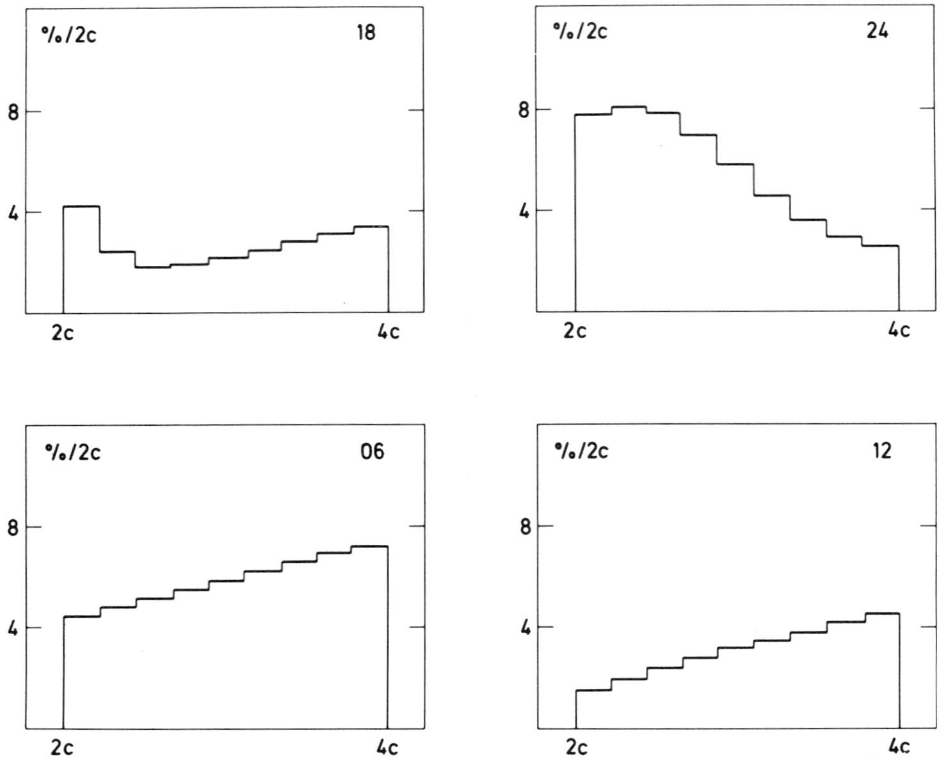


**FIGURE 15** DNA distribution (first 24 hr). Estimated absolute DNA "histograms" under model III, assuming identical synthesis rates in all subphases. It is important to notice the qualitative difference between the diurnal patterns of cell age distribution (Fig. 13) and DNA content distribution (here). The stack plot in Fig. 16 provides more complete information.

Assume that a specific proportion $p_i$ of the required amount of DNA is synthesized during the ith ficticious subphase, $i = 1, \ldots, M$, $p_1 + \ldots + p_M = 1$. (For given transit time $T_i$ of the ith subphase, this means that in this phase, the average synthesis rate is $p_i/T_i$.) An "expected histogram" of the DNA distribution at time t is now constructed in the following way. Let $q_0 = 0$, $q_i = p_1 + \ldots + p_i$, $i = 1, \ldots, M$, i.e., $q_M = 1$. Under the specified assumptions cells in subphase $i + 1$ contain a DNA content between $1 + q_i$ and $1 + q_{i+1}$, so that the expected number of cells with DNA content in $(1 + q_i, 1 + q_{i+1})$ is the expected number of cells in the $(i + 1)$ st subphase. Using the Poisson process $Y(t)$ this is, however,

$$\int_0^\infty \zeta(t-u)\, P\left\{Y(t) - Y(t-u) = i\right\} du$$

$$= \int_0^\infty \zeta(t-u) \left[\int_{t-u}^t \lambda(v)\, dv\right]^i \frac{\exp\left[-\int_{t-u}^t \lambda(v)\, dv\right]}{i!}\, du$$

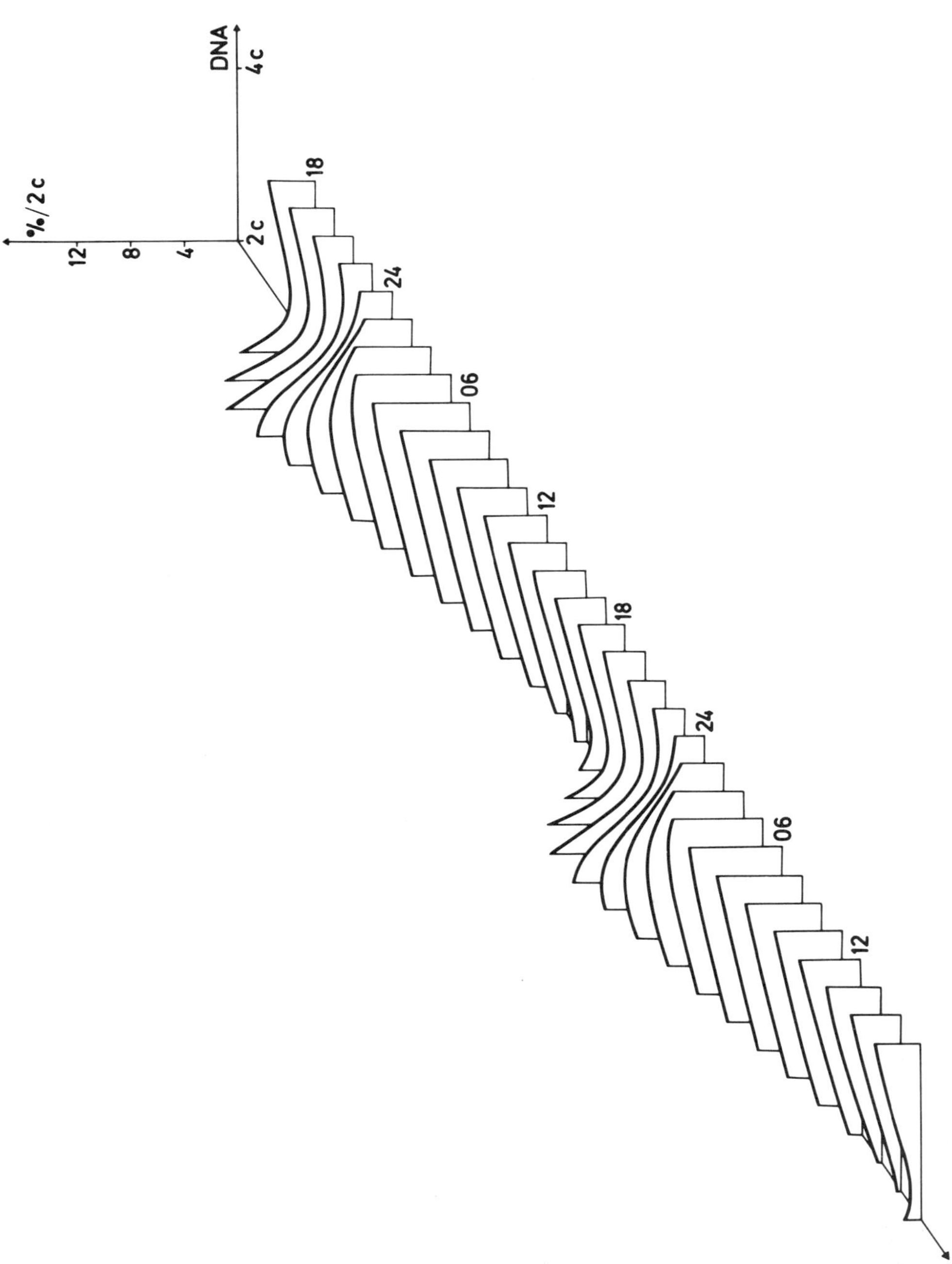


**FIGURE 16** Three-dimensional plot of DNA distributions in S phase estimated under model III.

In practice we replace $\zeta$ (t - u) by $\gamma(t)^{-1}\gamma(t-u)\,\iota(t-u)$ to obtain the expected *relative* number of cells with DNA content in $(1 + q_i, 1 + q_{i+1})$ at time t, which we shall denote $\Delta_{i+1}(t)$. We note that as required $\Delta_1(t) + \ldots + \Delta_M(t) = \alpha'(t)$, the expected relative number of cells in S phase at time t. The $\Delta_i(t)$ are explicit functions of the parameters of the model, so that "expected DNA histograms" may be estimated corresponding to each time of the day.

The histograms may be compared to estimated distributions of DNA content in S phase cells derived from, e.g., flow cytometry measurements. In Fig. 15 we have plotted such histograms under the simple assumption (White, 1980; Macdonald, 1981, p. 294) that $p_1 = \ldots = p_M = M^{-1}$, and Fig. 16 illustrates the diurnal variation in a three-dimensional plot. In further studies it would presumably be more natural to assume that the $p_i$ values are at first rather small, then larger, then small again, to reflect the biologically plausible assumption that the synthesis rate increases continuously from its $G_1$ phase value of zero and again decreases continuously to the $G_2$ phase value of zero (Dörmer et al., 1975).

## DISCUSSION AND CONCLUSION

### Cell Kinetic Conclusions

Our conclusion from the present study is that a considerable variation in S-phase influx (engineering interpretation: a DNA "pump") combines with a definite but less drastic diurnal variation in transition intensities (i.e., in DNA synthesis rate). Although the former variation is apparent already from a preliminary evaluation of the data (influx being also "directly" measured; see Fig. 7), the significance of the variation in transition intensity could only be assessed after detailed cell kinetic modeling and statistical analysis. The conclusion is in accordance with the results by Hopper and Brockwell (1978) based on earlier experiments from our laboratory.

Among the specific results from this work we should also like to point out the significant diurnal variation in cells per millimeter, and we want to emphasize that although the precision in the experiment only allowed for the identification of one significant harmonic term in both relative influx and transition intensity, the modeling contains as many terms as statistically and numerically feasible.

### Cell Kinetic Models Including Circadian Variation

There have recently been several papers on generalization of classic cell kinetic models to periodically varying intensities.

The recent review by Macdonald (1981, Sec. 3) concentrates on asymptotic properties (i.e., quasi-steady-state properties) of models that may essentially be viewed as branching processes ("stochastic models") or expected values of branching processes ("deterministic models"). Briefly, Klein and Valleron (1977), Klein and Guiguet (1978), and Klein and Macdonald (1980) developed a model allowing circadian rhythms using the device of hypothetical subphases introduced into cell kinetics by Takahashi (1966, 1968) and based on Kendall's (1948) multiphase birth process. The same model was used by Aarnæs et al. (1981) to study diurnal rhythms in hairless mouse epidermal basal cell proliferation. A similar approach in discrete time was suggested by Hopper and Brockwell (1978) and Hopper (1980), and applied to preliminary hamster data from our laboratory. A definitive general branching process formulation, with very general asymptotic results, has recently been obtained by Jagers and Nerman (1983) (cf. Jagers, 1983).

In this chapter we have, however, chosen to concentrate on a model for the S phase only, allowing, on the other hand, for description of transient phenomena not so easily captured by the quasi-steady-state assumption. The approach connects to the generalizations to time-inhomogeneous regimens of the classic integral equations describing the flow of cells through a stage in the cell cycle, given by Hartmann et al. (1977) and Hartmann and Møller (1978). The general model is a stochastic interpretation of this paradigm, conceiving the influx and efflux rates as intensities in relevant Poisson processes. Some general comments on the relationship between stochastic and deterministic population models were recently given by Kurtz (1981).

By choosing to base the parameterization of the transit time distribution upon subphase transition intensities, we follow the abovementioned line of work by Klein et al. Besides the lack of data necessary for the fit of a full model, we have two more fundamental reasons for restricting attention to the S phase. The first is that the long $G_1$ phase (about 4 days) must be supposed to damp diurnal variations in mitotic efflux; in the concise formulation of Izquierdo and Gibbs (1972, p. 407): ". . . synchrony of a dividing cohort of cells is lost before they divide again." In combination with the dramatic $G_1$-S flux one is lead to a hypothesis of a "switch" in late $G_1$ which at certain times of the day releases a number of cells from a $G_1$ pool into synthesis. Rather than attempt an ingenious choice of subphases, this is conveniently modeled by assuming a time-inhomogeneous Poisson-process input into S phase.

The second reason is that for the self-renewing epithelial tissue under consideration, about half of the cells are in fact lost (to either death or differentiation) during the cell cycle, the total number of cells being almost constant. To build a full model one will have to postulate exactly where in the cycle the cells are lost, and only then can it be assessed how the stability assumptions for growing populations are to be interpreted for the proliferating epithelial tissue.

In most of the abovementioned models, cell loss (death or differentiation) is assumed to take place just after mitosis. Indeed, this seems to be more of a mathematical convention than based on empirical evidence. [An exception is Hopper and Brockwell (1978), who assumed random loss over the whole cell cycle.] As we shall see below even our much less restrictive assumption of no cell loss during S phase may be questioned.

### Statistical Analysis and the Origin of the Diurnal Variation

Møller and Keiding (1982) summarized the discussion in the literature on the question of whether there is a diurnal variation in both influx and transition intensity. We shall here add some technical comments to some of the earlier results.

Brown and Berry (1968), studying Syrian hamsters, observed a considerable diurnal variation in labeling index and remarked that this could hardly be believed to be inherited through the $G_1$ phase from the diurnal variation in the mitotic index. From their observations on mean grain count, Brown and Berry were led to conclude that the variation originates in late $G_1$ and only there; see Keiding et al. (1982) for further discussion.

Grube et al. (1970) used double labeling with two different isotopes to directly measure influx into and efflux from the S phase of epidermal cells in hairless mice. A clear diurnal variation in both was present: efflux was almost equal to influx about 4 hr earlier, and the hypothesis of constant S-phase transit time seemed confirmed.

In their studies on Chinese (1972) and Syrian (1974) hamsters, Izquierdo and Gibbs again concluded that the source of the diurnal variation is at the $G_1$-S transition and not during S phase. Their claim of no diurnal variation of the S-phase transition was based on the comparison of two percentage labeled mitoses (PLM) curves. The standard methodology for interpreting PLM curves relies, however, fairly critically on the assumption of no circadian variation anywhere in the cell cycle, making this technique less suitable for interpreting these systems (Møller et al., 1974) unless the theory is modified (Macdonald, 1981). In addition, some of the more basic biological assumptions behind the applications of PLM curves to estimate the S-phase transit time may be violated (Clausen et al., 1981).

In her double-labeling experiments on the hairless mouse epidermis, Tvermyr (1972) was led to assume that both sources of variation were important. However, Clausen et al. (1979), working with the same experimental objects, found only little evidence of variation in S-phase influx. Clausen et al. used a deterministic cell kinetic model to fit flow cytometry-determined $G_1$, S, and $G_2$ fractions (mitoses being destroyed by the preparational method), autoradiographically recorded labeling indices and mitotic indices. As carefully explained by these authors, this means that the $G_1$-S flux is rather indirectly determined (in fact only after two successive applications of the continuity equation). It is conceivable that a considerable artificial damping of the $G_1$-S flux could result from this indirect and presumably suboptimal estimation procedure, which did not allow a statistical assessment of the accuracy of the estimated flux.

In their follow-up paper based on the same data, but using various modifications of the Klein-Valleron stochastic model, Aarnæs et al. (1981) did not either find it necessary to assume diurnal variation in $G_1$ transit time (or equivalently, in $G_1$-S flux). Again the evidence is somewhat indirect, and since their model contains only one subphase in $G_1$, the coefficient of variation of the $G_1$ transit time is large, thus allowing for absorbing as apparent random variation a considerable part of a possible diurnal variation.

#### Cell Loss and Extension to the $G_2$ Phase

The topic of this chapter is the evidence from double labeling with the same isotope concerning the influx and transition intensity of the S phase. We did, however, also register the accumulation of mitoses before the Colcemid block, and the following preliminary results from these measurements shed some light on the question of where in the cell cycle to expect the cell loss (Møller et al., 1982a).

By implementing the analysis of Thorud et al. (1980) of stathmokinetic data, the $G_2$-M flux (the mitotic rate) was estimated as a sum of three harmonic terms and a linear trend, in particular, the mean mitotic rate was 0.27% per hour ±0.02. In contrast to this, the S influx and S efflux under our model are assumed to have equal means, estimated as 0.55% per hour. That is, the flux seems to have been almost halved, with the theoretical consequence that almost all cell loss takes place in $G_2$. Before too much significance be attached to this indication, some possible pitfalls should be mentioned. First, there may be as yet unknown difficulties with the stathmokinetic method (like incomplete Colcemid block), and second, the exact value of the $S_0$-correction has considerable influence on the apparent difference: in fact, half of the $LI_1$ values have been reduced by 1.9%, increasing $3\iota(t)$ by 1.9%/2, or the influx by 1.9%/2/3 = 0.32% per hour. A different value of the correction could lead to quite a different impression of the difference. But on the other hand, estimation of the correction based on the data from the present experiment itself yielded a value of 2.0% in good agreement with the externally estimated correction.

As stated above, the existing literature on diurnally varying cell kinetics contains surprisingly few discussions of where in the cell cycle to expect cell loss. With but few exceptions, the same seems to be true for the specific literature on cell kinetics for epithelial tissues. We therefore hope (in forthcoming work) to achieve further insight by generalizing the modeling approach of this chapter to also accommodate $G_2$ transition and differentiation, so that further statistical evidence can become available.

### Smith-Martin Modeling

In contrast to the full stochastic process modeling of the cell cycle quoted above, which leads to some form of branching process, a simpler line of thought has attracted some interest since the paper by Smith and Martin (1973) (see also Chap. 5). According to this, the "intermitotic" period is composed of an A state of random exponentially distributed duration and a B state of deterministic duration. Cells are released from the A state to the B state (which is usually assumed to contain at least the conventional S, $G_2$, and M phases and perhaps also the late part of $G_1$) by a "transition probability," which might be assumed to vary according to changes (including cyclical changes) in environmental "initiators."

The basic empirical foundation of the Smith-Martin line of thought is the so-called $\alpha$, $\beta$ plot, and there has recently been quite convincing criticism of the theory based on deeper statistical analysis of this very plot (Castor, 1980; Nelson and Green, 1981).

In more general terms, however, our tentative conclusions of a "switch" in late $G_1$, strongly regulated by external factors, is broadly compatible with these ideas.

## ACKNOWLEDGMENTS

We are grateful to John Hopper, Philip Hougaard, and Søren Johansen for discussions at various stages of this work. The experiments were supported by the Danish Medical Research Council.

## REFERENCES

Aarnæs, E., Thorud, E., and Clausen, O. P. (1981). Model analysis of circadian rhythms in mouse epidermal basal cell proliferation. *J. Theor. Biol.* 88:355-370.

Brown, J. M., and Berry, R. J. (1968). The relationship between diurnal variation of the number of cells in mitosis and of the number of cells synthesizing DNA in the epithelium of the hamster cheek pouch. *Cell Tissue Kinet.* 1:23-33.

Brown, M. (1969). An invariance property of Poisson processes. *J. Appl. Prob.* 6:453-458.

Castor, L. N. (1980). A $G_1$ rate model accounts for cell cycle kinetics attributed to "transition probability." *Nature* 287:857-859.

Clausen, O. P. F., Thorud, E., and Aarnæs, E. (1981). Evidence of rapid and slow progression of cells through $G_2$ phase in mouse epidermis: A comparison between phase durations measured by different methods. *Cell Tissue Kinet.* 14:227-240.

Clausen, O. P. F., Thorud, E., Bjerknes, R., and Elgjo, K. (1979). Circadian rhythms in mouse epidermal basal cell proliferation. *Cell Tissue Kinet.* 12:319-337.

Dörmer, P., Brinkmann, W., Born, R., and Steel, G. G. (1975). Rate and time of DNA synthesis of individual Chinese hamster cells. *Cell Tissue Kinet.* 8:399-412.

Grube, D. D., Auerbach, H., and Brues, A. M. (1970). Diurnal variation in the labeling index of mouse epidermis: A double isotope autoradiographic demonstration of changing flow rates. *Cell Tissue Kinet.* 3:363-373.

Hartmann, N. R., and Christensen, I. J. (1981). Various applications of the continuity equation in the analysis of data measured by flow cytometry. In *Biomathematics and Cell Kinetics,* M. Rotenberg (ed.). Elsevier/North-Holland, Amsterdam, pp. 389-401.

Hartmann, N. R., and Møller, U. (1978). A compartment theory in the cell kinetics including considerations on circadian variations. In *Biomathematics and Cell Kinetics* A.-J. Valleron and P. D. M. Macdonald (eds.). Elsevier/North-Holland, Amsterdam, pp. 223-251.

Hartmann, N. R., Møller, U., and Lebeda, J. (1977). Analysis of cell kinetic data on the basis of mathematical models. In *Growth Kinetics and Biochemical Regulation of Normal and Malignant Cells,* B. Drewinko and R. M. Humphrey, (eds.). Williams & Wilkins, Baltimore, pp. 471-480.

Hopper, J. (1980). Applications of stochastic models in cell kinetics and epidemiology. Ph.D. Thesis, La Trobe University, Melbourne, Australia.

Hopper, J., and Brockwell, P. J. (1978). A stochastic model for cell populations with circadian rhythms. *Cell Tissue Kinet.* 11:205-225.

Izquierdo, J. N., and Gibbs, S. J. (1972). Circadian rhythms of DNA synthesis and mitotic activity in hamster cheek pouch epithelium. *Exp. Cell Res.* 71:402-408.

Izquierdo, J. N., and Gibbs, S. J. (1974). Turnover of cell-renewing populations undergoing circadian rhythms in cell proliferation. *Cell Tissue Kinet.* 7:99-111.

Jagers, P. (1975). *Branching Processes with Biological Applications,* Wiley, New York.

Jagers, P. (1983). Stochastic models for cell kinetics. *Bull. Math. Biol.* 45:507-519.

Jagers, P., and Nerman, O. (1983). Branching processes in periodically varying environment. Preprint 1983-12, Dept. Math., Chalmers Univ. Techn., Göteborg, Sweden.

Karlin, S. (1966). *A First Course in Stochastic Processes.* Academic Press, New York.

Keiding, N., Hartmann, N. R., and Møller, U. (1982). Circadian variation in influx and transition intensities in the S phase of hamster cheek pouch epithelium cells. Research Report 82/1, Statistical Research Unit, Danish Medical and Social Science Research Councils, Copenhagen.

Kendall, D. G. (1948). On the role of variable generation time in the development of a stochastic birth process. *Biometrika* 35:316-330.

Klein, B., and Guiguet, M. (1978). Relative importance of the phases of the cell cycle for explaining diurnal rhythms in cell proliferation in the tissues with a long $G_1$ duration. In *Biomathematics and Cell Kinetics* A.-J. Valleron and P. D. M. Macdonald (eds.). Elsevier/North-Holland, Amsterdam, pp. 199-210.

Klein, B., and Macdonald, P. D. M. (1980). The multitype continuous-time Markov branching process in a periodic environment. *Adv. Appl. Prob.* 12:81-93.

Klein, B., and Valleron, A.-J. (1977). A compartmental model for the study of diurnal rhythms in cell proliferation. *J. Theor. Biol.* 64:27-42.

Kurtz, T. G. (1981). Relationships between stochastic and deterministic population models. In *Biological Growth and Spread. Mathematical Theories and Applications,* Proceedings, Heidelberg, 1979, W. Jäger, H. Rost, and P. Taŭtu (eds.). *Springer Lect. Notes Biomath.* 38:449-467.

Macdonald, P. D. M. (1981). Measuring circadian rhythms in cell populations. In *The Mathematical Theory of the Dynamics of Biological Populations II,* R. W. Hiorns and D. Cooke (eds.). Academic Press, London, pp. 281-306.

Møller, U., Hougaard, P., and Keiding, N. (1982a). Diurnal variation in the mitotic rate of hamster cheek pouch epithelium cells (in preparation).

Møller, U., and Keiding, N. (1982). Circadian variations in influx and efflux of the S phase in a partially synchronized cell system. Double labeling with [$^3$H]thymidine in the epithelium of the hamster cheek pouch. *Cell Tissue Kinet.* 15:341-350.

Møller, U., Keiding, N., and Engel, F. (1982b). Subclassification of cells in S phase in a partially synchronized cell system. Methodological problems concerning pulse labeling of epithelial cells in the hamster cheek pouch. *Cell Tissue Kinet.* 15:157-168.

Møller, U., Larsen, J. K., and Faber, M. (1974). The influence of injected tritiated thymidine on the mitotic circadian rhythm in the epithelium of the hamster cheek pouch. *Cell Tissue Kinet.* 7:231-239.

Nelson, S., and Green, P. J. (1981). The random transition model of the cell cycle. A critical review. *Cancer Chemother. Pharmacol.* 6:11-18.

Smith, J. A., and Martin, L. (1973). Do cells cycle? *Proc. Natl. Acad. Sci. USA* 70:1263-1267.

Takahashi, M. (1966). Theoretical basis for cell cycle analysis I. *J. Theor. Biol.* 13:202-211.

Takahashi, M. (1968). Theoretical basis for cell cycle analysis II. *J. Theor. Biol.* 18:195-209.

Thorud, E., Clausen, O. P. F., Bjerknes, R., and Aarnæs, E. (1980). The stathmokinetic method in vivo. Time-response with special reference to circadian variations in epidermal cell proliferation in the hairless mouse. *Cell Tissue Kinet.* 13:625-634.

Tvermyr, E. M. F. (1972). Circadian rhythms in hairless mouse epidermal DNA-synthesis as measured by double labelling with H3-thymidine (H3Tdr.). *Virchows Arch. [Cell Pathol.]* 11:43-54.

White, R. A. (1980). A theory for the estimation of DNA synthesis rates by flow cytometry. *J. Theor. Biol.* 85:53-73.

# PART III

# REGULATION OF THE CELL DIVISION CYCLE

## EVENTS, SEQUENCES, PATHWAYS, AND TIMERS

In this and the following parts, we come to the real core of this book: the timing of the cell division cycle and the nature of the biochemical and molecular mechanisms by which it is regulated. This is clearly not a new area itself (Padilla et al., 1969, 1974; Mitchison, 1971; Jeter et al., 1978; Lloyd et al., 1982; Padilla and McCarty, 1982); in the ensuing 13 chapters, however, particular emphasis will be placed on the *time-measuring* aspects of these control devices and the identity of the "clocks"—whether they be sequences, pathways, or oscillators—that generate CDC periodicities.

In his overview (Chap. 9), J. M. Mitchison quite fittingly discusses the dissociation of the events that comprise the CDC and briefly reviews the progress made in the attempts of cell physiologists to dissect the CDC into a series of dependent sequences and to elucidate their control points, as has been done so elegantly in the yeast *Saccharomyces* (Hartwell, 1974; Pringle and Hartwell, 1982). He then analyzes two important sequences, the DNA-division cycle and the growth cycle (Mitchison, 1971, pp. 244-249), and their mutual interaction, particularly in his beloved fission yeast, *Schizosaccharomyces pombe* (also treated in depth by Peter A. Fantes in Chap. 14). Finally, he outlines some of the evidence for independent timers (i.e., "clocks"), the subject of Part IV (and also of Chaps. 25 and 26, dealing with periodicities in embryonic cell division cycles).

The following chapters provide in-depth treatments of many of these themes. Thus, in Chap. 10, Page R. Painter gives comprehensive coverage of the very important area of periodic enzyme synthesis in prokaryotic and eukaryotic cells [the original observations having provided the impetus for Goodwin's (1966) model for timed enzyme synthesis in bacteria and major support for Mitchison's (1971) postulate of a dissociable "growth cycle"]. The possible coupling of energy metabolism to the CDC is revisited (see Chap. 2) in prokaryotes by Robert K. Poole (Chap. 11): Could respiratory metabolism, representing the "ticking" of a CDC clock, be punctuated by identifiable, periodic events, reflecting its "chimes" striking the hour? The next two contributions, by Stephen Cooper (Chap. 12) and Neil H. Mendelson (Chap. 13), provocatively discuss, respectively, the "continuum model" as related to the clock model for the CDC and the "helix clock," which quite specifically describes the way in which spatial order could be used as a source of information about time and consequently used to establish a temporal order of events. Finally, Peter A. Fantes (Chap. 14) and John J. Tyson and Wilhelm Sachsenmaier (Chap. 15) give a detailed analysis of the temporal control of the CDC in *Schizosaccharomyces pombe* and of the nuclear division cycle in *Physarum polycephalum,* respectively; both contributions give strong support for a cell "sizer" (as opposed to a cell "timer") mechanism and provide, therefore, an appropriate juxtaposition to the consideration of CDC timing by autonomous oscillators that follows (Part IV).

## REFERENCES


Goodwin, B. C. (1966). An entrainment model for timed enzyme synthesis in bacteria. *Nature* 209:479-481.

Hartwell, L. H. (1974). *Saccharomyces cerevisiae* cell cycle. *Bacteriol. Rev.* 38:164-198.

Jeter, J. R., Jr., Cameron, I. L., Padilla, G. M., and Zimmerman, A. M., (eds.) (1978). *Cell Cycle Regulation.* Academic Press, New York.

Lloyd, D., Poole, R. K., and Edwards, S. W. (1982). *The Cell Division Cycle: Temporal Organization and Control of Cellular Growth and Reproduction.* Academic Press, London.

Mitchison, J. M. (1971). *The Biology of the Cell Cycle.* Cambridge University Press, London.

Padilla, G. M., Cameron, I. L., and Zimmerman, A., Eds. (1974). *Cell Cycle Controls.* Academic Press, New York.

Padilla, G. M., and McCarty, K. S., Sr. (eds.) (1982). *Genetic Expression in the Cell Cycle.* Academic Press, New York.

Padilla, G. M., Whitson, G. L., and Cameron, I. L. (eds.) (1969). *The Cell Cycle: Gene-Enzyme Interactions.* Academic Press, New York.

Pringle, J. R., and Hartwell, L. H. (1982). The *Saccharomyces cerevisiae* cell cycle. In *The Molecular Biology of the Yeast Saccharomyces: Life Cycle and Inheritance,* J. N. Strathern, E. W. Jones, and J. R. Broach (eds.). Cold Spring Harbor Laboratory, Cold Spring Harbor, New York, pp. 97-142.

# 9
# Dissociation of Cell Cycle Events

**J. M. Mitchison** / University of Edinburgh, Edinburgh, Scotland

## DNA-DIVISION CYCLE

The most conspicuous things that a cell does during the normal cell cycle is to double all its components and then undergo the two morphological splitting events of mitosis and cleavage. Of the synthetic processes, the best known is DNA synthesis, which is a periodic event occupying only a part of the cycle in nearly all eukaryotic cells and in slow-growing bacteria. It is the task of those who work with the cell cycle to try to understand the controls that operate on these events. The ultimate answers will have to be molecular, and good progress is being made, especially with the yeasts, in applying the tools of molecular genetics. But we still need information and analysis at the level of cell physiology in systems where the genetics is less well developed, both because it may take a considerable time to work out the molecular basis and because it is often the cell physiologist who poses the questions for the molecular biologist. If, for example, DNA synthesis had been found to be a continuous process, there would be little point in searching for the buildup of an initiating molecule present in the cytoplasm at the $G_1$ -S boundary.

For many years, the approach of physiologists to cell cycle controls has often been an attempt to dissect the events of the cycle into a series of dependent sequences (Mitchison 1969, 1971, 1974; Hartwell, 1974; Howell, 1974). Given a set of temporal events, there are two extreme cases. The first is that all the events are in a dependent sequence with direct causal connections. In this case, blocking an event will prevent the occurrence of all the subsequent events. The second is that the events are in an independent sequence with no causal connections between them. Blocking an event will therefore have no effect on the other events. In practice, the cell cycle seems to show dependent sequences but it is likely that there is more than one sequence, and events may lie on a series of parallel pathways (Pringle and Hartwell, 1982). Some time ago, I suggested that an important dependent sequence was the "DNA-division Cycle" (Mitchison, 1971, p. 244). This sequence included DNA synthesis, mitosis, and cleavage, together with the synthesis of histones and other proteins, including enzymes, that were associated with the main events. The evidence for the dependency came mainly from inhibitor blocks. Stopping DNA synthesis also stopped mitosis and division. This evidence has now been amplified by the use of cell cycle mutants (Hartwell, 1974; Nurse et al., 1976), and the concept of this dependent sequence is widely accepted.

The question of the control points of the DNA-division sequence is discussed in several chapters of this book and I only want to make two points here. There are obvious biological reasons why a normal cell maintains the sequence of DNA synthesis and division—it will not survive without a nucleus or a set of chromosomes. But there are cases where the events of the DNA-division cycle can be dissociated. Two or more rounds of DNA replication can happen without an intervening nuclear division. This is known to happen in *Tetrahymena* and other ciliates, in *Physarum,* and in mammalian cells (Mitchison, 1971, p. 249), and it presumably happens with spontaneous diploidization in *Schizosaccharomyces pombe.* The reverse situation of division without DNA synthesis is rare but there is the striking case of the enucleate activated sea urchin egg, which will go through several divisions in the absence of a nucleus (Harvey, 1956). Nuclear division without cell division, resulting in multinucleate cells, is much commoner. It occurs naturally in many fungi and it can be produced in yeast by cell cycle mutants, in mammalian cells by cytochalasin (Dickerman and Goldman, 1976), and in higher plant cells by caffeine (González-Fernández et al., 1966).

The existence of these cases of dissociation raises the question of exactly what we mean by dependency. If DNA synthesis has to be completed before mitosis can occur or mitosis has to be completed before another round of DNA replication can be started, then these cases are infringements of the rules of direct causal connection. It is more likely, however, that what is important in the control is the signal that starts mitosis rather than the one which marks the completion of mitosis. When cleavage is blocked by cytochalasin, the signal for cleavage may have been given but the cell is unable to execute division in the presence of the inhibitor. The signals might be given by an independent timer started earlier in the cycle, an "independent single timer sequence" (Mitchison, 1974). But this is all very speculative and it only serves to emphasize the second point I wish to make. This is our considerable ignorance about the nature and mechanism of dependency. It would be a considerable achievement if we could define the rate-limiting step in the cycle and understand its molecular basis. But it would still leave major questions to be answered about other events that are dependent on passing this step. There is good evidence in a number of systems that there is a cytoplasmic factor which can initiate DNA synthesis, but is there also a signal that DNA synthesis is complete and that the cell can go on to mitosis? $G_2$ and $G_1$ still remain major mysteries. Important things may be happening in them, for example, the changes of the chromosome cycle (Hittelman and Rao, 1978), but the dilemma is that cells can dispense with one or both of them, e.g., *Physarum, Amoeba,* and some mammalian cells (Liskay, 1977). It is probable that the solution of this dilemma will only come from a clearer understanding of the dependency relations between DNA synthesis and mitosis.

## GROWTH CYCLE

When I suggested that the DNA-division cycle was an important dependent sequence in the cell cycle, I also suggested that it could be dissociated from the "growth cycle" (Mitchison, 1971, p. 244). The growth cycle comprised most macromolecular synthesis and included most of the enzymes showing "step" and "peak" patterns of activity through the cycle. There were two important elements in this scheme. The first was that growth could be dissociated from the traverse of the DNA-division cycle. The second was that growth included a number of enzyme activity steps that marked the traverse of the cycle and could perhaps be controlled by a mechanism which was separate from DNA-division cycle. Since then, the first element has stood the test of time. One of the best demonstrations is that protein and RNA synthesis continue for two mass doubling times or more in cell cycle mutants of *S. pombe* blocked at various points in the DNA-division cycle (Nurse et al., 1976). The rate of synthesis increases in parallel with unblocked control cultures but then ceases to rise further at a point which is related to cell size (Elliott, Creanor, and Mitchison, unpublished observations). The state of the second element is much more equivocal, because of the present position of step enzymes. Ten years ago, it seemed as though a substantial proportion of the enzymes of the cell showed a step or peak rise in activity during the cycle and, by assumption, sharply periodic synthesis. The best-studied system was budding yeast and a review of that period (Halvorson et al., 1970) listed 30 enzymes, all of which showed periodic patterns. Today, many people would have doubts about these earlier results, partly because of the difficulties of analyzing age-fractionation data, especially in budding yeast, and partly because of the problems of perturbations in synchronous cultures. The activity of some enzymes is very sensitive to the handling procedures needed to set up a synchronous culture, and it is possible to generate spurious cell cycle patterns in asynchronous cultures (Mitchison, 1977). We have recently used a method of generating synchronous cultures from an elutriating rotor which causes minimal perturbation and have examined three enzyme activities in budding yeast (Creanor et al., 1983). These all showed continuous rises through the cycle even though they were classed as step enzymes from earlier work.

An important approach in recent years has been the analysis of the rate of synthesis through the cycle of polypeptides separated on two-dimensional gels. It is clear that periodic synthesis is rare among those proteins that can be separated in mammalian cells (Milcarek and Zahn, 1978; Bravo and Celis, 1980), in budding yeast (Elliott and McLaughlin, 1978), in *Physarum* (Turnock et al., 1981), and in bacteria (Lutkenhaus et al., 1979). In a recent careful study of budding yeast by Lorincz et al. (1983), about 900 proteins were examined. Only 20 of these (including three histones) showed a modulated pattern of synthesis, and none of them were among the abundant proteins. A point to note is that the method of normalizing the data would not have revealed changes in the overall rate of protein synthesis, for example, the "linear" pattern discussed below.

With the present information, we could construct a very simple model of cell cycle events. All the discrete events would fall in the DNA-division sequence and would include proteins associated with the S period, such as histones, and with mitosis, such as tubulin (Laffler et al., 1981). It would also include enzymes of DNA synthesis, many of which show step or peak patterns of activity (Mitchison, 1971, 1977), and enzymes associated with mitosis, such as histone kinase (Bradbury et al., 1974). It is, of course, an assumption that all the 20 modulated proteins found in the yeast gels mentioned above fall into these categories. Growth would then be a smooth uneventful process of exponential increase of all the other cell components. There would be no growth cycle or growth sequence, since there would be no discrete events or markers, but growth would control the DNA-division sequence through the attainment of a critical size for mitosis or DNA synthesis (see Chap. 14). Growth itself would be controlled by nutrients, permeability, and general metabolism, but there would not be specific cell cycle controls and, in particular, periodic controls which could be dissociated from the DNA-division cycle.

## GROWTH IN *Schizosaccharomyces pombe*

Despite the attraction of the simple model above, it does not fit all the facts, and I want here to give a brief survey of cell cycle growth patterns in the fission yeast *S. pombe* where they have been studied in considerable detail over the last 25 years. With the possible exception of labeled glucose incorporated into carbon skeletons (Creanor and Mitchison, unpublished observation), none of the measures of growth in *S. pombe* show exponential increase. The predominant pattern is a "linear" one in which there is a constant rate of growth for a cycle time followed by a more or less abrupt doubling in rate at a rate change point located in a characteristic stage of the cycle. This has been shown for total dry mass, measured by interferometry on single growing cells, with a rate change point toward the end of the cycle (Mitchison, 1957). It also occurs with three enzyme activities (sucrase and acid and alkaline phosphatase) with a rate change point at 0.2 of the cycle (Mitchison and Creanor, 1969). Carbon dioxide production shows the same linear pattern but with the rate change point at 0.75 of the cycle (Creanor, 1978a). Oxygen consumption differs in having a linear pattern but with two rate change points in each cycle, at 0.08 and 0.58 of the cycle (Creanor, 1978b). Total protein synthesis has a modified linear pattern in which the rate change takes place over an appreciable part of the cycle (Creanor and Mitchison, 1982). Fitting model curves to the data from synchronous culture suggests that the average single cell has an increasing rate of protein synthesis for the first 60% of the cycle followed by a constant rate for the remaining 40%. The pattern for ribosomal RNA seems to be approximately the same as that for protein (Elliott, unpublished observation), though earlier results by Fraser and Moreno (1976) gave a sharper rate change. Messenger RNA also has a linear pattern, though the evidence suggests that the rate change point is related to cell size (Fraser and Nurse, 1978, 1979; Barnes et al., 1979). The rate change point occurs at the same cell size in wild type cells and in the small *wee* mutants but it is much later in the cycle (0.81) in *wee* cells than in wild type (0.11).

The first point to emphasize about these results with *S. pombe* is that the patterns show periodicity when examined for the fine detail of rate doublings. The second point is that the times of rate change are not concentrated at one stage but are spread throughout the cycle. This clearly makes interpretation difficult, but we can at least see how far dissociation analysis helps to clarify the nature of the controls, bearing in mind that the experimental evidence is as yet only fragmentary.

We have seen that growth can influence and control the DNA-division cycle through a size control triggered at a critical size for mitosis or the initiation of DNA synthesis. How far can we identify control mechanism working in the reverse direction, so that growth is affected by the events of the DNA-division cycle? One obvious candidate is gene dosage, especially with linear patterns. Doubling the genes during the S period might double the rate of production of the gene products. This simple type of control has been suggested for *S. pombe* (Fraser and Moreno, 1976) and for some other cell systems. There is, however, growing evidence in *S. pombe* against it. A gene-dosage effect would concentrate the rate change points at the S period, which is short and occurs right at the end of the cell cycle. But in fact the rate change points are spread through the cycle and are not concentrated. Another approach is to exploit the fact that the S period in *wee* mutants is about a quarter of a cycle later than in wild type cells (see Chap. 14). A gene-dosage effect would be delayed by this amount in *wee* cells. But this does not happen with the patterns for protein synthesis (Creanor and Mitchison, 1982) or for messenger RNA (Fraser and Nurse, 1978). So gene dosage does not seem to be an important control in *S. pombe,* though an old problem remains as to how a cell regulates its growth when it moves from the haploid situation in $G_1$ to the diploid situation in $G_2$.

The second main event in the DNA-division cycle that might influence growth is mitosis. There is no question that this is an important control in most eukaryotic cells, where transcription ceases on condensed mitotic chromosomes. But this does not seem to happen in yeast, for reasons which are not clear. We have searched specifically for it in *S. pombe* but have been unable to find it; see the discussion in Creanor and Mitchison (1982). On the other hand, there is some evidence that mitosis may be connected in a subtler way with the patterns for protein and for ribosomal RNA (Elliott, Creanor and Mitchison, unpublished observation).

## INDEPENDENT TIMERS

We have considered so far interactions between growth and the DNA-division cycle, but there may also be other periodic controls which can continue to function when growth or division are blocked or perturbed. There is some evidence for such "independent timers," though it is not extensive, and we will consider four examples of it here.

As we have seen, carbon dioxide production in *S. pombe* shows a single rate change during the cycle. If a synchronous culture is treated with deoxyadenosine, DNA synthesis is blocked together with the division that would normally follow it. Yet there is a carbon dioxide rate change in the blocked culture at the same time as the rate change in controls (Creanor, 1978a). A very similar effect is shown with the division inhibitor, mitomycin C, though here there are two rate changes in the absence of two divisions. These independent cycles cannot be followed further since the effects of the inhibitors wear off.

The synthesis of some enzymes in cell cultures can be induced or derepressed by changing the growth medium. The maximum rate of increase in enzyme activity during induction is called the enzyme "potential." There has been a considerable amount of work in the last 18 years on changes of enzyme potential during the cell cycle using samples from synchronous cultures. In the case of *S. pombe,* sucrase potential was first measured by Mitchison and Creanor (1969), and there has been a later and more extensive study of sucrase and arginase potential by Benitez et al. (1980). Potential shows a stepwise pattern of increase through the cycle, which was originally interpreted as a delayed gene-dosage effect. It is clear that this interpretation is incorrect, since potential continues to increase after DNA synthesis has been blocked in cell cycle mutants. In particular, synchronous cultures of mutants blocked either in DNA synthesis or in mitosis show one or two further steps in potential with the normal timing observed in wild-type cells. This independent timer does not appear to be related to cell size. As with the carbon dioxide rate changes, it is impossible to carry on the measurements into later cycles, in this case because the large cells produced by the mutant block reach a critical protein/DNA ratio beyond which there is no further increase of potential. It should be remembered, however, that potential is only an indirect property of the cell cycle. It tells us what cells can do when induced and not what they are doing in the synchronous culture.

Another example of a timer that can be dissociated from the division cycle comes from the work of Cowan and Young (1978) on four mitochondrial enzymes in heat-synchronized *Tetrahymena*. Three of the enzymes show an exponential increase through the heat-shock cycle but the fourth (rotenone-insensitive NADH-cytochrome *c* reductase) has a step pattern of activity. When the heat shocks are stopped and the cells go into the shorter "free-running" cycles, the reductase steps continue for at least two cycles with the heat-shock cycle timing and are therefore out of phase and with a different frequency compared to the free-running divisions.

A final example comes from the fields of eggs and early embryos. They are particularly interesting since they break many of the normal rules of the cell cycle (see Chap. 26). Size control models are inappropriate, since eggs divide without growing and get smaller and smaller. This is not unknown in other systems, for example in *S. pombe* after nitrogen starvation (see Chap. 14), but there are usually only one or two divisions. Eggs also show the curious sharp change from early cycles, which are both short and constant, to later cycles, which are longer and more variable (Signoret and Lefresne, 1971). However, their interest here is that there are marked changes in the physical properties of the cell surface during the cycle which produce an apparent increase in the tension and stiffness of the surface at the time of cleavage, in sea urchin eggs (Mitchison and Swann (1955), and in amphibian eggs (Selman and Waddington, 1955). The reason for this change is not clear, but it has always been assumed to be part of the cleavage process and therefore of the DNA-division cycle from the point of view of control. However, it cannot be as simple as this, since the surface changes continue for at least six cycles, with the same timing as the controls, not only after the DNA-division has been blocked by inhibitors but also when the egg (or a fragment of it) is enucleate (Yoneda et al., 1978; Sawai, 1979; Hara et al., 1980). This is a very striking dissociation and has led Hara et al. (1980) to describe it in terms of a "cytoplasmic clock."

These independent timers must all share the property of being tied into or entrained by the normal DNA-division cycle, since they cause periodic events that occur at a particular time in the normal cycle (except perhaps for the *Tetrahymena* situation in which the normal unperturbed cycle has not been examined). It is possible that the three cases in growing cells represent damped residual oscillations which would rapidly die out. This can only be explored satisfactorily in a situation where an appreciable number of continuing oscillations can be observed. In view of the continuing oscillations in eggs, this explanation seems less probable here. It is more attractive to think in terms of a cytoplasmic clock which is set going by fertilization or activation. As Hara et al. (1980) have suggested, this clock may be the basic timer in egg development and trigger all cycle events, including the DNA-division cycle. A cell cycle timer or oscillator (Kauffman and Wille, 1975) is a much more attractive proposition for early embryos than it is for the normal growing cell, but we still have to search for its molecular mechanism and explain why it slows down so abruptly in later development and is perhaps superseded by other controls.

## SUMMARY

It is convenient to separate most of the cell cycle into two dissociable sets, growth and the DNA-division cycle. The latter cycle has three main events, DNA synthesis, mitosis, and cell division, and is a dependent sequence in the sense that an event will not normally occur unless the preceding one has been completed. There are, however, several cases in which these dependency relations break down. The DNA-division cycle also includes the periodic synthesis of a number of proteins associated with the main events, e.g., histones and enzymes of DNA synthesis. Most enzymes, however, do not appear to be synthesized periodically during growth, but there is evidence in fission yeast that most of the bulk components of growth follow linear or partially linear patterns with rate changes at characteristic but differing points of the cycle. Controls that may operate between growth and the DNA-division cycle include size controls that can trigger DNA synthesis or mitosis, and modulating effects of mitosis on growth. As well as these controls, however, there are some dissociation experiments which indicate the existence of independent timers that can control periodic events when the DNA-division cycle has been blocked or perturbed.


## REFERENCES


Barnes, A., Nurse, P., and Fraser, R. S. S. (1979). Analysis of the significance of a periodic, cell size-controlled doubling in rates of macromolecular synthesis for the control of balanced exponential growth of fission yeast cells. *J. Cell Sci.* 35:41-51.

Benitez, T., Nurse, P., and Mitchison, J. M. (1980). Arginase and sucrase potential in the fission yeast *Schizosaccharomyces pombe. J. Cell Sci.* 46:399-431.

Bradbury, E. M., Inglis, R. J., Matthews, H. R., and Langan, T. A. (1974). Molecular basis of control of mitotic cell division in eukaryotes. *Nature* 249:553-556.

Bravo, R., and Celis, J. E. (1980). A search for differential polypeptide synthesis throughout the cell cycle of HeLa cells. *J. Cell Biol.* 84:795-802.

Cowan, A. W., and Young, P. G. (1978). The formation of several mitochondrial enzymes during the cell cycle in heat-synchronized *Tetrahymena pyriformis* ST. *Exp. Cell Res.* 112:79-87.

Creanor, J. (1978a). Carbon dioxide evolution during the cell cycle of the fission yeast *Schizosaccharomyces pombe. J. Cell Sci.* 33:385-397.

Creanor, J. (1978b). Oxygen uptake during the cell cycle of the fission yeast *Schizosaccharomyces pombe. J. Cell Sci.* 33:399-411.

Creanor, J., Elliott, S. G., Bisset, Y. C., and Mitchison, J. M. (1983). Absence of step changes in activity of certain enzymes during the cell cycle of budding and fission yeasts in synchronous cultures. *J. Cell Sci.* 61:339-349.

Creanor, J., and Mitchison, J. M. (1982). Patterns of protein synthesis during the cell cycle of the fission yeast *Schizosaccharomyces pombe. J. Cell Sci.* 58:263-285.

Dickerman, L. H., and Goldman, R. D. (1976). The production of binucleate mammalian cell populations. In *Methods in Cell Biology,* vol. 14, D. M. Prescott (ed.). Academic Press, New York. pp. 81-86.

Elliott, S. G., and McLaughlin, C. S. (1978). Rate of macromolecular synthesis through the cell cycle of the yeast *Saccharomyces cerevisiae. Proc. Natl. Acad. Sci. USA* 75:4384-4388.

Fraser, R. S. S., and Moreno, F. (1976). Rates of synthesis of polyadenylated messenger RNA and ribosomal RNA during the cell cycle of *Schizosaccharomyces pombe. J. Cell Sci.* 21:497-521.

Fraser, R. S. S., and Nurse, P. (1978). Novel cell cycle control of RNA synthesis in yeast. *Nature* 271:726-730.

Fraser, R. S. S., and Nurse, P. (1979). Altered patterns of ribonucleic acid synthesis during the cell cycle: A mechanism compensating for variation in gene concentration. *J. Cell Sci.* 35:35-40.

González-Fernández, A., López-Sáez, J. F., and Giménez-Martin, G. (1966). Duration of the division cycle in binucleate and mononucleate cells *Exp. Cell Res.* 43:255-267.

Halvorson, H. O., Carter, B. L. A., and Tauro, P. (1970). Synthesis of enzymes during the cell cycle. *Adv. Microb. Physiol.* 6:47-106.

Hara, K., Tydeman, P., and Kirschner, M. (1980). A cytoplasmic clock with the same period as the division cycle in *Xenopus* eggs. *Proc. Natl. Acad. Sci. USA* 77:462-466.

Hartwell, L. H. (1974). *Saccharomyces cerevisiae* cell cycle. *Bacteriol. Rev.* 38:164-198.

Harvey, E. B. (1956). *The American Arbacia and Other Sea Urchins* Princeton University Press, Princeton, New Jersey, p. 141.

Hittelman, W. N., and Rao, P. N. (1978). Mapping G1 by the structural morphology of the prematurely condensed chromosomes. *J. Cell. Physiol.* 95:333-342.

Howell, S. H. (1974). An analysis of cell cycle controls in temperature sensitive mutants of *Chlamydomonas reinhardi.* In *Cell Cycle Controls,* G. M. Padilla, I. L. Cameron, and A. Zimmerman (eds.). Academic Press, New York, pp. 235-249.

Kauffman, S. A., and Wille, J. J. (1975). The mitotic oscillator in *Physarum polycephalum. J. Theor. Biol.* 55:47-93.

Laffler, T. G., Chang, M. T., and Dove, W. F. (1981). Periodic synthesis of microtubular proteins in the cell cycle of *Physarum. Proc. Natl. Acad. Sci. USA* 78:5000-5004.

Liskay, R. M. (1977). Absence of a measurable G2 phase in two Chinese hamster cell lines. *Proc. Natl. Acad. Sci. USA* 74:1622-1625.

Lorincz, A. T., Miller, M. J., Xuong, N.-H., and Geiduschek, E. P. (1983). The identification of proteins whose synthesis is modulated during the cell cycle of *Saccharomyces cerevisiae. Mol. Cell. Biol.* 2:1532-1549.

Lutkenhaus, J. F., Moore, B. A., Masters, M., and Donachie, W. D. (1979). Individual proteins are synthesised continuously throughout the *Excherichia coli* cell cycle. *J. Bacteriol.* 138:352-360.

Milcarek, C., and Zahn, K. (1978). The synthesis of ninety proteins, including actin, throughout the HeLa cell cycle. *J. Cell Biol.* 79:833-838.

Mitchison, J. M. (1957). The growth of single cells. I. *Schizosaccharomyces pombe. Exp. Cell Res.* 13:244-262.

Mitchison, J. M. (1969). Markers in the cell cycle. In *The Cell Cycle, Gene-Enzyme Interactions,* G. M. Padilla, G. L. Whison, and I. L. Cameron (eds.). Academic Press, New York, pp. 361-372.

Mitchison, J. M. (1971). *The Biology of the Cell Cycle.* Cambridge University Press, London.

Mitchison, J. M. (1974). Sequences, pathways and timers in the cell cycle. In *Cell Cycle Controls,* G. M. Padilla, I. L. Cameron, and A. Zimmerman (eds.). Academic Press, New York, pp. 125-142.

Mitchison, J. M. (1977). Enzyme synthesis during the cell cycle. In *Cell Differentiation in Microorganisms, Plants and Animals,* L. Nover and K. Mothes (eds.). VEB Gustav Fischer Verlag, Jena, Germany, pp. 377-401.

Mitchison, J. M., and Creanor, J. (1969). Linear synthesis of sucrase and phosphatases during the cell cycle of *Schizosaccharomyces pombe. J. Cell Sci.* 5:373-391.

Mitchison, J. M., and Swann, M. M. (1955). The mechanical properties of the cell surface. III. The sea urchin egg from fertilisation to cleavage. *J. Exp. Biol.* 32:734-750.

Nurse, P., Thuriaux, P., and Nasmyth, K. (1976). Genetic control of the cell division cycle in the fission yeast *Schizosaccharomyces pombe. Mol. Gen. Genet.* 146:167-178.

Pringle, J. R., and Hartwell, L. H. (1982). The *Saccharomyces cerevisiae* cell cycle. In *The Molecular Biology of the Yeast Saccharomyces. Life Cycle and Inheritance,* J. N. Strathern, E. W. Jones, and J. R. Broach (eds.). Cold Spring Harbor Laboratory, Cold Spring Harbor, New York, pp. 97-142.

Sawai, T. (1979). Cyclic changes in the cortical layer of nonnucleated fragments of the newt's egg. *J. Embryol. Exp. Morphol.* 51:183-193.

Selman, G. G., and Waddington, C. H. (1955). The mechanism of cell division in the cleavage of the newt's egg. *J. Exp. Biol.* 32:700-733.

Signoret, J., and Lefresne, J. (1971). Contribution a l'étude de la segmentation de l'oeuf d'Axolotl. I. Définition de la transition blastuléenne. *Ann. Embryol. Morphogen.* 4:113-123.

Turnock, G., Chambers, J., and Birch, B. (1981). *Eur. J. Biochem.* 120:529-534.

Yoneda, M., Ikeda, M., and Washitani, S. (1978). Periodic changes in the tension at the surface of activated non-nucleate fragments of sea-urchin eggs. *Dev. Growth Differ.* 20:329-336.

# 10

# Periodic Enzyme Synthesis and the Cell Division Cycle

**Page R. Painter** / University of California, Davis, California

**John J. Tyson** / Virginia Polytechnic Institute and State University, Blacksburg, Virginia

## INTRODUCTION

The operon model of Jacob and Monod (1961) has stimulated investigations in many areas of biology for more than two decades. One area where it has been particularly important is the study of oscillatory gene expression in microorganisms. This model for regulation of gene expression in bacteria both stimulated research that documented oscillatory phenomena in cells and provided a theoretical framework for the interpretation of experimental results. The model also stimulated a major area of research in mathematical biology which sought to define conditions for biological oscillations in negative-feedback systems and to describe quantitatively temporal patterns of gene expression. Though there have been difficulties in the past reconciling experimental evidence with theoretical predictions, refinements in the mathematical description of operon dynamics have recently led to quantitative predictions that are remarkably similar to experimental data. Therefore, it is now appropriate to reappraise theories of oscillatory repression as explanations for some periodic phenomena in cells.

## EXPERIMENTAL FOUNDATIONS

In both prokaryotic and eukaryotic cells, the synthesis of most enzymes continues, at approximately the same specific rate, throughout the cell cycle. The amount of an enzyme in a growing cell culture usually accumulates, like total protein, continuously and exponentially. However, a significant number of proteins exhibit periodic changes in activity. During synchronous growth of a cell culture, enzyme activity per milliliter of culture may show stepwise increases, peaks, or linear increase with periodic rate doubling

(Mitchison, 1969). Among bacteria, periodic enzyme synthesis has been reported in *Escherichia coli* (Masters et al., 1964; Kuempel et al., 1965; Ohki, 1978), *Bacillus subtilis* (Donachie, 1965; Masters and Donachie, 1966), and *Pseudomonas aeruginosa* (Kleber and Aurich, 1967). Periodic enzyme synthesis has been reported in a number of eukaryotic cells, including a budding yeast, *Saccharomyces cerevisiae* (Gorman et al., 1964), a fission yeast, *Schizosaccharomyces pombe* (Bostock et al., 1966), a green alga, *Chlorella sorokiniana* (Molloy and Schmidt, 1970; Vassef et al., 1973), an acellular slime mold, *Physarum polycephalum* (Braun and Behrens, 1969; Huettermann et al., 1970), and cultured Chinese hamster cells (Klevecz, 1969). Periodic enzyme synthesis has been reviewed by Schmidt (1974), Mitchison (1977), Halvorson (1977), and Klevecz and Forrest (1977).

Histones are examples of nonenzymatic proteins that are synthesized periodically during the cell cycle, specifically in S phase (Spalding et al., 1966). Histone mRNA sequences have been assayed with complementary DNA probes, and the mRNAs are found to be present preferentially in S phase, which suggests that histone synthesis is under transcriptional control (Rickles et al., 1982; Hereford et al., 1982).

Most reports of periodic gene expression have come from studies of synchronously dividing cell cultures. There are, however, some exceptions. In an early paper Torriani (1960) found that alkaline phosphatase (APase) is synthesized in periodic bursts after an asynchronous culture of *E. coli* uses up its supply of inorganic phosphate and must shift to glycerol-3-phosphate as a phosphate source. The bursts of APase synthesis seem to be correlated with fluctuations in $[P_i]$ that alternately repress and derepress APase synthesis. Masters and Donachie (1966) showed that periodic synthesis of ornithine transcarbamylase continued after the addition of FUDR, an inhibitor of DNA synthesis, which also inhibited cell division. Knorre (1968) observed damped oscillations in the specific rate of synthesis of $\beta$-galactosidase after an asynchronous culture of *E. coli* was shifted from glucose to lactose as the sole carbon source. The period of these oscillations (45 min) was very close to the cell generation time (49 min). Kleber and Aurich (1968) reported damped oscillations in the specific activity of carnitine dehydrogenase after an asynchronous culture of *P. aeruginosa* was shifted from glucose/$NH_4^+$ to carnitine as the sole carbon and nitrogen source. The period of oscillation (2 hr) was somewhat shorter than the mass doubling time (estimated to be 5 hr). In asynchronous cultures of *S. cerevisiae*, Bernhardt et al. (1965) found a large overshoot, followed by highly damped oscillations in the rate of synthesis of NAD-dependent glutamate dehydrogenase after the culture had been shifted from $NH_4^+$ to glutamic acid as the nitrogen source. In cultures of *S. pombe* that had been centrifuged in sucrose gradients to separate cells according to size, Mitchison (1977) showed that mixing the gradient before diluting cells into growth medium results in asynchronous growth; nonetheless, alcohol dehydrogenase and homoserine dehydrogenase show stepwise increases in enzyme activity with a period close to the cell doubling time. Recently, it was shown by Bliss et al. (1982) that periodic gene expression of the tryptophan operon of *E. coli* occurs in a mutant strain lacking normal feedback inhibition of the biochemical pathway for the synthesis of tryptophan. This occurs in asynchronous cultures, and the period is shorter than the cell doubling time by about 33%. As a general rule, it is fair to say that in synchronous cultures periodic enzymes show exactly the same periodicity as DNA synthesis and cell division, whereas in asynchronous cultures, if enzyme synthesis is still periodic, it shows a periodicity close to the interdivision time (±50%).

## THEORIES OF PERIODIC GENE EXPRESSION

Most models for periodic gene expression belong to one of two classes. Models in the first class attribute periodic gene expression to underlying cell cycle events, such as DNA replication. In these models, periodic gene expression is a marker for a stage of the cell cycle. In the second class of models, periodic gene expression is attributed to autogenous oscillations arising spontaneously in an unstable regulatory feedback loop. In this class of models periodic gene expression may give little useful information concerning the cell cycle. In addition, some models postulate a coupling between autogenous oscillations and cell age-specific events.

### Cell Age-Specific Models

The simplest cell age-specific model attributes stepwise increases in enzyme activity per milliliter of culture to periodic doubling of gene dosage as a result of synchronous DNA replication (Donachie, 1965). To see how this works, consider an unstable enzyme (half-life = ln 2/k) which is synthesized at a rate r proportional to the number of genes (per milliliter of culture). Then the enzyme activity per milliliter of culture E obeys the differential equation

$$\frac{dE}{dt} = r - kE$$

If the half-life of the enzyme is much shorter than the cell cycle time, then the enzyme rapidly approaches steady-state conditions, where the rate of synthesis is exactly balanced by the rate of degradation, i.e., $r = kE$ or $E = r/k$. Since k is constant and r doubles periodically with time as DNA in the synchronous culture is replicated, the enzyme activity per milliliter of culture would exhibit stepwise increases. This simple model can be tested by inhibiting DNA synthesis in the culture (the enzyme steps should be blocked), and by inhibiting protein synthesis in the culture (the enzyme activity should decay rapidly).

Several more complicated age-specific models attribute periodic enzyme synthesis to temporal restriction in gene expression. For example, to account for the stepwise synthesis of β-galactosidase in lactose-free synchronous cultures of *E. coli*, Hanawalt and Wax (1964) hypothesized that the β-galactosidase gene is available for transcription (i.e., derepressed) only for the brief period during which the *lac* operon genes are being replicated (see also Edelman and Edlin, 1974). To account for numerous step enzymes in budding yeast, Halvorson and his coworkers (Gorman et al., 1964; Halvorson et al., 1970) pictured RNA polymerase moving steadily along a chromosome and transcribing genes in sequence. In this way they attributed the temporal pattern of enzyme synthesis in *Saccharomyces* to the arrangement of genes on chromosomes. Although there are many surprising correlations between gene arrangements and the timing of enzyme steps, reviewed in Mitchison (1971, p. 171), it does not appear that gene transcription is limited to specific phases of the cell cycle by the linear reading of chromosomes by RNA polymerase (Donachie and Masters, 1969; Schmidt, 1974; Klevecz and Forrest, 1977). Tyson and Sachsenmaier (1979) have suggested a different mechanism for the temporal restriction of transcription during the cell cycle. They imagine a genetic regulatory system akin to the *lac* operon in bacteria and show that, under proper conditions, transcription of the structural genes can be derepressed during the period of time between replication of the operator region and replication of the gene coding for repressor, and then repressed during the period of time between replication of the repressor gene and the next replication of the operator. In this fashion gene transcription is periodically turned on and off once per cycle in concert with gene replication. However, at no time are the structural genes unavailable for transcription if challenged with sufficient inducer, nor does this model share the embarrassing feature of the linear reading model in which *all* genes are transcribed periodically. As in the other models considered in this section, enzyme periodicity in this model is driven by periodic gene replication. Consequently, if DNA synthesis were blocked, enzyme periodicity should disappear.

If transcription is restricted, by some mechanism, to a limited portion of the cell cycle, the pattern of enzyme synthesis will depend on the relative stability of mRNA and enzyme. If mRNA and enzyme are both unstable, the enzyme will show a peak of activity once per cycle. If mRNA is unstable and enzyme is stable, there will be a burst of enzyme synthesis causing a step pattern. If mRNA is stable and enzyme is unstable, then mRNA will increase stepwise and so will enzyme activity (see first paragraph of this section, and let r be proportional to mRNA level). If mRNA and enzyme are both stable, then the enzyme will show a linear increase with periodic rate doubling. Therefore, age-specific models can account, at least qualitatively, for the observed patterns of periodic enzyme synthesis.

However, a major question remains: whence comes the factor of two? Why should the magnitude of the steps or rate changes be two and not some other factor? In cell age-specific models the characteristic doubling of periodic enzyme events during synchronous cell division can ultimately be traced back to doubling in gene dosage during each cycle. Consider, for example, the case of unstable mRNA and stable enzyme. During the limited period of transcription, mRNA concentration (per milliliter of culture) builds up rapidly to a level proportional to the number of structural genes per milliliter of culture. Thus, enzyme is synthesized during this restricted period at a rate proportional to gene dosage. Since the duration of the phase of enzyme synthesis is constant from one cycle to the next but gene dosage doubles, the total amount of protein synthesized during each burst doubles from one cycle to the next. This argument is illustrated in Fig. 1. Similar considerations apply in the other cases.

### Autogenous Oscillators

Based on early observations of periodic enzyme synthesis, Kuempel et al. (1965) and Masters and Donachie (1966) argued that stepwise enzyme synthesis might be due to autogenous oscillations in the negative feedback system regulating enzyme synthesis. This hypothesis was supported by Goodwin (1966) using a mathematical model of gene expression, but this explanation has not been widely accepted, primarily for two reasons. First, it should be possible to show mathematically that negative-feedback genetic regulation can give rise to oscillations when biologically realistic estimates of parameters are used, but most attempts to do this have not been successful. Second, it should be possible to show mathematically that the period of such oscillations is near the observed period, which in most cases is the cell doubling time, but again most attempts to do this have failed. However, these failures may be due to inaccurate modeling of kinetic processes involved in gene expression and regulation, and more precise models may in fact give rise to oscillations with the desired period, as we shall discuss in later sections.

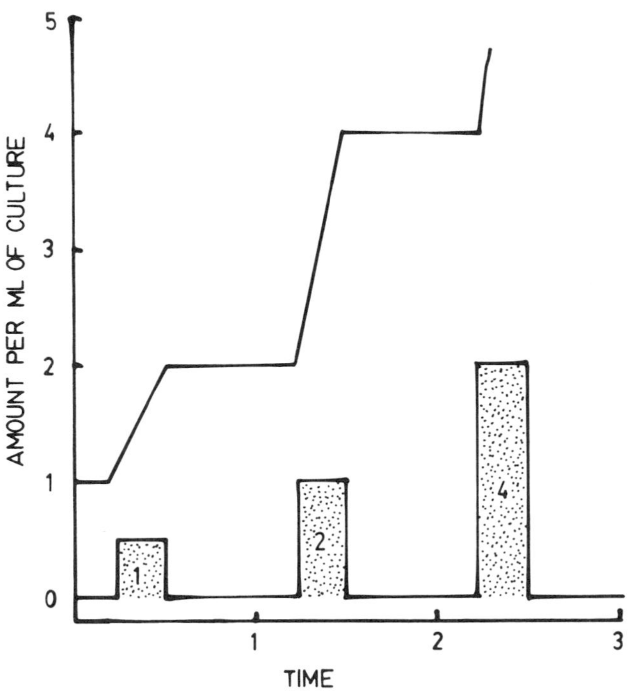


**FIGURE 1** Hypothetical cell age-specific model of a stable step enzyme coded by an unstable message. The amount of enzyme protein and mRNA (in arbitrary units) is plotted as a function of time (in units of the cell cycle time). Transcription is limited to 25% of the cell cycle, during which time mRNA builds up rapidly to a level proportional to gene dosage (number of genes per milliliter of culture). After transcription stops, mRNA concentration drops rapidly to zero. Thus, protein synthesis is restricted to that 25% of the cycle when mRNA is present in appreciable quantity. The amount of protein synthesized is proportional to the area under the mRNA curve (indicated by the stippling). Obviously, the amount of newly synthesized protein doubles from one cycle to the next, leading to a step pattern (assuming the enzyme is stable).

*Goodwin's Equations*

In his early work on negative-feedback biochemical systems, Goodwin extended traditional methods of biochemical kinetics to describe quantitatively a complex biological system, the repressible operon. The Goodwin equations in a generalized and widely studied form are

$$\dot{x}_1 = (1 + x_n{}^p)^{-1} - k_1 x_1$$
$$\dot{x}_i = x_{i-1}\, k_i x_i \qquad 1 < i \leq n \tag{1}$$

where the $k_i$ values are positive constants and the dot refers to a time derivative. Goodwin studied the three-dimensional case where $x_1$, $x_2$, and $x_3$ were interpreted, respectively, as the concentrations of mRNA, of enzyme, and of end product of the operon. The term $(1 + x_n{}^p)^{-1}$ describes end-product repression, and the positive constant p, the Hill number, describes the cooperativity of repression.

For the case $n = 3$ and $p = 1$, Goodwin (1965) published the results of analog computer simulations where $x_1$, $x_2$, and $x_3$ exhibit undamped oscillations about an unstable equilibrium. However, Griffith (1968) proved analytically that, with $n = 3$ and $p = 1$, the equilibrium is locally asymptotically stable. Later, Allwright (1977) and Banks and Mahaffy (1978) showed that the Goodwin Eqs. (1) with $p = 1$ are globally stable for all values of n. It therefore appears that the oscillations originally reported by Goodwin resulted from an artifact in the analog computer that introduced an error into the solution of Eqs. (1).

*Time Delays and Nonlinear Demand*

From his results, Griffith concluded that oscillations of operon activity could occur only if there is cooperative repression of such a high order as to be unlikely. (For example, with $n = 3$ the cooperativity parameter p must be greater than 8 for sustained oscillations to occur.) However, there are features of negative-feedback gene-control systems other than highly cooperative repression that can, in some cases, prevent them from approaching a stable equilibrium.

Introduction into Eqs. (1) of time delays in transcription and translation has been shown by computer simulations to lead to stable oscillations in particular cases where the cooperativity is greater than 1 (Landahl, 1969; Mackey and Glass, 1977). It was shown analytically by MacDonald (1977), Allwright (1977), and Banks and Mahaffy (1978) that time delays cannot destabilize the system with $p = 1$, but the results of MacDonald (1977) and Tyson and Othmer (1978) show that some systems with $p > 1$ can be destabilized by time delays (either discrete or distributed). In the case of discrete time delays, Mahaffy (1980) and Bliss et al. (1982) derived a criterion for local stability of the steady state.

Goodwin's equations have also been modified by replacing the term describing linear demand for end product, $k_n x_n$, by a nonlinear function such as the hyperbolic function,

$$\frac{k_n x_n}{K_n + x_n}$$

Where $K_n$ is a (suitably scaled) Michaelis constant for the enzyme involved in end-product utilization. Such terms have a destabilizing effect on the steady-state solution of Goodwin's equations, as shown numerically by Walter (1974) and analytically by Bliss et al. (1982). Nonlinear demand can destabilize the steady state *even for the case* $p = 1$, as shown by the simple example

$$\begin{aligned} \dot{x}_1 &= (1 + x_3)^{-1} - \frac{x_1}{10} \\ \dot{x}_2 &= x_1 - \frac{x_2}{10} \\ \dot{x}_3 &= x_2 - 10.701 x_3 (1 + x_3)^{-1} \end{aligned} \qquad (2)$$

The steady state of system (2) is located at $x_1 = 0.96668$, $x_2 = 9.6668$, $x_3 = 9.3447$. Linear stability analysis of the steady state leads to the characteristic equation

$$(z + 0.1)^3 + 0.0093447 = 0 \qquad (3)$$

which has roots $z = -0.31063$ and $z = 0.0053153 \pm 0.18241i$. Obviously, the steady state of system (2) is unstable and it can be shown, using Hopf bifurcation theory, that system (2) admits small-amplitude limit-cycle oscillations.

When the equilibrium of Eqs. (1) is unstable, Tyson (1975) showed (for n = 3) that there is a periodic solution. This result was extended to n-dimensional systems by Hastings et al. (1977). In locally unstable systems with time delays, criteria for the existence of periodic solutions for the one-dimensional case were found by Hadeler (1979) and in the two-dimensional case by an der Heiden (1979). Criteria for the existence of periodic solutions in the n-dimensional system with time delays were discovered by Mahaffy (1980). Numerical solution of cases with an unstable equilibrium by Walter (1974), Rapp (1975), and others were consistent with limit-cycle behavior of the Goodwin equations.

*Estimates of the Period of Autogenous Oscillations*

Both numerical (Walter, 1974) and analytical (Rapp, 1975; Tyson, 1979) studies of Goodwin's equations suggest that the period of oscillation is in general much longer than the half-life of the most stable component of the system. Since many well-studied step enzymes are known to be metabolically stable, their concentration decreases only as a result of cell growth, so their half-life is equal to (or slightly less than) the cell doubling time. Therefore, the period of an autogenous oscillator, like system (1), with a metabolically stable enzyme intermediate must be much longer than the cell doubling time, in contradiction to experimental results.

We must modify Goodwin's Eqs. (1) if we are to obtain autogenous oscillations with a period close to the cell cycle time. High cooperativity of feedback repression, time delays in transcription and translation, and nonlinear demand for end product have all been suggested (alone or in combination) as modifications to the classic Goodwin equations to bring the period in line with the enzyme half-life (i.e., the cell cycle time). To see how these factors may affect the period of oscillation, we will consider two simple cases.

*High Cooperativity and Time Delay*: Consider Goodwin's Eqs. (1) with n = 3 and p large, and introduce a transcriptional time delay $t_1$ and a translational time delay $t_2$. Then,

$$\begin{aligned} \dot{x}_1 &= H[x_3(t - t_1)] - k_1 x_1 \\ \dot{x}_2 &= x_1(t - t_2) - k_2 x_2 \\ \dot{x}_3 &= x_2 - k_3 x_3 \end{aligned} \tag{4}$$

where $H(x) = 1$ if $x \leq 1$, $H(x) = 0$ if $x > 1$. We assume that $k_2 = \ln 2/\tau$, $\tau$ = cell cycle time, i.e., enzyme is metabolically stable. Furthermore, we assume that $k_3 >> \tau^{-1}$, i.e., rapid turnover of end product on a time scale of one cell cycle. Then $\dot{x}_3 = 0$, and $x_3(t) = \dot{x}_2(t)/k_3$ on this time scale. Similarly, if mRNA also turns over rapidly on the time scale of a cell cycle, then $k_1 >> \tau^{-1}$ and

$$\begin{aligned} x_1(t) &= \frac{H[x_3(t - t_1)]}{k_1} \\ &= \frac{H[x_2(t - t_1)/k_3]}{k_1} \end{aligned} \tag{5}$$

This leaves us with only a single equation for enzyme $x_2$ as a function of time:

$$\dot{x}_2 = \frac{H[x_2(t - t_d)/k_3]}{k_1 - k_2 x_2} \tag{6}$$

where $t_d = t_1 + t_2$. It is convenient now to rescale the variables: let $x = x_2/k_3$ and $\hat{t} = k_2 t = (\ln 2)(t/\tau)$. Then Eq. (6) becomes

$$\frac{dx}{d\hat{t}} + x = RH[x(\hat{t} - \hat{t}_d)] \tag{7}$$

where $R = (k_1 k_2 k_3)^{-1}$. The parameter R has a simple interpretation: it is the ratio of the rate of transcription of mRNA when the operon is fully derepressed to the rate of transcription of mRNA under steady-state conditions. In bacteria, $R \simeq 10 - 100$. Equation (7) has been analyzed by Painter and Bliss (1981), who showed that it admits oscillations of period

$$T \simeq \ln\left(1 + \frac{R^2 \hat{t}_d}{(R - 1)}\right) \tag{8}$$

On the time scale $\hat{t}$, one cell cycle = ln 2 time units, so Eq. (7) admits oscillations with a period approximately equal to the cell cycle time if $R >> 1$ and $R\hat{t}_d \simeq 1$. In bacteria, $\hat{t}_d = (\ln 2)(t_1 + t_2)/\tau \simeq (0.7)(1 \text{ min})/(70 \text{ min}) = 0.01$. Thus, $R\hat{t}_d$ is indeed close to unity, and the period of oscillation can be expected to be close to the cell cycle time.

*High Cooperativity and Nonlinear Demand*: In this case we consider the Goodwin equations with $n = 3$, p large, and a constant demand for end product:

$$\begin{aligned} \dot{x}_1 &= H(x_3) - k_1 x_1 \\ \dot{x}_2 &= x_1 - k_2 x_2 \\ \dot{x}_3 &= x_2 - k_3 x_3 - \gamma \end{aligned} \tag{9}$$

The constant term $-\gamma$ represents utilization of the end product (e.g., an amino acid) for some major biosynthetic process (e.g., protein synthesis). Assuming that the enzyme and end product are stable molecules, we take $k_2 = k_3 = \ln 2/\tau$. Bacterial mRNA, on the other hand, is quite unstable, so we assume $k_1 >> \ln 2/\tau$. Reasonable values for these parameters are $k_1 = 1, \gamma = 1, 1 - 20, k_2 = k_3 \simeq 0.01$.

Equation (9) is a system of piecewise linear ordinary differential equations that can be solved analytically (Tyson, in preparation). For $k_1 = 1, \gamma = 1 - 20$, and $k_2 = k_3 = \ln 2/\tau << 1$, there exist sustained oscillations of period

$$T \simeq \tau \left(\frac{10}{\gamma}\right)^{0.41} \tag{10}$$

Thus, for $5 \leq \gamma \leq 20$, T varies from $1.3\tau$ to $0.75\tau$, which is quite consistent with experimental observations.

## THE TRYPTOPHAN OPERON

The tryptophan operon in *E. coli* is the best understood negative-feedback system which is known to be regulated by repressor-mediated feedback control. The operon contains

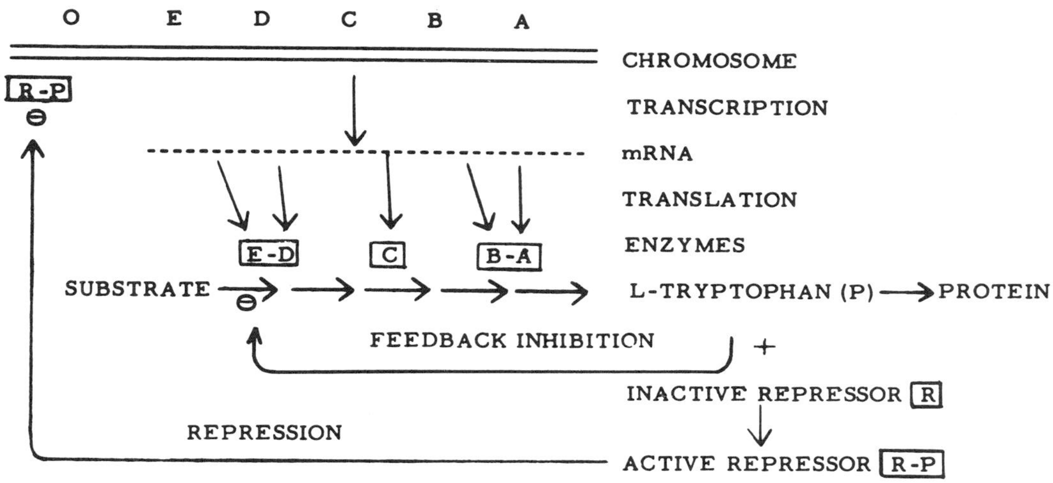


**FIGURE 2** The tryptophan operon. When the operator region (0) of the chromosome is free of active repressor (R-P), the structural genes (E-A) are transcribed to messenger RNA, which in turn is translated into proteins that catalyze the synthesis of tryptophan (P) from chorismic acid (substrate). The activity of the first enzyme of the pathway, anthranilate synthetase (E-D, or ASase), is inhibited allosterically by the end product, tryptophan. Furthermore, tryptophan binds to inactive repressor (R) to produce active repressor (R-P). The binding of R-P to O represses transcription of the structural genes. Since bacterial mRNA is unstable, the repression of transcription is soon manifested as a block of enzyme synthesis. The enzymes, being relatively stable, are eventually diluted by growth of the cells in culture. In mutant strains, lacking feedback inhibition of ASase, once the supply of tryptophan is insufficient to meet its demand for protein synthesis, the tryptophan pool is rapidly emptied, the repressor reverts to its inactive form, and transcription of the structural genes can begin again. In wild-type cells, with proper feedback inhibition of ASase, this instability in the regulation of transcription is masked because the rate of production of tryptophan can be matched to its demand by modulating the activity of the first enzyme of the biosynthetic pathway. Thus, in wild-type cells, large fluctuations of the tryptophan pool are avoided, so periodic repression and derepression of ASase synthesis do not occur.

five genes coding for polypeptides and the operator region where repressor binds to the chromosome. The polypeptides aggregate to form enzymes of the pathway for biosynthesis of tryptophan. When the repressor is not bound to the operator, the enzyme RNA polymerase initiates transcription near the operator and then transcribes continuously from genes E through A. While transcription is in progress, ribosomes initiate translation, following close behind the polymerase. The first enzyme produced by this coupled transcription-translation process, anthranilate synthetase, is the product of the E and D genes, and it catalyzes the rate-limiting step in the biosynthetic pathway.

Adequate levels of the amino acid tryptophan are essential for protein synthesis. Tryptophan also has two regulatory roles in the operon. First, it regulates gene expression by reversibly binding to the tryptophan repressor to convert it from an inactive form (aporepressor) to the active form (holorepressor) which then binds to the operator. This negative-feedback control is called repression. Second, tryptophan reversibly binds to anthranilate synthetase to convert this enzyme from an active form to an inactive form.

**TABLE 1** Values of Parameters Used in Numerical Solutions of Eqs. (11)

| Parameter | Units | Magnitude |
|---|---|---|
| $g$ | Operons per cell | 1.3 |
| $k_m$ | mRNA molecules/operon/sec | 0.07 |
| $k_p$ | Enzyme molecules/mRNA molecule/sec | 0.28 |
| $k_t$ | $\frac{\text{Moles of end product/enzyme molecule/cell}}{\text{Liters/cell}}$ | $10^{-8}$ |
| $k_1$ | $sec^{-1}$ | 0.016 |
| $k_2$ | $sec^{-1}$ | 0.0002 |
| $K_R$ | Molar | $3^{-1/2} \times 10^{-4}$ |
| $K_I$ | Molar | $10^{-4}$ (normal operon) |
| $K_I$ | Molar | $10^{-3}$ (feedback resistant) |
| $K_G$ | Molar | $10^{-6}$ |
| $k$ | $sec^{-1}$ | 0.0002 |
| $t_m$ | sec | 4 |
| $t_p$ | sec | 66 |

This negative feedback control is called feedback inhibition. The pattern of biosynthesis in the tryptophan operon together with the two feedback control loops is diagrammed in Figure 2.

In their quantitative description of the tryptophan operon, Bliss et al. (1982) attempted to (1) to identify major time delays in operon dynamics, (2) to describe the major nonlinearity in demand for tryptophan, and (3) to describe feedback inhibition as well as repression. Their equations are:

$$\begin{aligned} \dot{x}_1 &= gk_m R(x_3(t - t_m)) - k_1 x_1(t) \\ \dot{x}_2 &= k_p x_1(t - t_p) - k_2 x_2(t) \\ \dot{x}_3 &= k_t x_2(t) I(x_3(t)) - G(x_3(t)) - kx_3(t) \end{aligned} \tag{11}$$

where $x_1$, $x_2$ and $x_3$ are the concentrations of mRNA, enzyme, and end product of the tryptophan operon, respectively, $k_m$, $k_p$, and $k_t$ are rate constants, g is the concentration of the operon (gene dose), k is the specific growth rate, and $k_1$ and $k_2$ are constants describing first-order depletion as a consequence of both enzymatic degradation and exponential dilution by growth. The functions R and I describe, respectively, repression and feedback inhibition and are decreasing functions of $x_3$. The function G is an increasing function of $x_3$ that describes the demand for end product in protein synthesis. The time delay between initiation of transcription and initiation of translation is denoted by $t_m$, and the delay between initiation of translation and maturation of enzyme is denoted by $t_p$. Estimated values of these constants are listed in Table 1.

In the above study, R was assumed to be a Hill function $K_R{}^4/(K_R{}^4 + x_3{}^4)$, which is equal to 0.1 when $x_3 = 10^{-4}$ M, and I was assumed to be a Hill function $K_I{}^2/(K_I{}^2 + x_3{}^2)$, which is equal to 0.5 when $x_3 = 10^{-4}$ M in the normal operon. Values of $K_R$ and $K_I$ are listed in Table 1. The function G was defined by a detailed kinetic model for

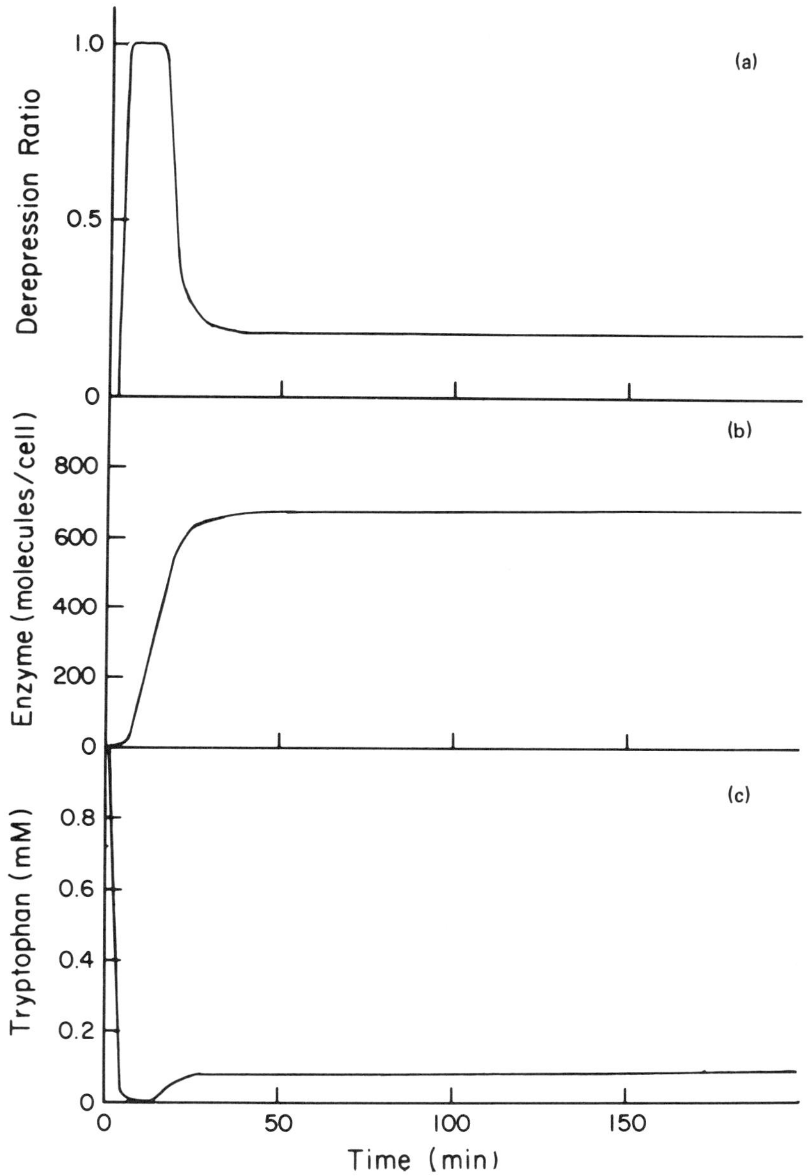


**FIGURE 3** Simulation of gene expression in the hypothetical example of a repressible operon with normal feedback inhibition. Equations (11) were solved numerically with the parameter values given in Table 1 ($K_I = 10^{-4}$ M). Initial conditions were chosen to simulate derepression at t = 0. (a) Derepression ratio expressed as the fraction of maximal gene activity. (b) Intracellular concentration of enzyme. (c) Intracellular concentration of end product.

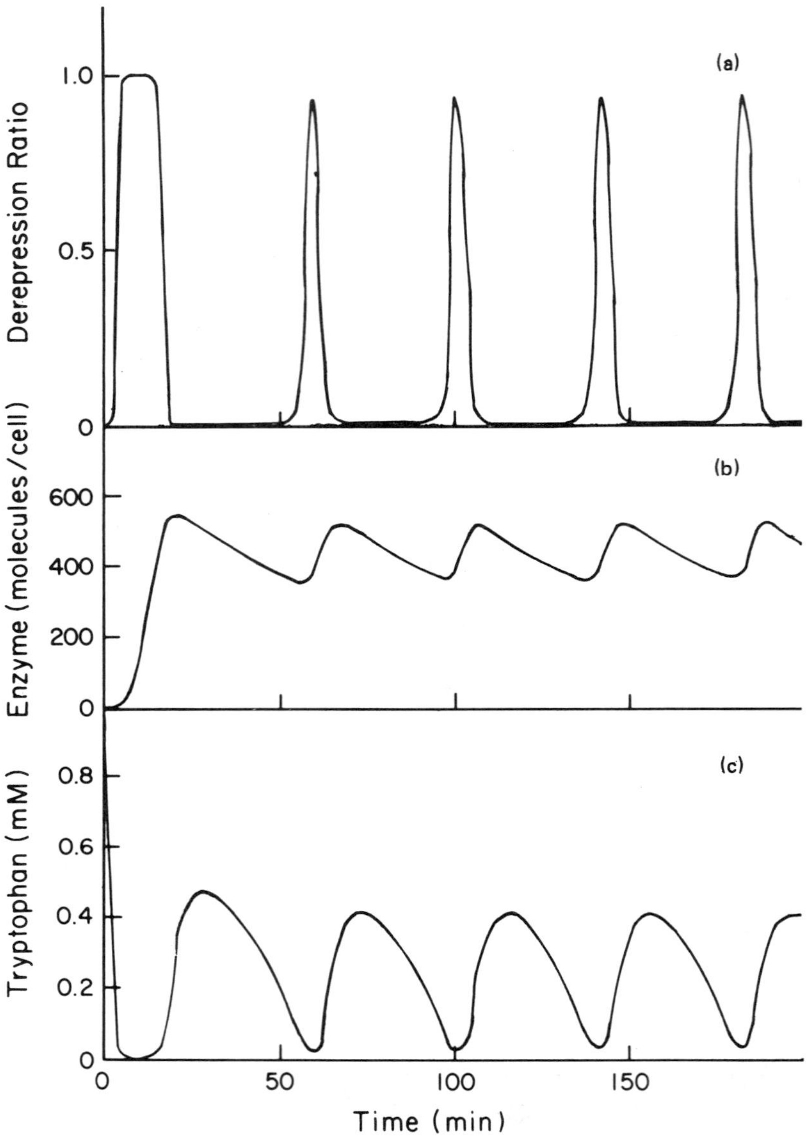


**FIGURE 4** Simulation of gene expression in the hypothetical example of a repressible operon with reduced feedback inhibition. Equations (11) were solved numerically with the parameter values given in Table 1 ($K_I = 10^{-3}$ M). Initial conditions were chosen to simulate derepression at t = O. (a) Derepression ratio expressed as the fraction of maximal gene activity. (b) Intracellular concentration of enzyme. (c) Intracellular concentration of end product.

protein synthesis as a function of transfer RNA concentration. However, G can be simplified to the Michaelis-Menten function $V_{max}x_3/(K_G + x_3)$, where $V_{max} = 4 \times 10^{-6}$ M/sec and $K_G = 10^{-6}$ M, with little effect on the solution of the above equations in this example, and this Michaelis-Menten function is used in the following.

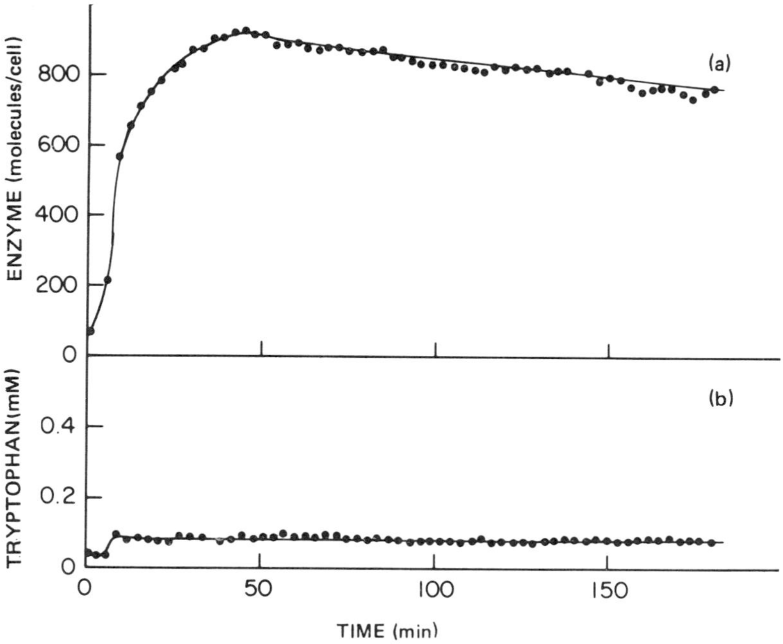


**FIGURE 5** Expression of the tryptophan operon of *E. coli* in strain W1485 tna⁻, which has normal feedback inhibition of tryptophan on anthranilate synthetase. At t = 0, the tryptophan operon was derepressed (as described in the text). (a) Intracellular concentration of anthranilate synthetase. (b) Intracellular concentration of tryptophan. (Data from Bliss, 1979.)

It is not possible to find an exact solution to Eqs. (11) using presently available techniques. However, these equations can be solved numerically. In doing this, initial conditions were chosen to simulate a derepression experiment where a culture is grown for many generations with a high concentration of tryptophan in the medium and where tryptophan is then removed from the medium (at time 0). With tryptophan in the medium, it was assumed that the intracellular concentration of tryptophan was $10^{-3}$ M. The initial values of $x_1$ and $x_2$ were then calculated by setting $\dot{x}_1 = \dot{x}_2 = 0$ in Eqs. (11). The dynamic behavior of the simulated normal operon is shown in Fig. 3. The operon components asymptotically approach their new equilibrium values, which are stable in this case.

Next, consider an example where feedback inhibition has been largely lost as a consequence of mutation. This can be simulated by changing the value of $K_I$ from $10^{-4}$ to $10^{-3}$ M, representing a 10-fold decrease in sensitivity to feedback inhibition. With this single change in the parameters used above, the dynamic behavior of the simulated operon (Fig. 4) is qualitatively different from the previous case. The solution oscillates about an equilibrium that appears to be unstable, and the period of these oscillations is about 47 min, which is significantly shorter than the cell doubling time of 58 min in this example.

Motivated by the above numerical predictions, Bliss (1979) carried out an experimental study of tryptophan operon expression following derepression. Results of this study are presented in Figs. 5 and 6. In these experiments, cells were grown in medium containing 50 $\mu$g/ml L-tryptophan to repress the tryptophan operon. During exponential growth in this medium, they were harvested by centrifugation, washed in ice-cold medium

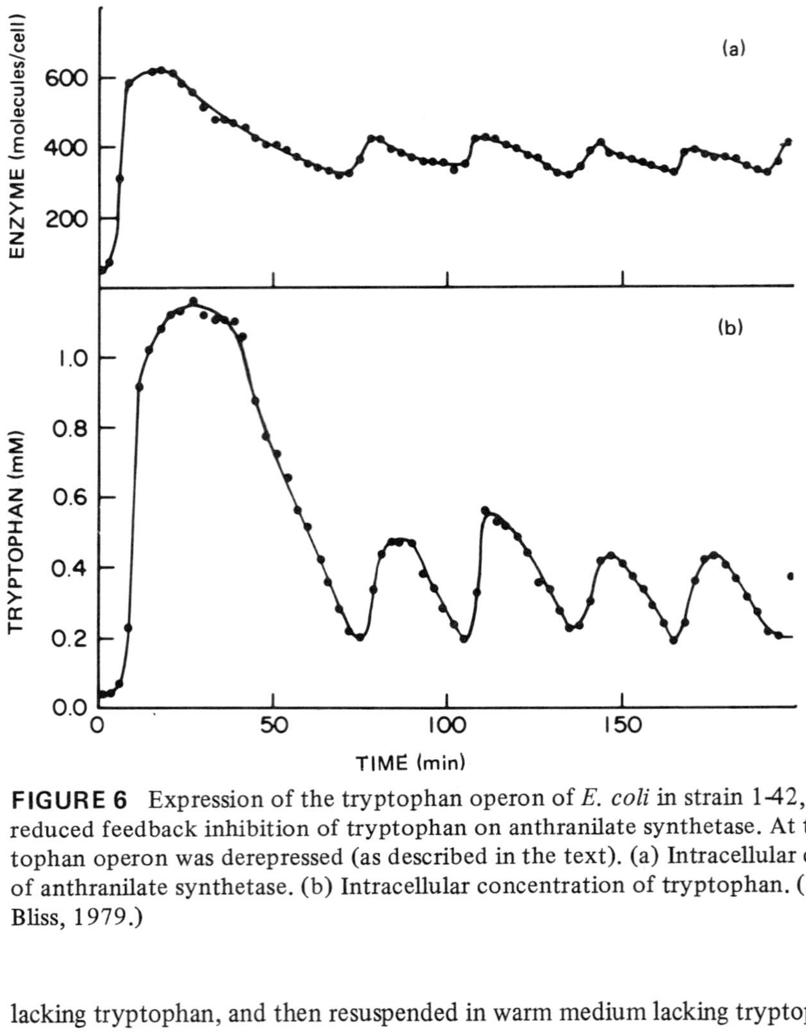


**FIGURE 6** Expression of the tryptophan operon of *E. coli* in strain 1-42, which has reduced feedback inhibition of tryptophan on anthranilate synthetase. At t = 0 the tryptophan operon was derepressed (as described in the text). (a) Intracellular concentration of anthranilate synthetase. (b) Intracellular concentration of tryptophan. (Data from Bliss, 1979.)

lacking tryptophan, and then resuspended in warm medium lacking tryptophan (at time designated 0). Figure 5 shows the response of *E. coli* strain W1485 *tna*⁻, and Fig. 6 shows the response of this strain's feedback-resistant derivative, 1-42, which has a $K_I$ for feedback inhibition that is much increased over that of the parent strain. As predicted by the foregoing equilibrium analysis and simulations, the tryptophan operon in the feedback-resistant strain appears to be unstable.

## COUPLING BETWEEN PERIODIC GENE EXPRESSION AND THE CELL CYCLE

Although it is possible for the period of oscillation in a negative-feedback system to be exactly equal to the cell cycle time, this seems very unlikely without a coupling mechanism. Goodwin (1966) was the first to propose such a mechanism for entrainment of the period of oscillation to the cell cycle time. He conjectured that addition of a pulse of one of the system components once per generation could, in some cases, entrain the oscillator to the cell cycle. To see if this conjecture is correct, we added a pulse of mRNA to our numerical study of system (11). These simulations showed that when the period of oscillation of system (11) is less than the cell cycle time, there is a range of pulse sizes, $P_{min} < P < P_{max}$, such that a pulse in this range does entrain the oscillator. Pulses smaller than $P_{min}$ and pulses larger than $P_{max}$ cannot entrain the oscillator. When the intrinsic period of system (11) is slightly less than the cell cycle time, $P_{max}$ is approximately twice $P_{min}$, but as the period decreases, the range of pulses giving entrainment narrows. When the period of system (11) is longer than the cell cycle time, entrainment by this mechanism does not appear to be possible, since the addition of a pulse usually lengthens the period (Painter, unpublished result).

The second coupling mechanism that we have studied is based on changes in gene potential during the cell cycle. Gene potential is defined here as the maximum possible rate of gene expression at a particular time in the cell cycle. It is well established in many bacterial systems that gene potential approximately doubles when a gene is replicated (Masters et al., 1964; Pato and Glaser, 1968). Changes in gene potential can be introduced into system (11) by changing the parameter g. Analytical studies (Painter and Bliss, 1981) and numerical studies (Painter, unpublished result) have shown that the period of oscillation increases as g increases, and in some cases there is a value of g, denoted $g_c$, which results in a period of oscillation equal to the cell cycle time. The dynamics of the DNA replication cycle with its consequent effect on gene potential can be incorporated into system (11) by doubling the value of g once per generation and by decreasing the value of g at constant specific rate k during each interval between doublings. Therefore, g oscillates (with a discontinuity in its path) over a twofold range during each generation. In numerical studies of system (11) in an unstable case where the effect of DNA replication was simulated, the period of oscillation can be entrained to the cell cycle time in some cases. This entrainment has only been observed when $g_c$ is in the range of oscillations of g, but the converse is not true, since entrainment does not occur in some cases when $g_c$ is near the maximum or minimum value of g (Painter, unpublished result).

In the above coupling mechanisms, the feedback oscillator plays a passive role and the cell cycle entrains the oscillator. In the third mechanism we have studied, it is the feedback oscillator that plays the active role. In this mechanism, derepression of the feedback system triggers DNA replication. The criterion for derepression used in our numerical studies is a fall in the concentration of end product $x_3$ below the unstable equilibrium level. Several minutes after derepression of the operon, replication of the operon was simulated by doubling the parameter g. During each interval between doublings, the value of g decreased at a constant specific rate k, the growth rate. Entrainment in this model occurs as a result of changes in g. For example, if the period of the oscillator is initially shorter than the mass doubling time, then the second doubling of g occurs before g has fallen to one-half of its value immediately after the first doubling. Therefore, the value of g following the second doubling will be higher than the value following the first doubling. As g increases, the period of the oscillator also increases. On the other hand, when the period of the oscillator is initially longer than the mass doubling time, values of g at the time g doubles sequentially decrease, resulting in shorter periods of oscillation. Numerical studies of system (11) with this type of coupling have shown that it can entrain the period of the feedback oscillator to the mass doubling time (i.e., to the cell cycle time) when the negative-feedback system remains unstable over the range of values of g that are encountered during entrainment.

## CONCLUDING REMARKS

The study of periodic enzymes has had a checkered history over the past 20 years. On the experimental side there was great enthusiasm at first as many periodic enzymes were discovered in bacteria, lower eukaryotes, and cultured mammalian cells. The seeming ubiquity of periodic enzyme synthesis was a major support of Mitchison's postulate of a "growth cycle," which is more or less dissociable from the "DNA-division cycle" (Mitchison, 1971, pp. 244-249). The significance of periodic enzyme synthesis was brought into question, however, at the end of the 1970s by two results. First, the "shaken gradient" controls (discussed in Experimental Foundations) led some people to believe that step enzymes were artifacts of the procedure used to synchronize cells and had more to do with nutritional perturbations of cells than with the cell cycle. Second, the technique of two-dimensional gel electrophoresis enabled investigators to study the synthesis of hundreds of individual proteins throughout the cell cycle, but few, if any, periodic enzymes were found in the organisms studied so far:

*E. coli* (Lutkenhaus et al., 1979)

*Caulobacter crecentus* (Milhausen and Agabian, 1981)

*S. cerevisiae* (Elliott and McLaughlin, 1978)

*S. pombe* (Dickinson, 1981)

*P. polycephalum* (Turnock et al., 1981)

*C. sorokiniana* (John et al., 1982)

HeLa cells (Milcarek and Zahn, 1978; Bravo and Celis, 1980)

Thus, a large fraction of the cell's most abundant proteins appears to be synthesized continuously throughout the cell cycle; periodic enzyme synthesis plays only a minor role in overall cellular protein synthesis.

These two results, however, do not imply that the early enthusiasm about periodic enzyme synthesis was misplaced. For one thing, two-dimensional gel electrophoresis would not be expected to reveal fluctuations of regulatory proteins that are present at very low concentrations. Second, the shaken-gradient controls may alternatively be taken to imply that periodic enzyme synthesis can be brought into synchrony in a population of cells by a nutritional perturbation even though the DNA-division cycle remains asynchronous. In this light, these experiments are further evidence for an oscillatory control system independent of periodic gene replication. Finally, the experiments of Bliss (1979) on the tryptophan operon show clearly that nutritional shifts can induce periodic enzyme synthesis in an asynchronous population of cells.

On the theoretical side, the original enthusiasm for the Goodwin model (Goodwin, 1965, 1966; Morales and McKay, 1967) waned as it became apparent that the conditions for sustained oscillations in negative-feedback systems are quite stringent (Griffith, 1968; Walter, 1969) and that the period and amplitude of Goodwin-type oscillations are much larger than experimental results (Tyson, 1979). However, the theoretical investigation of oscillations in negative-feedback gene control systems has been given new impetus recently by the development of improved models that give quantitative fit to experiments (Bliss et al., 1982) and by the development of new perspectives on the period and amplitude of autogenous oscillators (see Estimates of the Period of Autogenous Oscillations).

It is our conclusion that the study of periodic enzyme synthesis is an important approach to the understanding of cell division cycles and other oscillatory phenomena in cells. Furthermore, several different mechanisms can lead to periodic enzyme synthesis, and it is possible to determine experimentally which mechanisms are plausible explanations in a particular case. For example, when periodic enzyme synthesis results directly from replication of the enzyme gene or a regulatory gene, inhibition of DNA synthesis should block the steps or peaks of enzyme activity. On the other hand, when autogenous oscillations arise in an unstable metabolic regulatory system, then the steps or peaks should continue, at least transiently, after inhibition of DNA synthesis, or they should continue in a completely asynchronous culture and be observable, if there is some way to align the oscillation phases of individual cells.

In all of the models for periodic enzyme synthesis considered above, oscillations in enzyme activity result from periodic synthesis of some species of mRNA. This property of the models is due to the fact that control of gene expression in most bacterial systems studied to date is at the level of transcription. However, there is no a priori reason to believe that all cases of periodic enzyme synthesis are regulated at the level of transcription. For example, steps in enzyme activity could be due to activation of previously synthesized enzyme. This possibility can be investigated by the technique of density labeling newly synthesized protein (Mitchelson et al., 1978).

Another possibility not discussed above is that steps or peaks of enzyme activity could be controlled at the level of translation of the mRNA coding for the enzyme. This possibility can be addressed by measuring the concentration of the mRNA coding for the enzyme. If mRNA per milliliter of culture accumulates continuously while enzyme activity shows peaks or steps, then the process is not regulated at the level of transcription. It may be regulated at the level of translation, breakdown, or activation-inactivation of the enzyme.

Our understanding of periodic enzyme synthesis should soon benefit from recent technical advances in molecular biology, such as the development of immunoassays and recombinant DNA technology. The use of immunoassays both in the presence and abscence of protein synthesis inhibitors makes it possible to follow the synthesis and fate of an enzyme, and the use of DNA probes constructed by recombinant DNA technology makes it possible to follow the synthesis and degradation of mRNA coding for the enzyme and also to determine the time of replication of the enzyme gene. Studies using such methods to follow each of the components of gene expression will increase our insight into oscillatory phenomena in biology and will undoubtedly stimulate theoretical attempts at devising accurate and realistic models of periodic enzyme synthesis.

## ACKNOWLEDGMENTS

Parts of the work reviewed herein were supported by grants from the National Science Foundation (MCS-7919787) to JJT and the University of California, Davis (Faculty Research Grant) to PRP.

## REFERENCES

Allwright, D. J. (1977). A global stability criterion for simple control loops. *J. Math. Biol.* 4:363-373.

an der Heiden, U. (1979). Delays in physiological systems. *J. Math. Biol.* 8:345-364.

Banks, H. T., and Mahaffy, J. M. (1978). Global asymptotic stability of certain models for protein synthesis and repression. *Q. Appl. Math.* 36:209-221.

Bernhardt, W., Panter, K., and Holzer, H. (1965). Gedaempftes Oscillieren der Synthesegeschwindigkeit von DPN-Abhaengiger Glutamatdehydrogenase in Hefezellen. *Biochim. Biophys. Acta* 99:531-539.

Bliss, R. D. (1979). Analysis of the dynamic behavior of the tryptophan operon of *Escherichia coli.* Ph.D. thesis, University of California, Riverside.

Bliss, R. D., Painter, P. R., and Marr, A. G. (1982). Role of feedback inhibition in stabilizing the classical operon. *J. Theor. Biol.* 97:177-193.

Bostock, C. J., Donachie, W. D., Masters, M., and Mitchison, J. M. (1966). Synthesis of enzymes and DNA in synchronous cultures of *Schizosaccharomyces pombe. Nature* 210:808-810.

Braun, R., and Behrens, K. (1969). A ribonuclease from *Physarum*: Biochemical properties and synthesis in the mitotic cycle. *Biochim. Biophys. Acta* 915:87-98.

Bravo, R., and Celis, J. E. (1980). A search for differential polypeptide synthesis throughout the cell cycle of HeLa cells. *J. Cell Biol.* 84:795-802.

Dickinson, D. P. (1981). Polypeptide synthesis in cell cycle mutants of fission yeast. *J. Cell Sci.* 51:203-217.

Donachie, W. D. (1965). Control of enzyme steps during the cell cycle. *Nature* 205:1084-1086.

Donachie, W. D., and Masters, M. (1969). Temporal control of gene expression in bacteria. In *The Cell Cycle: Gene-Enzyme Interactions,* G. M. Padilla, G. L. Whitson, and I. L. Cameron (eds.). Academic Press, New York, pp. 37-76.

Edelmann, P. L., and Edlin, G. (1974). Regulation of the synthesis of the lactose repressor. *J. Bacteriol.* 120:657-665.

Elliott, S. G., and McLaughlin, C. S. (1978). Rate of macromolecular synthesis through the cell cycle of the yeast *Saccharomyces cerevisiae. Proc. Natl. Acad. Sci. USA* 75: 4384-4388.

Goodwin, B. C. (1965). Oscillatory behavior in enzymatic control. *Adv. Enzyme Regul.* 3:425-439.

Goodwin, B. C. (1966). An entrainment model for timed enzyme synthesis in bacteria. *Nature* 209:479-481.

Gorman, J., Tauro, P., LaBerge, M., and Halvorson, H. (1964). Timing of enzyme synthesis during synchronous division in yeast. *Biochem. Biophys. Res. Commun.* 15:43-49.

Griffith, J. S. (1968). Mathematics of cellular control processes. I. Negative feedback to one gene. *J. Theor. Biol.* 20:202-208.

Hadeler, K. P. (1979). Periodic solutions of the equation $x'(t) = -f(x(t),x(t-1))$. *Math. Meth. Appl. Sci.* 1:62-69.

Halvorson, H. O. (1977). A review of current models on temporal gene expression in *Saccharomyces cerevisiae.* In *Cell Differentiation in Microorganisms, Plants and Animals,* L. Nover and K. Mothes (eds.). VEB Gustav Fischer Verlag, Jena, Germany, pp. 361-376.

Halvorson, H. O., Carter, B. L. A., and Tauro, P. (1970). Synthesis of enzymes during the cell cycle. *Adv. Microb. Physiol.* 6:47-106.

Hanawalt, P., and Wax, R. (1964). Transcription of a repressed gene: evidence that it requires DNA replication. *Science* 145:1061-1063.

Hastings, S., Tyson, J., and Webster, D. (1977). Existence of periodic solutions for negative feedback cellular control systems. *J. Diff. Eq.* 25:39-64.

Hereford, L., Bromley, S., and Osley, M. A. (1982). Periodic transcription of yeast histone genes. *Cell* 30:305-310.

Huettermann, A., Porter, M. T., and Rusch, H. P. (1970). Activity of some enzymes in *Physarum polycephalum.* I. In the growing *Plasmodium. Arch. Microbiol.* 74:90-100.

Jacob, F., and Monod, J. (1961). Genetic regulatory mechanisms in the synthesis of proteins. *J. Mol. Biol.* 3:318-356.

John, P. C. L., Lambe, C. A., McGookin, R., Orr, B., and Rollins, M. J. (1982). Poly(A)+ RNA populations, polypeptide synthesis and macromolecule accumulation in the cell cycle of the eukaryote *Chlorella. J. Cell Sci.* 55:51-67.

Kleber, H. P., and Aurich, H. (1967). Gedaempfte Oszillationen in der Synthese der Carnitindehydrogenase bei Pseudomonas aeruginosa. *Hoppe-Seylers Z. Physiol. Chem.* 348:1727-1729.

Klevecz, R. R. (1969). Temporal order in mammalian cells. I. The periodic synthesis of lactate dehydrogenase in the cell cycle. *J. Cell Biol.* 43:207-219.

Klevecz, R. R., and Forrest, G. L. (1977). Metabolic regulations in the cell cycle. In *Growth, Nutrition, and Metabolism of Cells in Culture,* vol. III, G. H. Rothblat and V. J. Cristofalo (eds.). Academic Press, New York, pp. 149-196.

Knorre, W. A. (1968). Oscillations of the rate of synthesis of β-galactosidase in *Escherichia coli* ML30 and ML308. *Biochem. Biophys. Res. Commun.* 31:812-817.

Kuempel, P. L., Masters, M., and Pardee, A. B. (1965). Bursts of enzyme synthesis in the bacterial duplication cycle. *Biochem. Biophys. Res. Commun.* 18:858-867.

Landahl, H. D. (1969). Some conditions for sustained oscillations in biochemical chains with feedback inhibition. *Bull. Math. Biophys.* 31:775-787.

Lutkenhaus, J. F., Moore, B. A., Masters, M., and Donachie, W. D. (1979). Individual proteins are synthesized continuously throughout the *Escherichia coli* cell cycle. *J. Bacteriol.* 138:352-360.

MacDonald, N. (1977). Time lag in a model of a biochemical reaction sequence with end product inhibition. *J. Theor. Biol.* 67:549-556.

Mackey, M. C., and Glass, L. (1977). Oscillations and chaos in physiological control systems. *Science* 197:287-289.

Mahaffy, J. M. (1980). Periodic solutions for some protein synthesis models. *J. Math. Anal. Appl.* 74:72-105.

Masters, M., and Donachie, W. D. (1966). Repression and the control of cyclic enzyme synthesis in *Bacillus subtilis. Nature (Lond.)* 209:476-479.

Masters, M., Kuempel, P. L., and Pardee, A. B. (1964). Enzyme synthesis in synchronous cultures of bacteria. *Biochem. Biophys. Res. Commun.* 15:38-42.

Milcarek, C., and Zahn, K. (1978). The synthesis of ninety proteins including actin throughout the HeLa cell cycle. *J. Cell Biol.* 79:833-838.

Milhausen, M., and Agabian, A. (1981). Regulation of polypeptide synthesis during *Caulobacter* development: Two-dimensional gel analysis. *J. Bacteriol.* 148: 163-173.

Mitchelson, K., Chambers, T., Bradbury, E. M., and Matthews, H. R. (1978). Activation of histone kinase in G2 phase of the cell cycle in *Physarum polycephalum. FEBS Lett.* 92:339-342.

Mitchison, J. M. (1969). Enzyme synthesis in synchronous cultures. *Science* 165:657-663.

Mitchison, J. M. (1971). *The Biology of the Cell Cycle.* Cambridge University Press, London.

Mitchison, J. M. (1977). Enzyme synthesis during the cell cycle. In *Cell Differentiation in Microorganisms, Plants and Animals,* L. Nover and K. Mothes (eds.). VEB Gustav Fischer Verlag, Jena, Germany, pp. 377-401.

Molloy, G. R., and Schmidt, R. R. (1970). Studies on the regulation of ribulose-1,5-diphosphate carboxylase synthesis during the cell cycle of the eucaryote *Chlorella. Biochem. Biophys. Res. Commun.* 40:1125-1133.

Morales, M., and McKay, D. (1967). Biological oscillations in "controlled" systems. *Biophys. J.* 7:621-625.

Ohki, M. (1978). The cell cycle-dependent synthesis of envelope proteins in *Escherichia coli.* In *Bacterial Outer Membranes: Biogenesis and Functions,* M. Inouye (ed.). Wiley, New York, pp. 293-315.

Painter, P. R., and Bliss, R. D. (1981). Reconsideration of the theory of oscillatory repression. *J. Theor. Biol.* 90:293-298.

Pato, M. L., and Glazer, D. A. (1968). The origin and direction of replication of the chromosome of *Escherichia coli* B/r. *Proc. Natl. Acad. Sci. USA* 60:1268-1274.

Rapp, P. E. (1975). A theoretical investigation of a large class of biochemical oscillators. *Math. Biosci.* 25:165-188.

Rickles, R., Marashi, F., Sierra, F., Clark, S., and Wells, J. (1982). Analysis of histone gene expression during the cell cycle in the HeLa cells by using cloned human histone genes. *Proc. Natl. Acad. Sci. USA* 79:749-753.

Schmidt, R. R. (1974). Transcriptional and post-transcriptional control of enzyme levels in eucaryotic microorganisms. In *Cell Cycle Controls,* G. M. Padilla, I. L. Cameron, and A. Zimmerman (eds.). Academic Press, New York, pp. 201-233.

Spalding, J., Kajiwara, K., and Mueller, G. (1966). The metabolism of basic proteins in HeLa cell nuclei. *Proc. Natl. Acad. Sci. USA* 58:1535-1542.

Torriani, A. (1960). Influence of inorganic phosphate in the formation of phosphatases by *Escherichia coli. Biochim. Biophys. Acta* 38:460-479.

Turnock, G., Chambers, J., and Birch, B. (1981). Regulation of protein synthesis in the plasmodial phase of *Physarum polycephalum. Eur. J. Biochem.* 120:529-534.

Tyson, J. J. (1975). On the existence of oscillatory solutions in negative feedback cellular control processes. *J. Math. Biol.* 1:311-315.

Tyson, J. J. (1979). Periodic enzyme synthesis: Reconsideration of the theory of oscillatory repression. *J. Theor. Biol.* 80:27-38.

Tyson, J. J., and Othmer, H. G. (1978). The dynamics of feedback control circuits in biochemical pathways. *Prog. Theor. Biol.* 5:1-62.

Tyson, J. J., and Sachsenmaier, W. (1979). Derepression as a model for control of the DNA-division cycle in eukaryotes. *J. Theor. Biol.* 79:275-280.

Vassef, A. A., Flora, J. B., Weeks, J. C., Bibbs, B. S., and Schmidt, R. R. (1973). The effects of enzyme synthesis and stability and of deoxyribonucleic acid replication on aspartate transcarbamylase of *Chlorella. J. Biol. Chem.* 248:1976-1985.

Walter, C. (1969). Oscillations in controlled biochemical systems. *Biophys. J.* 9:863-872.

Walter, C. F. (1974). Some dynamic properties of linear, hyperbolic and sigmoidal multienzyme systems with feedback control. *J. Theor. Biol.* 44:219-240.

# 11

# Is Energy Metabolism in the Prokaryotic Cell Cycle Manifestly Coupled to a Clock?

**Robert K. Poole** / Queen Elizabeth College, University of London, London, England

## INTRODUCTION

The energy-transducing membranes of bacteria, mitochondria, and chloroplasts provide excellent examples of biological organization in space and time. The hierarchy of spatial organization (Lloyd et al., 1982) extends from, for example, the transmembrane distribution of small ions that constitute electrochemical gradients to some of the most complex enzymes known, such as terminal cytochrome oxidases and the energy-transducing ATP synthetases. These macromolecules are organized in the hydrophobic membrane such that they catalyze vectorial reactions and are in appropriate proximity with functionally related components.

This spatial hierarchy is accompanied by a temporal organization which, in bioenergetic systems, is particularly spectacular in its scope. Rates of electron transfer, particularly those to terminal electron acceptors such as oxygen, are among the fastest of all enzyme-catalyzed reactions ($1 \times 10^8\ M^{-1}\ sec^{-1}$) (Chance, 1975). The temporal hierarchy extends through the slower reactions of the "metabolic time domain" and epigenetic time to the cell division cycle (genetic) domain (Lloyd et al., 1982).

If we regard the cell division cycle as a clock, in the sense that it registers the passing of time, then the development of the respiratory "hardware" and its functions *are* coupled to this clock, since the *overall* increase during the cycle must be commensurate with other growth processes. The question addressed in this chapter, however, is more specific since, strictly, a clock is a machine with a bell. Is the inexorable development of respiratory metabolism (the ticking of the clock) punctuated by identifiable periodic events (the striking of the hour)? The analogy will be seen to be a surprisingly good one, since in those cases where overt periodicity has been observed, the periodic events recur at fairly regular fractions of the cell cycle time.

## TICKS AND CHIMES OF THE CELL CYCLE CLOCK

The varnished clock that click'd behind the door
Oliver Goldsmith, *The Deserted Village*

### Structure

Recent biochemical and biophysical investigations of energy-transducing membranes, pursued most intensively with certain electron-transferring complexes of the inner mitochondrial membrane, have yielded a view of the three-dimensional organization of these complexes. The components are highly ordered with respect to sidedness in the membrane (Eytan et al., 1975), intermolecular distances (Dockter et al., 1978) and the orientation of the chromophoric metal centers with respect to the membrane plane (Poole et al., 1980). Thus, all or any of these features could vary during the cell cycle, as could the stoichiometry of the components, their mobility in the membrane, and the degree of association with, and dissociation from, the membrane bilayer. Indeed, the fact that such measurements have usually been made on membranes derived from asynchronous cell populations casts some doubt on their usefulness, since such measurements are time averaged.

This degree of organization presumably imposes restraints on the order of assembly of the membrane and the constituent macromolecules and subunits. For example, the $H^+$-translocating ATPases of mitochondria and bacterial membranes consist of at least two major components (Nicholls, 1982). The catalytic site of ATP synthesis and hydrolysis is the so-called $F_1$ (or $BF_1$ in bacteria) and is rather peripheral to the membrane, being bound via the intrinsic $F_o$ (or $BF_o$) segment. The latter acts as a proton channel through the relatively proton-impermeable membrane. Uncontrolled integration of $F_o$ components in the absence of $F_1$ to block these channels may render the membrane "leaky" and uncoupled. Clearly, recent advances in the molecular biology of these components and, in *Escherichia coli,* the mapping of the *unc* operon and establishment of gene-polypeptide relationships (Downie et al., 1979) could fruitfully be accompanied by a description of the temporal relationships of these processes in the cell cycle.

Formally, the synthesis and development of the bacterial membrane may be regarded (Poole, 1980) as the assembly of previously synthesized and activated components (proteins and lipids) into a preexisting framework. The sequence may vary, however; incomplete proteins may be inserted with far-reaching consequences. For example, a mutant of *E. coli* that requires the presence of $\delta$-aminolevulinic acid for heme synthesis will nevertheless synthesize, and incorporate into the membrane, the cytochrome apoprotein in the absence of heme (Haddock and Schairer, 1973). (In fact, this mutant should provide a powerful tool for identifying the timing of incorporation of the spectroscopically "invisible" apoproprotein, since its presence in the membrane can be identified in a functional assay after reconstitution with heme in vitro.) An interesting consequence of the incorporation of this protein is the apparent failure of certain nonheme iron-sulfur clusters to assume their normal, fixed orientation in the membrane (Blum et al., 1980).

### Function

The overall capacity for electron transfer and energy transduction must double during the cycle. The course of development of these functions has been most frequently followed by measuring rates of oxygen consumption. These, together with the transmem-

**TABLE 1** Patterns of Oxygen Uptake Rates in Synchronous Cultures of Bacteria[a]

| Organism | Growth medium[b] | Cell cycle time (min) | Synchrony indices[c] $F_1$ and $F_2$ | Method of $O_2$ measurement[d] | Pattern observed | Approximate periodicity (fraction of cell cycle time) | Reference |
|---|---|---|---|---|---|---|---|
| *E. coli* B | Glucose | 50 | 0.27 (–) | Warburg manometry ± KOH | Continuous but $Q_{O_2}$ oscillated | Not applicable 1/2 | Maruyama (1956) |
| *E. coli* W1485 | Alanine | 90 | 0.69 (0.69) | Discontinuous | Oscillatory | 1/3 | Evans (1975)[e] |
| *E. coli* K12 (wild-type and ATPase-deficient mutant) | Glucose + casein | 44,70 | 0.65–0.88 | Continuous | Oscillatory | 1/2 | Poole (1977a) |
| *E. coli* A1002 | Glycerol + casein | 40 | 0.48 (0.47) | Continuous | Stepwise | 3/4 | Poole (1977a) |
| *E. coli* A1002 | Succinate | 90 | 0.3–0.6 | Continuous | Continuous (exponential) | Not applicable | Misri and Poole (unpublished); see Fig. 2 |
| *Alcaligenes eutrophus* H16 | Lactate | 76 | 0.72–0.82 (0.75–0.77) | Discontinuous (± CCCP) | Stepwise | 1/2 | Edwards and Jones (1977) |
| *Bacillus subtilis* NC113 | Glucose | 80 | 0.45 (–) | Discontinuous (± DCCD, Nbf-Cl, HQNO) | Stepwise | 1/2 | Edwards and McCann (1981) |

[a]Continuous flow centrifugation was used to prepare cultures except for Maryuma (filtration) and Misri and Poole (density gradient selection). Abbreviations: CCCP, carbonylcyanide-m-chlorophenylhydrazone; DCCD, NN′-dicyclohexylcarbodi-imide; HQNO, 2-heptyl-4-hydroxyquinoline-N-oxide; Nbf-Cl, 4-chloro-7-nitrobenzofurazan.
[b]All media are chemically defined media plus the tabulated components.
[c]Blumenthal and Zahler (1962).
[d]$O_2$ electrode except for Maruyama (1956).
[e]Timings from corrected figure provided by author.

brane movements of ions and solutes, the consequent membrane potential, and the levels of adenine and nicotinamide nucleotides may all provide evidence for the discontinuous manifestations of a clock.

## RESPIRATION DURING THE CELL CYCLE

### Measurement of Oxygen Uptake Rates in Synchronous Cultures

Table 1 presents a survey of published reports of bacterial respiration (oxygen uptake) during the cell cycle. For all three bacteria, centrifugation was used to select a homogeneous size and/or density class that was presumed to be also homogeneous with respect to age. Most of these results have been discussed previously (Poole, 1980).

Since then, Edwards and McCann (1981) have described a distinct stepwise pattern of oxygen uptake in *Bacillus subtilis.* A control experiment demonstrated that the steps were not an artefact of the continuous flow synchronization method. Two inhibitors of the $H^+$-translocating ATPase revealed oscillating sensitivity to inhibition of respiration during the cycle. An inhibitor of electron transfer, however, HQNO, exerted maximal effect only at 0.4 of a cycle, the time of the first increase in respiration rate.

The results are in clear contrast to the most recent data on *E. coli* (Misri and Poole, unpublished results) in which a continuous, probably exponential, increase was observed, with no evidence for coupling to a cell cycle clock.

### Comments on Methodology

#### *Polarographic Measurements*

The usual experimental procedure (described as discontinuous in Table 1) involves the periodic removal of samples from a synchronous culture and the monitoring of the disappearance of oxygen from solution when these samples are incubated in a closed reaction vessel. The "open" electrode system (Degn et al., 1980), with its ability to continuously record oxygen tension in a liquid sample over prolonged periods, was first applied, in a simplified form, to synchronous cultures by Poole (1977). More recently, we (Poole and Misri, unpublished results) have designed and constructed a modified apparatus (Fig. 1) that permits the simultaneous and continuous monitoring of uptake of both oxygen (polarographically) and any radioactively labeled substrate(s) by flow dialysis (Feldman, 1978) in a small (2-5 ml) synchronous culture. Further details of the apparatus will be published elsewhere. From a continuous recording of the oxygen tension in the liquid phase (see Fig. 2) and knowing the first-order rate constant for $O_2$ diffusion to the sample, $O_2$ uptake rates can be calculated at short time intervals (e.g., Fig. 2).

#### *Cell Synchronization*

There is considerable divergence in the degree of synchrony obtained by the size selection methods shown in Table 1 and reflected in the synchrony indices. It must be questioned whether the modest synchrony in some systems (and thus time averaging over the cycle) is obscuring true fluctuations in respiratory metabolism. The variability in synchrony is undoubtedly due in part to (1) differing degrees of success in isolating a homogeneous size class, and (2) the distribution of cell generation times (Koch, 1977; Lloyd et al., 1982). In addition, however, any sloppiness in the correlation between cell size and cell age may contribute to the asynchrony of age-related events when size selection forms the basis of synchronization. In such a case, improved resolution of size selection methods will not necessarily facilitate the detection of "chimes" that may be clocked independently

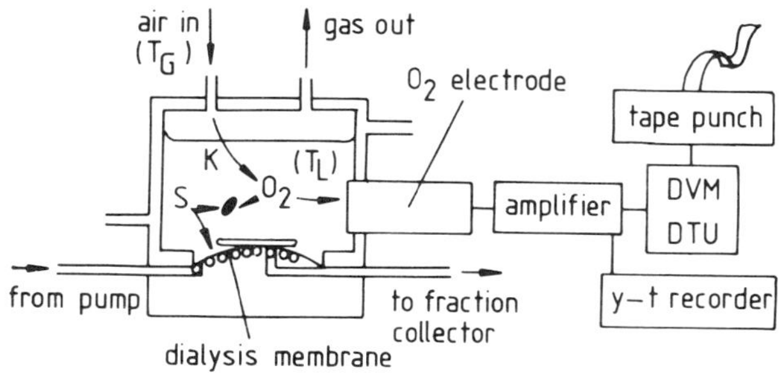


**FIGURE 1** Diagram of apparatus for continuous and simultaneous measurements of the uptake of $O_2$ and a growth substrate by small (3–5 ml) synchronous cultures. It is a modified open oxygen electrode (Degn et al., 1980). Respiration rate,

$$V_r = k(T_G - T_L) - \frac{dT_L}{dt}$$

where k is the rate constant for $O_2$ diffusion into the liquid phase and $T_G$ and $T_L$ are the $O_2$ tensions in gas and liquid phases, respectively. Uptake of a radioactively labeled solute S is measured by flow dialysis: the lower chamber consists of a fine spiral, which increases sensitivity and reduces response time. (From Feldman, 1978.)

of size. On the other hand, Koch (1977) has argued that periodic events of the cell cycle, such as chromosome initiation and cell division, are subservient to the major processes of RNA and protein synthesis and that "in balanced growth, the events of cell division would be expected to be more dependent on cell size than on cell age."

In most published experiments in which DNA synthesis has been measured in synchronous bacterial cultures, DNA synthesis has not been well phased with respect to the cell cycle, where time instead of cell size is the experimental variable. This is true even for cultures where cell division was well synchronized for many doublings after the synchronizing procedure (Kelley, 1974, cited by Koch, 1977). Chromosome replication is, however, certainly a discontinuous event, and in *E. coli* B/r, the timing between synthesis of a region of the chromosome and the synthesis of the same region in the daughter cell, is very precise (Newman and Kubitschek, 1978). We should be wary, therefore, of concluding too hastily, on the basis of some present synchronous culture experiments, that respiratory development is not coupled to a well-regulated clock.

### Are Growth Rates Limited by Respiration?

A continuous increase in respiration rates during the cell cycle is quite consistent with recent physiological data that suggest that respiration is rate-limiting for growth. The addition of the uncoupling agent 2,4-dinitrophenol (DNP) to chemostat cultures of *Paracoccus denitrificans* (Stouthamer and Bettenhausen, 1981) increased the maximum specific growth rate. The $Q_{O_2}$ at maximum specific growth rate was constant (10–10.8 mmol/g per hr) with various substrates, and equalled the highest value obtained after DNP addition. It was suggested that bacterial respiration may be analogous to mitochondrial state 3, in which the capacity of the respiratory chain is limiting under conditions of high substrate, oxygen and ADP levels (Nicholls, 1982).

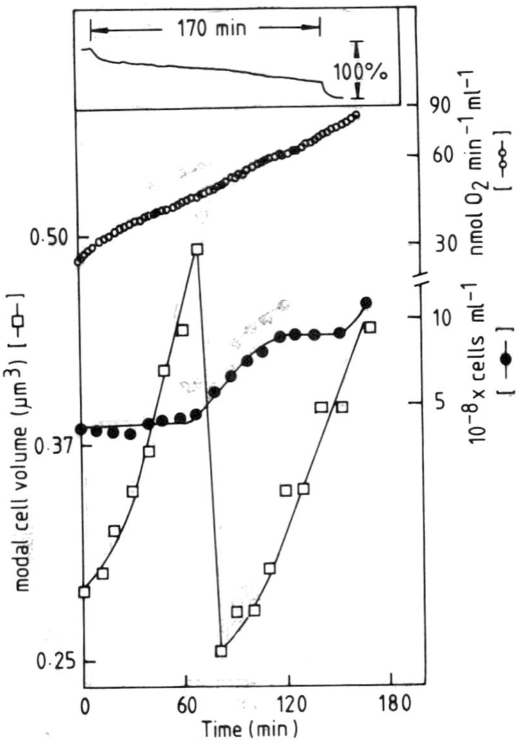


**FIGURE 2** Oxygen uptake in a synchronous culture of *E. coli* (succinate as carbon source) prepared by selection from a sucrose density gradient. The upper block shows the continuous recording from the open electrode; 100% refers to $O_2$ saturation. The calculated respiration rate (see legend to Fig. 1) is shown below. (Misri and Poole, unpublished data.)

**TABLE 2** Effects of Inhibitors on Growth and Division in *E. coli*

| Inhibitor and concentration (mM) | Target | Transition point[a] ($x_{TP}$) |
|---|---|---|
| KCN (0.3) | Respiratory electron transfer | 0.96 |
| CCCP (0.04) | Proton-motive force | 1.0[b] |
| Nbf-Cl (0.3) | ATP synthetase | 1.0[b] |
| D-chloramphenicol (0.6) | Protein synthesis | 0.73 |
| Nalidixic acid (0.1) | DNA synthesis | 0.58 |

[a]Calculated from

$$x_{TP} = 1 - \frac{\log_e (N/N_o)}{\log_e 2}$$

(Howell and Naliboff, 1973), where $N_o$ is the numbers of cells per milliliter culture at the time of adding the inhibitor and N the final population attained.

[b]No detectable increase in cell numbers observed after adding the inhibitor.

Similar conclusions have been drawn from batch cultures of *E. coli* grown in various media that supported specific growth rates from 0.2 to 1.3 $hr^{-1}$ (Andersen and von Meyenburg, 1980). Specific respiration rates ($Q_{O_2}$) were similar throughout at about 20 mmol $O_2$/g per hr, as was the apparent efficiency of oxidative phosphorylation.

That continued growth is absolutely dependent on respiration is also evident from the effects (Misri and Poole, unpublished results) of adding respiratory inhibitors to exponentially growing cultures of *E. coli* (Table 2). Transition points (Howell and Naliboff, 1973) are late in the cycle for such compounds, whereas inhibition of DNA or protein synthesis allows completion of the ongoing cell cycle and division.

All these observations are in accordance with the view of Koch (1977) that "ultimately the ability of the organism to produce protoplasm from the resources of the environment must limit growth and must time the initiation of chromosome replication, nuclear division and cell division." They also explain why, in rapidly growing cultures, the addition of uncouplers frequently does not stimulate respiration rates (Poole, 1977a; Burstein et al., 1979). It may be instructive to examine cell cycle patterns of respiratory development in cultures growing at submaximal rates, where respiration is not growth limiting and clocked patterns may be revealed. In summary, respiration in some bacteria may dictate rates of macromolecular synthesis and, ultimately, periodic cell cycle events, whereas in various eukaryotic microorganisms (see Chap. 2), the oscillations in respiratory metabolism are enslaved by the changing energy requirements of cell cycle progress.

### Respiratory Oscillations of Higher Frequency

> Would that life were like the shadow cast by a wall or a tree, but it is like the shadow of a bird in flight.
>
> Haggadah, *Palestinian Talmud* (fourth century)

None of the high-frequency oscillations (in the metabolic time domain) observed in various prokaryotes (Ried and Šetlík, 1971; Metzler, 1981) have yet been unequivocally associated with the cell cycle.

In *Klebsiella aerogenes*, undamped oscillations (period 2–3 min) were observed in oxygen tension and pyridine nucleotide fluorescence after a chemostat culture was subjected to an anaerobic shock; for a review, see Harrison (1976). Any association of these oscillations with the cell cycle was discounted on the grounds that the period appeared unrelated to growth rate. The recent finding that the period of cell cycle-dependent oscillations in *Acanthamoeba* (see Chap. 2) is also independent of cell cycle time (when modulated by temperature) suggests that it is premature to discount a cell cycle relationship in *Klebsiella.*

Disturbing the air supply to a late-exponential culture of *E. coli* (with glucose as carbon source) also induces oscillations in the fluorescence of flavins and pyridine nucleotides (Fig. 3). Culture reflectance also oscillated with a period equal to the prominent component of the fluorescence trace. Note that the preperturbation state of reflectance and fluorescence was rapidly restored on switching on the stirrer. Lloyd and Ball (1979), reporting similar oscillations in a yeast, have highlighted the distinctions between such high-frequency periodicities and cell cycle-related oscillations.

### Regulation of Respiration Rates in Growing Bacteria

Of the many levels at which respiration could in principle be governed (Harrison, 1976), only two, for which newer exponential evidence is available, are considered here.

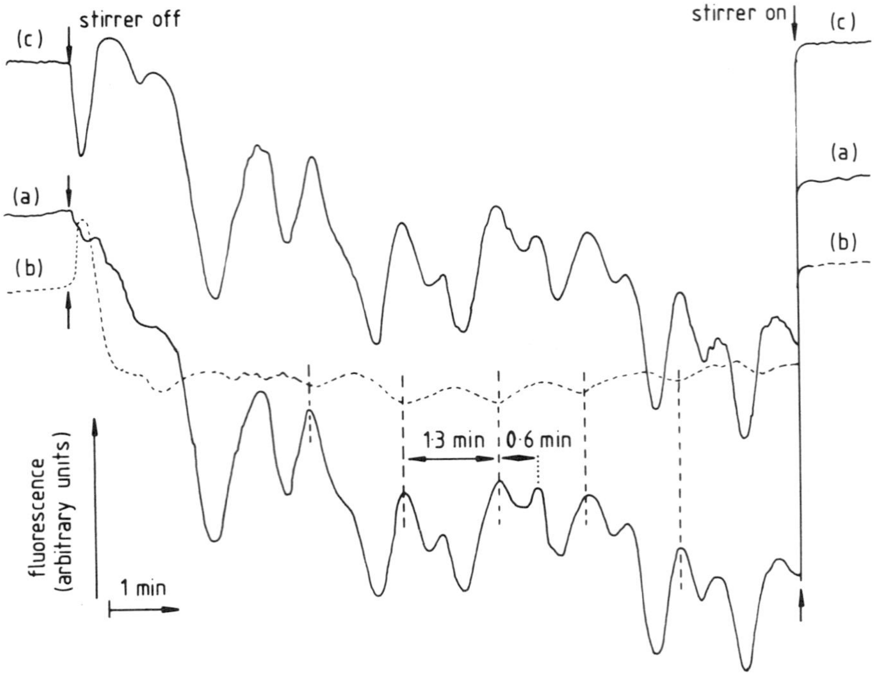


**FIGURE 3** Oscillations in flavin and pyridine nucleotide fluorescence following a sudden decrease in the $O_2$ transfer rate (by switching off the stirrer) in a late-exponential phase culture of *E. coli.* A Johnson Foundation compensated fluorometer/reflectometer utilized a bifurcated light guide which was immersed directly in the culture to measure fluorescence (a) and reflectance (b); (c) is corrected fluorescence [(a) – (b)]. (From Bashford, Misri, and Poole, unpublished data.)

### *Substrate Control of Respiration*

Although respiration rates could be regulated by rates of solute uptake, additional mechanisms for partitioning solute for anabolic and catabolic (i.e., respiratory) uses would have to operate. Discontinuous patterns of solute uptake have been observed during some bacterial cell cycles (Poole, 1980; Dietzel et al., 1978); these changes have not been identified as temporally related to other changes in respiratory metabolism; see, however, Ohki (1972).

The difficulties of measuring solute uptake rates in synchronous cultures without artefacts introduced by the synchronizing procedure are illustrated in Fig. 4. Frequent measurements of the uptake of succinate (carbon and energy source) by flow dialysis appear to show a pause in transport at about the same time as division (Fig. 4a). However, control experiments in which the sucrose gradient and cells were mixed up before removal of cells clearly show perturbations, which complicate interpretations (Fig. 4b,c). Similar results were shown for *Schizosaccharomyces pombe* by Creanor and Mitchison (1979), who also reported the disturbances to be different in two separate controls. Clearly, care is needed and control experiments are essential.

Kubitschek (1968a) has suggested that membrane-bound transport systems are incorporated and function just prior to cell division, resulting in a doubling of the linear

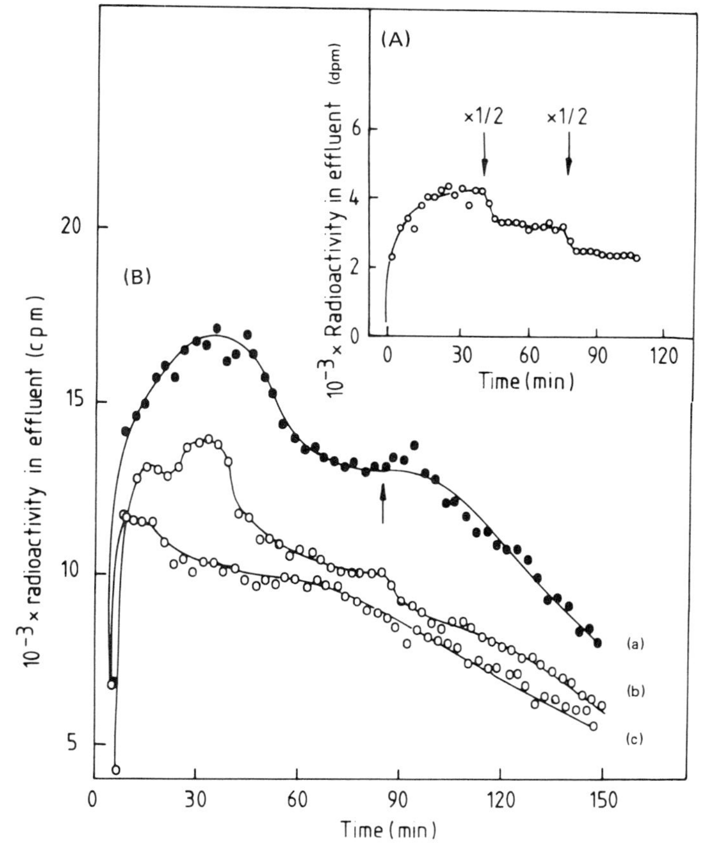


**FIGURE 4** (A) Performance of the flow dialysis cell (see Fig. 1) and (B) [$^{14}$C]succinate uptake by synchronous and asynchronous control cultures of *E. coli.* In (A), at zero time, 5μCi of [$^{14}$C]glucose was added to the upper chamber; at each arrow, the glucose concentration was halved by dilution. (B) The uptake of succinate in medium containing approximately 15 mM succinate is shown for a synchronous culture (a) and two asynchronous control cultures (b,c). The arrow in (a) marks the time of the midpoint of the step in cell numbers. (From Misri and Poole, unpublished data.)

rate of cell growth at this time. Measurements of transport capacity for many solutes support this idea (Kubitschek, 1968b; Kubitschek et al., 1971), but Koch (1975), on the basis of induction experiments with the *lac* permease, suggests that this carrier is synthesized, incorporated, and can start functioning at any time in the cycle.

### *Regulation of Adenylate Pools*

The finding that cell cycle-dependent respiratory oscillations in certain eukaryotic microorganisms may be explained by alternating availability of ADP (Poole, 1977b; Poole and Salmon, 1978). Lloyd and Edwards (see Chap. 2) have encouraged a search for similar mechanisms of respiratory regulation in bacteria. Although Edwards and Jones (1977) have shown that sensitivity to the uncoupler CCCP and respiration rates bear the same phase relationship in *A. eutrophus* as they do in the above eukaryotes, these fluctuations are also explicable in terms of oscillating $H^+$ conductance of the membrane and fluctuatons in the proportion of $F_1$ ATPase that is membrane bound (Edwards et al., 1978). It would be especially interesting to assay adenylate levels in this system.

Some independent lines of evidence suggest that ADP levels may not limit respiration in growing bacteria:

1. Under diverse growth conditions, the ADP content of *K. aerogenes* does not appear to fall low enough to give the equivalent of mitochondrial state 4 (Nicholls, 1982); the ATP/ADP ratio was poised below 3 (Harrison and Maitra, 1969). The caveat is that these measurements and others described by Harrison (1976) were made on chemostat cultures and must be time averaged over the cell cycle. However, very similar and constant values of the ATP/ADP ratio also occur through the cell cycle of *E. coli* (Misri and Poole, unpublished results).
2. The use of a mutant deficient in the $Mg^{2+}$-dependent $H^+$-translocating ATPase has shown that the respiratory oscillations (Table 1) observed after synchronization of *E. coli* by continuous flow centrifugation do not arise from respiratory control phenomena (Poole, 1977a).
3. As described above, there is evidence that rapidly growing bacteria respire at maximal capacity.

## ENZYME AND PROTEIN LEVELS DURING THE CELL CYCLE

Control experiments to determine the effects of certain cell synchronizing procedures (Mitchison, 1977) have cast doubt on some of the reports of periodic changes in enzyme activity in cell cycles; for catalogues, see Mitchison (1971) and Lloyd et al. (1982). Models to explain such periodicities, irrespective of their origin, are discussed by Painter and Tyson (Chap. 10). In this section, some of the better characterized periodicites in enzymes and proteins of respiratory metabolism are described.

### Bacterial ATPases

#### *In* E. coli

ATPase activity has been assayed following cell cycle analysis by two independent methods (Scott et al., 1980). In the first, small cells were selected from an exponentially growing culture using equivolumetric sucrose gradients; in the second, the high-resolution gradient was used to separate chilled cells into size classes, which were then analyzed directly for ATPase activity ("culture fractionation"). It was concluded that the activity of this enzyme does not increase continuously during the cell cycle of *E. coli,* based on the following findings. (1) Fluctuations in activity (on a per cell basis) after either method of cell cycle analysis were well outside the estimated limits of experimental error. (2) In control (mixed-up gradient) experiments, no fluctuations were observed. (3) The culture fractionation procedure was shown not to affect adversely the reproducibility of the assay (this being the only "control" situation available). (4) There was good correlation between the timings of maximum activities with respect to cell cycle age, when cycle analysis was carried out using the two independent methods or when repeat experiments were conducted with either method alone. However, it must be admitted that statistically it is far easier to *rule out* the possibility of continuous activity increase than it is to *demonstrate what the pattern is.* An oscillatory pattern that rises to two maxima per cell cycle was selected as the simplest solution compatible with the experimental data. The activity was also differentially sensitive to ruthenium red during the cell cycle (Scott et al., 1980).

#### *In* A. eutrophus

Marked oscillations in ATPase activity have also been observed by Edwards et al. (1978) during growth of *A. eutrophus,* synchronized by continuous-flow centrifugation. At the times of maximal and minimal activity, the enzyme was differentially sensitive to inhibitors of $BF_1$ or $BF_o$ components. Measurements of the $H^+$ permeability of the membrane revealed that all these oscillations could be accounted for by a scheme in which the $BF_1$ was associated with $BF_o$ to different extents during the cell cycle. However, in keeping with some reports of periodic changes in respiratory enzymes in other bacteria (Scott et al., 1980), control experiments have not yet shown that the discontinuities were not introduced by the synchronization procedure.

### Protein Accumulation and Synthesis

The frequently observed oscillatory and stepwise patterns of enzyme activity are in no way incompatible with the increasing number of reports that the majority of cell proteins are synthesized or accumulate continuously during the cell cycle; for an example in *E. coli,* see Lutkenhaus et al. (1979). If both kinds of experimental approaches are shown to be free of artefacts, attention will turn to modulations of activity, not synthesis.

As an alternative to the two-dimensional gel electrophoresis experiments, we have taken advantage of the ability to assay spectroscopically the amounts of certain electron transfer components in intact cells, without recourse to cell disruption and protein fractionation under disaggregating conditions, and to search for temporally structured changes in membrane composition. Cytochrome *o*, the major terminal oxidase of aerobically grown *E. coli,* was assayed with unequivocal specificity from low-temperature photodissociation spectra and was shown to accumulate continuously during the cell cycle (Scott et al., 1981). All the other cytochromes present in such cells, and detectable by numerical analysis of difference spectra, also behaved similarly (Scott and Poole, 1982). Four EPR-detectable centers assigned to membrane-associated protein all increased continuously during the cell cycle (Poole et al., 1981). Finally, the amount of the catalytic $F_1$ component of the ATPase, as measured by "rocket" immunoelectrophoresis in whole-cell digests, also accumulates in parallel with total cell protein, i.e., continuously (Smith, Boxer, Scott, and Poole, unpublished results).

Thus, we have been unable to detect any component of the respiratory chain of *E. coli* whose amount is increased in a discontinuous fashion through the cycle and which might be regarded as indicative of a cell cycle clock that times respiratory development. In addition, when the three-dimensional orientations within the membrane of (1) low-spin cytochrome heme (Poole et al., 1980) and (2) a ferredoxin-type Fe-S cluster (Poole et al., 1981) were determined in oriented "multilayers" of isolated membranes, no evidence was obtained for a cell cycle-specific rearrangement of these components.

Virtually nothing is known of the contributions of synthesis and turnover to the observed, continuous accumulation of membrane proteins. The analyses of Larrabee et al. (1980) and Mosteller et al. (1980) probably do not include such hydrophobic proteins. Nevertheless, a membrane-bound, ATP-stimulated endoprotease that could function in such turnover has recently been reported in *E. coli* (Voellmy and Goldberg, 1981).

## CONCLUSION

From the restricted experimental evidence available, no general picture of respiratory development has emerged. Continuous, peaked and stepwise patterns are observed with certain organisms under appropriate conditions. Some of these reports, however, are without adequate controls to test for the effect of experimental perturbations. For the discontinuous patterns, obligatory coupling with cell cycle progress is suggested by a few experiments where the timing of the periodicity, with respect to time elapsed during parasynchronous cell cycles, is relatively constant in successive cycles. Different synchronizing procedures (selection on the basis of size or age induction) may reveal further aspects of coupling. In particular, the ability to prepare cultures that exhibit good "free-running" synchrony over many generations (Kepes and Kepes, 1981) will allow (1) measurements to be made on a well-synchronized system several generations after synchronization, and (2) correlations between the degree of synchrony and respiratory periodicities to be drawn. The idea that bacterial growth rates are limited by respiration is supported by certain physiological and cell cycle considerations and forms a useful working hypothesis for future studies.

The cell cycle of aerobically grown *E. coli* is the most intensively studied in terms of energy metabolism. To date, the amounts of all respiratory chain components assayed, as well as respiration rates, increase continuously through the cell cycle. No information is yet available on the relative contributions of synthesis and turnover to these patterns. The only clearly defined periodic event in respiratory metabolism is ATPase activity. Interrelationships between oscillations of this sort, the metabolic oscillations described in bacteria, and the possibility of circadian rhythms remain to be discovered.

> For tribal man space was the uncontrollable mystery. For technological man, it is time that occupies the same role.
>
> Marshall McLuhan "Magic that Changes Mood," *The Mechanical Bride* (1951)

## ACKNOWLEDGMENTS

I am grateful to the many colleagues cited in the text and references for their collaboration in the experimental work, and especially to Ravi Misri for permission to use some of his unpublished results on the *E. coli* cell cycle. This work was supported in part by grants from the Science and Engineering Research Council.

## REFERENCES

Andersen, K. B., and von Meyenburg, K. (1980). Are growth rates of *Escherichia coli* in batch cultures limited by respiration? *J. Bacteriol.* 144:114-123.

Blum, H., Poole, R. K., and Ohnishi, T. (1980). The orientation of iron-sulphur clusters in membrane multilayers prepared from aerobically-grown *Escherichia coli* K12 and a cytochrome-deficient mutant. *Biochem. J.* 190:385-393.

Blumenthal, L. K., and Zahler, S.A. (1962). Index for measurement of synchronization of cell populations. *Science* 135:724.

Burstein, C., Tiankova, L., and Kepes, A. (1979). Respiratory control in *Escherichia coli* K12. *Eur. J. Biochem.* 94:387-392.

Chance, B. (1975). Freezing the biological burning. In *The President's Lectures,* University of Pennsylvania, Philadelphia, pp. 1-20.

Creanor, J., and Mitchison, J. M. (1979). Reduction of perturbations in leucine incorporation in synchronous cultures of *Schizosaccharomyces pombe* made by elutriation. *J. Gen. Microbiol.* 112:385-388.

Degn, H., Lundsgaard, J. S., Petersen, L. C., and Ormicki, A. (1980). Polarographic measurement of steady state kinetics of oxygen uptake by biochemical samples. *Meth. Biochem. Anal.* 26:47-77.

Dietzel, I., Kolb, V., and Boos, W. (1978). Pole cap formation in *Escherichia coli* following induction of the maltose-binding protein. *Arch. Microbiol.* 118:207-218.

Dockter, M. E., Steinemann, A., and Schatz, G. (1978). Mapping of yeast cytochrome *c* oxidase by fluorescence resonance energy transfer. *J. Biol. Chem.* 253:311-317.

Downie, J. A., Gibson, F., and Cox, G. B. (1979). Membrane adenosine triphosphatases of prokaryotic cells. *Ann. Rev. Biochem.* 48:103-131.

Edwards, C., and Jones, C. W. (1977). Respiratory properties of synchronous cultures of *Alcaligenes eutrophus* H16 prepared by a continuous-flow size selection method. *J. Gen. Microbiol.* 99:383-388.

Edwards, C., and McCann, R. J. (1981). Differential effects of inhibitors on respiratory activity of synchronous cultures of *Bacillus subtilis* prepared by continuous-flow centrifugation. *J. Gen. Microbiol.* 125:47-53.

Edwards, C., Spode, J. A., and Jones, C. W. (1978). The properties of adenosine triphosphatase from exponential and synchronous cultures of *Alcaligenes eutrophus* H16. *Biochem. J.* 172:253-260.

Evans, J. B. (1975). Preparation of synchronous cultures of *Escherichia coli* by continuous flow size selection. *J. Gen. Microbiol.* 91:188-190.

Eytan, G. D., Carroll, R. C., Schatz, G., and Racker, E. (1975). Arrangements of the subunits in solubilized and membrane-bound cytochrome *c* oxidase from bovine heart. *J. Biol. Chem.* 250:8598-8603.

Feldman, K. (1978). New devices for flow dialysis and ultrafiltration for the study of protein-ligand interactions. *Anal. Biochem.* 88:225-235.

Haddock, B. A., and Schairer, H. U. (1973). Electron transport chains of *Escherichia coli.* Reconstitution of respiration in a 5-aminolaevulinic acid-requiring mutant of *Escherichia coli. Eur. J. Biochem.* 35:34-45.

Harrison, D. E. F. (1976). The regulation of respiration rate in growing bacteria. In *Advances in Microbial Physiology,* vol. 14, A. H. Rose and D. W. Tempest, (eds.). Academic Press, London, pp. 243-313.

Harrison, D. E. F., and Maitra, P. K. (1969). Control of respiration and metabolism in growing *Klebsiella aerogenes.* The role of adenine nucleotides. *Biochem. J.* 112: 647-656.

Howell, S. H., and Naliboff, J. A. (1973). Conditional mutants in *Chlamydomonas reinhardtii* blocked in the vegetative cell cycle. I. An analysis of cell cycle block points. *J. Cell Biol.* 57:760-772.

Kepes, F., and Kepes, A. (1981). Long-lasting synchrony of the division of enteric bacteria. *Biochem. Biophys. Res. Commun.* 99:761-767.

Koch, A. L. (1975). Lag in adaptation to lactose as a probe to the timing of permease incorporation into the cell membrane. *J. Bacteriol.* 124:435-444.

Koch, A. L. (1977). Does the initiation of chromosome replication regulate cell division? In *Advances in Microbial Physiology,* A. H. Rose and D. W. Tempest (eds.). Academic Press, London, pp. 49-98.

Kubitschek, H. E. (1968a). Linear cell growth in *Escherichia coli. Biophys. J.* 8:792-804.

Kubitschek, H. E. (1968b). Constancy of uptake during the cell cycle in *Escherichia coli. Biophys. J.* 9:1401-1412.

Kubitschek, H. E., Freedman, M. L., and Silver, S. (1971). Potassium uptake in synchronous and synchronized cultures of *Escherichia coli. Biophys. J.* 11:787-797.

Larrabee, K. L., Phillips, J. O., Williams, G. J., and Larrabee, A. R. (1980). The relative rates of protein synthesis and degradation in a growing culture of *Escherichia coli*. *J. Biol. Chem.* 255:4125-4130.
Lloyd, D., and Ball, J. (1979). Perturbation of respiration in *Candida utilis*: Induction of metabolic oscillations. *J. Gen. Microbiol.* 114:463-466.
Lloyd, D., Poole, R. K., and Edwards, S. W. (1982). *The Cell Division Cycle. Temporal Organization and Control of Cellular Growth and Reproduction.* Academic Press, London.
Lutkenhaus, J. F., Moore, B. A., Masters, M., and Donachie, W. D. (1979). Individual proteins are synthesized continuously throughout the *Escherichia coli* cell cycle. *J. Bacteriol.* 138:352-360.
Maruyama, Y. (1956). Biochemical aspects of the cell growth of *Escherichia coli* as studied by the method of synchronous culture. *J. Bacteriol.* 72:821-826.
Metzler, H. (1981). The influence of organic substrates and inhibitors on the induction of a respiratory oscillation in the cyanobacterium *Anacystis nidulans*. *Arch. Microbiol.* 129:123-126.
Mitchison, J. M. (1971). *The Biology of the Cell Cycle.* Cambridge University Press, London, pp. 159-180.
Mitchison, J. M. (1977). Enzyme synthesis during the cell cycle. In *Cell Differentiation in Microorganisms, Plants and Animals.* L. Nover and K. Mothes (eds.). VEB Gustav Fischer Verlag, Jena, Germany, pp. 377-401.
Mosteller, R. D., Goldstein, R. V., and Nishimoto, K. R. (1980). Metabolism of individual proteins in exponentially-growing *Escherichia coli*. *J. Biol. Chem.* 255:2524-2532.
Newman, C. N., and Kubitschek, H. E. (1978). Variation in periodic replication of the chromosome in *Escherichia coli* B/r TT. *J. Mol. Biol.* 121:461-471.
Nicholls, D. G. (1982). *Bioenergetics. An Introduction to the Chemiosmotic Theory.* Academic Press, London.
Ohki, M. (1972). Correlation between metabolism of phosphatidylglycerol and membrane synthesis in *Escherichia coli*. *J. Mol. Biol.* 58:249-264.
Poole, R. K. (1977a). The influence of growth substrate and capacity for oxidative phosphorylation on respiratory oscillations in synchronous cultures of *Escherichia coli*. *J. Gen. Microbiol.* 99:369-377.
Poole, R. K. (1977b). Development of respiratory activity during the cell cycle of *Schizosaccharomyces pombe* 972h$^-$: Respiratory oscillations and heat dissipation in cultures synchronized with 2′-deoxyadenosine. *J. Gen. Microbiol.* 103:19-27.
Poole, R. K. (1980). Temporal diversity of bacterial respiratory systems: Membrane and respiratory development during the cell cycle. In *Diversity of Bacterial Respiratory Systems,* vol. 1, C. J. Knowles (ed.). CRC Press, Boca Raton, Florida, pp. 87-114.
Poole, R. K., Blum, H., Scott, R. I., Collinge, A., and Ohnishi, T. (1980). The orientation of cytochromes in membrane multilayers prepared from aerobically-grown *Escherichia coli* K12. *J. Gen. Microbiol.* 119:145-154.
Poole, R. K., and Salmon, I. (1978). The pool sizes of adenine nucleotides in exponentially growing, stationary phase and 2′-deoxyadenosine-synchronized cultures of *Schizocaccharomyces pombe* 972h$^-$. *J. Gen. Microbiol.* 106:153-164.
Poole, R. K., Scott, R. I., and Blum, H. (1981). Respiratory biogenesis during the cell cycle of aerobically-grown *Escherichia coli* K12. The accumulation of iron-sulphur clusters and their orientation in the membrane. *J. Gen. Microbiol.* 124:181-185.
Ried, A., and Šetlik, I. (1971). Oscillatory interaction of photosynthesis and respiration in algae. In *Proceedings of the II International Congress on Photosynthesis Research,* G. Forti, M. Avron, and A. Melandri (eds.). Dr. Junk N.V. Publishers, The Hague, pp. 2077-2087.

Scott, R. I., Gibson, J. F., and Poole, R. K. (1980). Adenosine triphosphatase and its sensitivity to Ruthenium Red oscillate during the cell cycle of *Escherichia coli* K12. *J. Gen. Microbiol.* 120:183-198.

Scott, R. I., and Poole, R. K. (1982). A re-examination of the cytochromes of *Escherichia coli* using fourth-order finite difference analysis: Their characteristics under different growth conditions and accumulation during the cell cycle. *J. Gen. Microbiol.* 128:1685-1696.

Scott, R. I., Poole, R. K., and Chance, B. (1981). Respiratory biogenesis during the cell cycle of aerobically grown *Escherichia coli* K12. The accumulation and ligand binding of cytochrome *o*. *J. Gen. Microbiol.* 122:255-261.

Stouthamer, A. H., and Bettenhausen, C. W. (1981). Influence of 2,4-dinitrophenol on the maximum specific growth rate and the respiration rate of chemostat cultures of *Paracoccus denitrificans*. *FEMS Microbiol. Lett.* 10:33-35.

Voellmy, R. W., and Goldberg, A. L. (1981). ATP-stimulated endoprotease is associated with the cell membrane of *E. coli*. *Nature* 290:419-421.

# 12
# Application of the Continuum Model to the Clock Model of the Cell Division Cycle

**Stephen Cooper** / University of Michigan Medical School, Ann Arbor, Michigan

## ANALOGIES AND CLOCKS

Physical analogies of natural processes have always been valuable educational and heuristic tools. I remember, many years ago, watching a demonstration of the principle of a chain reaction illustrated by a cage filled with set mousetraps. On each mousetrap were two ping pong balls. When a single ping pong ball was thrown into the cage it set off one trap, releasing two additional balls and this continued until all of the traps went off. The flurry of cascading ping pong balls was a memorable sight. But as clever as this model or analogy was, it was not the same as the real event. It stopped quite soon, and there were no disastrous consequences. One must always make whatever distinctions are necessary between the analogy and the ideas they represent so that incorrect ideas are not implied by the analogy.

In biology, whenever events that occur at relatively set intervals are observed, a clock analogy is quite often used to explain, or merely to discuss the event. The occurrence of cell divisions at relatively fixed intervals is one of those phenomena that are conveniently discussed by a clock analogy. But we must always be careful to note what is implied by the use of a "clock" to model the cell cycle.

I would like to note that two different types of mechanisms for clocks can be envisioned. One type of clock, illustrated in Fig. 1, is an "event"-determined clock. A sequence of events occur, each triggering the next, and after the series of events have taken place the final event takes place. The main characteristic of an event-based clock is that the events which occur are different. One could tell where one was in the sequence of events by observing which event was taking place. I suggest that it is this type of model that is implied by the current notions regarding the cell cycle.

**FIGURE 1** Illustration of an "event"-based clock. This illustration is by Rube Goldberg, a famous American cartoonist who is well known for his inventive approach to life's problems. As he described this invention, entitled "Professor Butts' Automatic Garage Door Opener," Professor Butts takes a drink of strange gin and evolves an invention for opening the garage door without getting out of the car. Drive auto bumper (A) against mallet (B) pushing it down and exploding cap (C) frightening rabbit (D) who runs toward his burrow (E) pulling string (F), which discharges pistol (G). The bullet penetrates can (H) from which the water drips into an aquarium (I). As the tide rises in the aquarium it elevates the floating cork upright (J), which pushes up end of seesaw (K) causing flea (L) to lose its balance and fall on gedunk-hound's tail (M), who wakes up and chases his tail round and round causing platform (N) to spin and turn on faucet (O). Water runs through hose (P) starting revolving lawn sprinkler (Q) on which rope (R) winds itself opening garage door. Of course, if you wish [Rube Goldberg added], you can drive right through the door and then there won't be any obstruction left to bother you in the future.

A different type of clock is illustrated by the more conventional cuckoo clock. A cuckoo clock moves forward in time by the occurrence of the same event over and over again. It is a "continuous" clock. The spinning of the gears, the swinging of the pendulum, the clicking of the escapement occur continuously and one could not tell where one was in the sequence of time merely by looking at the clock mechanism. When enough time has accumulated since the previous special event the clock starts another special event, that is, the singing of a little bird to announce that a particular time has arrived. Notice that no matter how long the bird sings, the clock continues on its way without missing a beat. The clockwork mechanism goes continuously independently of the event it started. The frequency of the cuckooing is determined by the speed of the clock, and is independent of the time the bird will sing. I suggest that this type of "continuous" clock is a more accurate reflection of the way in which cell divisions are "timed."

I do not wish to criticize or support the use of clocks as a model of cell division regulation, but merely indicate where they are useful as a model, where they are misleading, and how different aspects of the cell cycle can be regarding as being properly, or improperly, modeled by clocks.

## THE CONTINUUM MODEL

The continuum model of the cell cycle is illustrated in Fig. 2. There are two parts to the model. Is is proposed that the synthesis of some "hypothetical initiator" is occurring continuously during cell growth and division. When the amount of initiator that has accumulated reaches a certain amount or value, it initiates the replication-segregation sequence, which is composed of the replication of the genetic material of the cell and the subsequent cell division that follows this replication. The first part of the cycle, as we shall see in detail below, is not an event-oriented clock, but is a continuous clock. The clock moves forward in time by a succession of increments of the same event. The second part of the cycle, the replication-segregation sequence, has elements of an event-oriented clock—since, for example, each nucleotide added to the genetic material is an indication

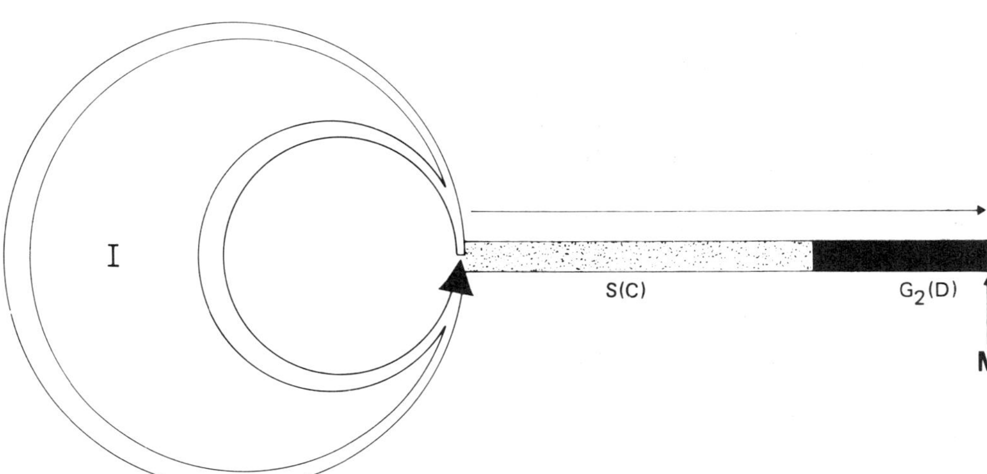


**FIGURE 2** The continuum model, illustrating the continuous synthesis of initiator (I), which occurs independently of DNA synthesis (S) or cell division (M) (Cooper, 1979).

of exactly where one is in the replication segregation sequence—but it should be noted that *the replication-segregation sequence does not control or alter the frequency of occurrence of cell divisions.* The frequency of initiation of DNA synthesis or the frequency of cell division is independent of the length of time required for the replication-segregation sequence. This is analogous to the fact that a cuckoo clock starts to sing precisely every hour whether the bird sings a short time at one o'clock or a long time at noon.

## DOES A "CLOCK" REGULATE CELL DIVISION?

### The Initiator Clock

The way in which the initiator clock regulates cell division is illustrated in Fig. 3. There is a continuous synthesis of "initiator" between starts of DNA replication (the S period in eukaryotes), and it is this rate of synthesis that determines the time between starts of DNA synthesis. If the rate of synthesis of initiator is low, there will be a relatively long time between starts of S periods. Since there is a constant period of time between the start of S and the subsequent cell division or mitosis (Cooper, 1979), if the time for replicatio and segregation is shorter than the time between starts of S periods, then the cell will exhibit a $G_1$ period. The slower the rate of synthesis of initiator, the less frequently initiation of DNA synthesis will occur, and the larger will be the $G_1$ period. As the rate of synthesis of initiator increases, the length of the $G_1$ decreases until one has no measurable $G_1$ period and the cell has only the S and $G_2$ periods. At this point the interinitiation time is equal to the replication-segregation period. As seen in Fig. 3, there is no sequence of different events (see Fig. 1) that determines the time between cell divisions. Rather, it is a continuous synthesis of a unitary substance, which is the "clock" that regulates the frequency of cell division. Thus, the clock analogy can hold, if one chooses the right kind of mechanism to run the clock.

### The Replication-Segregation Clock

The replication-segregation clock, in contrast, may be considered an event-based clock. Consider a simple view of DNA synthesis in which, upon initiation of DNA synthesis, nucleotides are added continuously and successively to a length of DNA. At each point in the synthesis one could see whether the first, second, hundredth, thousandth, or whatever nucleotide was being added. If the rate of addition of nucleotides to the chain was independent of the frequency of initiation of DNA synthesis, then one would have a situation that is a very close approximation of reality. As has been noted many times before (Cooper, 1979, 1981, 1982a,b), in both prokaryotes and eukaryotes there is a relatively constant period of time for DNA synthesis that does not vary with growth rate or interdivision time. Similarly, the time between the end of the S period and cell division (the classic $G_2$ period) is also relatively constant. This may be also visualized as an event-ordered clock in which a succession of events is triggered by the completion of DNA synthesis, and these events may be the events required for the final cell division.

Thus, the replication-segregation clock is an event-oriented clock, and may be so visualized, but as noted above, it is not the clock that determines how frequently the cell divides.

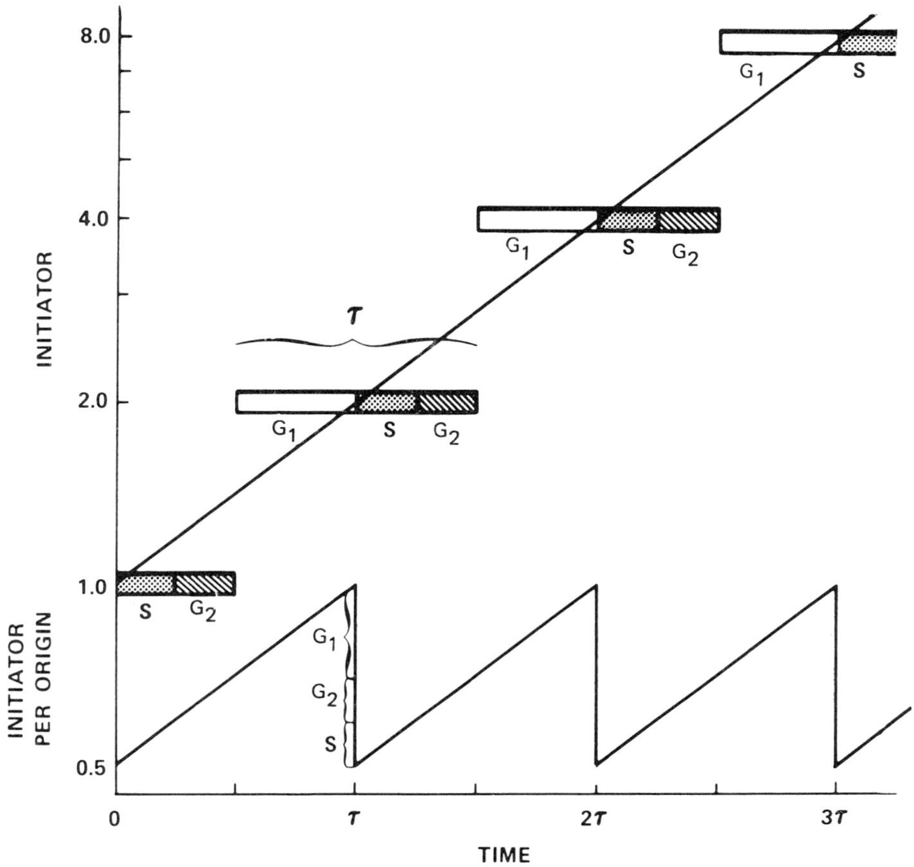


**FIGURE 3** A formal description of the continuum model demonstrating how the continuous synthesis of "initiator" can produce the observed cell cycles. The lower saw-tooth graph is the amount of initiator per origin, varying from 0.5 to 1.0. The upper bound is the critical amount required for the initiation of DNA synthesis. At the start the cells have just entered an S period and the cells are progressing through S and $G_2$. The amount of initiator per origin is increasing from 0.5 to 1.0 over one doubling time ($\tau$). When it reaches 1.0, another S period is initiated, but since at initiation there is a sudden doubling in the number of available origins of replication, the amount of initiator per origin decreases immediately to 0.5. If the rate of initiator synthesis was decreased and S and $G_2$ remained constant, then $G_1$ would increase. There is no $G_1$-specific synthesis of initiator but the synthesis is occurring continuously. As shown in the upper graph, this formal statement of the continuum model does not necessarily visualize any sudden appearance of disappearance of any particular molecules. (From Cooper, 1982a.)

### The Classic Clock

It is quite often the case that the ideas behind a certain model are not stated explicitly but are merely implicit in the discussion of the model. This is particularly true in biology, where the absence of mathematical rigor does not enable one to see all of the assumptions and implications of a model as clearly as in the physical sciences. I will therefore take the

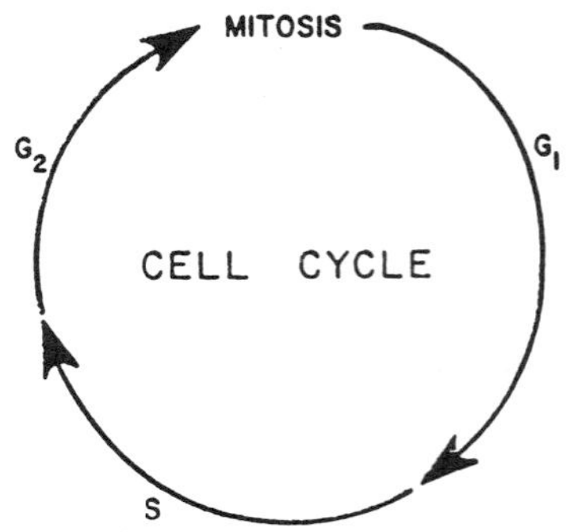


**FIGURE 4** The classic model of the cell cycle. (From Cooper, 1981.)

liberty of stating, as explicitly as possible, what I believe is implied by the current or classic (i.e., 1953-1979) model of the cell cycle. This is illustrated in Fig. 4. The classic model is an event-based clock in which the occurrence of a sequence of events is the basis for the regulation of the cell cycle. The faster a cell completes a sequence of events the more frequently will cell division occur. Most of the "events" that regulate the cell cycle have been proposed to reside in the $G_1$ period of the classic division cycle. The notion of "events" in $G_1$ derives from two types of experimental results. One is the observation that $G_1$ is the most variable period of the cell cycle. This observation can be made either within a single population, or for a given cell under different conditions. Within a cell population the variability of the $G_1$ period can be implied to mean that there is a great deal of variability in the time required to perform each of the events that occur in the $G_1$ period. The succession of $G_1$ events can be visualized as starting with the occurrence of mitosis. It is the mitosis that resets the stage for the occurrence of the $G_1$ events.

The second type of experimental evidence for $G_1$ events is the phenomenon of $G_1$ arrest. A wide number of experimental conditions that stop cell growth and division eventually lead to a population of cells that all have a $G_1$ content (2n rather than 4n) of DNA. For example, if an inhibitor of protein synthesis is added to a population of growing cells, after a period of time (usually over a day) all the cells are found with a $G_1$ content of DNA. It is usually suggested that they are arrested "in $G_1$," *at a point in $G_1$ where the cell requires additional protein synthesis for further progression through the cycle,* and the inhibitor of protein synthesis thus experimentally "defines" the $G_1$ event that requires protein synthesis. If a different inhibitor or condition (e.g., starvation, temperature in a temperature-sensitive mutant) is used to stop cell growth, then the different inhibitors may define different "events" in $G_1$. For example, if upon release of inhibitor A it is observed that DNA synthesis resumes in 5 hr, whereas upon release from inhibitor B DNA synthesis resumes in 10 hr, it is explained by stating that inhibitor B defines an event which occurs in $G_1$ and 10 hr prior to the start of the S period, and inhibitor A defines an event occurring 5 hr later and only 5 hr prior to the start of the S period. Irrespective of the timing of the "event," the finding that two different temperature-sensitive mutants are able to complement each other when fused has also been used to define two distinct events that occur during the $G_1$ period. See Cooper (1982a) for a complete and detailed analysis.

The classic clock is therefore an event-determined clock, with the events primarily in the $G_1$ period, and it is the rate of occurrence of these events which is the determinant of the frequency of cell division.

### The Clock Analogy Reexamined

We can now compare these two models of the cell cycle, the classic clock model and the continuum model using the language and images of clocks. The classic model is an event-determined model. A sequence of events, which it is believed can eventually be determined and identified, are occurring between cell divisions. Experimentally one examines this model by searching, for example, for proteins which are synthesized during one part of the cell cycle and not during another part of the cell cycle. The finding of such a protein would be a step toward understanding the sequence of events that occur between cell divisions.

The continuum model postulates that the regulation of cell division does not occur by an event-determined clock, but rather by a continuous clock mechanism as illustrated by the cuckoo clock analogy. There are elements of an event clock in the continuum model (the replication-segregation sequence is analogous to the singing of the bird), but these events do not determine the frequency of cell division.

## APPLICATION OF THE CONTINUUM MODEL

### Is the G(O) a Stopped Clock?

In eukaryotic tissues most of the cells are not continuously growing but are rather found in a resting state in which all of the cells are found to have a $G_1$ DNA content. These cells, which are therefore in a special kind of $G_1$ arrest, have been called G(O) cells and are thought to be "out of the cycle." Let us compare the classic clock model of the cell cycle and the continuum model with regard to how each explains the finding of G(O) cells. (Again, let me emphasize that I am taking the liberty of explicitly stating what is generally only implicitly contained in the literature. I leave it to the reader to decide whether my translation of the implicit to the explicit is accurate.)

The current model is illustrated in the top of Fig. 5. When cells that are growing normally are placed in a condition, either natural (as in tissues) or artificial (as when an inhibitor of growth is added to growing cells), the cells from each of the phases of the cell cycle are found with a $G_1$ content of DNA. If the cells are placed under these conditions of growth arrest for an extended period of time, it is considered that a separate type of $G_1$ is produced, the "$G_1$" or what is generally considered the G(O). Note that all of the cells are considered to be in the same state, and upon release from this state they are sometimes found, experimentally, to be synchronized, which is what would be expected if the cells were all in the same phase of the cell cycle.

The current model must be contrasted with the continuum model, which is illustrated in the bottom half of Fig. 5. As noted in Fig. 3, the amount of initiator per origin (formally taken as the operative amount of material that determines whether initiation will occur) varies continuously during the division cycle. Thus, at each point in the division cycle of growing cells, one may characterize the cells by the amount of initiator they contain per origin present. The continuum model proposes, quite simply, that all of the treatments, chemicals, or situations that lead to $G_1$-arrest [or G(O)] stop cell growth by stopping the synthesis of the hypothetical initiator. If one considers that the initiator synthesis requires general protein synthesis, then it is simple to consider the large number or inhibitors, or mutants, or starvation conditions, that lead to a general cessation of protein synthesis, and all of these lead to the production of cells with a $G_1$ DNA content. According to the continuum model the population of $G_1$ cells is produced because cells

| | ENTRANCE TO G1–ARREST ("G1") | EXIT FROM G1–ARREST ("G1") |
|---|---|---|
| CURRENT MODEL | G1 → "G1" | "G1" → G1 → S → G2 → M |
| | M → G1 → "G1" | "G1" → G1 → S → G2 → M |
| | G2 → M → G1 → "G1" | "G1" → G1 → S → G2 → M |
| | S → G2 → M → G1 → "G1" | "G1" → G1 → S → G2 → M |
| CONTINUUM MODEL | $G1_1 \equiv$ "$G1_1$" | "$G1_1$" $\equiv G1_1 \rightarrow S_S$ |
| | $M_M \rightarrow G1_M \equiv$ "$G1_M$" | "$G1_M$" $\equiv G1_M \rightarrow G1_1 \rightarrow S_S$ |
| | $G2_2 \rightarrow M_2 \rightarrow G1_2 \equiv$ "$G1_2$" | "$G1_2$" $\equiv G1_2 \rightarrow G1_M \rightarrow G1_1 \rightarrow S_S$ |
| | $S_S \rightarrow G2_S \rightarrow M_S \rightarrow G1_S \equiv$ "$G1_S$" | "$G1_S$" $\equiv G1_S \rightarrow G1_2 \rightarrow G1_M \rightarrow G1_1 \rightarrow S_S$ |

**FIGURE 5** Comparison of the current model for $G_1$ arrest with the continuum model. At the left the two models are compared for entry into a state of $G_1$ arrest (designated "$G_1$") and at the right the exit of the cells from the resting state is compared. Exit from the resting state could be due to a large number of causes from stimulation by addition of serum to shifting to the permissive temperature for a temperature-sensitive mutant. The subscripts in the continuum model denote the initiator per origin for those cells that enter arrest. When a cell enters a $G_1$-arrest state, the value for initiator per origin is fixed (since initiator synthesis stops, and initiation of DNA synthesis also stops) and the cells progress through the replication-segregation sequence as shown by the large letters. (From Cooper, 1982a.)

that have started DNA synthesis, or are in $G_2$ or M, are not prevented from completing DNA synthesis, the preparations for cell division, or the cell division itself. Thus, the cells are all found with a $G_1$ DNA content, but the cells are *not all in the same state.* They are in different states as determined by the amount of initiator per origin that cells have. This, in turn, is determined by the amount of initiator per origin found in the cells when they were diverted from growth to a resting state. Since the primary effect (I postulate) of the myriad conditions that lead to cessation of growth is the cessation of protein synthesis, and this in turn leads to a cessation of initiation of DNA synthesis [i.e., no new origins are produced), the ratio of initiator to origin remains the same. When cells are released from their resting state they are not synchronized; for a complete discussion of this point, see Cooper (1982a)] and are proposed to enter growth in sequence, depending on where they were when they were diverted to a resting state.

Whereas the classic clock model visualizes that a particular resting state is produced by causing the cells to come to rest at a particular point determined by an unique event that occurs during the division cycle, the continuum model proposes that resting cells are not actually arrested at any particular point in the division cycle. *Resting cells have a $G_1$ DNA content, but that does not mean they are "in $G_1$."*

To pursue the clock analogy further, the classic model would visualize that growing cells are a collection of clocks (let us imagine them the old-fashioned kind, with hands, rather than the digital ones that do not tick or tock as real clocks are supposed to do) which, if stopped by a particular treatment, all move to the same point in the mechanism, for example, 3:00. It is as though something was inserted in the mechanism which allowed the clocks to proceed until they arrived at 3:00, at which time they stopped. The continuum model, in contrast, holds that the cells, represented again by a collection of clocks, all may stop and appear to be resting at 3:00, but in reality they are stopped at different points in the mechanisms that move the hands, and the hands on the face do not accurately represent the clockwork mechanisms beneath the surface of the clock face.

I would like to parenthetically note an additional aspect of the classic clock model for the division cycle. As usually illustrated, the progression of events is placed on a circle with the events going around the circle (see Fig. 4). Events on the circle are sometimes discussed as occurring at particular times as though the events were lying on a clock face. The progression of events is thus viewed as though a clock hand was sweeping around the circle and determining the occurrence of different events during the division cycle. This problem is clearly one that is derived from our anthropomorphic view of clocks and the way in which the cycle is drawn. The continuum model, which is not circular (see Fig. 1), does not lend itself to being described by the simple clock mechanism.

### How Does Variability of Cell Cycles Arise?

To return to the other type of evidence for cell events determining the cell cycle, let us consider the variable $G_1$ as evidence for $G_1$ events. The variability of the $G_1$ period, and the associated variability of the interdivision times, have been interpreted to mean that the main determinant of the frequency of cell division is the rate with which cells pass through the different events occurring in $G_1$. The continuum model, as has been noted before (Cooper, 1979, 1982a), proposes that the variable of $G_1$ is a result or a symptom of the rate of growth varying rather than the varying rate of passage through $G_1$ events determining the rate of cell growth and division. The continuum model proposes that it is the time between initiations of DNA synthesis and the time for the replication-segregation sequence that determines the size or duration of the $G_1$ period, and not vice versa.

### Statistical Clocks

The transition-probability model of Smith and Martin (Cooper, 1982b) is a "random" clock mechanism for determining cell interdivision times. The rate at which a particular random transition occurs is the determinant of the frequency of cell division. The main evidence for this model is a detailed analysis of cell interdivision times using time lapse photography. When statistical variation is applied to the continuum model (Cooper, 1982b) the predictions of the continuum model are precisely those observed in support of the transition-probability model. Thus the continuum model can explain data that have been used to support a $G_1$-event-based model and a probabilistic model of the cell cycle.

## THE FUTURE OF THE CONTINUUM MODEL

The continuum model is a way of looking at the cell cycle from a different perspective. It is a different world view of the cell cycle, and when properly applied allows new, and generally simpler, interpretations of data. One example of such a reanalysis is the critique of the use of thymidine to measure mitogenesis (Guiguet and Cooper, 1982). The continuum model allowed an explanation of the dose-dependent kinetics observed for mitogenesis, explaining both the apparent threshold observed and the apparent inhibition of mitogenesis at higher mitogen concentrations. I believe that the continuum model can be applied to good effect in many other cases. I look forward to further analysis of cell phenomena using the continuum model, and in particular I look forward to experimental support of the continuum model such as has been presented by Okuda and Kimura (1982).

## REFERENCES

Cooper, S. (1979). A unifying model for the G1 period in prokaryotes and eukaryotes. *Nature* 280:17-19.

Cooper, S. (1981). The central dogma of cell biology. *Cell Biol. Int. Rep.* 5:539-551.

Cooper, S. (1982a). The continuum model. Application to G1-arrest and G(O). In *Cell Growth,* C. Nicolini (ed.). Plenum Press, New York, pp. 315-336.

Cooper, S. (1982b). The continuum model. Statistical implications. *J. Theor. Biol.* 94: 783-800.

Guiguet, M., and Cooper, S. (1982). A critique of the use of DNA synthesis as a measure of the effect of mitogens on lymphocytes. *Biosci. Rep.* 2:91-98.

Okuda, A., and Kimura, G. (1982). Effects of serum deprivation on the initiation of DNA synthesis in the second generation of rat 3Y1 cells. *J. Cell. Physiol.* 110:367-270.

# 13
# The Helix Clock
## Linking Spatial and Temporal Order

**Neil H. Mendelson** / University of Arizona, Tucson, Arizona

The purpose of this chapter is to present a specific model describing the way in which spatial order may be used as a source of "information" about time, and consequently used to establish temporal order of events. There are many examples of the relationship between spatial and temporal order in the biological world. The annual rings produced by the growth of certain trees (Stokes and Smiley, 1968), the annual lines of growth produced in the shell of bivalve mollusks (Ansell et al., 1978), the increasing number and size of chambers constructed during growth of *Nautilis* (Thompson, 1961), the growth of horn (Thompson, 1961), the developmental stages through which multicellular organisms pass during morphogenesis (O'Rahilly, 1967), the stages of cell division in eukaryotic cells (Hartwell et al., 1974), and even the particular cell morphology of certain bacterial cells as they progress through the cell cycle (Higgins and Shockman, 1976), all may be used to gauge time, at least in a relative sense. In the case of the helix clock a particular geometry serves as the basis for the assembly of structure that passes through an ordered series of intermediates to reach a natural limit. A system that can recognize these ordered intermediates can use the helix clock to order other events, such as the start of DNA replication, the start of cell division, or the start of cell growth. Although the helix clock pertains directly to prokaryotic cell growth and the relative ordering of events that go on during the cell cycle, the idea of using structure to provide information about temporal order may find broad application in other biological systems.

### GOALS AND SCOPE

The scope of this chapter will include, first, a brief discussion of evolutionary considerations focused primarily on the question, "why is temporal order needed in the cell cycle?", followed by a description of the helix clock model. Application of the helix clock model to the bacterial cell cycle in terms of a cell wall clock, and as a means to segregate daughter genomes, will be discussed prior to speculations about the reading of the clock, and the role of biomechanical factors in regulation. The experimental system upon which the helix clock theory is based, helical macrofibers of *Bacillus subtilis* (Mendelson, 1978, 1982b, 1982c; Mendelson and Karamata, 1982), will be considered only incidentally in the context of stress and strain deformations arising during growth as factors in regulation and timing.

## EVOLUTIONARY CONSIDERATIONS

The characteristic order of events that occur during a cell cycle from cell birth to division is remarkable in two respects: (1) in the complexity of processes that are coordinated during cell growth, and (2) in the precision with which successive generations reiterate the process. During steady-state growth of a bacterial population, for example, a newborn cell replicates its genome, segregates daughter genomes to positions destined to become isolated from one another by cell division, doubles its size, and produces by the precise location of the division septum two virtually identical-size daughter cells, each the equivalent of the initial cell from which it was derived. The phasing relationships between DNA replication, genome segregation, cell growth, and cell division are closely regulated, but the clock to which they are referenced is not yet understood. Both the characteristic order of the cell cycle and the need for a cell division process can be interpreted, however, in evolutionary terms, particularly in terms of selection and its efficiency.

### Why Temporal Order?

The evolutionary argument is based upon a few generally accepted basic assumptions concerning genes, organisms, and selection. First, one must agree that essential cell processes are regulated by genes. During the cell cycle these genes must replicate and pass to progeny cells. Accurate replication of DNA is required to assure that daughter genomes carry functional forms of the essential genes. Equally important at the cellular level is the accuracy of the genome segregation system. The products of DNA replication must eventually come to reside in independent entities for two reasons: (1) to assure genetic continuity from cell generation to generation, and (2) to assure efficient selection. These evolutionary objectives are achieved by the highly ordered sequence of events that characterize the cell cycle and by cell division. The essential genetic component of the cell cycle, therefore, is concerned with the replication of the genome and the distribution of the two products. Cell division per se compartmentalizes the daughter genomes, thereby facilitating selection, because the genomes reside in individual entities with phenotypic properties determined largely by the genes they harbor. If one measures success in evolutionary terms by the relative quantitative contributions of a given entity to the progeny population, then selection is most efficient when the genes governing the entity reside together and the entity is separated from all other entities and competing genes. These considerations imply that selective pressure must have led to the evolution of an ordered cell cycle, just as it did to the evolution of mechanisms that assure the accuracy of base sequences produced by DNA replication. The clock responsible for coordinating cell cycle events may be interpreted similarly: it is the organizing principal required for evolution of a system with at least two orders of complexity, replicating genetic material compartmentalized in growing cells.

## THE HELIX CLOCK MODEL

The helix clock theory was developed during the study of helical growth in *Bacillus subtilis.* Earlier work on the helical growth system led to the idea that even normal rod-shaped bacterial cells might grow by the assembly of their cell surfaces according to a helical plan (Mendelson, 1976). The rules of geometry governing helically organized cylindrical surfaces were explored, therefore, to establish the parameters that a helical growth system must obey. Use of the geometry to define a clock was recognized when it became apparent that the system always passes through an ordered series of intermediates to reach a natural limit structure. Using geometrical intermediates that arise during growth to provide information about time obviously need not be restricted to the particular case of helical geometry. The helix clock represents, therefore, one of a family of systems, any one of which may find application as a biological clock.

As described previously, the geometrical relationships of the helix clock are developed using an open-ended cylinder of constant diameter and examining the properties of wrapping string on the surface of the cylinder (Mendelson, 1982a). Three basic properties define the system: (1) the cylinder diameter, (2) the pitch angle assumed by the string(s) wrapped on the cylinder surface, and (3) the number of parallel strings that lie side by side on the surface. The diameter of each string is assumed to be constant. Any cylinder diameter may be used to illustrate the clock principle. The pitch angle assumed by the string(s) is measured relative to the cylinder axis as follows: the cylinder is drawn in projection as a rectangle, one side of which corresponds to cylinder length, the other to cylinder circumference. The rectangle is oriented such that the side corresponding to the cylinder circumference lies on a horizontal; that corresponding to the cylinder length lies vertical. In this projection a helically wrapped element on the surface of the cylinder will appear as a diagonal intersecting both sides of the rectangle. Pitch angle is defined as the angle formed by this diagonal line relative to the horizontal line that represents cylinder circumference. A pitch angle of 90° corresponds, therefore, to an element that lies parallel to the longest axis of the cylinder. The number of strings that lie parallel to one another refers to the fact that a cylinder may be covered by helically wrapping a single string on the surface, or by wrapping two or more strings that lie side by side on the cylinder surface. In either case, the entire surface of the cylinder may be covered by the string.

The three parameters defined above are related by the following equation:

$$\text{Cylinder diameter} = \frac{(\text{number of strings} \times \text{width of each string})}{\text{sine of pitch angle}} \quad \frac{1}{P_i} \tag{1}$$

From Eq. (1) it can be seen that if any one of the three parameters, cylinder diameter, pitch angle, or string number, is held constant, the other two must vary as a function of one another. There are therefore three relationships that may be defined: (1) the case where cylinder diameter is held constant and therefore pitch angle varies as a function of strand number, (2) the case where pitch angle is held constant and therefore cylinder diameter varies as a function of strand number, and (3) the case where strand number is

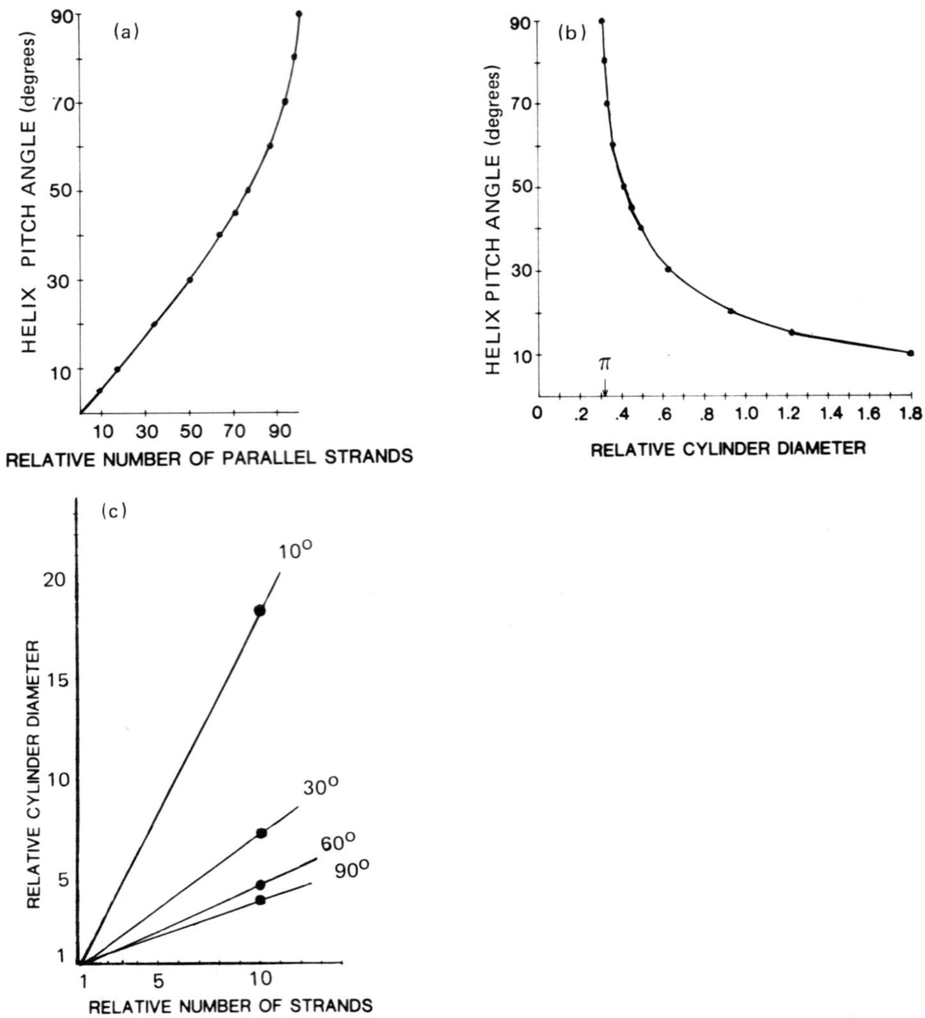


**FIGURE 1** The geometry of helical cylindrical surfaces. The model consists of a cylinder, the surface of which is covered by a wrapping of strings. (a) The helix pitch angle defined by the strings as a function of the number of parallel strings that lie side by side to cover the surface. Cylinder diameter is held constant. Zero degree pitch angle corresponds to an orientation perpendicular to the cylinder length. (b) The helix pitch angle defined by the string(s) as a function of the cylinder diameter. The number of strings is held constant. (c) The cylinder diameter as a function of the number of strings for several selected pitch angles. (From Mendelson, 1982a, 1982c.)

held constant, and cylinder diameter varies as a function of pitch angle. These three relationships are shown in Fig. 1. Relationship (1) is of primary concern, for it represents the geometrical basis of the helix clock.

### Using Form to Time Events

The relationship shown in Fig. 1a represents the situation where a cylinder of contant diameter increases in length by virtue of the addition of strings to the cylinder surface such that all strings lie parallel to one another on the cylinder surface. The orientation of the initial string is nearly perpendicular to cylinder length. The pitch angle of this initial string is defined only by the width of the string itself. When a second string of the same diameter and length as the first is inserted parallel to the first, the pitch angle of the two-string array changes as shown. Continued addition of new strings to the array causes a constant shift in pitch angle of the array until eventually all strings come to lie parallel to the long axis of the cylinder. At this point the pitch angle of the string is 90°, and the product of string number times string diameter equals the cylinder circumference. Once the limit structure of 90° has been reached, the addition of new strings into the array will cause an increase in cylinder circumference, and consequently in cylinder diameter, which by definition is being held constant. The relationship shown in Fig. 1a represents, therefore, a series of structural intermediates corresponding to a range of states with fixed boundaries: starting with the insertion of the first string, and ending with the insertion of the last string that brings the pitch angle of the entire array to 90°.

The relationship shown in Fig. 1a holds for strings of any diameter, so long as all the string have the same diameter, for strings of any length, so long as the strings are all longer than the cylinder circumference, and for strings of any composition. The rules of geometry governing parallel packing of strings in helical orientation on the surface of a cylinder thus appear to be a versatile means to establish temporal order. If one defines a helix clock cycle as running from 0 to 1, knowing the pitch angle of the entire string array at any given time is sufficient to determine which point in the cycle has been reached and therefore the relative age of the system. The conversion of this information to absolute time requires a knowledge of the rate at which strings are inserted into the array and whether that rate remains constant throughout the entire cycle.

So far as the topology of string packing is concerned the simplest situation involves the addition of strings to the array via a single insertion site, referred to in relation to bacterial growth as "a string-packing machine." A single machine inserting strings at constant rate of equal length such that the ends of all the strings in the array are aligned produces an ideal helix clock cycle. In a system of this sort, pitch angle information is sufficient to determine the relative length of the cylinder as well as the age of the system. Other, more complicated geometrical arrangements are also compatible with helix clock requirements. For example, a single reorienting string array may receive inputs from more than a single string-packing machine. The rate at which a system of this sort runs will be influenced by the number of such machines, whether they function simultaneously, sequentially, or both, and the rate at which each machine functions. The spatial orientations of multiple machines influences the manner of cylinder surfaces growth but does not distort the properties of the helix clock so long as the products of all machines pack parallel to one another in a single string array.

The geometry of the helix clock makes available 90° degrees of string reorientation within a single clock cycle. When the limit is reached the clock ceases to function, hence the system is essentially self-limiting and is suitable therefore only to measure a given amount of growth. Continuity of timing beyond the limits of a single clock system requires the establishment of new clock systems. A given length of cylinder may therefore possess a series of tandem helix clock systems, each independent in terms of its progress through the clock cycle. That such clocks may retain their independence yet actually be physically linked to one another will be discussed later in conjunction with a discussion of DNA segregation vis a vis the helix clock. The linking of multiple clocks to one another provides a means to maintain a hierarchy among them, thereby assuring the reading of the clocks in proper order.

The helical wrapping of string elements on the surface of a cylinder may assume either right- or left-handed orientation. Using the rectangular projection of a cylinder a right-handed orientation will appear as a diagonal line extending from the lower left to the upper right of the rectangle, whereas a left-handed orientation will appear as a diagonal line extending from the lower right to the upper left of the rectangle. The principle of the helix clock applies equally to either orientation. Once a clock is initiated, the first helically wrapped string must assume either one or the other orientation. All subsequent strings must, however, pack parallel to the first string in the same helix hand as the initial string. The properties of right and left hand reorienting helical string arrays are perfectly symmetrical, however, and therefore the ordered series of intermediates through which each passes are exactly equivalent in terms of a timing principal. The physical linkage of tandem clocks is readily achieved when all clocks are of the same hand. Tandem clocks of different helix hands can be linked although the geometry is more complex. In either case the independence of each clock system in terms of passage through the helix clock cycle is not influenced by which hand the clock assumes.

Although the helix clock is developed in terms of a 90° reorientation of a string array on the surface of a cylinder, the basic principle is applicable to any portion of the 90° spectrum, as well as to other geometrical forms. So long as string like elements of sufficient length assemble on the surface of a structure by parallel packing, and growth consists of an increase in the number of elements added to the array, then the shifting pattern of structural intermediates is adequate to serve as a timing device for a quantum event such as a single cell cycle. The progressive nature of the structural intermediates generated during a single helix clock cycle provides a basis for timing by reference to any portion of the 90° spectrum shown in Fig. 1a. Thus, depending upon how the clock is read, only a fraction of the entire geometrical range of intermediates available may be required to establish the temporal order of a cell cycle.

## APPLICATIONS OF THE HELIX CLOCK MODEL

The bacterial cell cycle is a dynamic process of structural change with time that results in the production of two equivalent daughter cells, both able to reiterate the process in precisely the same manner as did the parent cell from which they were derived. The phased relationship between the DNA replication cycle and the cell division cycle must be precisely regulated to achieve the observed continuity from cell generation to generation. To understand this relationship and the required temporal order of events that must be followed during the cell cycle one must consider first the basic properties of each cycle by itself.

The DNA replication cycle involves the following temporal order of events: (1) processes leading to the initiation of replication, (2) initiation of replication, (3) symmetrical bidirectional replication from origin to terminus of the template DNA, (4) termination of replication, and (5) separation of sister termini. Under steady-state growth conditions the rate of DNA replication is constant, hence a fixed amount of time is required for replication of the entire genome. When cells grow rapidly, a second round of replication may initiate at the chromosome origins of the sister molecules prior to the completion of the initial replication. Depending upon growth conditions, then, a replicating DNA molecule may consist of a series of waves of replication forks traveling from the origin toward the terminus down both left and right arms of the initial and daughter DNA templates.

Both the initiation of DNA replication and the replication process itself are known to be under genetic control. The replication process is believed to take place in association with a membrane-bound structure consisting of a number of gene products. Other regions of the DNA molecule have also been found associated with the cell surface. In particular the origin of replication, and also the DNA terminus, have been shown to copurify with cell surface components suggesting that they too are linked to the cell surface in vivo. A number of models of the bacterial cell cycle have ascribed particular significance to the attachment of the origin and terminus to the cell surface. Origin attachment is thought to provide a means to separate the products of replication prior to cell division, whereas the terminus attachment has been implicated in the regulation of cell division. These ideas will be incorporated into a model of the bacterial cell cycle based upon the helix clock theory.

The primary control of DNA replication within the cell cycle appears to be by control of the initiation of DNA replication. The phased relationship, therefore, between replication and division is thought to be regulated by control of the time during growth that DNA replication begins. Initially it was believed that all other cell cycle phenomena, including cell division, could be timed relative to this event. The dissociation of DNA replication and cell division found in certain mutants and in drug inhibition studies makes it unlikely, however, that DNA replication is the cell cycle timer. Rather, it appears more likely that another central clock exists that governs both the DNA replication cycle and the cell division cycle. The helix clock appears to meet the requirements of such a clock.

The cell division cycle in bacteria consists of cell growth, and the compartmentalization of the elongated structure by construction of a division septum. Maturation and cleavage of the septum results in the production of sister cell poles and the separation of sister cells from one another. Generally, growth is restricted to a single dimension. Thus, a rod-shaped cell grows solely by elongation, divides by the construction of a septum at a specific location, and produces two daughter cells of similar dimensions and shape. Many aspects of growth and division are known to be under genetic control. The division process per se can be considered to consist of seven different phenotypic processes of which only two are pertinent to this discussion: (1) regulation of cell division initiation, governed by genes called *div* I genes, and (2) regulation of the location of the cell division site, governed by genes called *div* IV genes. These findings suggest that specific gene products are needed for the initiation of cell division and for the positioning of the septum along the length of the cell. Little is known, however, about the mechanism by which these products act or about the relationship of regulation of the process at the gene or physiological level.

In rapidly growing bacterial cells events that require a relatively long time to be completed, such as the replication of the entire genome or the construction of a division septum, may become rate-limiting processes in growth. Multifork DNA replication, mentioned earlier, provides a means to bypass the need to wait for an entire genome to be replicated in a single cycle by providing newborn cells with partially replicated chromosomes which only need to be completed during a single cell cycle. The same strategy apparently pertains to septum formation. If cell division is already in progress in a newborn cell, then only the completion of division is required during a single cell cycle. Fast-growing cells must therefore contain multiple fork replicating DNA structures, and multiple divisions in progress, and it is essential that each replication fork must be somehow linked to its proper division septum in order to maintain viability. The segregation of sister DNA molecules produced when a given replication fork reaches the terminus must be coordinated with the proper cell division as well.

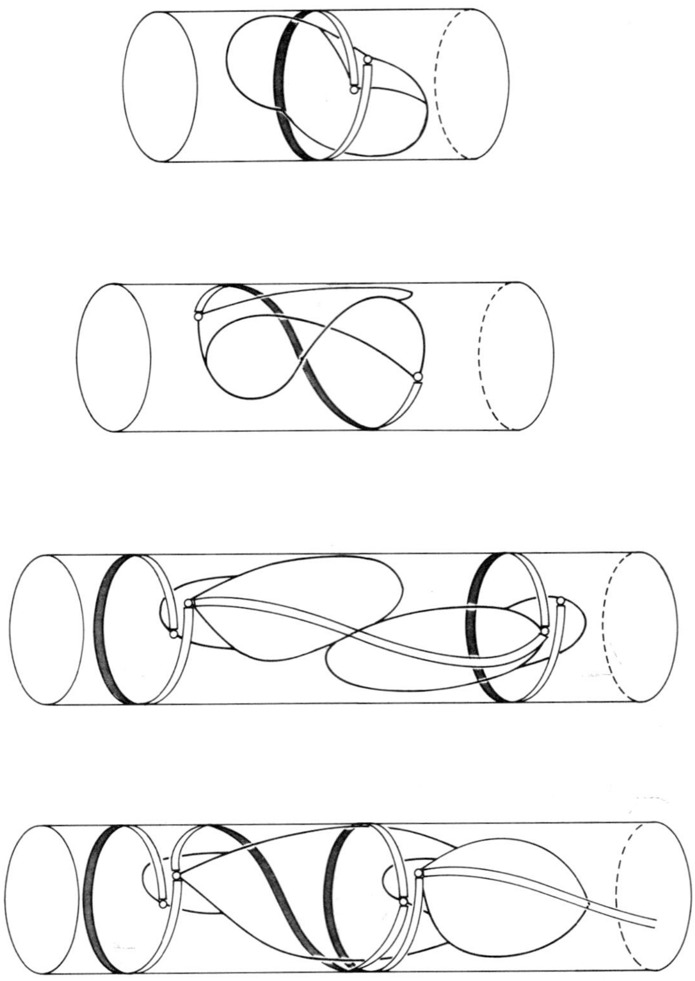

**FIGURE 2** Helical DNA segregation model. The open cylinders represent a portion of the length of a rod-shaped cell. Two systems are shown within the cylinder: cell surface DNA-segregating string(s) and replicating DNA. In the upper drawing a single cell surface DNA-segregating string is shown as a ribbon helically wrapped around the cylinder. A circular DNA molecule in the process of bidirectional replication is shown attached to the DNA-segregating string. The replication origins of both daughter DNA molecules are attached to the ends of the DNA-segregating string via connections (o). The chromosome terminus is attached to a point midway along the length of the DNA-segregating string. This drawing represents an early stage of DNA replication and segregation just after the initiation of both processes. In the second drawing DNA replication has progressed, and the position of the DNA-segregating cell surface string has become reoriented so that the two ends of the string and the daughter DNA origins are spatially separated. The position of the DNA terminus attached midway along the length of the DNA-segregating string remains fixed at the position it occupied in the first drawing. The reorientation of the cell surface DNA-segregating string is caused by the parallel packing of other cell surface strings into a single array as defined by the helix clock model. These additional strings are not shown. The third drawing depicts the situation when second rounds of replication have begun in both daughter molecules prior to the completion of the initial round. New cell surface DNA-segregating strings are associated with each of the second rounds, and the appropriate chromosome origins are attached to the ends of these strings. The chromosome terminus remains attached to the initial DNA-segregating string, which is still at its original position. Each new cell surface DNA-segregating string is shown attached to

### Bacteria—A Wall Clock

The helix clock theory has been used to construct a model of bacterial growth that appears to satisfy the requirements discussed above. In this model the initiation of DNA replication and the initiation of a helix clock cycle are linked temporally and structurally. One assumes that at some point associated with the cell surface a "machine" exists that serves two processes: (1) the insertion of string elements into the cell surface and (2) the replication of DNA. At the start of a helix clock cycle a single cell surface string is forced into the cell surface and assumes the configuration of a helically wrapped element oriented nearly perpendicular to the long axis of the cell cylinder. A DNA molecule becomes attached to this string such that the origin of replication is associated with one end of the string. The initiation of DNA replication results in the attachment of sister origins to both ends of the cell surface string. If the DNA terminus becomes attached to the same cell surface string to which the DNA origins are fixed, at a place along the cell surface string midway between the points of attachment of the origins, one then has in place appropriate markers for the subsequent positions of the daughter genomes and the cell division that will compartmentalize them in sister cells. The top panel in Fig. 2 illustrates a DNA molecule undergoing symmetrical bidirectional replication in association with the initial cell surface string of a helix clock cycle.

### Genome Segregation

Assuming the "machine" packs strings into the cell surface parallel to the initial string while DNA replication progresses, a configuration as shown in the second panel of Fig. 2 will be reached. Only the initial cell surface string is shown. The change in orientation of this string is due to reorientation as per the rules of the helix clock. As a result of this reorientation the ends of the string, which originally were close to one another, now be-

---

the ends of the original segregating string, which has continued to reorient in accord with the helix clock rules. The new DNA-segregating strings assume the same orientation as that shown in the first drawing. This orientation corresponds to the start of the helix clock. The bottom drawing illustrates the configuration found when a third round of replication begins at just about the time that the initial round is being completed. Only the left half of the entire system represented in the first three drawings is included. New cell surface DNA-segregating strings are once again associated with each new pair of replicating forks. The daughter origins always attach to the ends of these new segregating strings, and the new segregating strings themselves attach to neighboring, previously existing segregating strings is shown. The second segregating string from the right of the drawing volve the DNA origin attachment sites. Whether this corresponds to a portion of the segregating string itself or to some other component attached to it remains undefined. The chromosome terminus is shown still attached to the initial DNA-segregating string at its initial location along the cylinder length. When the terminus is released from this site, cell division will follow at this location. The free terminus will move to its new attachment place in the center of the length of the next-oldest DNA-segregating string, which in the bottom drawing has the same configuration as in the second drawing. In the bottom drawing the insertion of new cell surface DNA-segregating string between two existing segregating strings is shown. The second segregating string from the right of the drawing illustrates that new DNA-segregating strings become attached to neighboring DNA-segregating strings that are one generation different in age. Due to space limitations, the contour length of DNA shown in all of the drawings are reduced approximately 600-fold compared with the length of the DNA-segregating strings. (From Mendelson, 1982c.)

come redistributed, drawing with them the sister DNA origins. Thus the segregation of daughter DNA, according to the helical model, is due to string reorientation, a mechanism related to though not exactly the same as the idea proposed years ago, that cell surface growth between the points of attachment of the sister origins is responsible for DNA segregation (Jacob et al., 1963). Due to the symmetry of the reorientation process the point of DNA terminus attachment remains fixed at a place along the length of the cell cylinder throughout the helix clock cycle. This site therefore can serve to mark the place where cell division will occur.

The third panel in Fig. 2 introduces another order of complexity. A second round of DNA replication is shown to have initiated at the sister DNA origins prior to the completion of the first round of replication. For each new pair of replication forks associated with the sister origins, a new cell surface string is shown. Each pair of forks, therefore, has its own helix clock system which will progress through the clock cycle independently, yet the clock systems are linked to one another as shown. The initial cell surface string of the first clock cycle is shown having progressed nearly to completion of the helix clock string reorientation process. The original DNA terminus is still attached to the center of this string, as the replication forks of the first wave of DNA replication have not yet reached the terminus. Three tandem clock systems are in operation simultaneously in a cell of this configuration. The first cell division to occur, however, is governed by the oldest clock, and the site of that division is marked by the terminus attachment site on the initial DNA-segregating cell surface string.

The bottommost panel in Fig. 2 depicts yet a third wave of multiple forks and their cell surface helix clock string systems shown for only the left half of the system illustrated in the preceding three panels. Three and a half clock systems of the total seven tandem clocks are shown. The general rules followed are: (1) new clocks insert between older ones that differ in age by one clock generation, (2) each new clock is association with a pair of replication forks, (3) though structurally linked each clock progresses through its cycle independent of neighboring clocks, (4) the origins always attach at the ends of the first cell surface string inserted at the start of a helix clock cycle. This position corresponds to the juncture between old and new clock systems. (5) The DNA terminus always becomes attached at a place midway between the points of origin attachment on the same cell surface string. (6) The cell division site is determined by the place the DNA terminus is attached, and division starts only after the terminus dissociates from this position upon completion of DNA replication. The model does not define the mechanism by which the mass of replicated DNA moves to the place where the origins lie following completion of DNA replication.

Other specific details concerning the helical segregation model and its relationship to bacterial wall polymers will be considered elsewhere. In the context of this chapter one needs only to focus on the fact that the properties of the cell surface and its mode of assembly during growth are key elements in the model. Though little attention is given in this chapter to the properties of the machine that packs strings into the cell surface and carries out DNA replication, the assembly of particular gene products into this machine is obviously required for the start of any helix clock cycle. Gene products that regulate cell division must also play a role in the attachment of the DNA terminus to the proposed cell surface DNA-segregating string, as well as in events that go on at that place upon completion of DNA replication, release of the daughter termini, and the initiation of cell septation. The coordination of cell surface growth and DNA replication events is visualized as resulting from the transfer of information, perhaps by allosteric interactions between the components of the hypothetical machine.

At the geometrical limit of a helix clock cycle or the relative time within the cycle corresponding to the initiation of cell division, sister DNA termini detach from the cell surface string and septum construction begins. The production of a septum involves the insertion of cell surface elements in new orientation such that the building structure no longer lies in the plane of the cylinder but rather is directed perpendicular to it. So far as building materials are concerned those used for septum and lateral wall are probably identical. It would seem logical, therefore, to assign a role to the same string-packing machine responsible for cylinder elongation in septum formation. If so, this machine must reside at the place where termini attach to the cell surface string. The fact that septa lie perpendicular to the long axis of the cell cylinder suggests that string packing to form septa must resemble the orientation found at the start of a helix clock cycle. For reasons described elsewhere (Mendelson, 1982a) it is assumed that septal assembly differs from the initial stages of a cylinder elongation clock cycle in the helix hand used for assembly. Thus, if a cylinder surface is assembled as a right-hand helix, the septum at division is visualized as being of left-hand orientation. The signal to begin division and change helix hand is traceable to the machine responsible for string packing, and interpretations of the way in which this machine is made to stop assembly in the old direction and begin in the new one bears upon the question of how cells read the helix clock.

## READING THE CLOCK

The information provided by an ordered series of structural intermediates must be interpreted by cells in some manner to regulate the start of events, such as the initiation of DNA replication, the initiation of cell division, or even the beginning of growth in daughter cells. In the context of the helix clock the structural intermediates consist of a changing series of pitch angles that arise during reorientation of a surface string array. The question then is, how do cells sense this information and use it to control cell cycle events? From a theoretical point of view there are two possible answers: either cells respond directly to a particular structural intermediate uniquely related to each cell cycle process, or cells respond to some other factors, most likely physical states of the cell surface that are ramifications of and consequently absolutely coupled to each structural intermediate.

### Biomechanics

The study of bacterial helical macrofibers has shown that stress and its corresponding strain are important factors in regulation of cell shape and growth (Mendelson, 1982c). The deformation of a cylinder into a helical cell is thought to be caused by the introduction of stress into the cell surface during growth. The dynamics of macrofiber growth involves turning and writhing motions that lead to folding, and it is evident that growing structures are under great tension. Repeated growth and folding cycles result in the building of large multistranded helical structures. Folding appears to relieve some of the tension that develops during growth, but the folded structures are not totally relaxed. The digestion of macrofiber wall polymers by enzymes that reduce the chain length of the polymers results in specific relaxation motions indicating that helical morphology is at least maintained by a stressed condition of the wall polymers. The final state of relaxation prior to total dissolution of the wall still appears to retain some helical structure, however.

The principles of stress and strain may be applied to the question of how cells interpret a series of structural intermediates as follows. One must assume that the cell surface of normal cells consists of a strained structure. As the cell surface grows in terms of the helix clock the insertion of new strings and concomitant reorientation of the entire string array introduces additional stress into the cell surface. The initiation of cell cycle events is governed by the degrees of stress in the surface. In biochemical terms the start of DNA replication, for example, may require a given amount of stress to which the strain response consists of the assembly of a replication complex of proper configuration to combine with the DNA origin of replication, and the commencement of DNA replication. Similarly, the start of cell division might require yet another degree of stress to which the required gene products respond, leading to the redirection of cell wall assembly and the formation of a division septum. As a general principal, therefore, any system that passes through an ordered series of structural intermediates may generate a corresponding series of stress states that can serve to govern cell cycle processes. Reading the clock in biochemical terms can translate, therefore, into specific strain responses associated with the start of various processes.

### Speculations About Eukaryotic Cells

The structural differences between eukaryotic and prokaryotic cells does not preclude the possibility that similar systems operate in more complex cells. If one assumes that the physical state of the eukaryotic cell surface carries information useful for establishing temporal order, then there must be a means by which that information is transferred into the cell nucleus where processes such as gene replication occur. Usually, information transfer between the eukaryotic cell surface and the cell interior is thought of in terms of chemical messengers, but in the context of stress and strain vis-a-vis reading the helix clock, the idea that information may be transferred through the cell cytoskeleton in the form of strain and conformational alterations should be considered. A system of this sort would enable cells to respond not only to growth-related cell surface changes but possibly also to environmental factors such as cell-cell interactions, positional information of cells within tissues, or even the presence of various molecules attached to or embedded in the cell surface. All these ideas are predicated upon the assumption that physical forces can play a role in configuring cellular structures, which in turn govern biochemical processes.

## SUMMARY

The idea that cell cycle events may be timed using information produced by the assembly of an ordered series of structural intermediates has been developed in the context of a helix clock. The geometry governing the helix clock has been applied to the bacterial cell cycle and found particularly suited to explaining the mechanism of genome segregation and the positioning of cell division along the length of the cell. A consideration of stress and strain in relationship to structural intermediates has provided a basis for interpreting the means by which cells might read the clock. Speculations concerning the use of solid-state information by eukaryotic cells have focused on the transfer of information from the cell surface to the nucleus via the cell cytoskeleton. Conformational states of macromolecules and macromolecular assemblies which regulate biochemical processes are presumed to be influenced by physical forces resulting from the building of other structures. In the context of the helix clock spatial and temporal order are linked because the building of structures upon which the timing principle is based involves always passing through an ordered series of intermediates. The system consists therefore of (1) building of structure by assembly of elements that pass through an ordered series, (2) development of stress states corresponding to the structural intermediates, (3) influence of the stress state upon other cell components, (4) strain reactions affecting the conformation of target components, and (5) initiation of biochemical reactions driven by the altered conformation of the target components.

## ACKNOWLEDGMENTS

I wish to acknowledge discussions with Prof. F. R. Salemme and J. R. Mendelson concerning the helix clock and helical models of DNA segregation. Support by research grants from the National Institute of General Medical Sciences, and the Cell Biology Program of the National Science Foundation is gratefully appreciated.

## REFERENCES

Ansell, A. D., Parulekar, A. H., and Allen, J. A. (1978). On the growth of *Nuculana minuta* (Muller) (Bivalvia: Nuculamidae). *J. Moll. Stud.* 44:71-82.

Hartwell, L. H., Culotti, J., Pringle, J. R., and Reid, B. J. (1974). Genetic control of the cell division cycle in yeast. *Science* 183:46-51.

Higgins, M. L., and Shockman, G. D. (1976). Study of a cycle of cell wall assembly in *Streptococcus faecalis* by three-dimensional reconstruction of thin sections. *J. Bacteriol.* 127:1346-1358.

Jacob, F., Brenner, S., and Cuzin, F. (1963). On the regulation of DNA replication in bacteria. *Cold Spring Harbor Symp. Quant. Biol.* 33:329-347.

Mendelson, N. H. (1976). Helical growth of *Bacillus subtilis*: A new model of cell growth. *Proc. Natl. Acad. Sci. USA* 73:1740-1744.

Mendelson, N. H. (1978). Helical *Bacillus subtilis* macrofibers: Morphogenesis of a bacterial multicellular macroorganism. *Proc. Natl. Acad. Sci. USA* 75:2478-2482.

Mendelson, N. H. (1982a). The helix clock: A potential biomechanical cell cycle timer. *J. Theor. Biol.* 94:209-222.

Mendelson, N. H. (1982b). Dynamics of *Bacillus subtilis* helical macrofiber morphogenesis: Writhing, folding, close-packing and contraction. *J. Bacteriol.* 151:438-449.

Mendelson, N. H. (1982c). Bacterial growth and division: Genes, structures, forces and clocks. *Microbiol. Rev.* 46:341-375.

Mendelson, N. H., and Karamata, D. (1982). Inversion of helix orientation in *Bacillus subtilis* macrofibers. *J. Bacteriol.* 151:450-454.

O'Rahilly, R. (1967). Normal development of the human embryo. In *Normal and Abnormal Embryological Development,* C. H. Frantz (ed.). National Research Council, Washington, D.C., pp. 3-15.

Stokes, M. A., and Smiley, T. L. (1968). *An Introduction to Tree-ring Dating,* University of Chicago Press, Chicago, p. 75.

Thompson, D'A. (1961). *On Growth and Form,* J. T. Bonner (ed.). Cambridge University Press, London, p. 346.

14

# Temporal Control of the *Schizosaccharomyces pombe* Cell Cycle

**Peter A. Fantes** / University of Edinburgh, Edinburgh, Scotland

## INTRODUCTION

### General

The concept of the cell division cycle in eukaryotes as a cyclic process gained popularity with the discovery (Howard and Pelc, 1953) that the major event of DNA replication occurs within a discrete part of the cycle. The reasonable constancy of the position and duration of the S phase showed the existence of temporal organization within the interdivision period.

The use of a clock face (e.g., Fig. 1) to represent the progress of cells through the $G_1$, S, $G_2$, and mitotic phases was very likely responsible in part for introduction of the idea of an internal "clock." Interest in the hypothetical clock developed and was studied by the use of agents that (in general) delayed division, often with the implicit assumption that the clock motor rather than merely the hands was affected.

Studies on several systems, notably *Physarum,* concentrated on control of the whole intermitotic period. Some workers (Kauffman and Wille, 1975) favored the idea of a purely time-related control for mitosis, while others considered the process in terms of the assembly of a mitotic structure (Sachsenmaier et al., 1972) or accumulation of a critical level of mitotic activator (Rusch et al., 1966). In a limit-cycle model for *Physarum* (Kauffman and Wille, 1975), a particular class of oscillator was proposed to control mitosis and explain the clocklike behavior of the organism's division cycle. A different type of "clock" has been proposed for the control of cycle time in mammalian cells, which includes a subcycle of fixed length. Cycle time is quantized, and so only certain values are possible (Klevecz, 1976).

The start of the cycle began to be studied extensively, initially in mammalian cells and later in other organisms, when it was realized that cells usually arrest in the $G_1$ phase when they cease proliferation. Detailed analysis of resting mammalian cells led to the formulation of $G_0$ as a distinct state (Baserga, 1968). A different approach was used to define a unique stage in $G_1$, the restriction point, at which cells arrested after a variety of treatments that stopped proliferation (Pardee, 1974).

The kinetic analysis of slowly proliferating cell populations led Burns and Tannock (1970) to suggest that a random probabilistic event might be rate limiting for the cycle. Extension of this theory to more rapidly growing cells culminated in the transition probability theory. This theory, in both its original and revised forms (Smith and Martin, 1973; Brooks et al., 1980), includes a timer element as a major part of the cycle control. It is part of current thought, irrespective of models, that the period between the $G_1$-S boundary and mitosis is nearly constant both for individual cells within a population, and for different populations of cells, though direct evidence for these beliefs, especially the first, is scarce (Fantes and Nurse, 1981).

The fission yeast *Schizosaccharomyces pombe* has been the object of many studies into the control of cell division over the years, and this interest has expanded recently. The sum of the evidence, reviewed by Nurse and Fantes (1981) and Fantes (1981a), suggests strongly that the normal cell cycle in *S. pombe* is not controlled by a straightforward clock or timing mechanism of any kind. Nor does it seem likely that any significant part of the cycle is subject to timer control, with one possible exception, which will be described in due course. Much of the experimental evidence on which this chapter is based is presented elsewhere (Nurse and Fantes, 1981), and I shall only mention those parts that are relevant to the temporal control of the cell cycle.

The idea that a cell's behavior during the cycle may depend on its size is not new, with reports as least as early as 1924 (Wilson, 1924). Since cell size appears to be of particular importance to the cycle of *S. pombe,* I should perhaps explain what I mean by this vague term. It is deliberately chosen as imprecise for two reasons: first, measurements of cell volume, dry mass, protein content, and so on, have all been made in different studies. The diversity of observations suggests use of a general term to refer to them all, since in cases where they have been compared, they seem to correlate well. Second, we do not yet have any idea of which parameter the cell considers important.

### *S. Pombe* Cell Cycle

Before discussing the control mechanisms of the *S. pombe* cell cycle, it is necessary to outline the major events of the cycle in a wild-type strain under a set of "standard" conditions. Observations from a variety of sources on cells growing at near-maximum rate provide the picture of the cycle shown in Fig. 1. Cells are born with a length of about 8 $\mu$m, and increase in length only for about three-quarters of the cycle (Mitchison, 1957; James et al., 1975; Miyata et al., 1978). At about this time, the nucleus undergoes division (Mitchison, 1957) without breakdown of the nuclear membrane (McCully and Robinow, 1971). The daughter nuclei move to opposite ends of the cell, then return to take up positions about one-quarter cell length from the ends. Shortly after this a septum is laid down centrally across the cell, causing physiological separation of the daughter cells. The septum is cleaved, the newly formed ends of the daughters round off, and a new cycle starts. The length of the cell at the time of mitosis is about 14 $\mu$m, and there is little increase during the last quarter of the cycle (Mitchison, 1957; Fantes, 1977).

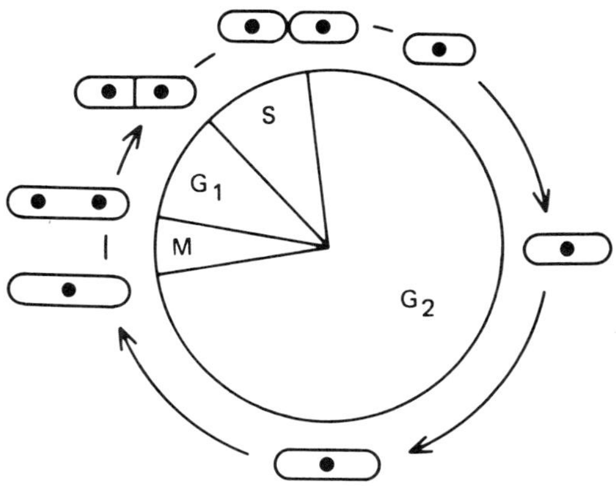


**FIGURE 1** The division cycle of *S. pombe*. The nuclear cycle is shown as a clock face, the size of each sector being roughly proportional to the length of the corresponding phase. The outer cycle shows the approximate appearance of cells at corresponding times. Note that cell separation occurs about 0.2 cycle after mitosis, so that the nuclear and cell cycles are out of phase.

Rounding out of the new ends accounts for the difference in total length of the two newborn daughters and the division length of the mother cell. The nuclear DNA is normally replicated very early in the cell cycle, the short S phase occurring simultaneously with cell separation (Nasmyth et al., 1979). The $G_2$ phase follows, and occupies the remainder of the cycle.

The *S. pombe* cell cycle is therefore quite similar to that of higher cells in the existence of $G_1$, S, $G_2$, and M phases, and in the symmetrical mode of division. The main differences are that the nuclear and cell cycles are about a quarter of a cycle out of phase, so that $G_1$ occurs at the end of the cell cycle but of course at the start of the nuclear cycle. *S. pombe* can therefore be used as a simple model eukaryote, and investigation of its cell cycle controls may have relevance to those of higher cells.

### Events of the Cycle

At this point I wish to make a distinction between the types of events that are associated with the cell cycle. Some events are essential for completion of the cycle. Examples of this are the synthesis of a round of nuclear DNA by the various polymerases, ligases, and so on; the assembly and subsequent disassembly of the mitotic spindle; and the segregation of daughter chromosomes into new nuclei. Other essential events have been defined genetically, in the sense that mutants unable to carry out certain steps in the developmental pathway have been isolated as conditional lethals (Simchen, 1978). In such cases, the nature of the defect is often not yet understood at the biochemical or ultrastructural level, although the major cycle event affected may be known. Many mutants unable to undergo mitosis fall into this class, and have been characterized in terms of physiology and gross nuclear morphology. By far the most common type of conditional cell cycle mutants are temperature sensitive, or more accurately heat sensitive. The temperature-sensitive phenotype is thought to result from temperature sensitivity of the mutant gene product, in general. It is, however, possible in some cases that the synthesis of the gene product is temperature sensitive. Conditional cell cycle mutants have been used in a variety of ways to investigate the cell cycle: an extensive review (Pringle, 1978) describes these in detail.

In *S. pombe,* conditional mutants in 28 genes specifically required for the cell cycle have been described. Mutants in all genes but one have been isolated as heat-sensitive lethals: one gene, *ben* 4, is defined by cold-sensitive mutants only (Roy and Fantes, 1982). In each case the observable event affected by the genetic lesion is in DNA replication, mitosis, or septum formation. Further details of the various mutants can be found elsewhere (Nurse et al., 1976; Fantes, 1979; Nasmyth and Nurse, 1981; Roy and Fantes, 1982): relevant properties of the genes mentioned in this chapter will be described at the appropriate place.

A second class of events is associated with the cycle. These are events that are coupled to the cell cycle under normal conditions but that are not, or do not appear to be, required for completion of the cycle itself. In this chapter I shall be concerned exclusively with the class of events required for the completion of the cycle, and to some extent with those events that have only been defined genetically. A separate chapter (9) by Mitchison discusses other events that are associated with the cycle and show periodic behavior.

## COORDINATION OF GROWTH AND DIVISION

### General Considerations

Any cell population that shows exponential growth over a substantial period of time must maintain a balance between growth, that is, the increase in biomass of the culture, and the rate of cell proliferation. The reason for this is straightforward: if the doubling time for biomass were greater than that for cell number, then division would continually run ahead of growth, and the mean cell size (total mass per number of cells) would decrease. Conversely, if biomass increased at a faster rate than cell number, cells would on average become larger with time. It is clear that in microorganisms at least this does not happen, provided the medium is replenished and the growth conditions do not alter. The various means by which growth and division might in principle be held in balance have been discussed elsewhere (Fantes et al., 1975; Fantes and Nurse, 1981) and will only be outlined here. The main conclusion to be drawn is that the most likely method is the operation of an active control that couples growth and division together. The coupling could be brought about in one of a number of ways. (1) Cells could grow until they have attained a particular critical size, which is monitored by the cell, and then divide. (2) Cells could grow by a fixed amount during the cycle, starting at birth, and then divide. (3) Cells could spend a constant time proceeding from birth to division, irrespective of either birth or division size. This third possibility does not maintain an element of active control between growth and division and is therefore not likely to operate. Modifications of this model, which result in active control, can include a growth rate that is varied by the cell so that (model 3a) the amount of growth during the cycle is constant or (3b) the final cell size at division is constant. These modifications imply that cells can in some way "look ahead" and adjust their growth rate in such a way that a particular target amount of growth or final cell size will be attained at some later stage.

### Cell Size and Cycle Time Homeostasis in *S. pombe*

Certain features of *S. pombe* make it highly suitable for investigating the relationship between division and growth: these are the mode of growth by linear extension and the symmetrical division mode. This has allowed studies on single cells growing and dividing in the field of view of the light microscope. Analysis of such experiments (Fantes, 1977; Miyata et al., 1978) has shown that there is substantial variation around a mean for both cell length at division and the interdivision time. This natural variation, whatever its cause (Fantes and Nurse, 1981; Brooks, 1981), has allowed the three possibilities outlined in the previous section to be examined. The results obtained in several studies (Fantes, 1977; Miyata et al., 1978; 1979a) support model (1), and are inconsistent with the other models.

Cells of *S. pombe,* whether larger or smaller than average at birth, divide at an almost constant mean size, irrespective of past history. In fact about three-quarters of the size deviation at birth is corrected within a single cycle, on average (Fantes, 1977). This is not quite in agreement with model (1), though close to it. One implication is that the amount of growth during the cycle is strongly dependent on initial size, in contradiction to model (2), which predicts a constant growth increment. For similar reasons model (3a) can be excluded. Model (3b), where the cell somehow targets its growth trajectory to a particular final value despite a fixed cycle time, cannot hold either. This is because the cycle time is observed to be strongly inversely correlated with birth size, ruling out the possibility of a constant time mechanism. *S. pombe* cells grow with a rate independent of birth size. Operation of a system close to that described by model (1) can explain the homeostasis of both cell size at division and cycle time (Fantes, 1977).

There are two situations where this simple model does not hold. One concerns the behavior of very large cells. Rather than undergoing extremely short cycles and attaining the normal size at the first division, very large cells undergo a minimum cycle (Fantes, 1977; Miyata et al., 1978). Completion of a minimum cycle does not bring a very large cell back to the normal size at the first division, since growth continues at a near-normal rate in these cells. Return to the normal division size can, however, occur in subsequent cycles. This emphasizes the point that any descriptive model of this type only holds true over a limited range of conditions. The behavior of very large cells does not contradict the validity of the model, provided it is only applied within limits.

A more serious objection to the model comes from a study (Miyata et al., 1979b) where oversized cells, made by blocking DNA synthesis with hydroxyurea, were transferred to a starvation medium free of the drug. Under these conditions growth is severely limited, but in most cells two successive divisions were possible. After release of the DNA synthesis block, the size of individual cells at the first and second divisions were measured, as was the duration of the intervening cycle. A negative correlation was still found between cell size at birth and the length of the subsequent cycle (Miyata et al., 1979b). No increase in cell length was observed during the cycle, and it is clearly impossible to argue that attainment of a critical size controlled division in these cells. Since the cycle times were strongly dependent on the cell size at birth, a minimum cycle cannot be invoked as an explanation, for this would predict a constant interdivision time. It seems as if under these unusual conditions cell size at birth is setting the cycle time (Miyata et al., 1979b). This explanation cannot, however, hold in general, since, as shown in the next section, cells in midcycle can adjust the time of division if the growth conditions change.

## CONTROLS OVER MITOSIS

### Nutritional Control

The studies outlined in the previous section show that the length of the whole cycle is controlled in some way, as is cell size at division. But the proposed cell size control need

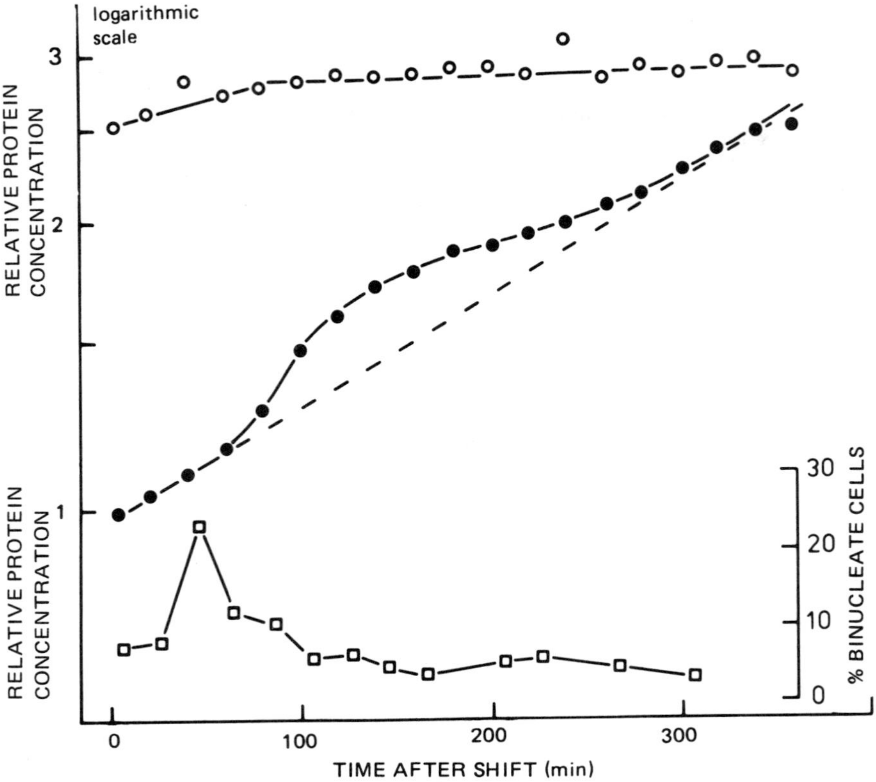


**FIGURE 2** Cells of wild-type strain 972 h⁻ were transferred from growth on minimal medium to medium containing all the components except a nitrogen source, at 25°C. The shift was done at zero time, and samples were taken at intervals to estimate cell number/ml culture (●); protein content/ml culture (o); and the percentage of binucleate cells (□). The broken line shows the increase in cell number/ml in a portion of the culture kept in the original medium. Abscissa shows time after shift (minutes); left ordinate is a logarithmic scale for relative cell number and protein content/ml culture; right-hand ordinate is a percentage of binucleate cells. Experimental procedures are described by Fantes and Nurse (1977).

not act directly over division, or even over a process that occurs shortly before division. Some midcycle event might be subject to a cell size control, and completion of this event might start the cell on a process that culminated in division (Fantes and Nurse, 1981). A similar situation is found in bacteria (Donachie, 1981) and budding yeast (Carter, 1981).

To answer this question for *S. pombe,* the relation between cell size and the growth rate was investigated by using different growth media. In this way different states of exponential growth with a range of doubling times can be achieved. There is a correlation between mean cell size and growth rate such that, in general, the faster cells grow, the larger they are (Fantes and Nurse, 1977). To investigate the basis of size variation with growth rate, a series of shift experiments was carried out, where cells growing exponentially under one set of nutritional conditions were transferred to new conditions (at constant temperature), and the patterns of growth and division followed (Fantes and Nurse, 1977). The conclusion to be drawn from these and other similar experiments (see Fig. 2) is as follows. Cells behave in these shifts (to a first approximation) as if there were within each cell a mass monitoring system, and a mechanism that initiated mitosis when a particular size had been attained. The actual size at initiation is not invariant, however; rather, it is dependent on the prevailing nutritional conditions, and is reset rapidly following environmental changes. The observations supporting this view (Fantes and Nurse, 1977) are as follows: when small cells growing at a low rate are shifted to better conditions where the division size would be larger during exponential growth, the first observable change is an abrupt and transient inhibition of mitosis in the population. This is followed shortly afterward by a transient cessation of cell division. When nuclear and cell division resume, the cells dividing after the shift have a larger size characteristic of the new conditions, sometimes after a short perturbed period. Cells appear to recognize the new conditions rapidly and respond by delaying mitosis in a cohort of cells which were close to mitosis before shift until they are large enough to undergo mitosis at the new larger size. Even more striking is the rapid stimulation of mitosis and cell division following a transfer of large fast-growing cells to poorer conditions. This can be accounted for by the cells' rapidly reducing their critical size at the time of shift. Cells in midcycle, which would not have entered mitosis for some time under the original conditions, are as large or larger than the new (smaller) critical size, and are accelerated into mitosis.

These inhibitions and accelerations of mitosis are largely independent of the precise nutritional conditions used to bring them about. Mitotic inhibition after shift-up occurs when the nitrogen, carbon, or phosphorus source is changed (Fantes and Nurse, 1977), while accelerations are observed after replacing a good nitrogen source with a poorer one, or even after complete nitrogen starvation of an exponential culture (Fig. 2) (Egel and Egel-Mitani, 1974).

The occurrence of delays and accelerations to mitosis has profound implications for our view of the whole cell cycle. First, it is clear that it is the length of the $G_2$ phase that is altered in cells following nutritional shift. This implies a flexibility in the length of a cycle phase often considered to be rather invariant in some cell types (Brooks, 1981). Second, *S. pombe* cells appear to be developmentally competent to enter mitosis some time before they actually do so. The main factor preventing a cell that is growing on a good medium and that is half-way through the cycle from entering mitosis appears to be completion of some event that can be described (to a first approximation) as attainment of a critical cell size. All events before the size control step may already have been completed in cells in midcycle, and these cells are awaiting a signal from the mass monitor before entering mitosis. Mitosis may therefore not occur as the culmination of a series of events in $G_2$, each of which takes a fixed time: the control mechanism has a greater degree of flexibility than this idea suggests.

### Genetic Control

Two types of mutants that affect mitosis have been isolated. In one type, entry into or progress through mitosis is conditionally defective, so that at one (permissive) temperature mitosis is possible, but transfer to another, usually higher temperature (restrictive), prevents progress through mitosis (Nurse et al., 1976; Fantes, 1979; Nasmyth and Nurse, 1981; Toda et al., 1981; Roy and Fantes, 1982). It seems likely that most mutants defective in mitosis have lesions in processes required for mitosis, though which have no regulatory influence over the timing of entry into mitosis. The genes defective in such mutants are analogous to "structural genes" of a metabolic pathway, which are typically defective for one enzyme of the pathway.

To continue with the comparison of the developmental pathway leading to mitosis with a metabolic pathway leading to an end product, specific mutations affecting regulation of the mitotic pathway might be expected. Such mutants do exist, and their regulatory nature is shown by the alteration in timing of mitosis, most convincingly in an advancement. This is because delays to mitosis might be caused by either a change in the regulatory system or inefficient function at the permissive temperature of one of the structural gene products required for mitosis, and the two are hard to distinguish. An advancement to mitosis might in principle have one of two consequences. Suppose that we have a mutant in which mitosis is normally timed at a permissive temperature but advanced at another restrictive temperature. On shift from permissive to restrictive temperature, mitosis will be advanced in the cycle that includes the shift. If mitosis in the next cycle (i.e., the first complete postshift cycle) is also advanced relative to the first mitosis after shift, then the cycle will be shortened, compared with the wild-type cycle. A succession of shortened cycles will ensue, and presuming that growth continues as in the wild type, division will run continually ahead of mitosis, and the mean cell size of the population will fall. Mutants of this type would be expected to be temperature-sensitive lethals, since continued growth and viability would not be maintained in cells of extremely small size. In fact, no such mutants have been obtained, despite an extensive screening that should have identified them (Thuriaux et al., 1978; Nurse and Thuriaux, 1980). A different type of "advanced mitosis" mutant has, however, been obtained (Nurse, 1975; Thuriaux et al., 1978; Nurse and Thuriaux, 1980; Fantes, 1981a). Such mutants are designated *wee* mutants because of their most striking property, which is a reduced cell size at division. Some, though not all, of these mutants are temperature sensitive, showing nearly normal cell size at 25°C but about 60% of the wild-type size when grown at 35°C (Nurse, 1975; Thuriaux et al., 1978). On shifting cells from 25 to 35°C, mitosis is indeed stimulated in the culture, some cells undergo mitosis ahead of schedule, and cell length at the first postshift division is much reduced (Nurse, 1975; Thuriaux et al., 1978). In subsequent cycles at 35°C this reduced cell length is maintained, and no further reduction occurs. The cells continue to divide with the same doubling time as the wild type, keeping a balance with the mass doubling time, and maintaining a constant cell size at division. The effect of temperature shift on such mutants is rather similar to that following nutritional shiftdown in wild-type cells, where a transient stimulation of mitosis is followed by attainment of a new steady state of growth with reduced cell size (see Nutritional Control). The transient stimulation of mitosis in temperature-shifted temperature-sensitive *wee* mutant cells show directly that mitosis is advanced relative to early cycle events, but after attainment of a new steady state of growth the intermitotic time is normal. This latter situation exists in *wee* mutants that are not temperature-sensitive: division and growth are in balance, but with a different proportionality from that of wild-type cells. Can non-temperature-sensitive *wee* mutants be regarded as altered in mitotic timing, despite the intermitotic period as a whole being unchanged? The time of mitosis can be considered advanced in *wee* mutants if the basis of time measurement is taken as the time in the wild-type cell cycle at which cells are the same size as are *wee* cells in mitosis. Since *wee* cells are just over half the wild-type size, the advancement is nearly a whole cycle. A more accurate calculation, based on the sizes and timing of mitosis in *wee* 1 and wild-type cells, and on the time of mitosis (Thuriaux et al., 1978) suggests an advancement of 0.81 of the cycle. A different way of considering *wee* cells is to regard them as "phase-shift" mutants, with the same intermitotic periodicity as wild-type cells, but with a 0.81 cycle phase advancement.

### Sizer or Timer?

Environmental conditions clearly affect cell size, as does the genetic constitution of the strain. The behavior of cells after shifts between "large cell" and "small cell" conditions can be readily explained if we assume that (1) cells are able to sense their size, or some parameter closely related to size, (2) mitosis is initiated when a critical size is attained, (3) the actual magnitude of the critical size is dependent on the environmental conditions, and on the alleles present at the loci which control cell size, *wee* 1 and *cdc* 2. This model is phenomenological rather than mechanistic; in other words, it provides a good description of the behavior of cells without attempting to explain by what means cells monitor their size, what molecular interactions set the correct size for mitosis, and so on. It is also possible to construct descriptive models where time rather than size is of prime importance to the cell. I do not intend to consider such models in detail, which has been done elsewhere (Fantes et al., 1975). It is important, I believe, to realize that the actual behavior of cells can be described by either system. Occam's razor suggests that the cell size or "sizer" model is preferable, containing fewer ad hoc assumptions, but only further analysis, probably at the molecular level, will decide which is correct. Having dealt with much of the material so far in terms of cell size, I shall also treat certain aspects in terms of the cell cycle as a temporally controlled pathway of cell development.

### Control Models and Mechanisms

Whether cell size or time in the cycle is important from the control aspect, any useful model should be able to explain the majority of observations. Some properties of *wee* and related mutants will now be summarized, as will certain simple molecular models.

One class of *wee* mutants maps at the *wee* 1 locus. The properties of *wee* 1 mutants suggest that they are "loss of function" mutants (Nurse and Thuriaux, 1980), implying that they have lost a nonessential function that normally inhibits mitosis. The wild-type *wee* $1^+$ function acts to inhibit mitosis, causing mitosis to occur at a greater size than in the absence of functional *wee* 1 product. In terms of mitotic timing, mitosis can be thought of as delayed in the presence of, or advanced in the absence of, functional *wee* $1^+$ gene product (see Genetic Control). It is not clear precisely what role the *wee* $1^+$ product plays in the inhibition of mitosis. It might be involved in the mass monitoring system itself, perhaps as the type of inhibitor suggested by Pritchard et al. (1969) for bacterial cells. If this were the case, then the *wee* $1^+$ product should be synthesized once every cycle, during a short period following mitosis. However, the dominance relationships between haploid and diploid strains of *S. pombe* (Nurse and Thuriaux, 1980) are not consistent with this simple but appealing idea, unless unjustifiable assumptions are made about the detailed biochemical kinetics involved. An alternative role for *wee* 1 is as part of a system that monitors the environment, particularly the nutritional environment, of the cell. *Wee* 1 cells do not show the rapid and substantial effects on mitosis of nutritional shifts which *wee* $1^+$ cells do (Fantes and Nurse, 1978). This may indicate that *wee* 1 cells are unable to sense their nutritional status, and "consider" themselves to be always in poor medium (Nurse and Thuriaux, 1980). This would explain their small size and the lack of effect on mitosis following nutritional shift.

The other known *wee* mutants are alleles of the *cdc* 2 gene (Nurse and Thuriaux, 1980). The *cdc* $2^+$ gene can be mutated to give two distinct phenotypes. Loss of function produces *cdc* $2^-$ cells, which are defective in mitosis; certain other changes result in an advanced mitosis and reduced cell size. These conclusions are borne out by a detailed study of the sizes and other properties of strains bearing the various *cdc* 2 alleles, both in haploids and in combinations in diploid strains (Nurse and Thuriaux, 1980). As with *wee* $1^+$, the actual role of *cdc* $2^+$ and its product in the control of mitosis is uncertain, though it is clear that the *cdc* 2 function has a positive regulatory role as well as being absolutely required for mitosis. The accumulation of a critical amount of functional *cdc* $2^+$ product, or of a substance whose synthesis is mediated by functional *cdc* $2^+$ product, might act as a rate-limiting step for entry into mitosis (Nurse and Thuriaux, 1980). It may be that the rate of synthesis of *cdc* $2^+$ product, or of the activity of this product, are in turn regulated by the *wee* $1^+$ product, an intriguing possibility to which I shall return.

## CONTROLS OVER S PHASE

### Entry to the Cycle and Commitment

The function of at least seven *cdc* genes is required for DNA replication to be completed (Nasmyth and Nurse, 1981). Of these, the enzymatic defects associated with *cdc* 17 (Nasmyth, 1977) and *cdc* 22 (Dickinson, 1981) have been defined. Among other mutants biochemical functions have been tentatively assigned to *cdc* 21 and 23 (Nasmyth and Nurse, 1981). Although nothing is known of the molecular lesion in *cdc* 10 mutants, this gene is of particular interest with regard to overall control of the cycle. *S. pombe* cells of appropriate mating types may undergo conjugation, nuclear fusion, and meiosis, followed by the formation of sexual spores. The first signs that cells are taking this developmental pathway rather than the vegetative mitotic cycle occur when cells of appropriate mating types are starved of nitrogen and mixed (Egel, 1971). Agglutination of the cells, followed by breakdown of the intervening cell walls, leads to cell fusion and nuclear fusion. This process normally takes place when the nitrogen-starved cells are in the $G_1$ phase, as is true for the related yeast *Saccharomyces cerevisiae* (Hartwell, 1974). In an elegant series of experiments, Nurse and Bisset (1981) showed that cells arrested at the step in the cycle controlled by *cdc* 10 were able to conjugate, while cells arrested by other mutational lesions in DNA replication were unable to do so. This suggests that the *cdc* 10 gene may control a key step in the *S. pombe* cycle and once cells have passed it and entered the mitotic pathway they cannot switch to the conjugation-meiosis pathway. The transition points of *cdc* 10 mutants are very early in the cell cycle (Nurse et al., 1976), and by analogy with the *S. cerevisiae* cycle it is possible that the *cdc* 10 step is the earliest in the mitotic cycle.

Of great interest is the observation that *cdc* 2 mutants, previously thought to be defective in a late $G_2$ event, were able to conjugate when arrested at the restrictive temperature (Nurse and Bisset, 1981). This lead to the discovery that *cdc* 2 mutants are defective in a $G_1$ event in addition to the $G_2$ event controlling mitosis.

### A Cryptic Size Control

Two major controls determine the timing of S phase in *S. pombe.* Cells must have completed mitosis before S phase can be started, and normally this takes place after a short $G_1$ period (Nasmyth et al., 1979). There is an additional requirement, only manifest in small cells, which appears to be the attainment of a critical cell size (Nurse and Thuriaux, 1977). Cells that are small for either genetic or physiological reasons exhibit this requirement, and in a range of different situations the size at which S phase occurs is fairly

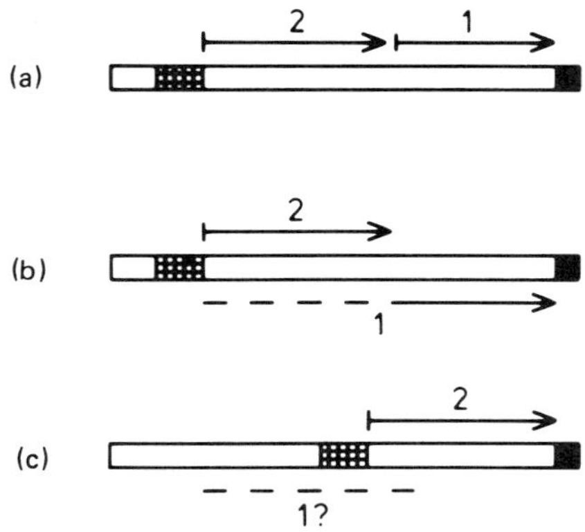


**FIGURE 3** Pathways leading to mitosis. Possible relationships between pathways 1 and 2 in wild-type (a, b) and *wee* cells (c) are shown. Each pathway is shown by an arrow: where the temporal extent of the pathway is uncertain, this is indicated by a broken line. Mitosis is shown by a shaded block; S phase by cross-hatching. See text for details.

constant (Nurse and Thuriaux, 1977; Fantes and Nurse, 1978). The possibility of nutritional or genetic modulation of the critical size cannot be excluded, because the best test for such a modulation, a nutritional or similar shift, has not been carried out to examine the behavior of S phase under such conditions. In wild-type sized cells under normal growth conditions the cell size requirement is not observed, presumably because such cells when newborn are already larger than the critical size.

In *wee* cells, expression of the size requirement delays the S phase by about one-quarter of the cycle, from being almost simultaneous with cell division in the wild type to about 0.25 in the cycle. The $G_2$ phase is correspondingly shortened by a quarter of the cycle.

### Control of the Cycle in *wee* Cells

The important question of what mechanism controls the time of cell division, and cell size at division in *wee* cells, has not yet been addressed here. One possibility is that the normal cell size control over mitosis operates, with a reduced critical size. This does not account for the loss of nutritional sensitivity in *wee* 1 cells, however: an additional postulate would be required to explain this. A different suggestion is that the mitotic size control that does not operate in *wee* 1 cells, or that at any rate the critical size is so small that its operation is never manifest. This implies that an alternative control would need to operate, otherwise cells would simply divide as fast as possible, faster than they were growing, and cell size would fall rapidly with time. A mechanism that has been suggested (Fantes and Nurse, 1978; Nurse and Fantes, 1981) includes the cell size control over S phase, and some kind of timer element that ensures that mitosis follows a constant time after S phase (Fig. 3). Several lines of evidence favor this idea (Fantes and Nurse, 1978). First, the duration of the $G_2$ phase in *wee* 1 cells is relatively constant (at fixed temperature) over a range of growth rates, suggesting that S phase and mitosis are separated by a constant period. The mean protein content of cells in S phase was also fairly constant in all conditions examined, consistent with a nutritionally independent critical size for S phase. Second, short delays to the start of S phase in *wee* cells caused a corresponding delay in the time of cell division, suggesting a temporal coupling between S phase and mitosis. Similar delays to S phase in wild-type cells did not affect the timing of mitosis. Third, when the rate of cellular growth was reduced by a method not involving alteration of the nutritional conditions, *wee* 1 cells accumulated partly in $G_1$, but wild types accumulated in $G_2$. This is again consistent with the absence of a cell size requirement for mitosis in *wee* 1 cells, since about half the cells arrested in $G_1$ when growth was limiting.

The nature of the "timer" element in this model is unclear: it may be that a certain set of processes need to take place during $G_2$ and these occupy a minimum time. Certainly it seems impossible to shorten the interdivision time of cells below a certain minimum when very oversized cells are allowed to divide (see Genetic Control). The shortening of the cycle would be expected to occur in $G_2$, since this phase represents the majority (three-quarters) of the cycle, and substantially shortened periods have been reported in some experiments (Mitchison and Creanor, 1971; Fantes and Nurse, 1978).

## PATHWAYS LEADING TO MITOSIS

### At Least Two Pathways Exist

Evidence presented in the preceding section implies the existence of two pathways, either of which can control the time of mitosis and cell size at that cycle stage. The normal pathway (now referred to as pathway 1) culminates in entry to mitosis when a certain cell size has been attained (Fig. 3a, b). It is not known what initiates this pathway, or at what cycle stage initiation occurs. Progress along this pathway is affected by the prevailing conditions, so that, for instance, nutritional shiftdown can speed up its completion. The other pathway (pathway 2) apparently controls mitosis in *wee* 1 cells. It starts around the time of S phase, and its initiation may be controlled by the same processes that initiate S phase (Fig. 3). In *wee* 1 cells pathway 2 ends in mitosis (Fig. 3c), and the beginning and end of the pathway appear to bear a constant temporal relationship to one another (see Controls over Mitosis: A Summary).

It is likely that both pathways are present in *S. pombe* cells in general, with the possible exception of *wee* 1 cells (see below). Both pathways must be completed before mitosis can occur (again with the possible exception of *wee* 1 cells), and the one finishing later will be rate limiting and actually determine the timing of mitosis. In wild-type cells, this is pathway 1, and in *wee* 1 cells, pathway 2. The relationship between the pathways is not certain (Fig. 3). Possibly pathway 1 could follow (in late $G_2$) after the completion of pathway 2 (Fig. 3a). This would explain the nutritional sensitivity of wild-type cells, since pathway 1 is nutritionally sensitive. In *wee* 1 cells pathway 1 is shortened drastically or even absent, accounting for the advancement of mitosis and the absence of nutritional sensitivity (Fig. 3c). An alternative is that the two pathways might run in parallel and partially overlap in time during $G_2$ (Fig. 3b). The only constraint on this possibility is that pathway 1 should finish later than pathway 2 in wild-type cells, so that the former actually controls the time of mitosis. There is one line of evidence that the two pathways operate in parallel rather than sequentially. In one study (Fantes and Nurse, 1978), *wee* $1^+$ cells carrying a termperature-sensitive defect in *cdc* 10, a gene required for initiating DNA replication, were subjected to a pulse of one-quarter of a cycle at the restrictive temperaure during growth at the permissive temperature. This pulse delayed the onset of DNA replication and presumably also the start of pathway 2, yet no delay to division was observed (Fantes and Nurse, 1978). If pathway 1 followed on from pathway 2, then a substantial delay would have been expected. In a similar experiment with a double mutant *wee* 1.6 *cdc* 10.129, a similar pulse did delay division, demonstrating that delaying the initiation of DNA replication also delayed the start of pathway 2 (Fantes and Nurse, 1978).

### Elements of the Pathway

This section is rather different from what has been discussed earlier. The material to be discussed is not directly concerned with the temporal control of the whole cell cycle, or even of the major events of DNA replication and mitosis. Rather, some experiments concerned with how cells carry out the processes signaled by control elements will be described. The reason for inclusion of this material is that some surprising results have emerged which have significance to the temporal controls as well as to our view of how the cell cycle functions as an integrated developmental process, particularly at the genetic and molecular levels of organization.

#### *Role of* cdc *2*

The *cdc* $2^+$ gene is unique among genes required for the cell cycle in *S. pombe.* As previously mentioned, the gene product has a rate-limiting role in controlling entry to mitosis, and is also implicated in entry to S phase. By analogy with metabolic pathways, we might expect a rate-limiting process to be located at the beginning of a developmental pathway. The role of *cdc* 2 in controlling entry into S phase, and the observation that cells can mate when arrested at the *cdc* 2 block in $G_1$, suggest a function early in $G_1$ for the *cdc* $2^+$ gene product. This is supported by evidence from *S. cerevisiae,* where comparable "start" mutants are defective in early $G_1$ events, and no mutants that are blocked earlier have yet been identified (Carter, 1981). Unfortunately, there are no such clear "landmark" events in mitosis comparable to the mating ability in $G_1$, which can be used in a similar way to investigate whether *cdc* $2^+$ acts early or late in mitosis. This is because mitosis itself is a very rapid process (Mitchison, personal communication), and the classical mitotic phases of prophase and metaphase are not normally visible (McCully and Robinow, 1971). However, an ultrastructural study of the terminal phenotypes of several mutants defective in mitosis (King and Hyams, 1982) and comparison with what is known of the wild-type process (McCully and Robinow, 1971) is consistent with the *cdc* $2^+$ gene controlling an early event.

Using a different approach, an attempt has been made to investigate the functional sequence of some events in mitosis in *S. pombe* (Fantes, 1982a) (Fig. 4). In this study, the *cdc* $2^+$ gene product was shown to act before a step in mitosis that was sensitive to the antimicrotubule agent benomyl. Furthermore, three other mutants (defective in *cdc* 1, 13, or 25) controlled steps that could only occur simultaneously with the benomyl-sensitive step, that is, after completion of the *cdc* 2-controlled process. One other mutant, defective in *cdc* 27, controlled a step upon which the benomyl-sensitive step was dependent (Fig. 4). The relationship between *cdc* 2 and *cdc* 27 could not, unfortunately, be determined in this study. Nevertheless, the observations are consistent with the *cdc* $2^+$ gene controlling an early step in mitosis, perhaps in pathway 1 referred to above. That the *cdc* 2-controlled step occurs close to mitosis is clear, both from the transition point (see following section) and from the observation that cells arrested at the *cdc* 2 stage have synthesized all the proteins required for mitosis, and can complete mitosis when the block to mitosis is removed under conditions where protein synthesis was inhibited almost completely (Fantes, 1982a).

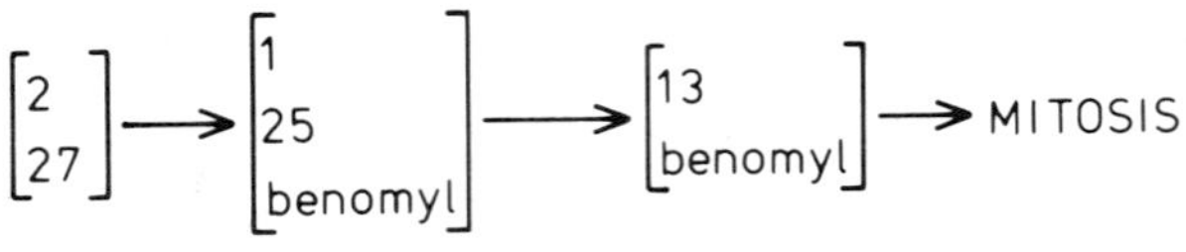


**FIGURE 4** Dependency relations in mitosis. Each number represents a step in the cell cycle controlled by the corresponding *cdc* gene, each of which is required for mitosis. Benomyl-sensitive steps are also shown. Steps bracketed together are mutually interdependent. An arrow implies the existence of a dependency of the event(s) on the right on the event(s) on the left of the arrow. (Data from Fantes, 1982a.)

### *The* wee *1 Gene as a Pleiotropic Regulator*

The *wee* 1 gene, like the *cdc* 2 gene, plays a rate-limiting role in the normal pathway (pathway 2) leading to mitosis. The wild-type *wee* $1^+$ gene product delays or inhibits mitosis, which can only occur when a critical cell size has been attained. In this section I shall address the question of how the *wee* $1^+$ gene product exerts its inhibitory effect on mitosis.

The time at which a particular gene product completes its function in the cell cycle can in principle be estimated if a temperature-sensitive mutation of the gene is available which shows first cycle arrest (Pringle, 1978). The mutant is grown at the permissive temperature, and a sample of cells shifted to the restrictive temperature. Cells that at the time of shift have completed the process controlled by the mutant gene are able to complete the entire cycle, but cells that have not completed the function cannot divide. The stage in the cycle at which a cell gains the ability to divide when challenged with restrictive conditions is called the transition point or execution point (Hartwell, 1974; Nurse et al., 1976). Knowledge of the transition point is informative about the time of completion of a particular function, but tells us nothing about its temporal extent. In *S. pombe,* mutants defective for mitosis usually have transition points shortly before mitosis itself (Nurse, et al., 1976; Fantes, 1979; Nasmyth and Nurse, 1981). Thus, a number of processes required for mitosis are completed shortly before the event itself. It follows that the times of completion of these processes are controlled and coordinated in some way.

Because the presence of a *wee* 1 lesion can be considered to advance the whole process of mitosis, the effect on the transition points of various mutants defective for mitosis of a defective *wee* 1 gene was investigated (Fantes, 1982b). The transition points of mutants of *cdc* 1 and *cdc* 27, and of some *cdc* 2 mutants, were advanced in *wee* 1 strains (Fig. 5). The effect was gene rather than allele specific for both the *wee* 1 and *cdc* mutations, apart from differences observed between the different *cdc* 2 alleles. The extent of the advancement for *cdc* 1 and *cdc* 27 mutations was about 0.4 of the cycle, while the advancement in *cdc* 2 mutants varied between 0 and 0.2 of the cycle (Fantes, 1982b) (Fig. 5). Double mutants of *cdc* 1 with *wee* 2 (= *cdc* 2w) (Nurse and Thuriaux, 1980) mutants had transition points close to those of the corresponding *cdc* 1 $wee^+$ strains, showing that *wee* 2 mutations do not affect the timing of the *cdc* 1 process in the same way as *wee* 1 mutations. Some transition points were unaffected by the presence of a *wee* 1 mutation.

The advancements described here suggest that the normal *wee* $1^+$ gene is involved in regulating the time of action of several *cdc* gene products in the cell cycle. This regulation might be mediated by repression of expression of the *cdc* genes, or by inhibition of

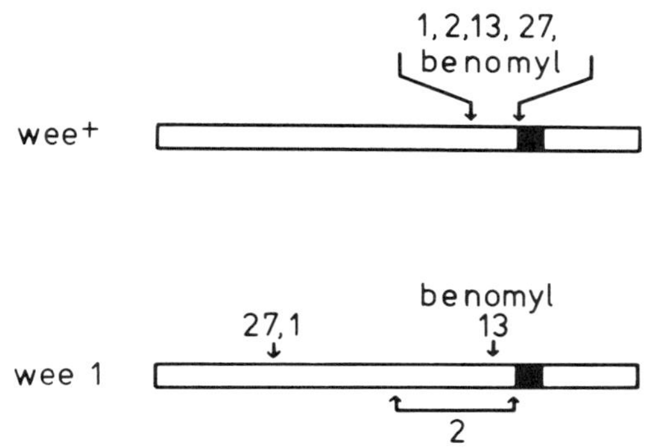


**FIGURE 5** Timing of events required for mitosis. The cell cycles of *wee*$^+$ and *wee* 1 strains are shown, with mitosis shown as a shaded block. The transition points of mutants defective in *cdc* 1, 2, 13, and 27 are shown, as is the transition point for the antimicrotubule agent benomyl. Note the substantial advancement of *cdc* 1 and *cdc* 27 transition points in a *wee* 1 background, and the lesser advancement of *cdc* 2 transition point. (Data from Fantes, 1982b.)

the function of the gene product. The differential effect of *wee* 1 mutations on the different *cdc* mutants may reflect different degrees of inhibition of synthesis or function of the *cdc* gene products in the normal cell. The extent of the advancements is not large, up to 0.4 cycle, but since the *wee cdc* double mutants have reduced cell size, completion of the various *cdc* functions occurs at a substantially reduced size compared with normal. Calculation suggests that a *wee* 1 cell completes the *cdc* 1 function at a cell size 43% of that of the wild-type when it completes the *cdc* 1 function. If, as seems likely, the size of a *S. pombe* cell is of great significance to its behavior in the cycle, the presence of a wild-type *wee* 1$^+$ gene may inhibit the rate of some process associated with *cdc* 1 very substantially.

### *A Paradox*

As mentioned in the previous section, several functions required for mitosis controlled by *cdc* genes are dependent on completion of an event controlled by the *cdc* 2$^+$ gene. In particular, the *cdc* 1 process is thought to be dependent on the *cdc* 2 step (Fig. 4). However, in *wee* 1 strains the *cdc* 1 step is apparently completed at around 0.3 in the cycle, while the *cdc* 2 step is not completed until between 0.5 and 0.7 (Fig. 5). This seems to imply that an event is preceding its cause, and some explanation must be offered for this apparent contradiction. One possibility is that the *wee* 1$^+$ product, in addition to coordinating the timing of processes controlled by the various *cdc* genes, also maintains the normal dependency relationships between them. Loss of *wee* 1$^+$ function allows the breakdown of these dependencies.

### *Other Control Elements*

Two other types of genetic element have been observed that may be involved in the mitotic control. Mutants defective in the *cdc* 25 gene have three properties that are of interest. First, *cdc* 25 mutants are typical temperature-sensitive lethals unable to complete mitosis at 35°C (Fantes, 1979). Interestingly, their cell size at 25°C is considerably greater than the wild type, though this may have a more trivial explanation, i.e., that the mutants are partially defective in mitosis even at 25°C. Second, the mutant phenotype of *cdc* 25 strains is almost completely suppressed by the presence of a *wee* 1 mutation: indeed, this property has been used to obtain a large number of *wee* 1 mutants, some with partial *wee* phenotype (Fantes, 1981b). The suppression of gene rather than allele specific, suggesting an interaction between systems rather than directly between gene products. Third, *cdc* 25 mutants appear to affect the efficiency of some nonsense suppressors, which implicates *cdc* 25 as playing a role in the translational machinery (Nurse et al., 1979). This might provide a link between the nutritional status of the cell and the mitotic control: nutritional conditions affect growth rate, which in turn is linked to the rate of protein synthesis. Some signal generated at the translational level might affect the activity of the *cdc* 25 product, which could in turn be sensed by the *wee* $1^+$ product, which controls entry into mitosis. This admittedly tenuous series of links nevertheless accounts for the suppression of *cdc* 25 phenotype by *wee* 1 mutations, and for the lack of nutritional sensitivity of *wee* 1 mutants.

A new class of mutants has recently been isolated, which shows altered nutritional sensitivity as regards mitosis. These mutants do not decrease their cell size by much after nitrogen starvation, in contrast to the wild type, where cells a quarter the normal size are generated. This increased starvation size is suppressed by the presence of a *wee* 1 mutant (Young, P. G., personal communication).

### Controls over Mitosis: A Summary

At this point it seems useful to recapitulate some of the previous arguments, and present a simplified version of the sequence of events in the nuclear division cycle of a wild-type *S. pombe* cell, starting with the completion of mitosis and the start of $G_1$. During the short $G_1$ the gene products of *cdc* $2^+$ and *cdc* $10^+$ are required, and completion of their functions starts the cell on the route to mitosis. The S phase follows, and around this time a sequence of events (pathway 1) is initiated. This sequence ends at about 0.5 in the cycle, after which time the cell is competent to undergo mitosis. However, mitosis is not signaled until the cell has attained a certain size, which depends on the prevailing nutritional conditions and perhaps the *cdc* 2 allele. In *wee* 1 cells this size requirement is either abolished, or is set so low that cells have always attained it by the time they are otherwise competent to enter mitosis. Pathway 2, which includes the nutritionally dependent cell size control, probably requires the function of genes *cdc* $1^+$, $2^+$, $25^+$ and $27^+$. The processes controlled by *cdc* 1, 2, and 27 are normally completed at about the same time, perhaps because of regulation of each process by the *wee* $1^+$ gene product. Alternatively, some processes, such as that controlled by *cdc* $1^+$, may normally (in *wee* $1^+$ cells) only take place after completion of the *cdc* $2^+$ (and *cdc* $27^+$) steps in the cycle. At the time of completion of the *cdc* $2^+$ function, all the proteins required for mitosis have probably been synthesized and all the subsequent events in mitosis take place at the level of cytoplasm or nucleoplasm.

## CONCLUDING REMARKS

Ideas about how the *S. pombe* cell cycle is controlled have progressed a long way since observations of excess division delay in response to treatments that affected mitosis led to the idea of division proteins (Mitchison, 1971). These ideas were applied to *S. pombe* (Herring, 1972; Polanshek, 1977) following the original studies on *Tetrahymena*, summarized by Mitchison (1971). A major conceptual step was the realization that growth and division were normally kept in balance (Mitchison, 1971; Mitchison and Creanor, 1971). Generation of unbalanced growth by blocking cycle progress caused the production of oversized cells, while release of this block was followed by shortened compensatory cycles (Mitchison and Creanor, 1971). Experiments such as these were carried out in other systems at about the same time, and gave similar results. This groundwork allowed more detailed analyses of the mechanism of cycle homeostasis to be carried out at the level of whole cell cycles. The next major conceptual step resulted from the isolation of cell size or *wee* mutants, which allowed the formulation of ideas about the molecular basis of cycle control in a more precise way. After a period of searching for molecular needles in haystacks, it now at least appeared that there was indeed a needle! Subsequent experiments involving different growth conditions confirmed the idea that cell size was an important factor in the cycle of *S. pombe* cells.

The analysis of mutants known to be defective in cycle controls has led to some unexpected observations, which perhaps suggest the way forward. The two most striking series of observations are concerned with the two genes that affect cell size at mitosis. Both *cdc* 2 and *wee* 1 mutants have very pleiotropic phenotypes: *cdc* $2^+$ function is required for initiating both S phase and mitosis: activity of the gene product is rate limiting for mitosis; and the $G_1$ step controlled by *cdc* 2 is a switch point between the mitotic and conjugation-meiosis pathways. Particularly interesting here is the recent finding that in *S. cerevisiae* the *cdc* 28 gene, previously thought to control only a key $G_1$ event, "start,'" also codes for a function required for mitosis (Piggott et al., 1982). The parallel with *cdc* 2 in *S. pombe* is striking, and similarity between the two systems may be more than superficial: using recombinant DNA techniques and yeast transformation, the precise functions of the two genes have been shown to be very similar (Beach et al., 1982).

The *wee* 1 gene appears to be involved in either cell mass monitoring or the sensing of the nutritional environment, or perhaps both. It certainly affects the cell size at which mitosis occurs, and can be considered to control the time of mitosis. In addition it appears to play an important role in coordinating the timing of several gene-controlled processes required for mitosis and in mediating the action of *cdc* 25.

Given the complexity of the behavior of these control genes, the reader may feel that cell cycle control has become too complex to analyze, or that the study of it has become obsessed with particular elements to the exclusion of all else. I have some sympathy with this view, but believe that our overall understanding will only progress by examining putative control elements at every level: physiological, genetic, and, following the development of recombinant DNA technology and yeast transformation systems, at the molecular level. It is of utmost importance, however, not to lose sight of the questions cells pose to us: what controls the how, why, and particularly the when of cell division.

## ACKNOWLEDGMENTS

I would like to thank Murdoch Mitchison, Paul Nurse, and Paul Young for invaluable discussions on the cell cycle over the past few years. I am very grateful to those who assisted in preparing the manuscript: in particular Joyce Hunter, and Judy Fantes, who with Jim Piper introduced me to the word processor facility at the MRC Unit, Western General Hospital, Edinburgh. Much of the recent experimental work was supported by the Science Research Council.

## REFERENCES


Baserga, R. (1968). Biochemistry of the cell cycle: A review. *Cell Tissue Kinet.* 1:167-191.

Beach, D., Durkacz, B., and Nurse, P. (1982). Functionally homologous cell cycle control genes in budding and fission yeast. *Nature* 300:706-709.

Brooks, R. F. (1981). Variability in the cell cycle and the control of proliferation. In *The Cell Cycle,* P. C. L. John (ed.). Cambridge University Press, London, pp. 35-61.

Brooks, R. F., Bennett, D. C., and Smith, J. A. (1980). Mammalian cell cycles need two random transitions. *Cell* 19:493-504.

Burns, F. J., and Tannock, I. F. (1970). On the existence of a G0-phase in the cell cycle. *Cell Tissue Kinet.* 3:321-324.

Carter, B. L. A. (1981). The control of cell division in *Saccharomyces cerevisiae.* In *The Cell Cycle,* P. C. L. John (ed.). Cambridge University Press, London, pp. 99-117.

Dickinson, J. R. (1981). The *cdc* 22 mutation in *Schizosaccharomyces pombe* is a temperature-sensitive defect in nucleoside diphosphokinase. *Eur. J. Biochem.* 119: 341-345.

Donachie, W. D. (1981). The cell cycle of *Escherichia coli.* In *The Cell Cycle,* P. C. L. John (ed.). Cambridge University Press, London, pp. 63-83.

Egel, R. (1971). Physiological aspects of conjugation in fission yeast. *Planta (Berl.)* 98: 89-96.

Egel, R., and Egel-Mitani, M. (1974). Premeiotic DNA synthesis in fission yeast. *Exp. Cell Res.* 88:127-134.

Fantes, P. A. (1977). Control of cell size and cycle time in *Schizosaccharomyces pombe. J. Cell Sci.* 24:51-67.

Fantes, P. A. (1979). Epistatic gene interactions in the control of division in fission yeast. *Nature* 279:428-430.

Fantes, P. A. (1981a). Cell division control: Why clocks are inadequate. In *International Cell Biology 1980-1981,* H. G. Schweiger (ed.). Springer-Verlag, Berlin, pp. 846-850.

Fantes, P. A. (1981b). Isolation of cell size mutants of a fission yeast by a new selective method: Characterization of mutants and implications for division control mechanisms. *J. Bacteriol.* 146:746-754.

Fantes, P. A. (1982a). Dependency relationships between events in mitosis in *Schizosaccharomyces pombe. J. Cell Sci.* 55:383-402.

Fantes, P. A. (1983). Control of timing of cell cycle events in fission yeast by the *wee* $1^+$ gene. *Nature* 302:153-155.

Fantes, P. A., Grant, W. D., Pritchard, R. H., Sudbery, P. E., and Wheals, A. E. (1975). The regulation of cell size and the control of mitosis. *J. Theor. Biol.* 50:213-244.

Fantes, P. A., and Nurse, P. (1977). Control of cell size at division in fission yeast by a growth modulated size control over nuclear division. *Exp. Cell Res.* 107:377-386.

Fantes, P. A., and Nurse, P. (1978). Control of the timing of cell division in fission yeast: Cell size mutants reveal a second control pathway. *Exp. Cell Res.* 115:317-329.

Fantes, P. A., and Nurse, P. (1981). Division timing: Controls, models and mechanisms. In *The Cell Cycle,* P. C. L. John (ed.). Cambridge University Press, London, pp. 11-33.

Hartwell, L. H. (1974). *Saccharomyces cerevisiae* cell cycle. *Bacteriol. Rev.* 38:164-198.

Herring, A. J. (1972). A study of induced delay in the division of yeast *Schizosaccharomyces pombe.* Ph.D. thesis, University of Edinburgh.

Howard, A., and Pelc, S. R. (1953). Synthesis of desoxyribonucleic acid in normal and irradiated cells and its relation to chromosome breakage. *Heredity (Suppl)* 6: 261-273.

James, T. W., Hemond, P., Czer, G., and Bohman, R. (1975). Parametric analysis of volume distributions of *Schizosaccharomyces pombe* and other cells. *Exp. Cell Res.* 94:267-276.

Kauffman, S. A., and Wille, J. J. (1975). The mitotic oscillator in *Physarum polycephalum. J. Theor. Biol.* 55:47-93.

King, S. M., and Hyams, J. S. (1982). Interdependence of cell cycle events in *Schizosaccharomyces pombe.* Terminal phenotypes of cell division cycle mutants arrested during DNA synthesis and nuclear division. *Protoplasma* 110:54-62.

Klevecz, R. R. (1976). Quantized generation time in mammalian cells as an expression of the cellular clock. *Proc. Natl. Acad. Sci. USA* 73:4012-4016.

McCully, E. K., and Robinow, C. F. (1971). Mitosis in the fission yeast *Schizosaccharomyces pombe*: A comparative study with light and electron microscopy. *J. Cell Sci.* 9:475-507.

Mitchison, J. M. (1957). The growth of single cells. I. *Schizosaccharomyces pombe. Exp. Cell Res.* 13:244-262.

Mitchison, J. M. (1971). *The Biology of the Cell Cycle,* Cambridge University Press, London, pp. 201-249.

Mitchison, J. M., and Creanor, J. (1971). Induction synchrony in the fission yeast *Schizosaccharomyces pombe. Exp. Cell Res.* 67:368-374.

Miyata, H., Miyata, M., and Ito, M. (1978). The cell cycle in the fission yeast, *Schizosaccharomyces pombe.* I. Relationship between cell size and cycle time. *Cell Struct. Funct.* 3:39-46.

Miyata, H., Miyata, M., and Ito, M. (1979a). The cell cycle in the fission yeast, *Schizosaccharomyces pombe.* III. The cycle duration in synchronized cells produced by hydroxyurea. *Cell Struct. Funct.* 4:81-89.

Miyata, H., Miyata, M., and Ito, M. (1979b). The cell cycle in the fission yeast, *Schizosaccharomyces pombe.* IV. Relationship between cell size and cycle time of non-growing cells due to a nutritional shift-down. *Cell Struct. Funct.* 4:209-214.

Nasmyth, K. A. (1977). Temperature-sensitive mutants in the structural gene for DNA ligase in the yeast *Schizosaccharomyces pombe. Cell* 12:1109-1120.

Nasmyth, K., and Nurse, P. (1981). Cell division cycle mutants altered in DNA replication and mitosis in the fission yeast *Schizosaccharomyces pombe. Mol. Gen. Genet.* 182: 119-124.

Nasmyth, K., Nurse, P., and Fraser, R. S. S. (1979). The effect of cell mass on the cell cycle timing and duration of S-phase in fission yeast. *J. Cell Sci.,* 39:215-233.

Nurse, P. (1975). Genetic control of cell size at cell division in yeast. *Nature* 256:547-551.

Nurse, P., and Bisset, Y. (1981). Gene required in G1 for commitment to cell cycle and in G2 for control of mitosis in fission yeast. *Nature* 292:558-560.

Nurse, P., and Fantes, P. A. (1981). Cell cycle controls in fission yeast: A genetic analysis. In *The Cell Cycle,* P. C. L. John (ed.). Cambridge University Press, London, pp. 85-98.

Nurse, P., Fantes, P. A., Munz, P., and Thuriaux, P. (1979). Control integrating growth rate and the initiation of mitosis in fission yeast. *Heredity* 42:282.

Nurse, P., and Thuriaux, P. (1977). Controls over the timing of DNA replication during the cell cycle of fission yeast. *Exp. Cell Res.* 107:365-375.

Nurse, P., and Thuriaux, P. (1980). Regulatory genes controlling mitosis in the fission yeast *Schizosaccharomyces pombe. Genetics* 96:627-637.

Nurse, P., Thuriaux, P., and Nasmyth, K. (1976). Genetic control of the cell division cycle in the fission yeast *Schizosaccharomyce pombe. Mol. Gen. Genet.* 146:167-178.

Pardee, A. B. (1974). A restriction point for control of normal animal cell proliferation. *Proc. Natl. Acad. Sci. USA* 71:1286-1290.

Piggott, J. R., Rai, R., and Carter, B. L. A. (1982). A bifunctional gene product involved in two phases of the yeast cell cycle. *Nature* 298:391-393.

Polanshek, M. M. (1977). Effects of heat shock and cycloheximide on growth and division in the fission yeast, *Schizosaccharomyces pombe. J. Cell Sci.* 23:1-23.

Pringle, J. R. (1978). The use of conditional lethal cell cycle mutants for temporal and functional mapping of cell cycle events. *J. Cell Physiol.* 95:393-405.

Pritchard, R. H., Barth, P. T., and Collins, J. (1969). Control of DNA synthesis in bacteria. *Symp. Soc. Gen. Microbiol.* 19:263-297.

Roy, D., and Fantes, P. A. (1982). Benomyl resistant mutants of *Schizosaccharomyces pombe* cold-sensitive for mitosis. *Curr. Genet.* 6:195-201.

Rusch, H. P., Sachsenmaier, W., Behrens, K., and Gruter, V. (1966). Synchronization of mitosis by the fusion of plasmodia of *Physarum polycephalum. J. Cell Biol.* 31: 204-209.

Sachsenmaier, W., Remy, U., and Plattner-Schobel, R. (1972). Initiation of synchronous mitosis in *Physarum polycephalum*: A model of the control of cell division in eukaryotes. *Exp. Cell Res.* 73:41-48.

Simchen, G. (1978). Cell cycle mutants. *Ann. Rev. Genet.* 12:161-191.

Smith, J. A., and Martin, L. (1973). Do cells cycle? *Proc. Natl. Acad. Sci. USA* 70:1263-1267.

Thuriaux, P., Nurse, P., and Carter, B. L. A. (1978). Mutants altered in the control coordinating cell division with cell growth in the fission yeast, *Schizosaccharomyces pombe. Mol. Gen. Genet.* 161:215-220.

Toda, T., Yamamoto, M., and Yanagida, M. (1981). Sequential alterations in the nuclear chromatin region during mitosis of the fission yeast *Schizosaccharomyces pombe*: Video fluorescence microscopy of synchronously growing wild-type and cold-sensitive *cdc* mutants by using a DNA-binding fluorescent probe. *J. Cell Sci.* 52:271-287.

Wilson, E. (1924). *The Cell in Development and Heredity*, 3rd ed. New York, Macmillan.

# 15

# The Control of Nuclear Division in *Physarum polycephalum*

**John J. Tyson** / Virginia Polytechnic Institute and State University, Blacksburg, Virginia

**Wilhelm Sachsenmaier** / Institute for Experimental Cancer Research, University of Innsbruck, Innsbruck, Austria

## INTRODUCTION

According to Mitchison, the cell cycle is "the period between the formation of the cell by the division of its mother cell and the time when the cell itself divides to form two daughters" (Mitchison, 1971, p. 1). Passage through the cell cycle is marked by the occurrence of a number of events that take place only during restricted portions of the cycle and that are repeated periodically from one cycle to the next. Some of these marker events are DNA synthesis, nuclear division, the synthesis of or capacity to synthesize certain enzymes, and the susceptibility to metabolic inhibitors or ionizing radiation. Under ideal conditions, this sequence of events is repeated indefinitely as the cell culture grows exponentially. Thus, the cell cycle is the fundamental unit of temporal organization in the history of a population of well-nourished cells.

How the cell controls this orderly progression of events has been explained in essentially four different ways:

1. Dependent sequences of events (Pringle and Hartwell, 1981). In this view, the cell cycle is a causal sequence of events that closes on itself, e.g.,

$$A \to B \begin{array}{c} \nearrow\ C \to D\ \searrow \\ \searrow\ E \to F\ \nearrow \end{array} G \to H \to A \to \ldots$$

   where the arrows represent causal relationships.
2. Cell size control (Fantes et al., 1975). In this view, cells must attain a minimum "size" before certain events (e.g., DNA synthesis) can occur. Size could be measured by cell

volume, surface area, protein content, RNA content, and so on. There may be one or more size "check points" during the cycle to coordinate growth and division.

3. Control by an autonomous clock (Kauffman and Wille, 1975). In this view, the cell has an internal clock that ticks off inexorably, triggering various periodic events of the cell cycle but remaining independent of the events it controls.
4. Random transitions (Smith and Martin, 1973). In this view, the cell does not cycle at all but proceeds fitfully from one holding state to another. Release from the holding states is conceived to be a random process, occurring with a certain constant probability per unit time.

In this chapter we shall assess these four views of cell cycle control with respect to the nuclear division cycle in *Physarum polycephalum* (Sachsenmaier, 1981b; Tyson, 1982). *Physarum* is an acellular slime mold (class Myxomycetes). In the diploid plasmodial stage of its life cycle, cell division does not occur. The cell grows as a large flat plasmodium containing hundreds of millions of nuclei in a common cytoplasmic environment. The cytoplasm is criss-crossed by a network of actomyosin-containing veins that pump nuclei, organelles, macromolecules, and cytosol vigorously throughout the plasmodium. Within this well-mixed cytoplasmic environment the nuclei all divide synchronously (about two or three times per day depending on growth conditions) and they all replicate their DNA synchronously during the first few hours after nuclear division. Furthermore, the activities of certain enzymes fluctuate periodically during the cell cycle (e.g., thymidine kinase, ornithine decarboxylase, histone kinase, DNA polymerase, RNA polymerase B, deoxyadenosine kinase), and some of these fluctuations are known to be caused by periodic enzyme synthesis (Tyson, 1982). This natural synchrony makes *Physarum* an important model organism for studies of the control of the cell cycle.

The *Physarum* cell cycle has no "$G_1$" period, that is, there is no gap between mitosis and the onset of DNA synthesis. Daughter nuclei begin DNA synthesis immediately after separation (telophase). This lack of $G_1$ is not uncommon among lower eukaryotes (Mitchison, 1971, p. 72) or unknown among mammalian cell types (Robbins and Scharff, 1967). It indicates, in *Physarum,* a close coupling between nuclear division and DNA synthesis. DNA synthesis is necessary before nuclei can divide, and the onset of nuclear division (though not its completion) is necessary before DNA synthesis can occur.

## SIZE CONTROL

Coordinating nuclear division, DNA synthesis, and other cell cycle events with cell growth by monitoring cell size can be accomplished by placing a single cell size restriction any place in the cell cycle. For instance, a cell size requirement could be placed at the onset of nuclear division, at the onset of DNA synthesis, or at some other event, as long as there are causal links between "secondary" events of the cycle and the "primary" controlled event. Thus, in the budding yeast, *Saccharomyces cerevisiae,* the "start" event seems to depend on the cell's reaching a minimum size (Johnston et al., 1977); in fission yeast, *Schizosaccharomyces pombe,* there seem to be size requirements at both nuclear division and the onset of DNA synthesis (Nurse and Thuriaux, 1977); in mammalian cells size controls seem to be manifested at the onset of DNA synthesis only (Killander and Zetterberg, 1965); and in *Physarum,* size controls seem to operate at nuclear division (Tyson, 1982; Loidl and Sachsenmaier, 1982). In what follows, we shall speak as if size controls mitosis, with DNA synthesis following automatically thereafter (possibly with some delay).

## Evidence for Size Control of Nuclear Division

### *Cell Fusion*

Besides natural mitotic synchrony, *Physarum* has another remarkable property that ideally suits it for cell cycle studies: if two genetically compatible plasmodia come in contact, their surface membranes over the region of contact disperse and their cell contents become mixed. The resulting fused plasmodium shares properties of the two original plasmodia. For instance, if a plasmodium due to undergo mitosis at 9 p.m. is fused at noon with a plasmodium due to undergo mitosis at 4:30 p.m., the resulting fused plasmodium will undergo mitosis at approximately 7:30 p.m. (Rusch et al., 1966). If the early-dividing plasmodium is twice as large as the late-dividing plasmodium, then nuclei of the fused plasmodium will divide considerably earlier, say about 6:00 p.m. (Rusch et al., 1966).

Such observations, which have been made by many different investigators, suggest that there is an intracellular substance that accumulates as plasmodia progress through the cell cycle and is "averaged out" on fusion of plasmodia at two different stages in the cycle.

### *Protein/DNA Ratio at Division*

The protein/DNA ratio in batch cultures of (asynchronous) microplasmodia is remarkably constant over a fourfold range of ploidy levels (Mohberg et al., 1973) and over a fourfold range of growth rates (achieved by varying nutritional conditions) (Plaut and Turnock, 1975). For a twofold range of growth rates of macroplasmodia (achieved by varying incubation temperature), the protein/DNA ration at mitosis does not change significantly (Exenberger, 1983). Furthermore, the protein/DNA ratio at mitosis in (synchronous) macroplasmodia seems to be tightly controlled: after perturbation of the mitotic cycle with heat, cycloheximide, or ultraviolet (UV) irradiation (Tyson et al., 1979), or with ionizing radiation (Exenberger, 1983), the protein/DNA ratio at mitosis is quicly restored to the normal level. This suggests some feedback between size, as measured by protein content, and DNA synthesis. In fact, from a more abstract argument, we can conclude that there *must* be some feedback between plasmodial size and intermitotic times. Sensitive measurements of protein accumulation (by the method of isotope dilution) indicate that *Physarum* plasmodia increase in protein content exponentially (Turnock et al., 1981). That is, the rate of protein synthesis increases continuously in proportion to the amount of protein in the cell. Now, if the timing of mitosis is controlled by some clock that is unaffected by plasmodial size, and if the period of the clock is subject to random variations (±5% is not unreasonable for the variation in intermitotic times in *Physarum*), then the variation in size (i.e., protein/DNA ratio) at division must increase from one generation to the next without bound. This happens because under exponential growth the protein/DNA ratio will drift randomly away from any preset value. Since such drift is not observed in exponentially growing plasmodia, there must be some direct coupling between growth and the division timing mechanism.

## Testing Models for Size Contrcl of Nuclear Division

### *Two Simple Models*

A simple model to account for the constancy of the protein/DNA ratio at mitosis and for the intermediate time of mitosis in fused plasmodia assumes the existence of a cytoplasmic substance whose cytoplasmic concentration changes monotonically as the plasmodium progresses through the cell cycle. When the concentration reaches a critical value, mitosis and a new round of DNA synthesis are triggered. The substance may be an activator, synthesized at a rate proportional to the mass of the cytoplasm, or an inhibitor, degraded at a rate proportional to the mass of the cytoplasm. During nuclear division or early in S phase, an aliquot of activator, proportional to the number of genome equivalents in the cell, is destroyed (or an aliquot of inhibitor is produced). Under steady-state growth conditions, plasmodia will enter mitosis at a constant nucleocytoplasmic ratio, and upon fusion plasmodia will adopt an intermediate activator (or inhibitor) concentration and thus an intermediate time until the next division.

Another simple model (Sachsenmaier et al., 1972) envisions an activator substance that accumulates in the cell in direct proportion to the size of the plasmodium (e.g., the total protein content). This substance binds tightly to sites on the nuclei, the number of sites per nucleus being fixed. Until all the nuclear sites are bound, the free cytoplasmic activator concentration is small, but after all the sites are filled the concentration of free activator in the cytoplasm increases abruptly (just like the titration curve of a strong acid with a strong base). This abrupt and uniform change in free activator concentration throughout the plasmodium triggers division in all the nuclei simultaneously. As before, under steady-state growth conditions, the trigger will be pulled whenever the nucleocytoplasmic ratio reaches a critical level. Since time until the next mitosis is essentially determined by the fraction of unfilled sites in a plasmodium, fused plasmodia will enter mitosis at an intermediate time because the fraction of unfilled sites will be the average of the fractions of unfilled sites in the individual plasmodia before fusion. (This assumes that activator binding to nuclear sites is reversible and sufficiently fast to allow the fraction of filled sites in two subpopulations of nuclei to come into equilibirum within the time necessary for plasmodial fusion; Sachsenmaier, in preparation).

*UV Irradiation Experiments*

Sudbery and Grant (1975) have shown that the timing of mitosis in plasmodia exposed to UV radiation can be used to distinguish between these two models. UV irradiation at $M_2$ (the second synchronous mitosis after inception of a macroplasmodium) causes a delay of $M_3$ and the destruction of a fraction of the nuclear population during the next S phase; the proportion of nuclei destroyed increases with increasing dose (Sudbery and Grant, 1975; Sachsenmaier et al., 1970). The intermitotic time from $M_3$ to $M_4$ is shortened because of the abnormally small nucleocytoplasmic ratio at the end of this S phase. Sudbery and Grant (1975) showed that the simple concentration model predicts not only a shortening of $M_3$-$M_4$ but a significant shortening of $M_4$-$M_5$ and $M_5$-$M_6$ as well. On the other hand, the nuclear sites-titration model predicts a shortening of $M_3$-$M_4$ only; $M_4$-$M_5$, and so on, should be of normal intermitotic duration. Since in *Physarum* it is possible to measure intermitotic times quite accurately over many generations, it is possible to test the two models by measuring the duration of $M_4$-$M_5$ with respect to control (unirradiated) plasmodia. The experiments of Sudbery and Grant (Table 1) are inconsistent with the simple concentration model and consistent with the nuclear sites-titration model.

However, Fantes et al. (1975), Sudbery and Grant (1975), and, recently, Wheals and Silverman (1982) have shown that the simple concentration model can be modified slightly to account for the results of the UV irradiation experiments. Suppose the cytoplasmic trigger substance is a very unstable inhibitor whose rate of production is at all times proportional to the total number of nuclei in the plasmodium (Fantes et al., 1975; Sudbery and Grant, 1975). Then the cytoplasmic inhibitor concentration will at all times be directly proportional to the nucleocytoplasmic (NC) ratio. Since the NC ratio falls by

**TABLE 1** Effect of UV Irradiation on *Physarum* Intermitotic Times[a]

| | Intermitotic times (irradiated/control) | | |
|---|---|---|---|
| | $M_3$-$M_4$ | $M_4$-$M_5$ | $M_5$-$M_6$ |
| Observations | 0.62 ± 0.03 | 0.95 ± 0.04 | 1.00 ± 0.05 |
| Predictions of simple concentration model | Given | 0.77 ± 0.02 | 0.87 ± 0.02 |
| Predictions of nuclear sites-titration model | Given | 1.00 | 1.00 |

[a]Plasmodia were irradiated (200 $J/m^2$) 4 hr after $M_2$. At $M_3$ about 20% of the nuclei in each irradiated plasmodium were lost, causing a decrease in the next intermitotic period by 38%. The two following intermitotic periods, however, were of the same length as the controls, within experimental error. The predictions of the models for $M_4$-$M_5$ and for $M_5$-$M_6$ were made assuming that $M_3$-$M_4$ was 0.62 ± 0.03.
*Source*: Adapted from Sudbery and Grant (1975).

a factor of 2 from one mitosis to the next, so does the inhibitor concentration. Presumably, when the inhibitor has been sufficiently diluted by cytoplasmic growth, its inhibitory effect is reduced and the nuclei divide synchronously. Alternatively, Wheals and Silverman (1982) imagine a very unstable activator, synthesized at a rate proportional to cytoplasmic mass and degraded by first-order kinetics with a rate constant proportional to the NC ratio. In this case the activator concentration will at all times be inversely proportional to the NC ratio, and nuclear division is triggered when activator concentration reaches a critical level. Both models can account for the fusion and UV experiments described so far.

### *Cycloheximide Pulse Experiments*

Cycloheximide is a potent inhibitor of protein synthesis in *Physarum*, but its effects can be nullified simply by transferring a plasmodium several times to fresh medium without the drug (Tyson et al., 1979). If 90 min treatments with cycloheximide (10 μg/ml medium) are initiated at various times during the mitotic cycle, one observes short delays (approximately 3 hr) of the next mitosis for treatments early in the cycle, and increasingly longer delays for treatments late in the cycle (maximum 7 hr delay for treatment started 45 min before metaphase) (Tyson et al., 1979; Scheffey and Wille, 1978).

The delay of mitosis caused by cycloheximide pulses is an awkward observation for the unstable inhibitor model since, if the inhibitor were a protein, it would be rapidly degraded during the cycloheximide treatment and the next mitosis would be advanced instead of delayed. The pattern of increasing delay as cycloheximide pulses are started later in the cycle is consistent with the unstable activator model of Wheals and Silverman (1982) and with Sachsenmaier's nuclear sites-titration model (Tyson et al., 1979). The titration of activator by nuclear sites seems to provide a more sensitive trigger for nuclear division than the unstable activator model (in which activator is degraded by nuclear sites) because the free cytoplasmic activator concentration in the former model increases abruptly by many orders of magnitude after the titration is complete, whereas cytoplasmic activator concentration in the latter model increases smoothly by a factor of only 2 during the intermitotic period. It is hard to see how small changes in activator concentration at the end of the intermitotic period can be sensed by all nuclei simultaneously unless some positive allosteric or cascade mechanism is appended to the model to amplify the final small change that is to trigger synchronous nuclear division.

Cycloheximide experiments suggest that the unstable activator is a protein. But it could belong to some other class of molecules whose synthesis is stringently coupled to protein synthesis. Johnston and Singer (unpublished) have suggested that cytoplasmic size is measured by accumulating portions of ribosomal precursor RNA (those segments not destined to end up in a ribosome). Binding to nuclear sites could be mediated by base pairing. These "coupons" could serve as the activator if their synthesis were inhibited by cycloheximide concordantly with protein synthesis inhibition. An obvious experiment, to repeat the protocol of the cycloheximide experiments using an inhibitor of RNA synthesis such as actinomycin C, has not yet been done.

### *5-Fluorodeoxyuridine Experiments*

In combination with uridine (100 μg Urd/ml medium), 5-fluorodeoxyuridine (5 μg FdUrd/ml medium) inhibits DNA synthesis in *Physarum* without blocking RNA synthesis. If DNA synthesis in a plasmodium is slowed down by continuous treatment with FdUrd + Urd, mitosis is delayed until its nuclear DNA is completely replicated (Sachsenmaier and Rusch, 1964). It is not clear, however, whether the putative mitotic activator is synthesized during the inhibitor treatment (but not effective since the DNA is not fully replicated) or whether activator synthesis is delayed until the completion of DNA synthesis. In *Physarum* one can easily test for the presence of activator in an inhibitor-treated plasmodium simply by fusing the treated plasmodium with an untreated $G_2$-phase plasmodium (Sachsenmaier et al., 1972). If the treated plasmodium has a full complement of activator, then the division-competent nuclei of the $G_2$-phase plasmodium will divide on schedule with a control fusion, but if the treated plasmodium has little or no activator, the $G_2$ nuclei will be considerably delayed with respect to the control. On performing this experiment, Sachsenmaier et al. (1972) found that the $G_2$ nuclei were delayed and divided at a time consistent with the hypothesis that the inhibitor-treated plasmodia did not synthesize activator until after completing DNA replication.

Tyson and Sachsenmaier (1979) have suggested a mechanism whereby activator synthesis can be turned off at the beginning of S phase and turned on at the end of S phase. Their mechanism assumes that the activator gene is inducible. Another possible explanation of the synthesis of activator only during $G_2$ phase relies on a possible connection between replication and transcription in *Physarum* (Pierron and Sauer, 1980). We could imagine the activator gene being replicated at the end of S phase with a burst of mRNA synthesis at the time of replication. This mRNA would support activator synthesis during $G_2$ but would have to be destroyed somehow during mitosis.

### *Evidence for a Miminum Intermitotic Period*

If nuclear division is delayed by perturbing a plasmodium with heat shocks, cycloheximide pulses, or UV irradiation, the next intermitotic time after the delayed mitosis is usually shortened (Kauffman and Wille,1975; Tyson et al., 1979; Sudbery and Grant, 1975). The more severe the perturbation, the longer is the delay and the shorter is the subsequent intermitotic period, down to a limit that seems to represent the minimum time needed to complete DNA synthesis and those processes in $G_2$ necessary for nuclear division (Tyson et al., 1979).

When plasmodia are growing as rapidly as possible, the intermitotic period cannot be further compressed, and any delay of the completion of DNA synthesis should cause a corresponding delay in the next mitosis. This effect was noticed by Sachsenmaier and

Rusch (1964), when they treated plasmodia with pulses of FdUrd + Urd to inhibit DNA synthesis for various lengths of time. (The inhibitory effect of FdUrd can be reversed by transferring a plasmodium to medium containing 50 μg thymidine per milliliter medium in place of FdUrd). They found that the delay of mitosis was approximately equal to the length of the FdUrd pulse, i.e., to the delay of the completion of S phase.

Kauffman and Shymko (1982) have presented evidence that delay of the completion of DNA synthesis by FdUrd + Urd treatment is not the only factor contributing to delay of the next mitosis. Using a pigmentless mutant of *Physarum* (strain 898/878), Kauffman and Shymko showed that a 3 hr treatment with FdUrd + Urd initiated 1 hr after mitosis delays the completion of DNA synthesis only by 2 hr, yet the next mitosis is delayed by 9 hr. Since this effect is *not* observed in the wild-type $M_{3C}$ strain of *Physarum* (Kauffman and Shymko, 1982), it seems reasonable to assume that some side effect of the inhibitor treatment causes the excess delay in the mutant strain.

## LIMIT CYCLE MODELS

### General Description

More than 50 years ago, Rapkine (1931) hypothesized that autonomous periodic biochemical fluctuations might provide the signal for mitosis and cell division in proliferating cells. He went so far as to pinpoint the signal in cyclic changes in oxidation state of intracellular SH groups. Sel'kov (1970) formulated a mathematical model of these ideas, consisting of two ordinary differential equations: one for reduced sulfur (-SH) and one for oxidized sulfur (-SS-). Under certain conditions these equations had limit cycle solutions, i.e., temporally periodic variations in [SH] and [SS] which recover their original form and period even after severe perturbations (Fig. 1). Sel'kov maintained that such periodic changes in thiol compounds could "set the rhythm of cell division." Within this framework of ideas, the cell cycle is governed by an autonomous limit-cycle oscillation (if not in thiol compounds, then in some other metabolites), quiescence corresponds to a stable steady state (no temporal variation in the controlling substances), and carcinogenesis occurs when the control system switches to some aberrant oscillatory solution (Gilbert, 1981).

Klevecz (1976) has advanced a somewhat different form of the oscillator hypothesis. He cites evidence that there is a "quantal" cycle of 3-4 hr period, and that cells are competent to divide (or perform other essential cell cycle tasks) only at certain phases of this quantal cycle. The quantal cycle is thought to be a limit-cycle oscillation in the intracellular concentrations of some unknown substances. Edmunds and Adams (1981) have adopted this idea to account for the entrainment of the cell division cycle to a circadian pacemaker, and they have suggested a specific mechanism (the "chronogene") for inserting and deleting time segments in cell cycles.

For *Physarum* Kauffman and his collaborators (Kauffman and Wille, 1975; Tyson and Kauffman, 1975) have suggested that an autonomous limit-cycle oscillator controls the timing of nuclear division and DNA synthesis. There are two major differences between Kauffman's model and the cell size models of the last section. First, cell size models belong to a class of oscillators called "relaxation" oscillators because at the end of the cycle the control variable is abruptly reset to its initial value. In limit-cycle models, the time-dependent variables change smoothly. (Though limit-cycles may approach relaxation oscillations arbitrarily closely, Kauffman takes pains to insist that his limit-cycle model is not working in the "extreme relaxation regime.") Furthermore, if perturbed,

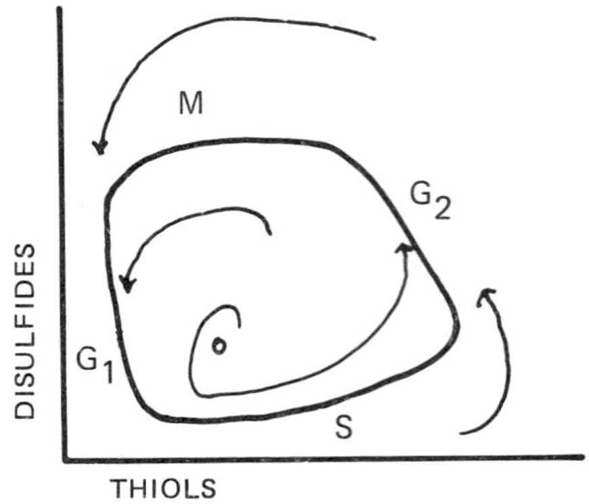


**FIGURE 1** A limit-cycle model of the cell cycle. The concentrations of thiol groups (-SH) and disulfide groups (-SS-) are imagined to oscillate indefinitely around a closed limit cycle in the phase plane. Different portions of the limit cycle are identified with different phases of the cell cycle. Perturbations away from the limit cycle are damped out as the system returns to the limit cycle along trajectories like the four curves with arrowheads. Inside the limit cycle is a resting point (o). If a perturbation lands the control system on the resting point, then oscillations in [SH] and [SS] are eliminated and, presumably, the cell cycle is halted indefinitely. (Adapted from Selkov, 1970.)

relaxation oscillators tend to recover rapidly, whereas smooth limit-cycle oscillators may take a long time to reestablish the undistrubed rhythm (see Fig. 1). In particular, if a perturbation lands the control system on the steady state within the limit cycle, then the oscillator never recovers.

The second major difference is that Kauffman's limit-cycle oscillator runs independently of the events it is timing, whereas in cell size models the discontinuous event of nuclear division or gene replication plays a fundamental role in generating periodicity.

### Testing the Limit-Cycle Hypothesis

As we just mentioned, perturbations exist that can suppress limit cycle oscillations by landing the control system on the steady state inside the closed orbit. Winfree (1974) has shown how this property of limit cycles can be used to test if a control system behaves like a limit-cycle oscillator. Limit cycles are suspected if one can find a critical perturbation of the cell cycle (by heat shock, inhibitor pulses, or plasmodial fusion) that abolishes cell cycle periodicity for a period longer than the normal cycle time under conditions that are otherwise normal. That is, the perturbed plasmodium should double in mass one or more times without concomitant nuclear division or DNA synthesis, or with temporally disorganized nuclear division and DNA synthesis.

Kauffman and his coworkers have presented evidence from fusion experiments (Kauffman and Wille, 1975) and heat shock experiments (Wille et al., 1977) that purports to establish just such a situation of skipped divisions. However, Tyson and Sachsenmaier (1978) have presented evidence, also from fusion and heat shock experiments, that belies limit-cycle control of the nuclear division cycle in *Physarum*. Though the reader is referred to the original article for our detailed critique of limit-cycle models, our basic objections can be summarized quite simply (Tyson, 1982). Referring to Fig. 2, we consider the fusion of a plasmodium shortly before mitosis with a plasmodium shortly after

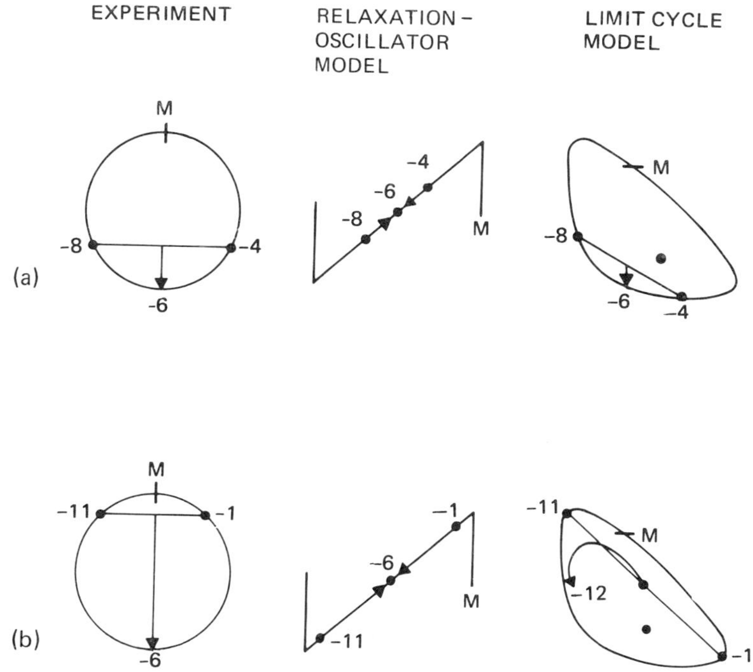


**FIGURE 2** Divergent predictions of a continuous limit-cycle model and a relaxation oscillator model. (a) If plasmodia 4 and 8 hr from metaphase are fused, the mixed plasmodium undergoes synchronous nuclear division about 6 hr after fusion. This result is predicted by both models of the mitotic control system. (b) If a plasmodium about 1 hr before metaphase is fused with a plasmodium that has recently gone through metaphase, the mixed plasmodium undergoes nuclear division about one-half to three-quarters of a cycle after fusion. This is consistent with the relaxation oscillator model, but the limit-cycle model predicts a full-cycle delay. (From Tyson, 1982.)

mitosis. According to a relaxation oscillator model, such as the nuclear sites-titration model, the fused plasmodium should undergo nuclear division after about a half-cycle delay. According to a limit-cycle model like Kauffman's, nuclei of the fused plasmodium should divide either immediately or after a full cycle delay. The uniform opinion of all investigators who have done this experiment is that the nuclei divide after about a half-cycle delay. The long argument in Tyson and Sachsenmaier (1978) merely shows that no amount of twisting, stretching, speeding up, or slowing down of the limit cycle and its nearby trajectories (short of making the limit cycle an "extreme relaxation oscillator") can account for this discrepancy without introducing equally disastrous errors elsewhere.

There is another way to express this difficulty with the limit-cycle hypothesis (Sachsenmaier, 1981). A *continuous* limit-cycle model predicts that if the phases of the two plasmodia to be fused are sufficiently close, then the phase of the fused pair *must* be close to the nearly common phase of the two original plasmodia. This is a consequence of the "continuity" assumption. On the other hand, relaxation oscillator models (which have a discontinuous jump at some phase in the cycle) predict that if the two plasmodia to be fused are at phases on opposite sides of the jump, then no matter how close they may be in phase, the phase of the fused pair must be much different from the nearly com-

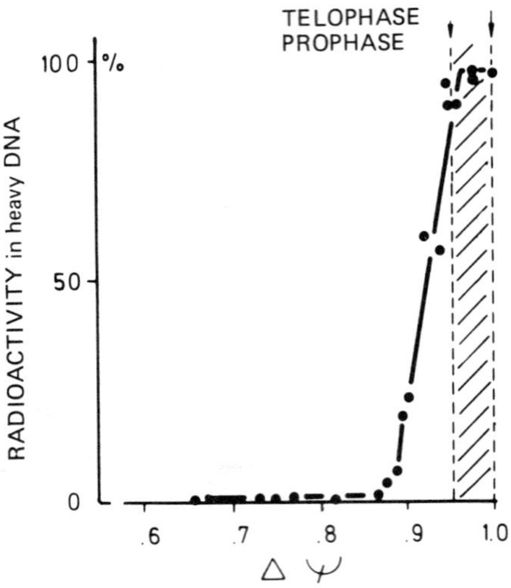


**FIGURE 3** Commitment of nuclei to mitosis. Two sets of macroplasmodia (A) and (B) are cultivated with different mitotic schedules. One plasmodium (B) is prelabeled with [$^{14}$C]thymidine throughout the S period following the first synchronous mitosis ($M_1$) after the start of the culture. At various time intervals during the $G_2$ period preceding mitosis ($M_3$), plasmodial pieces $\simeq$ 2 cm$^2$) are excised and fused with equal-sized pieces of an unlabeled plasmodium (A) that is just finishing mitosis and is about to enter the S period. At the same time, 5-bromo-2′-deoxyuridine (1.6 × $10^{-4}$ mol) (BrdUrd) is added to the medium in order to shift the density of newly synthesized DNA. The mixed plasmodia are harvested after the completion of the S period in control B cultures ($\simeq$ 5 hr after mitosis $M_3$) and the nuclear DNA is analyzed by ultracentrifugation in a preparative CsCl density gradient. Radioactive DNA is shifted from light density (= 1.709) to heavy density ( = 1.734) only if $G_2$ nuclei (B) enter mitosis and the ensuing S period unhindered during the exposure to BrdUrd. The results of two experiments with a total of 18 fused pairs of plasmodia (A/B) are shown. Abscissa: phase difference between A and B at the time of fusion; standard cycle = 1.0 (8.5 hr). Ordinate: percentage radioactivity shifted from light density to heavy density in the CsCl gradient, corresponding to the percentage of $G_2$ nuclei in the mixed plasmodium, which had become insensitive to the retarding influence of the "young" A plasmodium and proceed on schedule into mitosis and the following S period. (From Loidl and Sachsenmaier, 1982.)

mon phase of the two original plasmodia. Loidl and Sachsenmaier (1982) have studied this situation carefully, fusing a plasmodium just after mitosis (i.e., early S phase) with a plasmodium shortly before the expected onset of mitosis. Instead of finding a small advance in phase of the old nuclei and a small delay in phase of the young nuclei (as expected from a continuous limit-cycle model), they found a large delay of the old nuclei and a large advance of the young nuclei, consistent with a relaxation oscillator model with a discontinuous jump close to the time of mitosis.

The experiments of Loidl and Sachsenmaier also imply that the irreversible commitment of nuclei to mitosis practically coincides with the first visible signs of the onset of nuclear division (Fig. 3). When does the discontinuity in the control system occur? Clearly, it cannot occur before commitment. On the other hand, it cannot occur later than about 20 min after telophase because, were it to occur later, a plasmodium in telophase, when fused to a plasmodium in late $G_2$, would not be able to delay mitosis of the

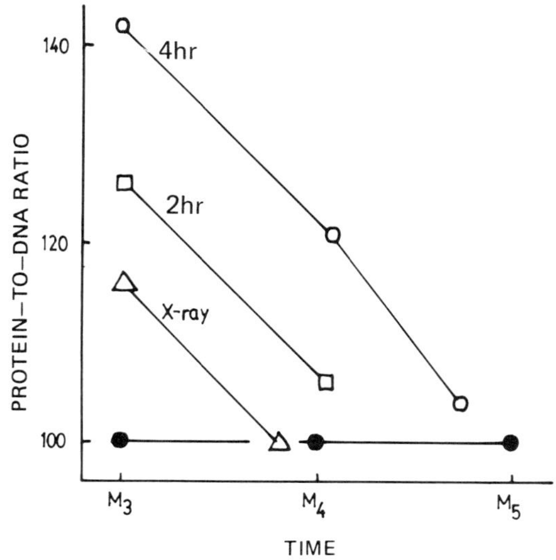


**FIGURE 4** Response to heat shock and X-irradiation. Abscissa: Intermitotic time, expressed as a fraction of the control intermitotic time. Ordinate: Protein/DNA ratio at nuclear division, expressed as percentage of the control ratio. Key: ●, control, o, 4 hr heat shock (at 37°C, started 1 hr before $M_3$); □, 2 hr heat shock (at 37°C, started 1 hr before $M_3$); △, X-irradiation (600 rad, 50 kV, at $M_2$). (Adapted from Tyson and Sachsenmaier, 1979, and Exenberger, 1983.)

late $G_2$ nuclei. It is impossible to be more precise because of the 30 min time necessary for fusion of two plasmodia.

Our objection to the heat shock experiments of Kauffman and Wille (Kauffman and Wille, 1975; Wille et al., 1977) is that long exposures to lethal temperatures have a drastic effect on overall plasmodial growth, so a basic assumption (that plasmodia grow normally after the perturbation) is not satisfied. Furthermore, there is no clear evidence of heat shock causing skips in or temporal disorganization of the DNA-division cycle. In a case studied carefully by many investigators (Brewer and Rusch, 1968; Kauffman and Wille, 1975; Tyson and Sachsenmaier, 1978; Wolf et al., 1979), heat shocks *of any duration* applied shortly before metaphase seem to block the separation of nuclei at telophase but DNA synthesis proceeds normally in these undivided nuclei to produce polyploid nuclei.

How do discontinuous, relaxation oscillator models account for the timing of nuclear divisions after heat shocks? The first mitosis after heat shock is generally delayed with respect to the control, and the amount of delay depends on the phase of the cycle at which the heat treatment is started. Presumably, the heat shock interferes with some heat-labile component of the nuclear division mechanism; histone kinase has been suggested by Bradbury et al. (1974). At the delayed mitosis the plasmodial protein/DNA ratio is larger than normal because the plasmodium has grown abnormally large during the lengthened intermitotic period. Subsequent mitoses are shorter than normal because the plasmodium does not have to double its size in order to bring the protein/DNA ratio back to its normal value. Since there is a limit to how short the intermitotic period may become, a plasmodium may require more than one short cycle to restore its protein/DNA ratio to the normal premitotic value (Fig. 4).

The shortening of intermitotic times after a delayed mitosis is a common reature of any perturbation that produces an abnormally large protein/DNA ratio at the delayed mitosis. Recent data of Exenberger (1983) on the response of plasmodia to X-irradiation shows the same pattern as the response to heat shock (Fig. 4).

## OTHER MODELS

### Dependent Sequences

In budding yeast, *Saccharomyces cerevisiae,* a complicated network of dependent sequences of events (also called "functional sequences") has been established by analyzing a host of conditional lethal mutants that, at the lethal temperature, are blocked at some point in the cell cycle (Pringle and Hartwell, 1981). A similar analysis of the nuclear division cycle in *Physarum* has not been possible because it has proven difficult to isolate cell cycle mutations that are expressed in the plasmodial stage of the cell cycle (Dee, 1982).

To view the cell cycle as a dependent sequence of events is not inconsistent with the other models of *control* of the cell cycle. The other models (cell size, limit cycles, random transitions) are viewed as triggers that set off a sequence of events terminating in division. Thus, the dependent sequence can be thought of as the "nuts-and-bolts" of the cell cycle, as distinct from a control system that sets the apparatus in motion. Of course, if the dependent sequence closes on itself,

A → B → C → D
↑_____________|

it would be self-perpetuating and in no need of an external triggering mechanism. However, even in *S. cerevisiae,* this does not seem to be the case. Johnston et al. (1977) and others (Hartwell and Unger, 1977; Tyson, C. B., et al., 1979) have presented a good deal of evidence that the dependent sequence in this organism, which starts at an event called "start," is correlated to cell growth by a minimum cell mass necessary for start to be executed. That is, it appears that a cell size mechanism triggers the first event in the sequences leading to bud emergence, DNA synthesis, nuclear division, nuclear migration, and finally cell separation. It is to be expected that a similar situation holds in *Physarum,* where cell size controls the onset of mitosis, and DNA replication events are causally connected to nuclear division events.

### Random Transitions

The evidence for random processes as essential components of the cell cycle comes mainly from observations of the distribution of cell cycle times in exponentially growing, asynchronous populations of cells, and particularly from measurements of the correlations of cell cycle times between related cells (Smith and Martin, 1973; Minor and Smith, 1974; Shields and Smith, 1977; Brooks et al., 1980). So far, the only attempt to perform analogous experiments with *Physarum* has been made by Sachsenmaier (1981a) and Exenberger (1983). They cultured macroplasmodia under identical conditions, determined the time of $M_2$ and $M_3$, and plotted the data as an $\alpha$ curve. From these data we calculate that, if *Physarum* has a random-exiting phase in its cell cycle, then the probability of exiting this phase per unit time (i.e., the first-order rate constant for exiting) is approximately 60 $\tau^{-1}$ (where $\tau$ = cycle time = 8.2 hr in Sachsenmaier's experiments). This is 10 times larger than the transition probabilities we calculate from the data in Smith

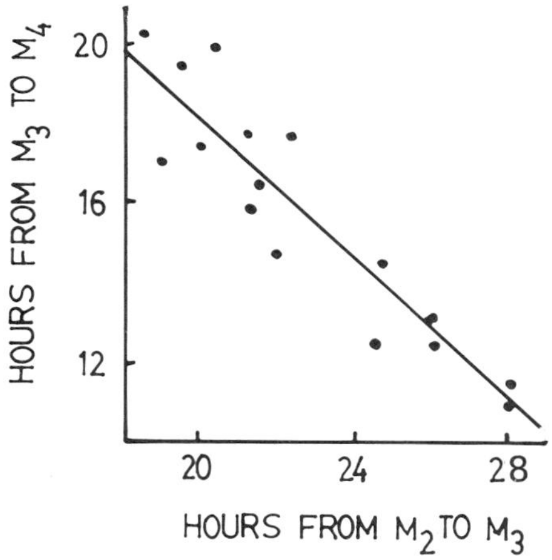


**FIGURE 5** Negative correlation of intermitotic times after an artificial delay of $M_3$ induced by cycloheximide treatment. Plasmodia were treated with a 2 hr pulse of cycloheximide (10 $\mu$g/ml medium) at various times between $M_2$ and $M_3$. As the treatment was begun later in the cycle, it caused increasing delays of $M_3$ and, subsequently, a marked shortening of the $M_3$-$M_4$ intermitotic period. In the graph, the intermitotic period from $M_3$ to $M_4$ is regressed against the intermitotic period from $M_2$ to $M_3$. The data points are highly negatively correlated; product-moment correlation coefficient = -0.92. (Data from Tyson et al., 1979.)

and Martin for cultured animal cell populations (slopes of $\alpha$ curves are approximately $6\,\tau^{-1}$, where $\tau$ = cycle time for the cell culture under consideration). This is just another way of saying that the coefficient of variation (CV) of intermitotic times in replicate *Physarum* plasmodia is much smaller than the CV of interdivision times in animal cell populations. Whereas the latter is typically 20%, the CV for intermitotic times in *Physarum* is typically 5%. Thus, if a random-exiting phase is present in *Physarum,* the transition probability is so large that the effects of the random-exiting phase on intermitotic times are scarcely noticeable.

The lack of large variations of nuclear cycle times in *Physarum* plasmodia does not necessarily contradict the transition probability concept for dividing *uninuclear* cells. If the cycle variation observed in such cells indeed results from random events involving only few molecules or structural elements, these may become deterministic in a large multinuclear plasmodium by increasing the number of the participating elements. Moreover, the *Physarum* nuclear cycle does not include a $G_1$ period as opposed to the division cycle of most higher eukaryotes. This could mean that those events specifically associated with the "A state" in $G_1$ of other cells are missing in the *Physarum* nuclear cycle. If this is true, one would have to conclude that the A state does not constitute an obligatory component of the cell cycle, at least in lower eukaryotes.

An aspect of division intervals in *Physarum* that has not been closely investigated is the correlation of two intermitotic periods, say $M_2$-$M_3$ and $M_3$-$M_4$. Are these intervals negatively correlated, as cell size models would predict, or are they uncorrelated, as random transition models would predict? After artificial delays, intermitotic periods are highly negatively correlated (Fig. 5), but what is the correlations between unperturbed cycle lengths? Our experience with *Physarum* cultures suggests some negative correlation, but this has never been quantified.

## CONCLUSION

We conclude that the evidence from studies of the cell cycle in *Physarum polycephalum* indicate that the major events of growth, DNA replication, and nuclear division in this organism are controlled by a system that monitors some "nucleocytoplasmic" ratio and triggers mitosis and a new round of DNA synthesis each time this ratio falls below a critical level. At present, the simplest model for monitoring the nucleocytoplasmic ratio that is consistent with all the data on *Physarum* (fusion, heat shock, UV irradiation, cycloheximide, and so on) is the nuclear sites-titration model originally proposed by Sachsenmaier et al. (1972). In this model an unstable activator (most likely a protein) accumulates to a level proportional to the amount of cytoplasm and binds to nuclear sites whose number is proportional to the DNA content of the plasmodium. When all the sites are occupied, the mitotic sequence is triggered. The trigger is fully effective, of course, only if the machinery for mitosis and DNA synthesis is fully operational. If some part of the machinery is out of commission, for physiological or genetic reasons, then some aspects of the DNA-division cycle may be delayed or aborted. Details of the machinery (the functional sequence of events) in *Physarum* are currently missing because of a paucity of cell division cycle mutants in the plasmodial stage of the life cycle of *Physarum*.

Furthermore, we conclude that, at present, there is no convincing evidence for the involvement of either limit-cycle oscillators or random transitions in controlling the progress of *Physarum* plasmodia through the cell cycle.

There is a great deal of evidence in organisms other than *Physarum* that cell size plays an important role in regulating the timing of periodic events in the cell cycle. Some of this evidence may be found in this book (Chap. 14) or in reviews published elsewhere (Tyson, 1982; Johnston et al., 1977; Mitchison, 1977). The only evidence that argues persuasively against size control of the cell division cycle is the observation in asynchronous cell cultures that mother-daughter interdivision times are usually uncorrelated, whereas cell size models predict significant *negative* correlation between mother and daughter cycle times (product moment correlation coefficient $\simeq -0.5$). This discrepancy has received much attention, and at present most investigators agree that there is a component of clonal heterogeneity in cell populations that tends to increase the correlation of mother-daughter generation times, bringing the observed correlation close to zero or even significantly positive (Shields and Smith, 1977; Brooks et al., 1980; Miyamoto et al., 1973).

The transition probability model, though it has had some success in explaining the statistics of interdivision times among related and unrelated cells (Brooks et al., 1980), cannot account adequately for the many experiments that indicate a direct role for cell size on cell cycle progress. Brooks (1981) has given some attention to the implications of cell size homeostasis for the transition probability model. He shows that a certain amount of homeostasis arises naturally, as long as growth is not exponential. However, his mechanism of cell size regulation takes several generations to restore cell size to its natural value after a perturbation (see Brooks' Table 1, ignoring the first generation, which is unreasonably extreme). The evidence, on the other hand, indicates that adjustments in cell size after perturbations or nutritional shifts are made very rapidly, usually within one generation (Tyson et al., 1979; Sudbery and Grant, 1975; Tyson and Sachsenmaier, 1978; Hartmann, 1928; Prescott, 1956; Frazier, 1973; Fantes and Nurse, 1977, 1978; Maaloe and Kjeldgaard, 1966).

Like the transition probability model, the limit-cycle model of the cell division cycle is a timing mechanism that runs independently of cell size and, therefore, has difficulty accounting fully for the experiments linking cell size directly with progress through the cell cycle. Furthermore, most of the evidence put forward for the limit-cycle hypothesis is circumstantial. The only well-documented, direct evidence for autonomous cytoplasmic limit cycle oscillations in control of the cell division cycle is the periodic cortical contractions observed in eggs (Bell, 1962), which continue with the same period as the division cycle even in the absence of any nuclear or centriolar material (Hara et al., 1980). The most direct and provocative evidence for the quantal cycle hypothesis is the bimodal phase response curve to serum pulses observed by Klevecz et al. (1978). However, without knowledge, for example, of the effects of serum pulses on macromolecular synthesis, it is difficult to assess the implications of these observations for size control models. Furthermore, the quantal cycle hypothesis predicts damped oscillations in the distribution of cell generation times (Klevecz et al., 1980), but evidence for such oscillations is meager (Murphy et al., 1978).

Hybrid models have been suggested. For instance, one could imagine a minimum cell size requirement for entry into the A state (the random exiting state) of the random transition model (Nurse, 1980). In some as yet unpublished calculations, Tyson has shown that such a model can account nicely for the distribution of the interdivision times among unrelated cells and sister cell pairs but that it still predicts significantly negative correlations of mother-daughter cycle times.

In general, then, though the cell size hypotesis is consistent with everything we know about the control of nuclear division and DNA synthesis in *Physarum,* it has not carried the field among students of the cell cycle in other organisms. This uncertainty about the importance of such fundamental notions as randomness and determinancy in the control of cell cycle events will spice the lives of cell cycle biologists for several years to come.

## REFERENCES

Bell, L. G. E. (1962). Some mechanisms involved in cell division. *Nature (Lond.)* 193: 190-191.

Bradbury, E. M., Inglis, R. J., and Matthews, H. R. (1974). Control of cell division by very lysine rich histone (F1) phosphorylation *Nature* 247:257-261.

Brewer, E. N., and Rusch, H. P. (1968). Effect of elevated temperature shocks on mitosis and on the initiation of DNA replication in *Physarum polycephalum. Exp. Cell Res.* 49:79-86.

Brooks, R. F. (1981). Variability in the cell cycle and the control of proliferation. In *The Cell Cycle,* P. C. L. John (ed.). Cambridge University Press, London, pp. 35-61.

Brooks, R. F., Bennett, D. C., and Smith, J. A. (1980). Mammalian cell cycles need two random transitions. *Cell* 19:493-504.

Dee, J. (1982). Genetics of *Physarum polycephalum.* In *Cell Biology of* Physarum *and* Didymium. H. C. Aldrich and J. W. Daniel (eds.). Academic, New York, pp. 211-251.

Edmunds, L. N., and Adams, K. J. (1981). Clocked cell cycle clocks. *Science* 211:1002-1013.

Exenberger, R. (1983). Ph.D. Thesis. University of Innsbruck, Innsbruck, Austria.

Fantes, P. A., Grant, W. D., Pritchard, R. H., Sudbery, P. E., and Wheals, A. E. (1975). The regulation of cell size and the control of mitosis. *J. Theor. Biol.* 50:213-244.

Fantes, P., and Nurse, P. (1977). Control of cell size at division in fission yeast by a growth-modulated size control over nuclear division. *Exp. Cell Res.* 107:377-386.

Fantes, P. A., and Nurse, P. (1978). Control of the timing of cell division in fission yeast. *Exp. Cell Res.* 115:317-329.

Frazier, E. A. J. (1973). DNA synthesis following gross alterations of the nucleocytoplasmic ratio in the ciliate *Stentor coeruleus*. Dev. Biol. 34:77-92.

Gilbert, D. A. (1981). The cell cycle 1981: One or more limit cycle oscillations? *S. Afr. J. Sci.* 77:541-546.

Hara, K., Tydeman, P., and Kirschner, M. (1980). Cytoplasmic clock with same period as division cycle in *Xenopus* eggs. *Proc. Natl. Acad. Sci. USA* 77:462-466.

Hartmann, M. (1928). Ueber Experimentelle Unsterblichkeit von Protozoen-Individuen. *Zool. Jahrb.* 45:973-987.

Hartwell, L. H., and Unger. M. W. (1977). Unequal division in *Saccharomyces cerevisiae* and its implications for the control of cell division. *J. Cell Biol.* 75:422-435.

Johnston, G. C., Pringle, J. R., and Hartwell, L. H. (1977). Coordination of growth with cell division in the yeast *Saccharomyces cerevisiae*. *Exp. Cell Res.* 105:79-98.

Kauffman, S. A., and Shymko, R. M. (1982). Mitotic delay following inhibition by 5′-fluorodeoxyuridine of S-phase in *Physarum* is not due to delay in termination of S-phase. *J. Cell Sci.* 53:143-154.

Kauffman, S. A., and Wille, J. J. (1975). The mitotic oscillator in *Physarum polycephalum*. *J. Theor. Biol.* 55:47-93.

Killander, D., and Zetterberg, A. (1965). Determination of DNA, RNA and mass content of age determined mouse fibroblasts in vitro and of intercellular variation in generation time. *Exp. Cell Res.* 38:272-281.

Klevecz, R. R. (1976). Quantized generation time in mammalian cells as an expression of the cellular clock. *Proc. Natl. Acad. Sci. USA* 73:4012-4016.

Klevecz, R. R., King, G. A., and Shymko, R. M. (1980). Mapping the mitotic clock by phase perturbation. *J. Supramol. Struct.* 14:329-342.

Klevecz, R. R., Kros, J., and Gross, S. D. (1978). Phase response versus positive and negative division delay in animal cells. *Exp. Cell Res.* 116:285-290.

Loidl, P., and Sachsenmaier, W. (1982). Control of mitotic synchrony in *Physarum polycephalum*. Phase shifting by fusion contradicts a limit cycle oscillator model. *Eur. J. Cell Biol.* 28:175-179.

Maaloe, O., and Kjeldgaard, N. O. (1966). *Control of Macromolecular Synthesis*. W. A. Benjamin, New York.

Minor, P. D., and Smith, J. A. (1974). Explanation of degree of correlation of sibling generation times in animal cells. *Nature* 248:241-243.

Mitchison, J. M. (1971). *The Biology of the Cell Cycle*. Cambridge University Press, London.

Mitchison, J. M. (1977). The timing of cell cycle events. In *Mitosis: Facts and Questions*, M. Little, N. Paweletz, C. Petzelt, H. Ponstingl, and D. Schroeter (eds.). Springer-Verlag, Berlin, pp. 1-19.

Miyamoto, H., Zeuthen, E., and Rasmussen, L. (1973). Clonal growth of mouse cells (strain L). *J. Cell Sci.* 13:879-888.

Mohberg, J., Babcock, K. L., Haugli, F. B., and Rusch, H. P. (1973). Nuclear DNA content and chromosome numbers in the myxomycete *Physarum polycephalum*. *Dev. Biol.* 34:228-245.

Murphy, J. S., D'Alisa, R., Gershey, L., and Landsberger, F. R. (1978). Kinetics of desynchronization and distribution of generation times in synchronized cell populations. *Proc. Natl. Acad. Sci. USA* 75:4404-4407.

Nurse, P. (1980). Cell cycle control—both deterministic and probabilistic? *Nature* 286: 9-10.

Nurse, P., and Thuriaux, P. (1977). Controls over the timing of DNA replication during the cell cycle of fission yeast. *Exp. Cell Res.* 107:365-375.

Pierron, G., and Sauer, H. W. (1980). More evidence for replication-transcription-coupling in *Physarum polycephalum. J. Cell Sci.* 41:105-113.

Plaut, B. S., and Turnock, G. (1975). Coordination of macromolecular synthesis in the slime mould *Physarum polycephalum. Mol. Gen. Genet.* 137:211-225.

Prescott, D. M. (1956). Relation between cell growth and cell division. II. The effect of cell size on cell growth rate and generation time in *Amoeba proteus. Exp. Cell Res.* 11:86-98.

Pringle, J. R., and Hartwell, L. H. (1981). The *Saccharomyces cerevisiae* cell cycle. In *The Molecular Biology of the Yeast Saccharomyces,* J. N. Strathern, E. W. Jones, and J. R. Broach (eds.). Cold Spring Harbor Laboratory, Cold Spring Harbor, New York, pp. 97-142.

Rapkine, L. (1931). *Ann. Physiol. Physicochim. Biol.* 7:382.

Robbins, E., and Scharff, M. D. (1967). The absence of a detectable G1 phase in a cultured strain of chinese hamster lung cell. *J. Cell Biol.* 34:684-685.

Rusch, H. P., Sachsenmaier, W., Behrens, K., and Gruter, V. (1966). Synchronization of mitosis by the fusion of the plasmodia of *Physarum polycephalum. J. Cell Biol.* 31:204-209.

Sachsenmaier, W. (1981a). Does the transition probability model apply to the nuclear cycle in *Physarum* macroplasmodia? In *Biology of Physarum–Current Problems,* L. Rakoczy (ed.). Jagiellonian University Press, Cracow, Poland, pp. 109-113.

Sachsenmaier, W. (1981b). The mitotic cycle of *Physarum.* In *The Cell Cycle,* P. C. L. John (ed.). Cambridge University Press, London, pp. 139-160.

Sachsenmaier, W., Donges, K. H., Rupff, H., and Czihak, G. (1970). Advanced initiation of synchronous mitosis in *Physarum polycephalum* following irradiation. *Z. Naturforsch.* 25b:866-871.

Sachsenmaier, W., Remy, U., and Plattner-Schobel, R. (1972). Initiation of synchronous mitosis in *Physarum polycephalum.* A model of the control of cell division in eukaryotes. *Exp. Cell Res.* 73:41-48.

Sachsenmaier, W., and Rusch, H. P. (1964). The effect of 5-fluoro-2′-deoxyuridine on synchronous mitosis in *Physarum polycephalum. Exp. Cell Res.* 36:124-133.

Scheffey, C., and Wille, J. J. (1978). Cycloheximide-induced mitotic delay in *Physarum polycephalum. Exp. Cell Res.* 113:259-262.

Sel'kov, E. E. (1970). Two alternative auto-oscillatory steady states in the metabolism of thiols–two alternative types of cell multiplication. *Biophysics* 15:1104-1112.

Shields, R., and Smith, J. A. (1977). Cells regulate their proliferation through alterations in transition probability. *J. Cell. Physiol.* 91:345-356.

Smith, J. A., and Martin, L. (1973). Do cells cycle? *Proc. Natl. Acad. Sci. USA* 70:1263-1267.

Sudbery, P. E., and Grant, W. D. (1975). The control of mitosis in *Physarum polycephalum.* The effect of lowering the DNA:mass ratio by UV irradiation. *Exp. Cell Res.* 95:405-415.

Turnock, G., Chambers, J., and Birch, B. (1981). Regulation of protein synthesis in the plasmodial phase of *Physarum polycephalum. Eur. J. Biochem.* 120:529-534.

Tyson, C. B., Lord, P. G., and Wheals, A. E. (1979). Dependency of size of *Saccharomyces cerevisiae* cells on growth rate. *J. Bacteriol.* 138:92-98.

Tyson, J. J. (1982). Periodic phenomena in *Physarum.* In *Cell Biology of* Physarum *and* Didymium, H. C. Aldrich and J. W. Daniel (eds.). Academic Press, New York, pp. 61-110.

Tyson, J. J., Garcia-Herdugo, G., and Sachsenmaier, W. (1979). Control of nuclear division in *Physarum polycephalum.* Comparison of cycloheximide pulse treatment, UV irradiation, and heat shock. *Exp. Cell Res.* 119:87-98.

Tyson, J., and Kauffman, S. (1975). Control of mitosis by a continuous biochemical oscillation: Synchronization, spatially inhomogeneous oscillations. *J. Math. Biol.* 1: 289-310.

Tyson, J. J., and Sachsenmaier, W. (1978). Is nuclear division in *Physarum* controlled by a continuous limit cycle oscillator? *J. Theor. Biol.* 73:723-738.

Tyson, J. J., and Sachsenmaier, W. (1979). Derepression as a model for control of the DNA-division cycle in eukaryotes. *J. Theor. Biol.* 79:275-280.

Wheals, A., and Silverman, B. (1982). Unstable activator model for size control of the cell cycle. *J. Theor. Biol.* 97:505-510.

Wille, J. J., Scheffey, C., and Kauffman, S. A. (1977). Novel behaviour of the mitotic clock in *Physarum. J. Cell Sci.* 27:91-104.

Winfree, A. T. (1974). Patterns of phase compromise in biological cycles. *J. Math. Biol.* 1:73-95.

Wolf, R., Wick, R., and Sauer, H. (1979). Mitosis in *Physarum polycephalum*: Analysis of time-lapse films and DNA replication of normal and heat-shocked macroplasmodia. *Eur. J. Cell Biol.* 19:49-59.

# PART IV

# TIMING OF THE CELL DIVISION CYCLE
## AUTONOMOUS OSCILLATORS AND ENVIRONMENTAL CONTROLS

As we already have seen in Part III, the consideration of mechanisms that control microbial cell cycles often includes the notion of timers and clocks in regulating cell cycle sequences that culminate in mitosis and cell division (Mitchison, 1974). In addition to various dependent pathways, whose blockage would result in the stoppage of all subsequent events of the cell division cycle (CDC), there exists the possibility of "independent timers," which would continue to function even if growth and cell division, for example, were inhibited (Chap. 9).

Inasmuch as mitosis is a periodic event of short duration relative to the total length of the CDC, it is not surprising that various types of *autonomous* biochemical and macromolecular oscillators have been proposed to underlie this and other "landmarks" comprising the CDC (Edmunds and Adams, 1981). These range from those of the relaxation type (in which a single continuous variable accumulates or declines in the cell, triggering some event in the CDC when it reaches a critical threshold and then resetting to a baseline so that the cycle starts again), to those exhibiting limit-cycle dynamics (Pavlidis, 1973; Winfree, 1980). In the latter, a central clock, characterized by a self-sustained oscillation of at least two continuously varying biochemical species, would maintain stable periodic behavior and coordinate the timing of the events comprising the CDC despite transient perturbations in phase.

In the opening chapter of this section, Ronald M. Shymko and coworkers (Chap. 16) provide an excellent marshaling of the evidence for the CDC as an oscillatory system and review the predictions and experimental testing of the various types of model in several cellular systems. (It should be read with reference to Chaps. 14 and 15, in which a strong case was made for a cell "sizer" in the timing of the CDC in yeasts and slime molds). In Chap. 17, Leland N. Edmunds and Danielle L. Laval-Martin summarize the experimental data implicating an autonomously oscillatory *circadian* oscillator (having a free-running period of approximately 24 hr under conditions held constant with respect to illumination, temperature, nutrients, and other variables) in the phasing, or "gating," of cell division in a number of microorganisms. Particular reference is made to the algal flagellate *Euglena,* in which the formal properties characterizing a bona fide circadian oscillator (entrainability, persistence, initiation, phase shiftability, singularity point, temperature compensation of the period) have now been amply documented. This theme is extended to rat and human (both in situ and in organ and cell culture) by Alexander G. Petrovic and co-workers (Chap. 18), who provide convincing evidence for the quantizement of cell generation time toward multiples of 24 hr in a manner analogous to that for the 3-4 hr quantal cycle ($G_q$) described by Klevecz (1976) for more rapidly dividing mammalian cell types.

The final three chapters particularly emphasize the role of environmental signals in the control of the CDC: Simone Puiseux-Dao (Chap. 19) discusses the importance of cell membranes, compartments, and the cytoskeleton; Sallie W. Chisholm and co-workers (Chap. 18) compare division patterns and the adaptive significance of cell cycle phasing in phytoplankton under natural conditions; and Kenneth J. Adams and co-workers (Chap. 21) undertake a quite detailed experimental and formal analysis of the photoperiodic control of the CDC in the algae *Euglena* and *Ceratium*.

## REFERENCES


Edmunds, L. N., Jr., and Adams, K. J. (1981). Clocked cell cycle clocks. *Science* 211: 1002-1013.

Klevecz, R. R. (1976). Quantized generation time in mammalian cells as an expression of the cellular clock. *Proc. Natl. Acad. Sci. USA* 73:4012-4016.

Mitchison, J. M. (1974). Sequences, pathways and timers in the cell cycle. In *Cell Cycle Controls,* G. M. Padilla, I. L. Cameron, and A. Zimmerman (eds.). Academic Press, New York, pp. 125-142.

Pavlidis, T. (1973). *Biological Oscillators: Their Mathematical Analysis.* Academic Press, New York.

Winfree, A. T. (1980). *The Geometry of Biological Time.* Springer-Verlag, New York.

# 16

# The Cell Cycle as an Oscillatory System

**Ronald M. Shymko** / City of Hope National Medical Center, Duarte, California

**Robert R. Klevecz** / The Beckman Research Institute of the City of Hope National Medical Center, Duarte, California

**Stuart A. Kauffman** / University of Pennsylvania School of Medicine, Philadelphia, Pennsylvania

## INTRODUCTION

In the past two decades, four distinct conceptual pictures of the cell cycle in eukaryotes have emerged. Each is based on a relatively distinct body of data and has spawned a relatively distinct research tradition. Briefly, these four views are as follows: (1) An image of the cell cycle as an interlocked and partially branched sequence of discrete events, linked in more or less complex causal chains. Central to this research tradition is the isolation of mutants that appear to block cells at specific points in the division cycle, coupled with attempts to resolve these into causal sequences. (2) A kinetic picture of the cell cycle with two or more discrete states and random transitions between them. This tradition has emphasized analysis of the distribution of cell age at mitosis in a population of presumptive identical cells. (3) "Sizer" models, based on the concept that a reliable ratio of cytoplasmic mass to nuclear content must be maintained, and hence that this ratio plays a critical role in the timing of cycle events. (4) Biochemical oscillator models, which in a variety of forms have been based on the view that essential variables wax and wane in concentration during the cell cycle and trigger events when they reach appropriate thresholds of concentration. This tradition has explored phenomena suggestive of smooth alterations in concentrations by external perturbations, leading to phase-resetting phenomena or to phase averaging when cells at different phases are fused.

These different classes of models are so unrelated at first glance, each addressing widely different aspects of cell behavior, that a coherent picture of the cell cycle has been difficult to achieve. Our aim here is to suggest a broader conceptual framework into which all four research traditions may fit. We will argue that an oscillator picture corresponds in essence to the most plausible microscopic organization of the cell cycle, and in addition provides a way of rationalizing much of the seemingly unrelated cell cycle data found in the literature.

## A GENERAL CAUSAL VIEW OF THE CELL CYCLE

To put the different views of the cell cycle in perspective, it is useful to start with a quite general framework. Some processes in the cycle have the approximate character of events; for example, the onset of DNA replication at a single origin, the termination of replication in a DNA region when two converging forks meet, or the onset of anaphase. One can initially idealize the cell cycle as comprising a large number of discrete events, each of which occurs or not at any given moment, depending upon the prior occurrence or absence of immediately causal events. This can be rephrased in terms of a general dynamical view of the cell cycle. The central concepts are these: (1) The cell's current state is specified at each moment by listing all the events occurring at that moment. (2) The state space available to the cell is composed of all the possible combinations of events that might occur together. Specifically, if there are N possible events each of which either occurs or does not occur at any moment, then there are $2^N$ possible cellular states. (3) The web of causal couplings among the events is such that each state changes to a well-defined successive state. Over a succession of moments, the cell undergoes a succession of transitions between its $2^N$ possible states, thus following a pathway through its state space. Note that the total number of logically possible states available to the cell is enormous. In practice most of these states are never seen; for example, in a cell's normal passage through the cycle it threads its way repeatedly through only a small subset of them.

This model emphasizes the important point that states other than those on the unperturbed cycle can be available to the cell, either as a result of specific signals or because of random fluctuations. Once off the normal cycle, it becomes possible for a cell to continue from the perturbed state on some path through the state space that eventually makes its way back onto the standard cycle, but that has now bypassed some of the normal events of the cycle. In contrast, in the more rigid version of a causal model, all cycle events are causally connected in a one-dimensional closed chain. Deviations from this path are not allowed, and the possible consequences of perturbations are more limited. Among our aims will be to show that some of these alternate paths are accessible to the cell under certain conditions.

## BRANCHED CAUSAL SEQUENCE MODELS

There is a large body of evidence that makes the branched causal sequence model of the cell cycle compelling, at least as a first approximation. The primary evidence comes from experiments on the budding yeast, *Saccharomyces cerevisiae.* Hartwell and his colleagues (1974) have isolated a number of temperature-sensitive cell division cycle (*cdc*) mutants that prevent normal cycling at the restrictive temperature. Many of these mutants have the property that cells arrest at a well-defined point in the cycle, without completing a critical event associated with the mutated gene. Often, further progress through the cycle is prevented, suggesting that passage through the cycle depends on the completion of a number of causally connected critical events.

There are, however, some mutations that prevent certain processes but do not impede later cycle events. Cells of the mutant genotype *cdc-4* are unable to replicate DNA but nevertheless continue budding on a periodic schedule not different from that in wild-type cells (Hartwell et al., 1974). This result is significant in that not only can events in the cycle beyond the site of action of the lesion continue to occur but also that normal cycling itself is unaffected. If this is true when a process as important as DNA replication is inhibited, then it would not be surprising if it were true for other processes as well, and one might suspect that a considerable number of such noncritical mutations have yet to be discovered.

This example points up some difficulties associated with the otherwise powerful *cdc* mutant approach to analyzing the cell cycle. The class of mutants that are readily selected and useful in revealing the logical structure of the cell cycle have the property that a cell's progress through the cycle is only affected at the critical site of action of the gene. That is, cells beyond that site progress normally around the cycle until they return to the site on the next cycle. In fact, this is the only possibility within the limited conceptual picture of the cell cycle as a closed, possibly branched sequence of causally related events, with each event dependent on the function of a single gene. Even in this restricted view, however, the mutant search will only recover mutants in processes that are necessary conditions for subsequent events. Thus, if events A and B must occur to induce event C, a mutant deleting event A will block progress through the cycle. But if either A or B can cause C, mutants in A will not block and not be recovered. Within its own framework, the *cdc* approach will readily uncover mutants only where the causal web is very simple and the deficient event measurable.

Beyond these limitations, however, are those implicit in the picture of the cell cycle as a closed sequence of strictly discrete events. The limitation can be described by an analogy to a grandfather clock, where the gears and escapements control the movements of hands, but the pendulum keeps time and gates the behavior of the system. Mutants that affect the amplitude and period of the pendulum are not phrasable in the *cdc* mutant approach. Indeed, the possible existence of a cellular pendulum is not phrasable in its framework. In contrast, the general view of the cell cycle as a continuous dynamical system encompasses the logical structure of the branched causal sequence but in addition allows for regulation of frequencies and amplitudes of continuous oscillations around closed orbits in the cell's state space.

## CONTINUITY IN THE CELL CYCLE

### General Considerations

Biochemical processes in the cell cycle depend on molecular concentrations and activity states, which in general vary continuously with time. In a branched causal picture of the cell cycle, this implies a continuous, rather than a stepwise, movement of the state of the cell through a continuous portion of its state space. In analogy with the idea of alternative branched pathways discussed earlier, a perturbation affecting the concentrations of reaction components will move the system off its normal trajectory, and from that point it will follow a different path, which may or may not return to the normal cycle after a time. Those events or processes that are strongly tied to the perturbed components now may be altered, or skipped altogether if, for example, a threshold requirement for initiation of such processes fails to be met as a result of the altered trajectory.

### Cell Fusion

Fusion experiments give strong evidence for continuously varying factors in the cell associated with cycle phase. If two cells at different phases in the cycle are fused, in most cases the two nuclei eventually undergo nuclear mitosis synchronously at some time between the times of expected division in the unfused cells (Mitchison, 1971). Therefore, phase of the cycle is in some sense averageable. Such observations, along with the corollary observation that mitosis is synchronous in multinucleate cells (Johnson and Rao, 1971), have led naturally to models that postulate that phase is strongly associated with some factor(s), probably in the cytoplasm, whose level(s) is averaged as a result of the fusion.

The synchronization and phase-averaging behavior is particularly well documented in the syncytial slime mold, *Physarum polycephalum.* In this organism nuclear mitosis is normally synchronous, with up to $10^8$ nuclei dividing within a few minutes of each other every 10-12 hr. Initiation of DNA replication is also synchronous, occurring within a few minutes of mitosis (Guttes and Guttes, 1964). Phase averaging and synchronous nuclear division after fusion can easily be demonstrated, and indeed, changing the mass ratios of the two plasmodia shifts the compromise phase toward that of the larger plasmodium (Rusch et al., 1966). Here again, such phenomena argue strongly that phase is averageable in much the same way that the concentration of a substance accumulating in the cytoplasm through the cycle is averageable. Notice that in a discrete model of the cell cycle there is no reason to expect fused cells to synchronize, or to do so smoothly at an intermediate phase according to their mass ratio.

It is important to keep in mind that not all events or processes of the cycle will be averageable. For example, fusing an S-phase HeLa cell with one in $G_1$ results in early initiation of S in the $G_1$ nucleus, indicating phase averaging. However, a $G_1/G_2$ fusion has no effect on initiation of S phase in the $G_1$ nuclei, yet mitosis in the $G_2$ nucleus is delayed (Rao and Johnson, 1970). Thus, in this case, S-phase processes occur relatively independently even while there is apparent phase averaging with respect to the timing of mitosis. Similar evidence for decoupling of S phase from mitosis can be found in *Physarum* (Kauffman and Shymko, 1982).

### Phase-Resetting Behavior

Further evidence for a continuous component in the cell cycle comes from phase perturbation experiments. Excess division delay, or division setback, in response to a variety of chemical and physical agents is a well-characterized phenomenon in unicellular eukaryotic organisms (Rasmussen and Zeuthen, 1962). When synchronized cultures of *Tetrahymena pyriformis* or *Schizosaccharomyces pombe* are briefly exposed to heat or cycloheximide, they show a pattern of increasing division delay as the time of exposure occurs later and later in the cycle (Zeuthen, 1971; Polanshek, 1977). The pattern is somewhat more complex in animal cells. Figure 1 shows the response of V79 Chinese hamster cells to serum pulses, heat shock, and ionizing radiation. Note the twice repeated pattern of phase shifts through the modal 8.5 hr cycle.

In most phase-response curves the phase shift varies smoothly with the time at which the perturbation is applied, with no more than one or two possibly discontinuous jumps. In a discrete causal picture of the cycle this behavior would not necessarily be expected. That is, distinct processes will in general have different responses to the perturbation, so that the delay or advance of the monitored event need not have any continuous variation through the cycle. That such curves are coherent over long periods in the cycle argues that the perturbation affects some continuous variables in a consistent way, and that those variables dominate the timing of a target event.

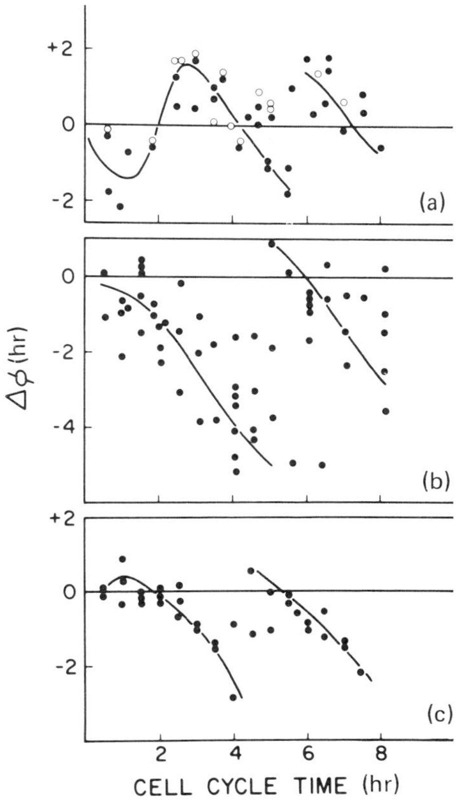


**FIGURE 1** Phase response of synchronous V79 cells to perturbation by serum, heat shock, and ionizing radiation. (a) Serum pulses. At intervals following mitotic selection, serum concentration in the medium was increased from 5 to 20%. Midpoints of first (●) and second (o) mitotic waves. (b) Heat shock. Midpoints of the first mitotic wave following synchronization and a 10 min 45°C heat shock are compared for each pair of heat-shocked and control cultures as described in (a). (c) Ionizing radiation. Synchronous V79 cells were exposed to 150 rad from a cobalt-60 source at 30 min intervals through the first synchronous cell cycle. Analyses of division advance or delay were determined by time-lapse videotape microscopy. (From Klevecz et al., 1980a.)

## RELAXATION OSCILLATOR MODELS

The simplicity of the results of fusion and phase-resetting experiments have led to a number of simple relaxation oscillator models of the cell cycle. These models postulate that a single continuous variable accumulates or declines in the cell, with some critical event in the cycle triggered when the control substance reaches a threshold level. Once the event is triggered or completed, the control substance is reset to a baseline, whereupon the cycle begins again.

In most of these models, cell division is given a special significance, with the control variables being reset as a result of division; i.e., if division does not occur, the reset does not occur. For example, Fantes et al. (1975) have discussed alternative models that satisfy general requirements for balanced growth, and for mitotic synchrony in multinucleate cells. In these models, trigger/reset of a cytoplasmic control factor at mitosis is assumed, with the factor coupled in some way to cell size or DNA content in order to achieve balanced growth. In assessing these alternative models for the mitotic cycle in *Physarum,* Sudbery and Grant (1975, 1976) have concluded that the most accurate model is one in which an unstable inhibitor of mitosis is diluted in the growing cytoplasm, resulting in nuclear division when the amount of inhibitor passes below a threshold.

Another form of this model has been proposed by Sachsenmaier (1976) and coworkers (1972) for the *Physarum* cycle; (see also Tyson et al., 1979). According to this model, a "division protein" begins accumulating after the completion of DNA replication, titrates against nuclear sites, and triggers mitosis at a critical threshold. Again, the effective level of division protein automatically resets as the number of nuclear sites halves at division.

In yeast, it has been proposed that both mitosis and the beginning of DNA synthesis are controlled by the cytoplasmic mass, with a separate threshold cell size for triggering each of these events (Fantes and Nurse, 1977; Nurse and Thuriaux, 1977). This model accounts for the behavior of metabolic mutants whose division cycle seems tightly coupled to their size and rate of growth.

Such models can account for many of the results of phase perturbation experiments. By postulating that the accumulating mitogen is heat labile, a pattern of increasing delays to heat shock through the cycle is predicted, as is observed in *Tetrahymena* (Zeuthen, 1971), *Physarum* (Brewer and Rusch, 1968), and in animal cells (Miyamoto et al., 1973; Klevecz et al., 1980a,b). Similarly, the shorter cycle times seen in *Physarum* after exposure to ultraviolet light is predicted by models based on cycle control by DNA/size ratio. Assuming that DNA is destroyed by the treatment, the DNA/mass ratio is effectively lowered in the subsequent cycle, leading to decreased cycle times as observed (Sudbery and Grant, 1975, 1976).

Fusion results can also be predicted. Fusing two cells at different phases averages the mitogen concentration and if further accumulation is not affected by the fusion, threshold is reached at a time intermediate to those in the unfused cells. Further, if the mass ratio of the two fused cells is shifted, as is possible for *Physarum,* the mean phase will shift toward that of the larger cell (or plasmodium), as observed (Rusch et al., 1966).

The relaxation oscillator model is attractive in its simplicity and capacity to account qualitatively or quantitatively for many aspects of cell cycle behavior. Nevertheless, it has significant faults.

The most serious problem arises from the assumption that mitosis (or DNA replication) is a fundamental component of the cycling of the "clock," so that cycling that does not include mitosis or DNA replication is impossible. Yet evidence in a number of systems suggests that this is not universally correct. For example, *Chorella* cells can pass through rounds of cell division without DNA synthesis, halving the DNA content at each cycle (Schmidt, 1966), implying in addition that they have completed previous rounds of DNA replication without cell division. The mutant blocking initiation of DNA replication in yeast of Hartwell et al. (1974) nevertheless undergoes rhythmic budding. Mano (1971) has shown that cell extracts that undergo neither mitosis nor DNA replication undergo biochemical oscillations with the period of the cells in question. And evidence in *Physarum,* after heat shock perturbation, suggests strongly that plasmodia can skip mitosis for up to two cycles yet "keep time" (Wille et al., 1977).

A somewhat different manifestation of cellular timekeeping comes from time-lapse microscopic observations on generation times in mammalian cells. Rather than showing a smooth distribution, generation times have a tendency to cluster at roughly 4

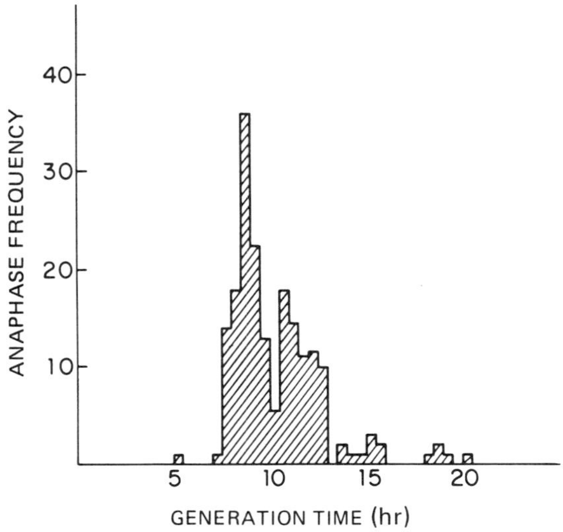


**FIGURE 2** Generation time distribution for unperturbed V79 cells, showing clustering of generation times at roughly 4 hr intervals. (From Klevecz, 1976.)

hr intervals (Klevecz, 1976; 1978). For example, rapidly growing V79 cells often show a bimodal distribution of generation times, with one peak at about 8.5 hr and another smaller peak at about 12 hr (Fig. 2). Such 4 hr subcycling is indirectly reflected in the phase-response curves for various perturbations, where the phase shift pattern repeats itself approximately every 4 hr (Fig. 1).

Results such as these imply that a cell can maintain its timekeeping ability independently of many cell cycle events. If so, there is now the question of how the timekeeping component is coupled to those events. A distinction must now be made between causally linked processes and that subset of processes which are functioning as "timers" under the conditions examined (Soll, 1979). The phenomenology implies that under some conditions, the timer variables are smoothly varying with respect to fusions and phase-resetting perturbations. Under different conditions, different processes may be rate limiting, and responses may differ, but phase averageability and smooth phase resetting must still be maintained.

One simple hypothesis, which essentially modifies the relaxation oscillator picture, would be that a mitogen that constitutes a *central clock* triggers some event, for example mitosis, at a critical threshold, and mitosis then unleashes a cascade of subsequent events that proceed independently of mitogen level, while postmitotic reaccumulation of the mitogen proceeds independently of those events. That is, the central clock gates mitosis but otherwise keeps time independently, in contrast to the usual relaxation oscillator model whose reset depends on mitosis. In modified forms of this hypothesis, different threshold levels of the central clock unleash different cascades–mitosis, DNA, replication, and others. This type of independent clock model has appeal in that it provides a way for coordination of all cellular processes, and provides a time sense for the cell which is reminiscent of the independently functioning circadian clocks found in higher organisms and unicellular eukaryotes (Hastings and Schweiger, 1976; Edmunds and Adams, 1981).

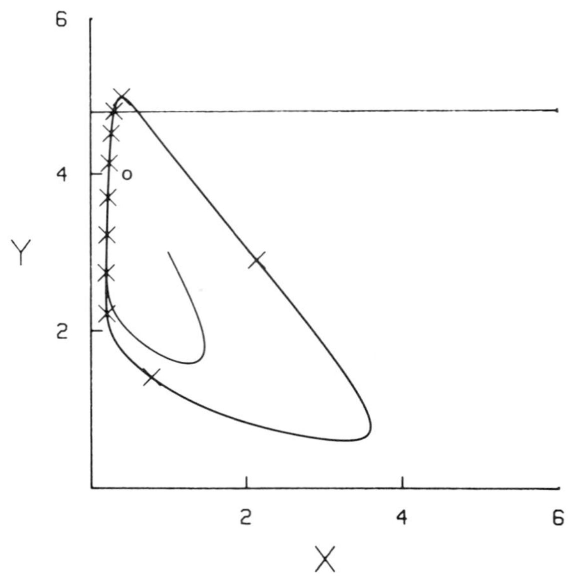


**FIGURE 3** Phase plane trajectory for a limit-cycle oscillator. Released at the point X = 1, Y = 3, the system moves in a clockwise direction to the stable trajectory surrounding the steady state (indicated by o). Crossing the threshold at Y = 4.8 is assumed to trigger gating between successive cell cycle stages. Crosses indicate intervals of 1/10 cycle. Note the rapid movement around the "fast arc" (Kauffman and Wille, 1975) in the lower right portion of the cycle. (From Klevecz et al., 1980a.)

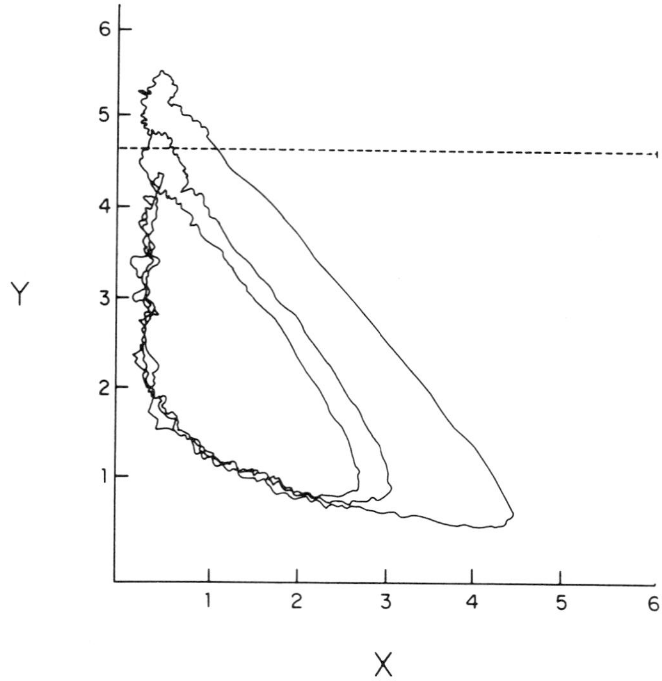


**FIGURE 4** The limit cycle of Fig. 3 with random noise added, showing subthreshold oscillations. (From Klevecz et al., 1980a.)

## LIMIT-CYCLE MODEL

The modified relaxation oscillator model described above has many desirable properties as a cell cycle model, but a more correct model must also take into account a more general property of cell cycles, namely, their stability. In spite of the complexity of processes occurring during the cell cycle, it is striking that cells that are perturbed from their normal cycle do typically return to that cycle after a time, even for rather large perturbations. This suggests that the overall dynamics of the cell cycle can be described rather simply as one in which most regions of the state space, as long as they are consistent with cell viability, relax smoothly back to the normal state cycle following perturbations.

A natural representation for this kind of behavior, which also has the property of continuity, comes from the concept of stable cycles in nonlinear dynamical systems. The simplest form for the stable cycle is termed a *limit cycle* (Pavlidis, 1973; Winfree, 1980).

The fundamental idea behind limit-cycle models is that cells contain a central clock, consisting of a self-sustained oscillation of two or more continuously varying biochemicals, which maintains a periodic behavior and coordinates the timing of cell cycle events. The limit cycle is the simplest system that can maintain an oscillation that will reestablish itself after some perturbation and return stably to the same period and amplitude (or more accurately to the same orbit in its state space). A minimum of two variables is required, with nonlinear interactions and energy or mass input to the system.

The typical properties of a limit cycle are shown in Fig. 3. The concentrations of two substances, X and Y, hypothetical inactive and active mitogens, are plotted at each instant in time in the XY state space of this oscillation. The normal cycling system progresses clockwise (in this model) around a closed orbit in the XY state space. This orbit is termed a limit cycle, because if the system is released on a state off the limit cycle it will follow a trajectory that winds inward or outward, approaching the limit cycle asymptotically. Inside the limit cycle is a unique unstable steady state, with the property that if the system is released at that state, it remains unmoved, but if perturbed off the steady state, it returns to the limit cycle.

In most formulations of this model, some cell cycle event, e.g., mitosis or initiation of DNA replication, is postulated to be triggered when one of the variables exceeds a threshold value, just as in the relaxation oscillator models. The difference now is that the continued oscillation of the system need not depend on the completion of the triggered event. This could happen for example, if some perturbation raises the threshold without affecting the limit-cycle dynamics, or if a perturbation or random fluctuation brings the oscillating variables within the cycle such that the system passes below threshold (Fig. 4). In other words, limit-cycle models admit the important possibility of subthreshold oscillations.

Therefore, the limit cycle has two important properties that mimic those of cell cycle dynamics in general. The first is the explicit possibility for skipping some events while cycling continues. If the stable cycle is extended to more dimensions, with thresholds for initiation of many cycle events, combinations of these dependent events might be skipped on any given cycle, depending on the perturbation (or mutation, as in the yeast *cdc* mutant experiments). The second property is that associated with the stability of the cycle. Typically, a perturbed cell returns to its normal cycle after the perturbation is removed. The limit cycle has the analogous behavior.

It should be emphasized that the relaxation oscillator and limit-cycle oscillator models are not distinguished merely by their kinetic behaviors, that is, the slow rise and rapid fall of the relaxation oscillator variable versus the more sinusoidal oscillations

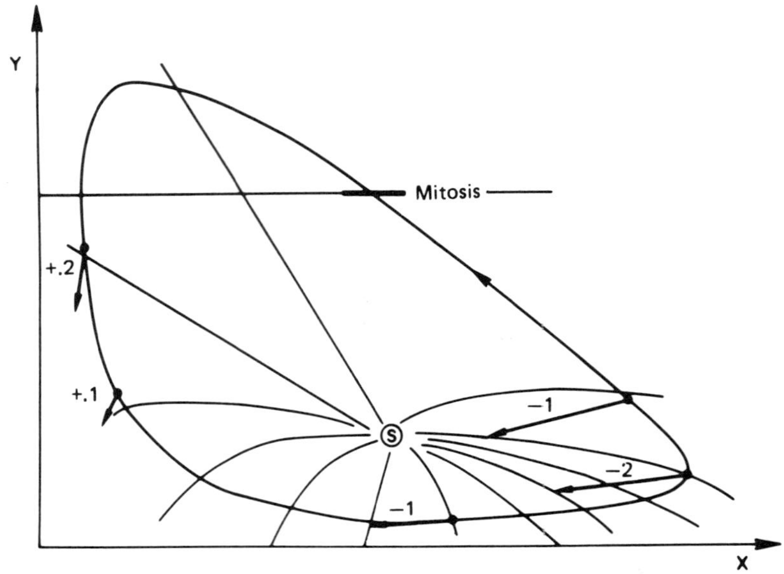


**FIGURE 5** Limit-cycle model for *Physarum,* showing the assumed destruction of X and Y by heat shock perturbation. Movement is counterclockwise in this model. Heat early in the cycle causes advances indicated by positive numbers, and delays are produced later in the cycle. Delay reaches a peak, then declines as mitosis is approached. (From Kauffman and Wille, 1976.)

possible in a limit-cycle oscillator. A limit cycle can easily be constructed that has rapid "relaxation" kinetics in some part of the cycle, making many of the predictions of both models similar.

### Predictions with Respect to Phase-Response

#### *Variable Phase-Resetting Curves with Both Advances and Delays*

Figure 5 shows the consequences of the simplest hypothesis about the effects of heat shocks in a simple limit-cycle model. Assume heat destroys X and Y in proportion to their concentrations and the duration of the shock. Then any shock drives the oscillating system off the limit cycle and toward the origin, X = Y = 0. After sufficiently strong shocks, the system will be found in a relatively small region of XY space near the origin, no matter when in the cycle the perturbation is applied. In this case, the perturbation has the effect of resetting the oscillator to a constant phase, that is, to a point at a constant time from the next threshold crossing. This is analogous to the complete destruction of mitogen in a relaxation oscillator, with reset to the beginning of the accumulation cycle.

An inherent feature of a limit cycle is that at some phases of the cycle, perturbations will cause a delay by driving the system in roughly the reverse direction to its normal progression around the cycle. At other phases, a shock should cause a phase advance by driving the system forward in the direction it was progressing (Fig. 5). Heat pulse experiments in V79 cells (Klevecz et al., 1980a,b) (Fig. 1), and in *Physarum* (Kauffman and Wille, 1975) have shown small phase advances for treatment given early in the cycle, with increasing delay at later times. In addition, the model for *Physarum* predicts a decreasing

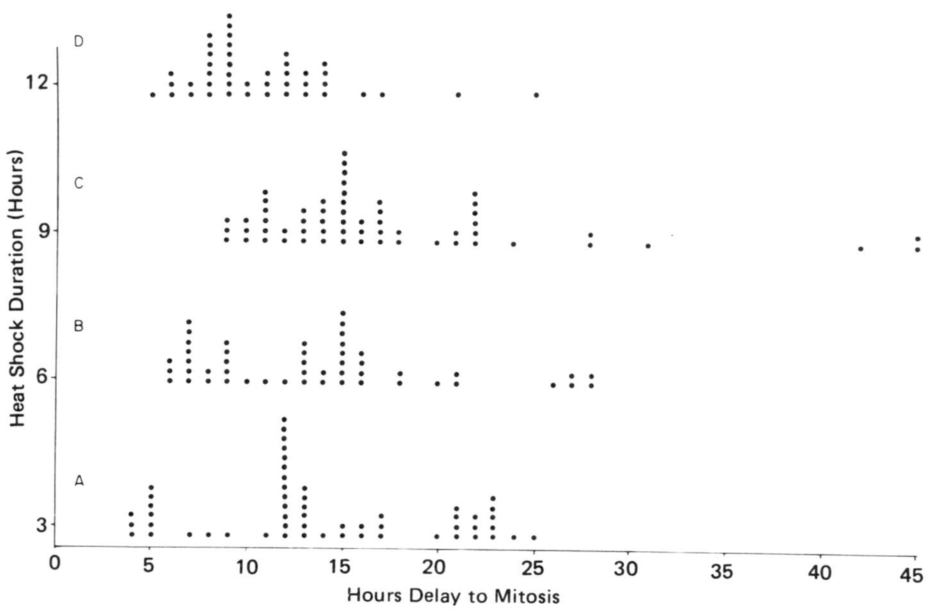


**FIGURE 6** Effects of 3, 6, 9, and 12 hr 37°C heat treatment on timing of mitosis in *Physarum*. The delayed mitoses cluster at 5, 12, and 22 hr post heat shock for 3 hr shocks, suggesting skipped mitosis with continuing timekeeping. Length and variability of delays increase with increasing shock duration to 9 hr, but 12 hr shocks show a decrease in median delay. (From Wille et al., 1977.)

delay as the time of heat shock approaches very near mitosis, which is observed (Kauffman and Wille, 1975). Such variable phase-response curves arise naturally in a limit-cycle model, but require ad hoc postulates in a relaxation oscillator model. For example, predicting the decreasing delay late in the cycle requires the assumption that a heat-labile mitogen accumulating through the cycle converts to a heat-stable form near mitosis (Mitchison, 1971). In an inhibitor-dilution relaxation oscillator, heat destruction of inhibitor should lead to phase advances, not delays; therefore, this model has even more difficulty in accounting for the observed phase-resetting curves.

### *Skipping of Mitosis After Perturbations*

In a two-dimensional limit cycle, if the system is driven near the steady state by some perturbation, it will spiral outward, eventually approaching the normal cycle. The time required before threshold is crossed depends on the exact nature of the limit cycle. One possibility is that there will be an unusually long delay in the event (e.g., mitosis) controlled by the threshold crossing, as several revolutions around the steady state occur before threshold amplitude is reached.

This topological property of a limit cycle leads to an unexpected prediction. If some perturbation at a particular time in the cycle drives the system near the steady state, then a stronger perturbation should drive it past the steady state. For sufficiently strong perturbations, the system will move completely across the limit cycle, where it will be near the normal cycle trajectory, and the phase shift will now be some fraction of a normal cycle period. That is, as the perturbation strength increases delay should first increase, perhaps to considerably more than one cycle time, but then decrease to less than one cycle time. Wille et al. (1977) have reported exactly this response in *Physarum* plasmodia after 37°C heat treatment for up to 12 hr (Fig. 6).

### Limit-Cycle Predictions with Respect to Fusions

The simplest hypothesis is that if two cells or two plasmodia are fused, all pertinent phase-determining variables freely mix, and the fused cell or plasmodium adopts a single state, which thereafter spirals onto the limit cycle. Under this idealized assumption, the limit-cycle model predicts synchronization in a straightforward way. As in the relaxation oscillator model, fused cells will tend to synchronize to an intermediate compromise phase and varying the mass ratio will shift the compromise phase toward the larger mass. The detailed behavior, however, can be more complex than in a relaxation oscillator model.

Mixing of variables within two fused systems places the composite system at a state that lies somewhere on the straight line joining the states of the original systems in the XY concentration diagram. For the limit cycle, in contrast to the relaxation oscillator, this fused state will in general not be on the normal cycle trajectory, but will be somewhere within the cycle. As in the case of phase perturbations, there is now the possibility for subthreshold oscillations of the fused system, with resulting skipping of cell cycle events or unusually long delays in the cycle.

Although subthreshold phenomena have not been directly observed, data consistent with this interpretation have been found in the early sea urchin embryo. This system exhibits almost complete mitotic synchrony during the first few cleavage divisions. This synchrony gives way to vegetal to animal pole mitotic waves later in cleavage. By the blastula stage local patches of cells divide synchronously, but neighboring patches are sharply out of synchrony with one another. If one supposes a limit-cycle oscillation triggers mitosis at some threshold, then this phase asynchrony across the surface of the blastula implies a spatial gradient of oscillatory variables across the blastula. This can lead to subthreshold oscillations in regions sandwiched between two other domains oscillating at full amplitude, but different phases. Such a synchrony would be expected to lead to an increase in the mean intermitotic time in the sea urchin blastula, but no increase in the minimum intermitotic time. This has in fact been observed by Parisi et al. (1978), who advanced this limit-cycle interpretation of their results.

Passage from synchrony to asynchrony coupled with an increase in mean intermitotic time is common in many early embryos; the interesting possibility is that maintenance of synchronous oscillations throughout the embryo may be difficult as cleavage proceeds and more cell membranes partition the volume, leading to the onset of phase asynchrony and subthreshold behavior. Thus, the oscillations hypothesis leads to new questions about the relation between tissue size, the onset of phase asynchrony and regulation of the fraction of cells undergoing mitosis.

#### *The Phase Singularity*

An important property of a limit cycle (Winfree, 1980) is that the steady state inside the cycle is a phaseless point, such that in its immediate vicinity are points that wind out to all possible phases on the limit cycle. Fusions of pairs of cells at all possible phases of the cycle, and all possible mass ratios, should include pairs in which the averaged limit cycle variables lie on or very near the steady-state singularity. Compromise phase for such pairs should be highly variable, reflecting the crowding of phase values near the steady state.

Tyson and Sachsenmaier (1978) have analyzed data from *Physarum* fusion experiments (Remy, 1968; Sachsenmaier et al., 1972; Sachsenmaier, 1976; Chin et al., 1972) in a search for evidence for unusual variability in the compromise phases. These authors concluded that the fusion data were more consistent with a relaxation oscillator model than a limit-cycle model for the cycle in *Physarum*. It should be noted, however, that one point with highly variable phase does appear in the data, for fusions of plasmodia at 4 and 9 hr from mitosis, resulting in compromise phases ranging between 4 and 11 hr from mitosis. The significance of this is difficult to judge at this point.

It must be said, however, that in general the variability predicted by the limit-cycle model is not easily observed. There are two possible reasons for this. First, since fusions of 1:1 mass ratios can be expected to place the fused system at the center of the line joining the prefusion states, a fusion experiment may fail to strike the singularity if the singularity is far away from the geometric center of the limit cycle. Although there must always be at least one pair of states for which instantaneous fusion places the fused system on the singularity (Winfree, 1980), if at least one of these states is on the fast arc of the limit cycle, as is typically the case (Fig. 3), the actual attainment of the singularity may be technically impossible.

Another possibility is that the assumptions behind the predictions for fusion experiments are not altogether accurate. For fused pairs to fall near the singularity requires that all relevant phase-determining variables diffusively mix to homogeneity in a time scale short with respect to the cycle time. If, for example, some variables mix much more slowly than others when two cells are fused, perhaps due to intranuclear versus cytoplasmic localization, averaging to a state near the singularity will be difficult or impossible. For example, the two may settle into coupled but rather different oscillations, creating a four-variable rather than two-variable limit cycle (Tyson and Kauffman, 1975).

## RANDOMNESS AND PROBABILITY IN THE CELL CYCLE

Generation times in a population of cultured cells typically fall in a right-skewed distribution, with a significant number of cells showing much longer generation times than the mode. The shape of the generation time distribution has been characterized as reciprocal normal, derived from the assumption that cycling rates are normally distributed (Kubitschek, 1962; 1971; Pardee et al., 1979), or negative exponential, based on the transition probability model (Burns and Tannock, 1970; Smith and Martin, 1973). In this second model, the cell cycle is divided into two parts, an indeterminate "A state," which cells enter after mitosis and leave randomly at a constant rate per unit time, and a second "B phase," which is traversed in a constant time $T_B$. The resulting generation time distribution has a right-skewed negative exponential tail. Most discussions of this model have been in terms of the so-called $\alpha$ curve (the plot of the fraction of cells undivided versus cell age) and the $\beta$ curve (the distribution of differences in sister cell generation times). In a semilogarithmic plot, both curves are predicted to have straight lines of equal negative slope, as is often observed in cultured cells (Minor and Smith, 1974; Shields, 1978).

Although the transition probability model does reasonably well in predicting some of the statistical properties of the cell cycle, it makes no predictions about phase-response curves or fusion results. In addition, there is another aspect of generation time variability, mentioned earlier, that is not accounted for in the transition probability or the reciprocal normal models. Our own observations on the distribution of generation times confirm the overall right-skewed shape of the distribution, but we see in addition that intermitotic

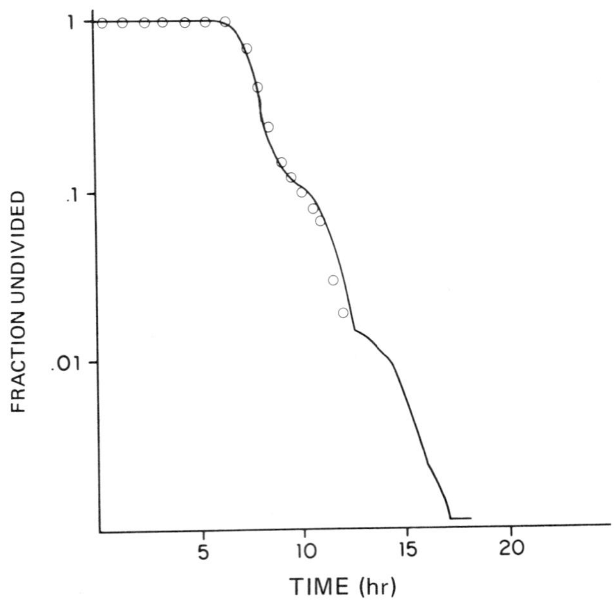


**FIGURE 7** $\alpha$ Curve for V79 cells along with a simulation (solid line) using a two-loop limit-cycle model with random noise added, resulting in subthreshold oscillations. Limit-cycle period was 4.25 hr. The undulations in the tail of the curve reflect the quantizement of generation times. (From Klevecz et al., 1980a.)

times tend to fall in clusters separated by slightly more than 4 hr (Klevecz, 1976; 1978). This clustering or quantizement is most easily seen in rapidly growing cells (Fig. 2).

We have shown (Klevecz et al., 1980a; Shymko and Klevecz, 1981) that both quantizement and the overall exponential envelope of the generation time distribution can be predicted by assuming that a limit cycle of nominal 4 hr period controls the gating of a cell between successive stages of the cell cycle. With random fluctuations added to the normal movement of the cells around the limit cycle, a population of cells moves around the cycle in a cloud whose size depends on the relative magnitudes of the fluctuations and the rate of return of the system to the normal cycle trajectory. If threshold crossing is required for gating into the next cycle stage, part of the cloud may fail to cross threshold and therefore those cells will fail to advance in their cell cycle. Furthermore, if the random fluctuations are large enough, the position of a cell within the cloud during one cycle will be essentially uncorrelated with its position one cycle later, so that its probability of achieving threshold is constant per cycle, or equivalently constant per unit time. In this system, we now have an explicit mechanism for generating an overall exponential cycle time distribution as in the transition probability model, but in addition, the cell cycle times are clustered at intervals corresponding to the limit-cycle period, which according to our experiments should be about 4 hr.

The $\alpha$ curve generated by this model, together with experimentally observed plot for V79 cells, is shown in Fig. 7. Note the undulations in the tail of the curve, reflecting the quantizement of generation times at 4 hr intervals.

As far as the exponential cycle time distribution is concerned, the limit-cycle period could be of any length. It is the timekeeping property of the limit cycle, along with the constant probability of exceeding threshold on each cycle, that results in the exponential distribution. The 4 hr period was chosen simply to be consistent with the generation time quantizement patterns.

As in the transition probability model, the random component (i.e., the 4 hr limit cycle with noise) could be placed anywhere in the cell cycle, and both quantizement and the exponential distribution would still result. However, since the quantizement interval, and by hypothesis the limit-cycle period, are a fraction of the total cell cycle time, an important question is whether the timekeeping behavior persists through more of the the cycle than $G_1$, where most of the variability in cycle time typically occurs (Pardee et al., 1978).

When V79 cells are heat shocked at 45°C for 10 min a characteristic pattern of resetting occurs (Fig. 1). The response curve for the modal mitotic delay appears as two parallel lines sloping downward to the right, with a small cluster of values between 4 and 6 hr of the cycle showing a constant 2 hr delay. Some of the 2 hr delays resulted from a splitting of the anaphase frequency histogram, suggesting that at these values of time and perturbing stimulus cells may be either slightly advanced or delayed, or quantally delayed by $\triangle\phi$ + 4 hr. This capacity to phase jump is further shown when on occasion the midpoint of a population shocked at 7 hr is delayed, not by 0.5-1.0 hr, but by 5 hr, or about a full subcycle.

The above results are consistent with a picture in which the perturbing agent affects the clock directly and cycling of the timekeeper continues through the entire cycle (in this case, repeating twice during the cell cycle). In particular, the repeated pattern in heat shock phase-response curves can be predicted by assuming that heat destroys most of the macromolecules constituting the limit-cycle oscillator (Klevecz et al., 1980a) (Fig. 5). To account for the greater variability in $G_1$ than later in the cell cycle, the threshold associated with the second loop of the limit cycle must be lower, or the fluctuations must be smaller, in order that threshold be more easily achieved in the second loop than in the first.

The implications of having two or more loops of a limit cycle during one cell cycle are quite different from those of having only one, as, for example, in the single-loop model of the *Physarum* mitotic cycle proposed by Kauffman and Wille (1975). If a single oscillator cycles twice per cell cycle, some additional marker is required to distinguish a cell in the first versus the second loop. If each loop of the limit cycle triggers passage into the next cycle stage, then position in the cell cycle could be identified by counting, in much the same way that the minute hand of an ordinary clock might be considered the timekeeper with the current "state" of the hour hand depending on its state during the previous revolution of the minute hand. This now suggests a way to relate the causal sequences of processes in the cycle, for example, representing the synthesis of structures in preparation for division, to an independent cellular oscillator. The causal sequence provides the counting mechanism for determining the progress of the cell through successive stages in the cycle, and the independent oscillator controls the timing of the individual events.

In the next section, we will discuss the possibility that cell size is an appropriate general parameter for representing the causally related events in the cell cycle. As we will show, the flavor of the global behavior of cycling cells can be captured by superimposing a size threshold constraint on a continuous limit-cycle timekeeper in the cell that gates the timing of cell cycle events.

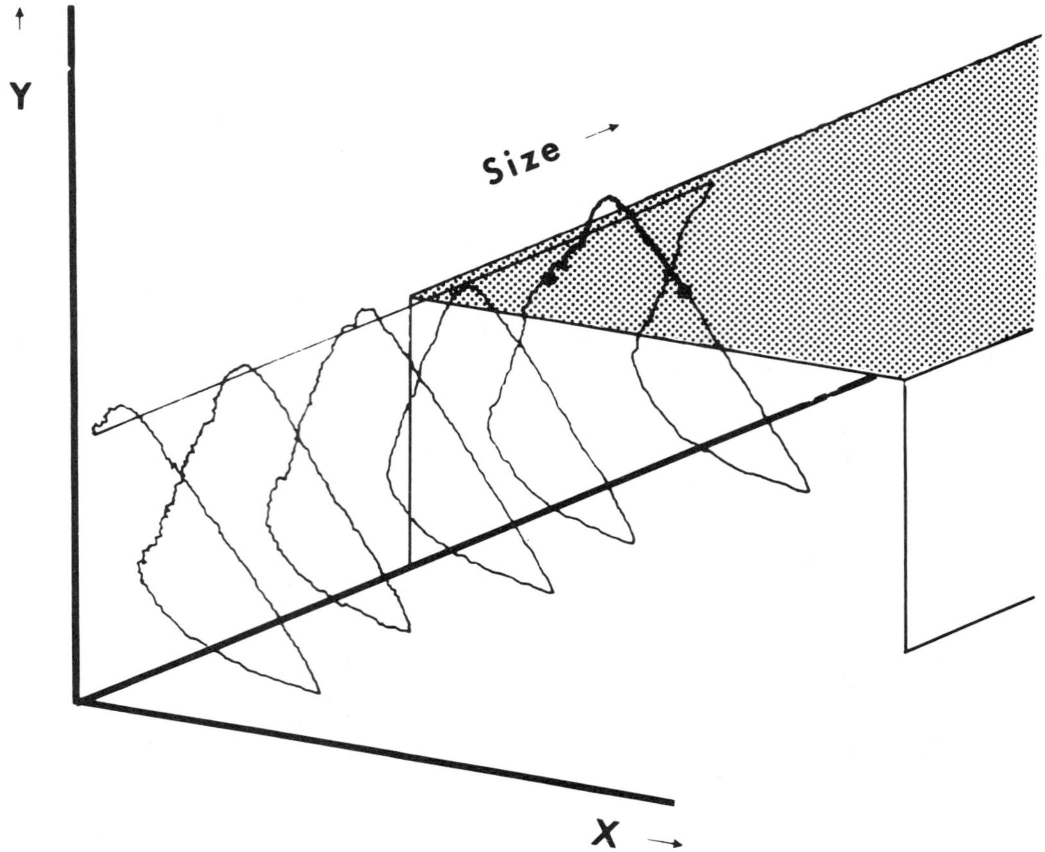


**FIGURE 8** Schematic representation of a limit-cycle model with a size threshold added. Limit-cycle oscillations occur in X and Y; size increases in the third dimension. Crossing of the Y threshold (stippled top of the box) is ineffective in triggering mitosis until the theshold size (open face of the box) is reached. (From Shymko and Klevecz, 1981.)

## CELL DIVISION GATED BY CELL SIZE

The idea that cell size has an effect on the timing of cell division is a recurrent theme in cell cycle studies. The evidence for such effects is strongest in nonmammalian cells, such as yeast, where initiation of DNA synthesis and nuclear division both appear to depend on the cell's attaining a critical size (Nurse, 1975; Nurse and Thuriaux, 1977).

In mammalian cells, evidence for a size effect in the cell cycle is mixed. In early experiments on mouse fibroblasts, Killander and Zetterberg (1965) concluded that there is a critical size requirement for entry into S phase. More recently, however, Yen and coworkers, using percentage of labeled mitoses methods (1975) and flow cytometry (Yen and Pardee, 1979), presented evidence suggesting that large cells, or cells with large nuclei, cycle faster than small ones, but concluded that this was not related to a critical size threshold for entry into S phase. Johnson et al. (1981), also using flow cytometry, could find no effect of nuclear size on cell proliferation. It appears, then, that cell size may play some role in regulating mammalian cell division, but it is most likely not the only controlling factor.

We have carried out a theoretical analysis to assess the consequences of a size threshold superimposed on a limit-cycle model of the cell cycle as described above (Shymko and Klevecz, 1981). In order to account for the quantizement effect, and phase perturbation results, we chose a two-loop model with a threshold crossing required for gating from one loop to the next, and further constrained the model by requiring that a critical size be reached before gating is implemented to exit from the second loop (Fig. 8). Then, if cell growth is rapid enough, size has no effect on the timing of mitosis, but with slow growth a cell remains within the second loop until the critical size is reached.

This kind of model, and in fact most models incorporating a size threshold, can be expected to have the qualitative property that sibling cell generation times are positively correlated, while those of mother and daughter cells are negatively correlated. The reason is that a cell that is larger than usual at division will have large daughters that more quickly reach the size threshold (if growth rates are unaltered) and therefore divide sooner on average than the daughters of a smaller parent. Conversely, if a cell randomly has a longer cycle time it will have a longer time to grow and will be larger than average at division. Its daughters will have a short cycle time as discussed above, and a pattern of negative correlation between cycle times in alternate generations will result. Positive correlations between sister cell generation times have been reported (McQuilkin and Earle, 1962; Minor and Smith, 1974), along with some reference to negative parent-daughter correlations (Killander and Zetterberg, 1965).

The issue of generation time correlations has been addressed by others in the context of the transition probability model. Brooks and coworkers (1980) were able to generate sister cell generation time correlations while retaining the random behavior inherent in the model, by proposing that a variable interval L begins in one cycle and is completed in the next cycle after division, after which it follows a purely random interval as in the simple transition probability model. The generation-spanning "memory" resulting from the L interval gives the positive sister-sister correlations, and the random interval maintains the desired properties of the $\alpha$ and $\beta$ curves.

Cell size provides a more direct way to give exactly the same relationship between sister cells as does the L interval because of the sharing of cell mass between dividing cells. Therefore, in addition to its previously described predictive power, the limit-cycle model with superimposed size threshold is able to account for that cell behavior addressed by the modified transition probability model, but now suggests a more physically based mechanism for that behavior.

Whether cell size does indeed play a strong role in cell cycle control in mammalian cells remains to be proven by experiment. A perhaps more important property of the model we have described here is that it provides a paradigm for a system that has both an independent timekeeper and a set of causally connected cell cycle events, in this case corresponding to cell growth, either of which can dominate the timing of cell cycle events, depending on the growth conditions.

## SUMMARY

The many different points of view on the cell cycle and the associated experimental emphases have led to a number of cell cycle models that are plausible and relatively simple in their own right but that seem to have little connection with one another. We have attempted to consider these models starting with a general framework with appropriate constraints in order to construct a minimal model incorporating as many of the essential elements of cell cycle control as possible.

Three properties of the cell cycle in most cells have emerged as fundamental. First, general considerations and direct evidence show that there must be some component to the cell cycle which is directly causal and perhaps irreversible. It is natural to associate such components with the progressive assembly of duplicates of cell structures that will ultimately be partitioned between two identical cells. In many of these structures, the process of assembly must be orderly and tightly controlled; otherwise the structures will not be complete at the time of cell division, and function in the newly divided cells will be impaired.

At the same time, since there are many processes in the cell that proceed in parallel, if the cell is to recognize a malfunction in any one of them in order that it can stop cycling until the deficit is remedied, it must contain a more central determinant of cell cycle phase. Therefore, a second requirement, based more on logical considerations than on direct evidence, is that the cell should have some global marker whose status coordinates the occurrence of cell cycle events. We have associated this marker with a continuously changing variable, or set of variables, probably residing in the cytoplasm, whose level can be averaged by fusion of cells at different phases, or can be altered by externally applied perturbations.

Finally, a cell must have the capability to accept some degree of deviation from the sequence of states it follows during a normal unperturbed cycle. This could result from an evolutionary selection for maximum flexibility in a constantly changing environment, or simply from an incidental ability to dispense on occasion with noncritical processes. In parallel with this flexibility, however, cells in general show a strong tendency to return to the normal cycle following a nonlethal perturbation, which implies that mechanisms generating such dynamically stable behavior must exist somewhere in the cell. We have argued that an appropriate model for this kind of stable metabolism is the limit-cycle model, which explicitly allows for deviation from the normal cycle while showing a global tendency to return to that stable cycle. More complex models are potentially more accurate, but the limit-cycle model is simple enough to be easily understood while incorporating much of the observed fundamental behavior of cycling cells.

Adding refinements to a simple limit-cycle model, such as having more than one loop of the limit cycle per cell cycle, or superimposing a size threshold in parallel with the limit-cycle control, preserves the general properties of the oscillator model while accounting for much of the detailed behavior of cells, such as the roughly exponential distribution of cell generation times along with cycle time quantizement, the tandemly repeated pattern of phase-response curves, and the intergenerational correlations between cell cycle times in cultured cells.

## ACKNOWLEDGMENTS

Supported in part by NIH grants GM31262 and AG03815, and by a Cancer Center Support (Core) Grant CA 33572.

## REFERENCES

Brewer, E. N., and Rusch, H. P. (1968). The effect of elevated temperature shocks on mitosis and the initiation of DNA replication in *Physarum polycephalum. Exp. Cell Res.* 49:79-86.

Brooks, R. F., Bennett, D. C., and Smith, J. A. (1980). Mammalian cells need two random transitions. *Cell* 19:493-504.

Burns, F. J., and Tannock, I. F. (1970). On the existence of a G0-phase in the cell cycle. *Cell Tissue Kinet.* 3:321-334.

Chin, B., Friedrich, P. D., and Bernstein, I. A. (1972). Stimulation of mitosis following fusion of plasmodia in the myxomycete *Physarum polycephalum. J. Gen. Microbiol.* 71:93-101.

Edmunds, L. N., and Adams, K. J. (1981). Clocked cell cycle clocks. *Science* 211:1002-1013.

Fantes, P. A., Grant, W. D., Pritchard, R. H., Sudbery, P. E., and Wheals, A. E. (1975). The regulation of cell size and the control of mitosis. *J. Theor. Biol.* 50:213-244.

Fantes, P., and Nurse, P. (1977). Control of cell size at division in fission yeast by growth-modulated size control over nuclear division. *Exp. Cell Res.* 107:377-386.

Guttes, E., and Guttes, S. (1964). Mitotic synchrony in the plasmodia of *Physarum polycephalum* and mitotic synchronization by coalescence of microplasmodia. In *Methods in Cell Physiology,* vol. I, D. M. Prescott (ed.). Academic Press, New York, pp. 43-54.

Hartwell, L., Culotti, J., Pringle, J. R., and Reid, B. J. (1974). Genetic control of the cell division cycle in yeast. *Science (N.Y.)* 183:46-51.

Hastings, J. W., and H.-G. Schweiger (eds.). (1976). *The Molecular Basis of Circadian Rhythms.* Dahlem Konferenzen, Berlin.

Johnson, R. T., and Rao, P. N. (1971). Nucleo-cytoplasmic interactions in the achievement of nuclear synchrony in DNA synthesis and mitosis in multinucleate cells. *Biol. Rev.* 46:97-155.

Johnson, T. S., Swartzendruber, D. E., and Martin, J. C. (1981). Nuclear size of G1/S transition cells measured by flow cytometry. *Exp. Cell Res.* 134:201-205.

Kauffman, S. A., and Shymko, R. M. (1982). Mitotic delay following inhibition by 5-fluorodeoxyuridine of S-phase in *Physarum* is not due to delay in termination of S-phase. *J. Cell Sci.* 53:143-154.

Kauffman, S. A., and Wille, J. J. (1975). The mitotic oscillator in *Physarum polycephalum. J. Theor. Biol.* 55:47-93.

Kauffman, S. A., and Wille, J. J. (1976). Evidence that the mitotic "clock" in *Physarum polycephalum* is a limit cycle oscillator. In *The Molecular Basis of Circadian Rhythms,* J. W. Hastings and H.-G Schweiger (eds.). Dahlem Konferenzen, Berlin, pp. 421-431.

Killander, D., and Zetterberg, A. (1965). A quantitative cytochemical investigation of the relationship between cell mass and initiation of DNA synthesis in mouse fibroblasts in vitro. *Exp. Cell Res.* 40:12-20.

Klevecz, R. R. (1976). Quantized generation time in mammalian cells as an expression of the cellular clock. *Proc. Natl. Acad. Sci. USA* 73:4012-4016.

Klevecz, R. R. (1978). The clock in animal cells is a limit cycle oscillator. In *Cell Reproduction,* E. R. Dirksen, D. M. Prescott, and C. F. Fox (eds.). Academic Press, New York, pp. 139-146.

Klevecz, R. R., King, G. A., and Shymko, R. M. (1980a). Mapping the mitotic clock by phase perturbation. *J. Supramol. Struct. Cell. Biochem.* 14:329-342.

Klevecz, R. R., Kros, J., and King, G. A. (1980b). Phase response to heat shock as evidence for a timekeeping oscillator in synchronous animal cells. *Cytogenet. Cell Genet.* 26:236-243.

Kubitschek, H. E. (1962). Normal distributions of cell generation rate. *Exp. Cell Res.* 26:439-450.

Kubitschek, H. E. (1971). The distribution of cell generation times. *Cell Tissue Kinet.* 4:113-122.

Mano, Y. (1971). Cell-free cyclic variation of protein synthesis associated with the cell cycle in sea urchin embryos. *J. Biochem. (Tokyo)* 69:11-25.

McQuilkin, W. T., and Earle, W. R. (1962). Cinemicrographic analysis of cell populations in vitro. *JNCI* 28:763-799.

Minor, P. D., and Smith, J. A. (1974). Explanation of degree of correlation of sibling generation times in animal cells. *Nature* 248:241-243.

Mitchison, J. M. (1971). *The Biology of the Cell Cycle.* Cambridge University Press, Cambridge, England.

Miyamoto, H., Rasmussen, L., and Zeuthen, E. (1973). Studies on the effect of temperature shocks on preparation for cell division in mouse fibroblast cells (L cells). *J. Cell Sci.* 13:889-900.

Nurse, P. (1975). Genetic control of cell size at cell division in yeast. *Nature* 256:547-551.

Nurse, P., and Thuriaux, P. (1977). Controls over the timing of DNA replication during the cell cycle of fission yeast. *Exp. Cell Res.* 107:365-375.

Pardee, A. B., Dubrow, R., Hamlin, J. L., and Kletzien, R. F. (1978). Animal cell cycle. *Ann. Rev. Biochem.* 47:715-750.

Pardee, A. B., Shilo, B.-Z., and Koch, A. L. (1979). Variability of the cell cycle. In *Hormones and Cell Culture,* G. H. Sato and R. Ross (eds.). Cold Spring Harbor Laboratory, Cold Spring Harbor, New York, pp. 373-392.

Pavlidis, T. (1973). *Biological Oscillators: Their Mathematical Analysis.* Academic Press, New York.

Parisi, E., Filosa, S., DePetrocellis, B., and Monroy, A. (1978). The pattern of cell division in the early development of the sea urchin *Paracentrotus lividus. Dev. Biol.* 65:38-49.

Polanshek, M. (1977). Effects of heat shock and cycloheximide on growth and division of the fission yeast, *Schizosaccharomyces pombe. J. Cell Sci.* 23:1-23.

Rao, P. N., and Johnson, R. T. (1970). Mammalian cell fusion: Studies on the regulation of DNA synthesis and mitosis. *Nature (Lond.)* 225:159-164.

Rasmussen, L., and Zeuthen, E. (1962). Cell division and protein synthesis in *Tetrahymena* as studied with p-fluorophenylalanine *C. R. Trav. Lab. Carlsberg* 32:333-358.

Remy, U. (1968). Ph.D. Thesis. Ruprecht-Karl-Universität, Heidelberg.

Rusch, H. P., Sachsenmaier, W., Behrens, K., and Gruter, V. (1966). Synchronization of mitosis by the fusion of the plasmodia of *Physarum polycephalum. J. Cell Biol.* 31: 204-209.

Sachsenmaier, W. (1976). In *The Molecular Basis of Circadian Rhythms,* J. W. Hastings and H.-G. Schweiger, (eds.). Berlin: Report of the Dahlem Workshop, 1975, Berlin, pp. 409-420.

Sachsenmaier, W., Remy, U., and Plattner-Schobel, R. (1972). Initiation of synchronous mitosis in *Physarum polycephalum. Exp. Cell Res.* 73:41-48.

Schmidt, R. R. (1966). Intracellular control of enzyme synthesis and activity during synchronous growth of *Chlorella.* In *Cell Synchrony–Studies in Biosynthetic Regulation,* I. L. Cameron and G. M. Padilla (eds.). Academic Press, New York, pp. 189-235.

Shields, R. (1978). Further evidence for a random transition in the cell cycle. *Nature* 273:755-758.

Shymko, R. M., and Klevecz, R. R. (1981). Cell division gated by oscillatory timekeeping and critical size. In *Biomathematics and Cell Kinetics,* M. Rotenberg (ed.). Elsevier/North Holland Biomedical Press, Amsterdam, pp. 329-348.

Smith, J. A., and Martin, L. (1973). Do cells cycle? *Proc. Natl. Acad. Sci. USA* 70:1263-1267.

Soll, D. R. (1979). Timers in developing systems. *Science* 203:841-849.

Sudbery, P. E., and Grant, W. D. (1975). The control of mitosis in *Physarum polycephalum.* The effect of lowering the DNA:mass ratio by UV-irradiation. *Exp. Cell Res.* 95:405-415.

Sudbery, P. E., and Grant, W. D. (1976). The control of mitosis in *Physarum polycephalum*: The effect of delaying mitosis and evidence for the operation of the control mechanism in the absence of growth. *J. Cell Sci.* 22:59-65.

Tyson, J., Garcia-Herdugo, G., and Sachsenmaier, W. (1979). Control of nuclear division in *Physarum polycephalum*: Comparison of cycloheximide pulse treatment, UV irradiation, and heat shock. *Exp. Cell Res.* 119:87-98.

Tyson, J., and Kauffman, S. (1975). Control of mitosis by a continuous biochemical oscillation: Synchronization, spatially inhomogenous oscillation. *J. Math. Biol.* 1: 289-310.

Tyson, J., and Sachsenmaier, W. (1978). Is nuclear division in *Physarum* controlled by a continuous limit cycle oscillator? *J. Theor. Biol.* 73:723-738.

Wille, J. J., Scheffey, C., and Kauffman, S. A. (1977). Novel behavior of the mitotic clock in *Physarum. J. Cell Sci.* 27:91-104.

Winfree, A. T. (1980). *The Geometry of Biological Time. Biomathematics,* vol. 8, Springer-Verlag, Heidelberg.

Yen, A., Fried, J., Kitahara, T., Strife, A., and Clarkson, B. D. (1975). The kinetic significance of cell size. I. Variation of cell cycle parameters with size measured at mitosis. *Exp. Cell Res.* 95:295-302.

Yen, A., and Pardee, A. B. (1979). Role of nuclear size in cell growth initiation. *Science* 204:1315-1317.

Zeuthen, E. (1971). Synchrony in *Tetrahymena* by heat shocks spaced a normal cell generation apart. *Exp. Cell Res.* 68:49-60.

# 17

# Cell Division Cycles and Circadian Oscillators

**Leland N. Edmunds, Jr.** / State University of New York, Stony Brook, New York

**Danielle L. Laval-Martin** / State University of New York, Stony Brook, New York, and Université Paris 7, Paris, France

## INTRODUCTION

The consideration of mechanisms that control microbial cell cycles (Mitchison, 1971; Padilla et al., 1974; Hartwell et al., 1974; Jeter et al., 1978; Pringle, 1981; Lloyd et al., 1982) often includes the notion of timers and clocks in regulating cell cycle sequences that culminate in mitosis and cell division (Mitchison, 1974; Klevecz, 1976; Wille, 1979; Gilbert, 1981; Edmunds and Adams, 1981; see also Chap. 16). Indeed, the cell division cycle (CDC) itself is a clock in the sense that it measures time—albeit imperfectly—under a given set of environmental conditions.[+] The duration of the CDC in the majority of microbes and of cultured cells ranges from approximately 10-30 min in certain prokaryotes (e.g., *Benekea natriegens, Escherichia coli*), 2-3 hr in the yeast *Saccharomyces cere-*

---

[+]Abbreviations: CDC, cell division cycle; $g$, average generation time (doubling time) of a population of cells; LL, continuous illumination; DD, constant darkness; LD: x,y, a repetitive light-dark cycle comprising x hours of light and y hours of darkness, whose period ($T$) is given by x + y; CR, circadian rhythm, whose free-running period is denoted by $\tau$ (taken here to be the time between onsets of cell division in a population of cells); $\overline{ss}$, average stepsize, or factorial increase in cell concentration (plateau to plateau) after a phased, or synchronized division step; $\phi$, phase of the rhythm; $\Delta\phi$, change in phase (phase shift); $\phi_R$, phase marker, or phase reference point (here, the onset of the division burst); PRC, phase response curve; CT, circadian time (CT 0 indicates the phase point of a free-running rhythm that has been normalized to 24 hr, which occurs at the onset of light in a LD: 12,12 reference cycle; ZT, environmental (Zeitgeber) time where ZT 0 = onset of light).

*visiae* or the ciliate *Tetrahymena pyriformis,* to 15-25 hr in cultured mammalian cells under ideal conditions, or even several days in some dinoflagellates (e.g., *Pyrocystis fusiformis*). The time domain of the CDC therefore, is of a higher order in the temporal hierarchy of living systems, several orders of magnitude greater than the ultrafast rapid, the metabolic, and the epigenetic domains (Lloyd et al., 1982; see also Chap. 2).

Inasmuch as mitosis is a periodic event of short duration relative to the total length of the CDC, it is not surprising that various types of biochemical and macromolecular oscillators have been proposed to underlie this and other events, or "landmarks," comprising the CDC (Edmunds and Adams, 1981; see also Chap. 16). These range from those of the relaxation type (Sachsenmaier, 1976; see also Chap. 15) to those exhibiting limit-cycle behavior (Kauffman and Wille, 1975; Gilbert, 1981; see also Chap. 16). Additionally, the role of "quantal" cell cycles in the clustering of the generation times of cultured mammalian cells has been considered (Klevecz, 1976, 1978). Most of these biological "clocks," however, have periods in the neighborhood of minutes to hours and thus fall between the higher frequency (ultradian) rhythmicities in intermediary metabolism (e.g., those of the glycolytic pathway) within the metabolic time domain and those having somewhat longer periods of about a day that are so ubiquitous in living systems. It is this latter type of periodicity and its relation to the CDC—particularly in unicellular organisms—to which we now turn.

## CIRCADIAN CLOCKS AND CELL DIVISION CYCLES

Circadian rhythmicities have been documented throughout the plant and animal kingdom at every level of eukaryotic (but not prokaryotic) organization (Brady, 1982). Their general properties are tabulated in Fig. 1. Typically they can be synchronized, or entrained, by imposed diurnal light or temperature cycles to precise 24 hr periods and can be predictably phase shifted by single light and temperature signals, yet they are able to "free-run" for long time spans as persisting rhythms under conditions held constant with respect to most environmental time cues, with a natural period close to but seldom exactly 24 hr. Furthermore, the free-running period $\tau$ is remarkably well compensated for changes in ambient temperature within the physiological range, as might be expected of an accurately functioning clock. Our question, therefore, concerns the extent to which the endogenous, light-sensitive, self-sustaining circadian oscillators presumed to underlie these rhythms may also interact with and control the CDC.

### Interaction of the Circadian Oscillators with the CDC in Unicellular Organisms

There is abundant evidence that the CDC of many unicellular algae, fungi, and protozoans, as well as mammalian cells in situ (and perhaps in cell culture) indeed do exhibit persisting circadian rhythms of cell division, or "hatching" (Edmunds, 1975, 1978). A number of these protistan systems are itemized in Table 1, including some quite recent experimental entries.

As is evident from this tabulation, the division of a number of algal cells occurs only at a certain phase of the circadian cycle, very often the times ("subjective" nights) in DD or LL corresponding to dark intervals in an environmentally synchronizing LD cycle. This phenomenon, commonly referred to as "gating," has been documented for a number of years in *Chlamydomonas* (Bruce, 1970; Bruce and Bruce, 1981), *Chlorella* (Pirson and Lorenzen, 1958; Hesse, 1972), *Euglena* (Edmunds, 1966, 1971) (to be considered in detail later), and *Gonyaulax* (Sweeney and Hastings, 1958; Hastings and Sweeney, 1964), and

PROPERTIES OF CIRCADIAN RHYTHMS

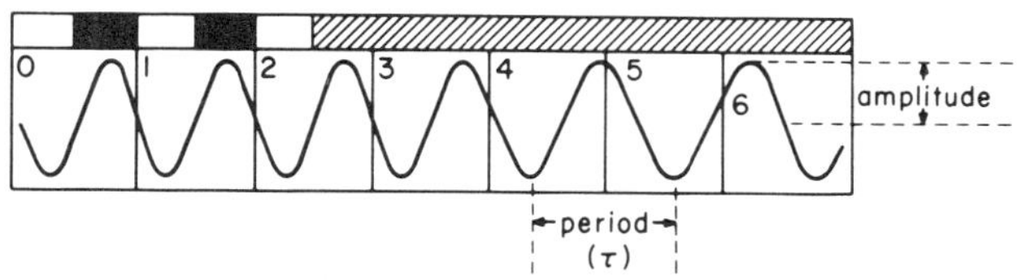


(a) ~24-HR PERIOD (τ)

(b) UBIQUITY (among eukaryotes, at every level of organization)

(c) ENTRAINABILITY (by light and temperature cycles and by social and other cues)

(d) PERSISTENCE (under "free-running" conditions held constant with respect to major environmental Zeitgeber)

(e) PHASE SHIFTABILITY (by single perturbations of light, temperature, or chemicals)

(f) MODULATION OF FREE-RUNNING PERIOD BY LIGHT INTENSITY (Aschoff's Rules) in diurnal and nocturnal animals

(g) TEMPERATURE COMPENSATION (temperature-independent free-running period within the physiological range)

(h) GENETIC BASIS (innateness and endogeneity)

**FIGURE 1** Properties of circadian (*sensu stricto*) rhythms. A generalized sinusoidal oscillation under entraining (LD) and free-running (dim LL) conditions is shown at the top over a time span of 7 days; white areas denote light intervals, black bars, darkness, and hatched areas, continuous dim illumination. Note that the period of the oscillation is longer in dim LL than it is in LD, where it exactly matches that of the synchronizing LD cycle. (Adapted from Edmunds, 1975.)

more recently in a number of marine phytoplankton (Chisholm, 1981; Chisholm et al., 1980; Weiler and Chisholm, 1976; see also Chap. 20). For example, Chisholm and Brand (1981) have examined the cell division patterns of 11 species of phytoplankton maintained in an LD: 14,10 cycle followed by 3 days of LL. All species except two diatom clones showed typical phased or synchronized division in the LD regime, although the timing of division relative to the LD cycle was variable. In six of the species (representing the algal classes Prasinophyceae, Xanthophyceae, Haptophyceae, and Dinophyceae), the division rhythm persisted throughout the 3 days of LL, suggesting endogenous control, although there was some evidence of damping. In others, the rhythm completely damped out after 1 day, or in the case of the two diatoms, was never evident in the first place.

**TABLE 1** Some Unicellular Systems for Which Persisting Circadian Rhythms of Cell Division (or "Hatching") Have Been Documented

| Organism | Reference |
|---|---|
| Algae | |
| *Ceratium furca* | Weiler & Eppley (1979); Chap. 21 |
| *Chlamydomonas reinhardi* | Bruce (1970); Bruce and Bruce (1981) |
| *Chlorella pyrenoidosa* Chick (211-8b) | Pirson and Lorenzen (1958); Hesse (1972) |
| *Emeliania huxleyi* (MCH, 451B) | Chisholm and Brand (1981) |
| *Euglena gracilis* Klebs (Z strain) | Edmunds (1966, 1971); Edmunds and Funch (1969) |
| $P_4$ZUL mutant | Jarrett and Edmunds (1970) |
| $P_7$ZNgL mutant | Mitchell (1971) |
| $W_6$ZHL mutant | Edmunds et al. (1976) |
| $W_n$ZIL mutant | Mitchell (1971) |
| $Y_9$ZNalL mutant | Edmunds et al. (1976) |
| *Gonyaulax polyedra* | Sweeney and Hastings (1958) |
| *Gymnodinium splendens* | Hastings and Sweeney (1964) |
| *Hymenomonas carterae* (Cocco 2) | Chisholm and Brand (1981) |
| *Olisthodiscus* sp. | Chisholm and Brand (1981) |
| *Prorocentrum* sp. | Chisholm and Brand (1981) |
| *Pyramimonas* sp. | Chisholm and Brand (1981) |
| *Pyrocystis fusiformis* Murray | Sweeney (1982) |
| Fungi | |
| *Candida utilis* | Wille (1974) |
| *Saccharomyces cerevisiae* (Y185 *rho*$^+$) | Edmunds (1980); Edmunds et al. (1979) |
| Protozoans | |
| *Paramecium bursaria* | Volm (1964) |
| *Paramedium multimicronucleatum* | Barnett (1969) |
| *Tetrahymena pyriformis* | |
| Strain W | Wille and Ehret (1968); Edmunds (1974) |
| Strain GL | Edmunds (1974) |

*Source*: Updated from Edmunds (1975).

The question arises as to whether other events comprising the CDC besides division itself can be controlled by a circadian oscillator. To experimentally approach this question, one needs an organism having a value of *g* greater than 1 day and several clearly distinguishable morphological landmarks. Recently, Sweeney (1982), working with the large (900 μm long) dinoflagellate, *Pyrocystis fusiformis* Murray, whose CDC has a minimum duration of 5 or 6 days, was able to isolate single cells or dividing pairs in capillary tubes kept in LD or DD and to determine five discrete morphological stages that could be correlated with the classic phases ($G_1$, S, $G_2$, M, and C) of the CDC. Cells changed from one stage to the next only during the night phase (subjective) of the circadian cycle, and cells in all stages displayed a circadian rhythm of bioluminescence. All morphological stages, therefore, not only division, appear to be phased by the circadian oscillator(s). Sweeney (1982) also interpreted these findings as being incompatible with a mechanism for circadian oscillations that invokes cycling in $G_q$, the quantal cycle Klevecz (1976) has hypothesized to be appended to the $G_1$ phase.

**Circadian Clock Control of the CDC in *Euglena***

The relationship between the CDC and circadian oscillators perhaps has been most intensively investigated in the algal flagellate, *Euglena gracilis* Klebs and, consequently, will serve as a basis for subsequent discussion (Edmunds, 1978; Edmunds and Adams, 1981). This organism can be grown on a variety of different, completely defined media, either photoautotrophically in the presence of $CO_2$, or organotrophically in the light or dark on carbon sources ranging from acetate and ethanol to lactic, glycolic, glutamic, and malic acids over a wide pH range. This versatility in growth mode, in conjunction with the fact that cell division can easily be synchronized by appropriate 24 hr (and other) lighting schedules (Cook and James, 1960; Edmunds, 1965) and temperature cycles (Terry and Edmunds, 1970), has made *Euglena* an important experimental organism for a variety of physiological and biochemical investigations (Buetow, 1968, 1982).

Furthermore, the temporal organization of *Euglena* has been extensively examined (Edmunds and Halberg, 1981; Edmunds, 1982). A large number of persisting circadian rhythms have been reported, a number of which are illustrated in Fig. 2. These studies have been aided by the fact that *Euglena* can be maintained in the "stationary" phase of growth (or infradian growth mode) for days or even weeks with little or no net change in cell concentration; circadian output can thus be monitored while divorced from the driving force of the CDC. Likewise, the fact that a number of photosynthetic mutants (or even completely bleached strains devoid of their chloroplast genomes) have been isolated which still exhibit light-entrainable circadian rhythms has effectively eliminated the problem of the dual use of imposed light spans and signals—as an energy source for growth, on the one hand, and as a timing cue for the underlying clock, on the other (Kirschstein, 1969; Jarrett and Edmunds, 1970; Mitchell, 1971; Edmunds et al., 1976). We will now review the major lines of evidence that implicate an endogenous self-sustaining circadian oscillator (see Fig. 1) in the control of the CDC in this unicell (and, by generalization, those microorganisms listed in Table 1).

*Entrainability*

Photoautotrophically grown cultures of wild-type *Euglena* (Z strain) can be routinely synchronized, or entrained—a key property of circadian rhythms (see Fig. 1)—by empirically chosen repetitive, 24 hr LD cycles so that cell division synchrony is confined entirely to the dark intervals (Cook and James, 1960; Edmunds, 1965). For example, in LD: 10,14 (Fig. 3a) a batch-cultured population doubles (factorial increase, or $ss \simeq 2.0$) at each step, the period $\tau$ of the rhythm in the population exactly matching that ($T$) of the imposed LD cycle, and, by inference, the length of the individual CDC also must average 24 hr (the rate of cell death is insignificant). Long-term synchronous cultures can be obtained using continuous or semicontinuous culture techniques by which the cell titer is maintained at a constant level without reaching limiting conditions (Terry and Edmunds, 1969), as occurs in batch cultures (due primarily to reduced light intensity caused by the mutual shading of the cells).

If one reduces the total duration of the light interval (for example, LD: 8,16) within a 24 hr framework, the amplitude of the cell division rhythm in the population is proportionately reduced ($\overline{ss} = 1.68$), and the average doubling time $g$ of the culture is lengthened to about 36 hr (Edmunds and Funch, 1966b). Not every cell divides during each division burst, and the culture is now no longer developmentally synchronous to the extent of a 1:1 mapping between the stages of the CDC in all constituent cells (Edmunds, 1978). Neverthe-

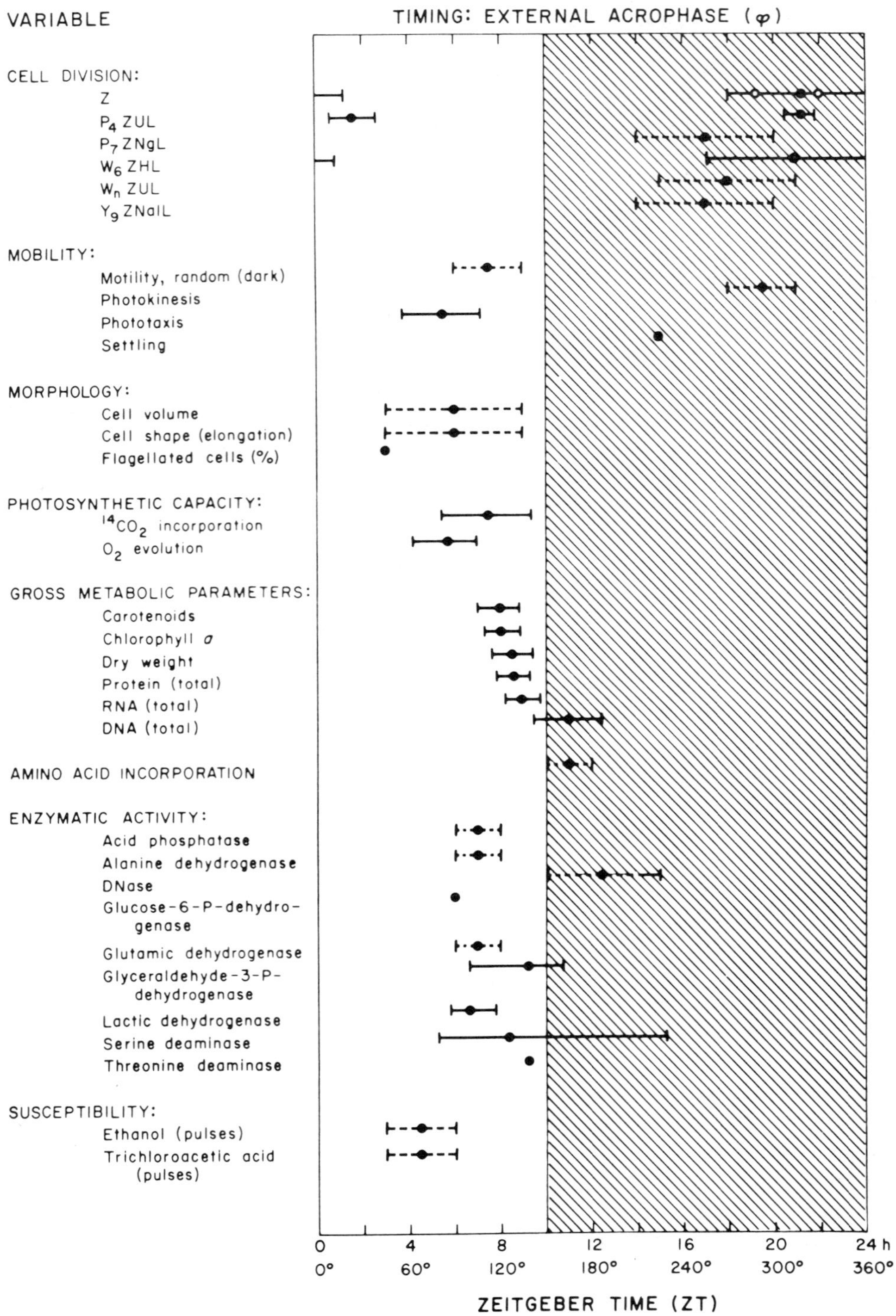


**FIGURE 2** Acrophase chart for *Euglena* showing the time relations among a number of circadian rhythms documented for this unicellular flagellate. The so-called external acrophase $\phi$ relates the peak of the fitted curve to the onset of light in a synchronizing LD cycle and to "subjective dawn" in DD or LL when the light would have come on had the

less, cell divisions, when they do occur, do so during the dark intervals only at intervals of 24 hr, and the culture continues to be synchronized in the sense of event simultaneity (Edmunds, 1978). This rhythmicity observed in LD: 8,16 (and in similar LD cycles) stands in sharp contrast to the asynchronous, exponential growth curve obtained in LL having a minimum doubling time of 12-14 hr (Edmunds, 1965).

Similarly, so-called skeleton photoperiods ($PP_S$) comprising the framework of a normal, full-photoperiod cycle (for example, *LD*: *3*,6,*3*,12) will also entrain the rhythm to a precise 24 hr period (Fig. 3b); in this case, divisions are confined to the main dark intervals, commencing at their onsets (Edmunds and Funch, 1969b). Entrainment by LD cycles having $T \neq 24.0$ hr (for example, LD: 10,10) may also occur, but only within certain limits (Edmunds and Funch, 1969b; Ledoigt and Calvayrac, 1979). Additionally, appropriate diurnal temperature cycles (for example, 18,25°C: 12,12 or 28,35°C: 12,12) will synchronize the cell division rhythm in cultures maintained in LL (Terry and Edmunds, 1970).

Finally, even more conclusive evidence for the role of a basic circadian oscillator in the control of the CDC has been obtained by utilizing photosynthetic mutants (all obligate heterotrophs) of *Euglena* (Edmunds, 1975, 1978). These studies have effectively circumvented the problem of the dual use of imposed light cycles and signals: as an energy source, or "substrate," for growth, on the one hand, and as a timing cue (Zeitgeber) for the underlying clock, on the other. Thus, the ultraviolet light-induced $P_4$ZUL mutant (Jarrett and Edmunds, 1970), although not entrainable to a 24 hr period by diurnal LD cycles when growing in the ultradian mode ($g < 24$ hr) at 25°C where a doubling of the population occurs every 10 hr, could be synchronized by a LD: 10,14 cycle if the temperature were lowered to 19°C, (yielding exponential growth with $g \simeq 24$-26 hr in DD or LL). In this infradian ($g > 24$ hr) growth mode (Ehret et al., 1977), divisions were set back or delayed 8-10 hr at 24 hr intervals. These results have been extended both to the naladixic acid-induced $Y_9$ZNalL photosynthetic mutant and to the white, heat-bleached, $W_6$ZHL strain (Fig. 4) that totally lacks chloroplasts (Edmunds, 1975; Edmunds et al., 1976), and are consistent with the data of Mitchell (1971) for the pale green nitrosoguanidine-mutagenized $P_7$ZNgL strain and the white, ultraviolet-bleached $W_n$ZUL mutant.

### *Persistence*

Although the synchronization of the cell division rhythm in *Euglena* by diurnal, full-photoperiod LD cycles is consistent with the notion that a putative circadian clock is entrained by the imposed light regime and, in turn, phases or "gates" cell division to the dark intervals every 24 hr (by acting, perhaps, on one or more key control points (Spudich and Sager, 1980) of the CDC, it does not demand it. Light (or darkness) could be acting by directly inhibiting (or promoting) division, and periodic shifts between light and dark would synchronize the culture (Campbell, 1957). A number of other observations, however, render this seemingly straightfoward hypothesis untenable.

One of the most basic tests for the existence of a circadian rhythm (*sensu stricto*) is to determine whether it will continue to free run for a number of cycles following transfer

---

synchronizer been continued. The bars extending to the left and right of the acrophase points give the 95% confidence interval, when calculated by cosinor analysis. Single points, or points with dashed lines and brackets, indicate subjective estimates of the acrophase from published data without benefit of statistical analysis. (Adapted from Edmunds and Halberg, 1981.)

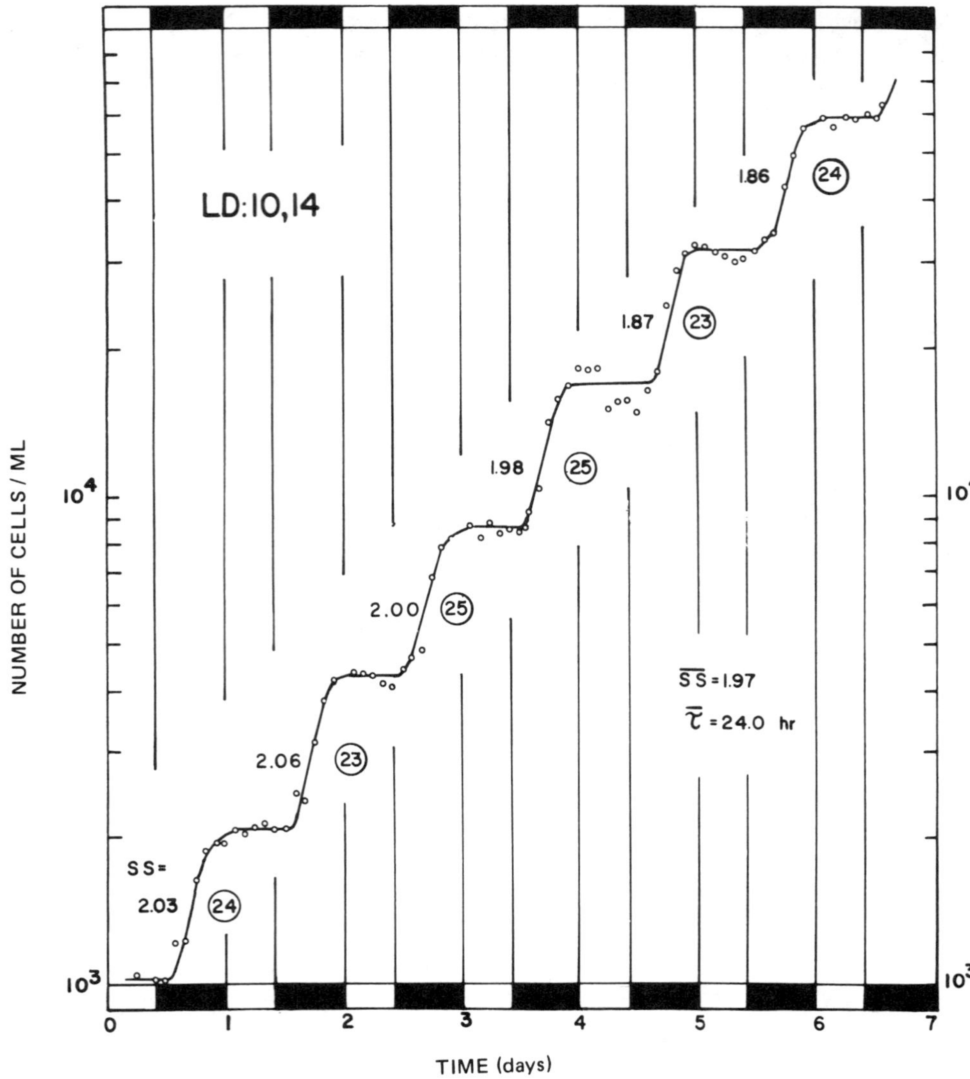


FIGURE 3a

**FIGURE 3** Entrainment of the cell division rhythm in populations of *Euglena gracilis* (Z strain) batch cultured photoautotrophically at 25°C by either a full-photoperiod, diurnal LD: 10,14 light cycle (a) or by a "skeleton" *LD*: *3*,6, *3*, 12 regime (b) comprising the framework of an LD: 12,12 cycle (see Fig. 5). Note the effects of the accidental reduction in the length of the second light pulse on the fourth day of the skeleton test cycle. Step-size (*ss*, ratio of number of cells per milliliter after a division burst to that just before the onsets of divisions) are given for successive steps; the estimated period $\tau$ of each oscillation is indicated by the encircled numbers (hours). The average period $\overline{\tau}$ of the rhythm in the culture were almost identical to that (*T*) of the synchronizing LD cycle in either regime; divisions were confined primarily to the main dark intervals, commencing at their onsets. Whereas a doubling of cell number ($ss \simeq 2.00$) usually occurred every 24 hr in the full-photoperiod regime, the amplitude of the rhythm was necessarily diminished in the skeleton LD cycle, corresponding to the reduction of the photofraction. (Modified from Edmunds and Funch, 1969b.)

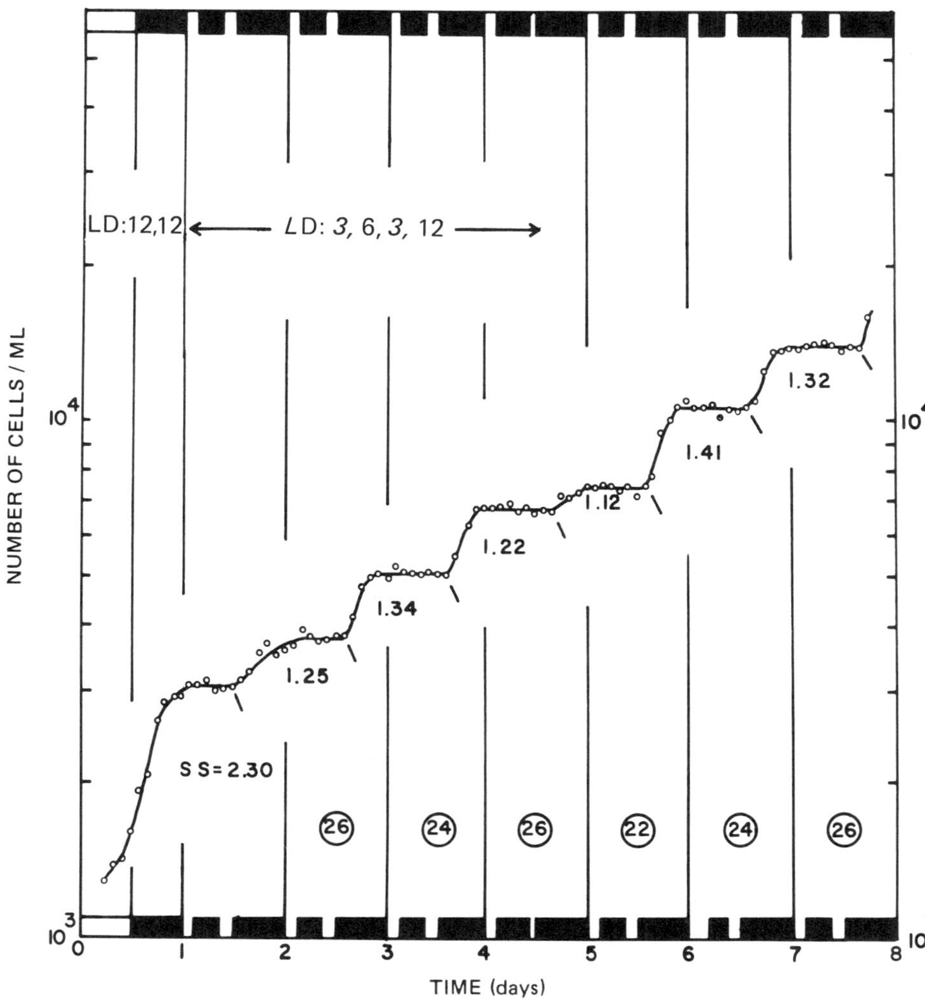


FIGURE 3b

of the organism to conditions held constant with respect to the major Zeitgeber (light and temperature); characteristically, the period $\bar{\tau}$ under such conditions only approximates 24 hr, as might be expected of an imperfect biological clock (see Fig. 1).

Indeed, rhythmic cell division has been found to persist for a number of days ($\bar{\tau}$ = 24.2 hr) in the autotrophically grown Z strain, batch cultured under dim LL (Edmunds, 1966). This series of experimental results was perhaps the most definitive but was restricted by the low light intensities (800 lux) that had to be used. (The rhythm damped out in more intense LL, and, of course, could not be observed in DD.) Nevertheless, although the division bursts were relatively small ($\bar{g} \simeq 5$ days), those cells that did divide did so during their subjective night at the times that they would have done so had the entraining LD cycle previously imposed been continued.

We have also observed free-running circadian rhythms of cell division in a variety of higher frequency (e.g., LD: 1,3 or LD: 1/3, 1/3) light cycles (Edmunds and Funch, 1969b;

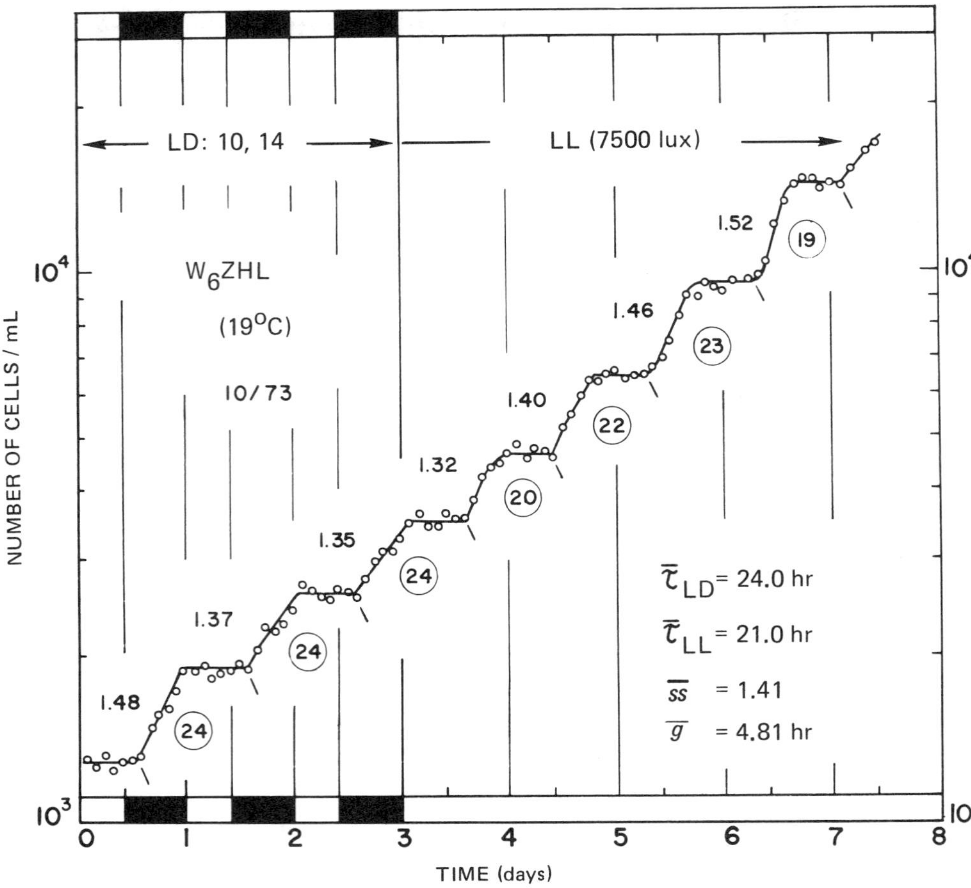


**FIGURE 4** Entrainment of the circadian rhythm of cell division and its persistence for at least 5 days under continuous illumination in the aplastidic, heat-bleached $W_6$ZHL mutant of *E. gracilis* (Z strain) batch cultured photoautotrophically at 19°C in an LD: 10,14 cycle and subsequently released into LL (7500 lux). Labels and notations as for Fig. 3. Note that although the period of the population rhythm was about 24 hr (i.e., circadian), the period of the average individual cell cycle approximated 48 hr, since less than half of the cells divided during any one 24 hr time span. (Adapted from Edmunds, 1975, 1978.)

Laval-Martin et al., 1979) and even "random" illumination regimes (Edmunds and Funch, 1969a), although we do not exclude the possibility that such short-period light regimens (which provide no information to the cells with regard to 24 hr periodicities) may modulate the period to some extent (see Chap. 21). Those high-frequency cycles (Figs. 5B and 6b) having symmetric photo- and scotophases (e.g., LD: 1,1 or LD: 3,3) and properly constructed random regimes (Fig. 6a) have proved particularly useful in that they afford an amount of light during a 24 hr time span identical to that received in a full-photoperiod LD: 12,12 (Fig. 5A) entraining reference cycle ($T = \tau = 24$ hr) yet elicit free-running circadian rhythms ($T << \tau \simeq 24$ hr) not only of cell division (Figs. 5B and 6), but also of random motility (Schnabel, 1968) and of photosynthetic capacity and chlorophyll content (Laval-Martin et al., 1979; Edmunds and Laval-Martin, 1981). Likewise, they have been employed (Edmunds et al., 1982) with advantage in the derivation of a phase-

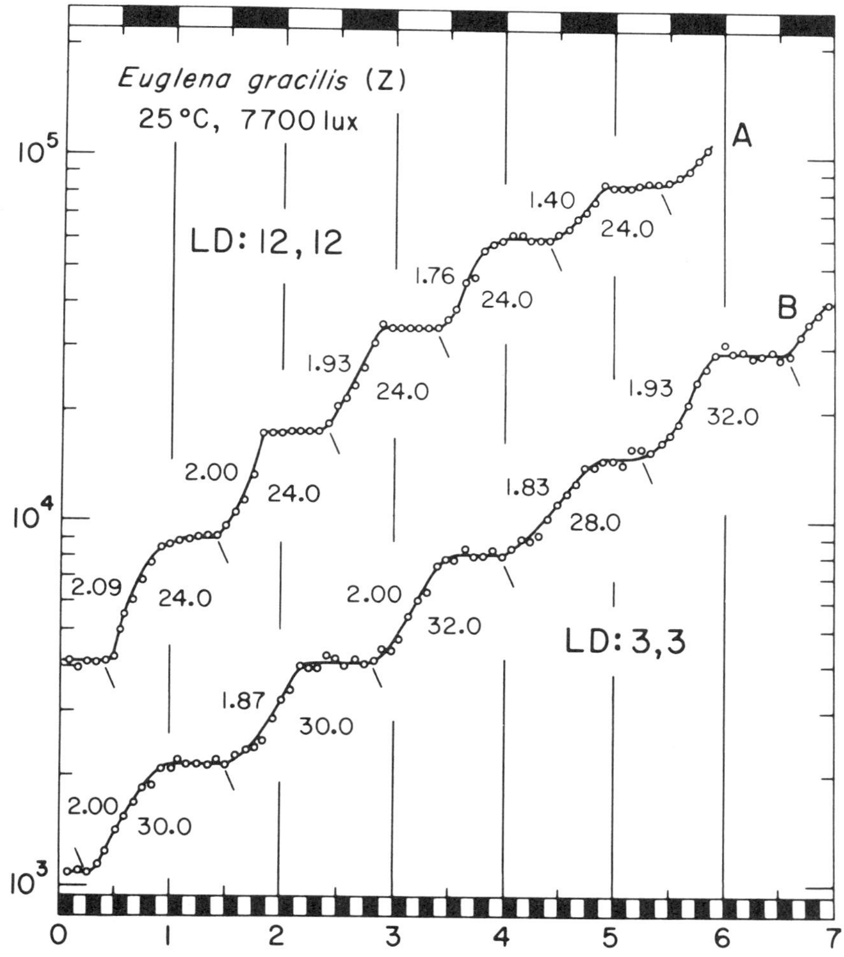


**FIGURE 5** Growth curves for synchronously dividing, photoautotrophic cultures of *E. gracilis* (Z) at 25°C in repetitive LD: 12,12 (curve A) or LD: 3,3 (curve B) light-dark cycles. Notations as for Figs. 3 and 4. Culture A was entrained by the diurnal regime to a precise 24 hr period; in contrast, culture B free ran with a circadian period $\tau$ here averaging 30.4 hr under the imposed higher frequency ($T = 6$ hr) LD: 3,3 cycle. Both cultures, however, received the same total duration of light per 24 hr time span, as reflected by their similar stepsizes ($ss \simeq 2.0$). (Reproduced by permission from Edmunds et al., 1982.)

response curve for light perturbations (see Phase Shiftability) and in studies on temperature compensation (see below). With these exotic light cycles, therefore, the total duration of light afforded for growth is held constant (reflected in the comparable stepsizes of the observed rhythmic growth curves), and the signaling information inherent in the light perturbations can be varied to manipulate the light-sensitive circadian oscillator.

Finally, persisting circadian rhythms of cell division have also been observed routinely in the $P_4$ZUL (Jarrett and Edmunds, 1970) and $W_6$ZHL (Edmunds, 1975; Edmunds et al., 1976) photosynthetic mutants batch cultured in the infradian growth mode at temper-

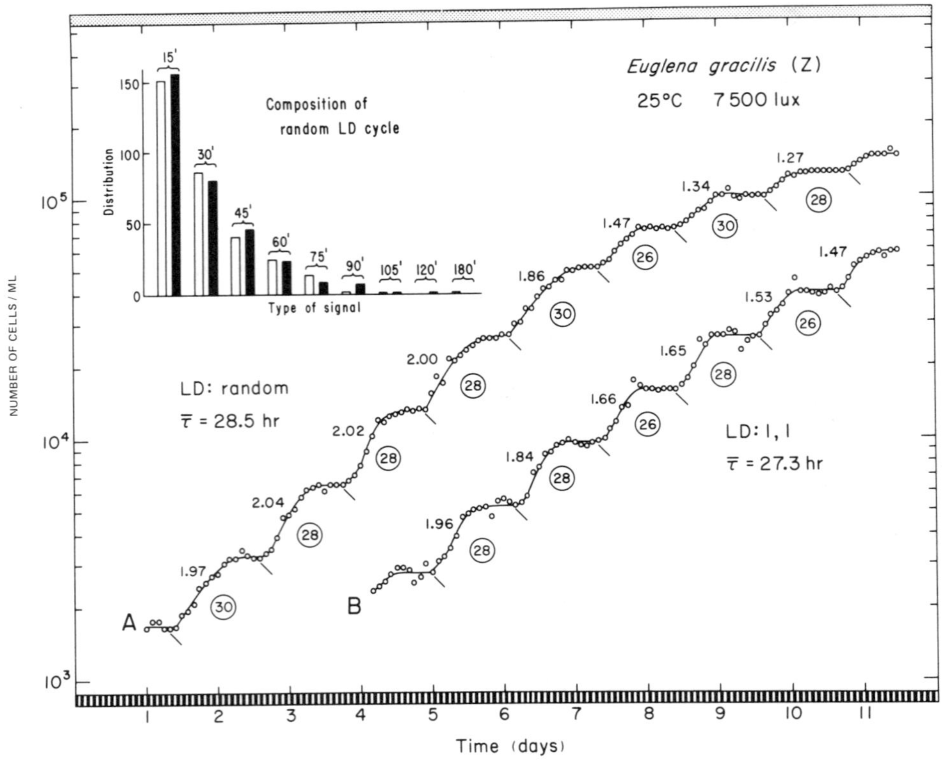


**FIGURE 6** Persistence of the free-running, circadian rhythm of cell division in cultures of *Euglena* (Z strain) grown photoautotrophically at 25°C under two exotic LD regimens, each of which furnished a total duration of 12 hr of illumination (cool-white fluorescent, 7500 lux) and 12 hr of darkness during a 24 hr time span but which provided no 24 hr time cue to the cells. The growth curve (A) at the top was obtained in a "random" illumination regime, the duration of whose signals varied between 15 and 180 min according to the distribution indicated in the inset. That at the bottom (curve B) was obtained from a culture exposed to a high-frequency LD: 1,1 cycle ($T = 2$ hr). The stepsizes (*ss*) and periods for the oscillations are indicated as in Fig. 3. The mean value of the freerunning period $\overline{\tau}$ for the rhythm obtained in the random LD regimen was 28.5 hr as read from the growth curve, while that in LD: 1,1 was 27.3 hr; these values corresponded, respectively, to those computed by time-series analysis of 28.2 and 26.6 hr.

atures less than 19°C in DD or LL (see Fig. 4). Similar observations have been made in semicontinuous culture of the $P_4$ZUL mutant: long trains of circadian oscillations were exhibited in DD or LL over a span of 10-14 days (Edmunds et al., 1971, 1974). Interestingly, we have observed (Edmunds et al., 1976) that both these mutants may gradually lose both their capacity to exhibit division synchrony in LL or DD and even to be entrained by LD cycles, reverting to random exponential growth ($g > 24$ hr). These properties could be restored, however, by the addition of certain sulfur-containing compounds (such as cysteine or methionine) to the medium at the onset of the experiment. If these substances were added at various times to an arrhythmic $P_4$ZUL culture in LL (after prior exposure to LD: 14,10), periodic division was likewise induced whose phase was that predicted on the assumption that the underlying clock had been running undisturbed (but unexpressed) throughout the experiment, merely having been uncoupled from division itself until the sulfur compounds were added. This hypothesis could be tested further by simultaneously monitoring another rhythm (such as motility)—presumably a manifestation of the same oscillator, but whose coupling to the clock would not be affected by these compounds (or lack thereof).

*Initiation*

It is a common feature of circadian rhythms that they damp out or are not observed in continuous bright illumination (in contrast to dim LL or to DD). This also holds true for the cell division rhythm in *Euglena*: in more intense LL (e.g., 3500-7500 lux), photoautotrophic cultures of the wild type (Z strain) at 25°C (Edmunds, 1965) revert to asynchronous, exponential growth ($\bar{g}$ = 11-15 hr). A single switch down in irradiance to 800 lux was sufficient to elicit a free-running circadian rhythm that persisted for at least two or three cycles, with the first $\phi_R$ being observed about 14 hr after the switch down (Edmunds, 1966). Two hypotheses could explain these observations: either (1) the endogenous oscillators in the individual cells comprising the population were absent or arrested at some specific phase point and subsequently initiated or released by the transition (primary arrhythmicity); or (2) the individual oscillators were all running in bright LL but were out of phase with each other (secondary arrhythmicity). The first explanation appears to be more likely (Edmunds, 1966), although this may not be the case for the initiation of rhythmicity (Jarrett and Edmunds, 1970) in arrhythmic cultures of the $P_4$ZUL mutant growing in DD by a single switch *up* in irradiance to LL (5000 lux).

More recently, we have explored this problem further in cultures of *Euglena* (Z strain) exposed to the higher frequency cycles discussed in the section Persistence. It is not necessary for the cultures to have been exposed to a prior entraining diurnal light cycle in order for rhythmicity to be exhibited: cells maintained in LD: 3,3 (Fig. 5B) or LD: 1,1 (Fig. 6B), or random illumination regimes (Fig. 6A) from the moment of inoculation all were rhythmic. Furthermore, if arrhythmic cultures growing exponentially in LL (7500 lux, 25°C) with a value of $\bar{g}$ of 11-15 hr were placed into any of these cycles, circadian periodicity was quickly induced, with $\tau$ values varying from 26.6 to 30.6 hr depending on the substrain and regimen used. For example, in some 15 different cultures transferred from LL to LD: 3,3, the $\phi_R$ values of the elicited rhythms were found to occur at intervals after the onset of the first dark pulse of the imposed cycle given by the expression 19.1 hr CT + n$\bar{\tau}$ (where n is an integer and all times have been normalized to 24 hr). Since the $\phi_R$ is known to occur at about CT 12 with reference to a LD: 12,12 cycle, one can infer from this empirical finding that all the oscillators comprising the population were stopped at approximately CT 16.9 in LL and were then set into motion from this phase point by the first transition (or, alternatively, that all the clocks were running in LL and were subsequently reset to CT 16.9 by the dark pulse). This value is somewhat larger than that of CT 12 found as an LL arrest point for many circadian systems; we note, however, that the LD: 3,3 regime itself may not be without effect (see Chap. 21).

*Phase Shiftability*

Another basic property of circadian rhythms (see Fig. 1) is that their phase can be reset (or shifted)—by the lengthening or shortening of the period of one or more oscillations—by single light (or dark) or temperature signals. Indeed, the twin notions of phase and period control are key elements of Pittendrigh's (1965) theory for the entrainment of circadian clocks by light cycles. It is also characteristic of circadian rhythms that both the

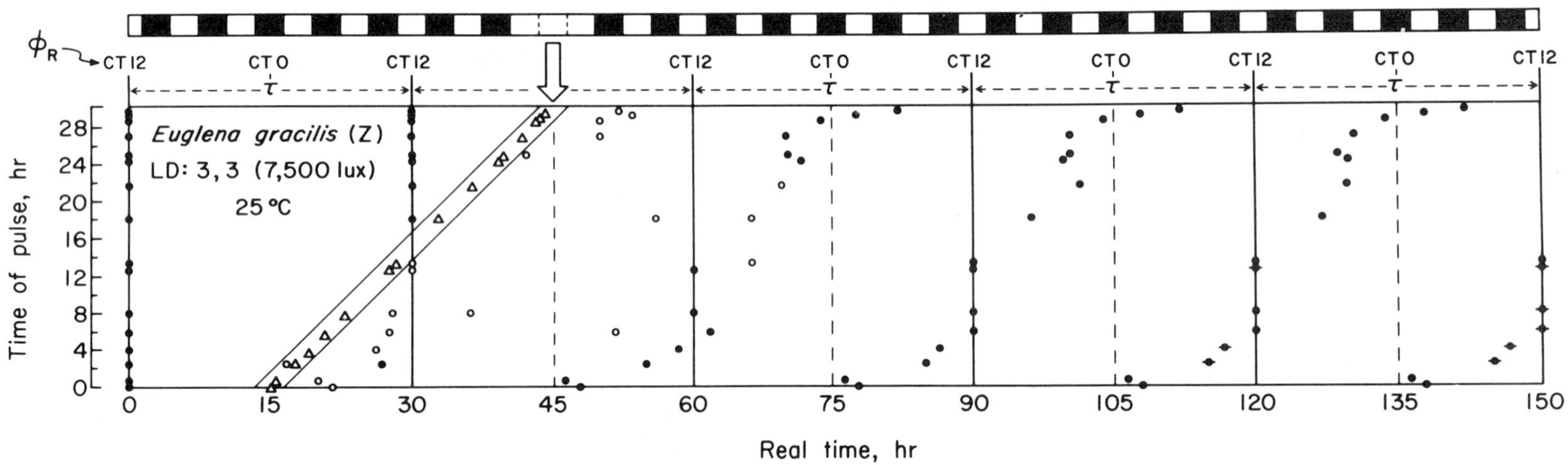


**FIGURE 7** The phase-shifting effects of 3 hr light pulses (700-7500 lux) imposed at different circadian times during the free-running rhythm of cell division in cultures of *E. gracilis* (Z) grown photoautotrophically in LD: 3,3 at 25°C. Sixteen in-phase cultures, whose phase reference points ($\phi_R$, taken to be the onset of cell division occurring at approximately CT 12 and corresponding to the onset of dark in a reference LD: 12, 12 cycle; see Fig. 5), are shown aligned vertically at real time 0. Each horizontal row of points indicates successive $\phi_R$ values determined for a given culture over a time span of 6 days (indicated on the abscissa in both real time and in circadian time). Each of these 16 cultures were perturbed at a specific circadian time by a *single* 3 hr light signal (whose midpoint is indicated by an open triangle, △, within the diagonal band between real time 16 and 45—CT 0-24) by replacing (open arrow) one of the usual dark intervals in the LD: 3,3 regimen with light of the desired intensity. Transients immediately followed the light perturbations, as indicated by the open circles (o) denoting the $\phi_R$ values, until a new steady state was achieved and the original, stable period $\tau$ of about 30 hr was attained once more (●). In some instances, toward the end of the run, this stable $\tau$ was projected to facilitate comparison (as indicated by those $\phi_R$ values denoted -●-). (From J. R. Malinowski, D. L. Laval-Martin, and L. N. Edmunds, Jr., unpublished, 1982.)

sign and magnitude of phase shifts engendered by the light signals are predictably dependent on the subjective circadian time at which the perturbations are applied. That light can reset circadian division rhythms also has been demonstrated in a general sort of way for several unicellular organisms, such as the dinoflagellate *Gonyaulax polyedra* (Sweeney and Hastings, 1958; Hastings and Sweeney, 1964; McMurry and Hastings, 1972).

Similarly, the phenomenon of phase perturbation is inherent in most, if not all models for cell cycle oscillators: The response of cells to external influences is often strongly dependent on the time of the CDC at which the agent is imposed. Thus, synchronous cultures of eukaryotic unicells, such as the ciliate *Tetrahymena pyriformis* (Zeuthen, 1974) or the fission yeast *Schizosaccharomyces pombe* (Zeuthen, 1974; Polanshek, 1977) and of animal cells, such as mouse fibroblast L cells (Miyamoto et al., 1973) all exhibit increased division setback (or "excess division delay") to heat shock and to a variety of chemical agents applied at progressively later times during their CDC. Perhaps more germane is the finding in *Schizosaccharomyces* (Smith and Mitchison, 1976) that the same agent can produce delays when imposed at one point of the CDC but advances when given at another. Indeed, Klevecz and coworkers have discovered that synchronous mammalian V79 cells perturbed by serum (Klevecz et al., 1978), heat shock (Klevecz et al., 1980b) or ionizing radiation (Klevecz et al., 1980a) at 0.5 hr intervals across their 8.5 hr CDC display a biphasic, 4 hr phase-response curve (PRC) comprising both advances and delays in subsequent cell divisions.

Despite these investigations, however, a detailed PRC for a *circadian* mitotic clock has been lacking. Recently, we have managed to fill this void by utilizing photoautotrophic cultures of *Euglena* free-running in a high-frequency LD: 3,3 cycle (see Persistence and Initiation) and displaying under the experimental conditions employed (Edmunds et al., 1982) a stable circadian period of 30.2 hr (±1.8 hr for over 156 monitored oscillations). At different times throughout the 30 hr division cycle, 3 hr light perturbations were imposed systematically on free-running cell populations by giving light during one of the intervals when dark would have fallen in the LD: 3,3 regimen (Fig. 7). Using the onset of division as the phase reference point, the net steady-state phase advance or delay ($\pm\Delta\phi$) of the rhythm was determined after transients, if any, had subsided (usually in one or two days) relative to an unperturbed control culture. Both $+\Delta\phi$ and $-\Delta\phi$ were found, with maximum values of approximately 11-12 hr being obtained at about CT 23 (the "breakpoint"); little, if any phase shift occurred if the light signal was given between CT 6 and CT 12 (Fig. 8, upper panel). The phase-shifting curve (Fig. 8, lower panel) obtained by plotting new phase ($\phi'$) *versus* old phase ($\phi$) was of the type 0 ("strong") variety (Winfree, 1980). Light perturbations, no matter when imposed, engendered new phases which mapped to a relatively restricted portion (CT 6 to CT 13) of the circadian cycle.

Several other interesting features of the PRC for *Euglena* emerge. In the first place, the CT at which all delay or maximum advance phase shifts are achieved correspond to the approximate position of the CDC during which division occurs (commencing about CT 12) and to the first few hours of the $G_1$ phase. These times, moreover, are exactly those when the free-running rhythm of photosynthetic capacity (Laval-Martin et al., 1979; Edmunds and Laval-Martin, 1981) in LD: 3,3 displays the lowest values; maximum values ($CO_2$ fixed per cell per hour) occur at the very time (CT 6-12) that 3 hr light signals are virtually ineffective in phase shifting the division rhythm. This observation demonstrates that although light is needed as a "substrate" for photosynthesis and the progression of the CDC in photoautotrophically cultured *Euglena,* it serves a quite distinct and separable function in phase shifting and entraining the circadian oscillator(s) hypothe-

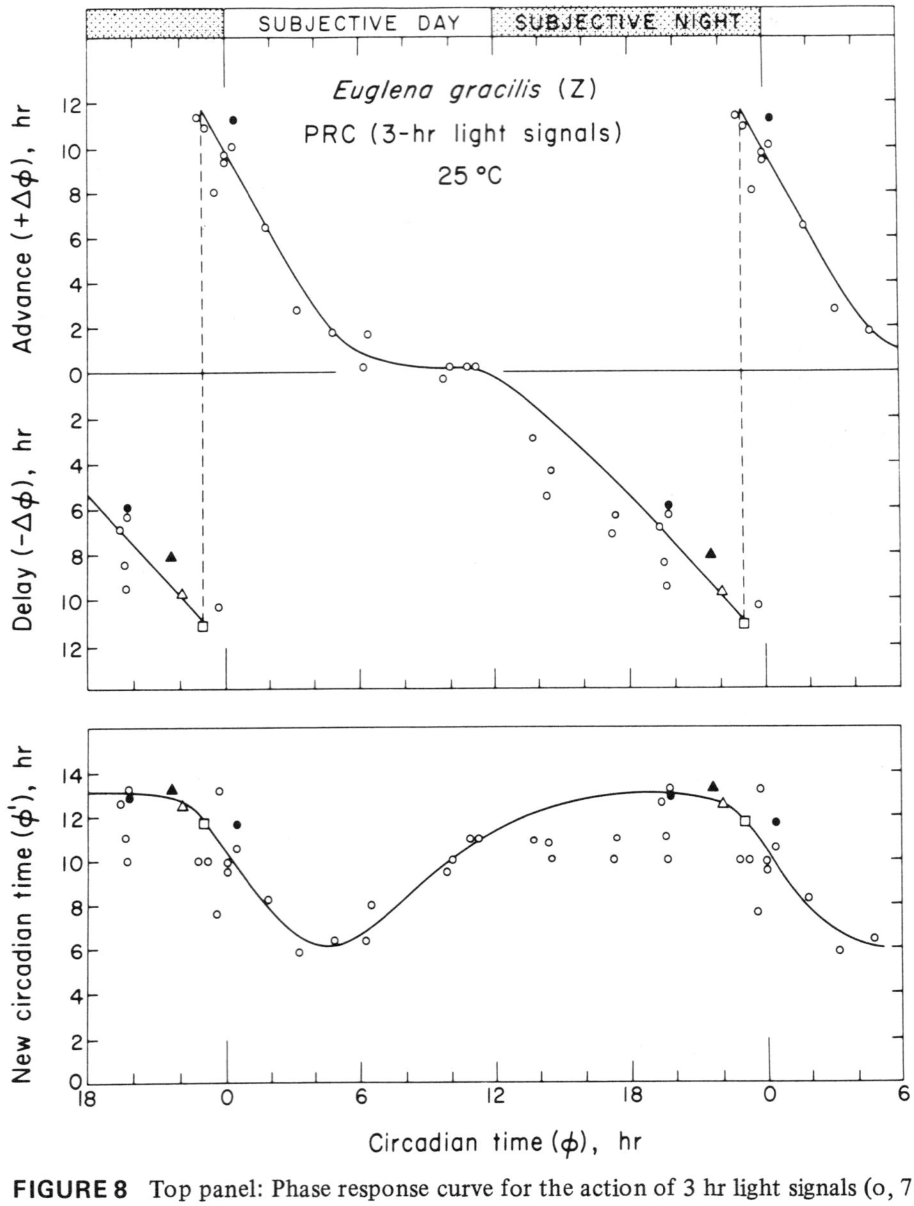


**FIGURE 8** Top panel: Phase response curve for the action of 3 hr light signals (o, 7500 lux; □, 3800 lux; △ 1500 lux; ●, 700 lux, ▲, 300 lux) on the free-running rhythm of cell division in photoautotrophic cultures of *Euglena* maintained in LD: 3,3. The steady-state phase shift ($\pm\Delta\phi$), determined as described in Fig. 7 but normalized to 24 hr, is double plotted as a function of the circadian time (modulo 24) at which the midpoint of the signal was given. The breakpoint inherent in this type of plot, indicated by the dashed line at CT 23, would disappear if the curve were drawn in monotonic form with all the phase shifts being treated as delays. Lower panel: The resultant new CT ($\phi' = \phi \pm \Delta\phi$) following the perturbation is plotted as a function of the CT ($\phi$) at which the signal was given; the smooth sinusoidal curve thus obtained indicates that the PRC is of the type 0 (strong) variety. (Modified from Edmunds et al., 1982, and Malinowski, J. R., Laval-Martin, D. L. and Edmunds, L. N., Jr., unpublished, 1982.)

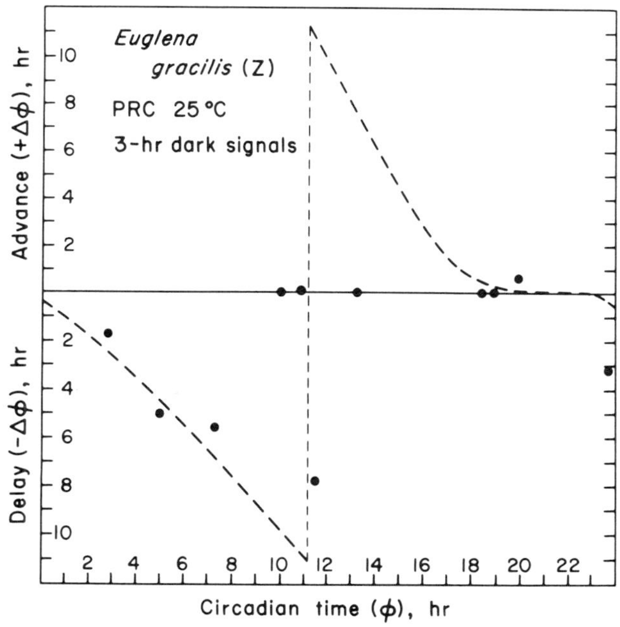


**FIGURE 9** Phase-response curve for the action of 3 hr *dark* perturbations on the free-running rhythm of cell division in cultures of *E. gracilis* (Z strain) growing photoautotrophically in LD: 3,3 at 25°C. The steady-state phase shift ($\pm\Delta\phi$), normalized to 24 hr, is plotted as a function of the circadian time at which the midpoint of the signal (replacing one of the usual light intervals in the LD: 3,3 cycle) was given. Other methodology similar to that described in Fig. 7. The dashed line denotes the PRC obtained for 3 hr light signals (Fig. 8) translated by 180°. (From J. R. Malinowski, D. L. Laval-Martin, and L. N. Edmunds, Jr., unpublished, 1982.)

sized to underlie the rhythm of cell division, whose rhythmic sensitivity to light is reflected in the PRC. Last, the phase shifts observed appear to constitute true developmental advances and delays: in the case of the former, a higher cell titer is attained before the same cell concentration is reached in the unperturbed control, but the converse holds for phase delays.

The PRC derived for the effects of white light pulses on the circadian rhythm of cell division in *Euglena* (Fig. 8) is qualitatively similar to those for many other organisms (Pittendrigh, 1965; Winfree, 1980). We note, however, that the breakpoint is displaced somewhat to the right; if the onset (as opposed to the midpoint) of the 3 hr light pulse were used in the calculations, however, or if the onset of the resultant 9 hr light span were taken, the PRC would then be shifted to the left and the discontinuity would lie at approximately CT 19-20—close to that found for the eclosion rhythm in *Drosophila*, for example (Pittendrigh, 1965). Finally, we caution that we have taken what is perhaps the simplest interpretation for the action of these light signals and have disregarded possible complications due to the interaction of the imposed LD: 3,3 cycle (free-running conditions) with the circadian oscillator (see Chap. 21 for a more detailed analysis).

In contrast to the phase-shifting effects of light pulses, the action of *dark* pulses on free-running circadian rhythms (usually in LL) has received far less attention. In animals, such as the tropical nocturnal bat *Taphozus melanopogon* (Subbaraj and Chandrashekaran, 1978) or the golden hamster, *Mesocricetus auratus* (Boulos and Rusak, 1982), the PRC for light or for dark perturbations have been reported to be "mirror images" of each other (i.e., 180° out of phase), although Ellis et al. (1982) find that dark signals in LL perturb the hamster circadian system in a somewhat more complex manner. Similarly, light and dark pulses seem to have opposite and mutually complementary influences in plants, such as *Bryophyllum* (Wilkins, 1960), and in the unicells *Gonyaulax* and *Acetabulaira* (Karakashian and Schweiger, 1976). Another generalization from these studies is that PRC for dark pulses seem to have smaller amplitudes than those of PRC for light signals of the same duration. We present here (Fig. 9) a PRC for 3 hr dark perturbations (replacing a normal 3 hr light interval in a manner analogous to that employed for the derivation of the light-pulse PRC described in Fig. 7) for the circadian rhythm of cell division in *Euglena* free running in LD: 3,3 and note that it, too, seems to follow these generalizations, with very little if any phase advance being observed.

### *Singularity Point*

Biological pacemakers may exhibit limit-cycle dynamics (Kauffman and Wille, 1975) (see also Chap. 16). Pavlidis (1968), among others, reformulated in terms of space-state topology the analytical model of Pittendrigh (1965) and coworkers for the entrainment of circadian oscillators by light cycles, whose dynamic behavior can be described by a set of two first-order differential equations involving two variables (which, if known at any given time allow the prediction of the behavior of the system at some later time). These two variables, termed the *state of the system,* can be represented as coordinates in a plane, and the behavior of the system can be described in terms of plane curves, or trajectories. The oscillatory motion of a biological clock, in turn, can be represented as a stable periodic trajectory that closes in on itself and is termed a *stable limit cycle.* Such a system, if disturbed (as, for example, by a light pulse or an inhibitor pulse), will always tend to return to an equilibrium configuration.

Theoretical and mathematical studies have further predicted that a circadian oscillator might be rendered arrhythmic—characterized by a phaseless, motionless state—by a critical pulse of a certain strength and duration given at a specific time (termed the *singularity point*) in the circadian cycle (Winfree, 1970); for extensive treatment, see Winfree (1980). As Winfree (1980) has pointed out, as stimulus strength is increased, the transition from type 1 (weak pulse) to type 0 (strong pulse) resetting is necessarily discontinuous at one special phase point, the breakpoint, corresponding to this unique singularity. This prediction has now been demonstrated for light perturbations in several circadian systems, including *Drosophila* (Winfree, 1970), *Kalanchoe* (Engelmann et al., 1978), *Sarcophaga* (Saunders, 1978), and *Culex* (Peterson, 1980), and most recently for critical pulses of anisomycin (an inhibitor of protein synthesis on 80s ribosomes) in the dinoflagellate *Gonyaulax* (Taylor et al., 1982).

We have now examined the threshold intensity of illumination for phase shifting the circadian rhythm of cell division in *Euglena* by 3 hr light signals (Malinowski et al., 1982, unpublished results). Perturbations having an intensity within the range of 700-7500 lux generated the same phase shift when imposed at a given CT (Fig. 10A). Light signals of lower intensity, however, elicited different responses, some quite dramatic. Thus, a 40-400 lux pulse given at CT 0.4 (the approximate location of the breakpoint at about CT 23) induced arrhythmicity (Fig. 10B), the population reverting to asynchronous, exponential growth. The intensity of this annihilating pulse and the CT at which

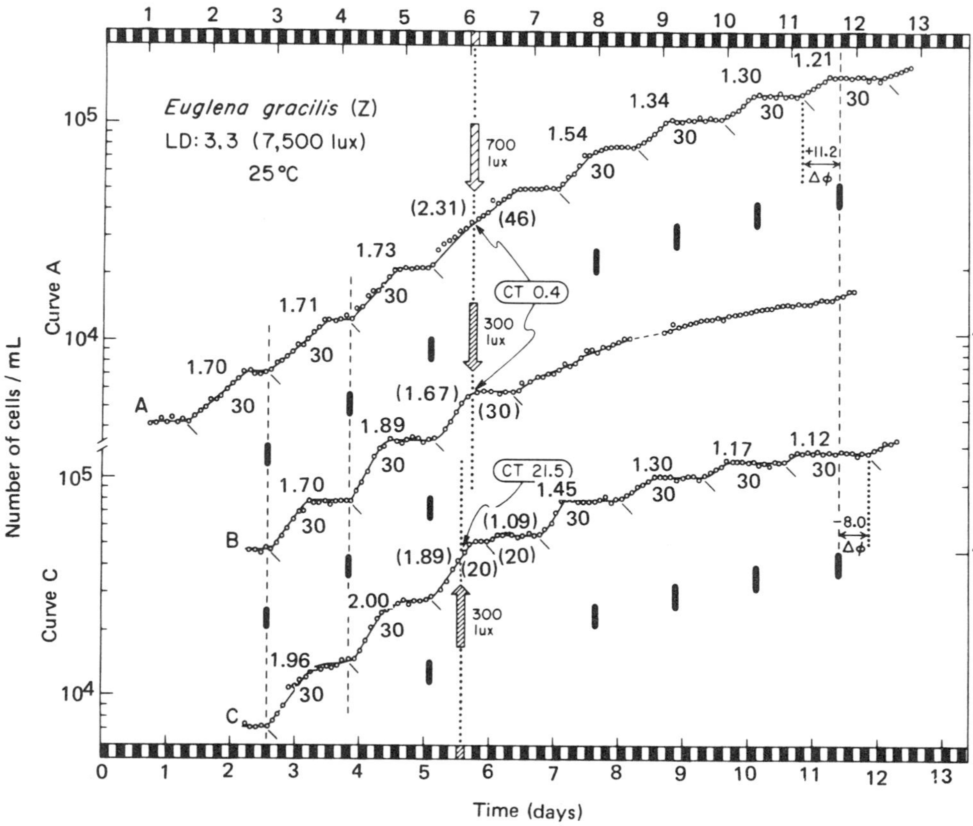


**FIGURE 10** Localization of a singularity point in the behavior of the circadian oscillator underlying the free-running rhythm of cell division in cultures of photoautotrophically grown *E. gracilis* (Z) perturbed by a "critical" 3 hr light signal. Three initially in-phase, synchronously dividing cultures were utilized, each displaying a stable $\tau$ of approximately 30 hr in an LD: 3,3 regimen at 25°C. At circadian time (CT) 0.4, culture A was perturbed by a 3 hr light signal of 700 lux, while culture B was perturbed by a 3 hr light signal of only 300 lux (each replacing one of the normal light intervals in the imposed LD cycle having an intensity of 7500 lux, as represented by the hatched arrows). Whereas the former signal elicited a phase shift ($\Delta\phi$) of + 11.2 hr (modulo 24) circadian time (equivalent to 14.0 hr real time) comparable to that seen for 7500 lux light signals (Fig. 8), the latter, weaker light pulse (curve B) induced total arrhythmicity, although cell division continued in the population. Nevertheless, if a third culture (curve C) were perturbed by a 3 hr signal having an identical intensity of 300 lux but imposed at a slightly different circadian time (CT 2.15), the cell division rhythm continued although its phase was delayed ($\Delta\phi$ = 8.0 hr CT, equivalent to −10.0 hr real time) in a manner observed for 7500 lux pulses (Fig. 8). (From J. R. Malinowski, D. L. Laval-Martin, and L. N. Edmunds, Jr., unpublished results, 1982.)

it was imposed were found to be quite specific: a 300 lux stimulus given at CT 21.5 (Fig. 10C) merely generated a phase delay of the same magnitude found for 7500 lux signals (Fig. 8). Different degrees of asynchrony were observed as one approached the boundaries (lux and CT) of the critical pulse. The location of the singularity point at close to CT 24 (CT 0) corresponds to the steepest part of the type 0 resetting curve (Fig. 8, lower

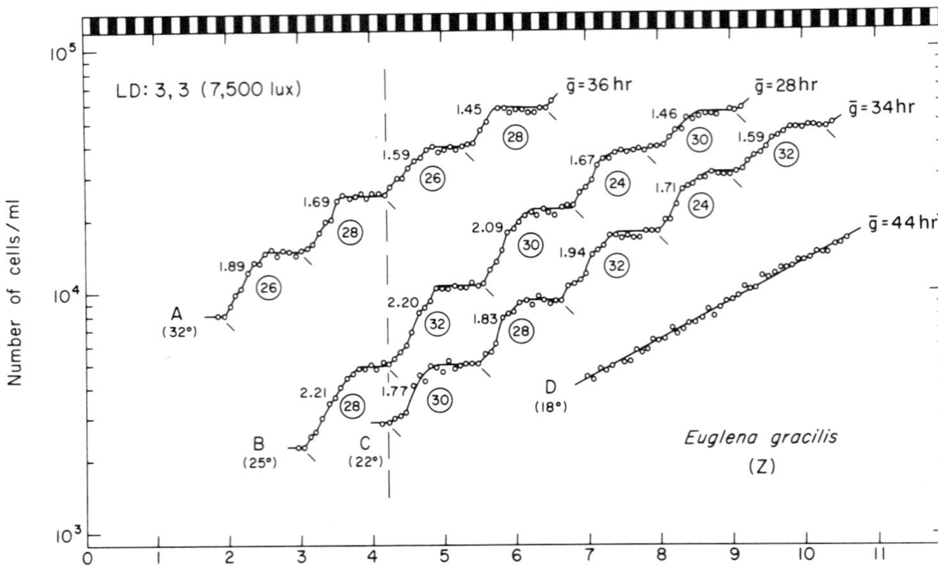


**FIGURE 11** Growth curves of *E. gracilis* (Z) photoautotrophically grown in LD: 3,3 at different temperatures. Notations as for Figs. 3 and 4. Only sections of the complete data trains are shown; to facilitate comparison, growth curves A, B, and C were aligned at one $\phi_R$ indicated by the vertical dashed line. The average generation (doubling) time $\bar{g}$ for each of the four cultures, as determined by linear regression, is given at the end of its growth curve. Note that no rhythm was obtained in LD: 3,3 at 18°C (curve D), despite the fact that an entrained rhythm ($T \simeq 24$ hr) was observed in a diurnal LD: 12,12 cycle (not shown). (From R. W. Anderson, D. L. Laval-Martin, and L. N. Edmunds, Jr., unpublished results, 1982.)

panel) and probably lies on the axis of *Euglena*'s three-dimensional phase-resetting surface.

The existence of this "critical pulse" and its corresponding singularity point not only further supports the hypothesis that a circadian oscillator regulates the CDC in *Euglena* but also suggests (though it does not demand) that the pacemaker may have limit-cycle dynamics. Inasmuch as the results were obtained with populations of cells, we cannot deduce the state of the oscillator(s) in individual cells. On the one hand, arrhythmicity in the culture may be the result of a dispersion of the phases of individual CDCs (cellular incoherence), entailing that the pulse differentially phase shift the cells which were synchronous at the time of the pulse; on the other hand, the critical pulse may have stopped each circadian clock comprising the population (but not the CDCs, which continue to run). To distinguish between these two alternatives for the *Drosophila* eclosion rhythm, Winfree (1970) gave a second light perturbation at different times after the first critical pulse. His finding of a unimodal distribution of phases suggested that reinitiation rather than resynchronization had occurred. We anticipate a similar situation to obtain for the cell division rhythm in *Euglena,* although at the moment this second pulse test is difficult to perform in our system, and proper controls are not feasible.

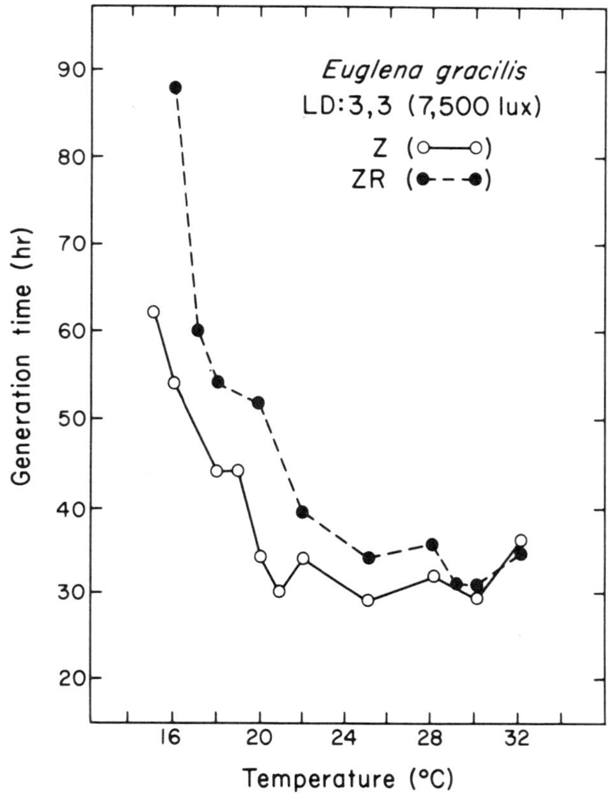


**FIGURE 12** (a) Influence of temperature on the average generation time $\bar{g}$ of two different strains (Z, wild type; ZR, DCMU-resistant) of *Euglena* maintained in LD: 3,3 (7500 lux). Growth curves obtained at each temperature during the exponential growth phase where light intensity had not become limiting (usually between 10,000 and 60,000 cells per milliliter) were fitted with straight lines by linear regression (least squares) to obtain their slopes.

*Temperature Compensation*

A final, and rather remarkable property of circadian rhythms (Fig. 1) is that their period —but not their amplitude—is only slightly affected by the ambient steady-state temperature over the physiological range (Sweeney and Hastings, 1960). In fact, this is just what one would anticipate in a functional biological clock measuring astronomical time. In contrast, the duration of the CDC (i.e., $g$) is commonly thought to be highly dependent on temperature, and indeed, this is true for *Euglena* also (Terry and Edmunds, 1970). A recent study (Anderson et al., 1982, unpublished results) of the effects of different constant temperatures ranging between 16 and 32°C on the generation time of the wild-type Z strain and a diuron (DCMU)-resistant line (ZR) photoautotrophically batch cultured in LD: 3,3 clearly illustrates this dependence (see Figs. 11 and 12a).

Despite this apparent paradox, several lines of evidence suggest that the period of the circadian oscillator hypothesized to underlie rhythmic cell division and to produce the gating effect observed in a population of cells is conserved. For example, Klevecz

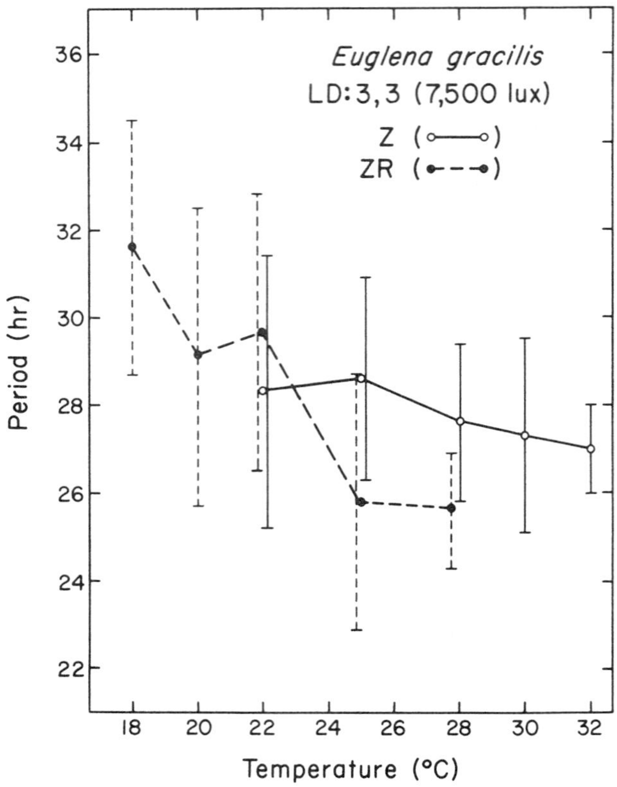


**FIGURE 12** (b) Temperature compensation of the period $\tau$ of the free-running circadian rhythm of cell division in two strains (Z, wild type; ZR, DCMU-resistant) of *Euglena* cultured in LD: 3,3. Periods were measured as the time between successive onsets of division ($\phi_R$) until the stationary phase of growth was reached (usually at a cell titer approximating 100,000 cells/ml). Errors bars denote ±1 standard deviation of the mean. Note that in strain ZR there was some lengthening of the period at lower temperatures. Although growth of the Z strain could be obtained at temperatures less than 22°C (18°C for ZR), rhythmicity disappeared. (From R. W. Anderson, D. L. Laval-Martin, and L. N. Edmunds, unpublished, 1982.)

and King (1982) have found that the $Q_{10}$ for cell division of the V79 line of Chinese hamster lung fibroblasts growing between 34 and 40°C was between 1.15 and 1.26, thus indicating that the mammalian CDC is temperature-compensated over a span of 6-7°C. We have observed previously (Edmunds and Adams, 1981) a similar compensation of the period of the free-running rhythm of cell division in cultures of the $P_4$ZUL photosynthetic mutant of *Euglena* grown in DD over a temperature rante of about 7°C (14-21°C).

In a more extensive and rigorous comparative study of temperature compensation of the free-running period in the Z and ZR strains of *Euglena* maintained in LD: 3,3 at different steady-state temperatures within the physiological range (Fig. 11), we have found a $Q_{10}$ of 1.05 in the former (Fig. 12b), indicating that it is virtually unaffected by changes in temperature over a 10°C range (22-32°C). The circadian clock was not as well compensated in ZR (Fig. 12), in which an average value for the $Q_{10}$ of 1.23 was observed over a temperature range of 18-28°C. The difference between the two strains was

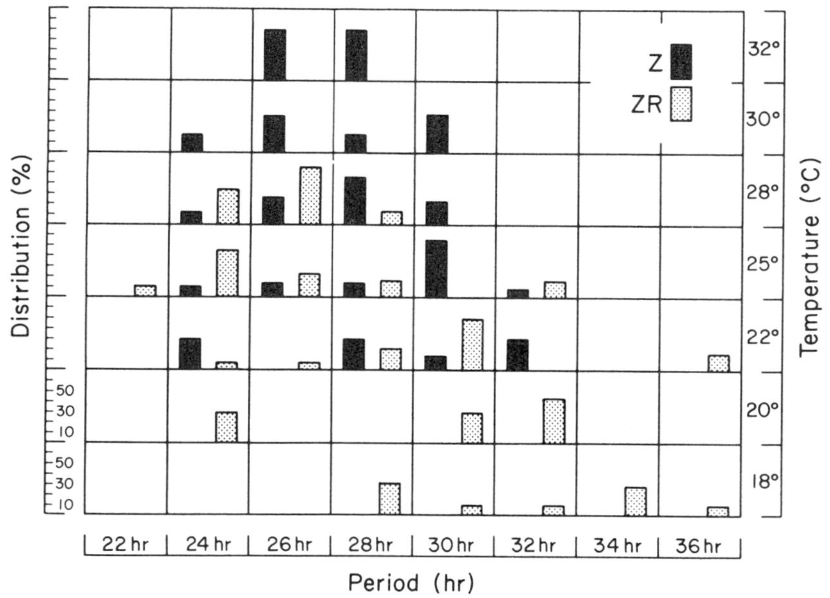


**FIGURE 13** The distribution of period lengths of the free-running rhythm of cell division in two strains of *E. gracilis* cultured photoautotrophically in LD: 3,3 at different steady-state temperatures. For each temperature used (right ordinate), the periods of all the oscillations observed in replicate experiments were pooled, and then the percentage distribution (left ordinate) of each particular period (abscissa) was calculated. A blank box in the grid indicates that no oscillation having that particular period was observed. (From R. W. Anderson, D. L. Laval-Martin, and L. N. Edmunds, Jr., unpublished results, 1982.)

also reflected in the distribution of their period lengths over the range of temperatures utilized (Fig. 13): the spread was considerably greater in ZR. Nevertheless, even the circadian oscillator of ZR was only modestly affected by temperature as compared to typical biochemical reaction rates and other biological processes, such as membrane transport. Finally, we note that a critical "permissive" temperature of 22°C was found for the Z strain (18°C for ZR): although slow exponential growth occurred at lower temperatures, synchrony was not seen.

## FINAL CONSIDERATIONS AND CONCLUSIONS

This overview has attempted to present a convincing case for the key role of circadian oscillator(s) in the control of the cell cycle in the algal flagellate *E. gracilis,* taken to be representative of other unicellular (and, perhaps, multicellular) systems. The *formal* properties of circadian clocks—entrainability, persistence, initiation, phase shiftability, and temperature compensation—were also found to characterize the circadian rhythm of cell division in this model microorganism. Our recent discovery of a singularity point at which the imposition of a critical light pulse generates an arrhythmic population (we do not yet know if the oscillators are stopped in individual cells comprising the culture or are merely dispersed in their phases) further suggests that the underlying oscillator may be of the limit-cycle type, although it does not demand it. Such oscillators are stable, under fixed conditions, and are independent of the starting concentrations of the components whose levels vary rhythmically with time.

According to our working hypothesis, mitosis would not be an essential part of the oscillator but would lie downstream from it: blockage of cell division should not stop the system from oscillating, at least at a subthreshold level. In this sense, then, cell division would be a "hand" of the underlying clock. We have recently tested this hypothesis in two ways. (1) Vitamin $B_{12}$ deprivation has been shown to completely block the CDC in *Euglena* (Bré et al., 1981). If the division rhythm (free running in LD: 3,3) was stopped due to low initial levels of $B_{12}$, and if this inhibition subsequently was released by re-addition of $B_{12}$ to the medium, the cell division rhythm started up again in phase with an unperturbed control (or in phase with an extrapolation of the stable period of the culture before $B_{12}$ deprivation), thereby suggesting that the putative underlying oscillator had continued to run unabatedly throughout the time span during which the CDC had stopped cycling (E. S. Monchak, D. L. Laval-Martin, and L. N. Edmunds, Jr., 1982, unpublished results). (2) If a pulse of lactate was given to a free-running culture, thereby temporarily accelerating the CDC (values of *g* approached 10-12 hr) and overriding circadian oscillator controls (Jarrett and Edmunds, 1970), the phase of the rhythm when it was finally restored after the supplemental substrate had been depleted was in phase with that of an unperturbed control, leading to a similar interpretation (J. E. Rhee, D. L. Laval-Martin, and L. N. Edmunds, Jr., 1982, unpublished results). These latter results are consistent with those found previously for the in-phase restoration of rhythmicity in the $P_4$ZUL mutant free running in LL by the addition of sulfur-containing compounds to the medium (Edmunds et al., 1976) (see Persistence).

Another basic question arises from the phase-shifting data reflected in the experimentally derived PRC (see Fig. 8) for light signals in *Euglena*: How are the observed shortenings and lengthenings of individual CDC generated by a master (circadian) clock at the biochemical or molecular level? The evidence reviewed in earlier sections *formally demands* that a clock of some sort predictably inserts time segments into, or deletes them from, the CDC. One way that the CDC might be "programmed" would be for a collection of timing loops of different lengths to couple together in various combinations to form a flexible timer, or "cytochron" (Edmunds and Adams, 1981), as figuratively diagrammed in Fig. 14. This scheme is sufficiently generalized to apply to any eukaryotic cell cycle (both its strength and its weakness), although it was originally devised to incorporate experimental findings in *Euglena* (Edmunds and Adams, 1981) (see also Chap. 21). Temporal loci, or control points, would exist along the cytochron track at which decisions would be made with respect to the addition or deletion of time loops.

On the view that this time "dilation" or "contraction" has an immediate molecular basis, we have hypothesized (Edmunds and Adams, 1981) the existence of "chronogenes" (our formal time segments) whose transcription would meter time and which would be inserted (or deleted) in varying number into the programmable cytochron by interaction with a circadian oscillator. Thus, the cytochron and the circadian clock are posited to be functionally independent (although not necessarily entirely separate as to mechanism). Of course, this hypothesis may be merely a needless multiplication of unexplained entities, but this is not the place for a detailed consideration of the molecular basis of circadian (Hastings and Schweiger, 1976) or cell cycle clocks (see Chaps. 9-15).

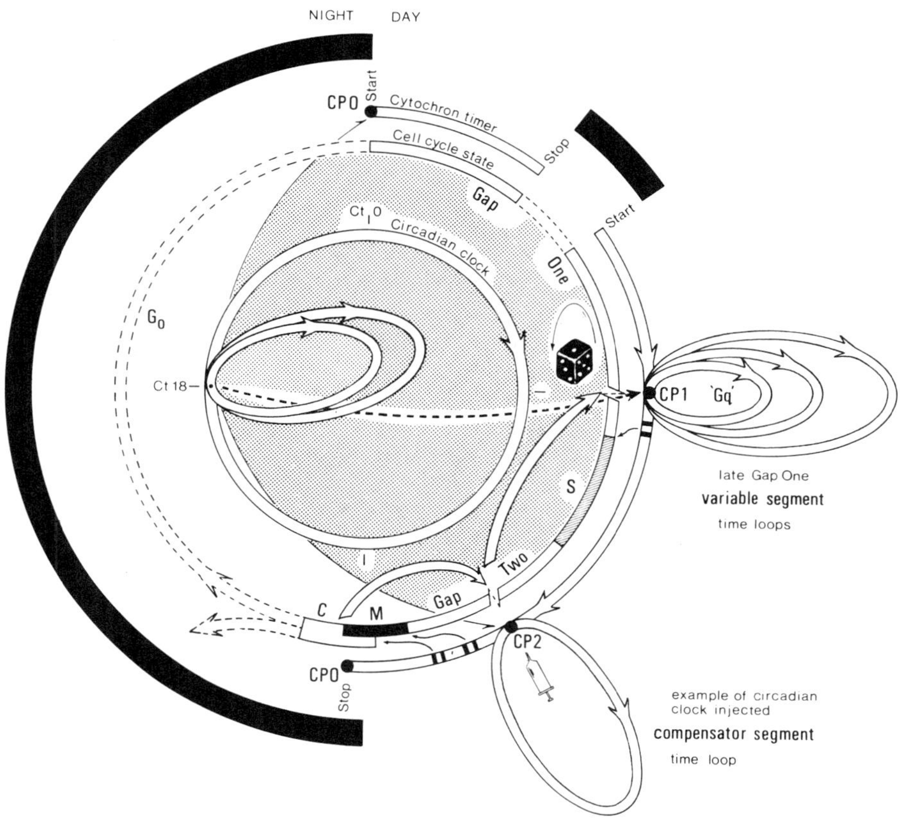


**FIGURE 14** A generalized model for the insertion and deletion of time segments in the track of the cytochron (cell cycle clock) hypothesized to program the events of the cell division cycle in *Euglena.* Cytochrons in each cell start up synchronously at dawn and meter time (unless interrupted by a dark pulse) around to the control point (CP 0), triggering sequentially (black bars) the events leading to chromatin replication S, mitosis M, and cytokinesis C. In photoautotrophic cultures in LD regimes, they stop (noncyclic mode) at CP 0 in the dark, having triggered M and C, leaving cells in an untimed $G_0$ state until dawn restarts the cycle. In LL, or on certain organic substrates, however, the cytochron is cyclic and runs on through CP 0 (i.e., there is no $G_0$). At CP 1, the addition of one or more variable-segment time loops by a random selector (dice symbol) generates variability in the duration of $G_1$ and disperses (desynchronizes) subsequent cell division across the next dark interval. A circadian clock, entrainable by LD Zeitgeber, can couple to the cytochron at a unique circadian time (CT 18) and inject (syringe symbol) a determinate compensator-segment time loop into the cytochron track anywhere within the arc CP 0-2 (stippled), so that the cell cycle is extended to a subsequent circadian cycle. Thus, divisions are phased in bursts or clusters at 24 hr intervals, the scatter within each cluster being generated by the $G_1$ variable-segment time loops, dawn synchronization of the cytochrons ensuring that each burst is confined to the dark interval in each LD cycle. The circadian clock can also apparently delete $G_1$ variable loops and reduce cell cycle variation under some regimes (dashed arrow). Finally, division is suppressed in cells approaching the infradian (stationary) phase of population increase, but multiple rounds of S raise the ploidy level. When conditions improve, ploidy is reduced by successive rounds of M and C. Anticlockwise arrows represent these temporary loop closures. (From Edmunds and Adams, 1981. Copyright 1981 by the American Association for the Advancement of Science.)

## ACKNOWLEDGMENTS

This work was supported in part by National Science Foundation research grant PCM-8204368 to L. Edmunds and is the story of our many students and coworkers in our laboratory. We acknowledge the recent unpublished results of R. W. Anderson, K. Goto, D. LeVine, E. S. Monchak, R. M. Palacios, and J. E. Rhee.

## REFERENCES

Barnett, A. (1969). Cell division: A second circadian clock system in *Paramecium multimicronucleatum. Science* 164:1417-1419.

Boulos, Z., and Rusak, B. (1982). Circadian phase response curves for dark pulses in the hamster. *J. Comp. Physiol.* 146:411-417.

Brady, J. (ed.) (1982). *Biological Timekeeping,* Soc. Exp. Biol. Seminar Series, No. 14, Cambridge University Press, Cambridge, England.

Bré, M. H., El Ferjani, E., and Lefort-Tran, M. (1981). Sequential protein dependent steps in the cell cycle. Initiation and completion of division in vitamin $B_{12}$ replenished *Euglena gracilis. Protoplasma* 108:301-318.

Bruce, V. G. (1970). The biological clock in *Chlamydomonas reinhardi. J. Protozool.* 17:328-333.

Bruce, V. G., and Bruce, N. C. (1981). Circadian clock-controlled growth cycle in *Chlamydomonas reinhardi.* In *International Cell Biology 1980-1981,* Proceedings of the Second International Congress on Cell Biology, Berlin, 1980, H. G. Schweiger (ed.). Springer-Verlag, Berlin, pp. 832-830.

Buetow, D. E. (ed.) (1968, 1982). *The Biology of Euglena,* vols. I, II, III. Academic Press, New York.

Campbell, A. (1957). Synchronization of cell division. *Bacteriol. Rev.* 21:263-271.

Chisholm, S. W. (1981). Temporal patterns of cell division in unicellular algae. In *Physiological Bases of Phytoplankton Ecology,* T. Platt (ed.). (*Can. Bull. Fish. Aquatic Sci.* Vol. 210), Canadian Government Publication Center, Hull, Quebec, pp. 150-181.

Chisholm, S. W., and Brand, L. E. (1981). Persistence of cell division phasing in marine phytoplankton in continuous light after entrainment to light:dark cycles. *J. Exp. Mar. Biol. Ecol.* 51:107-118.

Chisholm, S. W., Morel, F. M. M., and Slocum, W. S. (1980). The phasing and distribution of cell division cycles in marine diatoms. In *Primary Productivity in the Sea.* (Brookhaven Symposia in Biology, No. 31) P. Falkowski (ed.). Plenum, New York, pp. 281-300.

Cook, J. R., and James, T. W. (1960). Light-induced division synchrony in *Euglena gracilis* var. *bacillaris. Exp. Cell Res.* 21:583-589.

Edmunds, L. N., Jr. (1965). Studies on synchronously dividing cultures of *Euglena gracilis* Klebs (strain Z.). I. Attainment and characterization of rhythmic cell division. *J. Cell. Comp. Physiol.* 66:147-158.

Edmunds, L. N., Jr. (1966). Studies on synchronously dividing cultures of *Euglena gracilis* Klebs (strain Z). III. Circadian components of cell division. *J. Cell. Physiol.* 67:35-44.

Edmunds, L. N., Jr. (1971). Persisting circadian rhythm of cell division in *Euglena*: Some theoretical considerations and the problem of intercellular communication. In *Biochronometry,* M. Menaker (ed.). National Academy of Science, Washington, D.C., pp. 594-611.

Edmunds, L. N., Jr. (1974). Phasing effects of light on cell division in exponentially increasing cultures of *Tetrahymena* grown at low temperatures. *Exp. Cell Res.* 83: 367-379.

Edmunds, L. N., Jr. (1975). Temporal differentiation in *Euglena*: Circadian phenomena in non-dividing populations and in synchronously dividing cells. In *Les Cycles Cellulaires et leur Blocage chez Plusieurs Protistes,* M. Lefort-Tran and R. Valencia (eds.). Colloques Internationaux CNRS No. 240, Centre National de la Recherche Scientifique, Paris, pp. 53-67.

Edmunds, L. N., Jr. (1978). Clocked cell cycle clocks: Implications toward chronopharmacology and aging. In *Aging and Biological Rhythms,* H. V. Samis, Jr., and S. Capobianco (eds.). Plenum, New York, pp. 125-184.

Edmunds, L. N., Jr. (1980). Blue-light photoreception in the inhibition and synchronization of growth and transport in the yeast *Saccharomyces,* In *The Blue Light Syndrome,* H. Senger (ed.). Springer-Verlag, Berlin, pp. 584-596.

Edmunds, L. N., Jr. (1982). Circadian and infradian rhythms. In *The Biology of Euglena,* vol. III, D. E. Buetow (ed.). Academic Press, New York, pp. 53-142.

Edmunds, L. N., Jr., and Adams, K. J. (1981). Clocked cell cycle clocks. *Science* 211: 1002-1013.

Edmunds, L. N., Jr., Apter, R. I., Rosenthal, P. J., Shen, W.-K., and Woodward, J. R. (1979). Light effects in yeast: Persisting oscillations in cell division activity and amino acid transport in cultures of *Saccharomyces* entrained by light-dark cycles. *Photochem. Photobiol.* 30:595-601.

Edmunds, L. N., Jr., Chuang, L., Jarrett, R. M., and Terry, O. W. (1971). Long-term persistence of free-running rhythms of cell division in *Euglena* and the implication of autosynchrony, *J. Interdisc. Cycle Res.* 2:121-132.

Edmunds, L. N., Jr., and Funch, R. R. (1969a). Circadian rhythm of cell division in *Euglena*: Effects of a random illumination regimen. *Science* 165:500-503.

Edmunds, L. N., Jr., and Funch, R. (1969b). Effects of "skeleton" photoperiods and high frequency light-dark cycles on the rhythm of cell division in synchronized cultures of *Euglena. Planta* 87:134-163.

Edmunds, L. N., Jr., and Halberg, F. (1981). Circadian time structure of *Euglena*: A model system amenable to quantification. In *Neoplasms–Comparative Pathology of Growth in Animals, Plants and Man,* H. E. Kaiser (ed.). Williams & Wilkins, Baltimore, pp. 105-134.

Edmunds, L. N., Jr., Jay, M. E., Kohlmann, A., Liu, S. C., Merriam, V. H., and Sternberg, H. (1976). The coupling effects of some thiol and other sulfur-containing compounds on the circadian rhythm of cell division in photosynthetic mutants of *Euglena. Arch. Microbiol.* 108:1-8.

Edmunds, L. N., Jr., and Laval-Martin, D. L. (1981). "Free-running" circadian rhythms of photosynthesis elicited by short-period cycles of light and darkness in synchronously dividing and nondividing *Euglena.* In *Photosynthesis,* vol. VI, Proceedings of the Fifth International Congress on Photosynthesis, Halkidiki, Greece, 1980, G. Akoyunoglou (ed.). Balaban International Science Services, Philadelphia, pp. 313-322.

Edmunds, L. N., Jr., Sulzman, F. M., and Walther, W. G. (1974). Circadian oscillations in enzyme activity in *Euglena* and their relation to the circadian rhythm of cell division. In *Chronobiology,* L. E. Scheving, F. Halberg, and J. E. Pauly (eds.). Igaku Shoin, Tokyo, pp. 61-66.

Edmunds, L. N., Jr., Tay, D. E., and Laval-Martin, D. L. (1982). Cell division cycles and circadian clocks: Phase-response curves for light perturbations in synchronous cultures of *Euglena. Plant Physiol.* 70:297-302.

Ehret, C. F., Meinert, J. C., Groh, K. R., and Antipa, G. A. (1977). Circadian regulation: Growth kinetics of the infradian cell. In *Growth Kinetics and Biochemical Regulation of Normal and Malignant Cells,* B. Drewinko and R. M. Humphrey (eds.). Williams & Wilkins, Baltimore, pp. 49-76.

Ellis, G. B., McKlveen, R. E., and Turek, F. W. (1982). Dark pulses affect the circadian rhythm of activity in hamsters kept in constant light. *Am. J. Physiol.* 242:R44-R50.

Engelmann, W., Johnsson, A., Karlsson, H. G., Kobler, R., and Schimmel, M. (1978). Attenuation of the petal movement rhythm in *Kalanchoe* with light pulses. *Physiol. Plant.* 43:68-76.

Gilbert, D. A. (1981). The cell cycle 1981: One or more limit cycle oscillations? *S. Afr. J. Sci.* 77:541-546.

Hartwell, L., Culatti, J., Pringle, J. R., and Reid, B. T. (1974). Genetic control of the cell division cycle in yeast. *Science* 183:46-51.

Hastings, J. W., and Schweiger, H.-G. (eds.). (1976). *The Molecular Basis of Circadian Rhythms,* Dahlem Konferenzen, Berlin.

Hastings, J. W., and Sweeney, B. M. (1964). Phased cell division in marine dinoflagellates. In *Synchrony in Cell Division and Growth,* E. Zeuthen (ed.). Wiley Interscience, New York, pp. 307-321.

Hesse, M. (1972). Endogene Rhythmik der Produktionsfähigkeit bei *Chlorella* und ihre Beeinflussung durch Licht. *Z. Pflanzenphysiol.* 67:58-77.

Jarrett, R. M., and Edmunds, L. N., Jr. (1970). Persisting circadian rhythm of cell division in a photosynthetic mutant of *Euglena. Science* 167:1730-1733.

Jeter, J. R., Jr., Cameron, I. L., Padilla, G. M., and Zimmerman, A. M. (eds.) (1978). *Cell Cycle Regulation,* Academic Press, New York.

Karakashian, M. W., and Schweiger, H. G. (1976). Circadian properties of the rhythmic system in individual nucleated and enucleated cells of *Acetabularia mediterranea. Exp. Cell Res.* 97:366-377.

Kauffman, S., and Wille, J. J. (1975). The mitotic oscillator in *Physarum polycephalum. J. Theor. Biol.* 55:47-93.

Kirschstein, M. (1969). Das rhythmische Verhalten einer farblosen Mutante von *Euglena gracilis. Planta* 85:126-134..

Klevecz, R. R. (1976). Quantized generation time in mammalian cells as an expression of the cellular clock. *Proc. Natl. Acad. Sci. USA* 73:4012-4016.

Klevecz, R. R. (1978). The clock in animal cells is a limit cycle oscillator. In *Cell Reproduction: In Honor of Daniel Mazia,* E. R. Dirksen, D. M. Prescott, and C. F. Fox (eds.). Academic Press, New York, pp. 139-146.

Klevecz, R. R., and King, G. A. (1982). Temperature compensation in the mammalian cell cycle. *Exp. Cell Res.* 140:307-313.

Klevecz, R. R., King, G. A., and Shymko, R. M. (1980a). Mapping the mitotic clock by phase perturbation. *J. Supramol. Struct.* 14:329-342.

Klevecz, R. R., Kros, J., and Gross, S. D. (1978). Phase response versus positive and negative division delay in animal cells. *Exp. Cell Res.* 116:285-290.

Klevecz, R. R., Kros, J., and King, G. A. (1980b). Phase response to heat shock as evidence for a timekeeping oscillator in synchronous animal cells. *Cytogenet. Cell Genet.* 26:236-243.

Laval-Martin, D. L., Shuch, D. J., and Edmunds, L. N., Jr. (1979). Cell cycle-related and endogenously controlled circadian photosynthetic rhythms in *Euglena. Plant Physiol.* 63:495-502.

Ledoigt, G., and Calvayrac, R. (1979). Phénomènes périodiques, métaboliques et structuraux chez un protiste, *Euglena gracilis. J. Protozool.* 26:632-643.

Lloyd, D., Poole, R. K., and Edwards, S. W. (1982). *The Cell Division Cycle: Temporal Organization and Control of Cellular Growth and Reproduction.* Academic Press, London.

McMurry, L., and Hastings, J. W. (1972). No desynchronization among four different circadian rhythms in the unicellular alga, *Gonyaulax polyedra. Science* 175:1137-1139.

Mitchell, J. L. A. (1971). Photoinduced division synchrony in permanently bleached *Euglena gracilis. Planta* 100:244-257.

Mitchison, J. M. (1971). *The Biology of the Cell Cycle,* Cambridge University Press, London.
Mitchison, J. M. (1974). Sequences, pathways and timers in the cell cycle. In *Cell Cycle Controls,* G. M. Padilla, I. L. Cameron, and A. Zimmerman (eds.). Academic Press, New York, pp. 125-142.
Miyamoto, H., Rasmussen, L., and Zeuthen, E. G. (1973). Studies of the effect of temperature shocks on preparation for cell division in mouse fibroblast cells (L cells). *J. Cell Sci.* 13:889-900.
Padilla, G. M., Cameron, I. L., and Zimmerman, A. M. (eds.). (1974). *Cell Cycle Controls.* Academic Press, New York.
Pavlidis, T. (1968). Studies on biological clocks: A model for the circadian rhythms of nocturnal organisms. In *Lectures in Mathematics in the Life Sciences,* vol. 1, M. Gerstenhaber (ed.). American Mathematical Society, Providence, Rhode Island, pp. 88-112.
Peterson, E. L. (1980). Phase-setting a mosquito circadian oscillator. I. Phase-resetting surface. *J. Comp. Physiol.* 138:201-211.
Pirson, A., and Lorenzen, H. G. (1958). Ein endogener Zeitfaktor bei der Teilung von *Chlorella. Z. Bot.* 46:53-66.
Pittendrigh, C. S. (1965). On the mechanism of the entrainment of a circadian clock by light cycles. In *Circadian Clocks,* J. Aschoff (ed.). North-Holland, Amsterdam, pp. 277-297.
Polanshek, M. M. (1977). Effects of heat shock and cycloheximide on growth and division of the fission yeast, *Schizosaccharomyces pombe. J. Cell Sci.* 23:1-23.
Pringle, J. R. (1981). The genetic approach to the study of the cell cycle. In *Mitosis/Cytokinesis,* A. M. Zimmerman and A. Forer (eds.). Academic Press, New York, pp. 3-28.
Sachsenmaier, W. (1976). Control of synchronous nuclear mitosis in *Physarum polycephalum.* In *The Molecular Basis of Circadian Rhythms,* J. W. Hastings, and H.-G. Schweiger (eds.). Dahlem Konferenzen, Berlin, pp. 409-420.
Saunders, D. S. (1978). An experimental and theoretical analysis of photoperiodic induction in the flesh-fly, *Sarcophaga argyrostoma. J. Comp. Physiol.* 124:75-95.
Schnabel, G. (1968). Der Einflüss von Licht auf die circadiane Rhythmik von *Euglena gracilis* bei Autotrophie und Mixotrophie. *Planta* 81:49-63.
Smith, H. T. B., and Mitchison, J. M. (1976). Anaesthetics delay and accelerate division in the fission yeast *Schizosaccharomyces pombe. Exp. Cell Res.* 99:432-435.
Spudich, J., and Sager, R. (1980). Regulation of *Chlamydomonas* cell cycle by light and dark. *J. Cell Biol.* 85:136-145.
Subbaraj, R., and Chandrashekaran, M. K. (1978). Pulses of darkness shift in the phase of a circadian rhythm in an insectivorous bat. *J. Comp. Physiol.* 127:239-243.
Sweeney, B. M. (1982). Interaction of the circadian cycle with the cell cycle in *Pyrocystis fusiformis. Plant Physiol.* 70:272-276.
Sweeney, B. M., and Hastings, J. W. (1958). Rhythmic cell division in populations of *Gonyaulax polyedra. J. Protozool.* 5:217-224.
Sweeney, B. M., and Hastings, J. W. (1960). Effects of temperature upon diurnal rhythms. *Cold Spring Harbor Symp. Quant. Biol.* 25:87-104.
Taylor, W., Krasnow, R., Dunlap, J. C., Broda, H., and Hastings, J. W. (1982). Critical pulses of anisomycin drive the circadian oscillator in *Gonyaulax* towards its singulartiy. *J. Comp. Physiol.* 148:11-25.
Terry, O., and Edmunds, L. N., Jr. (1969). Semi-continuous culture and monitoring system for temperature-synchronized *Euglena. Biotechnol. Bioeng.* 11:745-756.
Terry, O. W., and Edmunds, L. N., Jr. (1970). Phasing of cell division by temperature cycles in *Euglena* cultured autotrophically under continuous illumination. *Planta* 93:106-127.

Volm, M. (1964). Die Tagesperiodik der Zellteilung von *Paramecium bursaria, Z. Vergl. Physiol.* 48:157-180.

Weiler, C. S., and Chisholm, S. W. (1976). Phased cell division in natural populations of marine dinoflagellates from shipboard cultures. *J. Exp. Mar. Biol. Ecol.* 25:239-247.

Weiler, C. S., and Eppley, R. W. (1979). Temporal pattern of division in the dinoflagellate genus *Ceratium* and its application to the determination of growth rate. *J. Exp. Mar. Biol. Ecol.* 39:1-24.

Wilkins, M. B. (1960). The effect of light upon plant rhythms. *Cold Spring Harbor Symp. Quant. Biol.* 25:115-129.

Wille, J. J., Jr. (1974). Light entrained circadian oscillations of growth rate in the yeast *Candida utilis.* In *Chronobiology,* L. E. Scheving, F. Halberg, and J. E. Pauly (eds.). Igaku Shoin, Tokyo, pp. 72-77.

Wille, J. J., Jr. (1979). Biological rhythms in protozoa. In *Biochemistry and Physiology of Protozoa,* vol. 2, 2nd ed., M. Levandowsky and S. H. Hutner (eds.). Academic Press, New York, pp. 67-149.

Wille, J. J., and Ehret, C. F. (1968). Light synchronization of an endogenous circadian rhythm of cell division in *Tetrahymena. J. Protozool.* 15:785-788.

Winfree, A. T. (1970). An integrated view of the resetting of a circadian clock. *J. Theor. Biol.* 28:327-374.

Winfree, A. T. (1980). *The Geometry of Biological Time.* Springer-Verlag, New York.

Zeuthen, E. (1974). A cellular model for repetitive and free-running synchrony in *Tetrahymena* and *Schizosaccharomyces.* In *Cell Cycle Controls,* G. M. Padilla, I. L. Cameron, and A. M. Zimmerman (eds.). Academic Press, New York, pp. 1-30.

# 18

# Temporal Organization of Rat and Human Skeletal Cells

## Circadian Frequency and Quantization of Cell Generation Time

**Alexandre G. Petrovic, Claudine L. Oudet, and Jeanne J. Stutzmann** /
Unité 213 de l'INSERM and Institut de Physiologie, Faculté de Médecine, Strasbourg, France

## INTRODUCTION

Cartilage and bone cells have been reported to exhibit seasonal and circadian rhythms of mitotic and metabolic activities (Simmons, 1962, 1964, 1968; Simmons et al., 1979; Hansson et al., 1974; Oudet and Petrovic, 1978, 1981a,b, 1982; Petrovic et al., 1981; Stutzmann et al., 1981). In conformity with our findings, the circadian periodicity of mitoses, both in situ and in organ culture, results not only from the generation time of individual cells, but also from the temporal organization of the cell populations (Stutzmann and Petrovic, 1979). Hence we were led to investigate the generation time of single isolated skeletal cells, as well as of cells forming associations at different levels of organization.

Indeed, an investigation in situ of the growth rate or renewal rate in a cell population presents a double inadequacy:

1. In the investigation in situ, one may estimate the *mean* duration of the generation time in the cell population as a whole, but we cannot detect the genuine intermitotic duration of an individual cell. Besides, in such an investigation, it is hard to evaluate the percentage of cells forming the mitotic compartment. Such a knowledge *may* be obtained in cell culture.
2. In the investigation in situ, it is arduous, and usually impossible, to make the distinction between the temporal organization of an individual cell itself and the effects of cellular interactions. In order to get some information in this field, we cultured *isolated* bone and cartilage cells.

## FEATURES OF SKELETAL CELLS DEALT WITH IN THE PRESENT STUDY*

There is a basic dissimilarity between the so-called primary and the so-called secondary cartilage (Petrovic and Charlier, 1967). In *primary* cartilages (epiphyseal cartilage of long bones, sphenooccipital synchondrosis, nasal septal cartilage, and so on), *chondroblasts* divide and synthesize cartilaginous matrix. In *secondary* cartilages (condylar cartilage of the mandible, fracture callus and so on), the cells which divide, the *prechondroblasts,* are not yet surrounded by cartilaginous matrix. As soon as the prechondroblast starts synthesizing the cartilaginous matrix, it stops dividing.

In this study, the generation time of the chondroblasts was investigated in the growing rat sphenooccipital synchondrosis and the generation time of the prechondroblast, in the condylar cartilage of the growing rat mandible. The generation time of the skeletoblast was investigated in the rat condylar cartilage and in human alveolar bone.

The histogenesis of cartilage and bone cells is a controversial problem. According to our experimental investigations in situ, in organ and cell culture, in transplantation, in both physiological and physiopathological situations, using animal and human tissues, there is a pluripotential stem cell (Holtzer, 1978) we name *skeletoblast* (Petrovic et al., 1974; Stutzmann and Petrovic, 1975, 1982; Petrovic and Stutzmann, 1981; Petrovic, 1982). The skeletoblast originates from the mesenchymal cell. Preferably, the skeletoblast differentiates into the preosteoblast-osteoblast, but it is also able under specific conditions to differentiate into prechondroblast-chondroblast type II. The commitment into either bone tissue or secondary cartilage tissue occurs as the effect of *local* unidentified determinants.

As for the chondroblast type I bypassing the skeletoblast stage, it probably derives directly from the mesenchymal cell. When the chondroblast type I originates from the skeletoblast at the "encoche d'ossification de Ranvier", it retains the ability to divide.

The skeletoblast is an elongaged cell forming a self-perpetuating compartment: in serial cultures, it may reach 60 consecutive divisions. Morphologically, it is a fibroblast-like cell. Its surface is rich in fibronectin and in cell transglutaminase. When the multiplication rate increases, the amount of fibronectin and cell transglutaminase decreases. Simultaneously, the skeletoblast becomes less flattened (Petrovic and Stutzmann, 1981). The prechondroblast type II and the preosteoblast are fairly round cells. Biologically they are each committed, but morphologically, especially in cell culture, they are not distinguishable. Both the preosteoblast and prechondroblast type II divide only a few times. Their surface is rather poor in fibronectin and in cell transglutaminase (Petrovic and Stutzmann, 1981).

The cytosolic $[Ca^{2+}]$ is basically lower in preosteoblast and in secondary prechondroblast (globular cells, relatively high rate of multiplication) than in skeletoblasts (flattened cells, relatively low rate of multiplication). When human skeletal cells are compared to corresponding rat cells, the cell volume appears to be slightly greater, but the amount of mitochondria moderately smaller.

## GENERATION TIME IN SITU

The generation time of the condylar prechondroblast and sphenooccipital synchondrosis chondroblast was estimated according to Baserga and Malamud (1969) and the calculations were carried out according to Quastler and Sherman (1959).

*For details, see Stutzmann and Petrovic (1982).

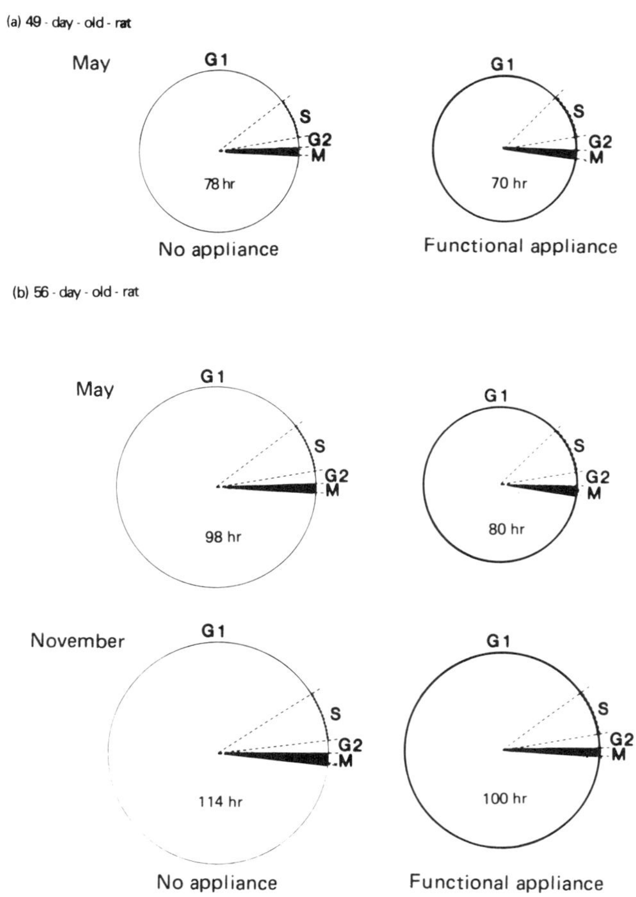


**FIGURE 1** Phases of the cell division cycle of the prechondroblast in the condylar cartilage of the rat's mandible in situ. The perimeters of the circles are proportional to the generation time in controls and in orthopedically treated rats. The length of the arcs representing the various phases of the cycle are proportional to their duration.

### Condylar Cartilage Prechondroblast

The generation time was estimated in control rats and in rats previously treated with the postural hyperpropulsor of the mandible, an orthopedic functional appliance known to increase the growth rate of the condylar cartilage (Charlier et al., 1968, 1969; Petrovic et al., 1975). All animals were injected with tritiated thymidine and sacrificed at appropriated time intervals. The number of tritiated thymidine-labeled mitoses was estimated on radioautograms. Only prechondroblast-labeled mitoses were investigated, in 49- and 56-day old rats.

#### *49-Day-Old Rat*

The experiment was performed in May. In controls, the generation time was 4680 min, i.e., 78 hr. In rats treated with the postural hyperpropulsor of the mandible, it was 4200 min, i.e., 70 hr (Fig. 1). The phases of the cell division cycle had very similar durations in both groups. In controls: $G_2$ = 117 min, S = 450 min, M = 45 min; in treated rats:

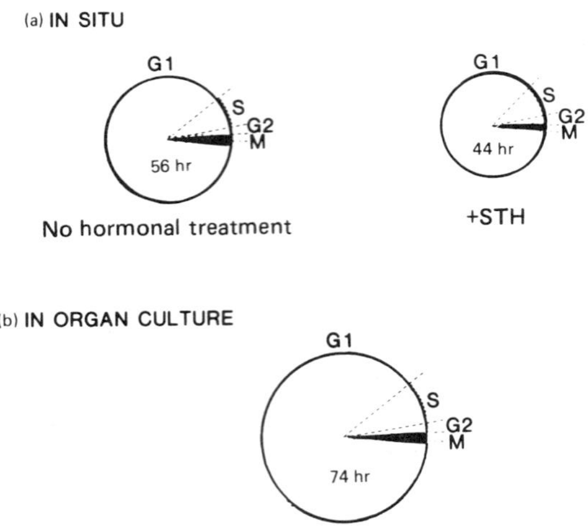


**FIGURE 2** Phases of the cell division cycle of the chondroblast of the rat sphenooccipital synchondrosis in situ (a) and in organ culture (b). The perimeters of the circles are proportional to the generation time in controls and in rats injected with STH.

$G_2$ = 108 min, S = 450 min, M = 45 min. This difference between the two cycle lengths primarily derives from the duration of phase $G_1$: in controls: $G_1$ = 4067 min, i.e., about 68 hr; in treated rats: $G_1$ = 3598 min, i.e., about 60 hr.

*56-Day-Old Rat*

The experiment was carried out in May and in November.

*May*: In controls, the mean generation time for prechondroblasts of the condylar cartilage was 5900 min, i.e., 98 hr. In rats treated with a postural hyperpropulsor, it was reduced to 4800 min, i.e., 80 hr. Here also, the variations in generation time are due to variations in the duration of phase $G_1$. The other phases of the cell division cycle had a very similar length in both groups. At this age, the following estimations were made: $G_2$ = 120 min, S = 450 min, M = 47 min. These values differ only slightly from the average duration observed in 49-day-old rats.

*November*: In controls, the mean generation time was 6800 min, i.e., 114 hr. In rats treated with the postural hyperpropulsor it was reduced to 100 hr (Fig. 1). Only the duration of the $G_1$ phase of the cell division cycle is responsible for this variation. So, the generation time of condylar cartilage prechondroblasts in situ depends, to some extent, on the age of the animals, on the season, and on the presence or absence of a growth-stimulating orthopedic treatment. It is interesting to emphasize that none of these factors, which modulate the generation time within a range of about 15 hr, modified the circadian periodicity of the number of tritiated thymidine-labeled cells (Oudet and Petrovic, 1978, 1981a,b).

### Sphenooccipital Synchondrosis Chondroblasts

The sphenooccipital synchondrosis was investigated in 18-day-old rats. Half of the rats had been injected, daily, with bovine-STH (NIH) for 3 days before sacrifice; the other half served as controls.

The mean generation time of chondroblasts in the sphenooccipital synchondrosis of controls reached 3336 min, i.e., about 56 hr. The phases of the cell division cycle were $G_2$ = 112 min, S = 450 min, M = 46 min, and $G_1$ = 2728 min (Fig. 2).

In rats injected with STH, the generation time was shortened to 2640 min, i.e., 44 hr, and the various phases were $G_2$ = 107 min, S = 450 min, M = 46 min, and $G_1$ = 2037 min. These generation times are obviously shorter than those of prechondroblasts observed in older rats (see above).

## GENERATION TIME IN ORGAN CULTURE

Petrovic and Charlier (1967) and Stutzmann and Petrovic (1970) have reported a discriminating difference of postnatal growth pattern in organ culture between primary and secondary cartilages. Following setting in organ culture, the decrease of the number of dividing cells in primary cartilage reaches only 40% of the chondroblasts, whereas it extends to almost 90% of the dividing prechondroblasts in secondary cartilage.

Hence, a detailed investigation on the possible persistence of circadian mitotic rhythms of cartilages has been performed by recording the number of tritiated methylthymidine-labeled chondroblasts in the cultured sphenooccipital synchondrosis of the 18-day-old growing rat (Stutzmann and Petrovic, 1979). Under these experimental conditions, the average generation time of the organ-cultured chondroblasts has been estimated to be about 74 hr. The distribution of the various phase duration was: $G_2$ = 119 min, S = 480 min, M = 42 min, and $G_1$ = 3780 min (Fig. 2).

There is, in organ culture, an obvious lengthening of the generation time as compared with the in situ experiment, the significant variation being due, once more, to the $G_1$ phase duration.

## GENERATION TIME IN CELL CULTURE

### Skeletoblasts and Prechondroblasts Originating from the Condylar Cartilage of the Rat Mandible

Explants from the condylar cartilage of the mandible of growing rats were introduced into a Rose chamber (Rose et al., 1958) containing TC 199 medium supplemented with 15% calf serum and maintained at 37°C. After several days, a monocellular layer develops and the original explant is removed. Time-lapse microcinematography (Dawson et al., 1965) enabled us to record cell populations in a Rose chamber containing skeletoblasts and/or prechondroblasts.

The proliferative cells have been identified by using morphological features, by detecting the nature of the intercellular substance (bone or cartilage) synthetized after transplantation into rat testis, or anterior chamber of the eye, as well as by measuring in time-lapse microcinematography, different biological parameters (mitosis duration, generation time, mitotic index).

#### *Skeletoblasts*

We selected 12 cultures of pure skeletoblasts and measured, in each chamber, at least 25 cell division cycles. In these experiments, we observed a maximal number of 43 skeletoblast generations. Under our experimental conditions, the generation time of skeletoblasts within population of a Rose chamber extended from 72 to 187 hr (Fig. 3). The histogram

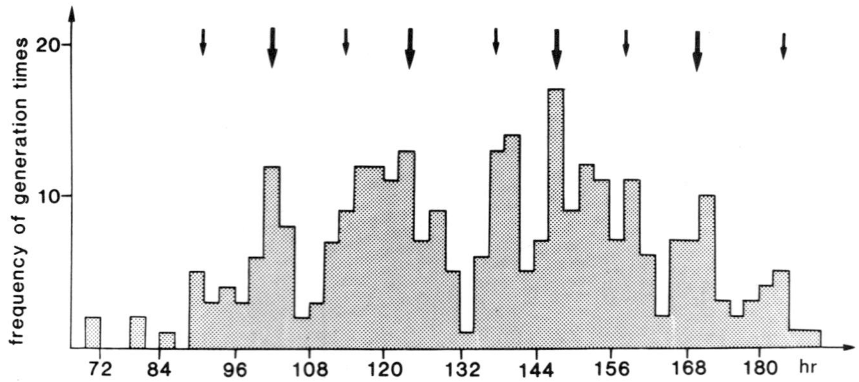


**FIGURE 3** Polymodal distribution of generation times in skeletoblasts originating from the condylar cartilage of the rat's mandible and cultures in a Rose chamber. In addition to the 24 hr rhythm of frequencies, there is an underlying 12 hr period.

of the cumulated generation times recorded in 12 populations of skeletoblasts, cultured in a Rose chamber, exhibits a polymodal distribution (Fig. 3). Distinct modes can be discerned at 102, 123.6, 147.6, and 171.6 hr.

The mean time interval between the four selected prevailing values is 23.2 hr, which could still be an expression of the fundamental circadian organization of motitic rate. Additionally, it is interesting to note that the time interval of 23.2 hr is almost an exact multiple (6) of 3.75 hr, i.e., the basic quantum for generation times proposed by Klevecz (1976).

A judicious observation of Fig. 3 shows that additional discrete "events" occur regularly in the histogram. These supplementary peaks are located at 90, 115.2, 136.8, 159.6 and 183.6 hr, respectively. The mean time interval between prevailing modes and supplementary peaks is almost exactly half the prevailing time interval, i.e., 11.4 hr, and, of course, three times the quantum $G_q$. Moreover, the first mode for generation times is observed after 102 hr, i.e., 30 × $G_q$ (if $G_q$ = 3.5).

In recorded cell division cycles, the mitoses lasted 45-55 min. The ratio of observed mitotic index to expected mitotic index (i.e., duration of mitosis to duration of cell division cycle) indicates that 72-87% of skeletoblasts belong to the compartment actually dividing. The mitotic index of skeletoblasts was 0.40-0.61%.

### *Prechondroblasts*

The maximal number of generations observed for prechondroblasts never exceeded nine.

The generation times extended from 19.2 to 55.2 hr in 15- and 49-day-old rats (Fig. 4) and had a tendency to increase with aging, e.g., up to 72 hr in 56-day-old rats (Fig. 4).

When generation times are cumulated from 6 populations in 15-day-old rats, from 3 populations in 40-day-old rats, and from 12 populations in 56-day-old rats, each histogram exhibits several modes separated by an average 8.2 hr time interval. The first mode corresponds to 24 hr, i.e., 6 × $G_q$ (if $G_q$ = 4) and the average increment is 2 × $G_q$. The recurrence of a 8 hr increment opens up the possibility of the maintenance of a 24 hr periodicity of generation time.

In 56-day-old rats, the histogram shows modes at 22.8, 32.4, 39.6, 49.2, and 56.4 hr. Accordingly, the basic temporal component of 8 hr may lead to various generation

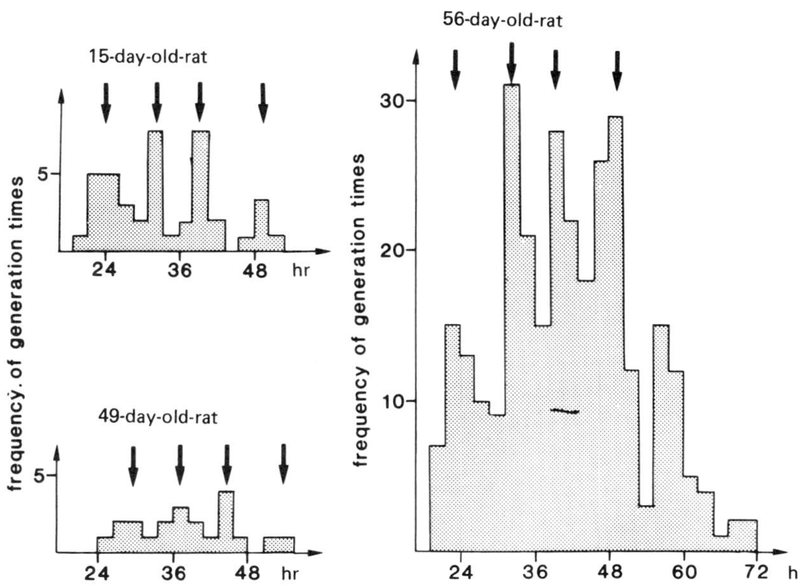


**FIGURE 4** Polymodal distribution of generation times in prechondroblast originating from the condylar cartilage of the rat's mandible and cultured in a Rose chamber. Whatever the age of the rat, there is an obvious 8 hr interval between prevailing values. Generation times skew to longer durations with the age of the animal.

times, from 8 to any multiple of 8 hr. However, taking into account that the sum of successive phases of the cell cycle exceeds 8 hr, the first actual possible generation time is 16 hr. In fact, as it will be reported below, in single isolated prechondroblasts, as well as in single isolated skeletoblasts, the circadian pattern appears to predominate.

It should be pointed out that the longest generation times observed in prechondroblasts coincide with the shortest generation times recorded in skeletoblasts. As reported, the prechondroblast originates from the skeletoblast; it is interesting to mention this aspect of the temporal organization between these two cell types.

Some data relative to cell division cycles should be mentioned: most mitoses lasted about 50 min. The ratio of observed mitotic index to expected mitotic index indicates that the number of cells actually belonging to the dividing compartment extended from 65 to 79% in 15-day-old rats, from 57 to 67% in 49-day-old rats, and only 51 to 63% in 56-day-old rats. The mitotic index of prechondroblasts was 1.22-2.74% in 15-day-old rats, 1.10-1.67% in 49-day-old rats, and 0.91-1.67% in 56-day-old rats. So, it appears that the age-dependent decreased mitotic activity is due to both the lengthening of the cell division cycle as detected in situ (see above) and the decrease of the percentage of cells actually dividing.

### Skeletoblasts and Preosteoblasts Originating From Human Alveolar Bone

Fragments of human alveolar bone were cultured in a Rose chamber in the same experimental conditions as the rat condylar cartilage explants. The outgrowing cells—skeletoblasts or preosteoblasts—were analyzed using time-lapse microcinematography.

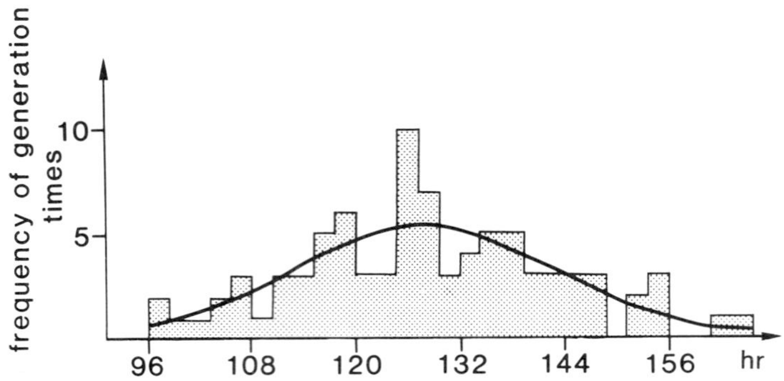


**FIGURE 5** Generation times of skeletoblasts originating from human alveolar bone and cultured in a Rose chamber. A prevailing value is observed at 126 hr. The distribution is normal around the mean. The normal distribution for a sample of same mean and standard deviation is drawn on the histogram.

*Skeletoblasts*

Twelve cultures of human skeletoblasts were carried out, and seven cell division cycles were measured in each Rose chamber.

Under our experimental conditions, the maximal number of human skeletoblast generations may reach 60. The generation time of skeletoblasts, within the cell population, extended from 96 to 168 hr. Interestingly, the shortest human skeletoblast generation time is a multiple of 24 hr. Figure 5 represents the histogram of the cumulated generation times (n = 84) recorded in the 12 populations of skeletoblasts cultured in a Rose chamber. No polymodal distribution could be substantiated: the observed frequency distribution seems to be normal ($\chi^2_{27}$ = 24.19; $p < 0.01$). Our findings do not allow us to rule out the existence of a polymodal distribution of the generation time frequency: indeed, the sample is not large enough to test statistically the existence of different modes. Nevertheless, the histogram displays, perhaps, some propensity to a multimodal distribution.

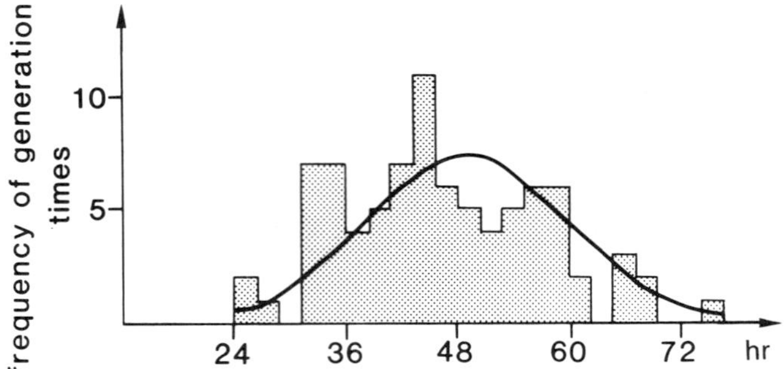


**FIGURE 6** Generation times of preosteoblasts originating from human alveolar bone and cultured in a Rose chamber. A prevailing value is observed at 43 hr. The distribution is normal around the mean. The normal distribution for a sample of same mean and standard deviation is drawn on the histogram.

In the recorded skeletoblast division cycles, the mitosis lasted 55 min. The observed mitotic index varied from 0.38 to 0.71%. The ratio of observed mitotic index to expected mitotic index indicates that 63-84% of skeletoblasts actually belong to the mitotic compartment.

*Preosteoblasts*

As for the skeletoblasts, 12 cultures of human preosteoblasts were carried out and 7 cell division cycles were measured in each Rose chamber.

The maximal number of human preosteoblast generations never exceeded nine. The generation time extended from 24 to 74 hr (one to three circadian cycles); i.e., it was shorter than the generation time of the human skeletoblasts, but highly similar to the generation time of rat prechondroblasts (see above and Fig. 6).

When analyzed statistically, the histogram of cumulated generation times recorded in the 12 selecting populations (n = 84) exhibited a normal distribution ($\chi^2_{18}$ = 24.47; $p < 0.05$) as in the human skeletoblast population.

In recorded preosteoblast division cycles, the mitosis lasted 55-60 min. The observed preosteoblast mitotic index was much higher than the observed skeletoblasts mitotic index: it varied from 0.80 to 1.83%.

The ratio of observed mitotic index to expected mitotic index indicates that 49-76% of the human preosteoblasts actually divide.

## GENERATION TIME OF SINGLE ISOLATED SKELETAL CELLS AND OF THEIR PROGENIES

In situ, in organ culture and in cell culture investigations, it is arduous, and usually impossible, to make the distinction between the temporal organization of an individual cell itself and the effects of cellular interactions. In order to get some information in this field of skeletal cell physiology, we have cultured single *isolated* cells.

Different skeletal cell samples were firstly cultured in a Rose chamber. Then, the monocellular layer was put into suspension. A single cell was drawn into a "capillary tube," a procedure derived from the technique of Sandford et al. (1961). Capillary tubes, each containing one cell, the culture medium, and an air bubble on each extremity, were sealed. After the single isolated cell had divided, one of the two daughter cells was removed by micromanipulation in order to measure the intermitotic duration in a single skeletal cell. Subsequently, the cell was allowed to divide as many times as possible. The generation time of each cell was scrutinized and recorded by an observer, the usual automatic time-lapse microcinematography being impractical in such a biological inquiry.

### Skeletoblasts and Prechondroblasts Originating from the Rat Condylar Cartilage

*Skeletoblasts*

Nine progenies from rat condylar cartilage skeletoblasts were followed. A considerable number of cells did not divide; an additional number did not survive. So, even after the longest culture (lasting 16 generations) the number of skeletoblasts in the capillary tube did not reach 100.

The generation time of the first single isolated skeletoblast fluctuated around 24.00 hr (Fig. 7). The *circadian* duration of the cell division cycle was observed only when the cell was a single one. But as soon as there were two or more interconnected skeletoblasts,

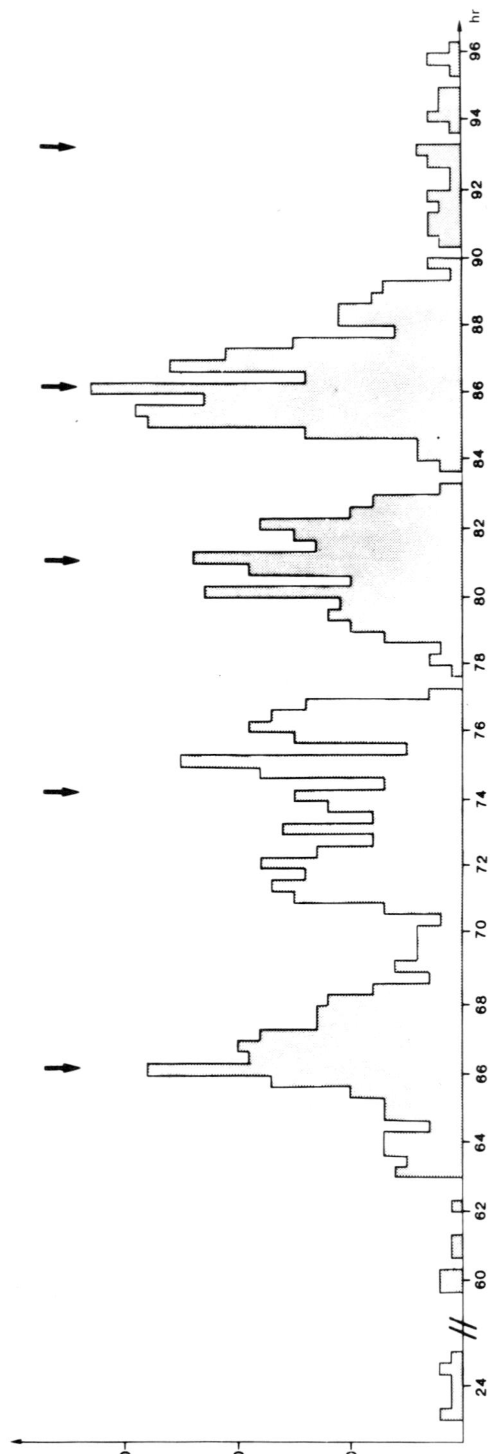


**FIGURE 7** Polymodal distribution of the generation times of progenies issued from a single isolated skeletoblast. The skeletoblast originated from the condylar cartilage of the rat mandible. Prevailing values occur every 6-7 hr. The number of generation times measured: n = 987.

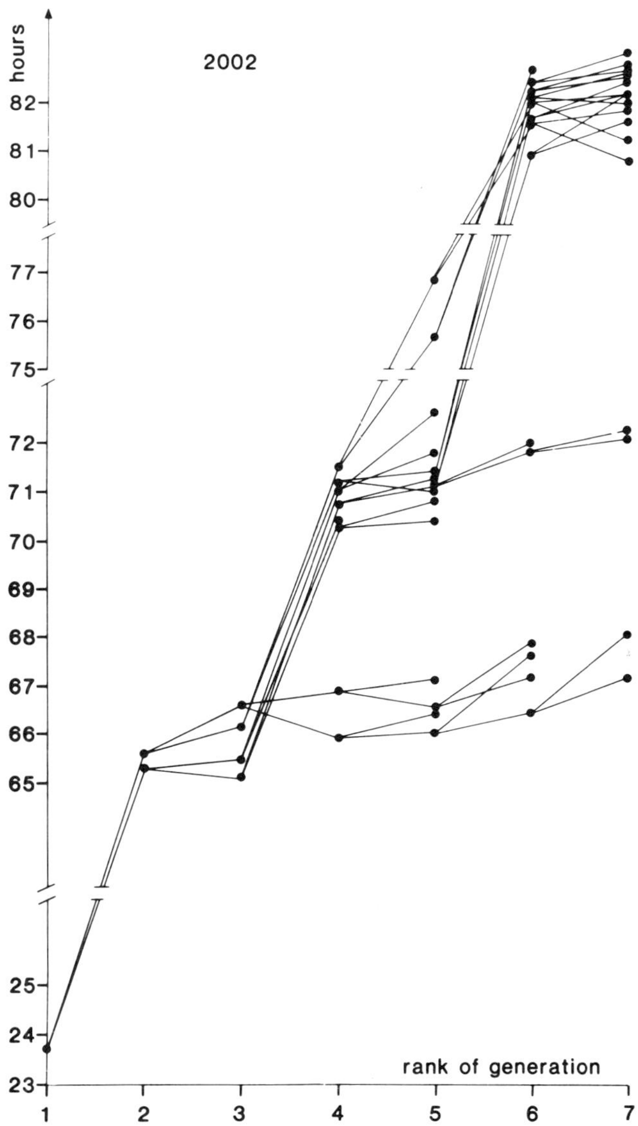


**FIGURE 8** Generation times in the progeny of a single isolated skeletoblast as related to the generation rank, the skeletoblast originating from rat condylar cartilage. Prevailing generation times appear to be quantized.

the duration of the cell division cycle jumped up to 60 hr and, skewing toward longer generation times, could reach up to 96 hr.

The distribution of the skeletoblast generation times observed in cells originating from the rat condylar cartilage exhibits distinct *discontinuities* (Fig. 7 and 8): the cell division cycle durations are distributed around prevailing values that are separated by time intervals of about 6.6 hr (Fig. 7). Additionally, in the rat condylar cartilage-derived skeletoblasts, an obvious tendency to intermediate periods of about 3 hr could be detected. The histogram does not display any tendency of the generation times to be normally distributed around each mode.

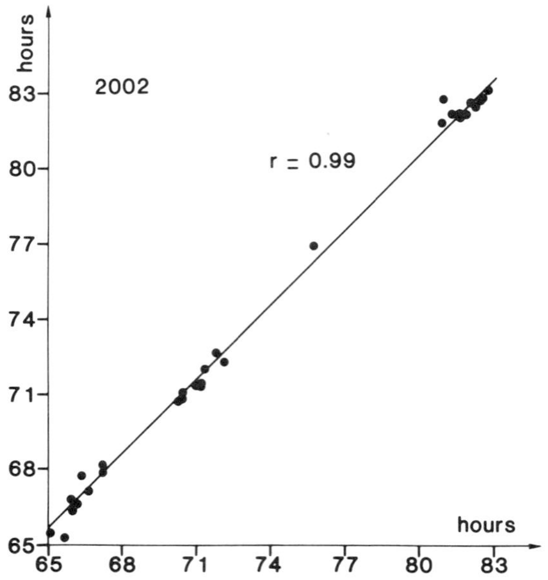


**FIGURE 9** Regression line for the generation times of sibling cells in the progeny of a single isolated skeletoblast originating from rat condylar cartilage. The degree of correlation is very high (r = 0.99). The different modes of the distribution appear clearly in this representation.

The restraining effect of the presence of two skeletoblasts on the cell division rate results obviously from signals operating after the formation of cell-to-cell junctions. Scrutiny of generation time modes (Fig. 7) shows that when the daughter cells of the single isolated skeletoblast divide, the first peak corresponds at least to 60 hr. The time interval between 24 and 60 hr, i.e., 36 hr, is precisely a multiple of a 6 hr quantum, which appears to be, in this specific case, a basic component of this cell's temporal organization.

The variations of the generation time in a same progeny account for the fact that, at a given moment, one randomly chosen skeletoblast may belong, for instance, to the fifth cell generation and another to the sixth cell generation after the division of the first single isolated skeletoblast. Figure 8 shows that a skeletoblast of the sixth generation can be observed after 421 hr-20 min, as well as after 468 hr. However, when the generation times were recorded for two sibling cells, a high degree of correlation was detected: r extends from 0.90 to 0.99 (Fig. 9). This feature was already reported for BHK 21 C13 (baby hamster kidney) cells by Minor and Smith (1974).

*Prechondroblasts*

Single isolated prechondroblasts originated from growing rats of different ages. As in skeletoblasts, the generation time of the first single isolated prechondroblast always oscillated around 24 hr. It skewed slightly during the first few generations. In some of the prechondroblasts, the generation time jumped up to 30 hr and, in a few cells, could reach, after two time discontinuities, 39 hr (Fig. 10). It is noteworthy that the leap in generation time occurs preferentially in prechondroblasts where the intermitotic interval is still close to 24 hr. This phenomenon was observed irrespective of the age of the rat (Fig. 11).

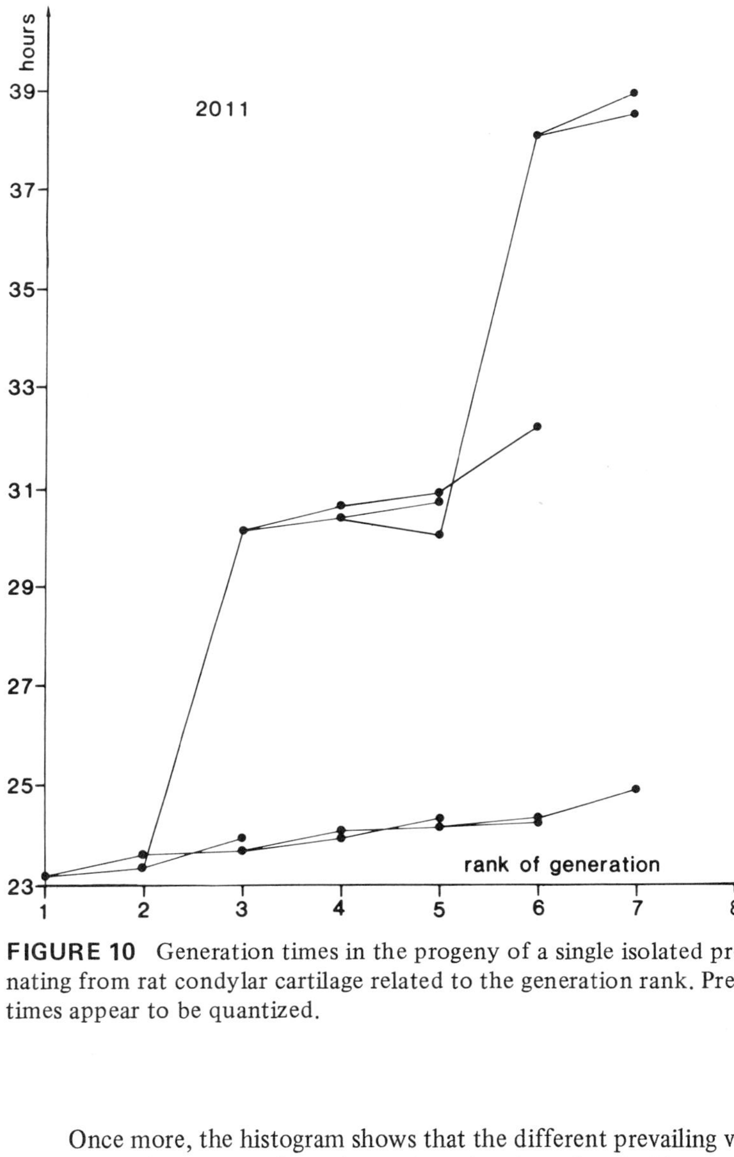


**FIGURE 10** Generation times in the progeny of a single isolated prechondroblast originating from rat condylar cartilage related to the generation rank. Prevailing generation times appear to be quantized.

Once more, the histogram shows that the different prevailing values for the generation time in the progeny from the single isolated prechondroblast seem to be quantized. The time interval between two modes is about 8 hr.

In prechondroblasts from rat, as in human preosteoblasts, cell-to-cell connections are looser than in skeletoblasts. So, the cell division restraining signal(s) reach the neighboring cells more directly in the skeletoblast population than in the prechondroblast or preosteoblast populations. This may account for the observation that, in comparison with the skeletoblast population, the leap toward longer generation times is delayed (it occurs only in the presence of numerous cells) and much less pronounced (it never exceeds 40 or 48 hr) in prechondroblast or preosteoblast populations.

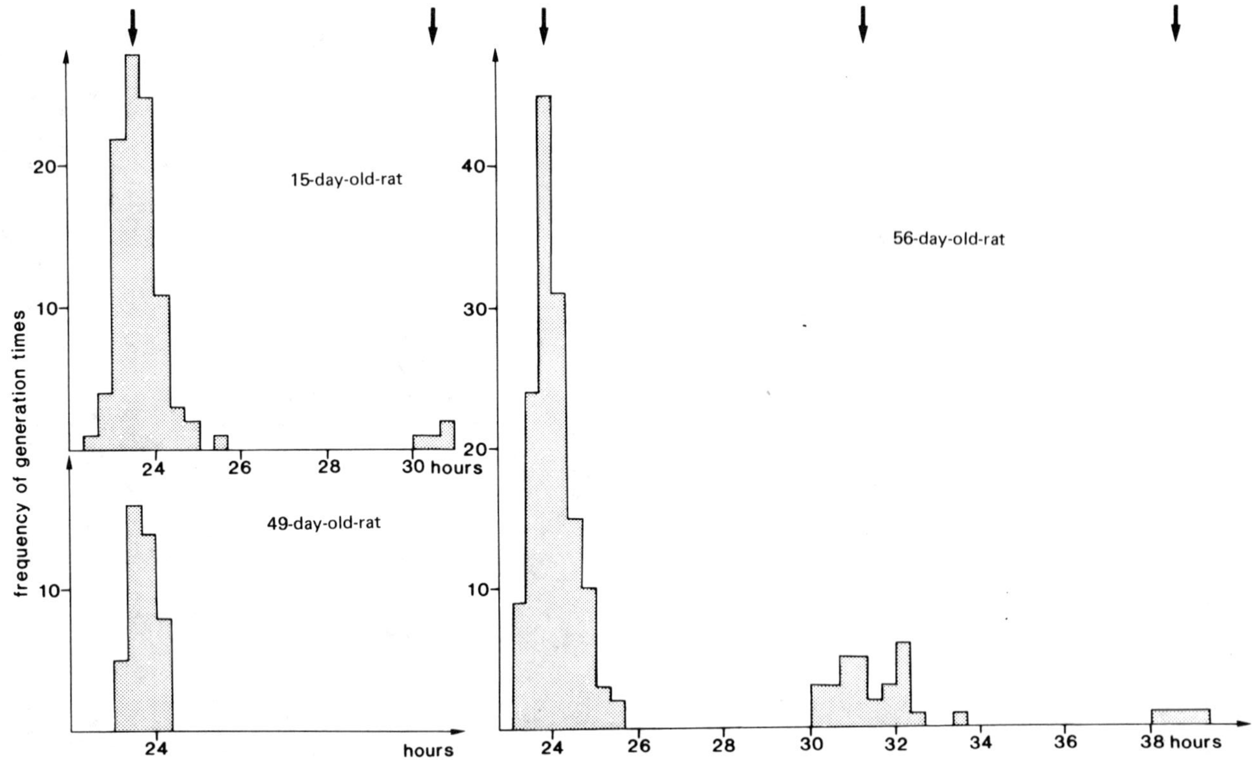


**FIGURE 11** Polymodal distribution of the generation times of progenies issuing from a single isolated prechondroblast. The prechondroblast originated from the condylar cartilage of the rat mandible. Most generation times are located around 24 hr. The following maxima occur, approximately, at an 8 hr interval: 31 and 39 hr. The number of generation times measured: n = 316.

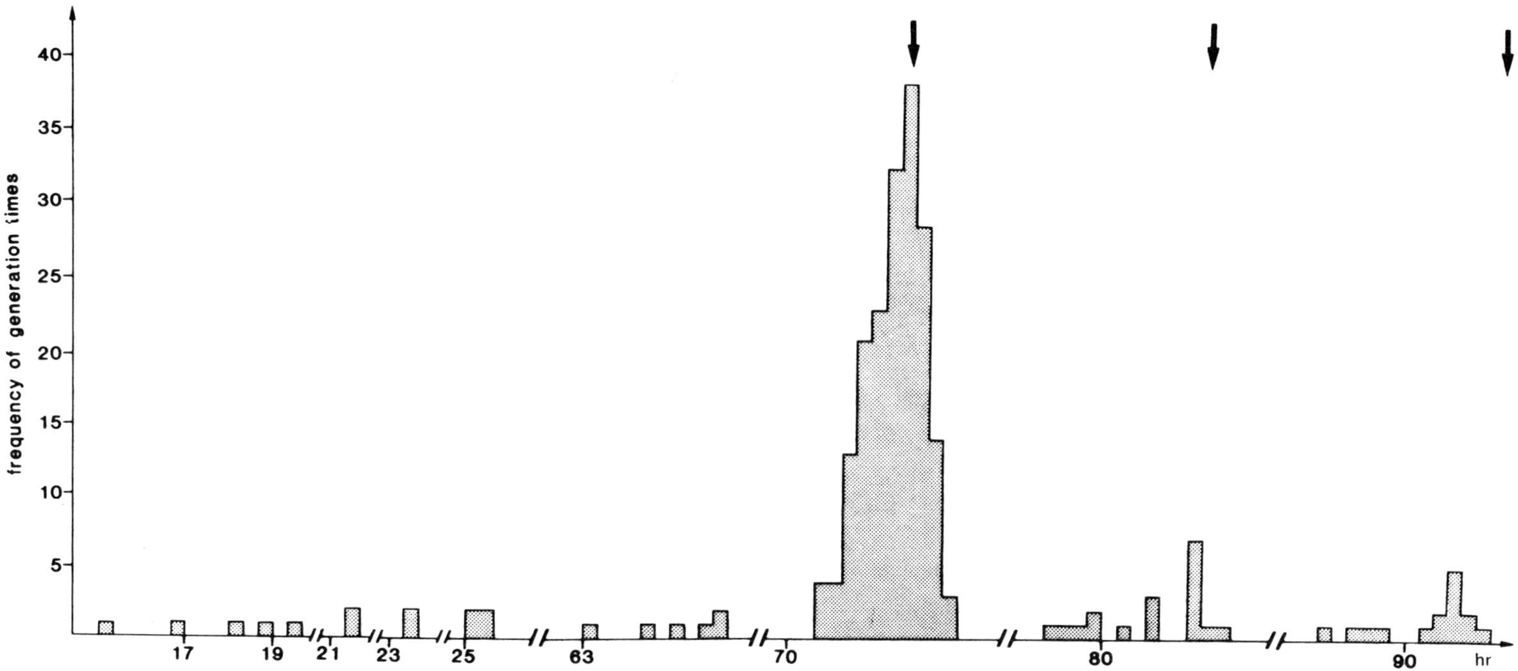


**FIGURE 12** Polymodal distribution of the generation times of progenies issuing from single isolated skeletoblast. The skeletoblast originated from human alveolar bone. Modes can be observed at about 73, 82, and 91 hr.

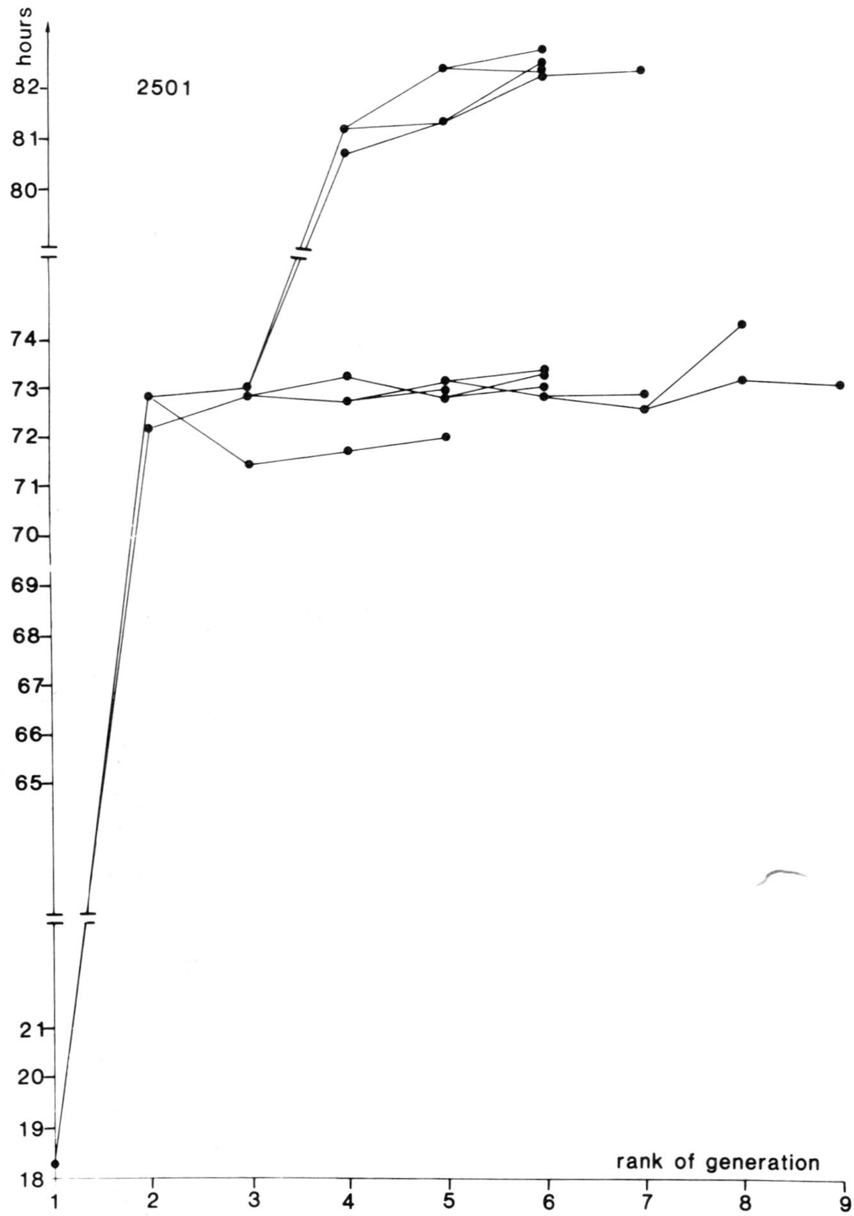


**FIGURE 13** Generation times in the progeny of a single isolated skeletoblast originating from human alveolar bone related to the generation rank. Prevailing generation times appear to be quantized.

## Skeletoblasts and Preosteoblasts Originating from Human Alveolar Bone

Single *isolated* skeletoblasts and preosteoblasts originating from the human alveolar bone were introduced into capillary tubes.

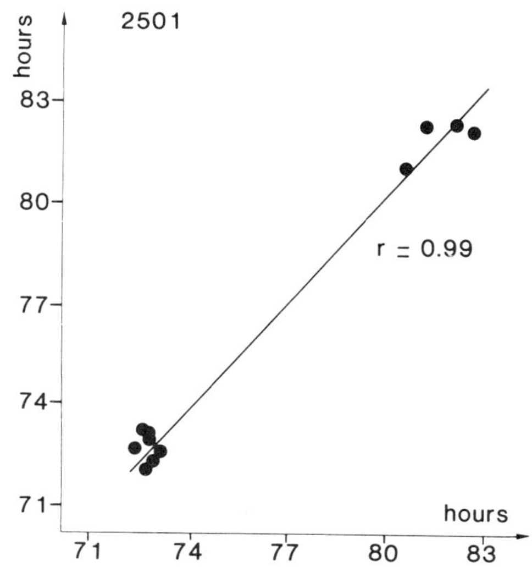


**FIGURE 14** Regression line for the generation times of sibling cells in the progeny of a single isolated skeletoblast originating from human alveolar bone. The degree of correlation is very high (r = 0.99). The different modes of the distribution appear clearly in this representation.

*Skeletoblasts*

Eight progenies of skeletoblasts from the human alveolar bone were followed. The histogram of cumulated generation times recorded in these eight progenies issuing from a single isolated skeletoblast exhibits a polymodal distribution. The generation time of a single isolated skeletoblast always oscillates around 24 hr, but, as soon as two or more skeletoblasts are associated with each other, the generation time lengthens. Obvious modes can be identified: around 64, 73, 82, and 91 hr. Most of the skeletoblast generation times (77.6%) are distributed around 73 hr (Fig. 12). It should be pointed out that, once again, 73 hr corresponds roughly to 3 X 24 hr. The average time interval between two modes is about 8.6 hr; intermediate "events" of 3 hr in the distribution of the human skeletoblast generation time could not be distinguished (Fig. 12).

When the generation time in the progeny of a single isolated skeletoblast is plotted against the generation rank (Fig. 13) an obvious quantizement of prevailing generation times can be discerned. As in the rat skeletoblast, the degree of correlation for the generation times of sibling cells is very high (Fig. 14).

*Preosteoblasts*

Seven progenies, each issuing from a single isolated preosteoblast, were followed. As for the skeletoblast, the histogram of cumulated generation times recorded in these seven progenies exhibits a multimodal distribution (Fig. 15). Once more, the generation time of a single isolated preosteoblast oscillates around 24 hr. However, in opposition to what was described for the skeletoblast, when two or more preosteoblasts are associated, the generation time continues to oscillate around 24 hr for several generations. Only after at least three to four cell generations does the generation time occasionally lengthen: it can jump up to 32 hr and even reach 41 hr for a few cells (Fig. 16). Here, too, an obvious

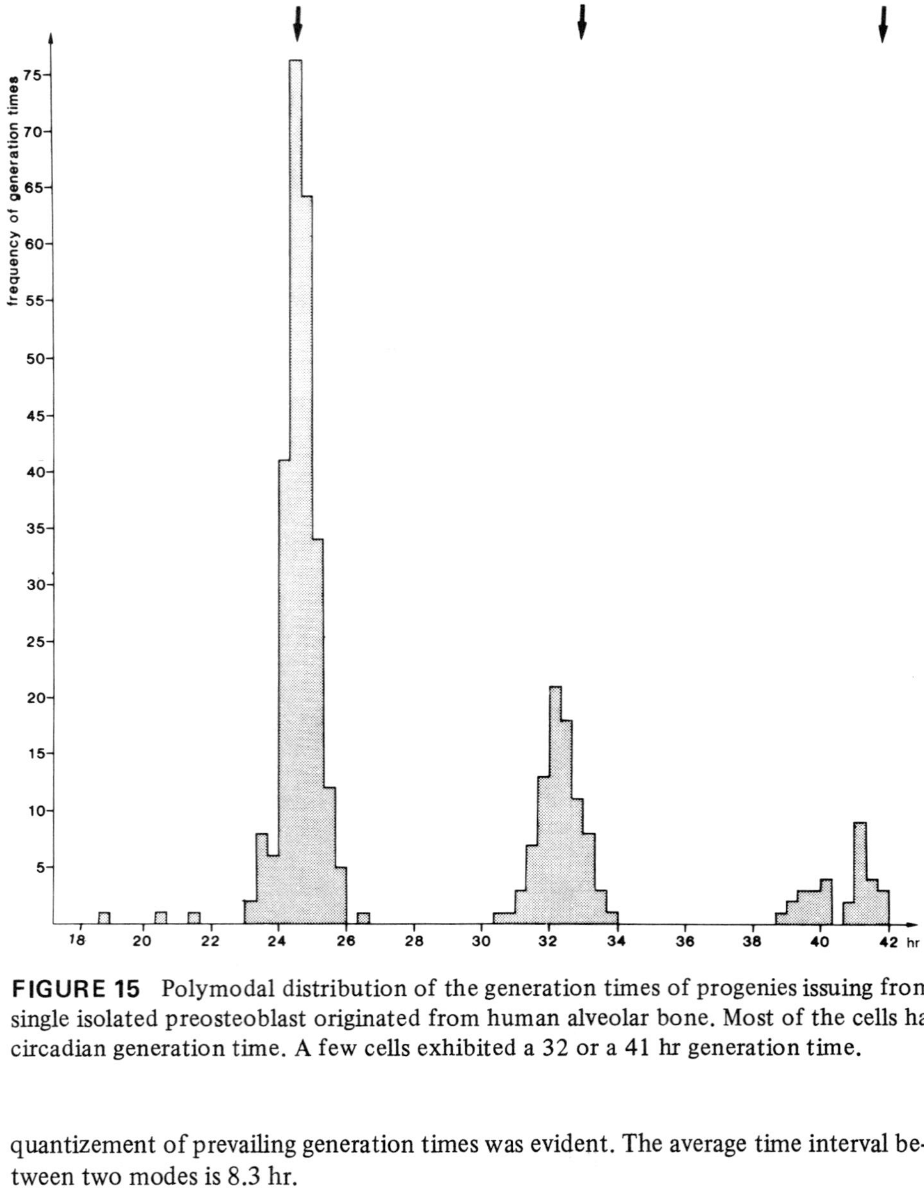


**FIGURE 15** Polymodal distribution of the generation times of progenies issuing from a single isolated preosteoblast originated from human alveolar bone. Most of the cells had a circadian generation time. A few cells exhibited a 32 or a 41 hr generation time.

quantizement of prevailing generation times was evident. The average time interval between two modes is 8.3 hr.

## SKELETAL CELL MULTIPLICATION-LINKED MEMBRANE EVENTS

It appears from the reported findings concerning skeletoblasts, preosteoblasts, and prechondroblasts that circadian rhythmicity is their usual endogenous pattern of multiplication, but that one basic "component" corresponds in fact to a 6-8 hr interval (i.e., 2 X $G_q$) and that, in a given cell population (in culture or in situ), the actual generation time is, for the most part, a multiple of this basic time component. Negative feedback intercellular processes acting on the mitogenic rate were detected, but the specific factors accounting for the overall lengthening of the generation time have not yet been identified.

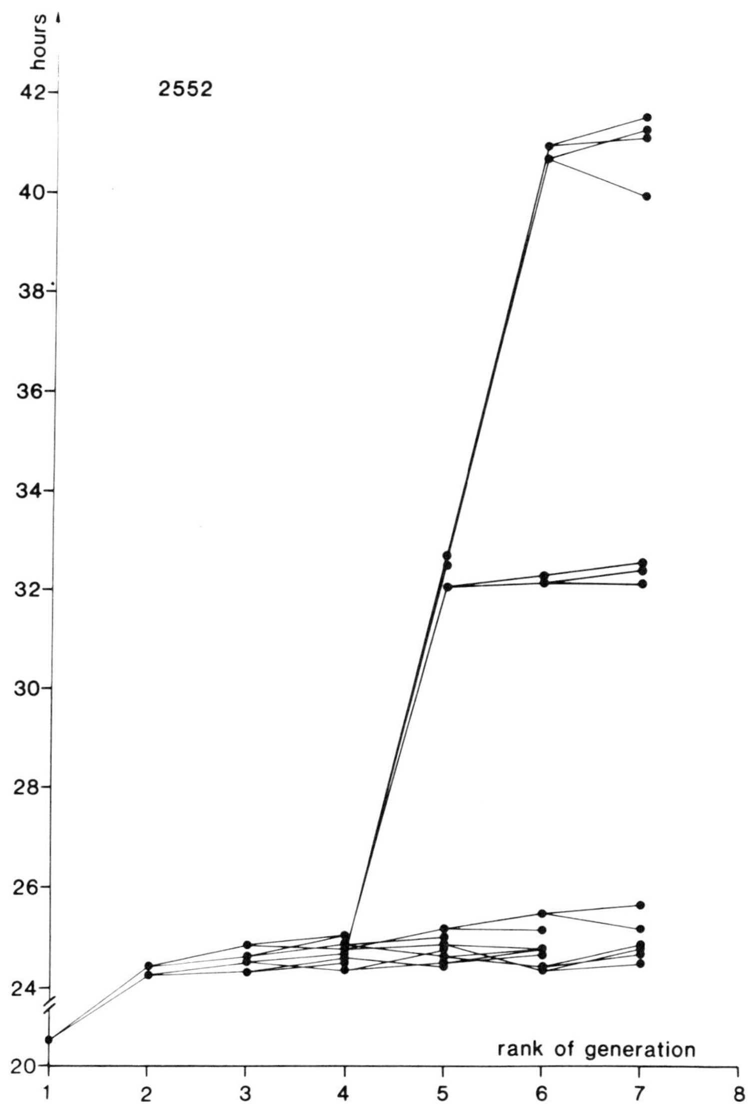


**FIGURE 16** Generation times in the progeny of a single isolated preosteoblast originating from human alveolar bone related to the generation rank. Prevailing generation times appear to be quantized.

Besides these phenomenological features of the temporal organization of skeletal cell multiplication, some enlightening data relative to physiological events at the cell membrane level in connection with skeletoblasts, preosteoblasts, and prechondroblasts multiplication were collected in our laboratory.

It is now well established that variations in ion fluxes and in mediators like cAMP and other unidentified endogenous growth-relating factors control cell multiplication, i.e., the initiation of DNA synthesis and of mitosis. However, the central problem in understanding the mechanism of action, i.e., the specific role of each factor and the discrimination between causal relationships and correlations, remains for the most part unanswered.

Recent investigations now in progress in our laboratory relative to the differences between the single isolated skeletoblast and two daughter cells have provided noteworthy information. In comparison with one single isolated skeletoblast, any one of the two daughter cells shows during the major part of $G_1$:

Consistent transmembrane ion flux variations, which were detected by using specific microelectrodes. The intracytoplasmic level of $Ca^{2+}$ displays a relative increase. The intracytoplasmic level of $K^+$ rises. The intracytoplasmic level of $Na^+$, especially in early $G_1$, is decreased. Such a low level of $Na^+$ acts to repress the mitogenesis-specific enzymes

A reduction in $H^+$ efflux, conducive to a persistent lowering (about 0.3) of the intracellular pH, which causes the discontinuation of the skeletoblast mitotic activity

A highly significant increase in two cell surface proteins, namely, the LETS protein (corresponding to the fibronectin) and the cell transglutaminase, detected and semiquantitatively evaluated by using histochemical immunofluorescence analysis. Glucosaminoglycans are also augmented.

The initiation of a new skeletoblast division involves a momentary surge of cytoplasmic $Ca^{2+}$: this transient phenomenon appears to be the actual triggering event inaugurating the skeletoblasts' new division process. The $Ca^{2+}$ surge initiates $Na^+$ influx to all appearances through calmodulin, and simultaneously, $[Na^+/K^+]$-ATPase activity is decreased. As a consequence, $H^+$ efflux is enhanced, leading to a long-lasting use of intracytoplasmic pH. The rise of pH, whatever the mechanism, is a critical condition for the onset of skeletoblast division. The deceleration of $Na^+$ efflux acts to derepress the mitogenesis-specific enzymes, which were repressed by the acceleration of the $Na^+$ efflux.

Investigations relative to preosteoblast and prechondroblast division provided us with additional information concerning cell membrane phenomena. The momentary surge of cytoplasmic $Ca^{2+}$ acts as the initiating event of a causal sequence leading to increased synthesis of endogenous cAMP. The transient rise of endogenous cAMP, in turn, is coupled to DNA synthesis. Indeed, under a threshold of cAMP concentration, there is no setting up of DNA synthesis. Moreover, the $Ca^{2+}$ surge alone, or the cAMP surge alone, is not conducive to preosteoblast or prechondroblast division. In these two cell types, also, the cytoplasmic $Ca^{2+}$ surge promotes $H^+$ efflux, i.e., contributes to the rise in intracellular pH.

It should be pointed out that, according to our experiments in cell culture (Stutzmann and Petrovic, 1982), cAMP plays two different roles in skeletal cell division. First, in all skeletal cells, cAMP is a critical factor in the causal chain inducing cell division. In this role, cAMP acts through a transient rise starting from a low basic level. Second, a persistently high level of endogenous cAMP (as well as the experimental addition of dibutyryl-cAMP) acts as an amplifier of the biologically detected, but not yet identified, intracellular signal causing the highly significant lengthening of the generation time when there are at least two skeletoblasts maintaining cell-to-cell specific junctions (Stutzmann and Petrovic, 1982).

It should be also reported that, irrespective of whether one is dealing with two or more skeletoblasts, the amount of LETS protein and of cell transglutaminase decreases during the recurrence of a new cell division.

The differentiation of the skeletoblast, a stem cell, into either a preosteoblast or a secondary prechondroblast, involves a substantial reduction of LETS protein and of cell transglutaminase, as well as a significant decrease in cytosolic $Ca^{2+}$. It is interesting to point out that skeletoblasts are rather flattened cells with a relatively low rate of multiplication, whereas preosteoblasts and secondary prechondroblasts are rather globular cells with a relatively high rate of multiplication (Petrovic, 1982). In this connection, it could be mentioned, that, according to Desmukh and Sawyer (1977), the rabbit articular chondrocyte switches its phenotype when cultured in calcium-free medium supplemented with dibutyryl-cAMP. Also, according to Benya and Shaffer (1982), fibronectin enhances the spreading of cultured chondrocytes and facilitates their "dedifferentiation" into flattened fibroblast-like cells. Moreover, we want to emphasize that under our culture conditions, we are dealing with skeletal cells that are genuinely different and not with experimentally induced cell modulations.

What is the first cause of skeletal cell division? Does it result from a necessary lapse of time, whatever the underlying process? Is it the accomplishment of a "program"? It should be mentioned that even after the inhibition of transcription, no consistent alterations in the period or phase of various rhythms were described (Matthews et al., 1976). However, the hypothesis of a gene-dependent circadian organization is far from refuted.

Our investigations concerning skeletoblast, preosteoblast, and prechondroblast division do not reveal the precise causal sequence of cellular and molecular events leading to the cell division. However, if the actual role of cell surface proteins is still a controversial problem, there is no doubt that skeletal cell division is triggered and mediated by modulations in ionic fluxes and by variations in other cell modulators. As a matter of fact, the following findings relative to the circadian fluctuations should be reported (Petrovic and Stutzmann, unpublished).

Between 1800 and 2400 hr, in comparison with observations made between 900 and 1600 hr, human alveolar preosteoblasts, collected during the first premolar extraction, have shown as a cell population:

Acceleration of $Na^+$ influx
Diminishment of [$Na^+/K^+$]-ATPase activity
Lowering of the cytoplasmic [$Ca^{2+}$]
Intensification of [$Ca^{2+}\cdot Mg^{2+}$]-ATPase activity
Amplification of $H^+$ efflux, leading to the rise of the cytoplasmic pH

A parallel study referring to skeletoblasts of the same origin has revealed, between 18.00 and 24.00 hr, a lessening in cell surface LETS protein and in cell transglutaminase activity and a reduction in the amount of basic endogenous cAMP.

In a previous paper (Petrovic, 1982) it was reported that, in culture, besides variations in ionic fluxes and in endogenous cell modulators, quantitative variations in cell-to-cell junctions between skeletoblasts seems to account, for the most part, for the modulation in generation time. Thus far, such structural variations could not be detected during the circadian cycle.

## CONCLUDING REMARKS

It is clear from our investigations that there is a basic circadian rhythm of mitotic activity in skeletal cells. However, in only the single isolated skeletal cell is the generation time approximately the same as the rotational period of the earth. In situ, the generation time of an individual cell is considerably longer than 24 hr; however, the cell population as a whole displays a circadian rhythmicity of the mitotic index. These findings strongly suggest that the circadian organization of cell division implicates an important endogenous component, and that the interactions between the cells of a population mask the endogenous circadian periodicity of the generation time.

The view of the existence of an ubiquitous 24 hr division cycle had been expressed by Harker (1958), but to the best of our knowledge has never been substantiated for mammalian, including human, cells. To evidence such a fundamental circadian temporal organization, animal and human cells had to be isolated and scrutinized individually.

The actual generation time in populations of skeletal cells is always somewhat greater than 24 hr. Indeed, the sequence of events leading to longer generation times should be considered as the outcome of signals restraining the mitogenetic activity and resulting from intercellular communications. In cells where strong interplays result from morphological peculiarities (for instance, in skeletoblasts), the circadian generation time is only expressed in the first division of the single isolated cell. The restraining effect on mitogenesis occurs as soon as two skeletoblasts coexist with cell-to-cell specific junctions. In cells whose intercytoplasmic connections are looser (for instance, in prechondroblasts and preosteoblasts), the 24 hr rhythm is maintained for several generations; in these circumstances, everything happens as if the generation time is lengthened when the cell density reaches a critical level, but this raises a question about the nature of the involved mechanism.

When the generationtime of individual cells increases by a small degree, it may display gradual, continuous changes but, as soon as the generation time becomes greater, the lengthening is converted into a stepwise, quantized pattern. In other words, the increase of the intermitotic duration appears to be discontinuous without being stochastic. Indeed, the generation time of skeletal cells is lengthened regularly toward multiples of 24 hr by the recurrence of 6-8 hr increments (both divisors of 24 hr). In a few cases, a more elementary quantum of 3-4 hr corresponding to the $G_q$ described by Klevecz (1976), was observed.

Identically, if no actual circadian periodicity could be detected in a malignant skeletoblast population culture, diversified but consistent ultradian mitotic index fluctuations were identified (Stutzmann et al., 1980): strikingly, the time interval between two successive peaks varied from 6 to 9 hr, which roughly corresponds to the elementary time component described above for normal skeletal cells. Such ultradian variations of bone sarcomatous cells were observed only at the clinical beginning of the cancer; later on, in most cases, significant fluctuations of the mitotic index could no longer be recognized.

The generation times of sibling cells in a progeny of skeletal cells are always highly correlated, which is probably another aspect of gene-coded phenomena. Klevecz (1976) also reported that the possible values for generation times in a cell strain are limited. Such facts argue in favor of the existence of a genetically determined component for generation times, as speculated by Edmunds and Adams (1981) in an attempt to establish models applicable to both mammalian cells and microorganisms. On the contrary, the close similarity of temporal pattern in the generation times of the progenies of rat prechondroblasts and human preosteoblasts, on the one hand, and on the other, those of rat condylar cartilage and human alveolar bone skeletoblast, argue in favor of a cell- and tissue-specific temporal pattern within animal species.

## ACKNOWLEDGMENTS

This work has been supported by grants INSERM (France): ATP 42 76 74, CRL 80 4 009, and ATP 76 79 108. Thanks are due for the technical assistance of Denise George, CNRS biologist and Martine Lecerf, INSERM biologist-biochemist.

## REFERENCES

Baserga, R., and Malamud, D. (1969). *Autoradiography. Techniques and Application.* Hoeber Medical Division, Harper and Row Publishers, New York.

Benya, P. D., and Shaffer, J. D. (1982). Dedifferentiated chondrocytes reexpress the differentiated collagen phenotype when cultured in agarose gels. *Cell* 30:215-224.

Charlier, J. P., Petrovic, A., and Herrmann, J. (1968). Déterminisme de la croissance mandilbulaire: effets de l'hyperpropulsion et de l'hormone somatotrope sur la croissance condylienne de jeune rat. *Orthod. Fr.* 39:567-579.

Charlier, J. P., Petrovic, A., and Herrmann-Stutzmann, J. (1969). Effect of mandibular hyperpropulsion on the prechondroblastic zone of young rat condyle. *Am. J. Orthod.* 55:71-74.

Dawson, K. B., Madoc-Jones, H., and Field, E. O. (1965). Variations in the generation times of a strain of rat sarcoma cells in culture. *Exp. Cell Res.* 38:75-84.

Desmukh, K., and Sawyer, B. D. (1977). Synthesis of collagen by chondrobytes in suspension culture: Modulation by Ca, 3′, 5′-cAMP, and prostaglandins. *Proc. Natl. Acad. Sci. USA* 74:3864-3868.

Edmunds, L. N., Jr., and Adams, K. J. (1981). Clocked cell cycle clocks. *Science* 211: 1002-1013.

Hansson, L. I., Stenstrom, A., and Throngren, K. G. (1974). Diurnal variation of longitudinal bone growth in the rabbit. *Acta Orthop. Scand.* 45:499-507.

Harker, J. E. (1958). Diurnal rhythms in the animal kingdom. *Biol. Rev.* 33:1-48.

Holtzer, H. (1978). Cell lineages, stem cells and the "quantal" cell cycle concept. In *Stem Cells,* B. Lord, C. Potten, and R. Cole (eds.). Cambridge University Press, Cambridge, England, pp. 1-27.

Klevecz, R. R. (1976). Quantized generation time in mammalian cells as an expression of the cellular clock. *Proc. Natl. Acad. Sci. USA* 73:4012-4016.

Matthews, H. R., Hardle, D. G., Inglis, R. J., and Bradbury, E. M. (1976). The molecular basis of control of mitotic cell division. In *The Molecular Basis of Circadian Rhythms.* J. W. Hastings and H. G. Schweiger (eds.). Dahlem Konferenzen, Berlin, pp. 395-409.

Minor, P. D., and Smith, J. A. (1974). Explanation of degree of correlation of sibling generation times in animal cells. *Nature* 248:241-243.

Oudet, C., and Petrovic, A. (1978). Growth rhythms of the cartilage of the manibular condlyle. Effects of orthopaedic appliances. *Int. J. Chronobiol.* 5:545-564.

Oudet, C., and Petrovic, A. (1981a). Spectral analysis, according to De Prins' method, of the nyctohemeral variations in the growth rate of the condylar cartilage of the mandible. *J. Interdiscipl. Cycle Res.* 12:167-175.

Oudet, C., and Petrovic, A. (1981b). Generation time in mandibular condylar cartilage cells of growing rat. Proceedings of the Fifteenth International Conference International Society for Chronobiology, Minneapolis, September 13-18. S. Karger (in press).

Oudet, C., and Petrovic, A. (1982). Daytime and seasons are source of variations for cartilage and bone growth rate. In *Factors and Mechanisms Influencing Bone Growth,* A. D. Dixon and B. Sarnat (eds.). Alan R. Liss, New York, pp. 481-498.

Petrovic, A. (1982). Postnatal growth of bone. Current trends, new approaches and innovative perspectives. In *Factors and Mechanisms Influencing Bone Growth,* Proceedings of the International Conference, UCLA, Los Angeles, January 5-12, A. D. Dixon and B. Sarnat (eds.). Alan R. Liss, New York, pp. 297-331.

Petrovic, A., and Charlier, J. P. (1967). La synchondrose sphéno-occipitale de jeune rat en culture d'organes: mise en évidence d'un potentiel de croissance indépendant. *C. R. Acad. Sci. [D] (Paris)* 265:1511-1513.

Petrovic, A., and Stutzmann, J. (1981). A cybernetic view of facial growth mechanisms. In *Long Term Treatment in Cleft Lip and Palate,* B. Kehrer, T., Slongo, B. Graf, and M. Bettex (ed.), H. Huber, Bern, Stuttgart, Vienna, pp. 15-56.

Petrovic, A., Stutzmann, J., and Oudet, C. (1975). Control processes in the postnatal growth of the condylar cartilage of the mandible. In *Determinants of Mandibular Form and Growth,* Monograph 4, Craniofacial Growth Series, J. A. McNamara, Jr. (ed.), Center for Human Growth and Development, The University of Michigan, Ann Arbor, Michigan, pp. 101-153.

Petrovic, A., Stutzmann, J., and Oudet, C. (1981). Turnover of human alveolar bone removed either in the day or in the night. *J. Interdiscipl. Cycle Res.* 12:161-166.

Petrovic, A., Stutzmann, J., Oudet, C., and Gasson, N. (1974). Kontrollfactoren des Kondylenwachstums–Wachstumshormon, Musculi pterygoidei laterales Vor und Fückschubgeräte des Unterkiefers. *Fortschr. Kieferorthopadie* 35:347-364.

Quastler, H., and Sherman, F. G. (1959). Cell population kinetics in the intestinal epithelium of the mouse. *Exp. Cell Res.* 17:420.

Rose, G. G., Pomerat, C. M., Schindler, M. D., and Trunnell, J. B. (1958). A cellophane strip technique for culturing tissue in multipurpose culture chamber. *J. Biophys. Biochem. Cytol.* 4:761-764.

Sandford, K., Covalevski, A. B., Dupre, L. T., and Earle, W. R. (1961). Cloning of mammalian cells by a simplified capillary technique. *Exp. Cell. Res.* 23:361-372.

Simmons, D. J. (1962). Diurnal periodicity in epiphyseal growth cartilage. *Nature (Lond.)* 195:82-83.

Simmons, D. J. (1964). Circadian mitotic rhythm in epiphyseal cartilage. *Nature (Lond.)* 202:906-907.

Simmons, D. J. (1968). Daily rhythm of $S^{32}$ incorporation into epiphyseal cartilage in mice. *Experientia* 24:363-364.

Simmons, D. J., Whiteside, L. A., and Whitson, S. W. (1979). Biorhythmic profiles in the rat skeleton. *Metab. Bone Dis. Relat. Res.* 2:49-64.

Stutzmann, J., and Petrovic, A. (1970). Particularités de croissance de la suture palatine sagittale de jeune rat. *Bull. Ass. Anat. (Nancy)* 148:552-562.

Stutzmann, J., and Petrovic, A. (1975). Nature et aptitudes évolutives des cellules du compartiment mitotique des cartilages secondaires de la mandibule et du maxillaire. Expériences en culture cytotypique et d'homotransplantation. *Bull. Assoc. Anat. (Nancy)* 59:523-524.

Stutzmann, J., and Petrovic, A. (1979). Persistence in organ culture of a growth rate circadian rhythm. In *Chronopharmacology,* Advances in Biosciences, vol. 19, A. Reinberg and F. Halberg (ed.). Pergamon Press, Oxford, pp. 89-100.

Stutzmann, J., and Petrovic, A. (1982). Bone cell histogenesis: The "skeletoblast" as a stem-cell for preosteoblasts and for secondary-type prechondroblasts. In *Factors and Mechanisms Influencing Bone Growth,* Proceedings of the International Conference, UCLA, Los Angeles, USA, January 5-12, A. D. Dixon and B. Sarnat (eds.). Alan R. Liss, New York, pp. 29-43.

Stutzmann, J., Petrovic, A., and George, D. (1980). Life cycle length, number of cell generations, mitotic index and modal chromosome number as estimated in tissue culture of normal and sarcomatous bone cells. In *Bone and Tumors,* Third Symposium CEMO, A. Donath and B. Courvoisier (eds.). Editions Médecine et Hygiène Geneva, pp. 188-194.

Stutzmann, J., Petrovic, A., and Malan, A. (1981). Seasonal variations of the human alveolar bone turn-over. A quantitative evaluation in organ culture. *J. Interdiscipl. Cycle Res.* 12:177-180.

# 19

# Environmental Signals and Rhythms on the Order of Hours

## Role of Cellular Membranes and Compartments and the Cytoskeleton

**Simone Puiseux-Dao** / Université Paris 7 and ERA C.N.R.S. 325, Paris, France

### INTRODUCTION

Life must be considered in four-dimensional space. The spatiotemporal organization of living systems includes cyclic phenomena whose periodicities vary from seconds or minutes (metabolic rhythms) to several hours (cell division cycles; circadian rhythms), a month (menstruation), or a year (bird migrations). For example, some of the metabolic rhythms have been extensively studied (Goldbeter, 1980); they involve some molecules assembled in at most two phases, a liquid-solid phase (such as a membrane) and a liquid phase. On the other hand, monthly or annual frequencies seem to imply elaborate structures such as entire organisms and their environment. We want to analyze rhythms on the order of hours: we shall work at the cellular level and look for the cellular assemblies which could be necessary for those rhythms to proceed.

### RHYTHMS ON THE ORDER OF HOURS: SOME EXAMPLES AND CHARACTERISTICS

#### The Cell Division Cycle

The rhythm of division in growing cell cultures shows a periodicity from about 1 hr (bacteria) or 3-4 hr (yeasts, *Tetrahymena*), to 12-26 hr (animal cells in culture). It is often entrained to become circadian by environmental conditions (such as light-dark cycles) as analyzed in *Euglena* (Edmunds, 1978; Edmunds and Adams, 1981), or pluricircadian (Chisholm, 1981).

The cell division cycle of eukaryotic cells has been divided into four steps: $G_1$, S, $G_2$, and M. The S phase, where DNA synthesis occurs, and cell division (M for mitosis)

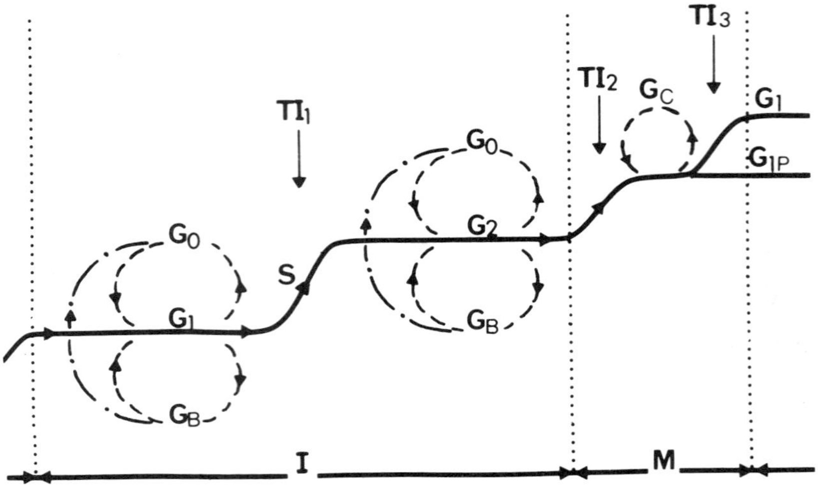


**FIGURE 1** Diagram showing the irreversible transitions of the cell cycle. $TI_1$, S phase; $TI_2$, chromosome division; $TI_3$, formation of the division membranes. During $G_1$ or $G_2$, the cells can enter dormancy ($G_0$) or they can return to a previous stage ($G_B$: Wounded or intoxicated cells that digest part of their contents). $G_0$ and $G_B$ are reversible stages, but the cells do not follow exactly the opposite path when they progress through the division cycle and when they pass to $G_0$ or $G_B$. This gives hysteresis cycles, $G_0$ and $G_B$, corresponding to the bifurcations of mathematical catastrophe theory. During mitosis, another hysteresis cycle can appear when the cells are treated with colchicine or by low temperatures. When the blockage is over, the cells can progress normally to $G_1$, or they can become polyploid ($G_{1p}$). I, Interphase; M, mitosis. (Modified from Puiseux-Dao, 1979).

both concern the behavior of the nuclear genome linked to replication and exist in eukaryotic cells, but also in prokaryotic cells. The $G_1$ and $G_2$ stages, related to cellular growth and the preparation for division, mainly comprise events such as transcription and translation and the behavior of proteins and various metabolites. The $G_1$ and $G_2$ phases (which do not concern DNA duplication in two copies equally distributed in the daughter cells) have long been considered as absent in prokaryotes, but $G_1$ is now described in relatively slow-growing cultures of prokaryotes where the generation time lasts longer than S plus M, which are called respectively, C and D in the case of bacteria (Cooper, 1979).

The cell cycle (Fig. 1) can be disturbed at any moment by changes in the environmental conditions, but the consequences are greater when S or M are involved, than when they occur in $G_1$ or $G_2$. Indeed, S and prophase and telophase of mitosis are *irreversible* transitions (Puiseux-Dao, 1979). Once DNA has replicated, it is impossible for the same cell to return to a state in which the quantity of DNA is reduced to that before replication. When chromosomes have divided, no reversion to a complement of polytene chromosomes in that cell is conceivable. The division of one cell into two daughter cells is irreversible (except if very peculiar experimental conditions permit the fusion of the young sister cells). And disturbances during these processes may give rise to stable anomalies that are then transmitted to the progeny: local mutations, polytene chromosomes, or abnormal number of chromosomes.

In contrast (Fig. 1), $G_1$ and $G_2$ are reversible steps (Puiseux-Dao, 1979) during which cell development can be arrested or the cell damaged. In the first case, the cells become dormant until convenient conditions are restored; in the second case, the cells can lose part of their material by amputation or digestion within vacuoles. Then they can either recover, if external factors become normal again, or they can enter a dormant stage. During mitosis, in eukaryotic cells, metaphase and anaphase are reversible events, as are $G_1$ and $G_2$. They can be interrupted, for example, by colchicine or cold, and start again when the environment allows the microtubules of the spindle to polymerize.

We want to point out that the irreversible transitions of S and M (taken in its entirety) have been shown to be less variable than those of $G_1$ and $G_2$ (see Chap. 6).

The dormant state has been called $G_0$; in general, it is assumed that cells enter $G_0$ from $G_1$ (Smith and Martin, 1973). However, the cell development can be reversibly stopped in $G_2$ or metaphase, with, for example, colchicine, or even in normal conditions (Gelfant, 1963) and the arrest could also be considered a $G_0$ stage (Fig. 1) (Puiseux-Dao, 1979).

### Cell Cycle Events That Can Cycle in the Absence of Cell Division

When analyzing the physiology of synchronized cells, one easily observes that the metabolic activities increase and decrease sequentially. For practically all of them, there is a maximum period and a minimum during each cell division cycle: this is the case for rRNA synthesis, ATP synthesis, and photosynthesis. When cultures stop dividing, some cellular activities can still cycle, for example, photosynthesis in stationary cultures of *Euglena* (Edmunds, 1975), enzyme potential in *Schizosaccharomyces* (Benitez et al., 1980), and budding initiation in *Saccharomyces* (Hartwell, 1971). This means that some of physiological oscillations of cell cycle events are not necessarily linked to DNA replication and cell division.

### The Case of *Acetabularia*

*Acetabularia* is a giant unicellular organism for which several circadian rhythms have been extensively studied (Schweiger and Schweiger, 1977; Vanden Driessche, 1980). The cell lives several months without any nuclear division. Therefore, the observed rhythms photosynthesis, chloroplast behavior, electrical polarity (Broda and Schweiger, 1981) or chloroplast migration (Broda et al., 1979) behave similarly to rhythms able to cycle in the absence of cell division. However, *Acetabularia* facilitates the analysis of the role of the nucleus: rhythms persist after the removal of the nucleus, though the nucleus controls the phase as shown by nuclear exchange where anucleate cells received a nucleus from a cell in the opposite phase—the phase of the nucleus became the phase of the new entire cell (Schweiger and Schweiger, 1977). Moreover, when *Acetabularia,* with opposite phases, were placed in the same dish or flask, each cell retained its own rhythm, suggesting that intercellular synchronization does not occur, at least for this type of cell.

## ONE NONOSCILLATING ENVIRONMENTAL SIGNAL IS ENOUGH TO INDUCE RHYTHMS ON THE ORDER OF HOURS

### Circadian Electrical Polarity of *Acetabularia*

When *Acetabularia* cells are transferred into the dark (for less than 6-8 weeks), they progressively enter a resting stage. In particular, they lose the rhythm of electrical polarity

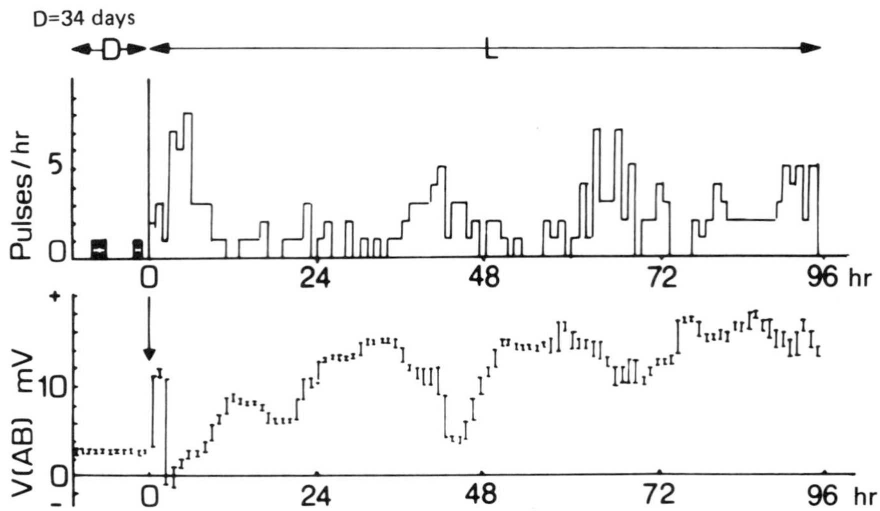


**FIGURE 2** Recovery of the electrical activity in *Acetabularia* after a dark treatment of 34 days (D). Continuous light ($\simeq$ 3-5 W·m$^{-2}$) was given at time 0. Abcissae: time in hours; ordinates: V[AB], potential difference between the apex and the basis in millivolts; pulses/hour, number of spontaneous action potentials per hour (records on five algae put together). The electrical activity in the dark was low and showed no periodicity. When continuous light was given, the electrical activity recovered with a circadian rhythm, after a short oscillation of 2-4 hr, depending on the algae. (Modified from Borghi et al., 1983.)

within about 8 days. As soon as these cells are illuminated again, changes in the electrical behavior takes place. In *continuous light,* after an oscillation lasting about 2-4 hr, electrical polarity recovers with a rhythm that becomes approximately circadian from the second day (Borghi et al., 1983) (Fig. 2). Therefore this rhythm, expected to be endogenous with experiments in the usual conditions—LD: 12,12 for several days, followed by continuous light (Broda and Schweiger, 1981)—can appear in cells completely deprived of any rhythm and placed in an environment without any periodicity. Such a possibility supports the existence of *endogenous rhythms* to a better extent than the previous experiments.

### Cell Division of Phytoplankton Algae

In general, phytoplankton algae are synchronized by dark-light cycles, the external conditions imposing the rhythm. However, some algae can be grown heterotrophically and escape the light requirement; this is the case with *Euglena.* And if a mutant, $P_4$ZUL, is cultured in continuous darkness in a medium containing glutamic and malic acids and then is transferred to continuous light (19°C; pH 3.5), this *single transition is sufficient to induce a division rhythm* with a period of 23 hr although no rhythm was detectable in the dark (Edmunds and Adams, 1981).

### Cell Division Cycle of Cultured Animal Cells

Synchronization of cultured animal cells has been extensively studied. In general, experiments start from quiescent cells induced to synthesize DNA and to divide by the

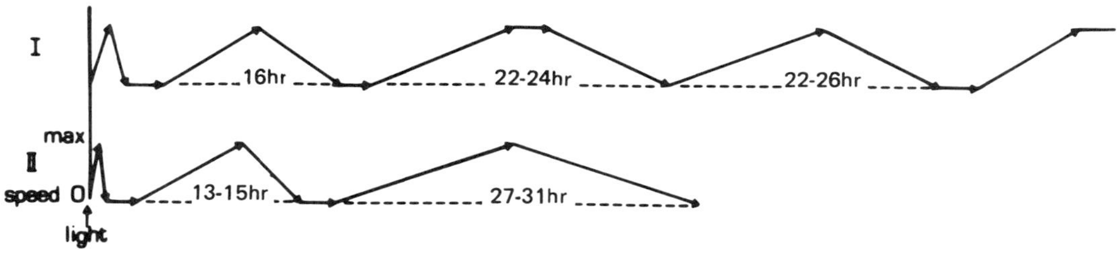


FIGURE 3 Recovery of the circadian rhythm of the electrical activity and cellular streaming in *Acetabularia* after a dark treatment of 5 weeks. I: Evolution of the electrical polarity (V[AB] as shown in Figs. 2 and 4). II: Evolution of the speed of cellular streaming as studied on five algae microcinematographed together. The two processes evolved in parallel. After a short period of 2-4 hr, a circadian rhythm was observed. Its first oscillation always lasted between 13 and 19 hr, and then the period varied between 17 to 31 hr, depending upon the groups of five algae. (Modified from Dazy et al., in press.)

addition of serum (Todaro et al., 1965) or of various other factors, such as hormones (Otto, 1982). Only a single cell cycle followed by a resting stage can be observed (Todaro et al., 1965), or several, at least two cycles (Teng et al., 1976); see also fibroblastic cells synchronized by isoleucine deprivation (McCormick and Bertram, 1982).

Although rapid damping of growth seems usual, we think that these cases are related to those described in algae and that the external signal plays a similar role as the dark-light transition in the preceding paragraphs. Lymphocyte stimulation by lectins also belongs to the same type of cell response (Wedner and Parker, 1976).

## TRANSMISSION OF THE EXTERNAL SIGNAL IMPLICATES THE CYTOSKELETON IN THE INDUCTION OF THE RHYTHMS ON THE ORDER OF HOURS

### The Cytoskeleton, Cellular Streaming, and the Circadian Electrical Polarity of *Acetabularia*

If *Acetabularia* cells are maintained in darkness (less than 6-8 weeks), not only does the electrical activity slow down and lose its rhythm, but cellular streaming also gradually decreases for the first week in the dark, then ceases more or less rapidly, during the next week, in general. When light is given again, cytoplasmic movements start in 1-3 min, then stop and become firmly established only after 2-4 hr of light. Even in *continuous* light, cytoplasmic streaming shows a rhythm with a periodicity of the same order as the periodicity of electrical polarity (Borghi et al., 1983) (Fig. 3). So *the timing of the recovery of the electrical behavior and cellular streaming* after prolonged dark periods is exactly the same. There are two steps: an immediate response, which is short, and stable activity, which starts after 2-4 hr of light and shows a rhythm that is roughly circadian, even in constant external conditions. Cytoplasmic streaming of *Acetabularia* is dependent on actin microfilaments and possibly on microtubules (Dazy et al., 1981; Koop, 1981).

### The Cytoskeleton and the Cell Division Cycle of Phytoplankton

With a phytoplankton alga, *Amphidinium carterae* (dinoflagellate), the effects of colchicine added to dark-light synchronized cultures at various times of the cell cycle have been analyzed. As usual, cell division was inhibited and this inhibition could be related to a decrease of microtubules observed under the electron microscope. Moreover, when the drug was given during $G_1$ (especially at the beginning), DNA synthesis did not occur as determined by fluorophotometric analyses (Durrand and Matthys-Rochon, submitted). The data suggest that the cytoskeleton, or at least the microtubules, could be necessary not only for mitosis, but also during $G_1$, for DNA replication to initiate.

### The Cytoskeleton and the Cell Division Cycle of Cultured Animal Cells

Microtubules (MT) seem to be involved in regulating the entry of quiescent fibroblastic cells into S phase. MT-disrupting drugs, such as colchicine, colcemid, or vinblastine, can enhance or inhibit the initiation of DNA synthesis after addition of a growth stimulating factor to quiescent cultures. The effect of colchicine is not a membrane effect since lumicolchicine has no similar action (Otto, 1982). Taxol, a drug able to stabilize MT, prevents DNA replication when colchicine stimulates entry into S phase (Crossin and Carney, 1981). MT-disrupting drugs are efficient when added during the first part of the lag period leading to S phase (Otto, 1982).

In a similar way, lymphocyte activation by lectins is colchicine- or vinblastine-sensitive, lumicolchicine being ineffective (Edelman et al., 1973; Greene et al., 1976). Cytochalasin B may also influence the process of lymphocyte stimulation (Belmont and Rich, 1981).

## A WORKING HYPOTHESIS

### What Could Link Light, Electrical Activity, and Cytoplasmic Streaming in Minutes or Less?

When *Acetabularia* cells are illuminated after a prolonged dark period, the electrical polarity and cytoplasmic streaming recover in a *few minutes* and evolve with exactly the same timing of oscillations, so they seem to be coupled. For this reason, some factors influencing these processes are under study. At the moment it is possible to settle the following points.

When light is switched on, the cytoplasm trembles at once, which is probably due to water transport: indeed, when the light energy is high ($\simeq$ 500 W·m$^2$), plasmolysis figures are then visible and even the cylindrical cell can break, in general just above the rhizoids (Durand, 1980). Moreover, tritiated water entry is enhanced when dark-treated cells are transferred to normal light as compared to cells transferred from the dark to dim light (Bonotto et al., unpublished data).

Calcium of the medium is necessary for streaming recovery (Durand, 1980; Puiseux- Dao et al., 1980, at least for the first short oscillation.

Blue light ($\simeq$ 450 nm) is much more efficient than red light ($\simeq$ 650 nm) in inducing cellular streaming recovery; the cells need at least 100 times more energy in red light to obtain the same effects as in blue light (Durand, 1980; Puiseux-Dao et al., 1980).

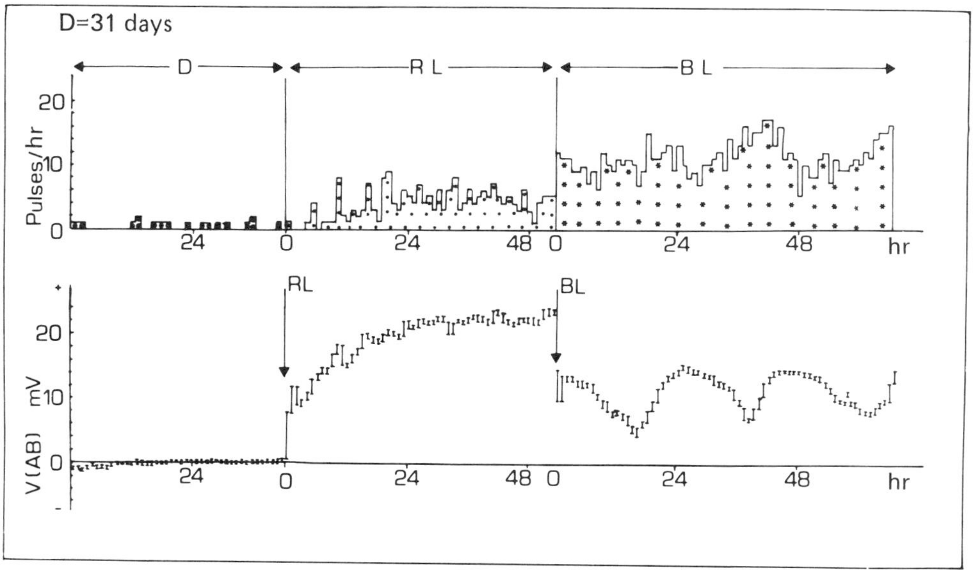


**FIGURE 4** Influence of the light wavelength on the recovery of electrical activity in *Acetabularia* after a dark treatment of 31 days. Same symbols as in Fig. 2. RL: red light; BL: blue light. Record of five algae put together. The electrical activity was low and showed no rhythm in the dark. When red light ($\simeq$ 650 nm, 3-5 W·m$^{-2}$) was given, the electrical activity increased, but had no periodicity. As soon as blue light ($\simeq$ 450 nm, 3-5 W·m$^{-2}$ replaced red light, a circadian rhythm of the electrical behavior appeared. (Courtesy of H. Borghi and A. C. Dazy, University of Paris, France.)

Red light can induce the recovery of electrical polarity, which, however, does not show any rhythm. If red light is replaced by blue light, the rhythmic nature of the recovery appears immediately. Streaming recovery also shows no rhythm in red light, although it does in white or blue light (Dazy et al., in press) (Fig. 4).

This last result should be related to data obtained by Clauss (1979), which showed that the photosynthetic activity in red light is low and constant, and that an additional irradiation with blue light of low intensity induces an increase of this activity with the appearance of a circadian rhythm after a first, shorter oscillation of 6 hr.

From these results, we can assume that *ions* (at least $Ca^{2+}$) and probably *water movements* are the cellular tools for the dark-light transition effects studied in *Acetabularia.* That ion and water transports can influence cell electrical activity is readily understood. But since microfilaments and microtubules are very sensitive to ion concentration, as are the associated kinases, it is conceivable that rapid changes in ion and water content can modify cellular streaming. Indeed, in *Acetabularia,* $K^+$, $Ca^{2+}$, and $Mg^{2+}$ control cytoplasmic movements (Sato, 1962), as does $Ca^{2+}$ in *Nitella* (Hayama et al., 1979a).

The effects of light on cell electrical potential have been investigated, particularly in Characeae by Tazawa and his collaborators (Kawamura and Tazawa, 1980; Tazawa and Shimmen, 1980a); in particular, a $K^+$ channel seems to be very rapidly activated by light (Tazawa and Shimmen, 1980b). In *Acetabularia,* very clearly the dark-light transition influences the electrical activity, but the light-dark transition modifies the electrical potential even more; $Cl^-$, $H^+$, and $K^+$ have been implicated (Saddler, 1970; Gradman, 1976).

It has also been shown, in Characeae, that water movements modify membrane polarization (Hayama et al., 1979b), as well as cytoplasmic streaming (Hayama and Tazawa, 1978).

Of course, ion and water transport in cells implicates membranes and compartments: the plasma membrane between the medium and the cell sap, the membranes of chloroplasts and mitochondria—organelles known to exchange various ions ($H^+$, $K^+$, $Ca^{2+}$), and the membranes of vacuoles. Here, for light activation, photoreceptors are necessary. Chloroplast pigments may be involved, as demonstrated by replacing *Chara* plastids by spinach chloroplasts in experimental algal cells (Tazawa and Shimmen, 1980a) but also phytochromes or phytochrome-like substances, as suggested by Weisenseel and Ruppert (1977) in *Nitella* or *Mougeotia* (Dreyer and Weisenseel, 1979).

### What Could Explain That the Rhythmic Machinery Once Set into Motion, Can Oscillate Without Any External Influence?

In giant *Acetabularia* cells, electrical polarity and cellular streaming and photosynthesis exist in anucleate growing cells; electrical polarity retains its rhythm (approximately circadian) (Novak and Sironval, 1976), as does photosynthesis (Schweiger et al., 1964). However, grafts and nucleus injection experiments have demonstrated a nuclear control on the phase of the photosynthetic rhythm (Schweiger et al., 1964); long-lived mRNAs (Schweiger and Schweiger, 1977; Vanden Driessche, 1980) are probably responsible for this control.

From the preceding paragraph, one can conclude that the environmental factor, light in the cases described, acts on *cytoplasmic structures,* such as membranes and the cytoskeleton, which is well correlated with the maintenance of rhythms in *anucleate* cells. In a first attempt, is it possible to explain rhythms without immediately involving the nucleus? In fact, the proposal of Burgoyne (1978), which is a combination of the Njus et al. (1974) and Schweiger and Schweiger (1977) hypotheses, with some changes, seems plausible. In the model, the rhythm-generating system is located on membrane(s), that are periodically refilled with newly synthetized proteins. The circadian rhythm would be due to assembly (comprised of synthesis of the implicated proteins and loading of the membrane) and disassembly (unloading and degradation of the proteins) of the concerned membrane (s) (Schweiger and Schweiger, 1977). The transport of a monovalent ion, $K^+$ according to Njus et al. (1974), rhythmically modified by the membrane composition, would change the cell ionic composition, which, in turn, would influence a number of syntheses, including the preceding membrane proteins. Such a feedback mechanism could result in the 24 hr period of circadian rhythms. Experiments showing that environmental factors can modify the protein composition of the plasma membrane support this hypothesis. For example, in *Paramecium primaurelia,* the surface antigen expression is regulated by temperature (Capdeville, 1979).

In our experiments, it is clear that such a process could explain how the external signal, a dark-light transition in this case, can induce a rhythmic activity in cells at rest following a prolonged dark treatment, even in otherwise constant external conditions. But, in our opinion, the model should implicate various ions—at least, $H^+$, $K^+$, and $Ca^{2+}$—from different sources (medium, plastids, mitochondria, vacuoles) and the assembly of cytoskeletal architecture able to assume the transport of adsorbed ions, synthesized molecules, and some organelles from one place to another, inside the cell, and so to transmit the metabolic responses from the vicinity of a membrane to other parts of the cell.

### What Could Explain the Variable Sensitivity of Rhythms to Inhibitors of Translation and Transcription?

One of the most disputed questions about circadian rhythms concerns their sensitivity to cycloheximide. In fact, the model of Schweiger and Schweiger (1977) implies that the pool of mRNA for the synthesis of the membrane proteins involved is not a limiting factor. It also lumps together protein synthesis and membrane loading, which could require an additional step, protein transport. But, it is easy to conceive that, depending upon the species or the type of cells in organisms, there can be a limiting step *at any level of the chain (i.e., transcription–translation–assembly) of the membranes and associated cytoskeleton. The molecule(s) with the shorter life span and the more limited pool size would become the most sensitive step of the mechanism.* In most cases, the rhythms are disturbed by translation inhibitors acting on the 80S ribosomes (Schweiger and Schweiger, 1977; Vanden Driessche, 1980). *But if assembly and disassembly of membrane(s) and cytoskeleton constituents concern long-lived proteins, no rapid sensitivity to cycloheximide or puromycin* would be detectable, and the feedback ionic mechanism giving the circadian rhythm would not influence protein synthesis but rather the assembly and disassembly processes. Moreover, as several types of membranes may be involved, some could be affected by protein synthesis inhibitors, others not, resulting in pathways still able to cycle, with or without modifications, in the presence of the inhibitors, while others could not. In Capdeville's experiments (1979), the replacement of a plasma membrane protein by another one takes a variable number of cell cycles, depending upon the mutant and the external conditions.

### The Hypothesis

In our opinion, the rhythms on the order of hours, including circadian rhythms, are mainly due to *physical processes of protein assembly and configuration.* Proteins imply a *nuclear control,* more or less delayed depending on the life span and pool size of the responsible mRNA and on the proteins themselves.

The assemblies concern membranes whose modifications induce extremely rapid changes of ion and water exchanges ($t \leqslant$ seconds), which in turn influence membrane enzymes (kinases, ATPases, nucleotide cyclases) and cytoskeletal structures very quickly ($t \simeq$ minutes). While membrane modifications can cycle by themselves with periodicities of some minutes–see, for example, electrical pulses of *Acetabularia* having a periodicity of about 15-20 min with a refractory period (Rogatykh et al., 1979)–the new ionic components and the products of reactions in the vicinity of the membranes would be transferred by cellular streaming to other parts of the cell. There, new molecule (enzyme mainly) assemblies and functioning would be induced. Among the synthesized molecules, membrane proteins would in return be transported by streaming to the related membranes to replace the aging or destroyed proteins and restore the decreasing specificity of the membrane to respond to the environmental conditions. This second phase, including cytoplasmic streaming and synthesis, would last several hours, frequently around 15-24 hr but much less for some protists. The adjustment between environmental periodic events and cellular periodic functioning would impose circadian rhythms. The role of the cytoskeleton, a structure that does not exist in prokaryotes, in controlling the transport would explain why those microorganisms do not show such rhythms, as would the absence of compartmentation (Wagner, 1977). In addition, the involvement of several ions, instead of $K^+$ alone, would allow understanding fo the contradictory results obtained with valinomycin, an antibiotic able to preferentially modify $K^+$ transport (Sweeney, 1974; Vanden Driessche, 1980).

In the experiments reported above, the external signal was a dark-light transition shown to modify the ion and water exchange at the level of membrane(s) containing a photoreceptor. But any substance that can induce a conformational change of a membrane with consequences on ion transports can replace the signal "illumination" with waves of "pleiotropic" effects on the cell metabolism organized in a network of reactions (Wagner, 1977). Peptidyl hormones, serum factors, and lectins should be such substances (Puiseux-Dao, 1981).

However, we still have to discuss the problem of nuclear control. Since the mechanism is based on proteins, any change in their genetic code may result in disturbance of the rhythms. And it has been shown that the plasma membrane protein composition can be modified by environmental conditions at the level of transcription by the basic phenomena of intergenetic and interallelic exclusion (Capdeville, 1979). Inherited alterations of lipid membrane components may also be involved. In addition, in *Acetabularia* grafts or nuclear implantations, when the nucleus imposes the phase of its rhythm to the cytoplasm, we can think that it does so by imposing the rhythm of waves of mRNA necessary for the concerned protein syhtheses. We must remark that this hypothesis, based on the "temporal inertia" linked to the necessary steps of the mechanism, does not exclude the possibility that the structure of nuclear DNA itself can induce a rhythmic transcription (Edmunds and Adams, 1981), but it does not demand it.

Here we put together cell rhythms related to the cell cycle and circadian rhythms, though they differ by their sensitivity to temperature. The period of circadian rhythms is known to be stable and temperature independent (Wagner, 1977; Edmunds, 1978), whereas the cell division cycle is temperature dependent. But, in our opinion, only the cyclic events of the cell division cycle are comparable to circadian rhythms, which can cycle even when DNA synthesis and mitosis do not occur. Normally, these events entrain DNA synthesis and cell division and would probably impose rhythmicity.

## REFERENCES

Belmont, J. W., and Rich, R. R. (1981). Calcium and magnesium and cytochalasin sensitive processes in lectin-stimulated lymphocyte activation. *Cell. Immunol.* 59:276-288.

Benitez, T., Nurse, P., and Mitchison, J. M. (1980). Arginase and sucrase potential in the fission yeast *Schizosaccharomyces pombe. J. Cell Sci.* 46:399-431.

Borghi, H., Puiseux-Dao, S., Durand, M., and Dazy, A. C. (1983). Morphogenesis, bioelectrical polarity and intracellular streaming in a giant cell, *Acetabularia mediterranea*: Studies on their recovery after prolonged dark period. *Plant Sci. Lett.* 31: 75-86.

Brōda, H., and Schweiger, H. G. (1981). Long-term measurement of endogenous diurnal oscillations of the electrical potential in an individual *Acetabularia* cell. *Eur. J. Cell Biol.* 26:1-4.

Broda, H., Schweiger, G., Koop, H. U., Schmid, R., and Schweiger, H. G. (1979). Chloroplast migration: A method for continuously monitoring a circadian rhythm in a single cell of *Acetabularia.* In *Developmental Biology of Acetabularia,* S. Bonotto, V. Kefeli, and S. Puiseux-Dao (eds.). Elsevier-North Holland, Amsterdam, pp. 163-167.

Burgoyne, R. D. (1978). A model for the molecular basis of circadian rhythms involving monovalent ion-mediated translational control. *FEBS Lett.* 94:17-19.

Capdeville, Y. (1979). Regulation of surface antigen expression in *Paramecium primaurelia*. II. Role of the surface antigen itself. *J. Cell. Physiol.* 99:383-393.

Chisholm, S. W. (1981). Temporal patterns of cell division in unicellular algae. In *Physiological Bases of Phytoplankton Ecology*, vol. 210, T. Platt (ed.). *Can. Bull. Fish. Aquat. Sci.*, Quebec, pp. 150-181.

Clauss, H. (1979). Auslösung der circadianen Photosynthese-Rhythmik bei *Acetabularia* durch Blaulicht. *Protoplasma* 99:341-346.

Cooper, S. (1979). A unifying model for the $G_1$ period in prokaryotes and eukaryotes. *Nature* 280:17-19.

Crossin, K. L., and Carney, D. H. (1981). Evidence that microtubule depolymerization early in the cell cycle is sufficient to initiate DNA synthesis. *Cell* 23:61-71.

Dazy, A. C., Borghi, H., Durand, M., and Puiseux Dao, S. The effects of blue and red light on the transcellular electrical potential, cytoplasmic streaming and rRNA transport in *Acetabularia acetabularia*. Proceedings of the XIth Intern. Seaweed Symposium, *Hydrobiologia* (in press).

Dazy, A. C., Hoursiangou-Neubrun, D., and Sauron, M. E. (1981). Evidence for actin in the marine alga *Acetabularia mediterranea*. *Biol. Cell* 41:235-238.

Dreyer, E. M., and Weisenseel, M. H. (1979). Phytochrome-mediated uptake of calcium in *Mougeotia* cells. *Planta* 146:31-39.

Durand, M. (1980). Action de la lumière, de la teneur en $Ca^{2+}$ du milieu et de la DNAase I sur les mouvements cytoplasmiques chez l'*Acetabularia mediterranea*. Diplôme d'Etudes Approfondies, University of Paris VII.

Durrand, A. M., and Matthys-Rochon, E. Application of flow cytometry on a dinoflagellate: Cell cycle kinetics and colchicine effect on DNA evolution (submitted).

Edelman, G. M., Yahara, I., and Wang, J. L. (1973). Receptor mobility and receptor cytoplasmic interactions in lymphocytes. *Proc. Natl. Acad. Sci. USA* 70:1442-1446.

Edmunds, L. N. (1975). Temporal differentiation in *Euglena*: Circadian phenomena in non-dividing populations and in synchronously dividing cultures. In *Les cycles cellulaires et leur blocage chez quelques protistes*, Colloque CNRS, vol. 240. CNRS, Paris, pp. 53-68.

Edmunds, L. N. (1978). Clocked cell cycle clocks: Implications toward chronopharmacology and aging. In *Aging and Biological Rhythms*, H. V. Samis and S. Capobianco (eds.). Plenum, New York, pp. 125-184.

Edmunds, L. N., and Adams, K. J. (1981). Clocked cell cycle clocks. *Science* 211:1002-1012.

Gelfant, S. (1963). Patterns of epidermal cell division. I. Genetic behavior of the $G_1$ cell population. *Exp. Cell. Res.* 32:521-528.

Goldbeter, A. (1980). Models for oscillations and excitability in biochemical systems. In *Mathematical Models in Molecular and Cellular Biology*, L. A. Segel (ed.). Cambridge University Press, London, pp. 250-291.

Gradman, D. (1976). Metabolic action potentials in *Acetabularia*. *J. Membr. Biol.* 29:23-45.

Greene, W. C., Parker, C. M., and Parker, C. W. (1976). Colchicine-sensitive structures and lymphocyte activation. *J. Immunol.* 117:1015-1022.

Hartwell, L. H. (1971). Genetic control of the cell division cycle in yeast. II. Genes controlling DNA replication and its initiation. *J. Mol. Biol.* 59:183-194.

Hayama, T., Nakagawa, S., and Tazawa, M. (1979a). Membrane depolarization induced by transcellular osmosis in internodal cells of *Nitella flexilis*. *Protoplasma* 98:73-90.

Hayama, T., Shimmen, T., and Tazawa, M. (1979b). Participation of $Ca^{2+}$ in cessation of cytoplasmic streaming induced by membrane excitation in Characeae internodal cells. *Protoplasma* 99:305-321.

Hayama, T., and Tazawa, M. (1978). Cessation of cytoplasmic streaming accompanying osmosis-induced membrane depolarization in *Nitella flexilis*. *Cell Struct. Funct.* 3: 47-60.

Kawamura, G., and Tazawa, M. (1980). Rapid light-induced potential change in *Chara* cells stained with neutral red in the absence of internal Mg-ATP. *Plant Cell Physiol.* 21:547-559.

Koop, H. U. (1981). Protoplasmic streaming in *Acetabularia. Protoplasma* 109:143-157.

McCormick, P. J., and Bertram, J. S. (1982). Differential cell cycle phase specificity of neoplastic transformation and mutation to ouabain resistance induced by N-methyl-N′-nitro-N-nitrosoaguanidine in synchronized C3H 10T 1/2 C18 cells. *Proc. Natl. Acad. Sci. USA* 79:4342-4346.

Njus, D., Sulzman, F. M., and Hastings, J. W. (1974). Membrane model for the circadian clock. *Nature* 248:116-120.

Novak, B., and Sironval, C. (1976). Circadian rhythm of the transcellular current in regenerating enucleated posterior stalk segments of *Acetabularia mediterranea. Plant Sci. Lett.* 6:273-283.

Otto, A., (1982). Microtubules and the regulation of DNA synthesis in fibroblastic cells. *Cell Biol. Int. Rep.* 6:1-18.

Puiseux-Dao, S. (1979). Mouvements cytoplasmiques et morphogenèse chez l'*Acetabularia mediterranea. Biol. Cell.* 34:83-90.

Puiseux-Dao, S. (1981). Cell-cycle events in unicellular algae. In *Physiological Bases of Phytoplankton Ecology,* T. Platt (ed.), *Can. Bull. Fish. Aquat. Sci.* 210:130-149.

Puiseux-Dao, S., Dazy, A. C., Durand, M., Marano, F., Brun, C., Hoursiangou-Neubrun, D., and Santa-Maria, S. (1980). Régulation par la lumière et la teneur en $Ca^{++}$ du milieu, des mouvements cytoplasmiques et du transport des RNA chez l'*Acetabularia. Biol. Cell.* 39:112.

Rogatykh, N. P., Melkumyan, V. G., and Zubarev, T. N. (1979). Action potential signals and cell morphogenesis in *Acetabularia.* In *Developmental Biology of Acetabularia,* S. Bonotto, V. Kefeli, and S. Puiseux-Dao (eds.). Elsevier-North Holland, Amsterdam, pp. 85-94.

Saddler, H. D. W. (1970). The membrane potential of *Acetabularia mediterranea. J. Gen. Physiol.* 55:802-821.

Sato, T. (1962). Effect of potassium, calcium and magnesium ions on the protoplasmic streaming in *Acetabularia calyculus. Bot. Mag. (Tokyo)* 75:384-390.

Schweiger, H. G., and Schweiger, M. (1977). Circadian rhythms of unicellular organisms: An endeavor to explain the molecular mechanism. *Int. Rev. Cytol.* 51:315-342.

Schweiger, E., Walraff, H. G., and Schweiger, H. G. (1964). Endogenous circadian rhythm in the cytoplasm of *Acetabularia*: Influence of the nucleus. *Science* 146: 658-659.

Smith, J. A., and Martin, L. (1973). Do cells cycle? *Proc. Natl. Acad. Sci. USA* 70:1263-1267.

Sweeney, B. M. (1974). The potassium content of *Gonyaulax polyedra* and phase changes in the circadian rhythm of stimulated bioluminescence by short exposure to ethanol and valinomycin. *Plant Physiol.* 53:337-342.

Tazawa, M., and Shimmen, T. (1980a). Direct demonstration of the involvement of chloroplasts in the rapid light-induced potential change in tonoplast-free cells of *Chara australis.* Replacement of *Chara* chloroplasts with spinach chloroplasts. *Plant Cell Physiol.* 21:1527-1534.

Tazawa, M., and Shimmen, T. (1980b). Demonstration of the $K^+$ channel in the plasmalemma of tonoplast-free cells of *Chara australis. Plant Cell Physiol.* 21:1535-1540.

Teng, M., Bartholomew, J. C., and Bissel, M. J. (1976). Insulin effect on the cell cycle: Analysis of the kinetics of growth parameters in confluent chick cells. *Proc. Natl. Acad. Sci. USA* 73:3173-3177.

Todaro, G. J., Lazar, G. K., and Green, H. (1965). The initiation of cell division in a contact-inhibited mammalian cell line. *J. Cell Comp. Physiol.* 66:325-334.

Vanden Driessche, T. (1980). Circadian rhythmicity: General properties—as exemplified mainly by *Acetabularia*—and hypotheses on its cellular mechanism. *Arch. Biol. (Bruxelles)* 91:49-76.

Wagner, E. (1977). Molecular basis of physiological rhythms. In *Integration of Activity in the Higher Plant.* D. H. Jennings (ed.). Cambridge University Press, London, pp. 33-72.

Wedner, H. J., and Parker, C. W. (1976). Lymphocyte activation. In *Progress in Allergy,* vol. 20, P. Kàllos, B. H. Waksman, and A. de Weck (eds.). Karger, Basel, pp. 195-203.

Weisenseel, M. H., and Ruppert, H. K. (1977). Phytochrome and calcium ions are involved in light-induced membrane depolarization in *Nitella. Planta* 137:225-229.

# 20

# Cell Cycle Controls in Phytoplankton

## Comparative Physiology and Ecology

**Sallie W. Chisholm, Daniel Vaulot,* and Robert J. Olson** / Massachusetts Institute of Technology, Cambridge, Massachusetts

### INTRODUCTION

Phytoplankton is the generic term for unicellular microalgae, most of which are obligate photoautotrophs, freely floating in the aquatic environment. The diversity of species belonging to this group of organisms (there are roughly 300 genera) is extraordinary in view of the simplicity of their growth requirements and the apparent uniformity of the aquatic environment in which they thrive. As photoautotrophs, these cells require only $CO_2$, light, vitamins, and essential inorganic nutrients to grow and reproduce. When grown on 24 hr photocycles, the cell division cycles of most species of phytoplankton become entrained; that is, division is phased in populations with doubling times greater than 1 day or synchronized in populations doubling about once per day. Although the latter might be considered induction synchrony in the strict sense of the term (Prescott, 1976; Zeuthen, 1974), it is a unique type of induction synchrony in that the periodic inducing stimulus (light) is a natural and ubiquitous one and is essential to cell survival. As such, the responses of the cells to this stimulus must have been shaped by natural selection.

A basic premise of this chapter, then, is that the cell division patterns of phytoplankton populations grown on light-dark cycles reflect an alignment of two periodic phenomena that is "optimized" for the long-term reproduction and survival of the population (Cohen and Parnas, 1976). Since the phase angle of this alignment, and its strictness, varies among phytoplankton species (Chisholm, 1981), we deduce that the requisites of the cell cycle and its controlling mechanisms must also vary. By analyzing in detail the similarities and differences between the growth patterns of various species under iden-

---

*On leave from Laboratoire d'Hydrobiologie, ERA 467 CNRS, U.S.T.L., Montpellier, France.

tical conditions, one should be able to make deductions about the regulation of their cell cycle progression and, ultimately, the adaptive significance of this regulation in the different species. A thorough descriptive review of the patterns of cell division in populations of phytoplankton entrained to photocycles can be found in Chisholm (1981). Our purpose here is to provide a brief review and to focus (admittedly, a bit myopically) on certain aspects of these patterns in a few species, which we believe reveal fundamental differences in cell cycle controls.

## PATTERNS OF MACROMOLECULAR SYNTHESIS OVER THE CELL CYCLE

Surprisingly little is known about the sequences, timing, and regulation of cell cycle events in phytoplankton, relative to the advances that have been made with other eucaryotic cell types. The most comprehensive recent review of the subject can be found in Puiseux-Dao (1981). Much work has been done on the cell cycle of *Chlorella* (Tamiya, 1964, 1966; Morimura, 1959; Setlik et al., 1975; Wanka and Mulders, 1967), but extrapolation from this genus to others is difficult because of the large number of daughter cells produced in one generation. In fact, it is a general source of frustration in compiling data for this type of discussion to find that much of the cell cycle research on algae has been done on genera that produce multiple daughter cells, such as *Platymonas* (Ricketts, 1977, 1979; Tanoue and Aruga, 1975), *Chlamydomonas* (Mihara and Hase, 1971; Lien and Knutsen, 1972; Kates et al., 1968), and *Scenedesmus* (Setlik et al., 1972). For simplicity here, we shall try to restrict ourselves to cell types that produce only two daughter cells. We recognize, however, that the species that produce multiple daughter cells may be of particular interest in terms of the optimization of growth rate under the constraints of strict cell cycle alignment (Bruce and Bruce, 1982).

Patterns of macromolecular synthesis in light-dark synchronized *Dunaliella* and *Euglena* show clear similarities. In both groups, mitosis occurs almost exclusively during the dark period in 24 hr photocycles except in very long photoperiods (Eppley and Coatsworth, 1966; Edmunds, 1965a; Marano et al., 1978). DNA synthesis begins in the middle of the light period and continues until the onset of mitosis, which in this case corresponds with the beginning of the dark period. In *Euglena* total RNA per cell increases continuously throughout the $G_1$ and S phases of the cell cycle, as do soluble protein, dry weight, and photosynthetic pigments (Edmunds, 1965b). In *Dunaliella,* increases in RNA and protein per cell are reduced during late S phase (Marano et al., 1978). Since most of the cellular RNA is ribosomal RNA and a significant fraction of this is in the chloroplasts (Puiseux-Dao, 1981; Cook, 1966b), consideration of the events surrounding plastid replication is important in interpreting patterns of biosynthesis in the whole cell (Ledoigt and Calvayrac, 1979; Puiseux-Dao, 1981).

Generalizations regarding macromolecular synthesis over the cell cycle in dinoflagellates are difficult to make. This group of species has several characteristics atypical of eucaryotes, including: (1) low acid-soluble protein content (and therefore histone content) in the chromatin (Rizzo and Nooden, 1973); (2) membrane-attached chromosomes and absence of a mitotic spindle (Bouligand et al., 1968; Kubai and Ris, 1969; Leadbeater and Dodge, 1967; Soyer, 1969); and (3) permanent condensation of chromosomes throughout the cell cycle (Loeblich, 1976). DNA synthesis has been reported to be continuous throughout the cell cycle in dinoflagellates (Dodge, 1965, 1966) as in procaryotic organisms, but more recent evidence has shown that a distinct S phase does exist in several species and it is synchronous for both nuclear and plastidal DNA (Loeblich, 1976, 1977; Galleron and Durrand, 1979). In *Cachonina niei,* synchronized on a LD: 12,12 cycle, for example, cell division begins at the end of the dark period and is completed within the first 4 hr of the light period. DNA synthesis begins later in the light period and is completed within the first 7 hr of the dark period (Loeblich, 1976, 1977). In *Amphidinium carteri,* cell division begins before or at the beginning of the dark period on most photocycles and is completed before the onset of light (Chisholm and Brand, 1981; Galleron and Durrand, 1979). The rate of DNA synthesis, determined as the rate of thymidine incorporation, is maximal immediately following cell division, suggesting a very short $G_1$ period (Galleron and Durrand, 1979). It should be noted that in *A. carteri* (Galleron and Durrand, 1979), *Prorocentrum micans* (Filfilan and Sigee, 1977), and possibly *C. niei* (Loeblich, 1976, Fig. 1b), residual DNA synthesis occurs outside the true S phase of the cell cycle, the nature of which is not as yet understood. This residual synthesis is probably what mistakenly was identified as a continuous S phase in the earlier studies.

## PATTERNS OF CELL DIVISION PHASING BY LIGHT-DARK CYCLES

In most species of phytoplankton, cell division is most pronounced during the dark period when cultures are maintained on 24 hr photocycles (Nelson and Brand, 1979; Chisholm, 1981). Both the degree of phasing and the timing of the onset of the division burst appear to vary between the species and can be influenced by photoperiod (Paasche, 1968; Eppley and Coatsworth, 1966) and other growth conditions (Chisholm et al., 1975), but a general trend of division onset in the dark predominates (Chisholm, 1981). For all these species there is a significant portion of the photocycle during which mitosis does not occur.

The dinoflagellates, in general, display some of the "tightest" cell cycle phasing, although heterotrophic forms appear to be exceptions (Weiler, 1978). In the well-characterized genera, *Gonyaulax* (Sweeney and Hastings, 1958, 1964), *Ceratium* (Weiler and Eppley, 1979), and *Pyrocystis* (Sweeney, 1982), mitosis is restricted to a 4-6 hr gate in the dark or at the dark-light transition, regardless of the mean generation time of the population. In all cases, the mechanism regulating this tight phasing has been attributed to coupling of the cell cycle to a circadian oscillator. We note here, however, that all three of these genera are typically slow-growing dinoflagellates with maximum growth rates of less than one doubling per day. In species with high intrinsic growth rates, such as *A. carteri* (discussed below), division patterns are more like those of other classes of phytoplankton, characterized by a fairly broad division burst but with a significant proportion of the burst occupying the dark period (Nelson and Brand, 1979; Chisholm and Brand, 1981; Chisholm, 1981). Typical examples are shown in Fig. 1. It has been our observation that cell populations belonging to species that display these types of patterns do not lose rhythmicity in division in the fast or "ultradian" growth mode, as suggested by Ehret and Wille (1970). The mechanism controlling the resultant patterns is most easily interpreted as one fitting a cell cycle "block point" model rather than a circadian clock (see below).

Diatoms (*Bacillariophyceae*) as a group are notably exceptional in their division patterns on photocycles (Nelson and Brand, 1979; Chisholm and Costello, 1980; Chisholm et al., 1980; Yoder et al., 1982). In this group, division patterns in populations grown on all but the shortest photoperiods are not phased, regardless of the growth rate (Chisholm and Costello, 1980). Population rhythmicities with a dominant 24 hr frequency have similar characteristics in both fast (ultradian) and slow (infradian) growing populations (Fig. 2). A typical pattern consists of a major burst of division at the beginning of the light period followed by one or more additional peaks that may extend into the dark period.

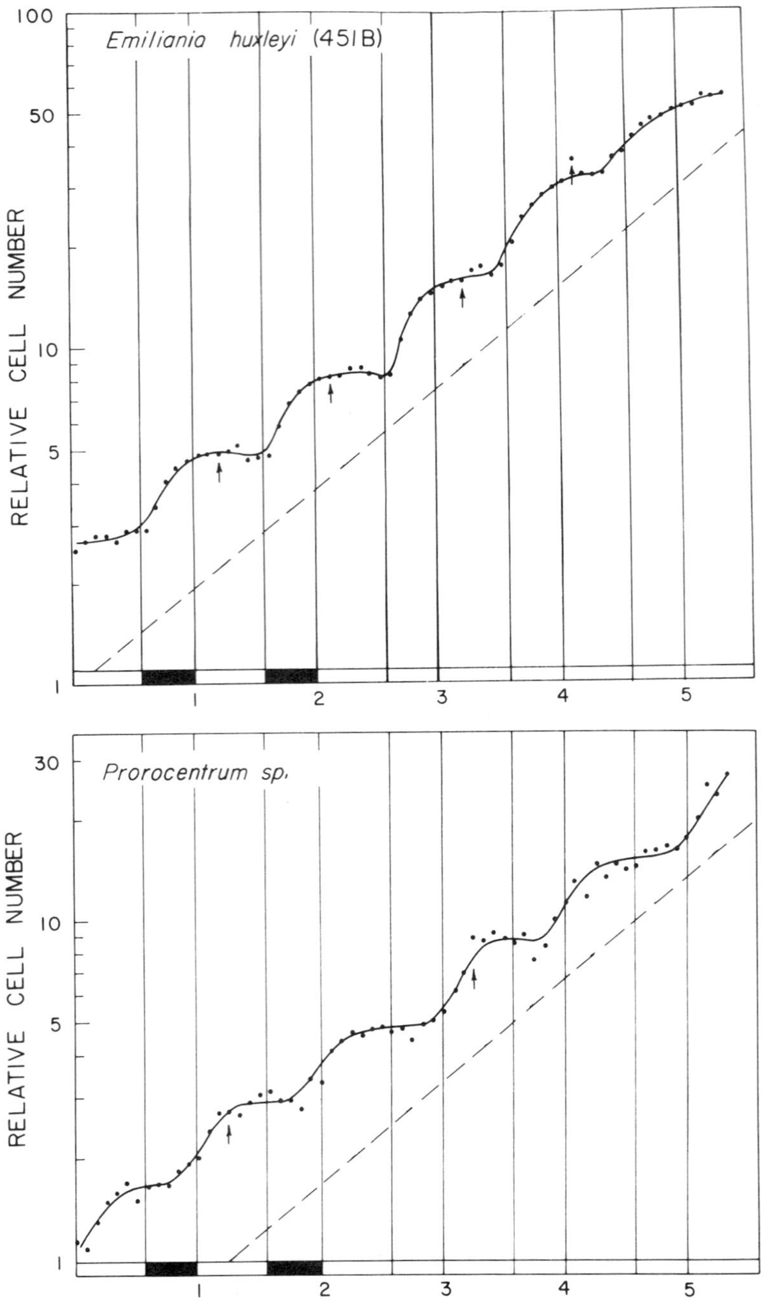


**FIGURE 1** Photocycle-induced cell division patterns of two species of marine phytoplankton grown in semicontinuous batch culture. Arrows indicate times when the cultures were diluted and dashed line has a slope of one population doubling per day. Note that division phasing persists for at least 3 days in continuous light and the average population growth rate does not change. (Adapted from Chisholm and Brand, 1981.)

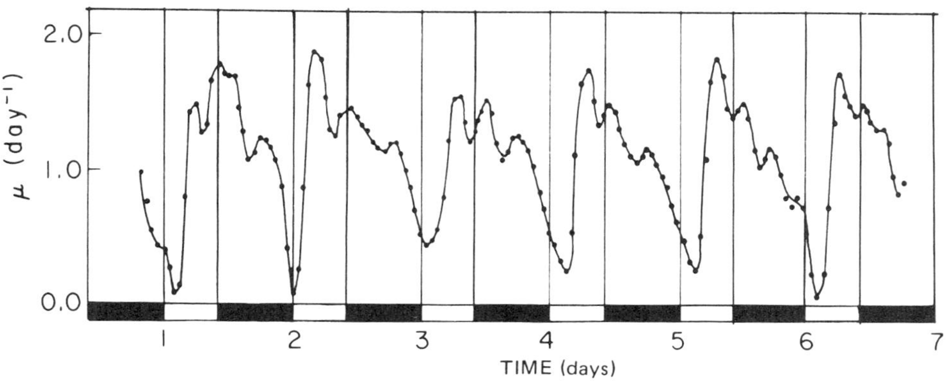


**FIGURE 2** Specific division rate as a function of time in the marine diatom *T. weissflogii* grown in a cyclostat on LD: 10,14. The specific division rate is the derivative of a smoothed function fitted to the cell concentration changes in the vessel (Corrected for dilution). The average growth rate of the population was 1.06 per day. (Adapted from Chisholm and Costello, 1980).

The precise timing and, to some extent, the magnitude of these peaks is conserved between photocycles for a given growth condition but assumes different "fingerprint" patterns when growth rate is limited by different factors, such as nutrients or temperature (Fig. 3). The latter is true, even if the average population growth rates of the cultures are the same, and regardless of whether they are greater than or less than one doubling per day. In all cases, some cells are dividing throughout the photocycle; i.e., there is no division gate.

It could be argued that the division rate periodicities observed in these diatom populations, which are calculated from rates of change of cell numbers, do not reflect underlying cell cycle events, but simply a rhythmicity in cytokinesis or cell separation. Recently, we have examined this possibility using flow cytometric techniques to distinguish between $G_1$ and $G_2$ cells in the population over the 24 hr photocycle (Olson et al., 1983). As can be seen in Fig. 4, the correspondence between the proportion of cells in $G_1$ and $G_2$ and the cell division patterns is quite good; thus, the periodicities we have seen in cell number do indeed reflect cell cycle events.

The resistance of the diatom populations to strict phasing by the light-dark cycle, regardless of the population growth rate, suggests that the cell cycle of these species is not coupled (or very loosely coupled) to an endogneous clock. We may also deduce that the duration of light exposure required for a cell to complete the cell cycle is much shorter than typical daily photoperiods, since in some cases two generations of cells are produced in a single photoperiod. Finally, we conclude that in general, cell cycle progression is not rigidly controlled in diatoms. Even when the average doubling time of the population is 1 day, there is no strictly preferred alignment of mitosis relative to the light-dark cycle. Thus, the change in the age distribution as a function of time in diatom populations entrained to light-dark cycles must differ distinctly from that of species in which division is phased or synchronized.

The division patterns displayed by diatom populations are difficult to interpret in terms of most existing models of cell cycle regulation. The multipeaked nature of the patterns suggested to us that generation times in these populations might be "quantized"

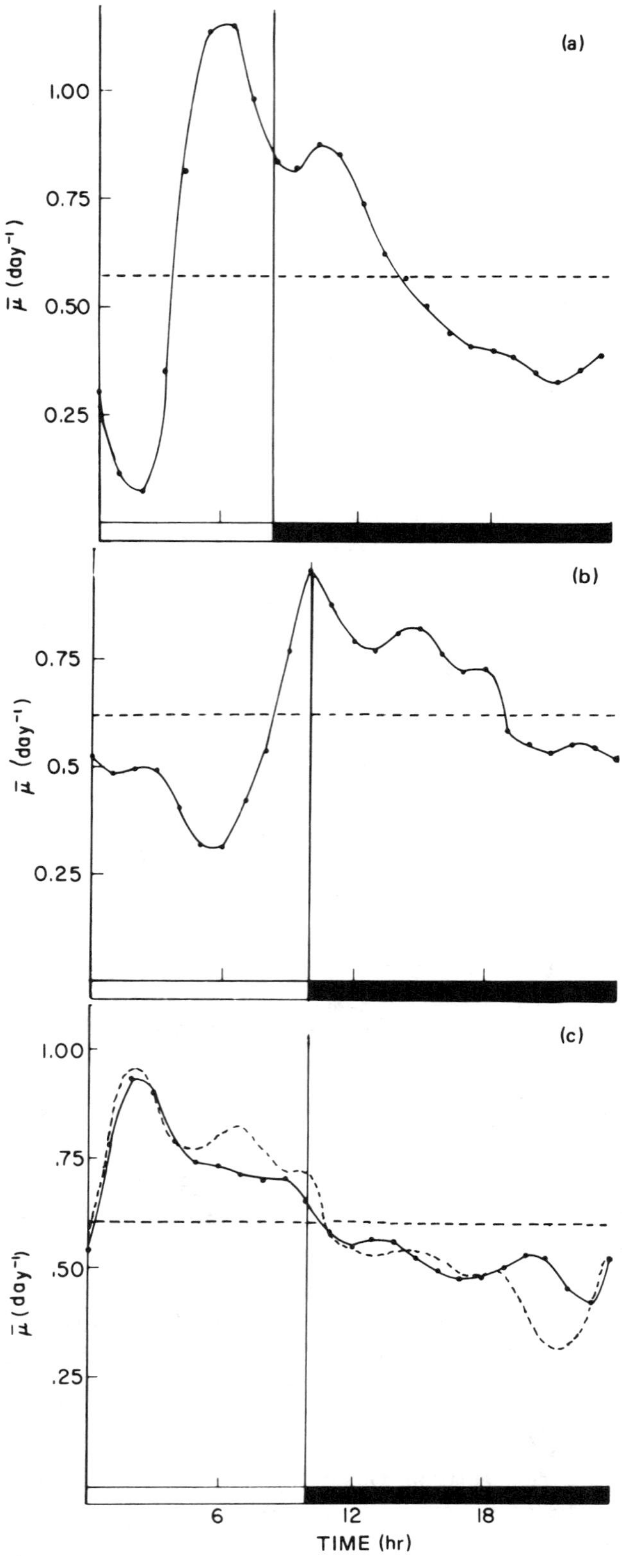
(a)
1.00
0.75
0.50
0.25
$\bar{\mu}$ (day$^{-1}$)
(b)
0.75
0.5
0.25
$\bar{\mu}$ (day$^{-1}$)
(c)
1.00
.75
.50
.25
$\bar{\mu}$ (day$^{-1}$)
6
12
18
TIME (hr)

and regulated in a manner similar to that proposed by Klevecz (1976) for mammalian cell lines. Using the Klevecz model as a basis, and superimposing a light requirement on the "$G_q$ loop," one can indeed simulate population division patterns similar to those observed experimentally with diatom populations (Chisholm et al., 1980). We have not been successful, however, in demonstrating directly that the distribution of generation times in these populations are truly quantized. Results of experiments using time-lapse video analysis to measure generation times of developing populations have yielded suggestive, but certainly not conclusive, evidence of quantization in two species of diatoms (Fig. 5). Since the concordance of our model with the data (Chisholm et al., 1980) does not at all imply uniqueness, we consider it still a hypothesis. We *can* conclude, however, that the cell division patterns of the diatom species examined thus far are not characteristic of species in which cell cycles are coupled to a circadian clock. We can also conclude (see below) that in most cases the diatom division patterns cannot be explained on the basis of the simple block-point theory of Spudich and Sager (1980).

Before completing this discussion it should be recognized that some species of diatoms can be synchronized using classic methods of induction synchrony, such as extended dark periods (Lewin et al., 1966), short (e.g., LD: 5,7) photocycles (Darley et al., 1976), and 24 hr photocycles with short photoperiods of high light intensity (Paasche, 1968; Eppley et al., 1967; Chisholm et al., 1980). Also, since diatoms have a silicified cell wall that is replicated only after mitosis and cytokinesis (Lewin et al., 1966), silicon starvation can be used to induce synchrony (Lewin et al., 1966; Busby and Lewin, 1967). In the absence of silicic acid in the medium, the cell cycle is blocked between cytokinesis and division (Lewin et al., 1966; Sullivan, 1977). When silicon is added to the medium, the cells begin synchronous development, but this synchrony decays completely in the subsequent cell cycle; i.e., exponential growth is resumed within a few hours after the synchronous cell wall formation (Busby and Lewin, 1967). This is consistent with our view that diatom populations express an unusual degree of nongenetic variability in generation times when grown in uniform environments (Chisholm et al., 1980).

## DIVISION PATTERNS OBSERVED IN SITU

What limited data exist on the division patterns expressed by phytoplankton species in nature confirm the observations from laboratory cultures; for detailed review see Chisholm (1981). Dinoflagellates divide almost exclusively during the late night-early morning hours (Swift and Durbin, 1972; Elbrachter, 1973; Heller, 1977; Weiler and Chisholm, 1976; Weiler and Eppley, 1979; Pollingher and Zemel, 1981), and diatoms exhibit a wide range of patterns, depending on the species and environmental conditions (Smayda, 1975; Lewin and Rao, 1975; Chisholm et al., 1978; Williamson, 1980). The only other group of

**FIGURE 3** Average specific division rate as a function of time for *T. weissflogii* populations grown in cyclostats: (a) on LD: 8,16 at 20°C, (b) on LD: 10,14 at 15°C, and (c) on LD: 10,14 at 20°C under phosphorus-limited conditions. The $\bar{\mu}(t)$ patterns in each case are an average of at least 5 days of data. The average population growth rates were similar in all three cultures and are indicated by the horizontal dashed line. The function described by the dotted line in (c) is the $\bar{\mu}$ (t) pattern from (b) shifted by 8 hr for comparison. (Adapted from Chisholm and Costello, 1980.)

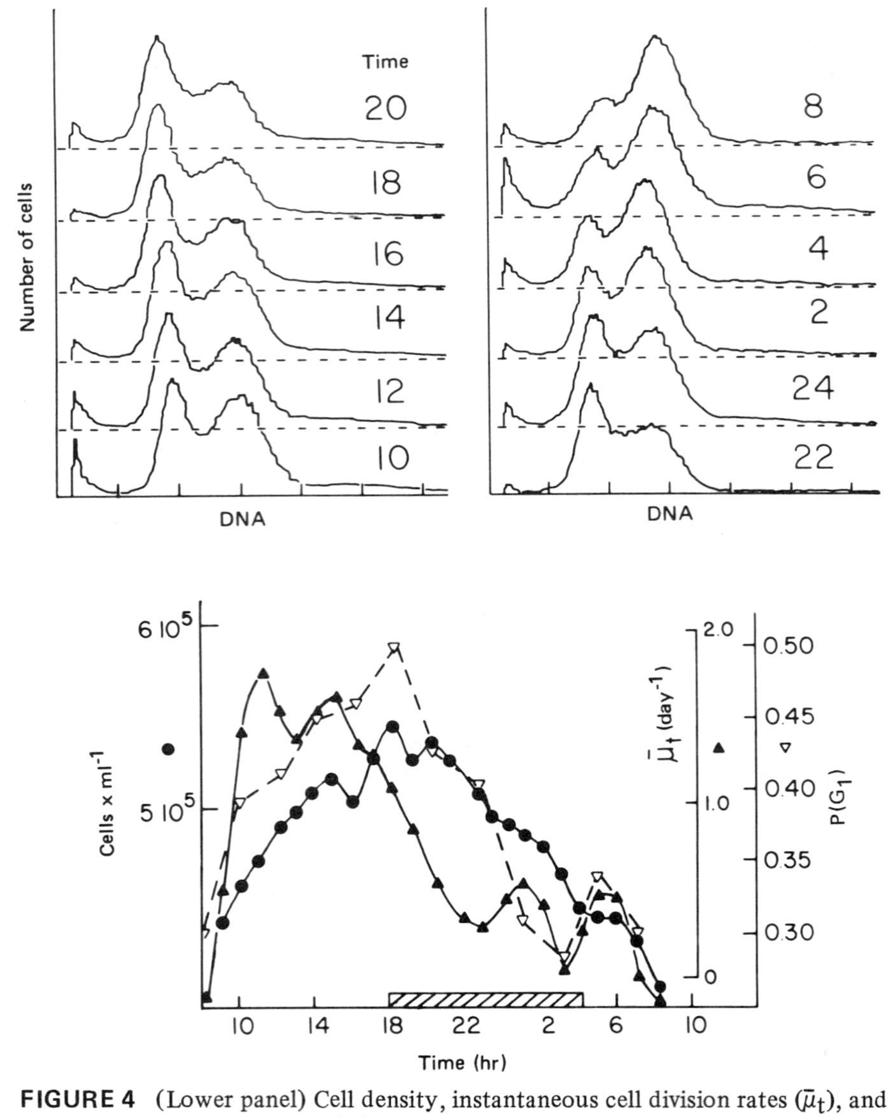


**FIGURE 4** (Lower panel) Cell density, instantaneous cell division rates ($\bar{\mu}_t$), and proportion of cells in $G_1$ phase [$P(G_1)$] over a 24 hr period for a nutrient-replete culture of *T. weissflogii* grown in a cyclostat on LD:14, 10. The average population growth rate was 0.98 per day. Values for $\bar{\mu}_t$ were obtained by averaging data from 5 consecutive days of hourly cell density samples (as described in Chisholm and Costello, 1980). (Upper panel) Relative amounts of DNA in cell populations taken from the culture shown in the lower panel at the times indicated. Hours 18-4 correspond to the dark period. The proportion of cells in $G_1$ at each time point was calculated from the DNA distributions as described in Olson et al. (1983).

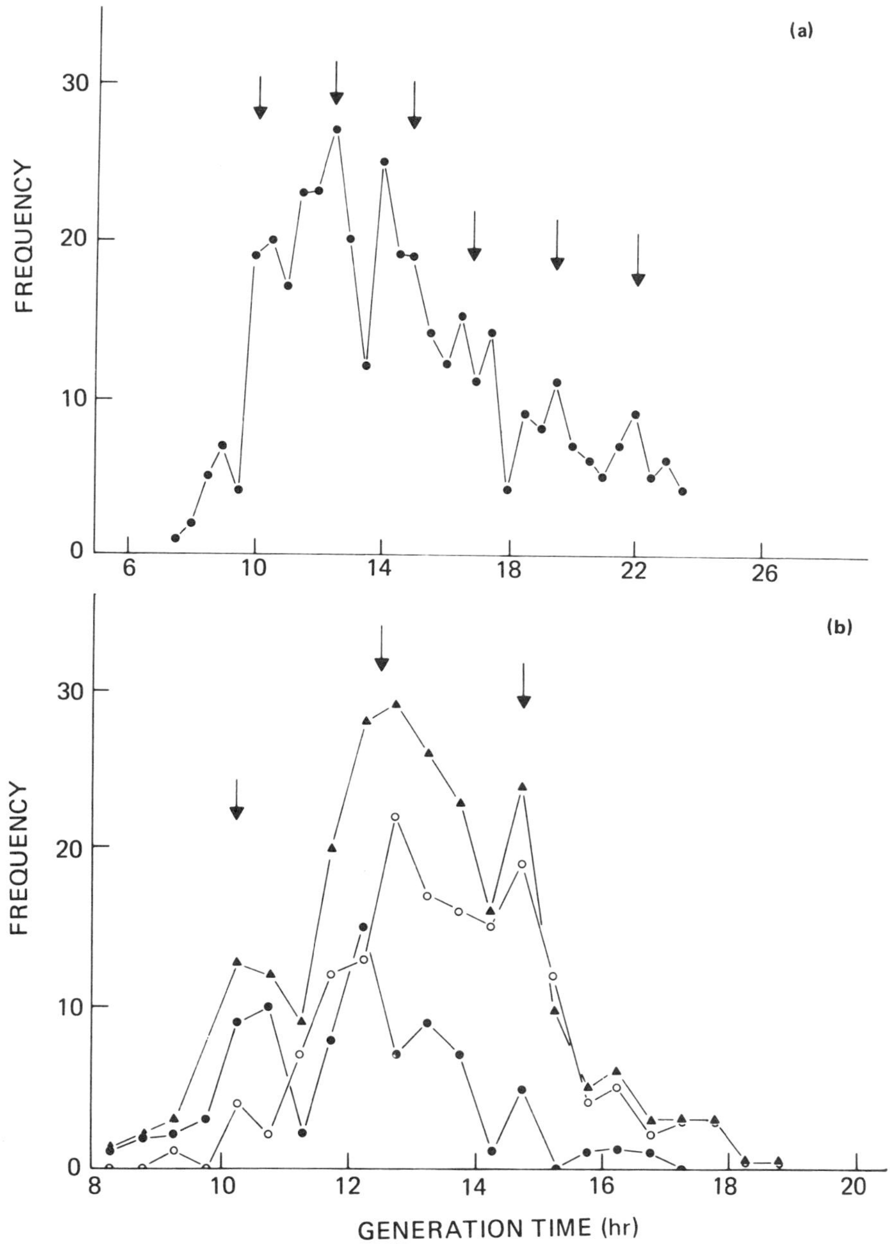


**FIGURE 5** Frequency distributions of diatom generation times, measured using a time-lapse video recording system. Light intensity 100 $\mu$Ein/m$^2$ per sec. (a) *Biddulphia aurita* grown at 18°C in a Palmer-Maloney cell (0.1 ml) filled with seawater medium and sealed with Vasoline. Data are pooled from several experiments. (b) *T. weissflogii* grown at 20°C on 1% agar made with enriched seawater. A thin layer of agar was laid down in a plastic culture dish which was sealed with Parafilm after inoculation. Data are from a single experiment in which two successive generations were followed. Solid circles = first generation observed, open circles = second generation observed, and triangles are the sum of the two. Arrows indicate hypothesized quantized generation times (see text).

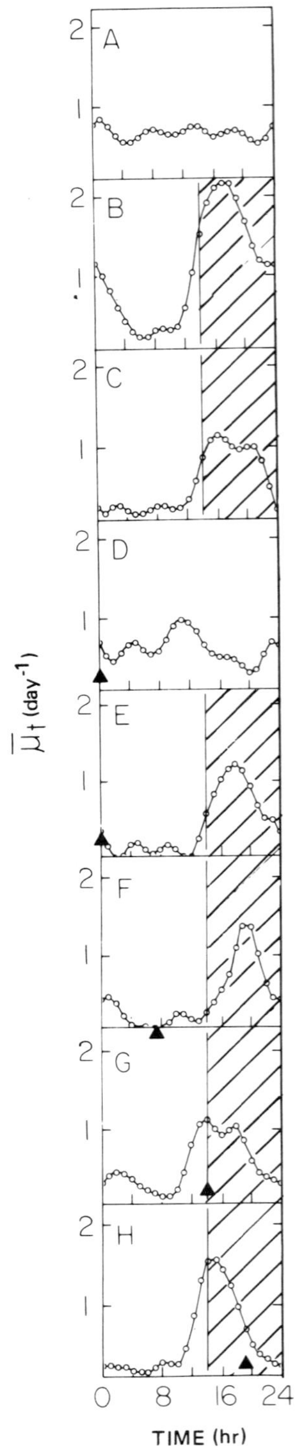


Figure 6

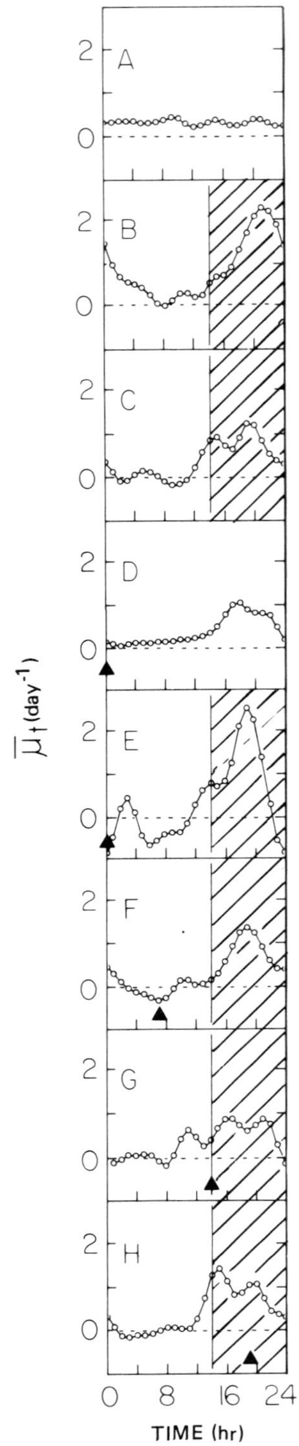


Figure 7

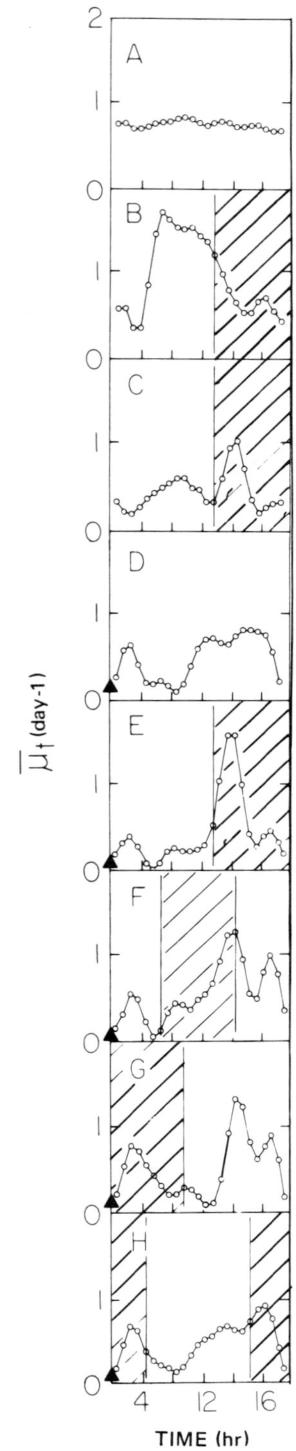


Figure 8

species that has been examined in this regard are members of the green algae in which cell division has been observed to be restricted to the dark period (Stayley, 1971; Simmer and Sodomkova, 1968), as is true of these species in laboratory cultures.

## CELL CYCLE PHASING BY NUTRIENT PULSES

The growth of phytoplankton cells in their natural environment is often (if not usually) limited by the supply of nitrogen (in the sea) or phosphorus (in freshwater). There is growing evidence that a major source of supply of these nutrients to the cells on a microscale is rapid remineralization by heterotrophic organisms, particularly in oligotrophic waters. Moreover, it is argued that the cells may experience this nutrient supply as micropatches (effectively pulses) of high concentration that are transported rapidly by the cells over a short interval relative to their generation times (McCarthy and Goldman, 1979; McCarthy, 1981; Lehman and Scavia, 1982). Cell cycle dependency of nutrient assimilation is reflected by the observation that pulses of limiting nutrients are known to synchronize populations of many cell types (Yoder et al., 1982; Hansche, 1969; Franke, 1970; Thomas et al., 1980). Since the individual phytoplankton cells can be subjected to periodic supplies of energy (light) and essential nutrients, and since the periodicity in nutrient supply may be irregular relative to the photocycle, it is of interest to us to understand to what degree these two stimuli compete in regulating the progression of the cell cycle.

To probe this question, we examined the division patterns of three taxonomically diverse phytoplankton species grown on 24 hr photocycles with periodic supplies of nitrogen (Olson and Chisholm, 1984). In these experiments nitrogen was supplied to cyclostat cultures (Chisholm et al., 1975) as an $NH_4^+$ pulse, once during each 24 hr period. The nitrogen supply was adjusted to be growth limiting and was introduced at various times during the light or the dark period. The responsiveness of the three species, *Hymenomonas carteri* (a coccolithophore), *A. carteri* (a dinoflagellate), and *T. weissflogii* (a diatom) to the competing periodic stimuli revealed distinct differences in terms of cell cycle phasing. In the coccolithophore, division was tightly phased by the light-dark cycle when nitrogen was supplied continuously, and only weakly phased by nitrogen pulses in continuous light. Not surprisingly, when growth conditions included both photocycles and nitrogen pulses the photocycle completely dominated as the phasing stimulus (Fig. 6).

---

**FIGURE 6** Average specific division rate as a function of time for *H. carteri* grown in a cyclostat at 20°C on the indicated LD cycle (where darkness is indicated by hatched areas) and with the indicated ammonium supply regime. (A) Control: continuous light and continuous supply of limiting ammonium. (B) LD: 14,10; nutrient replete. (C) LD: 14,10; continuous supply of limiting ammonium. (D) Continuous light; pulse of limiting ammonium at time indicated by arrow (From Olson and Chisholm, 1984).

**FIGURE 7** Average specific division rate as a function of time for *A. carteri.* Details as in Fig. 6.

**FIGURE 8** Average specific division rate as a function of time for *T. weissflogii.* Details as in Fig. 6.

In the dinoflagellate, division was strongly phased by both the photocycle and the nitrogen pulse. The photocycle alone tightly aligned division with the dark period, and the nitrogen pulse alone produced a regular division burst about 14 hr after the introduction of the pulse. When both periodic stimuli are supplied simultaneously, however, the photocycle almost totally dominated the nitrogen pulse as the phasing agent regardless of the phase angle between the two; i.e., the division burst was restricted to the dark period (Fig. 7).

Finally, in the diatom, typical multipeaked division rhythmicity was induced by both the photocycle and the nitrogen pulse when supplied independently, but when this species was grown with nitrogen pulse and photocycle imposed simultaneously, the nitrogen pulse completely dominated as the forcing stimulus. The main division burst always occurred 18 hr after the nitrogen pulse regardless of whether the pulse was given in continuous light or in any phase of the photocycle (Fig. 8).

The different responses of these three species to a periodic supply of nitrogen superimposed on the photocycle must reflect differences in the regulation of cell cycle progression. In other words, the phase angle of cell cycle alignment to the two stimuli should reflect an optimal "strategy" for the cells in "processing" these two growth-limiting substances. In an attempt to arrive at a more mechanistic explanation for the observed patterns the kinetics of nitrogen assimilation were examined in the three species by monitoring the changes in the intracellular pool sizes of $NH_4^+$ and free amino acids after the introduction of the nitrogen pulse (Wheeler et al., 1984). In all three species, the rate of $NH_4^+$ uptake and assimilation into free amino acids and the rate of disappearance of the free amino acid pool were essentially independent of the time in the photocycle at which the nitrogen pulse was introduced. The overall rate of metabolism of the nitrogen pulse in these three species, however, was inversely related to the relative strength of the nitrogen pulse in overriding photocycle entrainment. Large pools of free amino acids accumulated in the diatom (which has a large vacuole) and persisted for a significant portion of the photocycle, whereas in the other two species, assimilation of the $NH_4^+$ into free amino acids and the disappearance of this pool was much faster.

Although not conclusive, these results suggest that the assimilation of the limiting nutrient is the rate-limiting step for cell cycle progression in the diatom, whereas for the other two species the fulfillment of an energy requirement is controlling the progression. This is consistent with the observation that diatom cell cycles are not strictly phased by light-dark cycles except under very short photoperiods (Chisholm et al., 1980).

## MODELS OF CELL CYCLE ENTRAINMENT BY PHOTOCYCLES: CLOCKS VERSUS BLOCK POINTS

In order to explain the coupling between a periodic environmental forcing stimulus (the photocycle) and the growth patterns of algal cell populations, one has to hypothesize some kind of interaction between the forcing variable and the cell cycle. One of the most widespread theories assumes the existence of an internal cellular clock with a period close to 24 hr to which the cell cycle is coupled through gated division (Edmunds and Adams, 1981). A change in the environmental parameter (e.g., light-dark transition) resets the clock, aligning it with the forcing period, and as a consequence aligning the timing of mitosis with the same forcing period. Recently, Sweeney (1982) has shown that the timing of cell cycle stages other than mitosis also appear to be under circadian regulation. This view of cell cycle regulation involves coupled oscillators in the causal sequence:

Forcing stimulus → Clock → Cell cycle

An extensive vocabulary (Ehret and Wille, 1970) and elaborate conceptual models (Edmunds and Adams, 1981) have been developed to characterize the clock and its coupling, but no quantitative model has emerged to account for the growth dynamics of algal populations regulated by such a mechanism. The complexity of the conceptual models for the coupling mechanism is a strong deterrent for the formulation of useful quantitative models because of the multiplicity of parameters required.

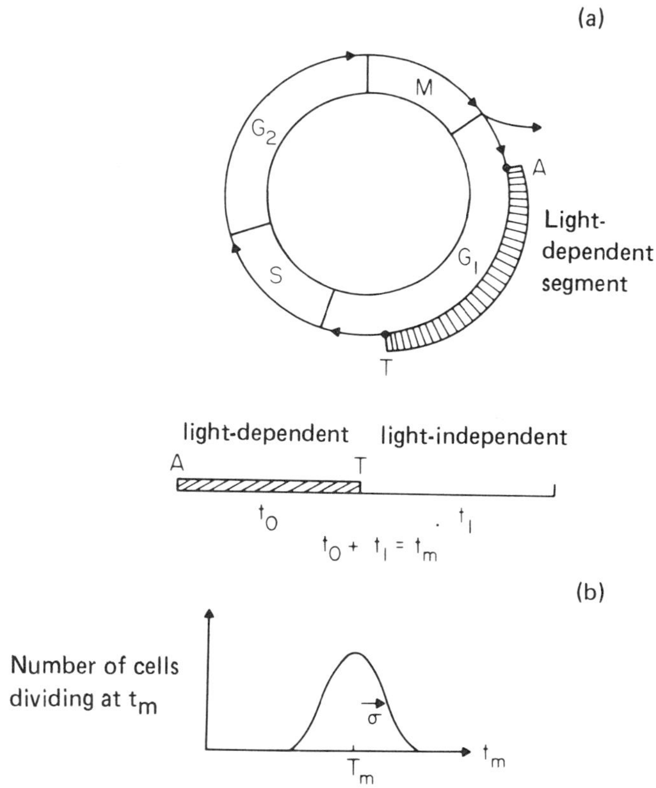


**FIGURE 9** The cell cycle with a light-dependent segment according to Spudich and Sager (1980). The cell cycle is viewed as a succession of four segments; $G_1$, S, $G_2$, and M; S corresponds to DNA synthesis and M to mitosis. Spudich and Sager postulated the existence of a light-dependent segment in $G_1$ from A to T (see text). (b) Stochastic model for entrained populations adapted form the Spudich and Sager model. Individual cell cycles are composed of two segments: one which is light dependent of length $t_0$ (the arrest point A has been placed at the beginning of $G_1$ for simplicity), one which is light independent of length $t_1$. The total length of the cycle is $t_m$ (= $t_0 + t_1$). $t_0$ and $t_1$ are random variables normally distributed among the population [$N(T_0\sigma_0)$], implying a normal distribution for $t_m$: $N(T_m, \sigma)$. For simplicity the variability has been concentrated in the light-independent segment ($\sigma_0 = 0$, $t_0 = T_0$); thus $\sigma = \sigma_1$ and $T_m = T_0 + T_1$. $T_0$ is called the mean light requirement and $T_m$ the mean generation time. The time evolution of such a stochastic population has been numerically simulated with a technique similar to the age transfer method used by Bronk et al. (1968).

The periodic nature of the cell cycle makes it analogous to a clock or, more appropriately, to an hourglass (Winfree, 1980). This invites the simple hypothesis that the cell cycle is directly entrained by the periodic forcing stimulus, reducing the causal sequence to:

Forcing stimulus → Cell cycle

Spudich and Sager (1980) have recently provided evidence that the regulation of the cell cycle in *Chlamydomonas reinhardtii* may be described by this simple model. By measuring the progression of *C. reinhardtii* through its cell cycle, both by its DNA content and by its light-scattering (cell volume) characteristics, they demonstrated the existence of a light-dependent segment in the cell cycle. This segment is bounded by two points: the primary arrest point A, where the cell cycle is blocked in the absence of light, and a transition point T, where a cell is committed to progress through the cycle to division (Fig. 9). After overcoming T, a cell does not require any more light energy to divide. The link between this segment of the cycle and photosynthesis was further demonstrated by the use of the photosystem II inhibitor DCMU, which mimicked the effect of darkness on division. One could easily extend this hypothesis and postulate the existence of other block points throughout the cell cycle that are overcome by the fulfillment of specific requirements, such as proteins and RNA (Alberghina and Sturani, 1981), cell size (Fantes, 1977), and mitogen levels (Cooper, 1982).

The relative simplicity of the block-point model, and its conduciveness to hypothesis testing, motivated us to build from it a quantitative model to stimulate phytoplankton population growth patterns under a variety of photocycle regimes. Our goal was to compare the results of simple "experiments" obtained using the block-point model with experimental results we have obtained with various species, and with results that would be predicted according to the clock-coupling hypothesis. We have not attempted to fit precisely experimental data by adjusting parameters, but have chosen parameters that are "realistic" based on our experience. The model is simple in construction, and is mechanistically similar to that described in Chisholm et al. (1980). In the model the cell cycle of each cell is composed of two segments (Fig. 9) similar to those described by Spudich and Sager (1980). Progression through the first segment (bounded by A and T) is light dependent, but once the cell has fulfilled its prescribed light requirement ($T_0$) and arrived at the transition point T, progression through the rest of the cycle is light independent. A stochastic component is inserted in the light-independent segment to allow for generation time variability. The precise form of this internal variability is not crucial to the present discussion. We have used a simple gaussian distribution, but recognize that other forms for the variability could be justified (Smith and Martin, 1973; Klevecz, 1976; Cooper, 1982). The evolution of the population itself is stimulated by following the progression of individual cohorts of cells along the cell cycle. This progression is regulated by the amount of light cells have received since the last round of division and, to a lesser degree, by the stochastic component (see Fig. 9 for details).

In deciding what "experiments" would be most useful to conduct with the model, we used the following conceptual framework. A typical entrainment experiment consists of submitting an algal population to a light regime characterized by three parameters: the photocycle length P, the photoperiod length L, and the light intensity I. Quite often two or more different photocycle regimes are applied in a row. For each regime one can distinguish two phases in the population response. (1) an initial transient response, which is dependent upon the initial conditions (and thus the previous regime) during which

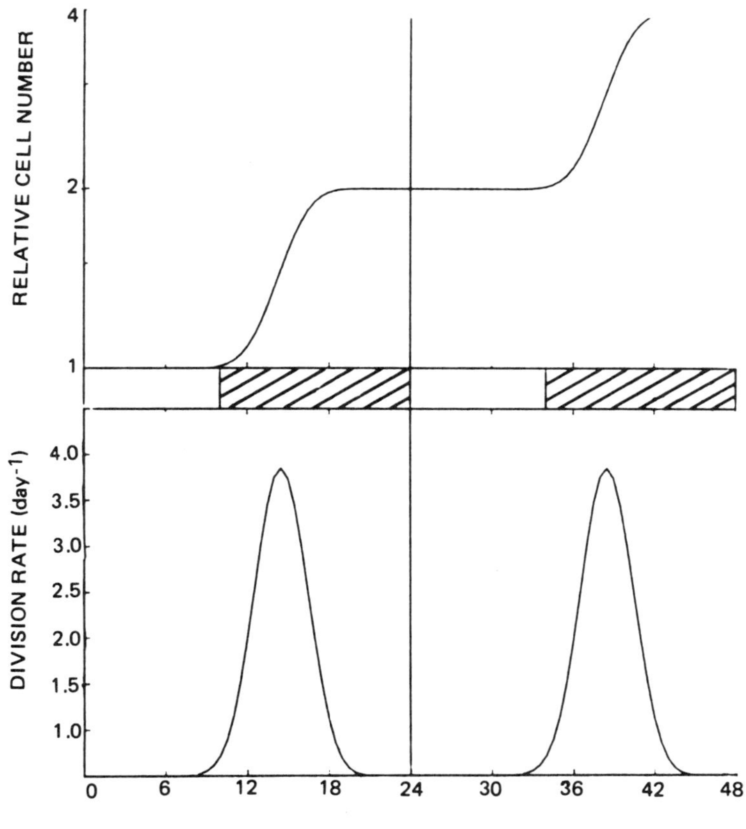


**FIGURE 10** The mechanism of entrainment according to the block-point hypothesis in a population growing at a rate near one doubling per day. Model parameters were $T_0 = 7$ hr, $T_1 = 8$ hr, $T_m = 15$ hr, and $\sigma = 2$ hr. The population was entrained to a LD: 12,12 photocycle for 10 days and its average growth rate is 0.68 per day. Cells from the generation born on the first day are synchronized by the imposition of the dark block and are released when the light is on again. They divide at the beginning of the next dark period.

the population adapts to the new conditions (e.g., Edmunds, 1966; Figs. 1 and 2); and (2) a long-term cyclic steady-state response, which is independent of the initial conditions. Of particular interest is the behavior of the population in the regime where P = L (continuous illumination), which is the "free-running" condition often used as diagnostic for clock control.

Unfortunately, in much of the literature describing population growth patterns on light-dark cycles, results are usually expressed as cell number as a function of time in batch cultures. Since an exponential trend is always present, precise frequency analyses are impossible; thus, comparisons with the model results are sometimes difficult. To ease this comparison, we have expressed the model results in two ways: the population division rate $\mu(t)$ as a function of time and the corresponding increase in cell number. The former is directly comparable with cyclostat data (neglecting mortality), and the latter with results from batch culture studies.

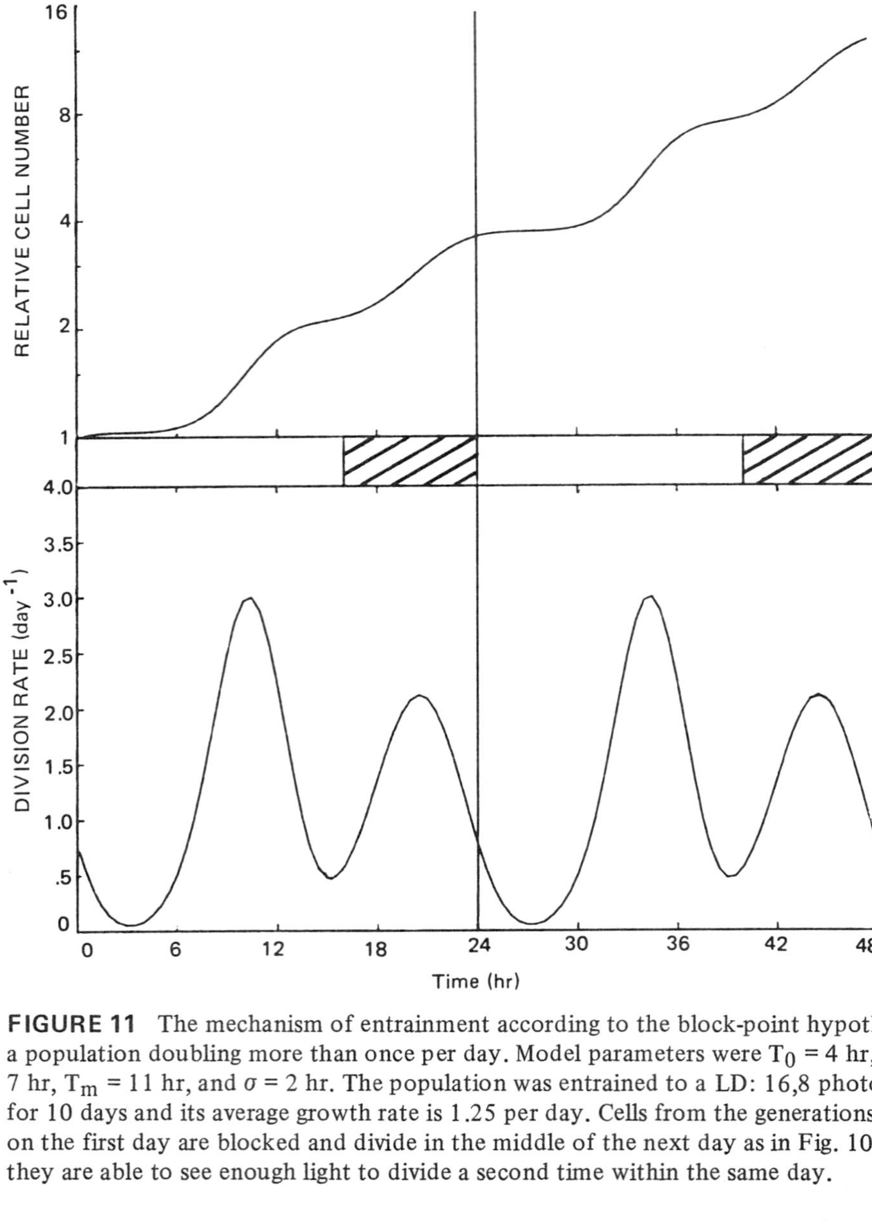


**FIGURE 11** The mechanism of entrainment according to the block-point hypothesis in a population doubling more than once per day. Model parameters were $T_0$ = 4 hr, $T_1$ = 7 hr, $T_m$ = 11 hr, and $\sigma$ = 2 hr. The population was entrained to a LD: 16,8 photocycle for 10 days and its average growth rate is 1.25 per day. Cells from the generations born on the first day are blocked and divide in the middle of the next day as in Fig. 10, but they are able to see enough light to divide a second time within the same day.

Let us consider first the steady-state population response to growth on 24 hr photocycles. We can see from the model simulation in Fig. 10 that the block-point model generates population growth patterns that are quite similar to "typical" experimental data from *Euglena* (Edmunds, 1965a) and various marine phytoplankton species (see Fig. 1). Cells from the generation born on the first day are synchronized by the imposition of the dark block and are released at the onset of light. They divide again at the beginning of the next dark period. These same results can easily be interpreted in the context of clock coupling, where it is hypothesized that at one of the transitions (presumably the dark-light transition), the clock is reset. Thus, if each cell divides at most once during

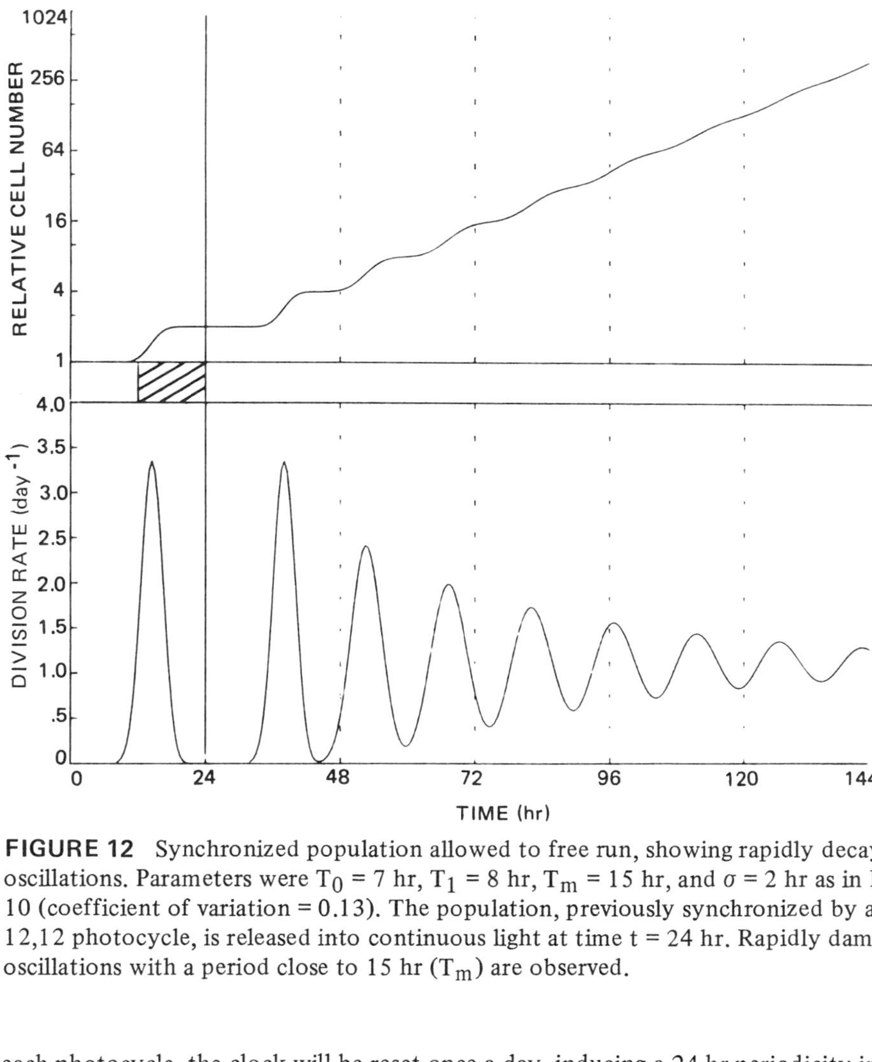


**FIGURE 12** Synchronized population allowed to free run, showing rapidly decaying oscillations. Parameters were $T_0 = 7$ hr, $T_1 = 8$ hr, $T_m = 15$ hr, and $\sigma = 2$ hr as in Fig. 10 (coefficient of variation = 0.13). The population, previously synchronized by a LD: 12,12 photocycle, is released into continuous light at time t = 24 hr. Rapidly damped oscillations with a period close to 15 hr ($T_m$) are observed.

each photocycle, the clock will be reset once a day, inducing a 24 hr periodicity in $\mu$. The maximum division rate should always occur at a fixed phase relative to the resetting signal.

Let us now extend the photoperiod so that the average population growth rate is greater than one doubling per day (Fig. 11). The model predicts a population response with a 24 hr period, but with two division bursts. In this case, the cell cycle is short enough and the light period is long enough that some cells are able to see enough light to divide a second time within the same 24 hr period, giving rise to a second division peak (Fig. 11). This multipeaked pattern is not unlike some we have observed for the diatom *T. weissflogii* and other fast-growing marine phytoplankton species (Chisholm and Costello, 1980). Moreover, Edmunds and Funch (1969) imply that such patterns might also be expressed by *Euglena* grown under long photoperiods of high light intensity. These types of experimental results are difficult to interpret with a clock-coupled cell cycle model because it is difficult to conceptualize how the cell division gating occurs. In theory (Ehret and Wille, 1970), the clock should be completely uncoupled from the cell cycle in this fast growth mode, and aperiodic exponential growth should ensue.

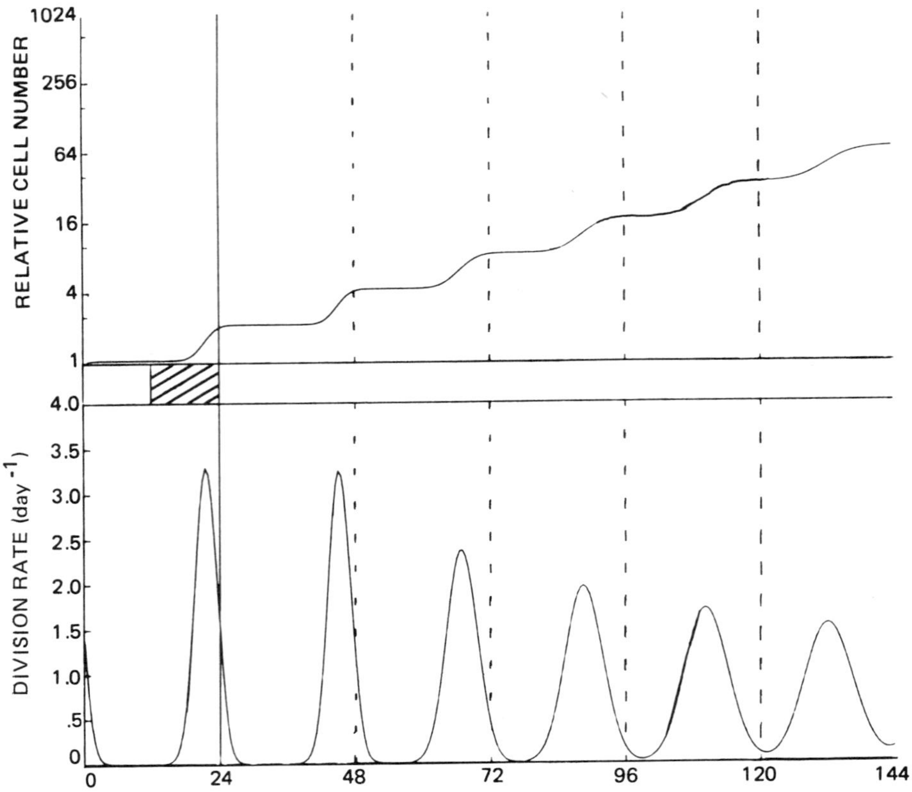


**FIGURE 13** Synchronized population allowed to free run showing slowly decay oscillations. Model parameters as in Fig. 12 except for $T_1 = 15$ hr, $T_m = 22$ hr (coefficient of variation = 0.09). The oscillations have a period close to 22 hr ($T_m$).

Consider now what the block-point model would predict for a population of cells synchronized as in Fig. 10 and released into continuous light (Fig. 12). Our results concur with the general conclusions of Bronk et al. (1968). The population growth rate exhibits decaying oscillations that have a period close to the mean generation time $T_m$ that would be expressed by the population in steady-state exponential growth under continuous light. $T_m$, in this example, is 15 hr. The loss of synchrony can be characterized by a damping time $T_d$ given by:

$$\frac{T_d}{T_m} = \frac{1}{2\pi^2}\left(\frac{T_m}{\sigma}\right)^2 \qquad (1)$$

where $T_m$ is the mean generation time defined above and $\sigma$ is the standard deviation of the generation times (Bronk et al., 1968). The more tightly the generation times are distributed, the longer the transient periodicity will last. For example, if the coefficient of variation (given by $\sigma/T_m$) is 13%, as in Fig. 12, $T_d/T_m$ is equal to 3, but if the CV is lowered to 9%, $T_d/T_m$ is doubled (Fig. 13). Thus, both the period length of the free-running oscillation and the time it takes for it to damp out are strongly dependent on the

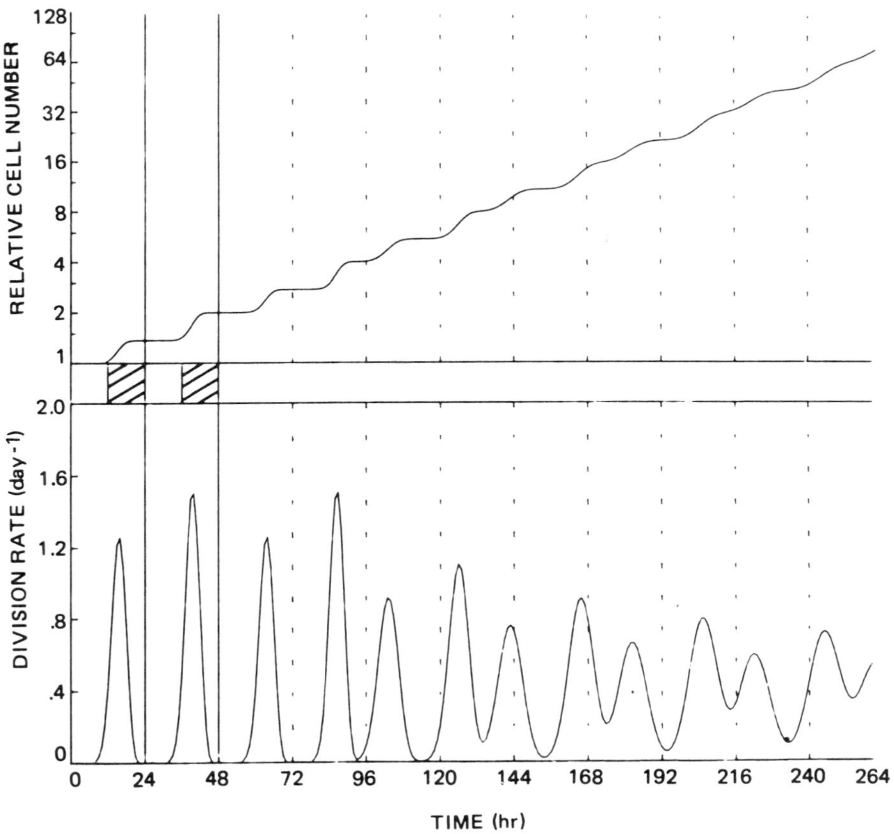


**FIGURE 14** Phase population allowed to free run. Model parameters were $T_0$ = 10 hr, $T_1$ = 30 hr, $T_m$ = 40 hr, and $\sigma$ = 3 hr. The population was previously phased by a LD:12, 12 photocycle and released into continuous light at time t = 48 hr. Damped oscillations of period 40 hr (= $T_m$) are observed. These oscillations have two components, 24 hr out of phase.

values of $T_m$ and the coefficient of variation of the generation times. For a species with a tight generation time distribution and a $T_m$ value close to 24 hr, the free-running patterns generated by the block-point model can appear as circadian oscillations (Fig. 13). We should point out here that population division rhythmicities under clock control are expected to damp out in free-running conditions, presumably because the free-running period lengths of the clocks of individual cells are not identical. The characteristics of decay of the rhythmicity, however, should differ from those of populations controlled by cell cycle block points. In theory, one could quantify these differences if the degree of variability in individual cellular clocks were measurable.

The situation gets more complicated when we consider entrained populations exhibiting phased cell division in which only a portion of the cells divide during each 24 hr period. This growth pattern can be simulated quite readily using the block-point model by choosing appropriate parameters (Fig. 14). When such population (with $T_m$ = 40 hr)

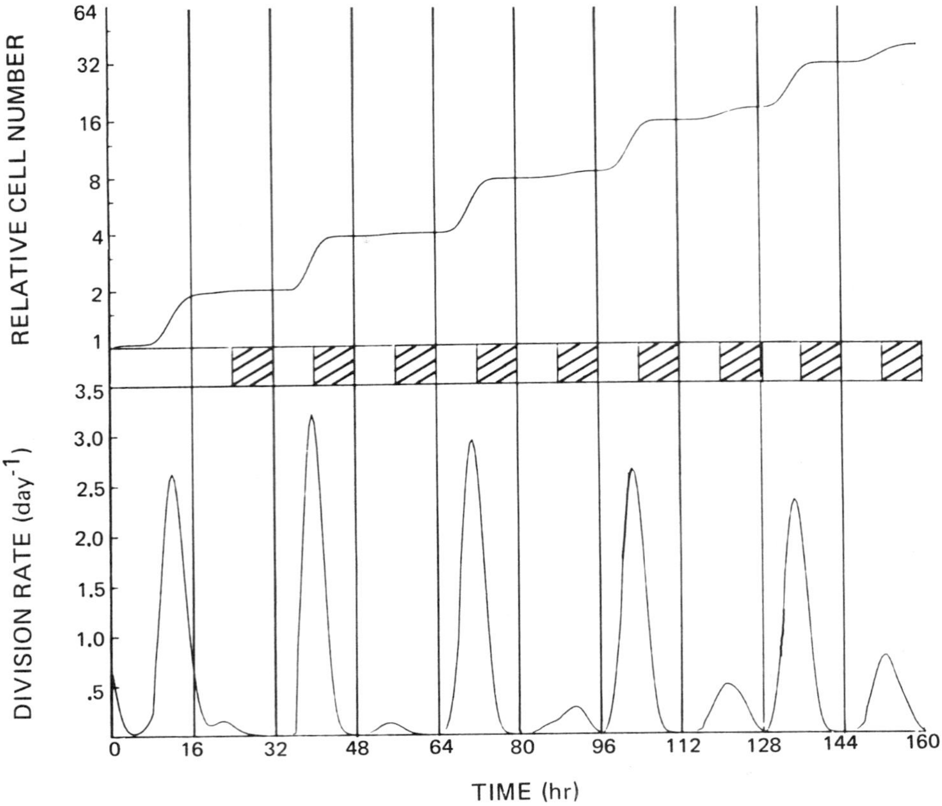


**FIGURE 15** Transients occurring when the light period length L is shorter than the light requirement $T_0$. Model parameters were $T_0 = 12$ hr, $T_1 = 4$ hr, $T_m = 16$ hr, and $\sigma = 2$ hr. An initially synchronized population was submitted to a LD photocycle at time t = 16 hr. An additional transient with period 32 hr (two photocycles) is observed, while the long-term behavior exhibits a 16 hr (one photocycle) periodicity.

is released into continuous light, we find that the overall free-running transient pattern has a 40 hr period as our theory would predict, but it contains two components dephased by 24 hr, which could easily be mistaken for a 24 hr circadian rhythm for several cycles.

Finally, consider an example where the light requirement is longer than one light period, but shorter than two light periods. In this case, each cell will need to be exposed to light during two consecutive days before fulfilling its light requirement. Here we find the period length of the major division burst during the initial transient to be twice that of the photocycle, or 32 hr in our example (Fig. 15). More generally, if the light requirement ($T_0$) is larger than (n - 1)L and less than nL, where L is the length of the light period and n is an integer, the transient will have a period of mean lenght nP. This same general response is illustrated by data from *Euglena* where, whenever the photoperiod is shorter than the light requirement (estimated as 10-14 hr; Chisholm, 1981), the period length of the division rhythm is always an integer multiple of the photocycle length (Table 1).

This modeling exercise serves to demonstrate that some of the qualitative properties of population growth that have been attributed to circadian regulation can be reproduced

**TABLE 1** Relationship Between the Period of the Entrained Division Rhythm and the Period of the Entraining Photocycle in *Euglena*[a]

| Light intensity (lux) | Photocycle length = P (hours) | Photoperiod length = L (hours) | Division rate period = $T\mu$ (hours) | $n = \frac{T\mu}{P}$ | Light received before dividing = n × L (hours) | Reference |
|---|---|---|---|---|---|---|
| 8000 | 24 | 14 | 24 | 1.0 | 14 | Edmunds (1965) |
| 8000 | 20 | 10 | 20 | 1.0 | 10 | Edmunds and Funch (1969) |
| 8000 | 16 | 8 | 33 | 2.1 | 16 | Edmunds and Funch (1969) |
| 8000 | 10 | 5 | 32 | 3.2 | 15 | Edmunds and Funch (1969) |
| 7500 | 6 | 3 | 30 | 5.0 | 15 | Edmunds et al. (1982) |

[a]Note than n is always close to or equal to an integer and that $n \times L \geq 10$ hr, the estimated required light exposure for *Euglena* under these conditions (see text).

using a block-point model. We do not assert that cell cycles are not clock coupled, but rather that long-term experiments with precise period length measurements are in some circumstances necessary to distinguish between these alternative hypotheses for cell cycle control. Moreover, the block-point model is useful for generating hypotheses that can be tested easily experimentally. It is likely that in its present form it will prove inadequate to describe precisely the growth patterns of many cell types under various conditions, but we see this approach as a fruitful methodology for probing the mechanisms underlying the observed division patterns. In the end, it is probable that a combination of the block-point model and the clock-controlled model will be required to adequately describe the growth patterns of many species (Bruce and Bruce, 1982).

## ADAPTIVE SIGNIFICANCE OF CELL CYCLE PHASING

In the absence of a precise understanding of the mechanisms that control cell cycle progression in the various groups of phytoplankton, it is difficult to postulate why cell cycle phasing might have (or have not) evolved in the different groups. If the timing of mitosis is regulated by a circadian clock, we might hypothesize that there is a selective advantage for each cell to divide at a certain time of day or for genotypically identical cells to divide simultaneously. It is equally plausible, however, that the timing of division per se has no adaptive significance and that natural selection has operated solely upon the coordination and timing of events leading up to division such that phased division is simply the end product of these precisely timed events (Sweeney, 1982). Even so, the clock coupled control mechanism implies that there is an optimal timing of events relative to the photocycle, the selective advantage of which is so strong that it is maintained even in the absence of the entraining stimulus.

If, on the other hand, the division patterns we observe are a result of a light-requiring block point in the cell cycle, any rhythmicity or phasing observed is simply a reflection of the periodic energy supply; i.e., it is a driven or forced rhythm, totally dependent on exogenous periodicity for its maintenance. In this case, the phase angle between the photocycle and the burst in cell division or the absence of division phasing can tell us something about the energetics of the cell cycle, but ecological interpretations can be misleading.

Recognizing the dangers involved in any discussion regarding "the adaptive significance of . . . ," we will attempt here to review briefly some hypotheses that have been put forward to explain photocycle-induced phytoplankton division patterns. The hypotheses fall into two categories: (1) those that consider the energetics of cell cycle progression and optimal alignment of biosynthetic (growth) and reproductive (division) processes, and (2) those that consider selective pressures imposed by the planktonic community (interspecific competition for resources, and mortality from zooplankton grazing) as important in dictating the timing of division. Let us consider the latter case first because it is rather easily dismissed for lack of supporting evidence.

The apparent asymmetry between the assimilatory and reproductive processes of phytoplankton and zooplankton has been noted for years (Wimpenny, 1938; Fleming, 1939). Grazing by zooplankton on phytoplankton ("assimilation") is generally believed to be most intense at night either because of vertical migration patterns or periodicity in actual grazing rates of individual animals (Hart, 1977; Singh, 1972; Arashkevich, 1978; Enright and Honeggar, 1977). *In general,* phytoplankton display a tendency to divide ("reproduce") at night and assimilate during the day. We can make reasonable hypotheses regarding the selective advantage of nocturnal grazing by zooplankton including (1) maximal caloric content and cell size of the phytoplankton, (2) maximal concentration of the phytoplankton crop, and (3) reduced vulnerability to visual predators while in the surface waters. Symmetrical arguments regarding the advantage of nocturnal division of phytoplankton, however, are difficult because dividing (large) cells are often selectively grazed upon (Richman and Rogers, 1969). Moreover, one could argue that there should be a selective advantage for phytoplankton cells to divide during the day (before the interval of most intensive grazing) to mazimize the probability of perpetuating the genotype.

Williams (1971) suggested that intraspecific synchrony combined with interspecific asynchrony in phytoplankton communities could allow temporal partitioning of the demand for limiting nutrients between species and thus reduce competition. This hypothesis has reappeared in several contexts (Eppley et al., 1971, Stross et al., 1973; Weiler and Chisholm, 1976) and is quite easy to support theoretically (Chisholm and Nobbs, 1976), but there is no evidence that such a mechanism is operational in nature.

Given that these synecological hypotheses (i.e., those involving selective pressures imposed by other species) regarding the adaptive significance of the timing of cell division are untenable at present, we might best focus our energies on examining the question in the context of optimal design and the autecology of individual species. Shuter (1979), following Rosen (1967), casts the framework elegantly:

> Given that the realized design of a particular biological structure is optimal in the context of natural selection, then it is also optimal in the sense that it minimizes some cost function which reflects differences in both the intrinsic cost (a measure of the energy required for production and maintenance) and the extrinsic cost (a measure of the selective advantage varying inversely with fecundity) associated with different designs. . . .

Accepting this premise, we conclude that the population growth rates expressed by unicellular algae under various environmental conditions reflect a minimization of this cost function for that condition. The various division patterns (and their degree of flexibility with changing conditions) displayed by different species of algae grown on light-dark cycles, then, must to some extent reflect "strategies" of cost minimization.

Cohen and Parnas (1976) have done a theoretical and experimental analysis of the optimal policy for the production and metabolism of storage materials in *Chlamydomonas reinhardi.* They hypothesized that storage materials are synthesized according to future requirements in algae grown on photocycles, and not simply as a result of an excess supply of energy over growth requirements during the day (Cook, 1966a). They were able to demonstrate that phototrophs produce storage materials during the day (even when energy limits growth) and divide at night because this maximizes the number of daughter cells produced per day. They also showed that cells adjust their biosynthetic and storage patterns according to the energy input. In their experiments, starch synthesis began earlier in the photocycle at low light intensities than at high light intensities, and cells subjected to darkness for 5 days began starch synthesis immediately upon illumination at rates higher than those existing prior to the extended dark exposure. These high levels of synthesis were maintained for at least 3 days, and the beginning of protein synthesis and cell division was delayed by 1 day in these cells.

Although there is a certain amount of teleology in the interpretation of these results, they do support the general tenet that the growth and division patterns of algae grown on light-dark cycles reflect an optimization of resource utilization based on past history designed to maximize long-term growth rate and/or survival of the population. If we compare the *Chlamydomonas* patterns with the division patterns of the diatom *Thalassiosira* (see Fig. 3), it is clear that the optimization strategy of these two species is distinctly different. The adaptive mechanisms dictating the predominance of daytime division in the diatoms are not at all clear at present. Nelson and Brand (1979) hypothesized that the energetic demands of silicification (which is tightly coupled to cell division) may favor daytime division in these cells, but evidence in support of this hypothesis is lacking.

We end this discussion with an anecdotal but intriguing (and testable) hypothesis regarding the selective advantage of the timing of mitosis in certain dinoflagellate species. It has been recognized for some time that dinoflagellate species are particularly sensitive to turbulence. Continuously stirred cultures generally die, or grow at reduced growth rates (White, 1976; Tuttle and Loeblich, 1975; Galleron, 1976). Recently, Pollinger and Zemel (1981) hypothesized that turbulence interferes specifically with the process of nuclear division in these species and showed that *Peridinium* cultures subjected to shaking during the premitotic and mitotic phases of the cell cycle suffered much higher mortality than those shaken during the other phases (i.e., during the light period of the photocycle). Since, on the average, turbulence is significantly reduced in lake surface waters during the night, there should be some selective advantage in nighttime division for these organisms. It is further noteworthy that the many dinoflagellate species actively migrate out of the mixed layer into even less turbulent deeper waters at night and have been observed to undergo division in the mud (Eppley et al., 1968).

## CONCLUSIONS

It should be clear from this review that the issues of cell cycle control mechanisms and the adaptive significance of division periodicity in phytoplankton populations are complex, and our information base is limited. In one sense these cells are not the organism of choice for cell cycle studies because well-developed genetic systems are lacking. They are, however, well suited for studies of population kinetics and analyses of cell lineages and the growth of single cells. Furthermore, the diversity of species with identical growth requirements but distinct growth patterns provides a unique system for comparative studies.

Based on our present knowledge, we must view the photocycle as playing a dual role in regulating cell cycle progression in these populations. By entraining the circadian clock, the photocycle can serve to temporally coordinate cellular processes that lead up to mitosis, which may or may not depend directly upon light energy. In some populations the photocycle also appears to regulate directly cell cycle progression in a manner that can be described using a simple cell cycle block-point model. Because of this dual role of the photocycle in regulating the growth of these organisms, deducing cell cycle control mechanisms from population division patterns can be quite difficult. As Bruce and Bruce (1982) have wisely pointed out, it is likely that multiple control mechanisms are operative within each species, which become relatively more or less important under different environmental conditions. From the point of view of the phytoplankton ecologist, it is important to understand which control mechanisms dictate population growth responses in a given environment and whether these mechanisms vary from species to species.

## ACKNOWLEDGMENTS

This work was supported in part by NSF grants OCE-8121270 and OCE-8118475, by funds awarded by MIT to S. W. Chisholm as Doherty Professor of Ocean Utilization, and to D. Vaulot as a Whitaker Fellow. D. Vaulot is a student in the MIT/Woods Hole Joint Program in Oceanography and is on temporary leave from CNRS, Laboratoire d'Hydrobiologie Marine, Montpellier, France.

## REFERENCES

Alberghina, L., and Sturani, E. (1981). Control of growth and of the nuclear division cycle in *Neurospora crassa. Microbiol. Rev.* 45:99-122.

Arashkevich, Y. G. (1978). Relationships between the feeding rhythm and the vertical migrations of *Cypridina sinuosa* (Ostracoda, crustaceae) in the western part of the equatorial Pacific. *Acad. Sci. USSR.* 17:466-469.

Bouligand, Y., Puiseux-Dao, S., and Soyer, M. O. (1968). Liaisons morphologiquement définies entre chromosomes et membrane nucleaire chez certains Péridiniens. *C. R. Acad. Sci. [D] (Paris)* 266:1287-1289.

Bronk, B. V., Dienes, G. J., and Pastkin, A. (1968). The stochastic theory of cell proliferation. *Biophys. J.* 8:1353-1338.

Bruce, V. G., and Bruce, N. C. (1982). Circadian clock-controlled growth cycle in *Chlamydomonas reinhardi.* In *International Cell Biology,* vol. 18, H. G. Schweiger (ed.). Springer-Verlag, New York, pp. 823-830.

Busby, W. F., and Lewin, J. (1967). Silicate uptake and silica shell formation by synchronously dividing cells of the diatom *Navicula pelliculosa* (Breb.) Helse. *J. Phycol.* 3:127-131.

Chisholm, S. W. (1981). Temporal patterns of cell division in unicellular algae. In *Physiological Bases of Phytoplankton Ecology,* T. Platt (ed.). *Can. Bull. Fish. Aquat. Sci.* 210:150-181.

Chisholm, S. W., Azam, F., and Eppley, R. W. (1978). Silicic acid incorporation in marine diatoms on light/dark cycles: Use as an assay for phased cell division. *Limnol. Oceanogr.* 23:518-529.

Chisholm, S. W., and Brand, L. E. (1981). Persistence of cell division phasing in marine phytoplankton in continuous light after entrainment to light:dark cycles. *J. Exp. Mar. Biol. Ecol.* 51:107-118.

Chisholm, S. W., and Costello, J. C. (1980). Influence of environmental factors and population composition on the timing of cell division in *Thalassiosira fluviatilis* (Bacillariophyceae) grown on light/dark cycles. *J. Phycol.* 16:375-383.

Chisholm, S. W., Morel, F. M. M., and Slocum, W. S. (1980). The phasing and distribution of cell division cycles in marine diatoms. In *Primary Productivity in the Sea,* P. Falkowski (ed.), Brookhaven Symposium on Biology, vol. 31, Brookhaven National Laboratory, Brookhaven, New York, pp. 281-300.

Chisholm, S. W., and Nobbs, P. A. (1976). Simulation of algal growth and competition in a phosphate-limited cyclostat. In *Modeling Biochemical Processes in Aquatic Ecosystems,* R. P. Canale (ed.). Ann Arbor Science, Ann Arbor, Michigan, pp. 337-355.

Chisholm, S. W., Stross, R. G., and Nobbs, P. A. (1975). Light/dark phased cell division in *Euglena gracilis* (Z) (Euglenophyceae), in $PO_4$ limited cultures. *J. Phycol.* 77: 367-373.

Cohen, D., and Parnas, H. (1976). An optimal policy for the metabolism of storage materials in unicellular algae. *J. Theor. Biol.* 56:1-18.

Cook, J. R. (1966a). Photosynthetic activity during the division cycle in synchronized *Euglena gracilis. Plant Physiol.* 41:821-825.

Cook, J. R. (1966b). Studies on chloroplast replication in synchronized *Euglena.* In *Cell Synchrony,* I. L. Cameron and G. M. Padilla (eds.). Academic Press, New York, pp. 153-168.

Cooper, S. (1982). The continuum model: Statistical implications. *J. Theor. Biol.* 94: 783-800.

Darley, W. M., Sullivan, C. W., and Volcani, B. E. (1976). Studies on the biochemistry and fine structure of silica shell formation in diatoms: Division cycle and chemical composition of *Navicula pelliculosa* during light/dark synchronized growth. *Planta* 130:159-167.

Dodge, J. D. (1965). Chromosome structure in the dinoflagellate and the problem of the mesocaryotic cell. *Excerpta Med. Int. Cong. Ser.* 91:339.

Dodge, J. D. (1966). The Dinophyceae. In *The Chromosomes of the Algae,* M. B. E. Godward (ed.). St. Martin's Press, New York, pp. 96-115.

Edmunds, L. N., Jr. (1965a). Studies on synchronously dividing cultures of *Euglena gracilis* Klebs (Strain Z). I. Attainment and characterization of cell division. *J. Cell. Comp. Physiol.* 66:147-158.

Edmunds, L. N., Jr. (1965b). Studies on synchronously dividing cultures of *Euglena gracilis* Klebs (Strain Z). II. Patterns of biosynthesis during the cell cycle. *J. Cell. Comp. Physiol.* 66:159-182.

Edmunds, L. N. Jr., (1966). Studies on synchronized dividing cultures of *Euglena gracilis* Klebs (Strain Z). III. Circadian components of cell division. *J. Cell. Physiol.* 67: 35-44.

Edmunds, L. N. Jr., and Funch, R. R. (1969). Effects on "skeleton" photoperiods and high frequency light/dark cycles on the rhythm of cell division in synchronized cultures of *Euglena. Planta* 87:134-163.

Edmunds, L. N., Jr., and Adams, K. J. (1981). Clocked cell cycle clocks. *Science* 211:1002-1013.

Edmunds, L. N., Jr., Tay, D. E., and Laval-Martin, D. L. (1982). Cell division cycles and circadian clocks. Phase-response curves for light perturbations in synchronous cultures of *Euglena. Plant Physiol.* 70:297-302.

Ehret, C. F., and Wille, J. J. (1970). The photobiology of circadian rhythms in protozoa and other eukaryotic microorganisms. In *Photobiology of Microorganisms,* P. Halldal (ed.). Wiley Interscience, London, pp. 369-416.

Elbrachter, M. (1973). Population dynamics of *Ceratium* in coastal waters of the Kiel Bay. *Oikos (Suppl.)* 15:43-48.

Enright, J. T., and Honeggar, H. W. (1977). Diurnal vertical migration: Adaptive significance and timing. Part 2. Test of the model: Details and timing. *Limnol. Oceanogr.* 22:873-886.

Eppley, R. W., and Coatsworth, J. L. (1966). Culture of the marine phytoplankter, *Dunaliella tertiolecta,* with light/dark cycles. *Arch. Mikrobiol.* 55:66-80.

Eppley, R. W., Holmes, R. W., and Paasche, E. (1967). Periodicity in cell division and physiological behavior of *Ditylum brightwellii,* a marine phytoplankton diatom, during growth in light/dark cycles. *Arch. Mikrobiol.* 56:305-323.

Eppley, R. W., Holm-Hansen, O., and Strickland, J. D. H. (1968). Some observations on the vertical migration of dinoflagellates. *J. Phycol.* 4:333-340.

Eppley, R. W., Robers, J. N., McCarthy, J. J., and Sournia, A. (1971). Light/dark periodicity in nitrogen assimilation of the marine phytoplankters *Skeletonema costatum* and *Coccolithus huxleyi* in N-limited chemostat culture. *J. Phycol.* 7:150-154.

Fantes, P. A. (1977). Control of cell size and cycle time in *Schizosaccharomyces pombe. J. Cell Sci.* 24:51-67.

Filfilan, S. A., and Sigee, D. C. (1977). Continuous DNA replication in the nucleus of the dinoflagellate *Prorocentrum micans* (Ehrenberg). *J. Cell Sci.* 27:81-90.

Fleming, R. H. (1939). The control of diatom populations by grazing. *J. Cons. Int. Explor. Mer.* 14:210-227.
Frank, E. K. (1970). A mathematical model of synchronized periodic growth of cell populations. *J. Theor. Biol.* 26:373-382.
Galleron, C. J. (1976). Synchronization of the marine dinoflagellate *Amphidinium carteri* in dense cultures. *J. Phycol.* 12:69-73.
Galleron, C., and Durrand, A. M. (1979). Cell cycle and DNA synthesis in a marine dinoflagellate, *Amphidinium carteri. Protoplasma* 100:155-165.
Hansche, P. E. (1969). A theoretical basis for the entrainment of chemostat populations. *J. Theor. Biol.* 24:335-350.
Hart, R. C. (1977). Feeding rhythmicity in a migratory copepod [*Pseudodiaptomus hessei* (Mrazek)]. *Freshwater Biol.* 7:1-8.
Heller, M. D. (1977). The phased division of the freshwater dinoflagellate *Ceratium hirundinella* and its use as a method of assessing growth in natural populations. *Freshwater Biol.* 7:527-533.
Kates, J. R., Chiang, K. S., and Jones, R. F. (1968). Studies on DNA replication during synchronized vegetative growth and gametic differentiation in *Chlamydomonas reinhardtii. Exp. Cell Res.* 49:121-135.
Klevecz, R. R. (1976). Quantized generation time in mammalian cells as an expression of the cellular clock. *Proc. Natl. Acad. Sci. USA* 73:4012-4016.
Kubai, D. F., and Ris, H. (1969). Division in the dinoflagellate *Gyrodinium cohnii* (Schiller). A new type of nuclear reproduction. *J. Cell Biol.* 40:508-528.
Leadbeater, B,, and Dodge, J. D. (1967). An electron microscope study of nuclear and cell division in a dinoflagellate. *Arch. Mikrobiol.* 57:239-254.
Ledoigt, G., and Calvayrac, R. (1979). Phénomènes périodiques, métaboliques et structuraux chez un protiste, *Euglena gracilis. J. Protozool.* 26:632-643.
Lehman, J. T., and Scavia, D. (1982). Microscale patchiness of nutrients in plankton communities. *Science* 216:729-730.
Lewin, J. C., Reimann, B. E., Busby, W. F., and Volcani, B. E. (1966). Silica shell formation in synchronously dividing diatoms. In *Cell Synchrony,* I. L. Cameron and G. M. Padilla (eds.). Academic Press, New York, pp. 169-188.
Lewin, J. C., and Rao, V. N. (1975). Blooms of surf-zone diatoms along the coast of the Olympic Peninsula, Washington. IV. Daily periodicity phenomena associated with *Chaetoceros armatum* in its natural habitat. *J. Phycol.* 11:330-338.
Lien, T., and Knutsen, G. (1972). Synchronous cultures of *Chlamydomonas reinhardii*: Synthesis of repressed and derepressed phosphatase during the life cycle. *Biochim. Biophys. Acta* 287:154-163.
Loeblich, A. R. (1976). Dinoflagellate evolution: Speculation and evidence. *J. Protozool.* 23:13-28.
Loeblich, A. R. (1977). Studies on synchronously dividing populations of *Cachonina niei,* a marine dinoflagellate. *Bull. Jpn. Soc. Physiol.* 25:119-126.
Marano, F., Amancio, S., and Durrand, A. M. (1978). Synchronous growth and synthesis of macromolecules in a naturally wall-less Volvocale, *Dunaliella bioculata. Protoplasma* 95:135-144.
McCarthy, J. J. (1981). The kinetics of nutrient utilization. In *Physiological Bases of Phytoplankton Ecology,* T. Platt (ed.) vol. 210. *Can. Bull. Fish. Aquat. Sci.,* Quebec, pp. 211-233.
McCarthy, J. J., and Goldman, J. C. (1979). Nitrogenous nutrition of marine phytoplankton in nutrient-depleted waters. *Science* 23:670-672.
Mihara, S., and Hase, E. (1971). Studies on the vegetative life cycle of *Chlamydomonas reinhardii* Dangeard in synchronous culture. I. Some characteristics of the cell cycle. *Plant Cell Physiol.* 12:225-236.

Morimura, Y. (1959). Synchronous culture of *Chlorella* I. Kinetic analysis of the life cycle of *Chlorella ellipsoidea* as affected by changes of temperature and light intensity. *Plant Cell Physiol.* 1:49-61.

Nelson, D. M., and Brand, L. E. (1979). Cell division periodicity in 13 species of marine phytoplankton on a light/dark cycle. *J. Phycol.* 15:67-75.

Olson, R. J., Frankel, S. L., Chisholm, S. W., and Shapiro, H. M. (1983). An inexpensive flow cytometer for the analysis of fluorescence signals in phytoplankton: Chlorophyll and DNA distributions. *J. Exp. Mar. Biol. Ecol.* 68:129-144.

Olson, R. J., and Chisholm, S. W. (1984). The effects of photocycles and periodic ammonium supply on three marine phytoplankton species. I. Cell division patterns. *J. Phycol.* (in press).

Paasche, E. (1968). Marine plankton algae grown with light/dark cycles. II. *Ditylum brightwellii* and *Nitzschia turgidula. Plant Physiol.* 21:66-77.

Pollinger, U., and Zemel, E. (1981). In situ and experimental evidence of the influence of turbulence on cell division processes of *Peridinium cinctum* Forma *westii* (Lemm.) Lefevre. *Br. Phycol. J.* 16:281-287.

Prescott, D. (1976). *Reproduction in Eucaryotic Cells.* Academic Press, New York.

Puiseux-Dao, S. (1981). Cell cycle events in unicellular algae. In *The Physiological Bases of Phytoplankton Ecology,* T. Platt (ed.), vol. 210. *Can. Bull. Fish. Aquat. Sci.,* Quebec, pp. 130-149.

Richman, S., and Rogers, J. N. (1969). The feeding of *Calanus helgolandicus* on synchronously growing populations of the marine diatoms *Ditylum brightwellii. Limnol. Oceanogr.* 14:701-709.

Ricketts, T. R. (1977). Changes in average cell concentrations of various constituents during synchronous division of *Platymonas striata* Butcher (Prasionophyceae). *J. Exp. Bot.* 28:1278-1288.

Ricketts, T. R. (1979). The induction of synchronous cell division in *Platymonas striata* Butcher (Prasinophyceae). *Br. Phycol. J.* 14:219-223.

Rizzo, P. J., and Nooden, L. D. (1973). Isolation and chemical composition of dinoflagellate nuclei. *J. Protozool.* 20:666-672.

Rosen, R. (1967). *Optimality Principles in Biology.* Butterworths, London.

Setlik, I., Berkova, E., Doucha, J., Kubin, S., Vendlova, J., and Zachleder, V. (1972). The coupling of synthetic and reproduction processes in *Scenedesmus quadricada. Arch. Hydrobiol. (Suppl.)* 41:172-213.

Setlik, I., Zachleder, V., Doucha, J., Berkova, E., and Bartos, J. (1975). The nature of the temperature block in the sequence of reproduction processes in *Chlorella vulgaris* Beijerinck. *Arch. Hydrobiol. (Suppl.)* 49:70-104.

Shuter, B. (1979). A model of physiological adaptation in unicellular algae. *J. Theor. Biol.* 78:519-552.

Simmer, J., and Sodomkova, M. (1968). Synchronization of natural populations of *Scenedesmus acuminatus* (Lagerh) Chod. induced by rhythmical changes in the environment. *Acta Univ. Carol. Biol.* 1967:251-254.

Singh, P. J. (1972). Studies on the food and feeding of the freshwater calanoid *Rhinediaptomus indicus* Kieffer. II. Diurnal variations in feeding properties. *Hydrobiology* 39:209-215.

Smayda, T. J. (1975). Phased cell division in natural populations of the marine diatom *Ditylum brightwellii* and the potential significance of diel phytoplankton behavior in the sea. *Deep-Sea Res.* 22:151-165.

Smith, J. A., and Martin, L. (1973). Do cells cycle? *Proc. Natl. Acad. Sci. USA* 70:1263-1267.

Soyer, M. O. (1969). Rapports existant entre chromosomes et membrane nucléaire chez un Dinoflagellé parasite du genre *Blastodinium* Chatton. *C. R. Acad. Sci. [D] (Paris)* 268:2082-2084.

Spudich, J., and Sager, R. (1980). Regulation of the *Chlamydomonas* cell cycle by light and dark. *J. Cell Biol.* 85:136-145.
Stayley, J. T. (1971). Growth rates of algae determined in situ using an immersed microscope. *J. Phycol.* 7:13-17.
Stross, R. G., Chisholm, S. W., and Downing, T. A. (1973). Cause of daily rhythms in photosynthetic rates of phytoplankton. *Biol. Bull.* 145:200-209.
Sullivan, C. W. (1977). Diatom mineralization of silicic acid. II. Regulation of $Si(OH)_4$ transport rates during the cell cycle of *Navicula pelliculosa. J. Phycol.* 13:86-91.
Sweeney, B. M. (1982). Interaction of the circadian cycle with the cell cycle in *Pyrocystis fusiformis. Plant Physiol.* 70:272-276.
Sweeney, B. M., and Hastings, J. W. (1958). Rhythmic cell division in populations of *Gonyaulax polyedra. J. Protozool.* 5:217-224.
Sweeney, B. M., and Hastings, J. W. (1964). The use of the electronic cell counter in studies of synchronous cell division in dinoflagellates. In *Synchrony in Cell Division and Growth,* E. Zeuthen (ed.). Interscience, New York, pp. 579-587.
Swift, E., and Durbin, E. G. (1972). The phased division and cytological characteristics of *Pyrocystis* spp. can be used to estimate doubling times of their populations in the sea. *Deep-Sea Res.* 19:189-198.
Tamiya, H. (1964). Growth and cell division in *Chlorella.* In *Synchrony in Cell Division and Growth,* E. Zeuthen (ed.). Interscience, New York, pp. 247-305.
Tamiya, H. (1966). Synchronous cultures of algae. *Annu. Rev. Plant Physiol.* 17:1-26.
Tanoue, E., and Aruga, Y. (1975). Studies on the life cycle and growth of *Platymonas sp.* in culture. *Jpn. J. Bot.* 20:439-460.
Thomas, K. C., Dawson, P. S. S., Gamborg, B. L., and Steinhuer, L. (1980). Relationships between cell size and efficiency of synchronization during nitrogen-limited phased cultivation of *Candida utilis. J. Bacteriol.* 141:10-19.
Tuttle, R. C., and Loeblich, A. R. (1975). An optimal growth medium for the dinoflagellate *Cryptothecodinium cohnii. Phycologia* 14:1-8.
Wanka, F., and Mulders, P. F. M. (1967). The effect of light on DNA synthesis and related processes in synchronous cultures of *Chlorella. Arch. Mikrobiol.* 58:257-269.
Weiler, C. S. (1978). Phased cell division in the dinoflagellate genus *Ceratium*: Temporal pattern, use in determining growth rates, and ecological implications. Ph.D. thesis, University of California, San Diego, La Jolla, California.
Weiler, C. S. and Chisholm, S. W. (1976). Phased cell division in natural populations of marine dinoflagellates from shipboard cultures. *J. Exp. Mar. Biol. Ecol.* 25:239-247.
Weiler, C. A., and Eppley, R. W. (1979). Temporal pattern of division in the dinoflagellate genus *Ceratium* and its application to the determination of growth rate. *J. Exp. Mar. Biol. Ecol.* 39:1-24.
Wheeler, P. A., Olson, R. J., and Chisholm, S. W. (1984). The effects of photocycles and periodic ammonium supply on three marine phytoplankton species. II. Ammonium uptake and assimilation. *J. Phycol.* (in press).
White, A. W. (1976). Growth inhibition caused by turbulence in the toxic marine dinoflagellate *Gonyaulax excavata. J. Fish. Res. Board Can.* 33:2598-2602.
Williams, F. M. (1971). Dynamics of microbial populations. In *Systems Analysis and Simulation in Ecology,* vol. I, B. C. Patten (ed.). Academic Press, New York.
Williamson, C. E. (1980). Phased cell division in natural and laboratory populations of marine phytoplankton diatoms. *J. Exp. Mar. Biol. Ecol.* 43:271-279.
Wimpenny, R. S. (1938). Diurnal variation in the feeding and breeding of zooplankton rhythms related to the numerical balance of the zoophytoplankton community. *J. Cons. Int. Explor. Mer.* 13:323-337.
Winfree, A. T. (1980). *The Geometry of Biological Time.* Springer-Verlag, New York.

Yoder, J., Martin, J., and Nill, A. (1982). Cell division periodicity and the nitrate environment of a marine diatom. *Limnol. Oceanogr.* 27:352-357.

Zeuthen, E. (1974). A cellular model for repetitive and free-running synchrony in *Tetrahymena* and *Schizosaccharomyces.* In *Cell Cycle Controls,* G. M. Padilla, I. L. Cameron, and A. U. Zimmerman (eds.). Academic Press, New York, pp. 1-30.

# 21
# Photoperiodic Control of Cell Division in *Euglena* and *Ceratium*

**Kenneth J. Adams** / North East London Polytechnic, Stratford, London, England
**Caroline S. Weiler** / Whitman College, Walla Walla, Washington
**Leland N. Edmunds, Jr.** / State University of New York, Stony Brook, New York

## INTRODUCTION

A remarkable feature of many unicellular algae is that although their biochemistry is subject to the vagaries of ambient illumination and temperature in their natural environment, they have become adapted to utilize several environmental cues to ensure that cell division can occur at a particular time of the day-night cycle, characteristic of each species (Chisholm, 1981). In most phytoflagellates, including both *Euglena* and *Ceratium*, mitosis and cytokinesis are usually confined to the hours of darkness. Furthermore, under short days many species extend their cell division cycle (CDC) over 2 days or more (even weeks, at low temperatures), the CDC being reduced in length by daily increments as the day length increases, until, at optimum temperatures, every cell is able to divide once per day. If the light intensity or temperature rises or falls during the diurnal cycle prior to cell division, the effect on the rate of progress through the CDC is compensated by either extending or curtailing the program so that cell division coincides with the dark phase of the day-night cycle. When conditions are optimal, however, many phytoflagellates can shrug off these CDC controls and complete several divisions per day, which suggests that this complex control system may have evolved to optimize the utilization of solar energy during the daylight hours.

Experimental investigation of the role of light in this control system in *Euglena* reveals that the mechanism has a photoperiodic element and appears to be based on two discrete but interactive endogenous clocks (Edmunds and Adams, 1981). A dawn-synchronized clock, the "cytochron," is apparently responsible for initiating and programming the events of the CDC, whereas a dusk-synchronized clock, the "circadachron," appears to stop the cytochrons in a proportion of the cells in populations grown under light-dark

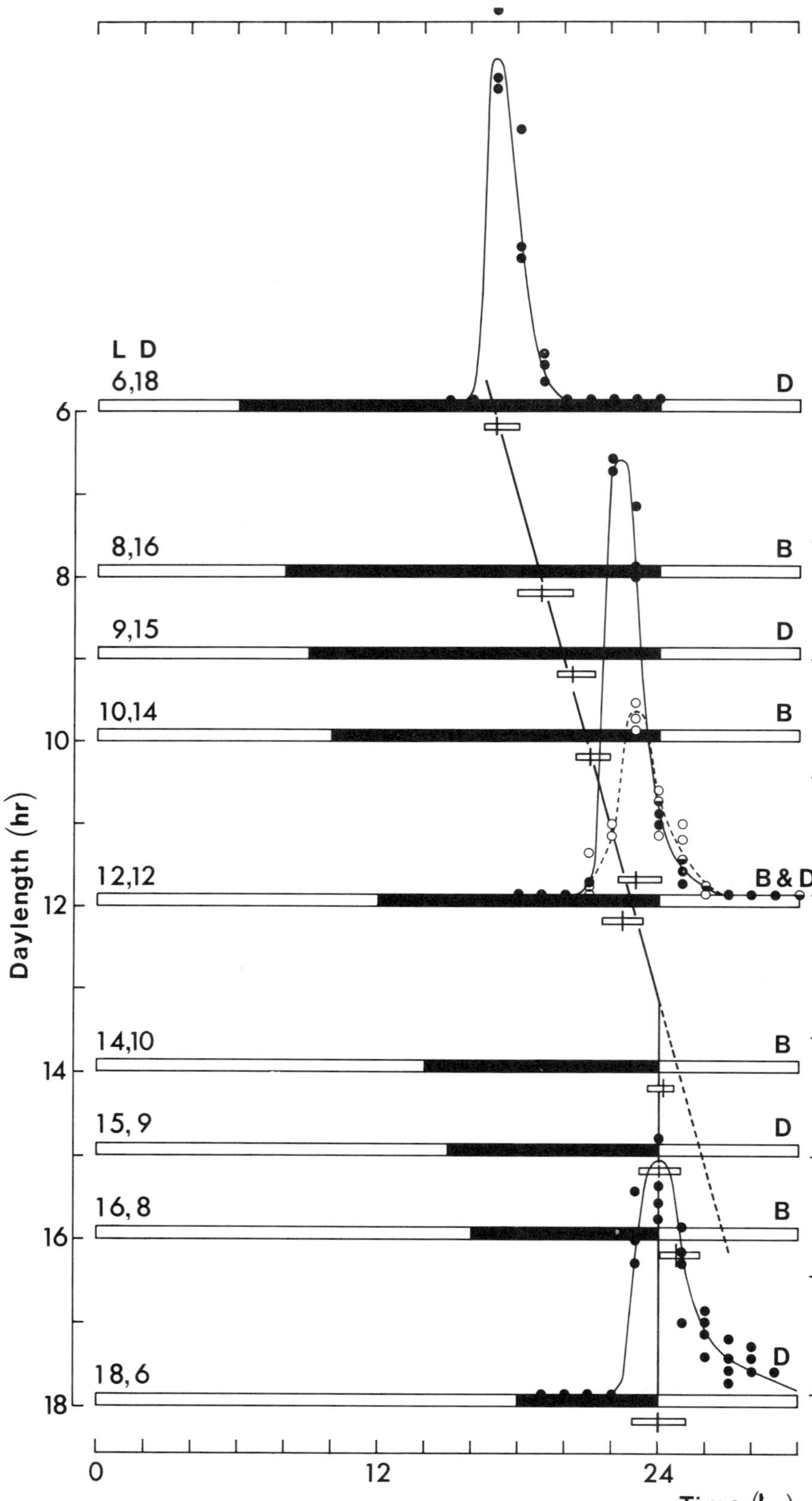

%
15
10
5
0
20
15
10
5
0
15
10
5
0
Recently Divided Cells/2 (D), Oslo data.
Binucleate Cells (B), San Diego data, and
L D
6,18
8,16
9,15
10,14
12,12
14,10
15, 9
16, 8
18, 6
D
B
D
B
B & D
B
D
B
D
6
8
10
12
14
16
18
Daylength (hr)
0
12
24
Time (hr)

(LD) regimes with photoperiods of less than about 12 hr, and thereby delays completion of their CDCs until the following night. A detailed study of the timing of cell division in the dinoflagellate *Ceratium furca,* reported here for the first time, also implicates a photoperiodic control mechanism, and the participation of two endogenous clocks, though the details of the way in which they interact to control cell division differ from that in *Euglena,* and have as yet to be fully elucidated. Under LD cycles in the range 6,18 to about 13,11 cell division appears to be gated by a dusk-synchronized clock some 11 hr post dusk. For day lengths in excess of about 13 hr the gating seems to be taken over by a dawn-synchronized clock, cell division occurring parallel to dawn astride the dark-light transition (Fig. 1). These alignments with the previous dawn and dusk on either side of the critical day length of 13 hr are an almost exact inverse of the pattern of eclosion versus photoperiod length in *Drosophila pseudoobscura* (Pittendrigh, 1966).

Since available evidence suggests that euglenoids and dinoflagellates parted company at around $10^9$ years before the present (Margulis, 1970; Tappan, 1980), the present-day occurrence of comparable photoperiodic control systems operated by a pair of endogenous clocks—too similar to be likely to have arisen independently—pushes back the origins of endogenous photoperiodic control mechanisms to an early unicellular stage soon after the advent of the eukaryotic level of organization. Furthermore, it suggests that the complex hierarchy of endogenous chronometers, evolved subsequently by higher organisms, may still nevertheless impinge at the cellular level on a very ancient, but "tried and trusted," device that masterminds the events of the CDC.

## THE ROLE OF LIGHT AND DARK

As most is currently known about the role of light and dark in controlling cell division in *Euglena gracilis,* we consider this in detail prior to comparing it with available data for *C. furca.* Perhaps not surprising for a cell with some seven times the amount of DNA (2.25 pg) (Edmunds, 1965b) in its 21 ± 2 pairs of chromosomes (Adams, unpublished results) as that of diploid *Drosophila,* as much as a higher plant or chicken (Shapiro, 1976), the mechanism restricting cell division to the hours of darkness in *Euglena* is as complex as that controlling dipteran eclosion, and that controlling the number of cells dividing according to photoperiod as sophisticated as that responsible for initiating diapause (Saunders, 1976).

**FIGURE 1** Day length dependence of cell division timing under steady-state LD regimes in *C. furca* (Ehr.) Clap. et Lachm. Entrained populations were scored for the percentage of binucleate (B) or of recently divided (hornless) (D) cells from hourly formalin-fixed samples. The B experiments were conducted on cells isolated from Southern Californian waters, grown at 20°C (Weiler and Eppley, 1979); the D experiments used a culture from the Oslofjord, Norway, grown at 18°C (Weiler unpublished results). Cool white fluorescent lighting gave $2.25 \times 10^{16}$ to $3.85 \times 10^{16}$ quanta/cm$^2$ per sec (B), and $1.20 \times 10^{16}$ quanta/cm$^2$ per sec (D), for the two sets of experiments over the band 400 to 700 nm. Data points and fitted curves given for four sets of data only, for clarity. Rest of curves summarized by their peak time planes and half-peak bandwidths only. Note the correlation with the previous dusk for daylengths $< 13$ hr and with the previous dawn for day lengths $> 13$ hr.

Unfortunately, most of the work on cell division control in *Euglena* has been carried out at 25°C, at which temperature, given sufficient illumination, cells can divide at least once per day under LD regimes, and in LL under really optimum conditions, twice a day. At the more natural temperatures of 15-20°C the CDC is extended over at least 2 days, and it is really under these conditions that the mechanism restricting cell division to the dark hours comes into its own.

Grown photoautotrophically at 25°C in LL *E. gracilis* becomes asynchronous, giving steady log-linear growth curves with doubling times in the range 10-15 hr, depending on medium composition, light intensity, aeration, and $CO_2$ or bicarbonate supply. Given an acceptable alternative carbon and energy source, such as ethanol, *Euglena* also adopts the asynchronous mode, immediately uncoupling the light-dark operated control system, switching to heterotrophic growth and dividing at a constant rate with doubling times as short as 8 hr—apparently ignoring any imposed LD regimes (Leedale, 1959; Edmunds, 1965a). Thus the photoperiodic control of cell division in *Euglena* appears to have evolved as an optional adaptation specifically for utilization during periods of growth on organically depleted media when photoautotrophic growth is the only means of competitive survival. Possibly this sophisticated adaptation has been developed to ensure that those stages that are most capable of photosynthesis, and are both flagellate and photo-

---

**FIGURE 2** Phased cell division in LD-cycled log-phase populations of *E. gracilis* var. *bacillaris* Cori strain (Cambridge Cult. Coll. No:1224/7). Cells grown autotrophically on modified Hutner and Provasoli (1955) medium, supplemented with vitamins $B_1$ and $B_{12}$, as batch cultures in 1 liter all-glass water-jacketed fermenters with automated cell counting at 6 min intervals by Fisher (MK1) Autocytometer (Adams, 1968). Photoperiods provided by three 100 VA tungsten mushroom lamps (2700 K), giving 7500 lux at the surface of the 10 cm diameter fermenter, switched suddenly on or off by time clock. Aeration continuous at 15 ml/min enriched with 4% $CO_2$ and dispersed through a glass sinter to give a $CO_2$ partial pressure in the medium of 29 mmHg. Cells kept in motion by 500 rpm motor-driven impeller to minimize settling rhythm effects. Temperature maintained at 25 ± 0.1°C throughout with thermostated water via water jacket. Mitosis and cleavage frequencies scored per 1000 cells.

Cell number counts for populations entrained on 24 hr cycles of (a) LD: 8,16 and 20 hr cycles of (b) LD: 8,12. L to D transition induced mobilization of settled cells (arrows). Cell number counts and age-gradient corrected, cumulative, entry curves (open data points) for the post mitotic cleavage phase c, (stippled), from LD: 10,14 entrained populations: (c) typical LD cycle with approximate doubling of the population, and (d) when LL given for one cycle. *Note*: mean cleavage duration of main wave approximately 85 min. Shoulder due to division of cells held over from previous day (arrows). Cell number counts and cleavage entry and exit curves for LD: 12,12 cycled populations from: (e) mid-log phase, and (f) late log phase. Pre-main wave cell number rise due to division of cells held over from previous cycle (between arrows). Main rise due to a majority of cells dividing once, and a minority dividing twice per LD cycle. Age gradient corrected entry and exit curves for both mitosis m, and cleavage c, given for the late log-phase population. *Note*: mean cleavage duration for mid-log phase is approximately 36 min right across the population. Mean mitotic duration for the late log-phase constant at approximately 55 min right across the population, but the cleavage duration is strongly correlated with cell cycle length, the last cells to divide stallin in cleavage until light-on. (Compare these curves obtained in 1967-1968 with those of Fig. 10 obtained in 1980, following further loss of ability by the Cori strain to be synchronized by LD cycles after 11 years of growth of stock cultures in LL.)

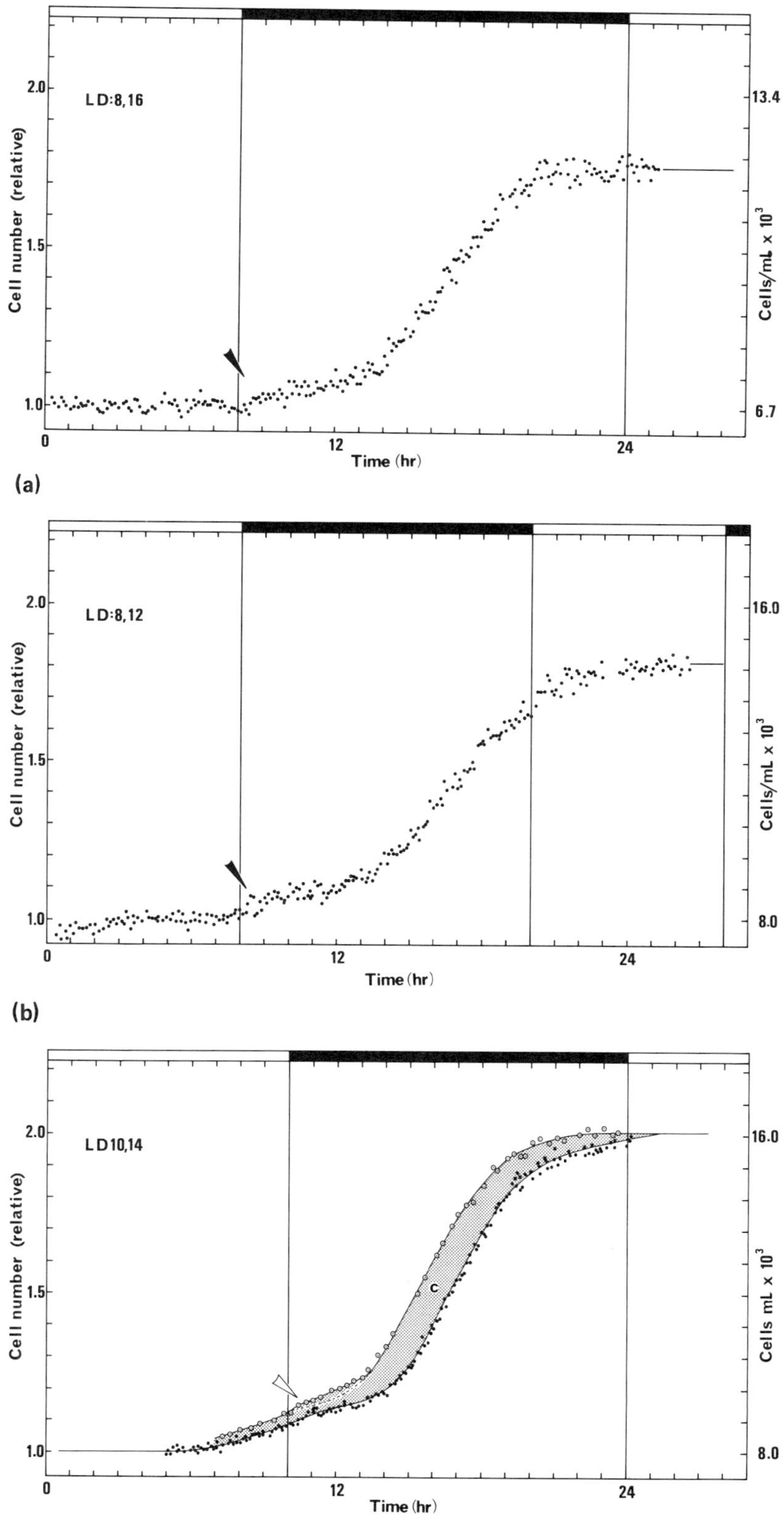
LD:8,16
Cell number (relative)
Cells/mL × 10^3
2.0
1.5
1.0
13.4
6.7
0
12
24
Time (hr)
(a)
LD:8,12
Cell number (relative)
Cells/mL × 10^3
2.0
1.5
1.0
16.0
8.0
0
12
24
Time (hr)
(b)
LD10,14
Cell number (relative)
Cells mL × 10^3
c
2.0
1.5
1.0
16.0
8.0
0
12
24
Time (hr)
(c)

FIGURE 2

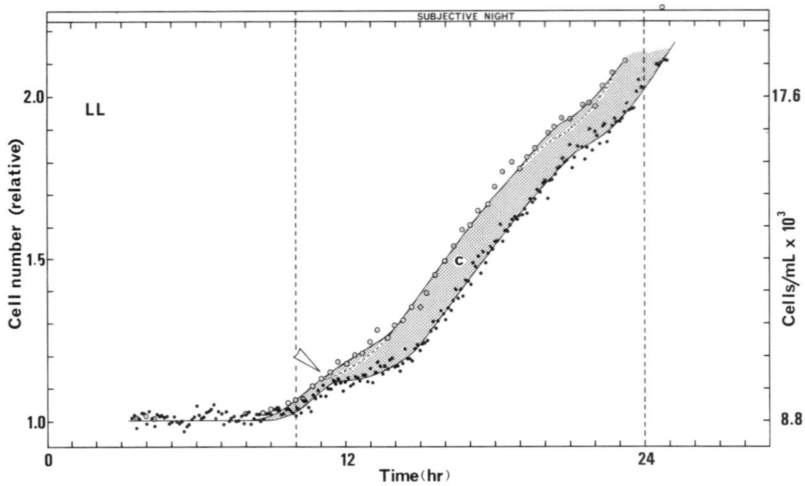


(d)

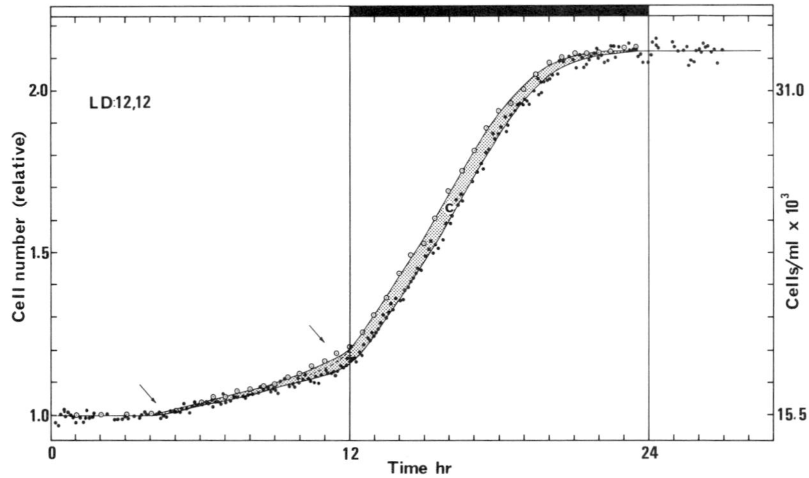


(e)

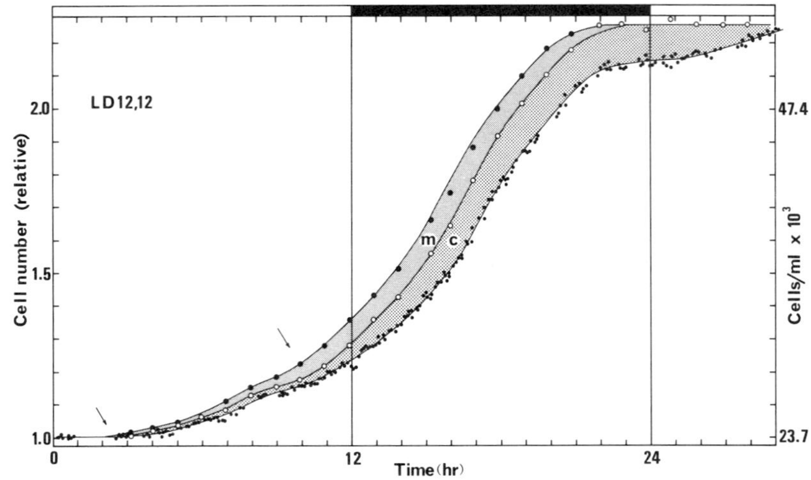


(f)

FIGURE 2 (Continued)

tactic, coincide with available light, and the dividing stages, which are either aflagellate or have two flagella pulling in different directions, are relegated to the dark hours.

At very low temperatures (3-4°C) when days are short and the CDC is extended to 20 days or more, *Ceratium* also abandons the strict LD cycle-entrained phasing of cell division and adopts the asynchronous mode (Weiler, unpublished results), though the advantages of such a change in strategy are far from clear. Indeed, it is not at all clear why *Ceratium* should align its tightly synchronized cell division waves parallel to either dawn or dusk in response to a change in photoperiod, either side of a critical 13 hr, over the more favorable temperature range of 15-20°C. It is, however, of interest to note that for the range of photoperiods examined (LD: 6,18 to LD: 18,6) with an isolate from the Oslofjord, at 18°C, there is considerably less variation in the timing of the onset of division with respect to midday (3.5 hr) than there is with respect to either lights-on (6.5 hr) or lights-off (5.5 hr). It may be, therefore, that it is most adaptive for this photosynthetic organism to time cellular events to maintain a constant phase relationship with the middle of the light period than it is with either dawn or dusk. A variation of 3.5 hr over a 12 hr change in photoperiod is clearly not perfect. Perhaps it is in some way adaptive for there to be deviations at the photoperiod extremes, or perhaps optimal growth conditions at these extremes (LD: 6,18 and 18,6) occur so infrequently in nature that there has been little selective pressure. *Ceratium* is most abundant in the Oslofjord, for example, in late summer and early autumn. Cells were rare in June 1980 when LD: 18,6 was approached, and by October, before the LD cycle was reduced to 6,18, water temperatures dropped to 3°C and the few cells to be seen had become arrhythmic.

### The Phasing of Cell Division

A striking difference between the CDC photocontrol strategies of cloned populations of *Euglena* and *Ceratium* grown autotrophically under experimental LD regimes ranging from 6,18 to 18,6 is apparent when the cell division waves associated with the dark periods are examined in detail. In *C. furca,* even though the CDC frequently extends over 3 days or more, cell division is confined to a narrow time zone only 4-5 hr wide during each dark period (see Fig. 1, also Weiler and Chisholm, 1976; Weiler and Eppley, 1979), whereas in *Euglena* the mitotic wave extends as a broad 8-10 hr wide band across each dark period (see Figs. 2 and 3). That this wide spread in *Euglena* is a deliberate component of the survival strategy is evident by comparison with dinoflagellate strategies where the variation in the timing of the CDC is quantized into almost exact multiples of 24 hr (Sweeney, 1969; Weiler and Eppley, 1979).

Work on the kinetics of cell division phasing in *Euglena gracilis,* first begun by Leedale (1959), has been pursued systematically by Edmunds and coworkers for a considerable range of LD (Edmunds, 1965a; Edmunds and Funch, 1969) and temperature regimes (Terry and Edmunds, 1970). High-resolution studies have, however, been lacking, and are reported here for the first time. Time courses for single cycles with automated cell counts at 6 min intervals are given for LD regimes ranging from 8,16 to 12,12 for *E. gracilis* var. *bacillaris* at a light intensity of 7500 lux and at 25°C in Fig. 2a-f.

In each case the main wave of cell division is preceded by either a small shoulder or a small steady rise in cell number, or both, during the latter part of the light period. These features are due to the complex interaction of the photoperiodic control mechanism together with interference from a L to D and D to L transition-influenced cell settling response, and will be analyzed in context. The main wave of cell division, however, rises rather sharply at 12-14 hr post dawn, irrespective of the duration or the intensity of the photoperiod, and terminates rather gradually some 8-10 hr later.

Under 24 hr cycles of LD: 8,16 (Fig. 2a), no cell divisions occur during the light period and only 70-80% of the cells divide each LD cycle. Following a small shoulder at the dusk transition, the curve of cell number remains flat until the main rise at around 12 hr post dawn. That the position of the main division wave is indeed timed from dawn, and not from dusk of the previous cycle, is indicated by the virtually identical curve for a population entrained to 20 hr cycles of LD: 8,12 (Fig. 2b). Note also that the small shoulder leveling off at around 9-10 hr post dawn occurs in the same place under both 24 and 20 hr cycles. It appears to arise from mobilization of a small fraction of cells settled on the fermenter walls as a result of being suddenly plunged into darkness rather than from a genuine increase in cell number.

On LD: 10,14 cycles (Fig. 2c), the population is able to double almost exactly in number each cycle, although at least 12% of the rise is the result of cells dividing in a small batch prior to the main wave. In this experiment the postmitotic cytokinesis phase frequencies were determined at half-hourly intervals from freshly fixed samples of 1000 cells. After correction for frequency depletion due to the rise in cell number since the start of the light period, and normalization to the same scale as the fitted cell number curve, they were plotted on top of it to give the cumulative entry curve for cytokinesis. Since the cell number curve is also the cumulative curve for cytokinesis exit, the gap between the two curves represents the mean time spent in cytokinesis for cells with any particular generation time. The duration of cytokinesis or "cleavage" is clearly around 85 min for the whole population (ignoring for the moment the small batch of cells dividing prior to the main wave). In Fig. 2d a population also entrained on LD: 10,14 has been subjected to LL from dawn of one cycle. Although the main wave is slightly depressed in slope, it clearly rises suddenly around 12-13 hr post dawn as in Fig. 2c, confirming that the main wave is actually timed from light-on and is not simply triggered by the dark onset. The pattern up to 19 hr post dawn is virtually the same as it would have been if the dark period had been present, but as the cells are now dividing in the light, their daughter cells can start their cell cycles immediately, so that by 19 hr post dawn some of them are ready to divide again and thus continue the division wave indefinitely. The small shoulder leveling off at about 12 hr post dawn is clearly visible (compare Fig. 2c) and cannot therefore be ascribed to any change in cell settling produced by the light-to-dark transition.

Although populations grown under LD: 10,14 cycles at a light intensity of 7500 lux (Fig. 2c) just about double every cycle, by increasing the light intensity of the 10 hr photoperiods it is possible to push the daily increment to 130%, some cells apparently dividing twice during the dark period. Such results have to be interpreted with caution, however, as under certain conditions (yet to be fully elucidated) *E. gracilis* can go through at least two rounds of S, thus raising the chromosome number to around 168, and this can be followed by two mitoses in rapid succession, giving rise to a complex cleavage figure as the cells divide into four (Adams, 1966).

In Fig. 2e a population has been subjected to LD: 12,12 cycles (at 7500 lux) with the result that the same pattern of cell division occurs but with the steeply rising main wave of division being somewhat greater in amplitude than in the 10,14 cycle of Fig. 2c. In this particular experiment $CO_2$-free air was flushed through the culture continuously during the sampled dark period, but identical results were obtained in a replicate in which (as in the other Fig. 2 examples) 4% $CO_2$ aeration was used throughout. Note, however, that in this experiment (as for its replicate) the mean duration of cytokinesis was only some 36 min as compared with the 85 min exhibited in Figs. 2c and d.

Although there is a tendency for cytokinesis to take longer as the stationary phase threshold is approached in batch cultures (Fig. 2f), and variations in the level of agitation can affect the delicate final cytokinesis stages—an increase causing them to part company more readily—the factors determining the variation in cytokinesis duration between experiments has yet to be fully elucidated. Fortunately, mitosis is remarkably stable in *Euglena,* with a mean duration of 51-60 min over a wide range of conditions, so that providing cytokinesis frequencies are available in addition to cell counts, the timing of the cell division wave can be estimated accurately from the cytokinesis entry curve. In Fig. 2f mitotic frequencies (prophase to telophase), in addition to those for cytokinesis, have been used to determine the mitotic entry and exit curves. Note that the mean duration of mitosis (approximately 55 min) is independent of position along the division wave, despite the strong positive correlation in this case for the mean duration of cytokinesis. Cytokinesis frequencies were determined on fresh samples immobilized by a trace of acetic acid. Mitotic phase frequencies were obtained by scoring samples of 1000 cells under oil immersion for each time plane, from permanent slides made with pelleted acetic/alcohol-fixed fresh samples, cold-stained in aceto orcein and suspension mounted (unsquashed) in euparal (Adams, 1966).

### Disentangling the Effects of LD Cycles

As a result of recent experiments and a reanalysis of existing data we are now able to demonstrate that the cell division waves obtained under daily LD regimes in *Euglena* have a complex structure, and that cell division can result from two (at least) different batches of cells in any one daily cycle. Most of the cells divide in the main wave, which always rises steeply from around 12 hr to about 20 hr post dawn. As the photoperiod is reduced in length, however, an increasing number of cells fail to be irreversibly committed to mitosis prior to a critical time plane some 8-9 hr post dusk, apparently being "frozen in their tracks" (irrespective of their stage in the CDC) and held over to a subsequent LD cycle. These cells divide during the subsequent cycle slightly ahead of, or merged with, the main wave of that cycle, and account for the rise up to the small shoulder indicated by the arrow in Fig. 2c and d. Note that the cells in this batch have what appears to be a cleavage duration only half that of the main-wave cells. As the cytokinesis fraction has been age gradient-corrected only for the current LD cycle in each case, this is an artefact that is caused by the failure of these cells to divide during the previous cycle, their frequency having been depleted by the increase in the number of those cells that did divide during the previous cycle. When the appropriate corrections are made the discrepancy disappears.

The cell division waves depicted in Fig. 2 clearly suggest a mechanism that meters time from the light-on time plane to give a rapidly rising and slowly tailing main division wave between 12 and about 20 hr post dawn, irrespective of the light or dark period duration (at 25°C). There appears, however, to be a minimum photoperiod of around 6 hr duration that must be satisfied during photoautotrophic growth before any cell can proceed toward cell division. Experiments with single photoperiods of 3.5 and 5 hr permit only 6 and 18%, respectively, of a population conditioned on a 10,14 cycle to divide in the subsequent dark period. These small fractions completing division are believed to be made up entirely of cells held over from the previous cycle.

In further experiments with *E. gracilis* var. *bacillaris* grown under LD: 14,10 and 16,8 cycles on 4% $CO_2$ (Adams, unpublished results), an even greater proportion of the population has been found to divide during the light periods than under the LD: 12,12

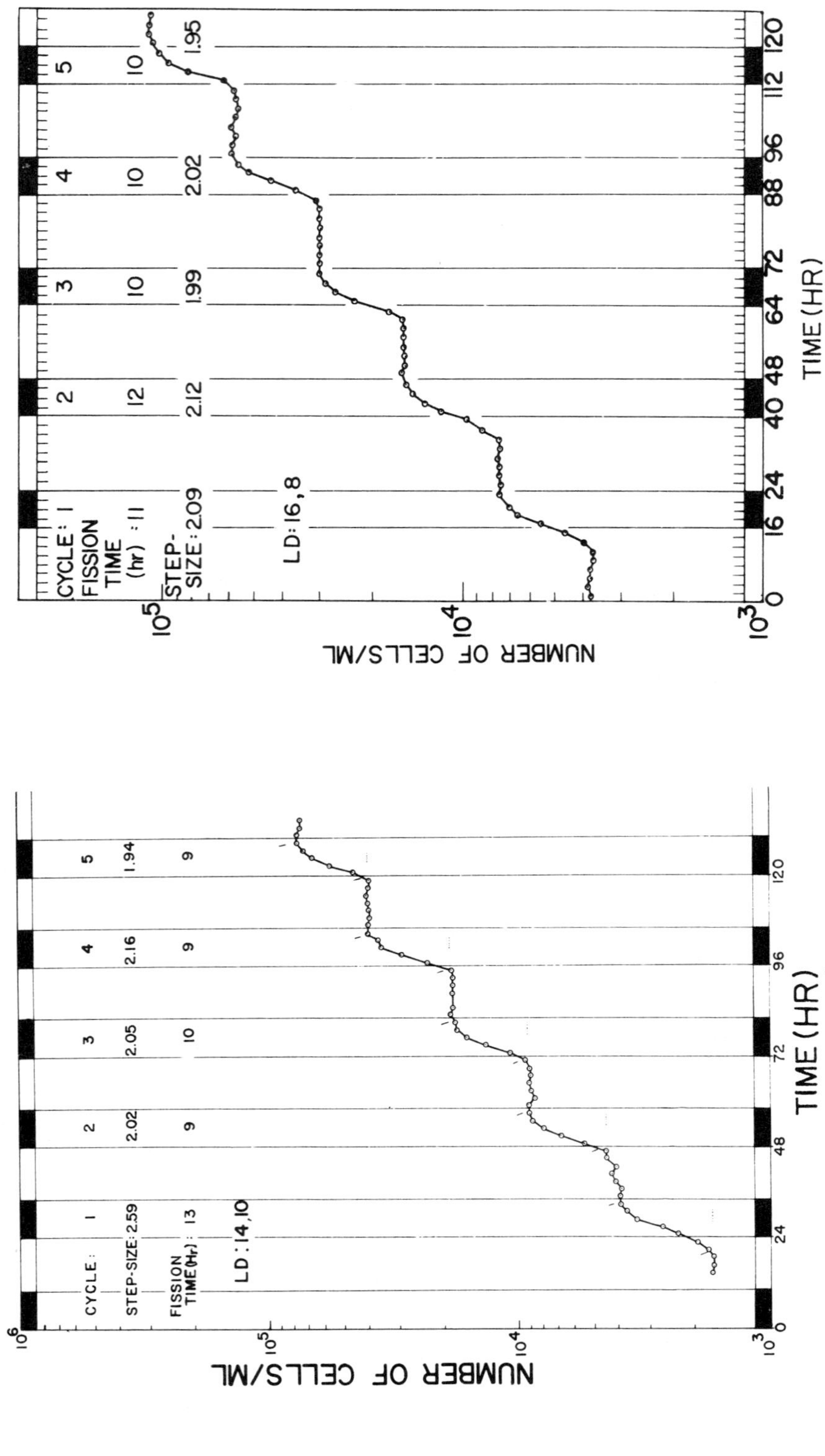
CYCLE: 1
FISSION TIME (hr): 11
STEP-SIZE: 2.09
LD: 16,8
2
12
2.12
3
10
1.99
4
10
2.02
5
10
1.95
NUMBER OF CELLS/ML
TIME (HR)
CYCLE: 1 2 3 4 5
STEP-SIZE: 2.59 2.02 2.05 2.16 1.94
FISSION TIME (Hr): 13 9 10 9 9
LD: 14,10
NUMBER OF CELLS/ML
TIME (HR)

(b)

(a)

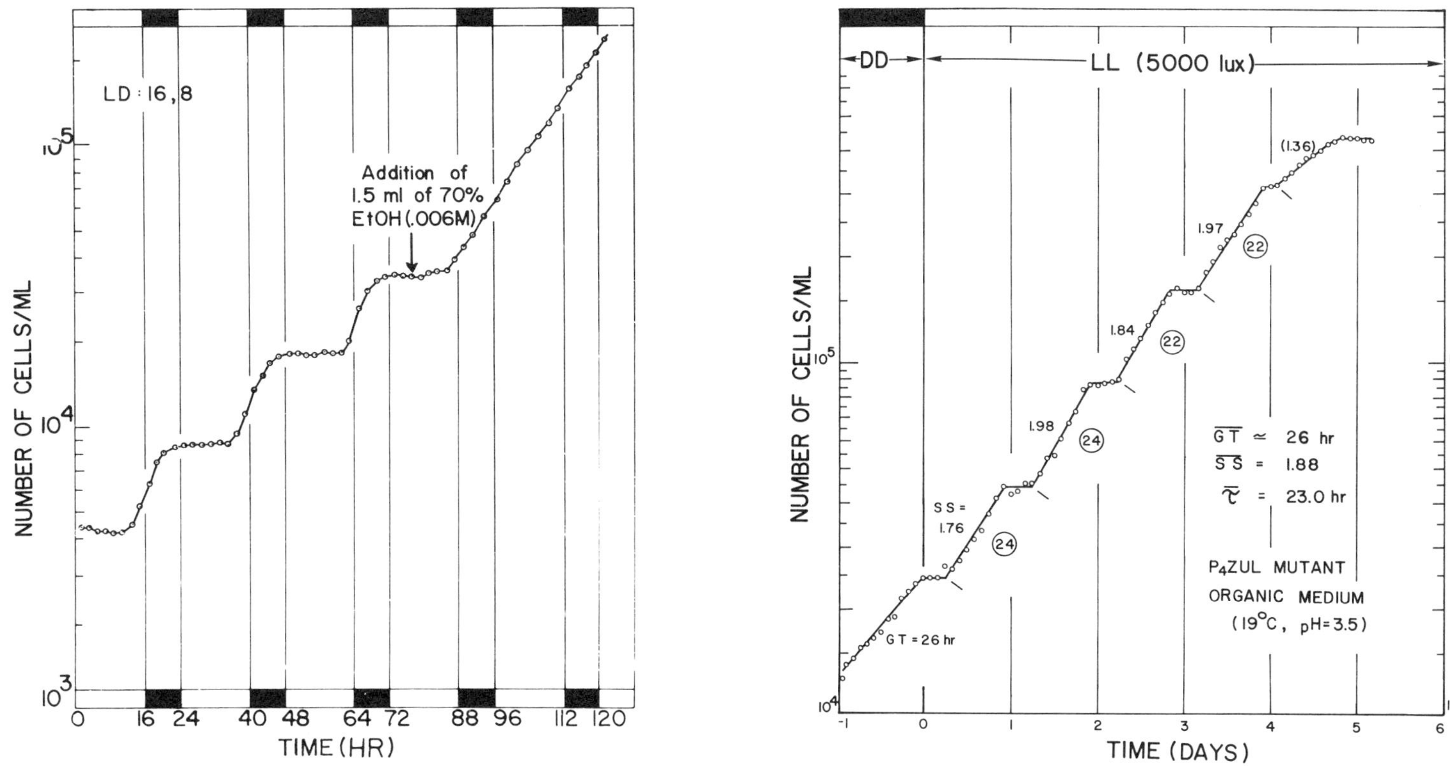


(c)

(d)

**FIGURE 3** Light-dark cycle induced cell division synchrony in autotrophic *E. gracilis* (Z), grown at 25°C and 3500 lux, with an approximate doubling of the population at each step under both (a) LD: 14,10 and (b) LD: 16,8 cycles. Addition of ethanol to a LD: 16,8 synchronized population, at 25°C, at a point 6 hr post light-on (c) results in continuation of the next division wave as log-linear growth, and abandonment of division phasing despite the continuing LD cycles. (d) Asynchronous growth of the photosynthetic mutant $P_4ZUL$ on organic medium at 19°C in DD is transformed into phased division at circadian intervals by a single DD to LL transition. *Note*: The log-linear as opposed to sigmoid rising curve segments as compared with the LD phased populations. (a, b, and c modified from Edmunds, 1965a, d from Jarrett and Edmunds, 1970.)

cycles depicted in Fig. 2e and f. This behavior is very different from that of *E. gracilis* strain Z grown on Cramer and Meyers medium (Edmunds, 1965a) at 25°C at an intensity of 3500 lux with aeration by room air at about 100 ml/liter per min. Under these conditions, after preconditioning, Edmunds (1965a) was able to obtain good synchrony with virtually no cell division during the light periods under LD: 14,10 and 16,8 cycles, the populations doubling almost exactly under both regimes (see Fig. 3 a and b). These differences may well be adaptive rather than strain specific, however. Gradual loss of rhythmicity in *E. gracilis* mutants is well known and can apparently be reversed by addition of a variety of sulfur compounds to the growth medium (Edmunds et al., 1976). Over the last 11 years var. *bacillaris* stocks have been maintained in LL at 20°C and there has been a gradual change in their response to imposed LD cycles. Cell division rhythms rapidly settle into steady-state patterns under LD regimes as before, but have become markedly biphasic with appreciable numbers of cells dividing during the light periods. The possible reason for this breakdown in ability to synchronize into a single cell division wave per LD cycle will be discussed later on in the context of the proposed dual clock-control mechanism.

## THE CASE FOR A CELL CYCLE CLOCK

What then are we to make of these various observations? Putting them all together we are able to assemble a purely descriptive model of the complex CDC controlling mechanism; nevertheless, it does eliminate many of the models that have previously been put forward and points up the value of detailed kinetic studies. The model that we have adopted to explain our observed kinetics is one involving a separate endogenous timing and sequencing mechanism, the cytochron, which periodically along its time track sends out signals that trigger a state change in the cell's biochemistry (Edmunds and Adams, 1981). Thus we see the cell's internal molecular milieu as confined by its substrate level feedback circuits to a steady-state biochemistry. In $G_1$, for example, the milieu can accumulate RNA, protein, chlorophyll, and paramylum, but cannot initiate the synthesis of DNA. If $G_1$ is extended in cells with long CDCs this is not achieved by triggering an additional Klevecz (1976) $G_1$ cycle, but by delaying the input of the signal from the clock to trigger the initiation of the S phase. Similarly, cells in $G_2$ can prepare for mitosis and cytokinesis but cannot trigger mitosis merely as a consequence of completing their preparations. Indeed, as we shall show, the cell cycle clock can bring them back from the brink of incipient mitosis when instructed to do so by our other hypothetical clock, the circadachron, and hold them in a kind of indeterminate extended $G_2$, although once initiated the mitotic process runs to completion. Similarly, Leedale (1957) has shown that *Euglena spirogyra* can be pulled back from early prophase by the onset of light, but again once past this stage mitosis is self-propelling. This may also apply to S, which involves the incorporation of nucleosomes as well as DNA synthesis, and possibly also to cytokinesis, although cells can apparently get "stuck" in the latter stages, maybe because they run out of available energy reserves (see Fig. 2f).

### The Concept of the Cytochron

We have called our cell-division-cycle clock the cytochron (Edmunds and Adams, 1981), and have elucidated quite a few of its properties. In LL it is fundamentally and effectively a cyclic clock: having completed programming of the CDC for one cell, its duplicates are immediately able to start up and program the CDCs of its daughter cells. Our cytochron becomes noncyclic, however, in the photoautrophic mode under LD regimes. Once it has triggered the terminal M and C stages of a cell division cycle its task is finished, but its duplicates in the daughter cells cannot start up and program the CDCs of the next generation without loss of time if M and C of the parent cell are completed in darkness. The daughter cell's biochemistry simply ticks over and "sits out" the dark period in a kind of untimed "$G_0$" phase of indefinite duration. As a result, no matter at what time during the night cells divide, be it early or late, their daughter cells are collected up in the $G_0$ state. Then at dawn the cytochrons in every cell start up in synchrony at the beginning of $G_1$. If dawn never comes they simply stay in $G_0$ indefinitely. If ethanol is introduced, however, they switch to the organotrophic mode and at the same time recyclize (as opposed to recycle) their cytochrons so that successive cytochron generations trigger one cell cycle after another without a break (Fig. 3c), just as they do in the photoautotrophic mode in LL (Fig. 2d).

Even in the photoautotrophic mode, however, if cells complete their cell cycles while still in the light pulse of a LD cycle, the cytochrons in the daughter cells can trigger entry into $G_1$ immediately after cell division is complete; however, since a minimum photoperiod of some 6 hr is apparently required before the cytochron can be allowed to run the full length of its track, if darkness ensues before the 6 hr are metered, the cytochron will stop and stay stopped until illumination is resumed.

In the case of the strain Z populations that double almost exactly in number each cycle under both LD: 14,10 and 16,8 regimes, it would seem that the energy input is finely enough balanced for all the cells to divide each night before the circadian clock can block them and hold them over to the latter part of the next light phase (Fig. 3 a and b). As the range of the generation-time distribution is apparently much greater in var. *bacillaris* populations, some cells are always available to be held over to the next cycle, and these cells may therefore continually feed into a subpopulation that recycles its cytochrons by virtue of their completion of cell division in the light, thus possibly accounting for the gradual rise in cell number during the light periods (Fig. 2c, e, and f).

So far we have ignored the question of the variability in the duration of the CDC in relation to the cytochron. Clearly, if we are not prepared to accept "sloppy timing" as the causal factor, another mechanism must be found, and since we postulate that the CDC phases are indeterminate in themselves, the variability has to be generated either by the cytochron or by some other timer that modulates the cytochron. Can we really justify our claim that the cells accumulate in a $G_0$ phase during the dark and start their $G_1$ phase in synchrony at dawn, thus implying that the spread of the cell division wave is due to variation in the duration of the CDC, and can we ignore the varying amounts of time spent in the dark by our $G_0$ cells? Apparently so, because if we compare the generation time distributions for asynchronous log-phase populations obtained by Cook and Cook (1962), with the range and form of the cell number curve for the putatively "dawn-synchronized" LD: 10,14 strain Z population of Fig. 4, the range is about the same and they are both skewed in favor of the shorter generation times. If we simply lined up all the cells in an asynchronous log-phase population at the start of $G_1$, the rising cell number curve would trace out the cumulative generation time distribution, and this seems to be just what the LD cycles achieve. It is as if each cell on completing division during the dark period has entered suspended animation after having been popped into liquid nitrogen, the entire population then being "thawed-out" at dawn to start the next cycle in synchrony. We conclude, therefore, that the spread of the rising cell number curve across the dark periods under LD regimes results from the same mechanism that generates the variability in generation time in asynchronous populations.

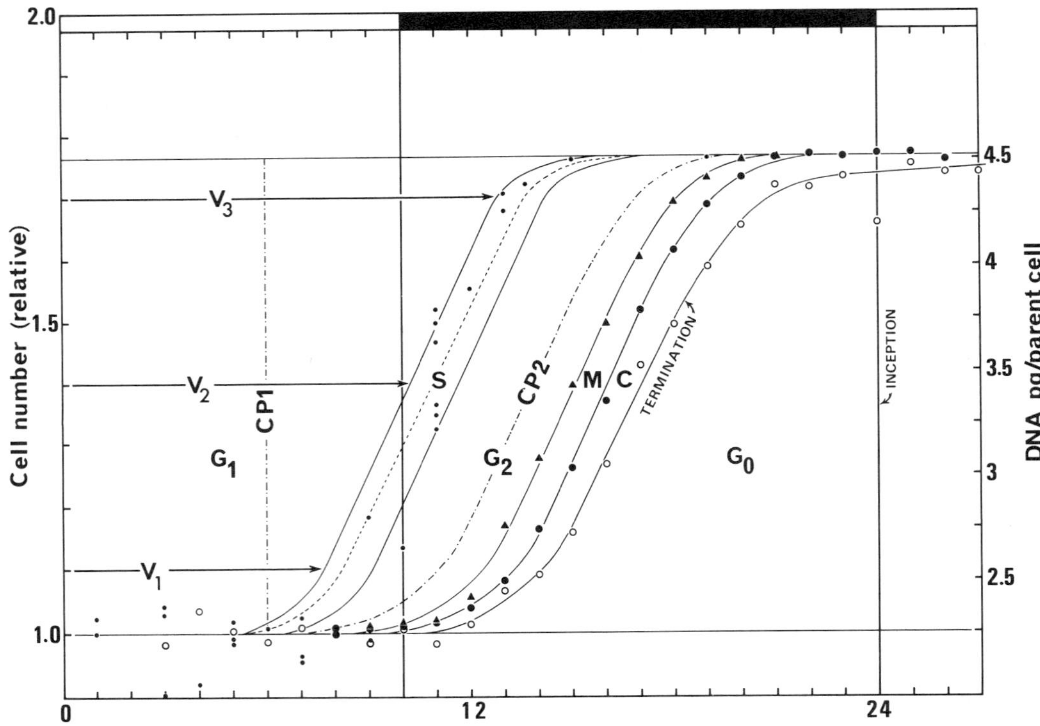


**FIGURE 4** Parameters of the cell-cycle phase distribution map for *E. gracilis* Z at 25°C under LD: 10,14 cycles in the autotrophic mode. Cell number (open circles), mitoses per 1000 cells, and cleavage counts per 1000 cells were scored for a typical cycle at a light intensity giving a step size of 1.76. After age-gradient correction the scores were used to determine the entry and exit curves for mitosis (triangles and solid circles) and cleavage (solid circles and open circles). The DNA content on a parent-cell basis is plotted for several replicate populations under LD: 14,10 cycles at different times during the cycle (small solid circles), normalized to the same height as the other curves. An eye-fitted curve through the DNA data gives an approximate time course for the stepwise increase during the S phase. Entry and exit curves for S have been drawn in on either side on the assumption that the variation in $G_2$ duration is minimal. The variation in the duration of the $G_1$ phase is "accounted for" by different varichron track lengths ($V_1$, $V_2$, and $V_3$), which start out at dawn to time the duration of the inception to S-entry gap. The essential light-driven early $G_1$ phase terminates at CP 1 and is equal to the shortest $G_1$ duration. The CP 2 control point occurs approximately 1 hr prior to mitosis, and the duration of mitosis is constant across the map at around 55 min. The duration of cleavage, however, is somewhat correlated with cell-cycle duration. On completion of division daughter cells enter the indefinite $G_0$ phase. They do not start $G_1$ until light-on induces inception of the CDC sequence. (DNA data from Edmunds, 1965b; other data from Adams and Edmunds, unpublished, 1970.)

Attempts have been made to fit the generation time distribution data for *Euglena* and several other organisms to a variety of theoretical curves, including the inverse-normal (Kubitschek, 1962; Castor, 1980), the log-normal (Schmid, 1967), the exponential (Smith and Martin, 1973), variants of the exponential (Brooks, 1976), and latterly, even the normal distribution (Hersh and Kitos, 1980). It is generally conceded, however, that the data are too noisy to distinguish unequivocally between continuous distributions that might fit one of these curves, on the one hand (Koch, 1980; Nurse, 1980; Smith et al., 1981), and on the other, discontinuous distributions that might be the result of repeated quantal cycles (Klevecz, 1976), or the random choice from a whole spectrum of genetically programmed timers (Edmunds and Adams, 1981). As in the case of mammalian cells, it would appear from the data for *E. gracilis* strain Z summarized in Fig. 4 that the majority of the variation in CDC duration is generated in the $G_1$ phase, bounded on the population map by the synchronous inception at dawn and the cumulative curve for entry into the S phase.

### The $G_1$ Varichron

We are now able to make an additional observation of relevance to all these models, however, as it seems that the mechanism generating $G_1$ variability, once set in motion, can run independently of the cell cycle, and presumably, therefore, of the cytochron. By extending all the cell cycles until this generator of variability has run down, the difference in CDC duration is effectively cancelled without affecting the relative distribution of individual cell cycles along the time axis. It is as if once the cytochron has committed a cell to $G_1$, an additional "hourglass" timer is started at the same time as the cytochron, the cytochron being unable to trigger S until the "sands" from this timer have run out. Given variable amounts of sand in the hourglass or "varichron" timers of each cell we have an analogue of the mechanism that generates $G_1$ variability. Imagine now an asynchronous population in which each daughter cell triggers its variable hourglass or varichron at the start of $G_1$. Suppose that we instantaneously stop all the cytochrons and keep them stopped for several hours, then simultaneously release them again. If the varichron timers have been running independently, some of them will have run out before the release point, and in the batch of cells involved the varichrons will no longer be able to hold up cytochron progress—CDC variability will have been lost for that generation. The population cell number curve, though remaining log-linear (because the population is still asynchronous) will steepen as these cells contribute their daughters to the population, but once this batch of cells has divided, the slope will abruptly decline to its former value. The batch of cells involved has not been synchronized; it has merely experienced a temporary reduction in its mean generation time. Such kinetics have been observed (Fig. 5) in the case of cell cycles extended by cold treatment (Adams, 1973), and perhaps most significantly when the cytochrons are stopped and restarted at circadian intervals by the circadachron (Fig. 3d). These kinetics begin to make sense with the realization that the CDC must be programmed by a clocklike timer (the cytochron) that can be stopped and started at any locus along its track. In order to discuss these concepts in some detail it is appropriate at this juncture to present our new ideas concerning the mechanism of interaction between *Euglena*'s circadian clock and its cytochron.

## THE CASE FOR A CIRCADIAN CLOCK

Thus far we have yet to present experimental evidence for our contention that a determinate sequence of cellular biochemistry is inadequate to explain the sequence of events of the CDC. Ironically, it was the discovery of an entrainable, free-running, circadian rhythm of cell division in *Euglena* that prompted a reexamination of the existing data from Edmunds and Funch (1969) and Adams (1973) that in turn gave rise to the concept of the cytochron, and latterly that of the varichron.

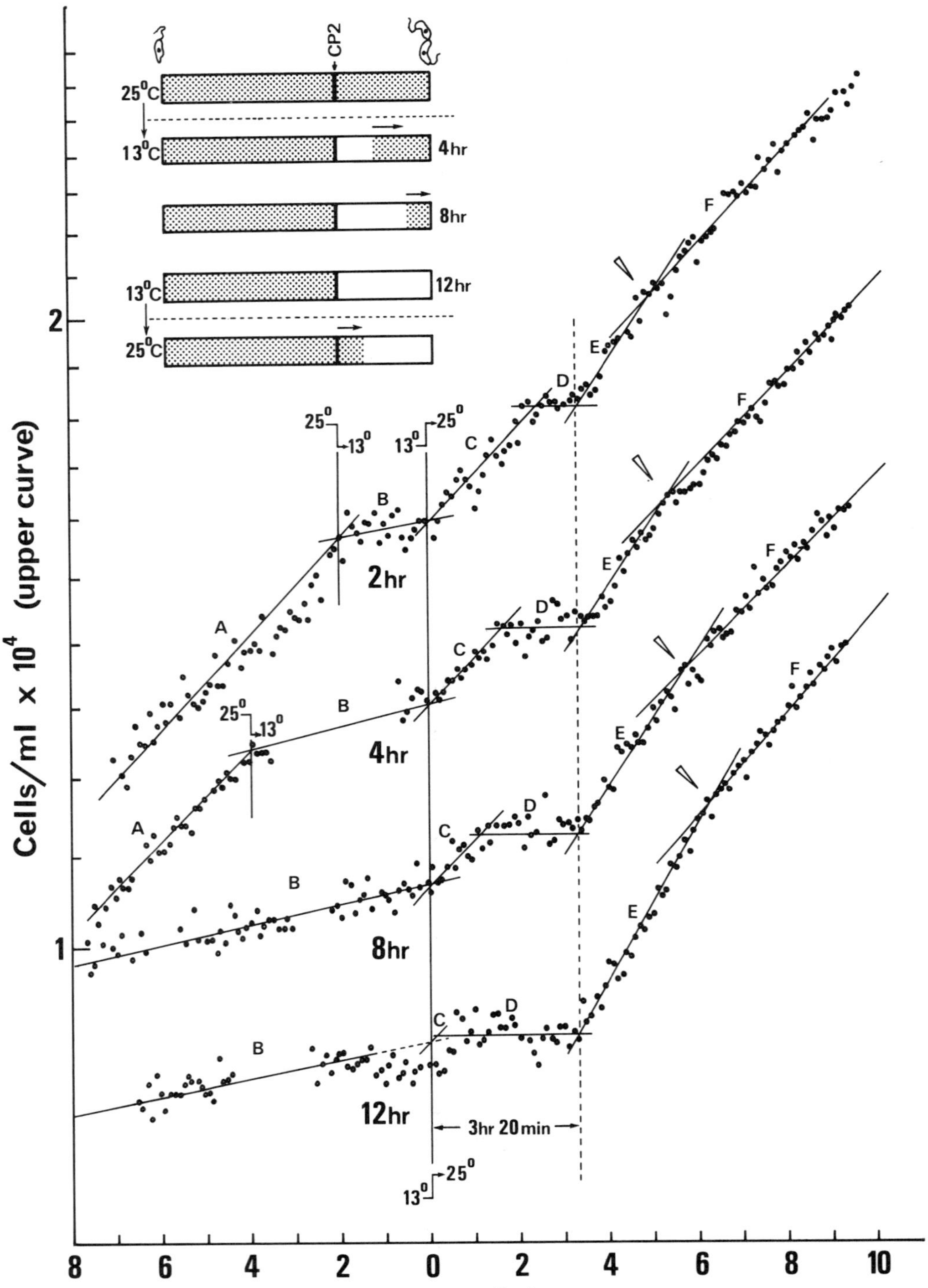


**FIGURE 5** Localization of a control point in the mid $G_2$ phase (CP 2) by analysis of cell number transients across low-temperature pulses. Automated cell counts at 6 min intervals. Replicate asynchronous log-phase populations of *E. gracilis* var. *bacillaris*, grown in

In 1964, Edmunds discovered that *E. gracilis,* once entrained to LD cycles, would exhibit a free-running rhythm of cell division with a period close to 24 hr when transferred to continuous dim (800 lux) light, although the number of cells dividing was small and the rhythm damped out to give asynchronous growth if the light intensity was raised (Edmunds, 1966). Later experiments demonstrated that both photosystem-2 defective (Jarrett and Edmunds, 1970) and aplastidic (Edmunds et al., 1976) mutants when grown on glutamate-malate medium at 19°C (but not at 25°C) can sustain vigorous free-running circadian rhythms of cell division in both DD and continuous bright (5000 lux) light. Furthermore, when grown asynchronously in DD and simply transferred to LL, they synchronize to give what is clearly an endogenously programmed, free-running, circadian rhythm of cell division (Edmunds et al., 1976). The cell number curves for these free-running circadian rhythms (Fig. 3d) are very different from those obtained with daily LD cycles (Fig. 3a and b), comprising log-linear, as opposed to sigmoid rising segments, interrupted at circadian intervals by steps. Furthermore, these rising segments are much steeper than the log-linear curves obtained for the same populations in the asynchronous mode, and the overall growth rate is the same as that of the asynchronous starter populations, despite the newly imposed circadian steps. Analysis of these growth patterns suggests that as expected on an organic substrate, the cytochrons must be operating in the cyclic mode and that the populations are therefore still in the asynchronous log phase. The rises and steps result not from synchronization but from the periodic extension of the cell cycles of every cell. They are not caused by localized CDC blocks or by the simple occurrence of permissive "division gates" at circadian intervals—both classic explanations of circadian cell division pulses. Every 24 hr synchronized circadian clocks in each cell of the population actually seem to stop all the CDCs, irrespective of their position in the CDC phase sequence, and then several hours later release them. It is this phenomenon that explains the log-linear nature of the rising segments and that has necessitated the concept of the cytochron.

If the CDC consisted of a continuous, determinate, chain reaction of rate-limiting biochemical events that gave rise to each of the classic phases in turn, any clocklike mechanism interfering with the sequence would be unlikely to be able to simultaneously

---

LL with 4% $CO_2$ in air as sole carbon source (slope A), were rapidly cooled to 13°C for 2, 4, 8, or 12 hr and then returned to 25°C. At 13°C all pre-CP 2 cells are apparently 'frozen' in their tracks, but post-CP 2 cells continue toward cell division, though more slowly (see CDC sequence charts top left), and contribute to the population rise (B), thus creating an increasingly wide phase gap, virtually all of them having divided after 12 hr at 13°C (see CDC charts top left). On return to 25°C, any residual post-CP 2 cells complete division at the same rate (slope C) as prior to the low-temperature pulse, whereas the now unfrozen pre-CP 2 cells have to traverse all the post-CP 2 CDC phases before dividing and contributing to the steep rise (E) at the end of the phase gap (D). The duration of this traverse locates the CP 2 control point at approximately 3 hr 20 min prior to CDC termination at 25°C. The rise (E) is log-linear with a doubling time of approximately 8 hr (equivalent to the shortest possible CDC at 25°C), and increases in amplitude with cold-treatment time, suggesting that contributing cells progressively lose their CDC variability during the cold pulses. Rise (E) inflexes abruptly (wedge symbol) to slope (F) which parallels the asynchronous starter population slope (A), and appears to be due to cells that were 'frozen' in the early stages of the CDC, and have retained their original CDC variability.

block all the reactions going on in cells of different cycle age. Blocks imposed at a limited number of key sites would no doubt suffice to freeze the sequence; they would, however, result in batches of cells in phase segments between the blocks accumulating in synchronous clusters behind each of them. On the other hand, if the biochemistry is indeterminate and the control positive, rather than negative, a clocklike programmer outside the substrate level biochemistry could induce or activate the enzymes required, say, for the S phase, in the appropriate place in the timed sequence, and the individual S phases would remain asynchronous even if the clocks in all the cells were stopped and later restarted. Thus, it is relatively easy to see how the stopping and starting of a CDC programming clock could leave the asynchronous status of the individual CDCs unaffected as apparently demanded by the observed kinetics, but extremely difficult to see how the CDCs of large numbers of cells in all possible stages of the cycle can be stopped simultaneously if such a clock does not exist.

Analysis of the log-linear segments of these free-running division rhythms has shown that their doubling times are equal to the *minimum* generation times that can be obtained in asynchronous populations grown under the same conditions. Thus, in addition to being frozen in their tracks during the steps, the cytochrons also appear to lose their variable length. While recognizing these phenomena, we have previously suggested that the circadian clock might insert extra "time loops" into the cytochron tracks that in effect stretch the CDC of each cell and thus create the steps; and that a "random choice" mechanism might similarly insert variable time loops to generate the variation in CDC length observed in both LD-cycled and asynchronous LL populations (Edmunds and Adams, 1981). Further analyses, however, have cast doubt on both these suggestions. Phase shifting experiments suggest that the width of the steps can be expanded or contracted by advancing or delaying the circadian clock (Fig. 6a-c). It therefore seems that the time gaps represented by the steps are actually metered by the circadian clocks; they stop the cytochrons at a certain circadian time and restart them, again at a definite circadian time. Similarly we have to conclude that the variation in CDC length that results in the typical generation time distribution, skewed in favor of the shorter generation times, is generated by a separate mechanism that runs in parallel rather than in series with the cytochron's time-metering track; hence our additional timer/clock, the varichron.

### The Concept of the Circadachron

The term *circadian clock* is useful in a general sense but restrictive when several types of clock, all with approximately 24 hr cycle times, have to be considered at once in the same cell, as will become apparent when we dissect the *Ceratium* cell division control mechanism. Furthermore, it seems probable that in *Ceratium* there are two circadian clocks that both respond to light signals and can therefore be expected to have different light-reset phase-response curves.

The clock that stops and starts the cytochrons in *Euglena* we have therefore called a *circadachron.* Recent analysis (KJA) of the skeleton-photoperiod and high-frequency LD cycle data of Edmunds and Funch (1969) has provided the clue as to why it has not been possible to achieve free-running circadian rhythms in wild-type cells in bright LL. Continuous bright light appears to "stop" the circadachron in the same way as it does in *Drosophila* (Pittendrigh, 1966) and *Gonyaulax* (Sweeney, 1969). The circadian clocks of these organisms have very similar properties to the circadachron of *Euglena* and are

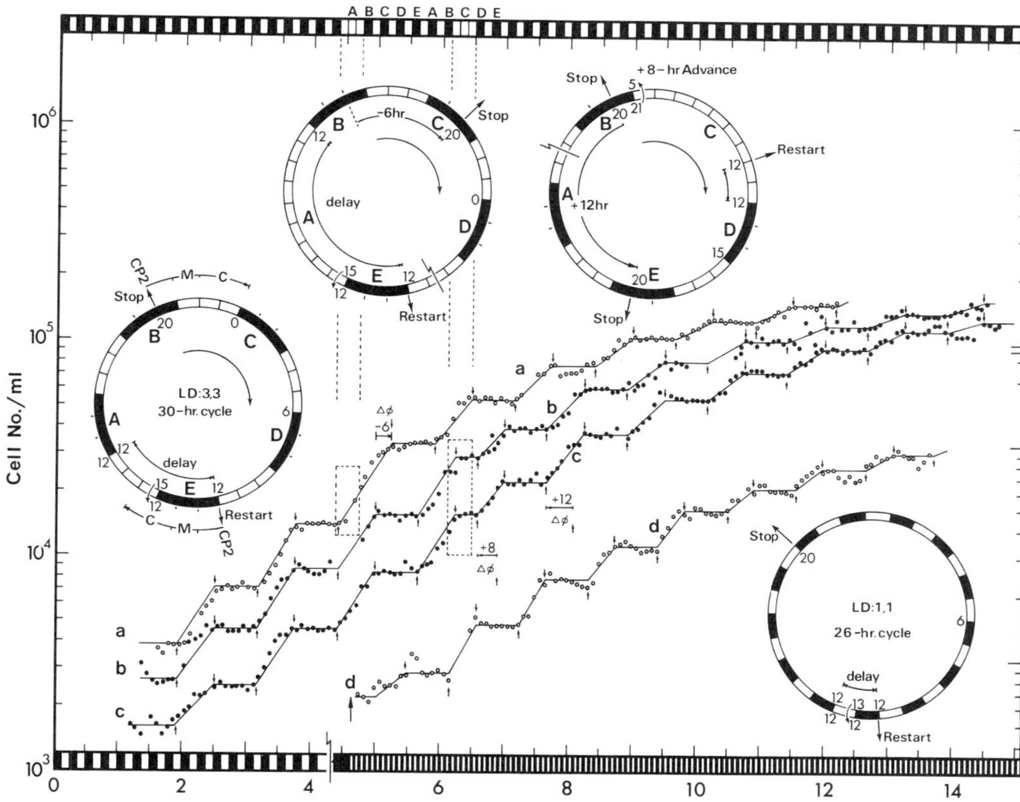


**FIGURE 6** Pseudocircadian cycles of cell division in *E. gracilis* Z obtained with high-frequency LD: 3,3 and 1,1 regimes, and their phase shifting by omission of single dark pulses. As the circadachrons have an endogenous circadian period of slightly in excess of 24 hr they reach CT 12 just after the onset of a dark pulse under 3,3 regimes (far left dial), and lock onto 30 hr repeating cycles of five light and dark pulses (A-E). On reaching CT 15 they are set back to CT 12 and held there for 3 hr, giving a total delay of 6 hr, and hence a 30 hr cycle. On reaching CT 20 they signal all cytochrons to stop, and the cell number curves (a, b, c) shoulder 4 hr later when all post CP 2 cells complete division. On first reaching CT 12 they signal the restarting of the cytochrons, the overt rise being evident 4 hr later. If one dark pulse (A) is omitted (top left dial and curve a) circadachrons are held at CT 12 for an extra 6 hr by the extended light pulse, resulting in a subsequent 6 hr phase delay ($-\Delta\phi$) of the overt rhythm. If one dark pulse (C) is omitted (top right dial and curves b and c) the extended light pulse confirms the potential advance shift ($+\Delta\phi$) at CT 21 (light-on) of about 8 hr. This pulls CT 12 into the light period, resulting in a 2 hr delay, but avoids the usual 6 hr delay resulting from a setback to CT 12. The steady-state overt rhythm is therefore phase advanced by +12 hr ($+\Delta\phi$). A population synchronized on LD: 12,12 cycles and released (arrow) into a LD: 1,1 regime, locks onto a 26 hr repeating cycle (dial, lower right and curve d). (An alternative interpretation of data from Edmunds et al; see Chap. 17.)

therefore circadachrons. In accordance with convention we define the circadian time that the circadachron restarts from at dusk as CT 12. Analysis of the full range of skeleton photoperiod and high-frequency LD data of Edmunds and Funch (1969) has facilitated the deduction of a set of formal rules that appear to govern the operation of the *Euglena* circadachron. Analysis of new phase-shift data obtained by Weiler, and reported here for the first time, suggests that *Ceratium* also has a dusk-synchronized circadachron with virtually identical characteristics to that of *Euglena.* In addition, however, *Ceratium* appears to have a dawn-synchronized clock that programs the timing of cell division unless preempted by the dusk-synchronized circadachron. As there is some evidence that the dawn-synchronized clock may be advanced by light-on signals, it is not yet at all clear whether this is a cytochron in the euglenoid sense, or whether *Ceratium* has both a photo- and a scoto-circadachron that interact to modulate the cytochron. This situation is reminiscent of the dawn and dusk oscillators of *Nasonia* (Saunders, 1976) and nocturnal rodents (Pittendrigh and Daan, 1976).

### *The* Euglena *Circadachron*

The formal rules that appear to govern the behavior of the *Euglena* circadachron have been derived following the examination of a wide range of experimental data on the basis of the assumption that the circadachron has a natural period very close to 24 hr, and that this period is remarkably stable and accurate for the conditions of any one experiment. The rules may be summarized succinctly as follows. (1) *Euglena*'s circadachron will stop if it reaches CT 12 during a light period or light pulse and will stay stopped until light-off. (2) If darkness prevails as the circadachron meters its time track through CT 12 it will run on, but should a light pulse intervene during its traverse of the CT 12 to CT 18 segment of the circadian cycle, it will be set back to CT 12 and held there until darkness again ensues. (3) If the circadachron reaches CT 12 in the light and is stopped and held there until dusk (or if it is set back to CT 12 and held there until dusk), then it cannot be set back again by a subsequent light pulse arriving on the CT 12 to CT 18 segment during the same circadian cycle. (4) The segment of the circadian cycle from CT 6 to about CT 12 is insensitive or "blind" to light signals. (5) In common with the circadachrons of other organisms, the *Euglena* circadachron is sensitive to light-on transitions (but blind to light-off) in the segment from about CT 18 to about CT 6, the responses in this segment being "patterned" advances, i.e., the magnitudes of the advances are nonlinear and are characteristic of the circadian time of the light-on transition. (6) The advance expected of a given light-on signal is negated unless it is confirmed by a light pulse in excess of some 4-5 hr. These features are summarized in the light-response resetting map and the phase-response curve of Fig. 7. The data for the *Euglena gracilis* mobility circadian rhythm from Feldman (1967) and Schnabel (1968) have been analyzed on the same basis and it would seem that the circadachron involved in control of mobility is identical to that responsible for the photoperiodic effects on cell division. The mobility data have therefore been plotted together with the few values available for the cell division rhythm.

Edmunds and Funch (1969) interpreted the various regular cell-division-rhythm periods observed over a range of high-frequency LD cycle regimes as being free-running rhythm variants, with differences in period ranging from 24 to 33 hr. This, unhappily, suggests that slight differences in culture conditions produce changes in the natural period of the circadachron of as much as 9 hr. A new analysis of this data has revealed, however, that if a stop/reset to CT 12 mechanism is assumed, these large differences in period can be accounted for and predicted, not on the basis of free running, but as a result of the synchronizing effect of the high-frequency cycles. Using this interpretation it is possible to reduce the estimate of the variation in natural period at 25°C to no more than the ±0.3 hr observed for the free-running mobility rhythm (Brinkmann, 1966; Schnabel, 1968). In accordance with this hypothesis the LD: 3,3 high-frequency cycle data of Edmunds and Funch (1969), and the LD: 1,1 data (Edmunds and Laval-Martin, unpublished data) can be interpreted as shown in Fig. 6. Although the dark pulses probably induce small delays in individual cells (as newly divided daughter cells collect briefly in $G_0$) the considerable variation in CDC length precludes the synchronization of the cytochrons into a single 24 hr time frame, so that the cytochrons are effectively asynchronous and free running. In contrast, the precise natural (endogenous) period of the circadachrons results in resonance with the regular high-frequency LD regimes. Furthermore, no matter whether the cells enter the 3,3 or 1,1 regimes from LL or from, for example, LD: 12,12, the circadachrons will be restarted synchronously from CT 12, either at the onset of the first dark pulse, if led in from LL, or from the beginning of the last dark pulse of a 12,12 regime. Once already synchronized, resonance with the high-frequency LD pulses maintains circadachron synchrony indefinitely.

Since the 3 hr light pulses are too short to be confirmed as light-on advancing signals, the circadachrons are able to meter time round from CT 18 to CT 12 oblivious to the high-frequency LD regimes. When, however, CT 12 falls during or just after a light pulse, the circadachrons will either be stopped and held, or reset back and held, for the duration of one of the light pulses (see the far left and bottom right dials of Fig. 6). For LD: 3,3 possible integral multiples of the 6 hr regime close to the natural circadian period are 24 and 30 hr. If the natural period was exactly 24 hr, or slightly less, then the circadachrons would be able to lock on to a 24 hr repeating cycle; if slightly more than 24 hr (say, 24.1 hr) the circadachrons would precess against the 30 hr cycle until CT 12 occurs just after the onset of a dark period. At the beginning of the next light pulse (since they would have run on through CT 12 in the dark, to in our model, CT 14.9) they would be collectively set back to CT 12, held there for the duration of the 3 hr light pulse, and then allowed to restart from CT 12 at the onset of the next dark pulse; a total delay of 6 hr is thus added to the natural period of 24 hr, the difference of 0.1 hr being lost in the reset. Since the data in Fig. 6 clearly fit a 30 hr repeating cycle, the natural period of these populations must slightly exceed 24 hr. (See also the LD: 3,3 data in Chap. 17). A similar argument applied to the 1,1 data of Fig. 6d predicts a 26 hr repeating, lock-on cycle, the good fit of the data again confirming a natural period of slightly in excess of 24 hr.

Thus, under these high-frequency LD regimes, populations with natural circadian periods of slightly in excess of 24 hr lock on to exact "pseudocircadian" repeating cycles equal to 24 hr plus the duration of the integral LD component. The slight inherent slowness of the natural cycle, together with any intercell sloppiness in timing, is taken up in the reset-back to CT 12. Although no diurnal periodic cues are supplied by these high-frequency cycles, the periodic coincidence between CT 12 and a light pulse at circadian intervals serves to maintain tight synchrony of the circadachrons, each being locked on to the pseudocircadian cycle characteristic of the particular LD regime.

A further complication can theoretically arise if the natural period of the circadachron in any one strain of *Euglena* has a value of around 24 hr with the slight spread of the "noise" in the system straddling 24 hr. Under a LD: 3,3 regime a population with a natural period of, say 24 ± 0.3 hr would give rise to an overt rhythm of 27 hr, half the

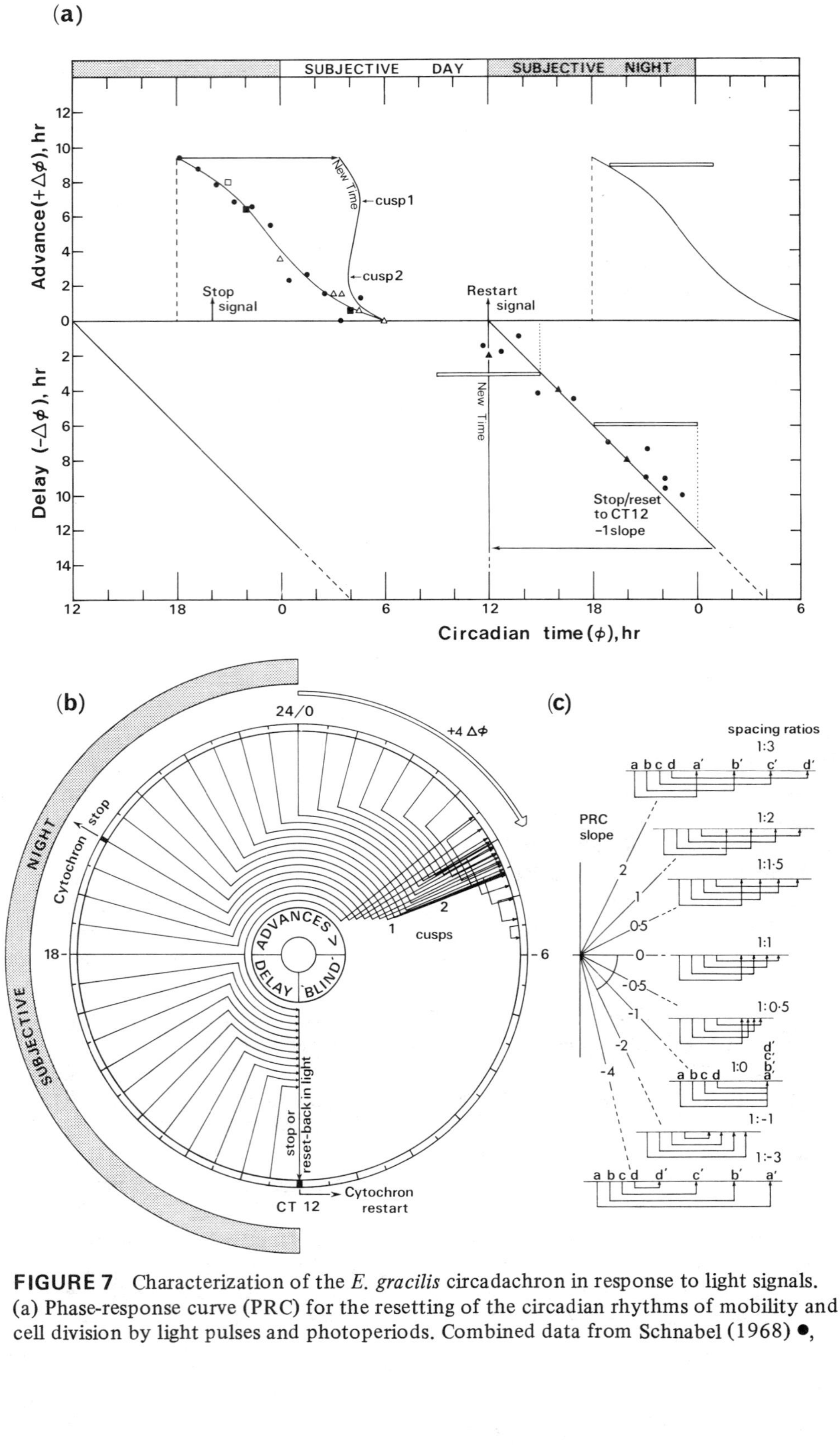


**FIGURE 7** Characterization of the *E. gracilis* circadachron in response to light signals. (a) Phase-response curve (PRC) for the resetting of the circadian rhythms of mobility and cell division by light pulses and photoperiods. Combined data from Schnabel (1968) ●,

population reaching CT 12 before the coincident dark pulse and half reaching it after its onset, the former cells having their period stretched a fraction of an hour to fit the 24 hr cycle, the latter being set back and held until they fit a 30 hr cycle. Even more complex "hybrid" overt rhythm periods can be envisaged for populations with periods very close to, but not exactly 24 hr, that have their variability straddling 24 hr. Virtually any period could be expected in such hybrids in the range 24-30 hr, depending on the relative proportions of the population being pushed in one direction or the other. Fortunately, the majority of the available data of Edmunds and coworkers suggests that the *E. gracilis* strain Z clone used has maintained a circadachron period somewhat in excess of 24 hr (at 25°C), with most of its variability also in excess of 24 hr, for many years.

Many of the high-frequency cycles result in more complex responses than those of the 3,3 and 1,1 regimes, giving, for example, bimodal and trimodal symmetries, accounting for pseudocircadian rhythms of 12 hr in the case of 4,8 cycles, for instance (Edmunds and Funch, 1969). For regimes in which the light pulses exceed 5 hr in length the CT 18 to CT 6 "advancing segment" is able to respond. Thus, in the case of a 10,10 cycle, the endogenous circadian period is actually shortened by the advance at light-on to give a very tight, stable, lock-on cycle of 20 hr (Edmunds and Funch, 1969).

Investigation of the cell number kinetics of these various LD cycles suggests that the circadachrons stop when they reach CT 12 during a light pulse, collect there in syn-

---

Feldman (1967) ▲, for mobility; from Adams ■, and from the LD: 3,3 data of Fig. 6 □, for cell division. Available values for *C. furca* △ are plotted for comparison. Data points plotted at the time planes of light-on for advances, and at light-off for delays. All delays assumed to result from stopping at CT 12 in the light, or resetting back to CT 12, the clock remaining at CT 12 in both cases until darkness ensues. Note the satisfactory fit to a -1 slope consistent with this assumption. Examples of resetting with 6 hr light pulses, shown as narrow bars: A 6 hr pulse starting at CT 9 only gives a 3 hr delay because the clock does not stop until it reaches CT 12. A 6 hr pulse starting just before CT 18 will reset the clock back to CT 12 and hold it there until the end of the pulse. A 6 hr pulse starting just after CT 18 will advance the clock at light-on of the pulse as defined by the advance segment of the PRC. (b) Resetting map for the light-induced resetting of the *Euglena* circadachron. Concentric arrows, clockwise for the advancing segment, and anticlockwise for the reset to CT 12 segment (delay), link the times that the circadachron is challenged by a light pulse with the times to which it resets. Note the double cusp in the advancing segment reset pattern caused by two nodes or "chronostats" centered on the -1 slope (see text for discussion of significance). (c) The relationship between PRC slope and resetting map spacing ratio. The times at which successive equally spaced pulses (a, b, c, d), fall on a segment of given slope, and are reset to (a′, b′, c′, d′), are linked by arrows. For a given slope (S) there is a simple relationship between the spacing of the time of occurrence of the light signals (set at unity) and the spacing of the times they reset the clock to. The spacing ratio is defined as 1:S + 1. For slopes within the segment 0 to -2, differences in timing between clocks in a population are reduced after being reset. All clocks are set to the same time at a slope of -1, and on either side the pattern opens out until the initial and final spacings are equal at slopes of 0 (spacing ratio 1:1) and -2 (spacing ratio 1:-1), and any timing differences are conserved. Note that for slopes of $< -1$ the sequence a, b, c, d is inverted to d′, c′, b′, a′, hence the negative ratio. For slopes of $>0$ or $< -2$ the differences in timing between clocks are compounded by the resets. Thus for slopes of +2 and -4, differences are magnified by a factor of 3.

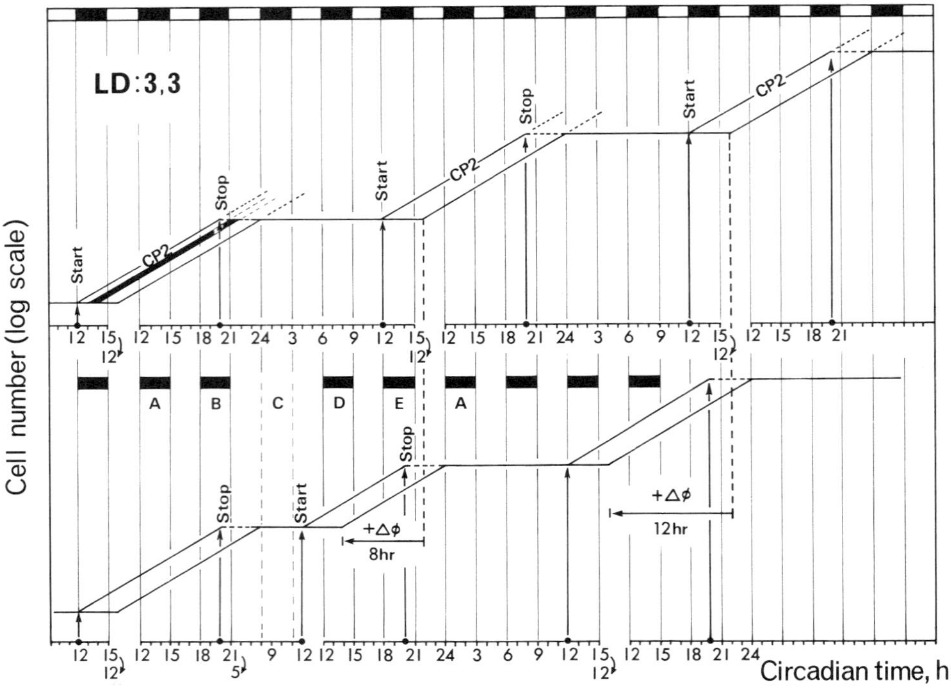


FIGURE 8 Mechanism of the phasing of cell division in autotrophic *E. gracilis* Z under high-frequency LD: 3,3 regimes on the assumption that a stop/reset to CT 12 occurs once every 30 hr cycle to lock the cells in synchrony with the LD regime (see also Fig. 6). In the top diagram light-on coincides with the circadachron reaching CT 15 and resets it back to CT 12, holding it there for the duration of the light pulse. It runs on again at the onset of the next dark pulse, unimpeded, until it reaches CT 15 again. The process is repeated every 30 hr cycle of five light and five dark pulses. On reaching CT 12 the circadachron releases all the cytochrons and 4 hr later the cells, having progressed through mitosis (black diagonal) and cytokinesis (clear diagonal), start to contribute to the population. At CT 20 the circadachron stops all the cytochrons in pre CP 2-phase cells, and 4 hr later the cell number curve inflexes as the post-CP 2 cells complete division. In the lower diagram, leaving out the dark pulse (C) results in a sufficiently long light pulse to confirm the potential reset at CT 21 (light-on), thus advancing the circadachrons to CT 5. This advances the CT 12 release point by +8 hr. Note, however, following through the subsequent sequence, that the final overt shift of the rhythm is actually +12 hr (see also Fig. 6b and c).

chrony, and then run on in unison from CT 12 at dusk, ignoring any subsequent light pulse during the same circadian cycle. On reaching CT 20 the circadachrons then generate a signal that stops all the cytochrons not irreversibly committed to triggering cell division, and freezes them in their tracks. They are then restarted as the circadachrons first reach CT 12 of the next circadian cycle (see Fig. 8).

With both *Euglena* and *Ceratium* it is possible to test whether the circadachron actually stops at CT 12 during a light period or is merely reset back to CT 12 at dusk. Under LD: 16,8 cycles *E. gracilis* var. *bacillaris* expresses the CT 20 shutdown in the early part of the light period. Should the circadachron run on through CT 12 in the light, as a result of the advance at light-on it should reach CT 20 again shortly after the start of the main cell division rise, and shut down cell division. As it clearly does not do so Adams, unpublished results), we conclude that the circadachron in *Euglena* stops when it reaches CT 12 during a light period and remains stopped until dusk.

### *Control Points—The Interclock Links*

The cell number kinetics for the high-frequency LD cycles also reveal that it takes 3-4 hr for the response to the CT 12 and CT 20 signals to become overt. This implies that progress toward cell division cannot be stopped by the circadachron past a point approximately 1 hr prior to mitosis. We have previously recognized loci along the cytochron track that appear to be natural control points (Edmunds and Adams, 1981) susceptible to both internal and external influence (CP 0, CP 1, and CP 2). The CP 0 control point potentially interconnects the inception and termination of the cytochron track and is the locus at which the switch that determines whether the cytochron is operating in the cyclic or noncyclic mode is postulated to occur. CP 1 lies in late $G_1$, and marks the end of the cytochron metered "essential-light-phase." CP 0 and CP 1 appear to correspond to the *Chlamydomonas* A point and T point of Spudich and Sager (1980), which suggests that the essential light phase of *Euglena* requires a constant photosystem-2 input to keep the cytochron running.

Evidence for the location of the CP 2 control point at 3-4 hr prior to termination of the CDC, i.e., approximately 1-2 hr prior to mitosis, at 25°C, is presented in Fig. 5 and in Adams (1973). Rapid cooling to 13°C can be seen to freeze the cytochrons in their tracks unless they have progressed beyond CP 2.

The maintained log-linear kinetics subsequent to the chill treatments suggests that rather than a general metabolic slowing down, the effect is due to a "safety-catch" mechanism that shuts down all cytochron progress as soon as possible adverse conditions are detected. The same effect, namely, a prevention of CDC progress toward cell division, has been observed when cells are transferred from LL to DD, when $10^{-3}$ M mercaptoethanol is added (Adams, 1966), and as a transient effect when $CO_2$ is flushed from the fermenter medium. All these observations suggest the existence of a control point at a locus 1-2 hr prior to mitosis in mid-$G_2$; hence the designation CP 2.

One possible explanation of these phenomena is that the cytochron track only extends as far as CP 2, the events of M and C being self-propelling once initiated, thus implying that CP 2 and CP 0 are one and the same control point. This would also explain why the circadachron appears to operate at the CP 2 control point. If it does indeed stop all cytochrons progressing along their tracks, and thus in turn prevents the further initiation of phase transitions, the terminal phases of M and C would escape if they have already been set in motion by the CP 2 signal at the end of the cytochron track (see Figs. 6 and 8).

Control points CP 1 and CP 2 in *Euglena* almost certainly correspond to the $G_1$ ("start") and $G_2$ potential arrest points of *Schizosaccharomyces pombe* and *Saccharomyces cerevisiae.* Furthermore, the recent discovery that the temperature-sensitive mutant *cdc 2* and *cdc 28* gene products are required to override *both* arrest points at permissive temperatures in the two yeasts (Beach et al., 1982) suggests that the cytochron normally stops at CP 1, and again at CP 2, unless these two gene products are released by the CDC control mechanism following assurance from "environmental probing" that it is (probably) safe to proceed.

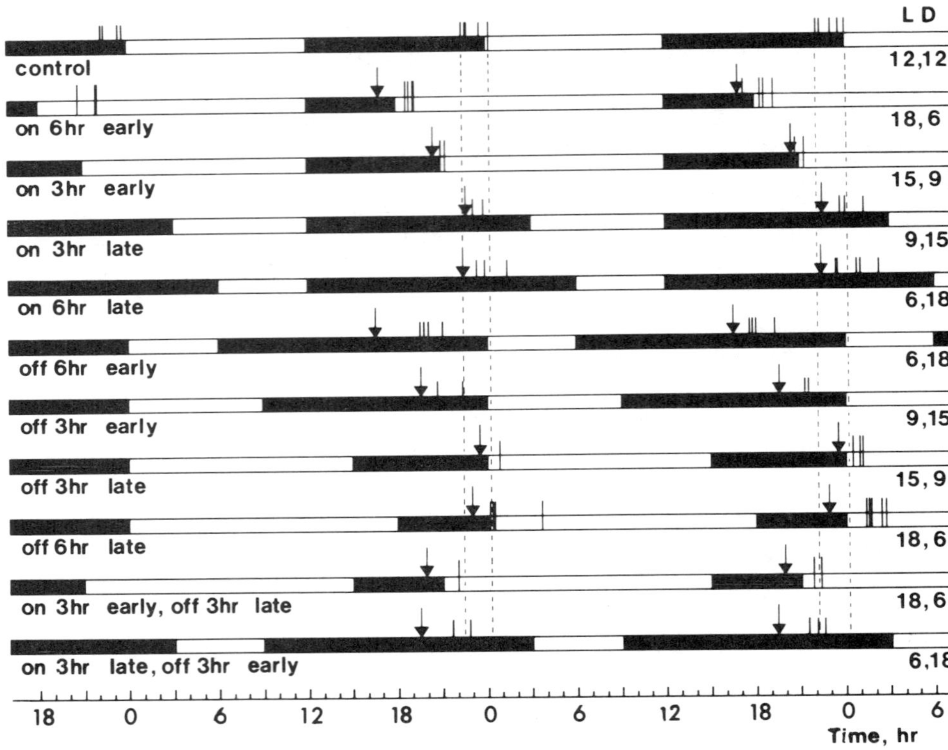


**FIGURE 9** Phase-shift transients in the timing of cell division in *C. furca* induced by subdividing 10 liter cultures preconditioned on LD: 12,12 for 3 weeks, and transferring them to the new LD regimes as specified at the end of each track, where "early" and "late" are in relation to the LD: 12,12 preconditioning and control cycles. The phase-shift experiment was repeated three times, with duplicate subcultures per treatment, on cells grown from an Oslofjord population at 18°C ± 0.5°C and 1.2 × $10^{16}$ quanta/$cm^2$ per sec (400 to 700 nm) throughout. Hourly samples were fixed and 300-500 cells scored per sample per treatment for the frequency of recently divided (hornless) plus partial horn-regenerated cells. The time planes at which the recently divided cell fraction reached its ½ max value are shown for each sample as vertical strokes. The dashed vertical lines encompass the range of values obtained for the LD: 12,12 control regime, during which the cells would be expected to divide unless influenced by the new LD regimes, for comparison. The vertical arrows indicate the steady-state ½ max value time planes that each subculture is expected to precess to after reaching steady-state under each appropriate new regime (from the same data as Fig. 1.)

### *The* Ceratium *Circadachron*

Since in *Ceratium* the cell-division waves have an almost negligible spread (Fig. 1), if the varichron mechanism of *Euglena* exists at all in *Ceratium,* it must be strictly quantized into 24 hr segments. Furthermore, again unlike the division wave of *Euglena,* the *Ceratium* division wave does not have a constant relationship with either dawn or dusk. For day lengths in the range 6-13 hr the division wave runs parallel to dusk, occurring some 11 hr after the onset of darkness, whereas in the range 13-18 hr it runs parallel to dawn astride the L to D transition (Fig. 1). This arrangement suggested that a dusk-synchronized circadachron might be responsible for the dusk-related segment up to a day length of 13 hr, and a cytochron might be the basis for the dawn-associated response for day lengths exceeding 13 hr.

Analysis of data from previously unpublished phase-shift experiments, designed and executed by Weiler (Fig. 9), has permitted one of us (KJA) to construct a 2-clock model based on separate dawn- and dusk-synchronized clocks. Weiler's experimental design is particularly useful for analyzing the effect of both the light-on and the light-off signals in photosynthetic organisms, since it does not necessitate maintaining the cultures in continuous darkness. The "light-on early" data confirms the predictions of the model that when LD regimes are adjusted so that a circadachron of the *Euglena* type is uncoupled from CDC control, the cell division wave is timed solely from the previous dawn (see Fig. 10, tracks n and o). It appears, therefore, that a dawn-starting cytochronlike clock programs the division waves to occur some 23 hr post-dawn as the "fall-back" position. A dusk-synchronized circadachron of the *Euglena* type can then preempt the cytochron by initiating cell division some 11 hr post dusk, presumably by cutting short the $G_2$ phase. Once the day length reaches 13 hr, both clocks would be expected to trigger cell division simultaneously (see Fig. 1 and Fig. 10, tracks c and d), whereas at longer day lengths the cytochron would preempt the circadachron. Iterative analysis of all the phase-shift data confirms that the presumed circadachron has virtually identical characteristics to that of its *Euglena* equivalent. It is insensitive to light in the CT 6 to CT 12 segment, has a stop/reset to CT 12 mechanism, and has a dawn-triggered advancing segment with patterned advances that, as in *Euglena,* pulls CT 12 into the light period and thereby collects all the circadachrons together for a synchronous restart from dusk. Furthermore, as in *Euglena,* the phase-shift data seem to rule out the possibility that the circadachron keeps on running past CT 12 in the light and resets to CT 12 at dusk. When dawn is given 6 hr early following a LD: 12,12 entraining regime (Fig. 10, tracks l to o), the division wave is advanced some 4 hr during the subsequent cycle, as if the dawn-synchronized cytochron has taken over the timing. With this arrangement dawn should coincide with the circadachrons at CT 18. Slow ones would be expected to be reset back to CT 12 and held there for the duration of the light period, while fast ones would be given a maximum advance of the order of 10 hr (see *Euglena* PRC, Fig. 7). Either way, if not stopped and held at CT 12 for the duration of the light period, it would trigger cell division several hours earlier than actually observed. Analysis of the data also suggests that the point in the CDC that the two clocks trigger cell division is approximately 2.5 to 3 hr prior to the recently divided cell stage scored by Weiler as a measure of the timing of cell division for the phase-shift experiments. Subtracting 1.5 hr to encompass the prophase to cell separation sequence, the division switch must occur about 1.5 hr prior to mitosis, agreeing well with the location of CP 2 in *Euglena* (see Fig. 10, tracks l and m). Using this information, the circadachron cell division control locus comes out at around CT 20.5 and that for the cytochron at 21.5 hr post dawn. Furthermore, the track length of the cytochron appears to be only 21.5 hr. The "on 6 hr early" data (Fig. 10, tracks l and m) is also of interest because the division wave in the subsequent dark period is actually 1.5 hr later than expected on the basis of a synchronous cytochron start from dawn. In addition, the division wave expected to occur during the initial extra 6 hr of light is actually advanced some 1.5 hr (see track l, left). The only logical explanation is that the cytochron must have been advanced by 1.5 hr at the dawn transition (see track m). If it has an advancing segment, can this be a cytochron in the euglenoid sense? Or does *Ceratium* have a "photocircadachron" in addition to its cytochron? The dilemma cannot be resolved until further

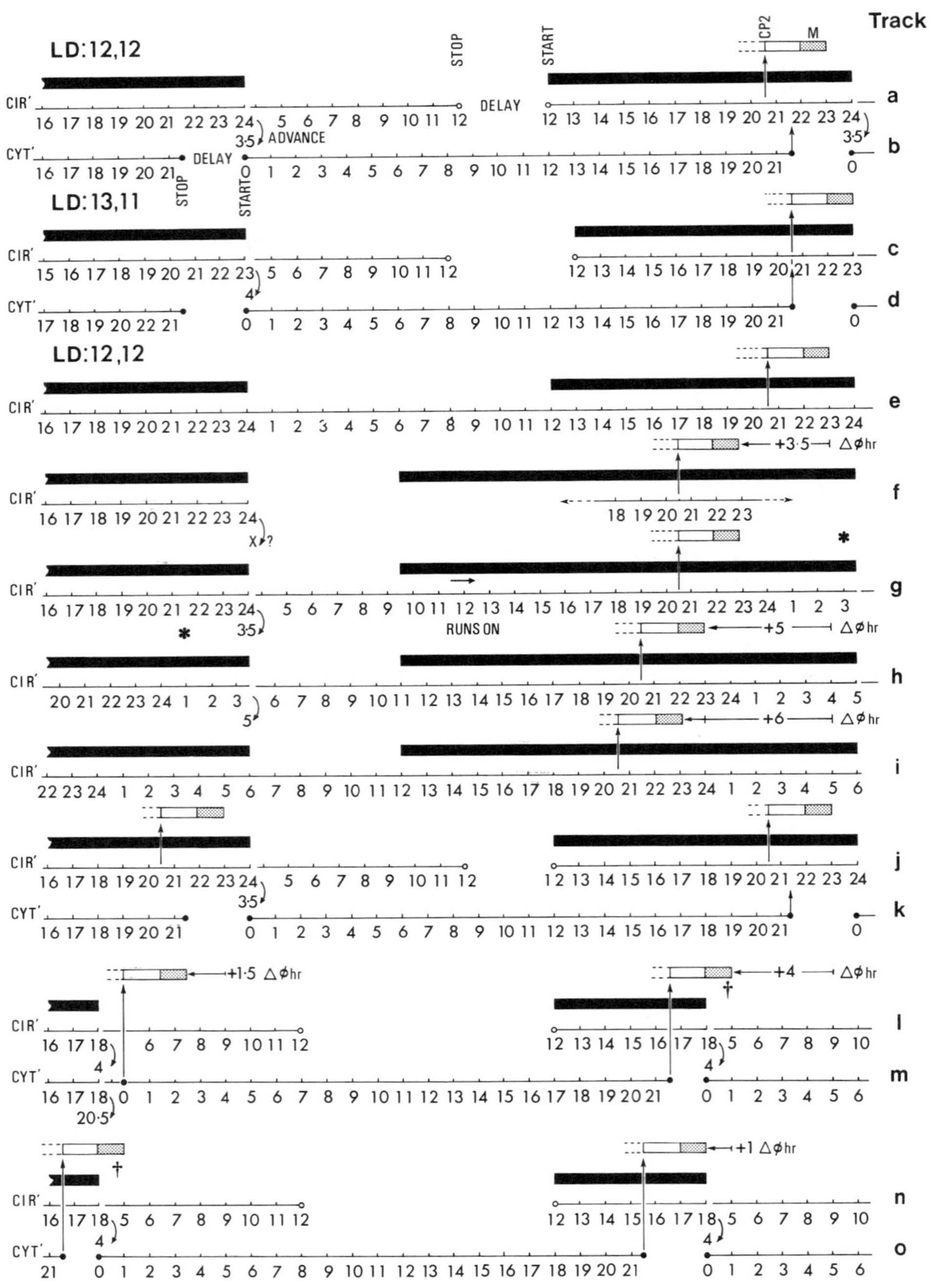


**FIGURE 10** A model for the control of the timing of cell division in *C. furca* based on the data of Figs. 1 and 9. In track (a) the dusk-synchronized circadachron (CIR') meters time from left to right across a LD: 12,12 cycle. At dawn it reaches CT 24 and is phase advanced +3.5 hr, then meters on to CT 12, stops, and remains at CT 12 until dusk. Restarting at dusk it runs on to CT 20.5 and triggers cell division at the CP 2 control point. Meanwhile the cytochron (track b), which has a cycle time of only 21.5 hr, starts up from

data are available. It does mean, however, that there may not be a single light reset phase-response curve for the circadian rhythm of cell division and that the kinetics must be separately unraveled to reveal the PRC for both the dawn- and the dusk-synchronized clocks. To avoid complicating the situation at the present time, the dawn-synchronized clock in *Ceratium* is referred to as the cytochron in Fig. 10, cytochron time (CYT′) being defined as running from CYT 0 when synchronized for a dawn start (Fig. 10, track b).

### *The Light Reset Phase-Response Curve*

Although reset phase-response curves for both light (Feldman, 1967; Schnabel, 1968) and temperature (Brinkmann, 1966) have been obtained for the mobility rhythm in *Euglena*, until recently, no data were available for the circadian rhythm of cell division. Edmunds et al. (1982) (see also Chap. 17) have published a PRC based on the, perhaps, simplest assumption that under LD: 3,3 cycles *Euglena* exhibits a free-running rhythm of cell division, and that the phase of the rhythm can be judged by inspection of the inflexion points of the cell number curves (although they noted that, indeed, there may be some interaction of the high-frequency cycle with the putative circadian clock).

If, however, a stop/reset to CT 12 lock-on mechanism exists, as we propose in this chapter, an alternative interpretation is inevitable. On this basis it should only be possible to replace five dark pulses with a 3 hr light pulse (giving together with the 3 hr pulses on

---

CYT 0 at dawn and meters through to CYT 21.5, its terminus and the locus of the division trigger, in this case preempted by the circadachron. It stops at CYT 21.5 until light-on at dawn recycles it. In tracks (c) and (d) the same pattern is repeated for a LD: 13,11 regime. As the circadachron is held at CT 12 for an extra hour, both clocks now trigger cell division simultaneously. For day lengths >13 hr the cytochron will preempt the circadachron, as the latter will be held at CT 12 progressively longer as the day length increases. The phase advances (see Fig. 7) were deduced as follows: In track (e) the circadachron is shown running unimpeded across a steady-state LD: 12,12 cycle. To show that it is advanced at dawn, and that CT 12 is pulled into the light period, dusk is brought forward 6 hr, thus "covering-up" CT 12 and allowing it to run on in darkness. This results in a 3.5 hr advance of the timing of cell division (track f, right). Writing in the hours in both directions from CT 20.5 (track f), we are able to deduce the unknown advance (X?) from the 3.5 hr mismatch at dawn (track g). By maintaining dusk at 6 hr early for subsequent cycles a series of advance values can be determined for other circadian times. At the end of track (g), dawn coincides with the circadachron at CT 3.5. Following the sequence (*) to the next cycle, since cell division is advanced a further 1.5 hr (track h), we deduce that the advance for CT 3.5 is +1.5 hr, taking it to CT 5. Finally, (track i), successive small advances push CT 6 to the dawn locus. Since CT 6 is at the start of the "dead zone" of the PRC (Fig. 7), the circadachron locks on to this position under what is now a steady-state LD: 6,18 regime. Slightly slow clocks are advanced to CT 6; slightly fast ones collect at CT 12 until dusk. If a LD: 12,12 entrained population (tracks j and k) is subjected to dawn 6 hr early (l and m), the circadachron is either advanced from CT 18 to around CT 4 (advance of 10 hr) or set back and held at CT 12 for the duration of the light period (not shown). Cell division is advanced immediately (track l, left) some 1.5 hr by the earlier dawn. Thus the cytochron must have been advanced by +2.5 hr at dawn, taking it to CYT 20.5. Being illuminated, it now recycles through CYT 21.5/0 and triggers cell division again 21.5 hr later. Following the sequence (†) to the next cycle (tracks n and o), the cytochron locks on to a steady state under the new 18,6 regime and takes over the timing of cell division.

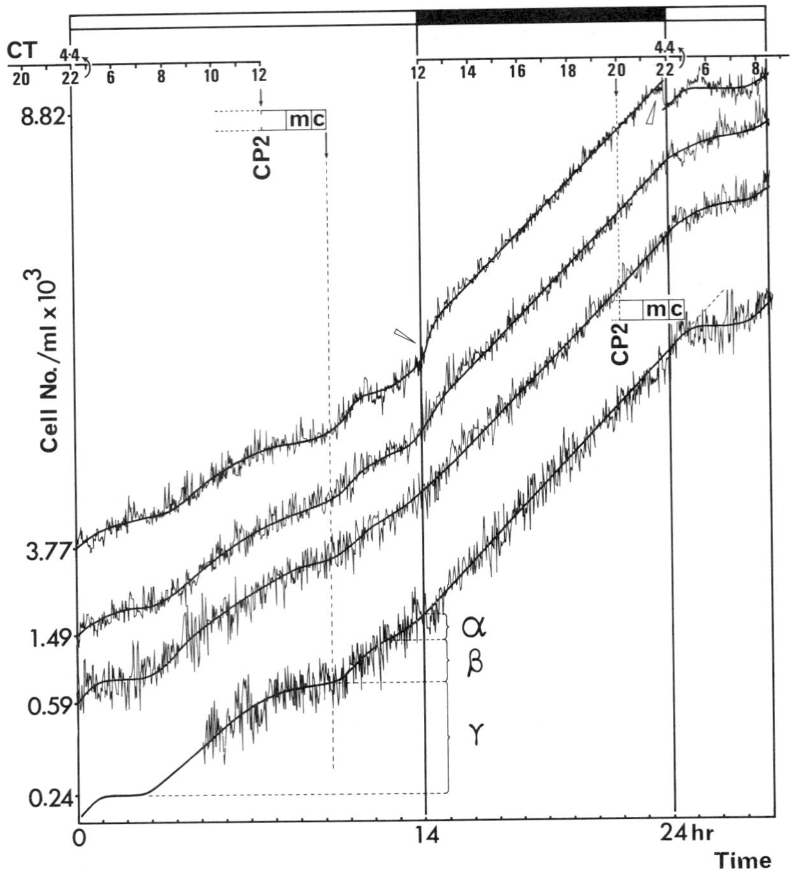


**FIGURE 11** Semilog plot of automated cell counts at 100 sec intervals for a *E. gracilis* var. *bacillaris* population under four successive LD: 14,10 cycles at 25°C, inoculated from a stock culture maintained in continuous light for 11 years. Note the breakdown in division synchrony with the occurrence of a second wave of division during the light periods. The main cell division wave rises at 12.5 hr post dawn with some cells dividing in the light (α). Circadachrons restart from CT 12 at dusk (top centre), meter time to CT 20, and then stop all cytochrons that have not reached the CP 2 control point in their tracks. Post-CP 2 cells add to the population rise for the next 2.5 hr. At light-on the circadachrons are advanced from CT 22 to CT 4.4. They then meter on to CT 12, and stop, at the same time releasing those cells held over from the previous CT 20 freeze. These cells after completing mitosis (m) and cleavage (c) begin to divide some 2.5 hr later, forming fraction (β). The large fraction (γ) appears to be due to cells held over from the previous cycle that are released early to complete their cell cycles by the dawn transition, instead of waiting to be released at CT 12 by the circadachrons. Note the mobilization of a small fraction of settled cells at dusk, and some resettling at dawn, at cell densities in excess of about 2000 per milliliter (open arrows).

either side, a 9 hr light pulse) per circadian cycle, since the circadachrons are tightly locked into the 30 hr cycle of five light and dark segments. With this interpretation of the data, all the advances and delays of the Edmunds et al. (1982) PRC have to be relegated to five clusters of data points corresponding to the five dark pulses that can be replaced in a 30 hr cycle. Although the data support the CT 12 stop/reset mechanism hypothesis, only one of the proposed advances (corresponding to dark pulse C replacement; see Fig. 6) can be used as an estimate for a value on the advancing segment. Data-point clusters corresponding to dark pulses A, B, and E (Fig. 6) can all be explained by the CT 12 stop/reset mechanism, but those associated with pulse D cannot be deciphered because of interaction with the LD regime.

On this basis only three advance values are available, to date, for the advancing segment of the PRC for the *Euglena* circadachron controlling cell division (Fig. 7), the other two coming from Fig. 11 and from a LD: 8,16 cycle experiment (Adams, unpublished results). When, however, these advance values plus the general information about the limits of the advancing segment (CT 18 to CT 6), the "dead zone" (CT 6 to CT 12), and the CT 12 stop/reset zone (CT 12 to CT 18) are considered, in comparison with the available data on the mobility rhythm PRC, a composite PRC can be plotted for light resetting that suggests a common circadachron control system for both mobility and cell division (Fig. 7).

The data of Feldman (1967) and Schnabel (1968) have been reworked on the assumption that pulses of light providing light-on signals over the CT 18 to CT 6 circadian cycle segment induce advances (plotted at light-on), and that periods of light intersecting the CT 12 to CT 18 segment stop the clock at CT 12, or reset it back to CT 12, and hold it there until darkness again ensues. The data have also been corrected for differences in circadian period. The "delay" data are clearly consistent with the -1 slope (Fig. 7a) expected of a stop/reset to CT 12 mechanism, the slope extending indefinitely in continuous light. Thus, *Euglena* does not appear to have a patterned delay segment (unlike *Drosophila*) and the PRC would not therefore be expected to have the traditional vertical step between the maximum advance and delay values at the breakpoint or singularity, which apparently lies at CT 18. The advancing segment data are sufficient to define a sigmoid curve peaking at around 10 hr maximum advance.

To bring out the functional importance of the change in PRC slope with circadian time the sigmoid advancing segment of the PRC has been differentiated in two ways: in Fig. 7a, by simply reflecting the advances parallel to the abscissa to delineate a "new-time curve"; and in Fig. 7b, by plotting a resetting map. Both plots reveal a pair of cusps with instantaneous slopes of -1 at their apices. Analysis of the spacing ratio (Fig. 7c) suggests that these cusps may serve as "chronostats," zones of the PRC where the clock is most effective at correcting daily time drifts against the slowly changing light-on and light-off time checks provided by the rotating earth as it wheels around the sun on its annual cycle. These cusps would be expected to occur at circadian times when the clock has to be at its most accurate. In photoperiodic organisms, for example, they appear to operate around the critical day lengths (Adams, in preparation). Cusp 1, (Fig. 7a and b) lies close to the proposed CT 20 cytochron-stop locus. This would become critical at day lengths of around 16 hr, when this locus occurs at the dawn transition (since dawn would cause an advance prior to the clock reaching the cytochron-stop trigger) thus permitting an uncoupling of the circadachron at daylengths in excess of around 16 hr.

Furthermore, in var. *bacillaris* kept in LL for many years, growth becomes biphasic with a second division wave occurring during the light periods (see segment $\gamma$ of Fig. 11).

Since this second wave is also present in LD: 8,16 entrained populations, its presence seems to suggest that the circadachron-cytochron link can become labile, and that light-on can restart the stopped cytochrons, thus preempting the normal restart by the circadachron at CT 12. This change may well account for the observations of apparent breakdown of synchrony in *Euglena* strains kept for long periods in culture. Possibly a lack of sulfur compounds in culture media in some way affects DNA modifications (e.g., methylation) and leads to altered gene expression.

Interestingly, the apex of cusp 2, (Fig. 7a and b) is close to the CT value at which Edmunds et al. (Chap. 17) have achieved desynchronization by a step-down in light level to a critical intensity. The variation in the slope of the advancing segment of the PRC lies between the limits of 0 and -2. In this region successive daily timing errors (e.g., a, b, c, d of fig. 7c) are continually minimized as a result of resetting (a′, b′, c′, d′), whereas slopes of < -2 would cause positive feedback in successive cycles, and are apparently only used in clocks (e.g., *D. pseudoobscura*), for a once-only, rapid, unbalanced "flip-over" to correct a major timing error.

The five advancing segment values obtained for *C. furca* from the phase-shift data are also plotted on the *Euglena* PRC (Fig. 7a). They suggest that the advancing segment of the *Ceratium* PRC is close to that of *Euglena,* at least in the region from CT 24 to CT 0.

## THE CYTOCHRON, VARICHRON, AND CIRCADACHRON COMPLEX

The cell division control mechanisms in both *Euglena* and *Ceratium* are clearly collections of clocklike time-metering and programming devices. In *Euglena* the cytochron in each cell programs the sequence of CDC phases but is apparently overridden by two other components of the collection. In each daughter cell the cytochron starts up from CP 0 to time a "minimum $G_1$ period." Apparently at the same time, however, a varichron timer also starts up and proceeds to meter a variable period within the range 6-12 hr. It seems a cell cannot enter S until the light-driven, cytochron-metered, minimum $G_1$ and the varichron period have both run to completion in that particular cell. Whether the varichron period is timed by a transition probability mechanism (Smith and Martin, 1973) or by random selection of a determinate timer, e.g., from a set of alternative $G_1$ duration-controlling genes, as in mating-type cassette selection in yeast (Leupold, 1980), is still a completely open question. Furthermore, it is obvious that at low temperatures and low light intensities, the duration of the light-driven early $G_1$ segment is increased. Control of the range of the varichron timing segments is also obviously modulated by the environment, but again, just how and to what extent is unknown. The elaborate stopping and restarting of the cytochrons by the circadachrons is probably just a back-up mechanism that supplements control by these other systems. With such a large genome and such a long evolutionary history we should perhaps expect the photosynthetic protists to have evolved a sophisticated and highly redundant control system, much as birds have a redundant series of navigational aids.

In contrast to *Euglena,* the division waves of *Ceratium* are tightly synchronized, any spread in the timing being accountable in terms of biological noise. Both *Euglena* and *Ceratium* do, however, utilize a circadachron to modify the cytochron output. If the light period exceeds 6 hr, the light-on transition induces an advance that pulls CT 12 into the light period, each circadachron stopping as soon as it reaches CT 12, so that all the circadachrons in the population are collected together for a synchronous start at dusk. With only 8 hr to run independently before reaching CT 20 they therefore reach this critical locus more or less in unison.

The reason *Ceratium* moves its division wave across the dark period in response to a change in photoperiod is unclear, although the mechanism appears to be yielding to analysis. Even more problematical, however, is that as in *Euglena* the CDC is extended over several days when photoperiods are reduced in length. Just how a given cell is selected to either divide or be held over until a subsequent day is a mystery. One possibility is that *Ceratium* has a varichron timer that is quantized in multiples of 24 hr segments, serving to split the population so that some cells divide every 1, 2, 3, or more days apart. If the varichron is light dependent, then it might stop and restart either side of each dark period until it has run full track. These ideas are purely speculative, however, and probably fruitless in the absence of more experimental data.

What does this all mean? It would seem to indicate that models which suggest that cells "divide when they reach a certain size" or "when they have accumulated sufficient reserves" per se, are at best simplistic. Both *Euglena* and *Ceratium* seem to have evolved several linked timing devices that work in conjunction with each other to ensure that as the day length decreases, a greater and greater proportion of the population receives two (or more) daily inputs of solar energy before being committed to cell division. In this way, the cells have a greater chance of receiving enough energy and fixing enough carbon to reach an optimum size before dividing. They divide when their clocks program them to, no matter how big, small, or well endowed with food reserves they are. That they do so without successive generations of daughter cells getting bigger and bigger (or smaller and smaller) is due to the natural selection of a "microcomputerlike clockshop" that on average maintains a balance between growth and cell division by adjusting the timing of cell division in relation to photoperiod, temperature, and nutrient status.

It would seem that the overall strategy, namely, the prevention of cell division until growth is likely to have reached a prescribed norm, is the same in both euglenoids and dinoflagellates. Furthermore, the similarity between their dawn- and dusk-synchronized clocks and the dawn and dusk oscillators of higher organisms suggests the possibility of a phylogenetic thread that reaches back in time to the earliest photosynthetic eukaryote. Indeed, selection for photosynthetic efficiency may well have provided the evolutionary drive for the development of the very first endogenous clock.

## ACKNOWLEDGMENTS

This work was supported by a Science Research Council Fellowship (Queen Mary College, London) and grants from the University of London Central Research Fund to K.J.A., and by National Science Foundation research grants PCM-7805832 and PCM-8204368 to L.N.E.

## REFERENCES

Adams, K. J. (1966). Mitotic cycle in algal unicells. Ph.D. thesis, University of London, England.

Adams, K. J. (1968). An automatic recording cell counter for monitoring growing populations of unicellular micro-organisms. *J. Scien. Instrum. (J. Physics [E]) (Series 2)* 1:861-863.

Adams, K. J. (1973). Characterisation of a cell division commitment threshold in the cell cycle of *Euglena gracilis. Br. Phycol. J.* 8:204.

Beach, D., Durkacz, B., and Nurse, P. (1982). Functionally homologous cell cycle control genes in budding and fission yeast. *Nature* 300:706-709.

Brinkmann, K. (1966). Temperature effects on circadian rhythms of *Euglena gracilis* under mixotrophic and autotrophic conditions. *Planta* 70:344-389.

Brooks, R. F. (1976). Regulation of the fibroblast cell cycle by serum. *Nature* 260:248-250.

Castor, L. N. (1980). A $G_1$ rate model accounts for cell-cycle kinetics attributed to "transition probability." *Nature* 287:857-859.

Chisholm, S. W. (1981). Temporal patterns of cell division in unicellular algae. *Can. Bull. Fish. Aquat. Sci.* 210:150-181.

Cook, J. R., and Cook, B. (1962). Effects of nutrients on the variation of individual generation times. *Exp. Cell Res.* 28:524-530.

Edmunds, L. N. (1965a). Studies on synchronously dividing cultures of *Euglena gracilis* Klebs (Strain Z). I. Attainment and characterization of rhythmic cell division. *J. Cell. Comp. Physiol.* 66:147-158.

Edmunds, L. N. (1965b). Studies on synchronously dividing cultures of *Euglena gracilis* Klebs (Strain Z). II. Patterns of biosynthesis during the cell cycle. *J. Cell. Comp. Physiol.* 66:159-182.

Edmunds, L. N. (1966). Studies on synchronously dividing cultures of *Euglena gracilis* Klebs (Strain Z). III. Circadian components of cell division. *J. Cell. Comp. Physiol.* 67:35-44.

Edmunds, L. N., and Adams, K. J. (1981). Clocked cell cycle clocks. *Science* 211:1002-1013.

Edmunds, L. N., and Funch, R. (1969). Effects of "skeleton" photoperiods and high frequency light-dark cycles on the rhythm of cell division in synchronized cultures of *Euglena. Planta* 87:134-163.

Edmunds, L. N., Jay, M. E., Kohlmann, A., Liu, S. C., Merriam, V. H., and Sternberg, H. (1976). The coupling effects of some thiol and other sulfur containing compounds on the circadian rhythm of cell division in photosynthetic mutants of *Euglena. Arch. Microbiol.* 108:1-8.

Edmunds, L. N., Tay, D. E., and Laval-Martin, D. L. (1982). Cell division cycles and circadian clocks. Phase response curves for light perturbations in synchronous cultures of *Euglena. Plant Physiol.* 70:297-302.

Feldman, J. F. (1967). Lengthening the period of a biological clock in *Euglena* by cycloheximide, an inhibitor of protein synthesis. *Proc. Natl. Acad. Sci. USA* 57:1080-1087.

Hersh, R. T., and Kitos, P. A. (1980). Is $G_1$ normally distributed? *J. Theor. Biol.* 86:117-122.

Hutner, S. N., and Provasoli, L. (1955). In *Biochemistry and Physiology of the Protozoa*, vol. II, S. N. Hutner and L. Lwoff (eds.). Academic Press, New York, p. 106.

Jarrett, R. M., and Edmunds, L. N., Jr. (1970). Persisting circadian rhythms of cell division in a photosynthetic mutant of *Euglena. Science* 167:1730-1733.

Klevecz, R. R. (1976). Quantized generation time in mammalian cells as an expression of the cellular clock. *Proc. Natl. Acad. Sci. USA* 73:4012-4016.

Koch, A. L. (1980). Does the variability of the cell cycle result from one or many chance events? *Nature* 286:80-82.

Kubitschek, H. E. (1962). Normal distribution of cell generation rate. *Exp. Cell Res.* 26:439-450.

Leedale, G. F. (1957). Division cytology of the Euglenineae. Ph.D. thesis. University of London, England.

Leedale, G. F. (1959). Periodicity of mitosis and cell division in the Euglenineae. *Biol. Bull.* 116:162-174.

Leupold, U. (1980). Transposable mating-type genes in yeast. *Nature* 283:811-812.

Margulis, L. (1970). *Origin of Eukaryotic Cells.* Yale University Press, New Haven, Connecticut.

Nurse, P. (1980). Cell cycle control—both deterministic and probabilistic? *Nature* 286: 9-10.

Pittendrigh, C. S. (1966). The circadian oscillator in *Drosophila pseudoobscura* pupae: A model for the photoperiodic clock. *Z. Pflanzenphysiol.* 54:275-307.

Pittendrigh, C. S., and Daan, S. (1976). A functional analysis of circadian pacemakers in nocturnal rodents. V. Pacemaker structure: A clock for all seasons. *J. Comp. Physiol.* 106:333-355.

Saunders, D. S. (1976). *Insect Clocks.* Pergamon Press, Oxford.

Schmid, P. (1967). Temperature adaptation of the growth and division process of *Tetrahymena pyriformis.* II. Relationship between cell growth and cell replication. *Exp. Cell Res.* 45:471-486.

Schnabel, G. (1968). The influence of light upon the circadian rhythm in *Euglena gracilis* in mixotrophic and autotrophic cultures. *Planta* 81:49-63.

Shapiro, H. S. (1976). Deoxyribonucleic acid content per cell of various organisms. In *Handbook of Biochemistry and Molecular Biology,* 3rd ed., vol. 2, *Nucleic Acids,* G. D. Fasman (ed.). Chemical Rubber Company Press, Cleveland, Ohio, pp. 284-310.

Smith, J. A., Laurence, D. J. R., and Rudland, P. S. (1981). Limitations of cell kinetics in distinguishing cell cycle models. *Nature* 293:648-650.

Smith, J. A., and Martin, L. (1973). Do cells cycle? *Proc. Natl. Acad. Sci. USA* 70:1263-1267.

Spudich, J. L., and Sager, R. (1980). Regulation of the *Chlamydomonas* cell cycle by light and dark. *J. Cell Biol.* 85:136-145.

Sweeney, B. M. (1969). *Rhythmic Phenomena in Plants.* Academic Press, New York.

Tappan, H. (1980). *The Paleontology of Plant Protists.* W. H. Freeman, San Francisco.

Terry, O. W., and Edmunds, L. N., Jr. (1970). Phasing of cell division by temperature cycles in *Euglena* cultured autotrophically under continuous illumination. *Planta* 93:106-127.

Weiler, C. S., and Chisholm, S. W. (1976). Phased cell division in natural populations of marine dinoflagellates from shipboard cultures. *J. Exp. Mar. Biol. Ecol.* 25:239-247.

Weiler, C. S., and Eppley, R. W. (1979). Temporal pattern of division in the dinoflagellate genus *Ceratium* and its application to the determination of growth rate. *J. Exp. Mar. Biol. Ecol.* 39:1-24.

# PART V

# CELL CYCLE CLOCKS AND CANCER
## PATHOLOGY AND CHRONOTHERAPY

The foregoing chapters have documented much of the experimental evidence for viewing the cell cycle as an oscillatory system. In this part, the role of cell cycle clocks in the pathology and treatment of cancer will be considered.

John J. Wille, Jr., and Robert E. Scott (Chap. 22) first discuss cell division cycle (CDC)-dependent, integrated control of cell proliferation and differentiation in normal and in neoplastic mammalian cells. Particularly germane to this book is their exciting speculation that neoplastic transformation might be attributable to defective control of the CDC. They suggest that the two-stage model of carcinogenesis could be explained by specific alterations in the kinetic parameters controlling the dynamics of a nonlinear, dynamic biochemical system comprising both a high-amplitude and a low-amplitude stable limit-cycle oscillation (see Chaps. 16 and 17) "nested" within the mitotic oscillator (Wille, 1979).

The remaining two contributions, by Lawrence E. Scheving (Chap. 23) and Ulla Møller (Chap. 24), stress the importance of chronobiology for basic cancer research and chemotherapy. There is no lack of awareness that the CDC is most significant for cancer treatment, if for no other reason than that studies of synchronized mammalian cells have shown that their sensitivity to a large number of cytostatic drugs (such as cytosine arabinoside), as well as to ionizing radiation, is highly dependent on the stage of the CDC (see Baserga, 1971). Indeed, some of these drugs may also lead to a partial synchronization of the CDC and have been so employed to obtain synchronized cell populations both in vitro and in vivo. This strategy can maximize the chance for survival and cure by applying the minimum dose of drug necessary to kill the phased, malignant cells at the time of their CDC when they are most susceptible. It is obvious, therefore, that the diurnal (and circadian rhythmicities observed in cell flux, and in the distribution and duration of CDC phases (see

Part II) must be taken into account in the design of an appropriate chemotherapeutic regimen.

But there is another, equally important consideration for maximizing the results of radio- and chemotherapy: host tolerance. It is unfortunate that in most of the earlier cancer work there is little mention of the role of rhythmic variations (particular circadian periodicities) in the susceptibility of the whole organism to the toxicity of the drug(s) being utilized for treatment. Both Scheving and Møller review the experimental evidence revealing that properly designed protocols (such as sinusoidally varying drug courses during the 24 hr day) can dramatically enhance survival and cure rates by concomitantly maximizing the tolerance of the host to the drug through a temporal "shielding" of normal, healthy tissues. Thus, "when" to treat must assume importance together with the "what" and "where" (Halberg, 1975).

## REFERENCES

Baserga, R. (ed.) (1971). *The Cell Cycle and Cancer.* Marcel Dekker, New York.

Halberg, F. (1975). When to treat. *Indian J. Cancer* 12:1-20.

Wille, J. J., Jr. (1979). Biological rhythms in protozoa. In *Biochemistry and Physiology of Protozoa,* vol. 2, 2nd ed., M. Levandowsky and S. H. Hutner (eds.). Academic Press, New York, pp. 67-149.

# 22

# Cell Cycle-Dependent Integrated Control of Cell Proliferation and Differentiation in Normal and Neoplastic Mammalian Cells

**John J. Wille, Jr., and Robert E. Scott** / Section of Experimental Pathology, Mayo Clinic/Foundation, Rochester, Minnesota

## INTRODUCTION

Neoplastic transformation of cells has been reported in virtually all vertebrates, some invertebrates, higher plants, and some insects. It may be inferred from this that carcinogenesis is a process of universal occurrence among complex multicellular organisms (Dawe, 1973). Many different theories of cancer have been proposed, such as the somatic mutation theory, the oncogene/virus activation theory, and the cell selection theory; and these have been extensively discussed in numerous reviews on cancer (Upton, 1982; Pitot, 1978; Foulds, 1975; Pardee, 1982). Very few studies have, however, considered the importance of the cell cycle mechanisms in oncogenesis (Baserga, 1971; Burton and Canham, 1973). In this chapter the emphasis will therefore be placed on the role of the cell cycle in regulating the integrated control of proliferation and differentiation in normal and neoplastic mammalian cells.

### Cell Cycle in Normal Mammalian Cells

The concept of the cell cycle has been periodically reviewed (Baserga, 1976; Mitchison, 1971; Prescott, 1976a,b; Pardee et al., 1978) and the relationship between the cell cycle and cell clocks has been well covered (Hastings and Schweiger, 1976; Edmunds and Adams, 1981; Wille, 1979; Kauffman and Wille, 1975; Klevecz, 1976). According to current dogma, the cell life cycle or cell cycle consists of a proliferative or mitotic cycle, subdivided into mitosis (M), pre-DNA synthetic period ($G_1$), the period of DNA synthesis (S), and premitotic period ($G_2$) (Howard and Pelc, 1953). Early autoradiographic studies documented this in proliferating cell cultures (Baserga, 1968).

After mitosis some cells permanently leave the proliferative compartment and undergo maturation, differentiation, senescence, and ultimately death. Other cells leave the proliferative cycle and enter into a quiescent phase from which they may be induced to synthesize DNA and divide and thereby reenter the proliferative cycle (Baserga, 1976). This state has been termed $G_0$ for lack of a more precise definition. Under most circumstances, $G_0$ cells have a $G_1$ content of DNA (Lajtha, Oliver, and Gurney, 1962). There are a few instances in which mammalian cells growth arrest in $G_2$ (mouse ear epidermis, kidney, and duodenum) and they are also considered to be in $G_0$ by some (Pederson and Gelfant, 1970). The microenvironmental conditions that induce cells into arrest at $G_0$ in $G_1$ include: high density (Todaro et al., 1965; Dulbecco, 1970), serum limitations (Bartholomew et al., 1976; Martin and Stein, 1976), suspension growth (Otsuka and Moskowitz, 1976; Stocker et al., 1968), and nutrient deprivation (Holley and Kiernan, 1973; Ley and Tobey, 1970). Recent studies have suggested that each of these conditions actually induces cells to arrest at distinct states (Yen and Pardee, 1978; Moses et al., 1980; Wille and Scott, 1982a), and we have given each of these states a specific designation: (1) arrest due to growth factor or serum deprivation ($G_S$), (2) arrest due to contact inhibition, i.e., density-dependent inhibition ($G_C$) (Holley and Kiernan, 1968), and (3) arrest due to nutrient deprivation, e.g., depriving cells of isoleucine or other amino acids ($G_N$) (Tobey and Ley, 1971) or of a variety of other low molecular weight compounds (Mitchison, 1971). Another arrest state complex, designated $G_D/G_D'$, has also been described at which the integrated control of proliferation and differentiation is mediated (Scott et al., 1982a).

The sequences of metabolic events that accompany the progress from quiescence to initiation of DNA synthesis have been termed the *pleiotypic response* (Hershko et al., 1970). Evidence has also been obtained for a multiphasic temporal sequence within $G_1$ for the commitment and progression to DNA synthesis, which are separately controlled by PDGF (platelet-derived growth factor) and insulin, respectively (Pledger et al., 1978). As part of the pleiotypic response, biochemical changes have been demonstrated in membrane-controlled transport of sugars and nucleosides and in membrane-associated receptors for various hormones (Pardee et al.,1974). Changes in the rates of protein synthesis and degradation, in the levels of cyclic nucleotides, in the rates of accumulation of cytoplasmic mRNAs, in the synthesis of polysome associated poly A + RNA, and in template availability of chromatin (Baserga, 1976; Pardee et al., 1978) have also been described.

The fact that the $G_1$ phase of the cell cycle is the most variable and is subject to preferential lengthening under conditions leading to the cessation of cell growth forms the basis for the conclusion that events that occur in the $G_1$ phase influence the regulation of cell proliferation, and thereby the control of the initiation of DNA synthesis (Baserga, 1976; Pardee et al., 1978).

Temperature-sensitive cell cycle mutants that show a variety of defects that influence $G_1$ growth arrest have also been isolated. The fact that such cell cycle mutants require different intervals to initiate DNA synthesis has been interpreted as evidence for the existence of different temporal states in $G_1$ (Ashihara et al., 1978; Burstin et al., 1974; Naha et al., 1975; Talavera and Basilico, 1978).

A number of hypotheses have been offered to account for the fact that cells exit from the proliferative cycle and arrest in the $G_1$ phase of the cell cycle. Pardee (1974) originally proposed that the decision to proliferate or not is made at a single state in $G_1$, called the restriction point (R). Earlier studies on the kinetics of restimulation of BHK cells arrested in $G_0$ by serum and nutrient deprivation tended to support this proposal

(Martin and Stein, 1976; Pardee et al., 1974). Later experiments on 3T3 cells (Yen and Pardee, 1978) and temperature-sensitive $G_1$ mutants of BHK 21 cells (Burstin et al., 1974), however, provided evidence for either temporally distinct arrest states within $G_1$ or at least several different metabolic states of the cell cycle. Additional recent studies (Yen and Pardee, 1978; Moses et al., 1980; Wille and Scott, 1982a) support the latter possibility but do not prove it. Despite all these recent studies there continues to be considerable controversy as to whether cells in $G_0$ or at $G_D/G_D'$, $G_C$, $G_S$, or $G_N$ have actually left the cycle and entered in a qualitatively distinct state (Edmunds and Adams, 1981; Wille, 1979; Klevecz, 1976, 1978) or whether they are merely undergoing a prolonged $G_1$. Evidence for the former is the fact that return of resting cells to the $G_1/S$ boundary following serum stimulation of growth may require an interval that is greater than the normal length of $G_1$, and evidence that such cells possess biochemical properties that are distinct from cycling $G_1$ cells (Baserga, 1976; Wu et al., 1982). There is, in fact, evidence that cells held for progressively longer times in $G_0$ continue to sink into progressively deeper states of arrest (Baserga, 1976; Wille and Scott, 1982a).

Interpretation of kinetic data on the age distribution of intermitotic times of exponentially growing mammalian cells in culture has also led to the probabilistic cell cycle model (Smith and Martin, 1973). According to this model, cells enter into an A state immediately after mitosis. In the A state cells do not progress toward completion of events associated with DNA replication, but have a fixed probability per unit time of leaving this state and entering a B state. The propensity of cells stochastically to leave the A state is termed the *transition probability.* The transition probability is dependent on cell type and environmental conditions. From this view, cells arrested in $G_0$ possess a very low transition probability, so that most cells will reside in the A state. Rapidly proliferating cells, by contrast, will have a very high transition probability and spend relatively little time in the A state.

Another model of proliferation control proposes that cell cycling is driven by an underlying biochemical oscillator, which triggers mitosis/cell division and gates the events occurring at the other cell cycle phases (Kauffman and Wille, 1975). Such a model was shown to predict much of the phase-dependent cycle resetting demonstrated by heat shock experiments and the results of some studies on the fusion of cells at different phases of the cell cycle (Kauffman and Wille, 1975). It also predicted the novel observation (Wille et al., 1977) that application of a critical stimulus to a cell in a specific cell cycle phase will induce the mitotic clock to stop because the oscillating system will transiently arrest at a nonoscillating steady state. The model further predicted that the same perturbation given at a noncritical phase merely resets the phase of the oscillating system. This model therefore predicts the existence of multiple nonoscillating steady states and further suggests that proliferating cells may leave the cell cycle and arrest at a variety of $G_0$-like states (Wille, 1979).

As will be discussed later, the most important aspect of cell cycle control mechanisms in vivo relates to the integrated control of stem cell proliferation and differentiation. The restriction point model (Pardee, 1974) and the probabilistic model (Smith and Martin, 1973) do not explain the effect of differentiation on cell cycle control mechanisms. The cell clock-biochemical oscillator model (Kauffman and Wille, 1975; Wille, 1979) by contrast has the potential to show how the mitosis/proliferation mode and the differentiation mode are controlled by different mechanisms of the same biochemical oscillator. The potential of this model to explain the integrated cell cycle-dependent control of cell proliferation and differentiation will be further developed in this paper.

**TABLE 1** Comparison of Growth Rate and Saturation Density of Some Normal Mouse Embryo Cells and Their Transformed Derivatives

| Cell line | Transforming agent | Generation time (hr) | Saturation density (cells/cm$^2$ X $10^5$) |
|---|---|---|---|
| Swiss 3T3 | | | |
| Normal[a,b] | – | 18 | 0.5 |
| | | 23 | 0.75 |
| Transformed | | | |
| SV101[c] | SV40 virus | 16 | 5 |
| SVR-c182[b] | SV40 T-antigen negative revertant | 20 | 1.5 |
| Swiss 3T3[a,d] | Spontaneous; high-density selection | 14 | 6 |
| C3H/10T½ | | | |
| Normal[e] | – | 16 | 0.3 |
| MCA-C3H[f] | Methylcholanthrene (several clones) | 22-27 | 0.7 |
| AKR-2B | | | |
| Normal[g] | – | 17 | 1 |
| MCA-AKR[g] | Methylcholanthrene | 16 | 3 |
| Balb/c 3T3 and 3T3 T proadipocytes | | | |
| Normal[h] | – | 22 | 0.4 |
| SV 3T3[i] | SV40 virus | 16-18 | 6 |
| DMBA-3T3[i] | Dimethylbenz (a) anthracene (four lines) | 14-19 | 4 |
| SST 3T3 T[j] | Smooth surface plates (six clones) | 22-33 | 0.6-1.5 |

[a]Todaro and Green, 1963.
[b]Risser and Pollack, 1974.
[c]Todaro, Green, and Goldberg, 1964.
[d]Stiles, Desmond, Chuman, Sato, and Sauer, 1976.
[e]Reznikoff et al. (1973b).
[f]Reznikoff et al. (9173a).
[g]Moses, Proper, Volkenant, Wells, and Getz, 1978.
[h]Aaronson and Todaro, 1968.
[i]DiPaolo, Takano, and Popescu, 1972.
[j]Wille et al. (1982).

### Cell Cycle in Neoplastically Transformed Cells

Detailed cell cycle parameters have been published for various established neoplastically transformed cell lines (Altman and Katz, 1976). However, most are of little use in understanding their role in oncogenesis because they cannot be compared to their normal stem cell of origin. For this and other reasons, many studies on carcinogenesis have employed quasinormal embryonic cells and their corresponding transformed cell derivatives. These include mouse embryonic cell lines derived from Balb/c (Aaronson and Todaro, 1968), Swiss (Moses et al., 1978), AKR (Todaro and Green, 1963), and C3H mouse embryos (Reznikoff et al., 1973b).

Neoplastic transformation of embryo cells grown in vitro has been successfully accomplished by viral infection with simian virus (SV40) and polyoma DNA tumor viruses, and by treatment with a variety of chemical carcinogens (Brown and Holley, 1979; Holley et al., 1976; DiPaolo et al., 1972; Gehley et al., 1982; Barrett and Ts'o, 1978; Bertram and Heidelberger, 1974). Ultraviolet light (Mondal and Heidelberger, 1976; Namba et al., 1978; Chan and Little, 1976) and X-irradiation (Terzaghi and Little, 1976) have also been shown to induce either initiation of carcinogenesis or complete transformation in vitro. Drugs have also been found to be active in the induction of transformed cells. These include ara-cytidine, 5-fluorodeoxyuridine, and bleomycin (Benedict et al., 1975).

Comparisons of the growth arrest properties of normal and transformed cells in culture have shown that transformed cells in general reach a higher saturation density in 10% serum than their normal counterparts, and in some cases have a more rapid proliferative rate during the exponential phase of culture growth. These comparisons are shown for four different mouse embryo cell lines in Table 1.

In many cases the altered growth properties of the transformed cell have been shown to be due to an inability to arrest proliferation in $G_1$ (Pardee, 1974; Yen and Pardee, 1978; Moses et al., 1980). Both the restriction point and the transition probability models alluded to above have attempted to account for these observations by assuming that transformed cells lack the capacity to undergo growth arrest at a certain restriction point or other state in the $G_1$ phase of the cell cycle. However, the situation is not quite this simple.

In some instances it has been shown that chemically transformed AKR and C3H/10T½ cells can arrest their growth in $G_1$ through exhaustion of low molecular weight nutrients (Moses et al., 1978). A benzo-a-pyrene transformed derivative of Balb/c 3T3 cells has also been shown to growth arrest in $G_1$ within 48 hr after transfer of a rapidly growing culture to lower serum-containing medium (Brown and Holley, 1979). We have also reported that some transformed cells can also $G_S$, $G_C$, and $G_N$ arrest (Wille et al., 1982).

Despite the above-described studies most of the data on transformed cells in culture are deficient in many respects. This is so because they do not account for many of the salient features of the oncogenic process in vivo in humans. Any proposed mechanisms for the pathogenesis of cancer must account for a wider spectrum of fundamental observations. These include the following facts: (1) stem cells are the targets of most carcinogenic agents (Stevens, 1967; Pierce et al., 1967; Damjanov et al., 1971) and stem cells regulate their proliferation and differentiation in a coordinated manner (Lajtha et al., 1978; Rheinwald, 1979; Iversen et al., 1968; Holtzer et al., 1982; Nadal-Ginard, 1978; Rovera et al., 1980; Yun and Scott, 1982), (2) neoplastic transformation of stem cells can result in the development of tumors with a variety of growth and differentiation potentialities (Evans, 1968; Ackerman and Rossi, 1974; Schiffer, 1968; Zimmerman et al., 1968; Scott et al., 1971), and (3) carcinogenesis is at least a two-stage process, including an initiation phase in which an irreversible change is fixed in stem cells and a promotion phase, which allows initiated cells to proliferate and form a neoplasm (Berenblum, 1981; Farber, 1973, 1982; Becker, 1981).

A new concept is therefore needed to account for these fundamental observations. In this regard, we have presented a hypothesis that suggests that the regulation of stem cell growth in vivo is regulated through the integrated control of proliferation and differentiation and that this integrated control process is mediated at two distinct but interrelated arrest states in the $G_1$ phase of the cell cycle, $G_D$ and $G_D'$. Our hypothesis further suggests that cancer results from defects that uncouple this integrated control system. In the following sections we present evidence for this hypothesis from our own studies on normal and transformed proadipocytes and from the published data of other investigators.

## INTEGRATED CELL CYCLE-DEPENDENT CONTROL OF PROLIFERATION AND DIFFERENTIATION

Our studies have focused on attempts to establish the mechanisms by which proadipocyte stem cells regulate the coupling of growth and differentiation in a coordinated fashion. We used a cloned murine proadipocyte stem cell line designated Balb/3T3 T (Diamond et al., 1977) and the related Swiss 3T3 L1 proadipocyte stem cell line (Green and Kehinde, 1975). These cells have many characteristics in common with normal

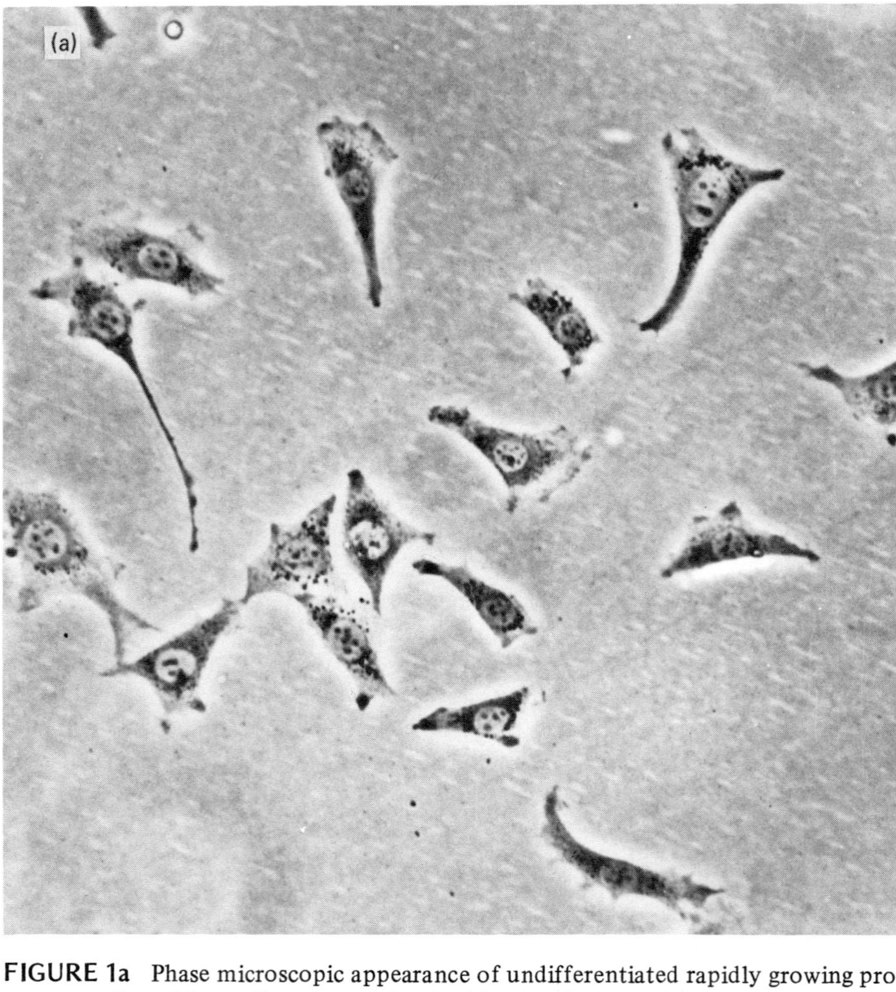


FIGURE 1a Phase microscopic appearance of undifferentiated rapidly growing proadipocytes cultured in medium containing 10% fetal calf serum, 400X.

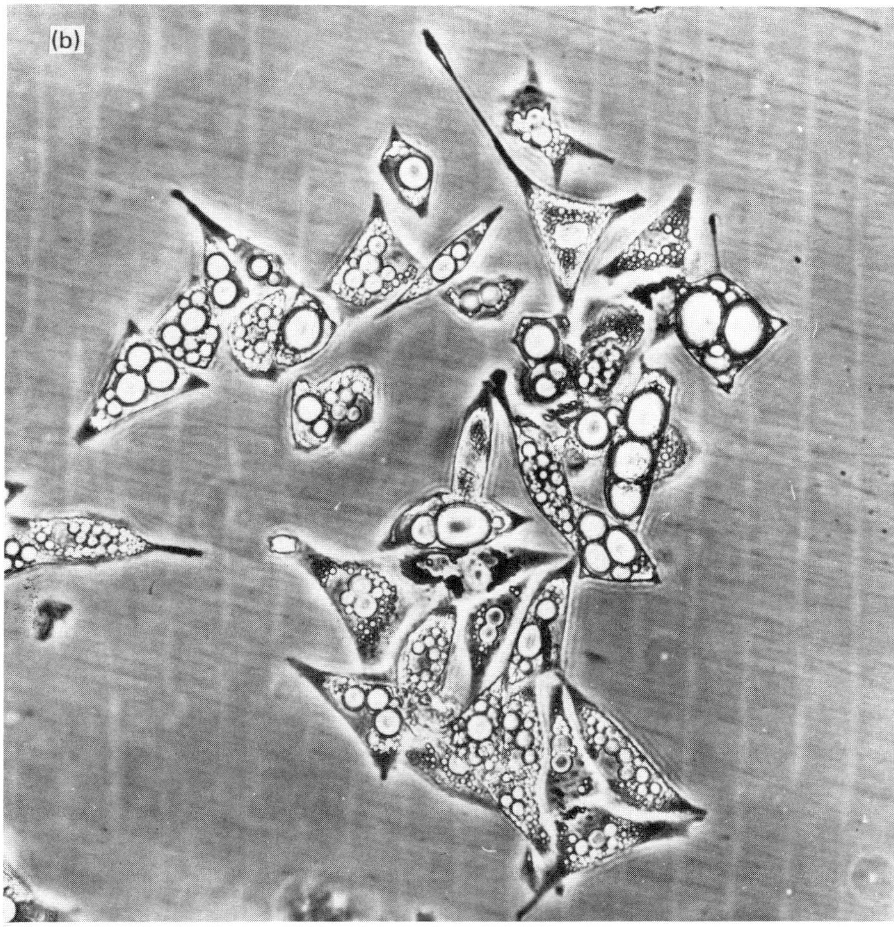


FIGURE 1b Differentiated adipocytes cultured in heparinized medium containing 25% human plasma, 400X.

stem cells; they can proliferate, they can remain quiescent, and they can differentiate under various environmental conditions. In contrast to many other differentiation systems which employ transformed cells—erythroleukemia cells (Friend et al., 1966), neuroblastoma (Sheppard and Prasad, 1973), and melanoma cells (Silagi, 1977)—Balb/3T3 T proadipocytes are nontransformed and noninitiated with respect to the two-stage model of carcinogenesis (Scott and Boone, 1981). Therefore, although 3T3 T proadipocytes are not normal stem cells, they represent the most physiologically "normal" cloned stem cell line available for study.

In this cell system, we have developed methods to induce rapid and synchronous differentiation under physiological conditions (Krawisz and Scott, 1982; Scott et al., 1982a). For example, the culture of low-density heparinized DMEM containing only human plasma has been shown to induce adipocyte differentiation of >90% of the cells within 5-10 days or less. Figure 1 illustrates the morphology of such undifferentiated and differentiated cells.

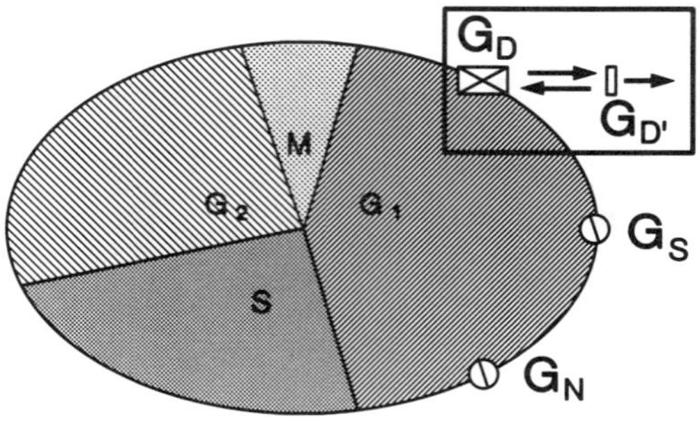


**FIGURE 2** Cell cycle model illustrating the topography of the $G_D$ and $G_D{}'$ arrest states relative to the $G_S$ and $G_N$ arrest states in the $G_1$ phase of the cell cycle.

The first discovery we made was that, prior to differentiation, proadipocytes must growth arrest at a distinct state in $G_1$, designated $G_D$. The $G_D$ state was shown to be distinct from other $G_1$ growth-arrest states, including those induced by contact inhibition ($G_C$) or serum deprivation ($G_S$) and by nutrient deprivation ($G_N$) (Scott et al., 1982a). Figure 2 illustrates the relative topography of $G_D$, $G_S$, and $G_N$ in the $G_1$ phase of the cell cycle in monolayer cultures of proadipocytes (Wille and Scott, 1982a), and Table 2 lists the characteristics of $G_D$-arrested cells that distinguish them from cells at $G_S$ and $G_N$. It was further demonstrated that proadipocytes can be arrested and maintained for long intervals at the $G_D$ state by culture in heparinized medium containing a specific human plasma fraction. Other plasma fractions have been isolated that induce only the nonterminal differentiation; still other plasma components induce only terminal differentiation. Table 3 lists these human plasma fractions and the process of growth arrest and differentiation that they control.

These studies led us to construct a model for the integrated control of cell proliferation and differentiation (Scott et al., 1982b). The model presented in Fig. 3 illustrates that three steps are involved in promoting the differentiation process and two steps antagonize differentiation and promote proliferation. For differentiation to occur, cells first undergo a process (arrow 1) that induces growth arrest at a distinct state in the $G_1$ phase of the cell cycle, designated $G_D$. Cells at $G_D$ can remain quiescent; they can be induced

**TABLE 2** Characteristics That Distinguish $G_D$-Arrested Proadipocytes from $G_S$- and $G_N$-Arrested Proadipocytes

1. $G_D$ cells can differentiate in the absence of DNA synthesis[a]
2. $G_D$ cells reside at a topographically distinct state in the $G_1$ phase of the cell cycle[b]
3. $G_D$ cells can be induced to proliferate by methyl isobutyl xanthine[c]
4. $G_D$ cells can be killed at high cell density by 8-bromo AMP[d]
5. $G_D$ cells have abundant and prominent cell surface microvilli[e]

[a]Yun and Scott (1982).
[b]Wille and Scott (1982a).
[c]Scott et al. (1983).
[d]Florine, Hoerl, and Scott (1982).
[e]Boman et al. (1983).

**TABLE 3** Human Plasma Fractions That Induce $G_D$ Arrest, Nonterminal Differentiation, and Terminal Differentiation[a]

| Human plasma fraction | Processes of adipocyte differentiation: $G_D$ arrest | Nonterminal differentiation | Terminal differentiation |
|---|---|---|---|
| HP | + | + | + |
| BaP | + | – | + |
| CEP | – | + | – |
| CEPH | + | + | – |
| BaPH | – | – | + |

[a]HP, whole human plasma; BaP, barium-adsorbed human plasma prepared at pH 7.8; CEP, human plasma components precipitated with barium at pH 7.8 then eluted with sodium citrate at pH 7.8; BaPH, BaP prepared at pH 8.6; CEPH, CEP prepared at pH 8.6.

to differentiation (arrows 2 and 3) or to reinitiate proliferation (arrow 5). The differentiation of $G_D$-arrested cells involves two processes: nonterminal differentiation (arrow 2) and terminal differentiation (arrow 3). Under appropriate conditions cells can be blocked in a state of nonterminal differentiation and they are designated to be at the $G_D'$ state. $G_D'$ cells can be induced to differentiate terminally (arrow 3), to lose the differentiation phenotype (arrow 4), and then to reinitiate proliferation (arrow 5). Under selected conditions, $G_D'$ cells can also be induced to proliferate without losing the differentiated phenotype (small arrow 5). The integrated control of proliferation and differentiation can, therefore, be regulated at $G_D$ and $G_D'$ because at both these state cells either can remain quiescent, can reinitiate, or can differentiate.

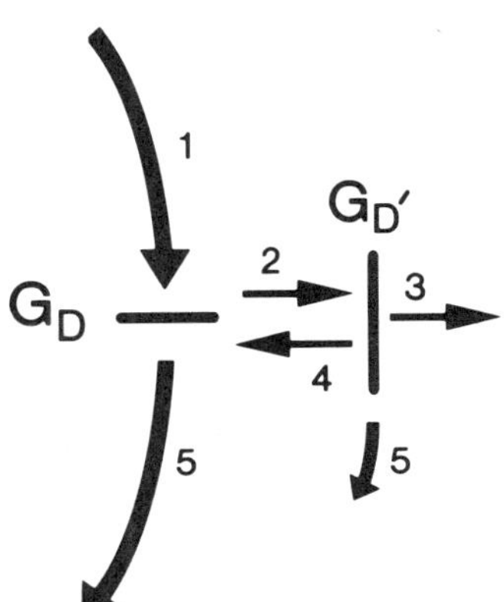


**FIGURE 3** Simple model for the integrated control of the coupling of growth arrest and differentiation. Cells first undergo $G_1$ growth arrest at $G_D$ (arrow 1). $G_D$-arrested cells can remain quiescent at the $G_D$ state or they can either differentiate (arrow 2) or reinitiate proliferation (arrow 5). If $G_D$-arrested cells differentiate, they first nonterminally differentiate (arrow 2) and can remain in the nonterminally differentiated state ($G_D'$). Such cells can subsequently terminally differentiate (arrow 3), they can lose the differentiated phenotype (arrow 4), then reinitiate proliferation (arrow 5), or they can reinitiate DNA synthesis without loss of the differentiated phenotype (small arrow 5).

These studies, therefore, show that the regulation of cell growth can be mediated by the integrated control of cellular proliferation and differentiation and that mediation occurs at two distinct yet interrelated states in the $G_1$ phase of the cell cycle, designated $G_D$ and $G_D'$. We have suggested that these growth control mechanisms mimic the physiological regulation of stem cell proliferation in vivo and that other $G_1$ regulatory processes, i.e., $G_C$, $G_S$, and $G_N$, are backup systems to prevent stem cells from growing under adverse conditions, such as, in too crowded a situation or in the absence of growth factors or nutrients, respectively.

## DEFECTIVE INTEGRATED CONTROL OF PROLIFERATION AND DIFFERENTIATION IN TRANSFORMED CELLS

The data presented above on the control of proadipocyte growth and differentiation provided us with the basis for additional studies to determine if defects in the integrated control of proadipocyte growth and differentiation exist in various disease states, especially in cancer (Scott and Florine, 1982). To study the mechanism of carcinogenesis in vitro, the use of standard mouse embryo cell lines is inadequate to test the hypothesis that neoplastic transformation results from defects in the coordinated control of stem cell proliferation and differentiation. This is so because such cells are not stem cells; i.e., they are not capable of physiological in vitro differentiation and they are initiated with respect to the two-stage mechanisms of carcinogenesis—they form tumors when implanted on plastic plates in syngeneic mice (Scott and Boone, 1981) or grow in soft agar under certain conditions (Todaro et al., 1979). For this reason, we have studied the effect of neoplastic transformation of noninitiated 3T3 T proadipocyte stem cells on their capacity to coordinately regulate growth and differentiation.

In particular, we tested to determine if neoplastic transformation is associated with selective modulation of the ability of a cell to $G_D$ arrest and differentiate with or without associated defects in growth control mediated at $G_C$, $G_S$, or $G_N$. To accomplish this goal, proadipocytes were transformed by the smooth surface tumorigenesis (SST) method (Scott and Boone, 1981; Boone and Jacobs, 1976) or by spontaneous events and then cloned. Table 4 summarizes the results of these studies and shows that neoplastic transformation of proadipocyte stem cells is associated with a complete defect in the ability to $G_D$ arrest and differentiate and with a partial, i.e., variable, defect in the ability to $G_C$ arrest. By contrast, significant defects were not observed in the ability of transformed proadipocytes to $G_S$ or $G_N$ arrest (Wille et al., 1982).

Based on these and other data we have proposed that the development of defects in the integrated control of stem cell proliferation and differentiation mediated at $G_D/G_D'$ may be associated with the initiation phase of carcinogenesis and that promotion of carcinogenesis may result from the development of defects in the backup/fail-safe growth control processes mediated most importantly at $G_C$ but perhaps also at $G_S$ or $G_N$.

Recent additional but somewhat preliminary results support this conclusion (Wille and Scott, 1982b). We have isolated approximately 60 clones of proadipocytes which have various combinations of defects in growth control mediated at $G_D/G_D'$, $G_C$, $G_S$, and $G_N$ and have then tested to determine which combination of cell cycle defects are required for initiation and complete transformation. Clones with only $G_D/G_D'$ defects are not tumorigenic but appear to be initiated. By contrast, clones with defects in both $G_D/G_D'$ control and $G_C/G_S$ control appear to be fully transformed (metabolic events at $G_N$ do not appear to be important in this regard).

**TABLE 4** Neoplastic Transformation Associated with Selective Defects in the Ability of Cells to $G_D$ Arrest and Differentiate[a]

| Cell designation | Number of cell lines used | Tumorigenicity (% animals with tumors) | % Cell population | | | | |
|---|---|---|---|---|---|---|---|
| | | | $G_D$ arrest | Differentiation | $G_C$ arrest | $G_S$ arrest | $G_N$ arrest |
| 3T3 T proadipocytes | 2 | 0 | 100 | 100 | 100 | 100 | 100 |
| SST and spontaneously transformed 3T3 T proadipocytes | 8 | 75 | 0 | 0 | 25 | 87.5[b] | 87.5[b] |

[a]SST designates smooth surface tumorigenesis. Tumorigenicity was assayed following inoculation of $\simeq 1 \times 10^7$ cells in suspension in the dorsal subcutaneous tissues of syngeneic mice. $G_D$ arrest was assayed by the ability of cells to undergo $G_1$ growth arrest at low cell densities when cultured in heparinized DMEM containing 25% human plasma and the ability of these cells to respond to exposure to methyl isobutyl xanthine with DNA synthesis. Differentiation was assayed morphologically by the appearance of fat cells. $G_S$ and $G_N$ arrest capability was assayed by the ability of cells to undergo $G_1$ growth arrest following culture in DMEM containing 0.5% FCS or in an isoleucine-deficient DMEM containing 10% dialyzed FCS, respectively; and to be stimulated to grow when subsequently refed growth factors or isoleucine, respectively. $G_C$ arrest was assayed as the ability to undergo $G_1$ arrest in DMEM containing 10% FCS at a designated saturation density of between 3 and $7 \times 10^4$ cells per $cm^2$.

[b]All transformed cells growth arrested under $G_S$- and $G_N$-inducing culture contitions, but 12.5% failed to shown an enrichment in $G_1$-arrested cell populations.

Our suggestion that normal stem cell growth is regulated by the integrated control of proliferation and differentiation and that transformation results from defects in this integrated control process is supported by studies on normal and leukemic myeloid development. Sachs (1978, 1980) has shown that growth and differentiation of normal myeloblasts are coupled and are regulated by specific proteins, designated MGI (macrophage-granulocyte inducing factor). He and Lotem also showed that MGI consists of two activities, MGI-1 and MGI-2 (Lotem and Sachs, 1982). MGI-1 causes growth and sustained viability of normal precursor myeoblasts and MGI-1 stimulates normal stem cells to produce their own differentiation-inducing factors, i.e., MGI-2. Thereby, the integrated control process is mediated. By contrast, leukemic myeloblasts can proliferate independently of MGI-1 and show variable defects in differentiation control processes. That is, they are unable to integrate the control of proliferation and differentiation.

Our results and those of Lotem and Sachs (1982) provide strong evidence that important aspects of carcinogenesis result from defects in the integrated control of cell proliferation and differentiation. We suggest that if the mechanisms that control or fail to control the integrated control process in normal and neoplastic stem cells can be established, an important step in understanding the pathogenesis of cancer will have been made. In the following section a brief discussion of the goal of future studies is presented.

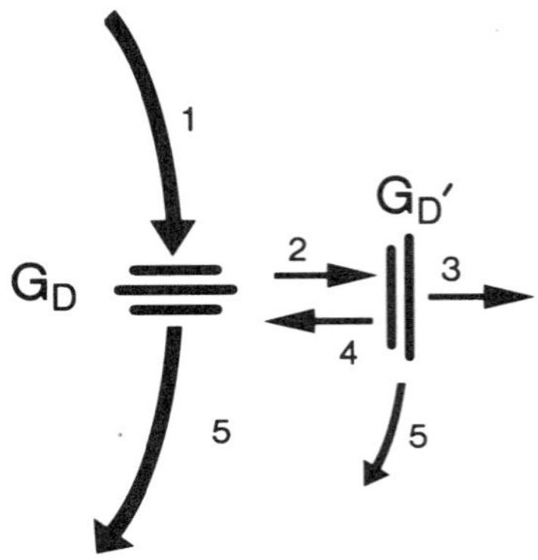


**FIGURE 4** More complex model for the integrated control of growth arrest and differentiation. Compared with the model presented in Fig. 3, this model shows that $G_D$ is complex and is composed of three substates (see text). This model also shows that the $G_D{}'$ arrest state is also complex. In particular, it suggests that two metabolic events are required to convert nonterminally differentiated cells to a terminal state of differentiation (see text).

## FUTURE DIRECTIONS

### Relevance of $G_D/G_D{}'$ Integrated Control Model for Normal Human Diploid Stem Cells

Although 3T3 T proadipocytes are useful cells to study the integrated control of proliferation and differentiation in vitro they are not the ideal stem cell system. It would be better to establish the mechanisms for the integrated control of proliferation and differentiation with normal human diploid epithelial cells, especially since epithelial stem cells give rise to the vast majority of cancers in humans. The best candidate for such studies in our view is human epidermal keratinocyte stem cells. Recent developments have made it possible to culture these cells free of contaminating fibroblasts and to carry out studies on their growth kinetics (Tsao et al., 1982). In this system, proliferative cultures of basal stem cells can be maintained in the absence of differentiation by culture in medium containing low calcium levels (less than 0.1 mM $Ca^{2+}$), and the entire program of terminal differentiation can be induced in these cultures by increasing the calcium concentration in the medium to 1.2 mM (Watt and Green, 1982). We are now using this system to establish how normal epithelial cells integrate the control of proliferation and differentiation. We suggest that this research focus should in the future provide important information on the mechanisms of epithelial stem cell transformation.

### Mechanisms That Control the Decision to Proliferate or Differentiate

In proadipocyte stem cells the decision to proliferate or to differentiate has been shown to be mediated at two distinct $G_1$ growth arrest states in the cell cycle, i.e., $G_D$ and $G_D{}'$. In the future it will be important to establish the specific biochemical or metabolic events that control these processes. Preliminary evidence suggest that $G_D$ and $G_D{}'$ are biologically complex arrest states. Figure 4 (compare to Fig. 3), for example, suggests that $G_D$ may consist of three substates. The first substate (shown by the top horizontal bar) can be induced when proadipocytes are cultured in differentiation-promoting medium in the presence of the tumor promotor phorbol myristate acetate (PMA). Cells at this state can be induced to move to the second substate (the middle horizontal bar) by removal of PMA. This second substate is distinct because therein cells are uniquely sensitive to the mitogenic effect of methyl isobutyl xanthine (MIX). Cells at this substate may then move to the third substate prior to differentiation, but this remains to be established. We have, however, shown that the third substate of $G_D$ (bottom horizontal bar) is arrived at when nonterminally differentiated cells are induced to lose the differentiated phenotype and arrest at a MIX nonsensitive state from which they can redifferentiate without traversing the cell cycle. A detailed description of the data on which these conclusions are based will be presented elsewhere.

Figure 4 also suggests that $G_D'$ is a complex state. The basis for this conclusion is derived from experiments that show that conversion from the nonterminal $G_D'$ state to the terminal state of differentiation involves two steps (vertical bars at $G_D'$). The first $G_D'$ substate (left vertical bar) appears to be selectively reversed by retinoic acid and the second $G_D'$ substate (right vertical bar) appears to be selectively reversed by phorbol myristate acetate. The metabolic basis for the $G_D$ and $G_D'$ substates and their physiological significance, however, remains to be established. One suggestion is that different gene cascade sequences might be involved in the mechanism coupling growth arrest and differentiation at $G_D$ and $G_D'$. This and other possible mechanisms need to be carefully studied in the future because carcinogenesis could be associated with defects in one of these critical processes.

### Relevance of Cell Clocks to the Integrated Control of Proliferation and Differentiation in Normal and Neoplastic Cells

As pointed out earlier, most models of cell cycle regulation do not adequately consider the question of how the integrated control of cell proliferation and differentiation can be mediated. Since this is an extremely important question we have attempted to determine if the cell clock-biochemical oscillator model (Wille, 1979; Klevecz, 1976; Kauffman and Wille, 1975) can account for multiple cell cycle arrest states and, in particular, for $G_D$ and $G_D'$-like arrest states. Further studies in this direction are needed. Figure 5 presents our current concept of how the biochemical oscillator model can explain important aspects of the integrated control process. A more detailed discussion of this topic will be presented elsewhere.

Our explanation of the integrated control process based on the biochemical oscillator model requires one primary assumption. That is, it must be assumed that normal stem cells can proliferate, differentiate, or remain quiescent because they possess a central biological clock that is governed by an underlying biochemical oscillator. Figure 5 (panel a) is a schematic representation of the hypothetical limit-cycle oscillator with its parameters optimized for a high-amplitude oscillation and, therefore, cell proliferation. This limit-cycle mode does not permit growth arrest or differentiation because in this metabolic state it is not possible to enter the $G_1$ quiescent modes (shaded circle). If, however, low-amplitude oscillations are generated (Fig. 5, panel b), arrest in $G_1$ can occur at various states. The large circle could represent $G_N$ and the smaller circle could represent $G_C$/$G_S$ (see legend to Fig. 5). The integrated control of proliferation and differentiation at $G_D$ and $G_D'$ is schematically depicted in Fig. 5 (panel c). Arrest in a $G_D$ state (inside circle) or at $G_D'$ (stippled area) can be achieved only when the system's trajectories are moving into the $G_1$ domain at the time or phase that the controlling biochemical parameters are varied. In addition, the rate of change of the parameters must be sufficiently great to

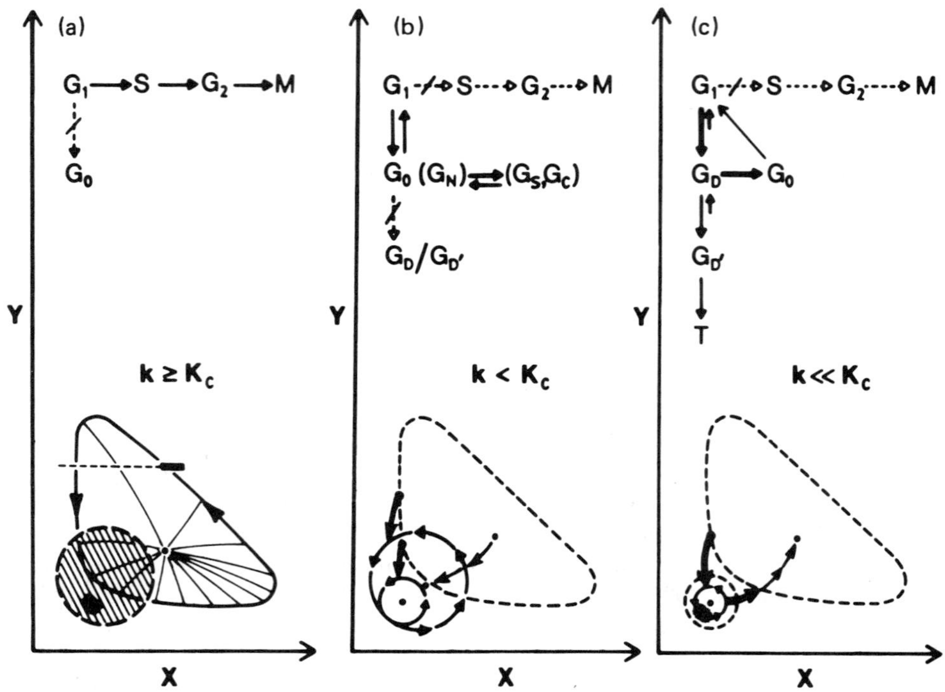


FIGURE 5 Phase portrait of a hypothetical XY biochemical oscillator for the control of cell proliferation and differentiation. It shows multiple critical points and limit cycles that result from parametric control of the system's dynamics. The top of each diagram (a, b, and c) shows the permissible (solid arrows) and nonpermissible (slashed line across dotted arrows) transitions between cell cycle phases and various growth arrest states predicted by the behavior of the underlying limit cycle dynamical system shown at the bottom of each diagram. $G_1$, pre-DNA synthetic phase; S, DNA synthetic phase; $G_2$, premitotic phase; M, mitotic phase; $G_0$, quiescent phase; $G_N$, nutrient-deficiency arrest state; $G_S$, serum-deficiency arrest state; $G_C$, density-dependent arrest state; $G_D$, predifferentiation arrest state; $G_D'$, nonterminal differentiation arrest state; T, terminal differentiation. (a) Proliferative mode of the oscillator is governed by constant kinetics ($k \geq K_c$). A sustained high-amplitude oscillation rotates in a closed roughly triangular path in a counterclockwise fashion (solid arrows) around an unstable focus (central dot). Lines emanating from the neighborhood of the central unstable focus and crossing the limiting trajectory are called isochrons. They mark equal intervals of time along the oscillation; where isochrons are closely spaced they indicate a slow angular velocity of rotation. Isochrons also define the locus of all states not on the path of the limiting trajectory that have the same ultimate phase as the state it intercepts on the path of the oscillation. Mitosis/cell division (dotted line with horizontal bar) is triggered when the system's trajectory passes a threhsold concentration of the variable Y. The region of state space which defines the $G_1$ phase coincides with an arc of the limiting trajectory where the system's variable have their minimal values (shaded circle). If dynamics governing the proliferative mode of the oscillator are modified, it is then possible to induce arrest at some state in $G_1$. The model further proposes that within $G_1$ is a volume of state space at which differentiation is regulated (dark stippled area). (b) The quiescent mode of the oscillator is governed by variable kinetics ($k < K_c$) when the control equations are modified in one or more parameters (Rosen, 1970). If the parameters of this hypothetical oscillator are slowly decreased, the high-amplitude oscillation (dotted closed path) is lost

move the trajectories through a repelling region associated with an anti-limit cycle (dashed circle) that surrounds the system's stable singularity. Once within the attractor region of this stable singularity, a very low amplitude oscillation defines $G_D$ arrest. If the $G_D$ oscillation changes slightly it may enter the domain of space where differentiation is triggered (stippled area) and where the $G_D'$ state exists. The $G_D'$ state can be maintained or reversed if a signal of sufficient strength causes the oscillation to be stabilized or increased to a higher amplitude. Upon collapse of the metabolic oscillation, $G_D'$ cells enter the T state of terminal differentiation, which represents the system's ultimate singularity.

It is predicted that positive effectors are necessary to drive the oscillating system off the high-amplitude limit cycle into the attractor region of the stable singularity, and other positive effectors are needed to drive the low-amplitude oscillation back out to the high-amplitude limit cycle. These predictions are entirely consistent with findings of studies on the regulation of the integrated control of proliferation and differentiation in myeloblasts and proadipocytes.

We further speculate that the two-stage model of carcinogenesis might find its explanation in specific alterations in the kinetic parameters controlling the dynamics of such an oscillator. If, indeed, the initiation step corresponds to the loss of ability for $G_D/G_D'$ arrest and differentiation but retention of the capacity to arrest at other $G_0$-like states, initiation of carcinogenesis could be mediated by changes in the system's

---

and subthreshold limit-cycle oscillations (solid closed paths and arrows) can develop. Two such potential oscillations are shown. They can be reached by vectors (thick arrows) either from the limiting trajectory as it spontaneously traverses the $G_1$ domain or by slow collapse of the high-amplitude oscillation to the system's stable singularity (see path of the unstable focus, thin arrows). The higher amplitude oscillation (larger circle) defines a temporally distinct $G_0$ arrest state which consists mostly of $G_1$ but which may include in its limiting path some component in S. Such a state could correspond to $G_N$ arrest (Yen and Pardee, 1978). The lower amplitude oscillation (small circle) lies entirely within the $G_1$ domain and may include the $G_S$ and $G_C$ growth arrest states. The model predicts that conversion from $G_N$ to $G_S/G_C$ can occur as can conversion from $G_C/G_S$ to $G_N$. It further suggests that the mechanisms to induce $G_D$ arrest are different than those required to induce $G_N$, $G_C$, or $G_S$ arrest. (c) The quiescent mode of the oscillator for the integrated control of cell proliferation and differentiation are also governed by variable kinetics ($k << K_c$). Quiescence at $G_D$ results when the kinetic parameters are rapidly varied at a critical phase of the high amplitude oscillation. This occurs when a vector (thick arrow, left) drives the limiting trajectory off the limit cycle in the $G_1$ domain with sufficient magnitude to pass through the repelling region of an anti-limit cycle (dashed circle). A distinct very low amplitude oscillation within the anti-limit cycle defines the $G_D$ arrest state. If in the $G_D$ state critical oscillations are further modified the model suggests that nonterminal differentiation occurs (dark stippled area). If additional critical oscillations become unstable the model suggests that the system collapses to the stable singularity, T (dot in small circle), which represents the state of terminal differentiation. The model, however, also predicts that: (1) it should be possible to maintain a stable nonterminal state of differentiation by the action of factors that sustain important low amplitude oscillations within the anti-limit cycle, and (2) it should be possible to reverse $G_D'$ arrest by perturbations that return the system to the high-amplitude limit cycle (thick arrow, right).

dynamics that abolish the anti-limit cycle. A promotional step leading to complete transformation could be mediated by changes in the system's dynamics that yield very high amplitude oscillations. We suggest that further investigation of this oscillator model to the phenomenology of carcinogenesis could aid in the quest for the actual molecular mechanisms involved.

## ACKNOWLEDGMENTS

The authors thank Bryan Hoerl, Michael Zschunke, and Peter Maercklein for excellent technical assistance and Ms. Karen Connelly for excellent secretarial assistance. We also thank Dr. K. Yun and Marjorie Wier for their invaluable help in preparing this manuscript.

## REFERENCES

Aaronson, S. A., and Todaro, G. J. (1968). Development of 3T3-like lines from Balb/c mouse embryo cultures: Transformation susceptibility to SV 40. *J. Cell. Phsyiol.* 72:141-148.

Ackerman, L. V., and Rossi, J. (1974). *Surgical Pathology.* Mosby, St. Louis.

Altman, P. L., and Katz, D. D. (eds.) (1976). *Biological Handbooks: Cell Biology,* vol. I. Fed. Am. Soc. Exptl. Biol., Bethesda, pp. 11-16.

Ashihara, T., Chang, S., and Baserga, R. (1978). Constancy of the shift-up point in two temperature sensitive mammalian cell lines that arrest in $G_1$. *J. Cell. Physiol.* 98: 15-22.

Barrett, J. C., and Ts'o, P. O. (1978). Evidence for the progressive nature of neoplastic transformation in vitro. *Proc. Natl. Acad. Sci. USA* 75:3761-3765.

Bartholomew, J. C., Yokota, H., and Ross, P. (1976). Effect of serum on the growth of Balb 3T3/A31 mouse fibroblasts and on SV-40 transformed derivative. *J. Cell. Physiol.* 88:277-286.

Baserga, R. (1968). Biochemistry of the cell cycle: A review. *Cell Tissue Kinet.* 1:167-191.

Baserga, R. (Ed.) (1971). *The Cell Cycle and Cancer.* Marcel Dekker, New York.

Baserga, R. (1976). *Multiplication and Division in Mammalian Cells.* Marcel Dekker, New York.

Becker, F. F. (1981). Presidential address: Recent concepts of initiation and promotion in carcinogenesis. *Am. J. Pathol.* 105:3-9.

Benedict, W. F., Baker, M. S., and Gardner, A. (1975). Malignant transformation with ara-C, FUdR, MTX, and bleomycin in 10 T 1/2 cells. *Proc. Am. Assoc. Cancer Res.* 16:40.

Berenblum, I. (1981). Sequential aspects of chemical carcinogenesis: Skin. In *Cancer: A Comprehensive Treatise,* vol. I, 2nd ed., F. F. Becker (ed.). Plenum, New York, pp. 451-484.

Bertram, J. S., and Heidelberger, C. (1974). Cell cycle dependency of oncogenic transformation induced by N′-methyl N′-nitro-N-nitrosoguanidine. *Cancer Res.* 34:526-537.

Boman, B. M., Maercklein, P. B., Hoerl, B. J., and Scott, R. E. (198 ). Cell surface characteristics of proadipocytes growth arrested at the predifferentiation $G_D$ state: Defects associated with neoplastic transformation. *Lab. Invest.* 48:199-204.

Boone, C. W., and Jacobs, J. B. (1976). Sarcomas routinely produced from putatively nontumorigenic Balb/3T3 and C3H/10T½ cells by subcutaneous inoculation attached to plastic plates. *J. Supramol. Struct.* 5:131-137.

Brown, K. D., and Holley, R. W. (1979). Epidermal growth factor control of proliferation of Balb 3T3 and benzo-(a) pyrene-transformed Balb 3T3 cells. *J. Cell. Physiol.* 100: 139-146.

Burstin, S., Meiss, H., and Basilico, C. (1974). A temperature sensitive cell cycle mutant of the BHK cell line. *J. Cell. Physiol.* 84:397-408.

Burton, A. C., and Canham, P. B. (1973). The behavior of coupled biochemical oscillators as a model of contact inhibition of cell division. *Theor. Biol.* 39:555-580.

Chan, G. L., and Little, J. B. (1976). Induction of oncogenic transformation in vitro by ultraviolet light. *Nature* 264:442-444.

Damjanov, I., Solter, D., Belicza, M., and Skreb, N. (1971). Teratomas obtained through extrauterine growth of seven-day mouse embryos. *JNCI* 46:471-475.

Dawe, C. (1973). Comparative neoplasia. In *Cancer Medicine,* J. F. Gallard and E. Frie (eds.). Lea and Febiger, Philadelphia, pp. 193-240.

Diamond, L., O'Brian, T. G., and Rovera, G. (1977). Inhibition of adipocyte conversion of 3T3 T fibroblasts by tumor promoters. *Nature* 269:247-249.

DiPaolo, J. A., Takano, K., and Popescu, N. C. (1972). Quantitation of chemically-induced neoplastic transformation of Balb/3T3 cloned cell lines. *Cancer Res.* 32: 2686-2695.

Dulbecco, R. (1970). Topoinhibition and serum requirement of transformed and untransformed cells. *Nature* 227:802-806.

Edmunds, L. N., Jr., and Adams, K. J. (1981). Clocked cell cycle clocks. *Science* 211: 1002-1013.

Evans, R. W. (1968). *Histological Appearance of Tumors.* Williams & Wilkins, Baltimore.

Farber, E. E. (1973). Carcinogenesis: Cellular evolution as a unifying thread. *Cancer Res.* 33:2537-2550.

Farber, E. E. (1982). Chemical carcinogenesis: A biologic perspective. *Am. J. Pathol.* 106:272-296.

Florine, D. L., Hoerl, B. J., and Scott, R. E. (1982). Growth arrest of proadipocytes at a distinct predifferentiation $G_1$ state associated with cytotoxic responsiveness to 8-bromo cyclic AMP. *Cell Differ.* 11:185-202.

Foulds, L. (1975). *Neoplastic Development,* vol. II. Academic Press, London.

Friend, C., Patuleia, M. C., and DeHarven, E. (1966). Erythrocytic maturation in vitro of murine (Friend) virus-induced leukemia cells. *Natl. Cancer Inst. Monogr.* 22: 505-522.

Gehley, E. B., Landolph, J. R., Heidelberger, C., Nagasawa, H., and Little, J. B. (1982). Induction of cytotoxicity, mutation, cytogenetic changes, and neoplastic transformation by benzo(a) pyrene and derivatives in C3H/10T½ clone 8 mouse fibroblasts. *Cancer Res.* 42:1866-1875.

Green, H., and Kehinde, O. (1975). An established proadipocyte cell line and its differentiation in culture. *Cell* 5:19-27.

Hastings, J., and Schweiger, H. G. (eds.) (1976). *The Molecular Basis of Circadian Rhythms.* Dahlem Konferenzen, Berlin.

Hershko, A., Mamont, P., Shields, R., and Tomkins, A. G. (1970). Pleiotypic response. *Nature [New Biol.]* 232:206-211.

Holley, R. W., Baldwin, J. H., Kiernan, J. A., and Messmer, T. O. (1976). Control of growth of benzo-[a]pyrene-transformed 3T3 cells. *Proc. Natl. Acad. Sci. USA* 73: 3229-3232.

Holley, R. W., and Kiernan, J. (1968). Contact-inhibition of cell division in 3T3 cells. *Proc. Natl. Acad. Sci. USA* 60:300-304.

Holley, R. W., and Kiernan, J. A. (1973). Control of the initiation of DNA synthesis in 3T3 cells: Low molecular weight nutrients. *Proc. Natl. Acad. Sci. USA* 71:2942-2945.

Holtzer, H., Pacifici, M., Tapscott, S., Bennett, G., Payette, R., and Dlugosz, A. (1982). Lineages in cell differentiation and cell transformation. In *Expression of Differentiated Functions in Cancer Cells,* R. P. Revoltella, G. M. Pontieri, C. Basilico, G. Rovera, R. C. Gallo, and J. H. Subak-Sharpe (eds.). Raven Press, New York, pp. 169-180.

Howard, A., and Pelc, S. R. (1953). The synthesis of deoxyribonucleic acid in normal and irradiated cells and its relation to chromosome breakage. *Heredity (Suppl.)* 6:261-273.

Iversen, O. H., Bjerknes, R., and Devik, F. (1968). Kinetics of cell renewal, cell migration and cell loss in the hairless mouse dorsal epidermis. *Cell Tissue Kinet.* 1:351-367.

Kauffman, S. A., and Wille, J. J., Jr. (1975). The mitotic oscillator in *Physarum polycephalum. J. Theor. Biol.* 55:47-93.

Klevecz, R. R. (1976). Quantized generation time in mammalian cells as an expression of the cellular clock. *Proc. Natl. Acad. Sci. USA* 73:4012-4016.

Klevecz, R. (1978). The clock in animal cells is a limit cycle oscillator. In *Cell Reproduction: In Honor of Daniel Mazia,* E. R. Dirksen, D. M. Prescott, C. F. Fox (eds.). Academic Press, New York, pp. 139-146.

Krawisz, B. R., and Scott, R. E. (1982). Coupling of proadipocyte growth arrest and differentiation. I. Induction by heparinized medium containing human plasma. *J. Cell Biol.* 94:394-399.

Lajtha, L. G., Lord, B. I., Dexter, T. M., Wright, E. G., and Allen, T. D. (1978). Interrelationship of differentiation and proliferation control in hematopoietic stem cells. In *Cell Differentiation and Neoplasia,* G. F. Saunders (ed.). Raven Press, New York, pp. 179-193.

Lajtha, L. G., Oliver, G., and Gurney, C. (1962). Kinetic model of a bone-marrow stem-cell population. *Br. J. Haematol.* 8:442-460.

Ley, K., and Tobey, R. (1970). Regulation of DNA synthesis in Chinese hamster cells. II. Induction of DNA synthesis and cell division by isoleucine and glutamine in $G_1$ arrested cells in suspension culture. *J. Cell Biol.* 47:453-459.

Lotem, J., and Sachs, L. (1982). Mechanisms that uncouple growth and differentiation in myeloid leukemia cells: Restoration of requirement for normal growth-inducing protein without restoring induction of differentiation-inducing protein. *Proc. Natl. Acad. Sci. USA* 79:4347-4351.

Martin, R. G., and Stein, S. (1976). Resting state in normal and simian virus transformed Chinese hamster lung cells. *Proc. Natl. Acad. Sci. USA* 73:1655-1659.

Mitchison, M. (1971). *The Biology of the Cell Cycle.* Cambridge University Press, London.

Mondal, S., and Heidelberger, C. (1976). Transformation of C3H/10T½ Cl 8 mouse embryo fibroblasts by ultraviolet irradiation and a phorbol ester. *Nature* 260:710-711.

Moses, H. L., Proper, J., Volkenant, M., and Swartzendruber, D. E. (1980). Growth arrest of AKR-2B cells maintained in the presence of epidermal growth factor 12-0-tetradecanoylphorbol-13-acetate: Evidence for two separate $G_1$ arrest points. *J. Cell. Physiol.* 102:367-378.

Moses, H. L., Proper, J. A., Volkenant, M. E., Wells, D. J., and Getz, M. J. (1978). Mechanism of growth arrest of chemically transformed cells in culture. *Cancer Res.* 38: 2807-2812.

Nadal-Ginard, B. (1978). Commitment, fusion, and biochemical differentiation of a myogenic cell line in the absence of DNA synthesis. *Cell* 15:855-864.

Naha, P., Meyer, A., and Hewitt, K. (1975). Mapping of the $G_1$ phase of a mammalian cell cycle. *Nature* 258:49-53.

Namba, M., Nishitani, K., and Kimoto, T. (1978). Carcinogenesis in tissue culture 29: Neoplastic transformation of a normal human diploid cell strain, WI-38, with Co-60 gamma rays. *Jpn. J. Exp. Med.* 48:303-311.

Otsuka, H., and Moskowitz, M. (1976). Arrest of 3T3 cells in $G_1$ by suspension culture. *J. Cell Physiol.* 87:213-220.

Pardee, A. B. (1974). A restriction point for the control of normal animal cell proliferation. *Proc. Natl. Acad. Sci. USA* 71:1286-1290.

Pardee, A. B. (1982). Principles of cancer biology: Cell biology and biochemistry of cancer. In *Cancer Principles and Practices of Oncology,* V. J. DeVitta, Jr., S. Hellman, and S. A. Rosenberg (eds.). Lippincott, Philadelphia, pp. 59-72.

Pardee, A. B., de Asua, L. J., and Rozengurt, E. (1974). Functional membrane changes and cell growth: Significance and mechanism. In *Control of Proliferation in Animal Cells,* B. D. Clarkson and R. Baserga (eds.). Cold Spring Harbor Laboratory, Cold Spring Harbor, New York, pp. 547-561.

Pardee, A., Dubrow, R., Hamlin, J., and Kletzien, R. (1978). The animal cell cycle. *Annu. Rev. Biochem.* 47:715-750.

Pederson, T., and Gelfant, S. (1970). G2-Population cells in mouse kidney and duodenum and their behavior during the cell division cycle. *Exp. Cell Res.* 59:32-36.

Pierce, G. B., Stevens, L. C., and Nakane, P. K. (1967). Ultrastructural analysis of the early development of teratocarcinomas. *JNCI* 39:755-761.

Pitot, H. L. (1978). *Fundamentals of Oncology.* Marcel Dekker, New York.

Pledger, W., Stiles, C., Antoniades, H., and Scher, C. (1978). An ordered sequence of events is required before Balb/c 3T3 cells become committed to DNA synthesis. *Proc. Natl. Acad. Sci. USA* 75:2839-2843.

Prescott, D. M. (1967a). *Reproduction of Eukaryotic Cells.* Academic Press, New York.

Prescott, D. M. (1976b). The cell cycle and the control of cellular reproduction. *Adv. Genet.* 18:99-177.

Reznikoff, C. A., Bertram, J. S., Brankow, D. W., and Heidelberger, C. (1973a). Quantitative and qualitative studies on chemical transformation of cloned C3H mouse embryo cells sensitive to post-confluence inhibition of cell division. *Cancer Res.* 33: 3239-3249.

Reznikoff, C. A., Brankow, D. W., and Heidelberger, C. (1973b). Establishment and characterization of a cloned line of C3H mouse embryo cells sensitive to post-confluence inhibition of division. *Cancer Res.* 33:3231-3238.

Rheinwald, J. G. (1979). The role of terminal differentiation in the finite culture lifetime of the human epidermal keratinocyte. *Int. Rev. Cytol. (Suppl.)* 10:25-33.

Risser, R., and Pollack, R. E. (1974). Biological analysis of clones of SV-40 infected mouse 3T3 cells. In *Control of Proliferation in Animal Cells,* B. D. Clarkson and R. Baserga (Eds.). Cold Spring Harbor Laboratory, Cold Spring Harbor, New York, pp. 139-150.

Rosen, R. (1970). *Dynamical Systems Theory in Biology,* vol. 1. Wiley Interscience, New York.

Rovera, G., Olashaw, N., and Meo, P. (1980). Terminal differentiation in human promyelocytic leukemia cells in the absence of DNA synthesis. *Nature* 284:69-70.

Sachs, L. (1978). Control of normal cell differentiation and the phenotypic reversion of malignancy in myeloid leukemia. *Nature* 274:535-539.

Sachs, L. (1980). Constitutive uncoupling of pathways of gene expression that control growth and differentiation in myeloid leukemia: A model for the origin and progression of malignancy. *Proc. Natl. Acad. Sci. USA* 77:6152-6156.

Schiffer, L. M. (1968). Kinetics of chronic lymphocytic leukemia. *Ser. Haematol.* 1(3): 3-23.

Scott, J. L., McMillan, R., Davidson, J. G., and Marino, J. V. (1971). Leukocyte labelling with $^{51}$Chromium. II. Leukocyte kinetics in chronic myelocytic leukemia. *Blood* 38:162-173.

Scott, R. E., and Boone, C. W. (1981). Promotion of smooth surface tumorigenicity by phorbol myristate acetate in BALB/3T3 cells and BALB/3T3 T proadipocytes. *JNCI* 66:733-736.

Scott, R. E., and Florine, D. L. (1982). Cell cycle models for the aberrant coupling of growth arrest and differentiation in hyperplasia, metaplasia, and neoplasia. *Am. J. Pathol.* 107:342-348.

Scott, R. E., Florine, D. L., Wille, J. J., Jr., and Yun, K. (1982a). Coupling of growth arrest and differentiation at a distinct state in the $G_1$ phase of the cell cycle: $G_D$. *Proc. Natl. Acad. Sci. USA* 79:845-849.

Scott, R. E., Hoerl, B. J., Wille, J. J., Jr., Florine, D. L., Krawisz, B. R., and Yun, K. (1982b). Coupling of proadipocyte growth arrest and differentiation. II. A cell cycle model for the physiological control of cell proliferation. *J. Cell Biol.* 94: 400-405.

Scott, R. E., Yun, K., and Florine, D. L. (1983). Differential mitogenic effects of methyl isobutyl xanthine and a tumor growth factor on $G_1$ arrested 3T3 T proadipocytes at the predifferentiation $G_D$ state and the growth factor deficiency $G_S$ state. *Exp. Cell Res.* 143:405-414.

Sheppard, J. R., and Prasad, K. N. (1973). Cyclic AMP levels and the morphological differentiation of mouse neuroblastoma cells. *Life Sci.* 12:431-439.

Silagi, S. (1977). Reversible suppression of malignancy and differentiation of melanoma cells. *Am. J. Pathol.* 89:671-681.

Smith, J. A., and Martin, L. (1973). Do cells cycle? *Proc. Natl. Acad. Sci. USA* 70:1263-1267.

Stevens, L. C. (1967). Origin of testicular teratomas from primordial germ cells in mice. *JNCI* 38:549-551.

Stiles, C. D., Desmond, W., Chuman, L. M., Sato, G., and Sauer, M. H., Jr. (1976). Growth control of heterologous tissue culture cells in the congenitally athymic nude mouse. *Cancer Res.* 36:1353-1360.

Stocker, M., O'Neill, C., Berryman, S., and Waxman, V. (1968). Anchorage and growth regulation in normal and virus transformed cells. *Int. J. Cancer* 3:683-687.

Talavera, A., Basilico, C. (1978). Requirement of BHK cells for the exit from different quiescent states. *J. Cell. Physiol.* 97:429-440.

Terzaghi, M., and Little, J. B. (1976). X-radiation-induced transformation in a C3H mouse embryo-derived cell line. *Cancer Res.* 36:1367-1374.

Tobey, R., and Ley, K. (1971). Isoleucine-mediated regulation of genome replication in various mammalian cell lines. *Cancer Res.* 31:46-51.

Todaro, G. J., DeLarco, J. E., Marquardt, H., Bryant, M. L., Sherwin, S. A., and Sliski, A. H. (1979). Polypeptide growth factors produced by tumor cells: A possible growth advantage for the producer cells. In *Cold Spring Harbor Conference on Cell Proliferation. VI. Hormones and Cell Culture,* S. Sato and R. Ross (eds.). Cold Spring Harbor Laboratory, Cold Spring Harbor, New York, pp. 113-127.

Todaro, G. J., and Green, H. (1963). Quantitative studies on the growth of mouse embryo cells in culture and their development into established cell lines. *J. Cell. Biol.* 17:299-313.

Todaro, G. J., Green, H., and Goldberg, B. D. (1964). Transformation of properties of an established cell line by SV 40 and polyoma virus. *Proc. Natl. Acad. Sci. USA* 51:66-73.

Todaro, G. J., Lazar, G. K., and Green, H. (1965). The initiation of cell division in contact-inhibited mammalian cell lines. *J. Cell. Comp. Physiol.* 66:325-334.

Tsao, M. C., Walthall, B. J., and Ham, R. G. (1982). Clonal growth of normal human epidermal keratinocytes in a defined medium. *J. Cell. Physiol.* 110:219-229.

Upton, H. C. (1982). Principles of cancer biology: Etiology and prevention of cancer. In *Cancer Principles and Practice of Oncology,* V. J. DeVitta, Jr., S. Hellman, and S. A. Rosenberg (Eds.). Lippincott, Philadelphia, pp. 33-58.

Watt, F. M., and Green, H. (1982). Stratification and terminal differentiation of cultured epidermal cells. *Nature* 295:434-436.

Wille, J. J., Jr. (1979). Biological rhythms in protozoa. In *Biochemistry and Physiology of Protozoa,* vol. 2, 2nd ed., M. Levandowsky and S. Hutner (eds.). Academic Press, New York, pp. 67-149.

Wille, J. J., Jr., Maercklein, P. B., and Scott, R. E. (1982). Neoplastic transformation and defective control of cell proliferation and differentiation. *Cancer Res.* 42:5139-5146.

Wille, J. J., Jr., Scheffey, C., and Kauffman, S. A. (1977). Novel behavior of the mitotic clock in *Physarum polycephalum. J. Cell Sci.* 27:91-104.

Wille, J. J., Jr., and Scott, R. E. (1982a). Topography of the predifferentiation $G_D$ growth arrest state relative to other growth arrest states in the $G_1$ phase of the cell cycle. *J. Cell. Physiol.* 112:115-122.

Wille, J. J., Jr., and Scott, R. E. (1982b). Oncogenic potential of cells expressing variant $G_1$ growth arrest properties (abstract). *J. Cell Biol.* 95:19a.

Wu, R. S., Tsai, S., and Bonner, W. M. (1982). Pattern of histone variant synthesis can distinguish $G_0$ from $G_1$ cells. *Cell* 31:367-374.

Yen, A., and Pardee, A. B. (1978). Arrested states produced by isoleucine deprivation and their relationship to the low serum produced arrested state in Swiss 3T3 cells. *Exp. Cell Res.* 114:389-395.

Yun, K., and Scott, R. E. (1982). Biological mechanisms of phorbol myristate acetate-induced inhibition of proadipocyte differentiation. *Cancer Res.* 43:88-96.

Zimmerman, T. S., Goodwin, H. A., and Perry, S. (1968). Studies of leukocyte kinetics in chronic lymphocytic leukemia. *Blood* 31:277-291.

# 23

# Chronobiology of Cell Proliferation in Mammals

## Implications for Basic Research and Cancer Chemotherapy

**Lawrence E. Scheving** / University of Arkansas College of Medicine, Little Rock, Arkansas

## BRIEF HISTORY OF INVESTIGATIONS INTO CELL PROLIFERATION RHYTHMS IN MULTICELLULAR PLANTS AND ANIMALS

### Plants

One of the first reports in the literature suggesting a circadian rhythm in cell proliferation was that by Killicott (1904). She reported that the highest mitotic index in the *Allium* root tip occurred at 2300 and the lowest at 0700. Karsten (1918) reported similar phenomena for the mitotic index of *Spirogyra,* with the highest activity during midday, and for *Zea mais* with the highest mitotic index occurring during the night. Stalfelt (1921) reported that the highest mitotic index in *Pisum sativum* occurred at midday and the minimum during the morning. With these and several other early studies, the botanists pioneered in exploring rhythms in cell division.

### Animals

Fortuyn-Van Leyden (1917, 1926) was the first to report on mitotic-index rhythms in animals, specifically on rhythmic variations in several tissues of the cat. In these studies she used "once-a-day" and "once-a-night" sampling and small numbers of animals; consequently her findings frequently were contradictory from one study to another. In spite of the problems, she was correct in her suggestion that cell proliferation in the cat was periodic and therefore in this respect was not unlike cell proliferation previously reported in plants.

In the 1930s and 1940s, there was renewed interest in this phenomenon, and several papers appeared. Carleton (1934), Ortiz-Picón (1934), and Cooper and Franklin

(1940) all reported a rhythm in cell division in the epidermis of the mouse. Although early studies were very limited as to the number of sampling times along the 24 hr scale and the total number of animals used, mititoc activity generally was reported to be highest in the epidermis at about midday and lowest at night. The work of Ortiz-Picón (1934) was, up to that time, the most extensive in that he sampled the epidermis at four time points, 1200, 1900, 2000, and 2400, and used six mice per time point. In reality, however, the above represents only three circadian stages, because the 1900 and 2000 time points are so close together.

More extensive investigations were made on rhythms in cell proliferation in the rat by Blumenfeld (1939, 1942). He found that the maximum mitotic index of the abdominal epidermis occurred between 0800 and 1000 and the minimum occurred between 2000 and 2400. In these same rats, he also determined the mitotic index in the submandibular gland and kidney. Although the times of peak mitotic index were different for these three tissues, it was of interest that the time of the lowest mitotic index was identical for all. Blumenfeld discussed at length the differences between the timing of the peaks of the ryhthms but failed discuss the similarity of phasing of the lowest values. We have found that the time at which the low point of the rhythm occurs each day is frequently more reliably predictable than that at which the peak of the rhythm occurs. The early reports invariably failed to comment on how their animals were standardized, and of course this makes it difficult to compare them.

### Humans

In 1938, Cooper and Schiff analyzed mitosis in the prepuce of 8-day-old infants and concluded that the highest mitotic index occurred at 0800 and the lowest at 2200; only 13 specimens were used to evaluate the entire 24 hr span, and no sampling was performed between 1245 and 0730. A similar study followed in 1939, when Broders and Dublin examined 14 infant foreskins and concluded that the highest mitotic activity occurred at night. We also carried out extensive studies on the human infant prepuce but were unable to demonstrate a circadian rhythm; as to why these data were not published, see Scheving (1981). There was, however, a somewhat irregular high-frequency fluctuation in the data from the infant prepuce suggesting to us that its mitotic activity might be characterized by an ultradian rhythm. The sampling interval of 2 hr that we used was too great to resolve an ultradian rhythm. Our observations, in retrospect, might prove to be more in line with what has been demonstrated in newborn rats, where the circadian rhythm in the mitotic index of the corneal epithelium does not manifest itself until about day 15 of postnatal life (Goodrun et al., 1974).

It is our opinion that until more extensive documentation with adequate sampling intervals is available, the question remains open as to whether a circadian or an ultradian rhythm characterizes cell proliferation in the infant prepuce.

Earlier anatomists and pathologists frequently were bewildered by the sometimes sparse numbers of mitotic figures that were found in histological preparations. This was especially puzzling because the epidermis was known to be continually shed and renewed. The question often asked was whether the paucity of mitotic figures frequently characterizing biopsied skin specimens (more often than not samples were obtained from the morning autopsies) could account for the renewal of all of the skin known to be lost during the course of a single day. Several investigators, including Frieboes (1920), Bostroem (1928), Cameron (1936), Levander (1950), and others, even postulated that epidermal cells may indeed have a mesodermal origin. It was in 1949 that Andrew and Andrew introduced the hypothesis that cell replacement in the human epidermis was accomplished

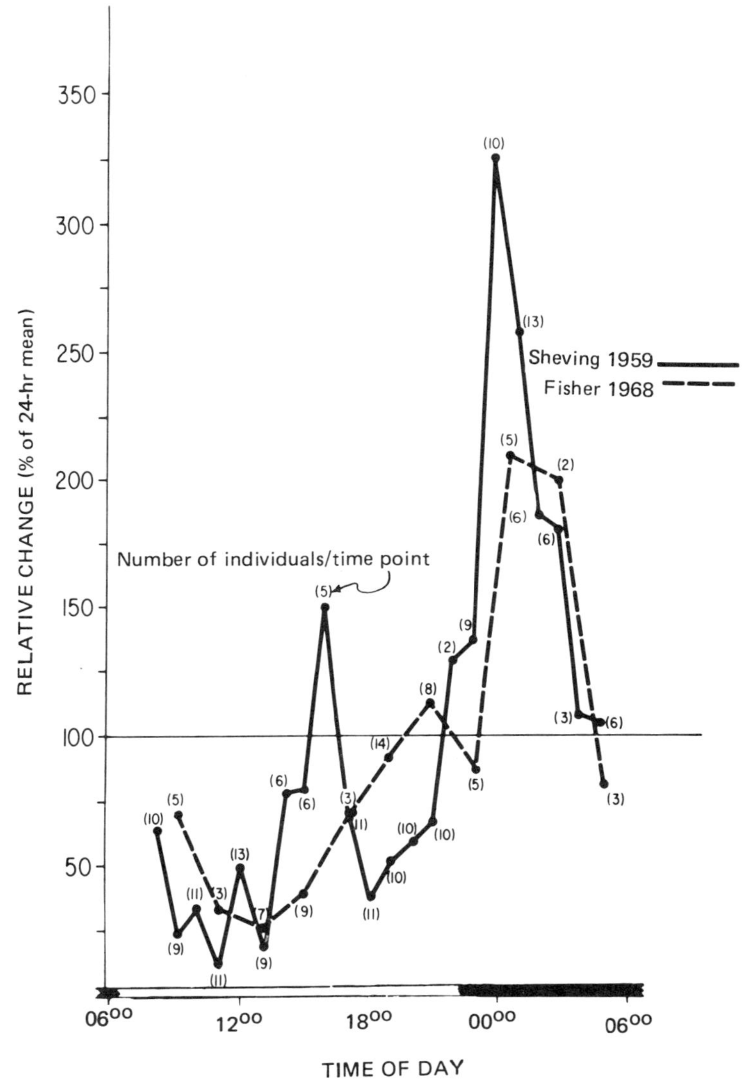


**FIGURE 1** Reproducibility of the mitotic rhythm in adult human epidermis. For details, see Scheving (1959) and Fisher (1968).

by lymphocytes that migrated through the skin and transformed into clear cells, which ultimately differentiated into epidermal cells. Andrew (a histologist) believed that this was the reason one normally did not find many mitotic figures in histological preparations obtained from human skin.

Such an explanation was difficult to accept. It seemed more logical to me, having background knowledge of what had been done during the 1920s through 1940s on rhythmicity in cell proliferation in plants and rodents, that the paucity of mitotic figures in the adult human epidermis might well be explained by the fact that mitosis occurred in daily cycles. Therefore a study was designed to test this hypothesis in humans; the result (Scheving and Gatz, 1955; Scheving, 1959) was the finding that the mitotic index in the epidermis

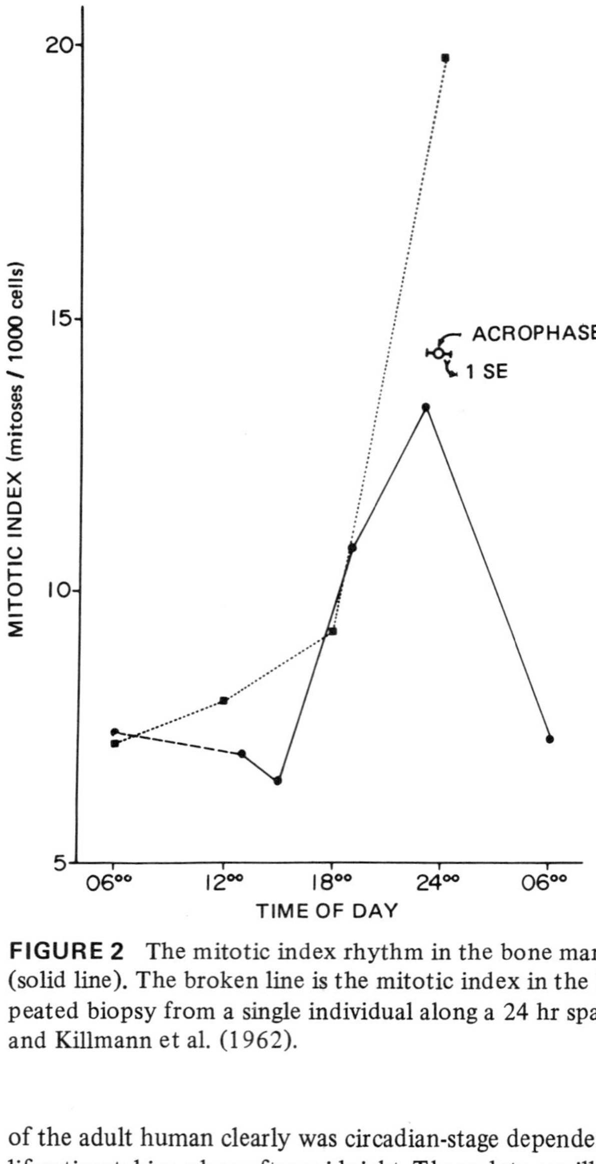


**FIGURE 2** The mitotic index rhythm in the bone marrow of a group of young men (solid line). The broken line is the mitotic index in the bone marrow obtained by repeated biopsy from a single individual along a 24 hr span. For details, see Mauer (1965) and Killmann et al. (1962).

of the adult human clearly was circadian-stage dependent with a great bulk of cell proliferation taking place after midnight. These data are illustrated in Fig. 1 along with data that confirmed them (Fisher, 1968). It is of interest that this finding was confirmed again recently (Zugula-Mally et al., 1979). Gelfant et al. (1982) claim to have found no variation in the mitotic index from the results of their study on human beings with psoriatic disease. They sampled at only four time points from both involved and uninvolved areas of their patients and pooled data obtained from uninvolved skin of the arm, elbow, knee, and back. Aside from the infrequent sampling intervals, their own data would seem to cast doubt on their interpretation of their results, and this has been pointed out (Rubin and Scheving, 1983).

In addition to the studies on skin and the extensive documentation of the unequivocal circadian variation in cell proliferation found in human circulating blood cells, there are, to the best of our knowledge, only two other reports in the literature showing a mitotic index rhythm in human tissues, and these were reported in bone marrow (Killmann et al., 1962; Mauer, 1965). The data of these two studies are illustrated in Fig. 2.

In spite of the evidence that has accumulated, cellular proliferation within mammalian tissues traditionally has been considered by most classic cell kineticists as if it were an asynchronous or random event. The objective of this chapter is to consider some of the more recent work and to demonstrate how an appreciation of rhythms can be used advantageously in carrying out basic research on growth and in the treatment of growth-related disorders, such as in cancer chemotherapy.

## MORE RECENT STUDIES ON CELL PROLIFERATION IN ANIMALS

### Techniques for Evaluation of Cell Proliferation

Up until the 1950s most studies on the rhythmic nature of cell proliferation were based on mitotic counts; these were very time consuming and laborious. For example, to prepare most tissues properly and to evaluate the mitotic activity with an adequate number of samples and at frequent enough intervals of sampling along a single 24 hr time scale could occupy the time of one technician for the better part of a year. Naturally there were attempts to make this task easier, and one involved the technique of arresting mitotic figures with colchicine over a fixed period of time. Such techniques, however, were sometimes fraught with pitfalls, which will be discussed later. In our opinion, the practice of sampling at less frequent intervals, in order to reduce the volume of work involved, was one of the main sources of the ambiguity and confusion that arose during that time as to the nature of the rhythmicity in many tissues (Scheving, 1976).

With the advent of radioisotopes, such as tritiated thymidine ([$^3$H]Tdr), it became possible to study cell proliferation by measuring the uptake of [$^3$H]TdR into DNA, which supposedly reflected the synthesis (S) phase of the mitotic cycle. This could be done either by preparing the tissue for autoradiography and then counting the labeled cells or by scintillation counting, which was less time consuming. Only recently has flow cytometry (FCM) made the task of evaluating cell proliferation in tissue easier. This technique, however, works only for certain tissues because it requires a single-cell suspension; this is frequently difficult to obtain for many heterogeneous tissues and is particularly difficult when solid tumors are involved. In this chapter we shall discuss briefly studies using all three techniques.

### Rhythms in the Mitotic Index of Rodents

All normal proliferating tissues examined to date in adult mammals have been shown to undergo circadian variation when (1) the uptake of [$^3$H]TdR into DNA is measured, reflecting the synthesis (S) phase, (2) the $G_2$ phase is determined, or (3) the mitotic index phase of the cell cycle (M) is monitored along the 24 hr time scale; in some tissues all three phases have been monitored simultaneously using FCM. Figure 3 illustrates chronograms (data collected and plotted along a time scale) of rhythms in the mitotic indices of the epithelium of the duodenum and tongue. The mitotic index in the epithelium of the tongue has a high-amplitude rhythm, frequently dropping to very low levels at one circadian stage. On the other hand, the overall mean of the mitotic index of the duodenum remains rather high during the 24 hr span, whereas the amplitude of the rhythm is low. Figure 4 illustrates data showing the rhythmic pattern of the mitotic index in five diverse tissues obtained from the same animals. In 1973, a review article by Scheving and Pauly

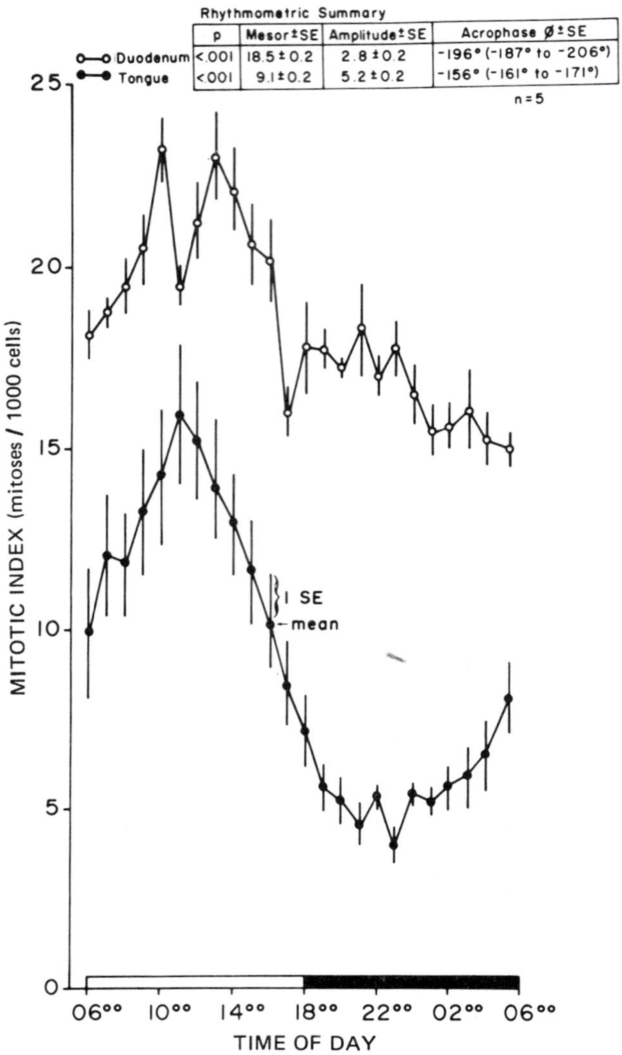


**FIGURE 3** Chronograms illustrating the mitotic indices in epithelium of the duodenum and tongue of the same animals (duodenum data from Scheving and Pauly, 1973; tongue data from Gasser et al., 1972a). Note that the rhythm in the tongue epithelium has a relatively high amplitude, whereas the mitotic index of the duodenal tissue represents a low-amplitude rhythm. The rhythmometric summary is explained in detail by Halberg et al. (1972). Essentially such analysis implies that the data showed a statistically significant fit to a 24 hr cosine curve. The abscissa illustrates the light-dark cycle to which the animals have been standardized for at least 2 weeks prior to the experiment (light from 0600 to 1800).

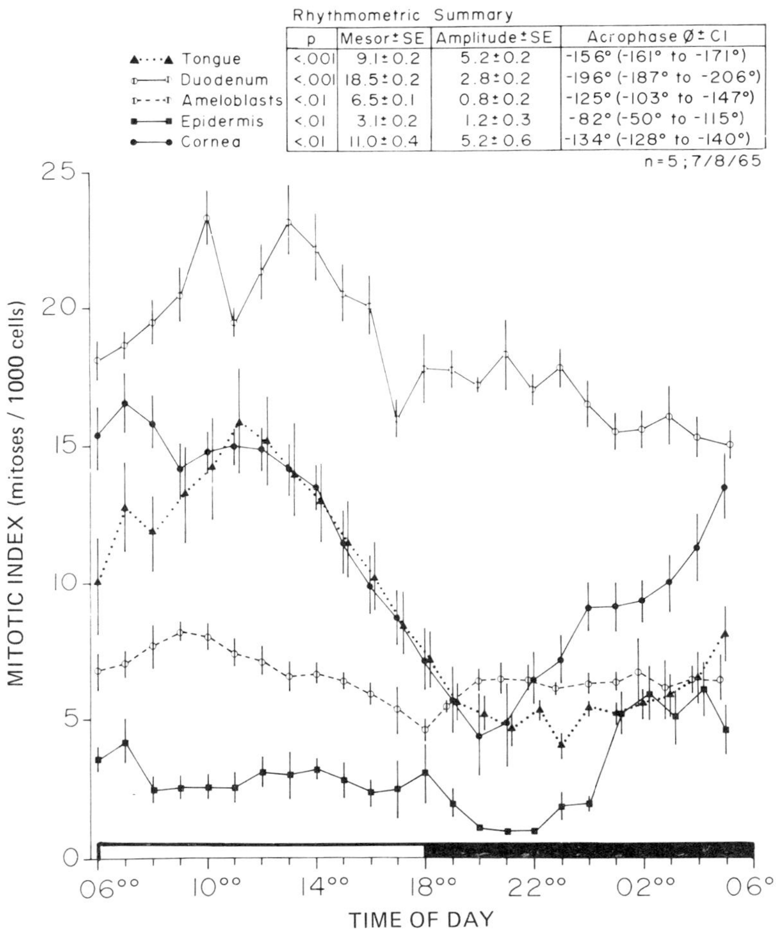


**FIGURE 4** A composite of chronograms illustrating the mitotic indices of five different tissues, all of which were obtained from the same animals. The duodenum and tongue data are the same as that illustrated in Fig. 3. The cornea data were previously published by Scheving and Pauly (1967), the epidermis by Scheving and Pauly (1977), and the ameloblasts by Gasser et al. (1972b). Note that the trough levels in all tissues occur between 1800 and 2200. Note also that the evaluations of the mitotic indices were made by five different technicians. The ameloblasts have the lowest amplitude rhythm, but it was a statistically significant rhythm as revealed by cosinor analysis and analysis of variance.

summarized time-series studies that had been published up to that time on the circadian variation in the mitotic index of 15 different tissues of the rodent. As documented, some of these studies were based on data obtained from two or only a few sampling time points along the 24 hr time scale.

The corneal epithelium has been extensively investigated over the past 20 years because it can easily be prepared as a whole mount and counts of mitotic figures can be made more easily than from tissues, which have to be sectioned and stained. Thus, its

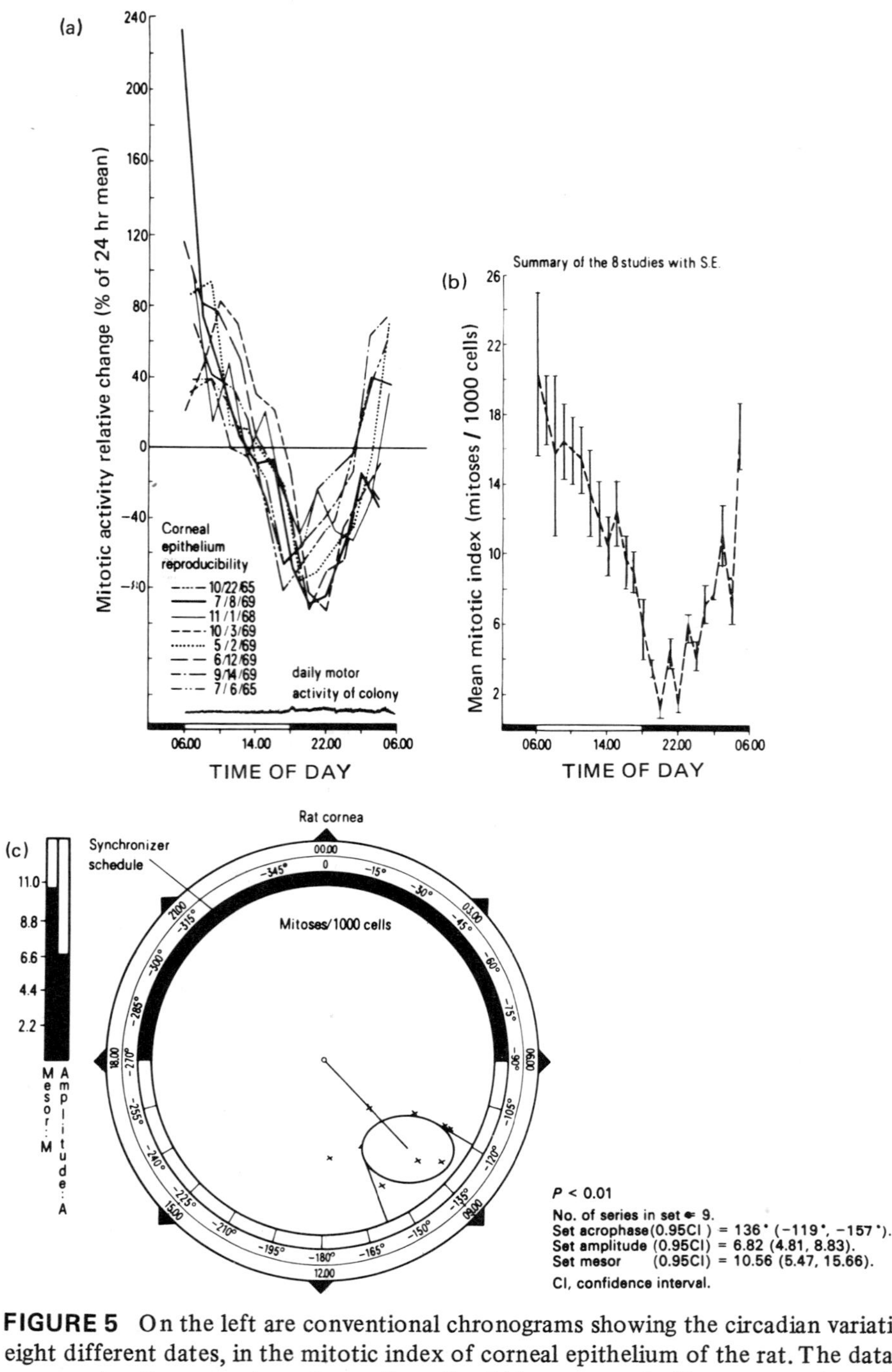


**FIGURE 5** On the left are conventional chronograms showing the circadian variation, on eight different dates, in the mitotic index of corneal epithelium of the rat. The data are expressed as the percentage change from the 24 hr mean to facilitate comparison. Also shown is the typical daily motor activity of the colony based on noises emanating from the colony. The chronogram on the right represents the summary of all eight studies expressed as absolute values ±SE. The polar plot on the bottom is a display of the data obtained by a fit of the eight time series to a 24 hr cosine curve. For a complete description of this method, see Halberg et al. (1972). For a complete description of (1) method of standardization, (2) mechanism of determining motor activity, and (3) manner of evaluating the mitotic index, see Scheving and Pauly (1967).

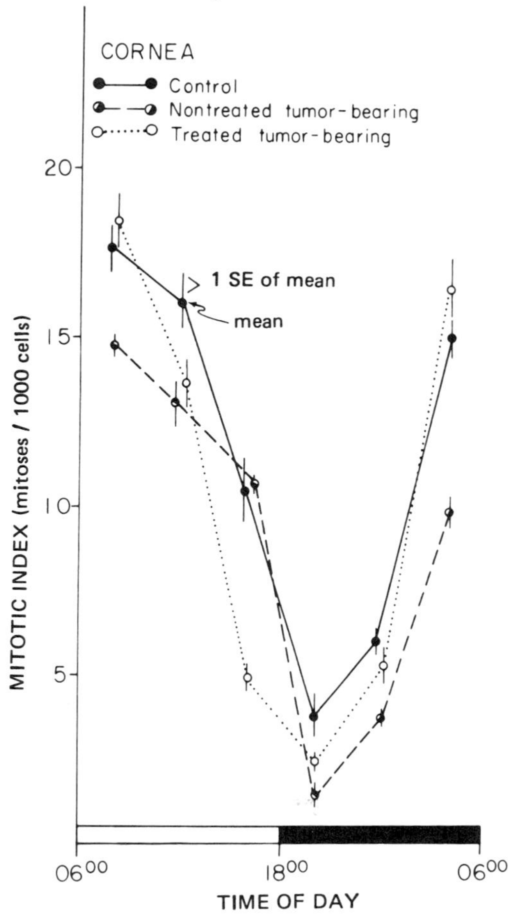


**FIGURE 6** The tumor-bearing mice are subgroups of mice that were inoculated at one of six different circadian stages (0800, 1200, 1600, 2000, 2400, and 0400) with 4.4 X $10^6$ leukemic cells; the mice were standardized to 12 hr of light alternating with 12 hr of darkness. Exactly 44 hr later one group of eight mice along with eight control mice (no leukemia) were killed at each time point and the mitotic index was evaluated. At these same times eight tumor-bearing mice were treated in a chronobiological manner previously described (Scheving et al., 1977b); they were killed 48 hr later. The point we wish to emphasize is that the rhythm persisted and the timing was remarkably stable in spite of the rather large tumor load and the harsh course of treatment. It is noteworthy that in a subsequent study we found that this same rhythm persisted until the day of death in leukemic mice; however, the overall levels of mitosis had decreased (all data previoulsy unpublished).

rhythmic pattern has been extensively documented. The rhythm of the mitotic index and of DNA synthesis in the cornea exhibits one of the larger amplitude changes over the 24 hr span of all the tissues in the body; it is illustrated in Fig. 5. Often the peak in the mitotic index is more than 20 times the trough value. The trough value frequently is zero or almost zero; thus we see what seems to be a "shutting off" of cell proliferation at one circadian stage. In the chronobiology literature relative to cell proliferation in unicellular organisms and insects, such a phenomenon has been called "gating." Edmunds

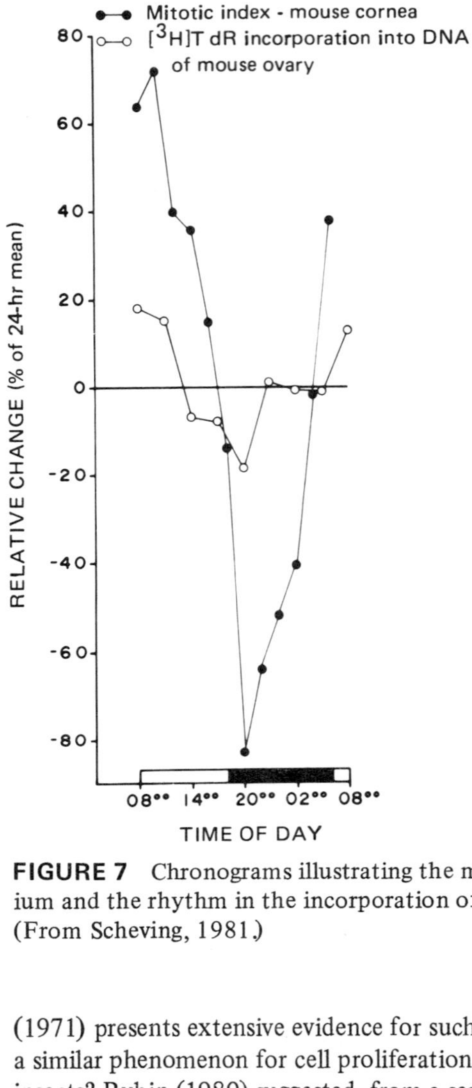


**FIGURE 7** Chronograms illustrating the mitotic index rhythm in the corneal epithelium and the rhythm in the incorporation of [$^3$H]TdR into the DNA of mouse ovary. (From Scheving, 1981.)

(1971) presents extensive evidence for such a gate in mitosis of *Euglena.* Are we observing a similar phenomenon for cell proliferation in the cornea of rodents to that in *Euglena* or insects? Rubin (1980) suggested, from a series of radiation studies using the corneal epithelium as a target organ, that gating could explain her interesting results pertaining to recovery of the mitotic index in the cornea from radiation; the interested reader is referred to her paper. The wide variation around the 24 hr mean illustrates the hazard of sampling this tissue at only one or two random times during the 24 hr cycle. The maximum and minimum values clearly do not occur at random but are predictable from one experiment to another as long as the animals used have been kept under the same standardized conditions. It is interesting to note that in mice bearing L1210 leukemia we found that the rhythm in the corneal epithelium persisted until the day of death (Fig. 6).

In our experience with rodents, two types of cell proliferation rhythms in tissues are encountered. One type is a high-amplitude rhythm with very low levels of mitotic activity, or none, occurring at certain circadian stages. The other type of rhythm, as exemplified by the uptake of [$^3$H]TdR into DNA and the mitotic index in the duodenum, is characterized by relatively high levels of activity over the whole 24 hr time scale, but a low-amplitude rhythm is still present. The two types of rhythms are illustrated in Fig. 3

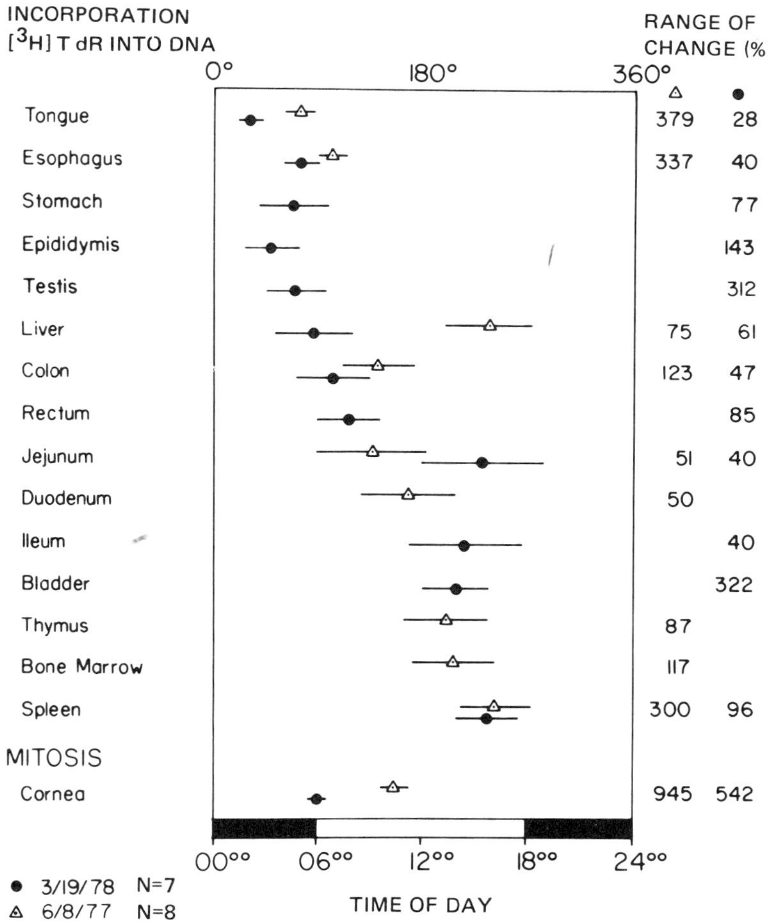


**FIGURE 8** An acrophase map, showing the approximate peak of the circadian rhythm in either the incorporation of [$^3$H]TdR into DNA or in the mitotic index of the corneal epithelium of mice. The acrophase is represented by a dot or triangle, and the bars extending out from the dots or triangles represent the confidence limits. The dark bar on the abscissa portion represents the 24 hr dark span of the environmental light-dark cycle. The percentage range of change reflects the change from the lowest to the highest time point means, when the lowest means equal to 100%; values rounded to the nearest integer. (From Scheving, 1981.)

using the rat duodenum (low amplitude) and tongue (high amplitude) and Fig. 7 for the mouse ovary (low amplitude) and cornea (high amplitude). In our opinion the majority of the tissues of the body exhibit a high-amplitude rhythm, but most do not show the almost complete absence of cell proliferation at certain circadian stages seen in the corneal epithelium (Scheving et al., 1978).

Figure 8 is an acrophase map that shows the approximate peak of the circadian cycle in the incorporation of [$^3$H]TdR into DNA in many tissues of the mouse as well as the acrophase of the mitotic index of the corneal epithelium (Scheving, 1981). The acrophase indicates the point along the 24 hr time scale when the data values are on the average highest as determined by the cosinor analysis (Halberg et al., 1972). It should be noted that this point is not necessarily identical to the time when the peak (value) was recorded.

### Why Confusion Exists Regarding Presence or Absence of a Rhythm

A number of tissues have been reported to be devoid of circadian rhythmicity, and a brief explanation of why some of these reports occurred should be mentioned. For example, the mitotic index of the uterine epithelium of the rat was reported to be devoid of circadian rhythmicity by Bertalanffy and Lau (1963). However, Krueger et al. (1975) clearly demonstrated that the mitotic index of the same tissue was characterized as a very high amplitude rhythm; the variation between the lowest and highest record means along the 24 hr time span represented an 1100% change. The reasons Bertalanffy and associates frequently reported the absence of rhythms in many normal tissues, including the cornea (Bertalanffy and Lau, 1962) and the digestive tract (Bertalanffy, 1960), have been discussed by us (Scheving et al., 1972). Their findings, however, are still being widely cited (without mention of the evidence to the contrary) as having demonstrated that certain tissues are devoid of a rhythm in cell proliferation. We believe that most of the negative reports regarding cell proliferation rhythms have been the result either of keeping animals in continuous light or of using colchicine to collect mitotic figures over too long a span of time, such as the 6 hr span that Bertalanffy and associates frequently used. Both of the above techniques are contraindicated for demonstrating rhythms (Scheving and Pauly, 1977; Scheving et al., 1972, 1983a).

## SYNCHRONIZATION OF CIRCADIAN RHYTHMS

Both the light-dark cycle and meal timing have been shown to influence and even to synchronize circadian rhythms in cell proliferation. However, the effect of either potential synchronizer varies between different organs and tissues. We first shall consider the synchronizing effect of the light-dark cycle with emphasis on the mitotic index in the corneal epithelium of the rodent.

### Effects of Light-Dark (LD) Regime on Cell Proliferation Rhythms

Figure 9 illustrates the mitotic index rhythm in the corneal epithelium of male rats standardized to an LD: 12,12 cycle, with food available ad libitum. Note that the trough occurs shortly after the time of transition from light to dark and the peak shortly after the transition from dark to light. The strong synchronizing effect of the light-dark cycle on this particular rhythm becomes evident when the light-dark cycle is reversed 180°. The profile of the rhythm in the mitotic index is completely phase shifted within 1 week and remains so a week later (Scheving and Pauly, 1973). It should be realized that not all rhythmic variables invert so rapidly; some have been shown to take 3 weeks or more. The incorporation of [$^3$H] TdR into DNA in the spleen is a good example of a rhythm in cell proliferation that takes about 3 weeks to invert completely when mice are subjected to 180° changes in the light-dark cycle.

### What Happens to Rhythms in Cell Proliferation in the Absence of the LD Cycle?

In the corneas of animals subjected to continuous darkness (DD) or to blinding, the colony rhythm of the mitotic index persists but a change in phasing is evident (Fig. 10).

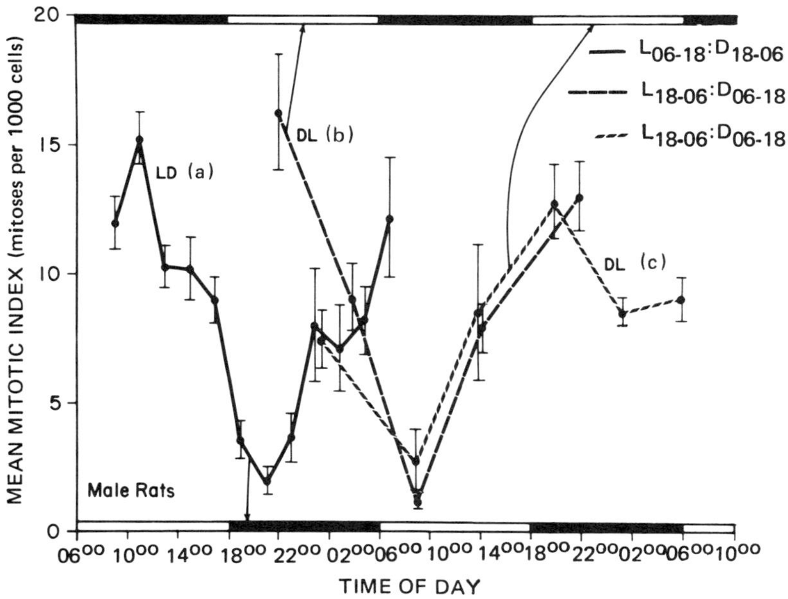


**FIGURE 9** The solid curve (a) illustrates a typical mitotic index rhythm in rats standardized at 12 hr of light alternating with 12 hr of darkness shown on the bottom abscissa. The broken line (b) right represents the inverted pattern 7 days subsequent to inverting the light-dark cycle (see horizontal axis at the top of the figure). The slanted broken line (c) accompanying it represents another study conducted a week later showing that the rhythm was still "locked on" to the light-dark cycle. In all cases n = 5. (From Scheving and Pauly, 1973.)

In contrast, continuous light (LL) has a more drastic effect on this rhythm in that there is a tendency for the amplitude of the colony rhythm to be reduced or even to flatten completely (Fig. 11). It is not known why LL has such an effect (Scheving et al., 1973, 1974c). On the one hand, the apparent loss of the colony rhythm may reflect depression of the sympathetic nervous system by light with a resulting cessation of circadian variation in various mitogens (such as epidermal growth factor, to be discussed later) from glands that may be involved in the control of circadian time structure. Chemical sympathectomy has been shown to suppress the colony circadian variation in DNA synthesis in the small intestine of rodents (Klein, 1979), and the circadian rhythm in the level of epidermal growth factor in the submandibular salivary gland is also dependent on activation of the sympathetic nervous system, being abolished by superior cervical ganglionectomy (Krieger et al., 1976). Alternatively, the lack of a colony rhythm in constant light may be the result of individual mice free running with different frequencies; thus, mice become desynchronized from one another and the end result is loss of the colony rhythm. For example, it has been shown, through longitudinal studies on guinea pigs in constant light, that the circadian rhythms in motor activity, temperature, and response to histamine persist in the individual animals (Scheving et al., 1973). In the histamine studies, however, the phasing of rhythms was different in the individual rodents. Thus, from such studies it is clear that the comparison of the averaged values obtained for a group of animals in LL at different time points over the 24 hr span may belie the extent of the rhythmic

RHYTHMOMETRIC SUMMARY

| | Noise-to signal SE/A | Mesor, M M ± SE | Amplitude, A A ± SE | Acrophase, Ø (.95 confidence limits) |
|---|---|---|---|---|
| A | .10 | 12.4 ± .5 | 7.7 ± .7 | −119°(−108°, −131°) |
| B | .15 | 11.6 ± .6 | 5.6 ± .8 | −179°(−161°, −197°) |

Acrophase reference = local midnight

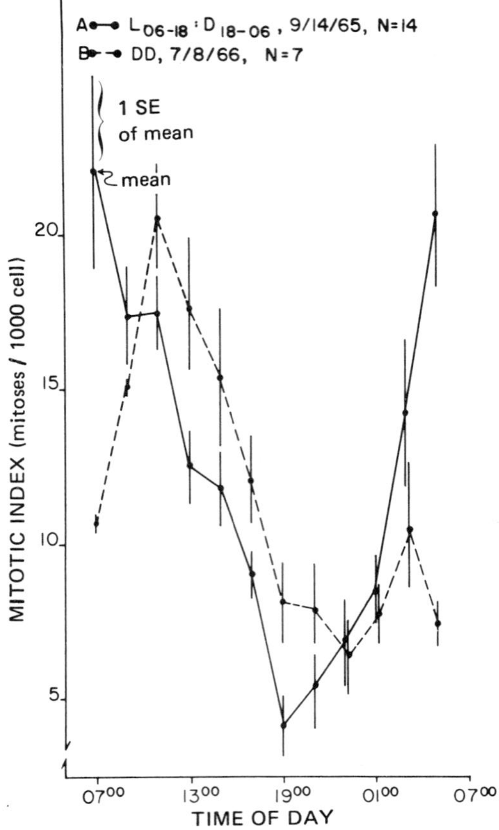


**FIGURE 10** Comparison of rhythm seen in the mitotic index of the corneal epithelium of rats maintained and artificially controlled, alternating periods of light and darkness (LD: 12,12) with that of animals maintained a constant darkness (DD). Note the phase shift in the colony rhythm in animals in DD (the acrophase in DD occurs at 1100 rather than the 0800 as the LD). (From Scheving et al., 1974c).

variation in the individual animals. The problem of carrying out longitudinal sampling for mitosis in individual animals makes it difficult to prove the latter. However, the author has seen some unpublished data obtained by the late Gordon Rosene by longitudinal sampling (repeated biopsy) on individual mice subjected to LL, which suggests the persistence of a mitotic index rhythm in the epidermis of the pinna. This study needs to be repeated since the data seem to be no longer available.

### Synchronization of Cell Proliferation Rhythms by Meal Timing

When feeding time was restricted to a single 4 hr span for mice standardized to an LD: 12,12 cycle, the phasing of the rhythms in DNA synthesis and in the mitotic index of

RHYTHMOMETRIC SUMMARY

| | Noise-to signal SE/A | Mesor, M M ± SE | Amplitude, A A ± SE | Acrophase, Ø (95 confidence limits) |
|---|---|---|---|---|
| A | .10 | 12.4 ± .5 | 7.7 ± .7 | −119° (−108°, −131°) |
| B | .39 | 12.3 ± .3 | 1.2 ± .4 | −80° |
| C | .38 | 12.9 ± .5 | 2.1 ± .8 | −54° |

Acrophase reference = local midnight

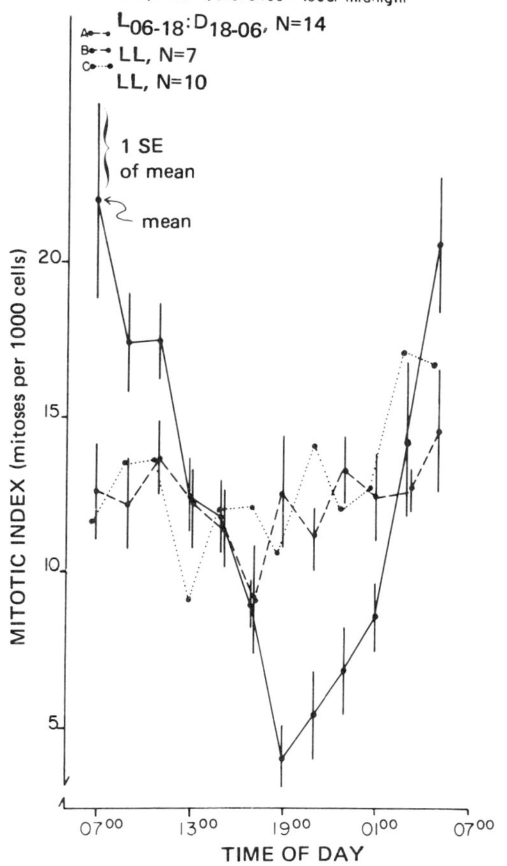


**FIGURE 11** Comparison of rhythms in the mitotic indices of the corneal epithelium of rats maintained on artificially controlled, alternating periods of light and darkness (LD: 12,12) with those of animals maintained under constant illumination (LL) at 600 lux (two studies). Note flattening of the curves and the reduction in amplitude in LL conditions. (From Scheving et al., 1974c.)

the corneal epithelium were relatively resistant to change from the pattern seen in ad libitum fed animals (Fig. 12). Therefore, for this tissue the lighting schedule appeared to dominate meal timing in the synchronization of cell proliferation (Scheving et al., 1974c; Philippens et al., 1977); this does not imply that the overall mean may not be affected by restricted meal timing. Although such an LD schedule synchronizes the rhythm in the corneal epithelium, it does not generate the rhythm, as is evidenced by the persistence of the rhythm in mice (as discussed above) that were either blinded or raised in continuous darkness (Figs. 10 and 11). The transition from light to dark may cause changes in the cells in the corneal epithelium or of local concentration of a mitogen, which leads to

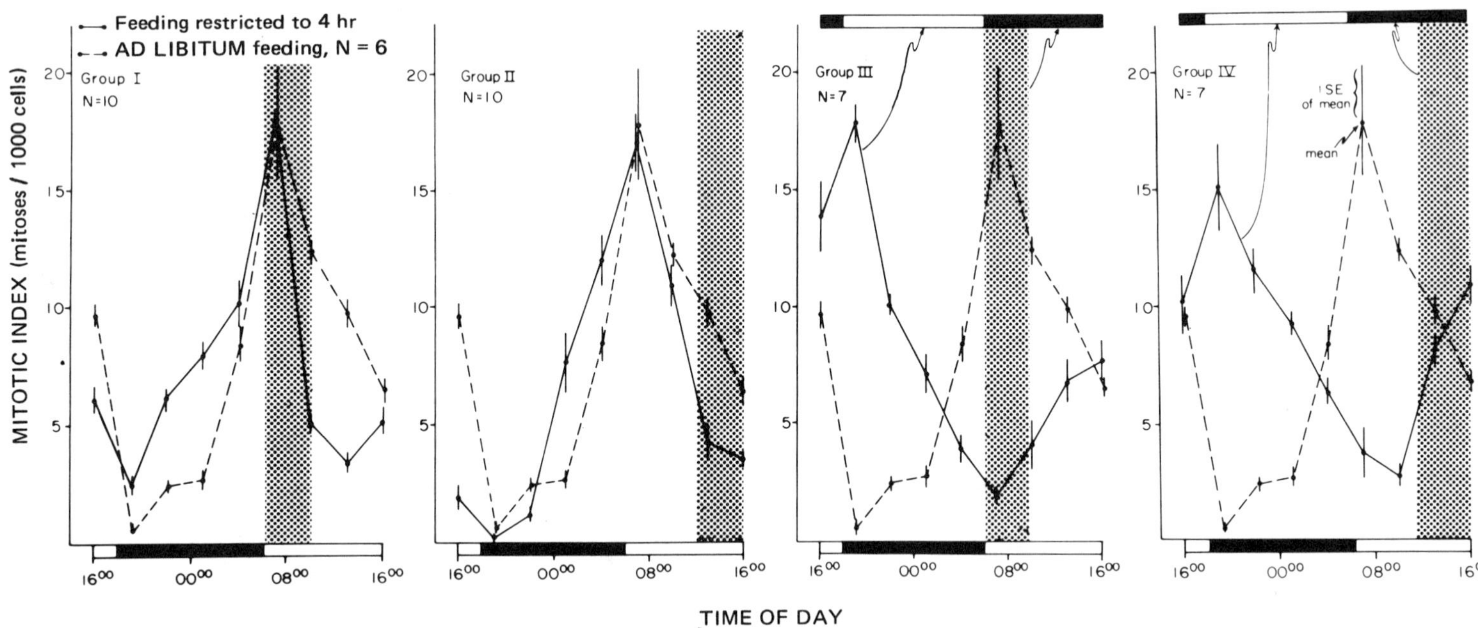


**FIGURE 12** Effect of feeding on the mitotic index rhythm in the rat corneal epithelium. The light-dark cycle to which the animals were subjected is depicted on the horizontal axis of each chronogram. Within each group of mice ($CD2F_1$), controls were fed ad libitum and the experimental mice were restricted in their access to food each day to 12 hr during the first week, to 8 hr the second week, to 6 hr the third week, and 4 hr during the fourth and fifth weeks (shaded areas). Note that the mitotic index rhythm remained remarkably synchronized to the light-dark cycle. In all instances, the time was local clock time (CST). For details of the study, see Scheving et al. (1974c).

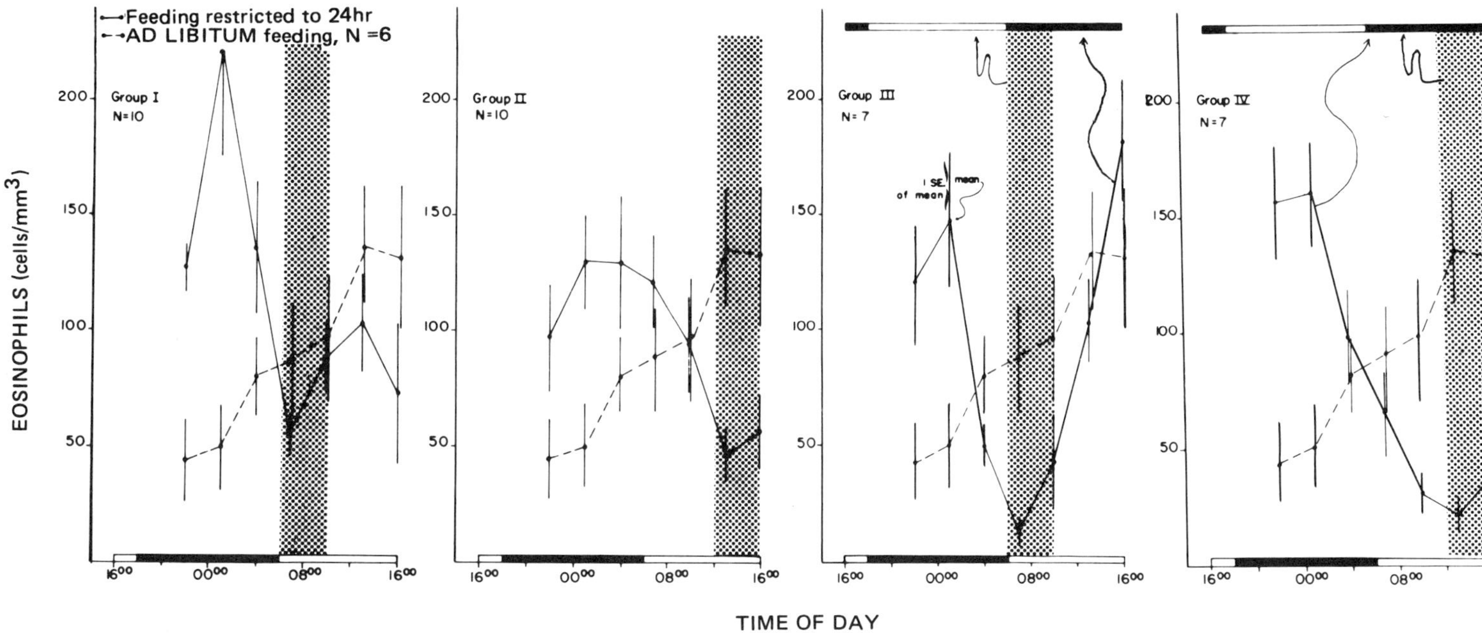


**FIGURE 13** The effect of restricted feeding schedules on the eosinophil levels in mice in the same animals described in Fig. 12 (Pauly et al., 1975). Unlike the case of the mitotic index rhythm of the corneal epithelium, the feeding schedule strongly synchronized the eosinophil rhythm; both tissues came from the same animals.

increased DNA synthesis and the ultimate division of the target cells. In other organs, such as some of those involved in hemopoiesis or digestion, the restricted feeding schedule appears to be able eventually to dominate the light-dark cycle in the synchronization of the rhythm in DNA synthesis; that is illustrated in Fig. 13 for eosinophil levels in the blood of mice (Pauly et al., 1975). Cell proliferation in the digestive tract is strongly dependent on meal timing (Scheving et al., 1976b; 1983a).

Normally, feeding, motor activity, and DNA synthesis in tissues increase about the time of the light-dark transition in animals fed ad libitum and synchronized to LD: 12,12. When feeding is restricted to a single 4 hr span sometime during the LD: 12,12 span, the time of peak DNA synthesis in such organs as the duodenum, esophagus, tongue, spleen, and bone marrow phase shifts to coincide with the time that food is presented. The shift is not immediate; it generally takes from several days to more than a week, depending on the interval when the feeding time occurs in relation to the LD: 12,12 cycle. Thus, the effect that meal timing has is itself circadian-stage dependent, at least in certain organs or tissues. For example, in comparison to both ad libitum feeding and to other restricted times of feeding, a 4 hr feeding span falling during the latter part of the light span dramatically increases the amplitude of the circadian rhythm in DNA synthesis of the duodenum, spleen, and bone marrow (Scheving et al., 1976b; 1983b). Moreover, in some organs, such as the esophagus, the shift of the rhythm is compelted following a week of restricted feeding, regardless of where in the 24 hr span the mealtime occurs. In other organs, such as the tongue, the extent of shift of the original monophasic rhythm depends on the time the feeding span occurs in relation to the LD cycle. For example, in mice fed late in the dark span, the DNA synthesis rhythm in the tongue (but not in the esophagus) became biphasic: one phase appeared to be cued by food presentation and the other by the onset of darkness (Scheving et al , 1983b). One *cannot generalize* about the effect of meal timing as a synchronizer of rhythms based on studies of one or two variables; even in the digestive tract the response varies from one region to another. Extensive data by Philippens et al. (1977) obtained on the cornea of rats and other tissues, especially in the liver, support this conclusion.

### Significance of Ability to Synchronize the Cell Proliferation Rhythm

The ability to shift some rhythms both within an organ and between organs, thereby increasing the synchronization of cellular proliferation in the intact animals simply by altering meal timing or the light-dark cycle, could provide a useful strategy for predictably shifting normal cells to a phase of the cell cycle where they would be more resistant to radiotherapy or chemotherapy. Moreover, the combination of environmental and hormonal challenges, involving the administration of a combination of mitogens, such as EGF, insulin, glucagon, and fibroblast growth factor (FGF), could be a powerful tool to accomplish this. Indeed, it has been reported that immunocytoma-bearing rats survive longer following adriamycin treatment when meal timing is restricted to the early part of the light span (Nelson et al., 1974; Halberg et al., 1977). It becomes an important task for scientists to determine how environmental, hormonal, and perhaps social factors can be used to manipulate cell proliferation rhythms in humans in order to shield the normal cells from the toxic effects of radiotherapy and chemotherapy.

## INVESTIGATIONS INTO THE MECHANISMS OF RHYTHMS IN CELL PROLIFERATION

### Effect of Ablation of Adrenal and Pituitary Glands on Cell Proliferation

In spite of much research into the mechanism of cell proliferation and its associated rhythmicity, it is not fully understood. We and many others have carried out ablation studies in attempts to shed light on what might be controlling the basic rhythmicity. Initially we investigated the effect of adrenalectomy, adrenomedullectomy, and hypophysectomy on rhythms in cell proliferation, using primarily the corneal epithelium of the rodent. In each case, the circadian patterns of cell proliferation persisted, but the amplitude was reduced when compared to the controls (Fig. 14) (Scheving and Pauly, 1967).

### Effect of Suprachiasmatic Nucleus on Cell Proliferation Rhythms

In response to reports that complete destruction of the suprachiasmatic nucleus (SCN) disrupted or abolished rhythms in motor activity, body temperature, and several endocrine functions, we evaluated the effects of lesioning the SCN both on the mitotic index rhythm of the mouse corneal epithelium and on DNA synthesis in the tongue and in several different regions of the intestinal tract. We found that this procedure does not abolish the mitotic index rhythm of the corneal epithelium but does have a profound effect in reducing its amplitude and in producing a phase advance of the rhythm as described by cosinor analysis (Powell et al., 1980) (Fig. 15). Similar results were obtained for the tongue and intestinal tract; this is demonstrated by data on the tongue (Fig. 16), where we found an acrophase advance and an amplitude reduction (Scheving et al., 1983b). We conclude that the SCN appears to function as a phase resetter and amplifier of the rhythm. The SCN, in our opinion, represents but one cog in an overall central control mechanism. When the SCN is damaged, circadian rhythms in cell proliferation are altered but are not destroyed (Scheving et al., 1983b).

### Effect of Epidermal Growth Factor on Cell Proliferation

More recently we have been carrying out comprehensive, chronobiologically controlled studies to evaluate the effect of various exogeneous peptides and hormones on the DNA synthesis and mitotic index rhythms in various tissues. From recent studies (L. A. Scheving et al., 1979, 1980; Yeh et al., 1981) of the effect that epidermal growth factor (EGF) has on [$^3$H]TdR incorporation into DNA of 20 different tissues, several conclusions have been reached. First of all, EGF has a rather striking effect on cell proliferation and we have hypothesized that EGF could be playing an important role in the control of rhythms, especially in cell proliferation in the digestive tract and cornea (L. A. Scheving et al., 1979). Figure 17 illustrates but one example of a tissue that responds to EGF, namely, the mouse rectum. Interestingly, in two of the tissues most responsive to EGF, namely, the cornea and the esophagus, one can dissociate the phasing of the DNA synthesis rhythms one from the other by changing the lighting schedule and feeding time, suggesting that a general systemic release of EGF is insufficient by itself to account for rhythmic variation in the tissues. Second, the different tissues exhibit a considerable variation in the response to EGF at 4, 8, and 12 hr after it is injected. For example, the cornea, tongue, and esophagus, which are the most rhythmic of all tissues studied, consistently respond to EGF with increased DNA synthesis at 4, 8, and 12 hr after injection; the glandular and nonglandular stomach, colon, and rectum respond only at 8 and 12 hr; the lungs and

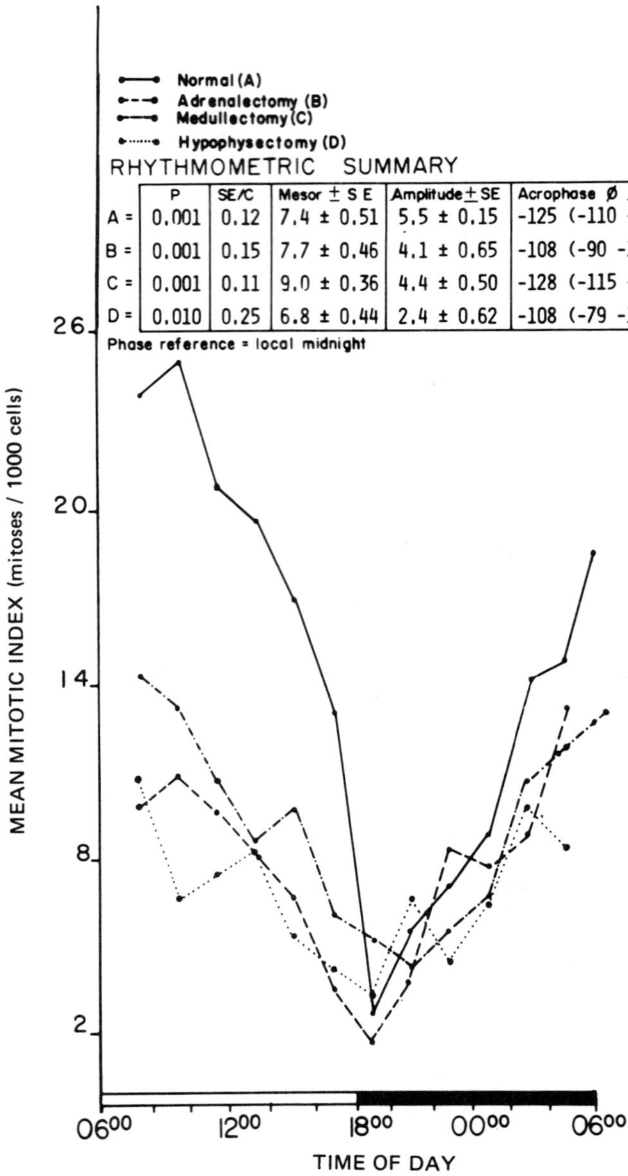


| | P | SE/C | Mesor ± S E | Amplitude ± SE | Acrophase ø ± SE |
|---|---|---|---|---|---|
| A = | 0.001 | 0.12 | 7.4 ± 0.51 | 5.5 ± 0.15 | -125 (-110 -140) |
| B = | 0.001 | 0.15 | 7.7 ± 0.46 | 4.1 ± 0.65 | -108 (-90 -125) |
| C = | 0.001 | 0.11 | 9.0 ± 0.36 | 4.4 ± 0.50 | -128 (-115 -141) |
| D = | 0.010 | 0.25 | 6.8 ± 0.44 | 2.4 ± 0.62 | -108 (-79 -136) |

**FIGURE 14** Comparison of mitotic index rhythm in corneal epithelium of normal rats with that of rats sampled 3 weeks after adrenalectomy, adrenomedulectomy, or hypophysectomy. Adrenalectomized rats died when deprived of physiological saline, and this was taken as evidence of complete adrenalectomy. For a more detailed description, see Scheving and Pauly, (1967).

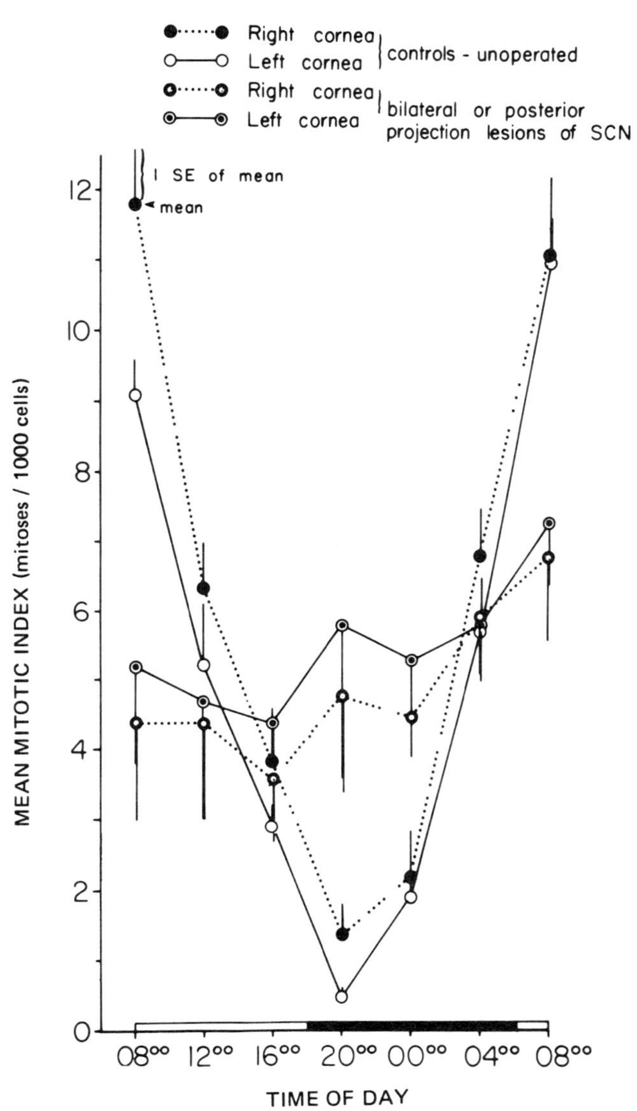


**FIGURE 15** Comparison of the rhythmic pattern of the mitotic index in SCN-lesioned and sham-operated mice (for description, see Powell et al., 1980).

aorta at 8 hr; and the liver and testis at 12 hr. Thus, the response to EGF is tissue specific and related to the time of day (circadian stage) it is administered. Third, there are certain circadian stages when the stimulatory effect of EGF appears to be maximally potentiated. For example, the injection of EGF at 1500 into mice standardized to an LD: 12,12 cycle (light 0600, dark 1800) results in a striking generalized increase in DNA synthesis 4 hr later (at 1900), particularly in such tissues as the parotid, thymus, small intestine, and rectum, which fail to respond normally to EGF at 4 hr after injection at other times (this can be seen in Fig. 17). Interestingly, the concentrations of various hormones, such as insulin, gastrin, and corticosterone, peak during this period and the rate of feeding is increased (L. A. Scheving et al., 1980). Fourth, the results indicate that EGF can also

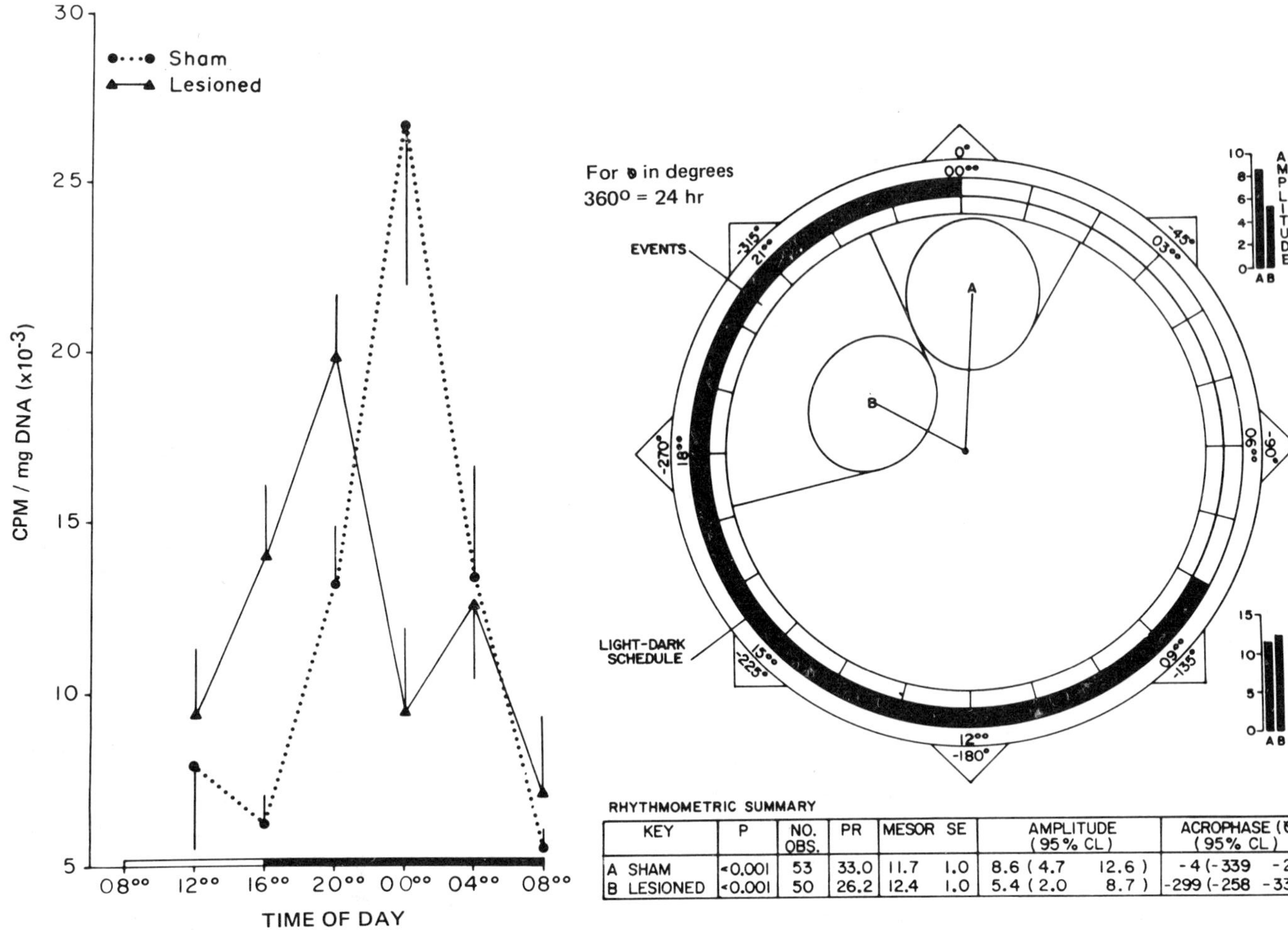


RHYTHMOMETRIC SUMMARY

| KEY | P | NO. OBS. | PR | MESOR | SE | AMPLITUDE (95% CL) | ACROPHASE (ϕ) (95% CL) |
|---|---|---|---|---|---|---|---|
| A SHAM | <0.001 | 53 | 33.0 | 11.7 | 1.0 | 8.6 ( 4.7 12.6 ) | -4 (-339 -29) |
| B LESIONED | <0.001 | 50 | 26.2 | 12.4 | 1.0 | 5.4 ( 2.0 8.7 ) | -299 (-258 -339) |

**FIGURE 16** Circadian rhythm in [$^3$H] TdR incorporation into DNA of tongue; rhythm was phase advanced by about 4 hr in SCN-lesioned animals. Dots and vertical bars equal mean and standard error (SE). Results of cosinor analysis are presented in the polar plot. (From Scheving et al., 1983b.)

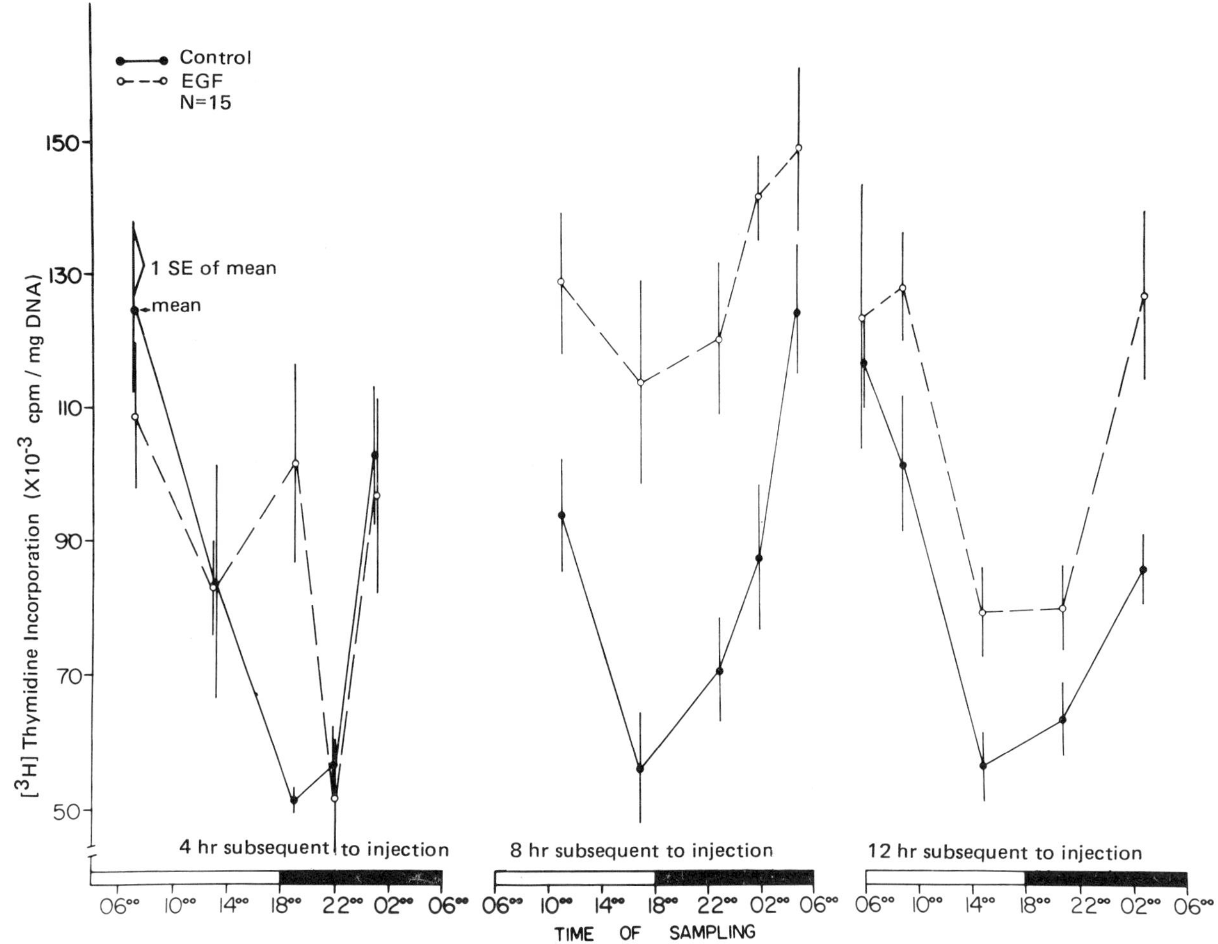


**FIGURE 17** Effects of EGF on [$^3$H]TdR incorporation into the rectum is plotted versus time of sacrifice for the different injection times. Mice were killed at either 4, 8, or 12 hr after injection for each injection time. (Modified from L. A. Scheving et al., 1980.)

inhibit DNA synthesis at certain times in several tissues, including the small intestine, thymus, bone marrow, and spleen.

Results also suggest that the DNA-synthesizing regions of the digestive tract of the mouse that are the most rhythmic exhibit the greatest overall response to EGF. Although the relationship between rhythmicity, response to exogenous EGF, and the potential to undergo transformation remains to be elucidated, the EGF cell-surface effector system has been implicated in the in vitro chemical and viral transformation of certain cell types; this is evidenced by the fact that transformed cells, in contrast to their normal counterparts, exhibit a decreased requirement for exogenous EGF to undergo cell proliferation (Cherington et al., 1979). Also, several tumor growth factors capable of interacting through the EGF cell-surface receptors have been identified in transformed cells, such as SV40 sarcoma.

EGF itself has been shown to be cocarcinogenic in the induction of certain skin tumors (Roberts et al., 1976), and recently it has been shown in rats that removal of the submandibular gland, a major source of EGF, results in a reduction in the number of dimethylebenzanthracene-induced tumors of the colon (Li et al., 1980). The same chemical appears to be most carcinogenic at a time when the glandular content of EGF is expected, from the work of Krieger et al. (1976), to be highest in male rodents. Moreover, it is of interest that induction of submandibular gland sarcomas using the same chemical was shown more than 20 years ago to be circadian-stage dependent (Chaudhry and Halberg, 1960; Halberg, 1964). Clearly EGF is involved in normal and perhaps in abnormal growth.

### Effects of Insulin and Glucagon on Cell Proliferation

Our interest in insulin and glucagon arose from the demonstration by Bucher et al. (1978) of the importance of EGF, insulin, and glucagon in supporting cell proliferation in liver regeneration of rodents. Insulin and glucagon were injected intraperitoneally (i.p.) at four different circadian stages into separate groups of female adult BALB/Cann mice that had been standardized to 12 hr of light alternating with 12 hr of darkness. Comparable control groups were injected with saline. Subgroups of seven mice that had been injected 30 min earlier with [$^3$H]TdR were killed 4, 8, 12 and 18 hr after each of the four injection times; a total of 336 mice were used. Incorporation of [$^3$H]TdR into DNA of the esophagus, stomach, duodenum, jejunum, caecum, colon, rectum, and spleen were subsequently determined (L. A. Scheving et al., 1982).

In summary, the results of the above studies demonstrate for the first time that both hormones affect the incorporation of [$^3$H]TdR into DNA in all the examined organs but in different ways and at different circadian stages. The effect of these hormones are complex, but several generalizations have emerged. (1) Insulin, when effective, tends to increase incorporation of [$^3$H]TdR into DNA in examined organs, whereas glucagon tends to decrease it. (2) Insulin is more effective in stimulating incorporation of [$^3$H]TdR into DNA when injected either at the end of the dark span or the beginning of the light span, as opposed to the end of the light span or the beginning of the dark span. (3) Insulin has its greatest effect on [$^3$H]TdR incorporation into DNA in the glandular stomach and rectum, whereas glucagon has its greatest effect on the colon and spleen. (4) The effects of both insulin and glucagon are different from those of epidermal growth factor, as revealed in a similar study done by us. Our results suggest that insulin, glucagon, and epidermal growth factor play important roles in the control of cell proliferation of various endodermally derived organs.

### Effect of ACTH on Cell Proliferation

Other extensive studies have shown that ACTH 1-17 also has a role in the control of rhythms in cell proliferation. For example, ACTH statistically significantly inhibits incorporation of [$^3$H] TdR into DNA in the stomach at 4, 8, and 12 hr after a single administration. All responses seen in the many tissues studied are remarkably circadian-stage dependent (Scheving et al., 1983, 1984a, 1984b).

### Conclusions Relative to Mechanism

In spite of much work over many years, scientists still cannot fully explain normal or abnormal of cell proliferation. As yet there is no satisfactory unifying concept of a control mechanism. Recognition of the underlying rhythmic nature of substances capable of influencing cell proliferation as described above may help us to better understand both normal and abnormal growth. We must learn how all the above hormones, peptides, and the SCN intermodulate with one another in time.

## WHAT VALUE IS THERE IN RECOGNIZING RHYTHMS FOUND IN CELL PROLIFERATION?

### Basic Research

From the studies involving EGF, insulin, glucagon, ACTH 1-17, and the SCN, it is apparent that it would be virtually impossible to obtain a complete picture as to what effect each of these has on cell proliferation without a chronobiological analysis. Biological investigators could benefit greatly by recognizing the oscillating nature of this or of any system being investigated. Sampling at the same time of day does not take care of the rhythm problem.

The fact that a single stimulus, such as a drug, can phase shift a rhythm makes suspect many studies that involve perturbation of a physiological system and then later test for its effects on some variable, such as the mitotic index, against unperturbed control animals. For example, it repeatedly has been shown that drugs can alter the rhythm in the mitotic index (Scheving and Pauly, 1971, 1973; Burns and Scheving, 1973; Scheving et al., 1974d). As an example, Fig. 18 illustrates the results of an investigation performed on adult mice standardized for 7 days prior to the beginning of the study to artificial light from 0600 to 1800, alternating with darkness from 1800 to 0600; food and water were available ad libitum. A total of 40 BD2F$_1$ mice were injected intraperitoneally with 2.6 mg 1-β-D-arabinofuranosylcytosine (ara-C) in 0.2 ml saline at 1800 and again at 2100 on the same day. Starting 2 hr after the second injection, subgroups of five mice from both the controls and the experimental groups were sacrificed at intervals indicated in Fig. 18 over a 36 hr period. The mitotic rate in the corneal epithelium in the saline-injected animals shows the expected phasing in the circadian rhythm. However, in the ara-C-injected animals, a phase delay of about 4 hr is apparent on the first day after the drug administration; on the second day after treatment, the usual phase relationship has been reestablished between the mitotic index rhythm and the synchronizer cycle of the alternating light-dark schedule (Scheving and Pauly, 1973).

The rationale for performing the above experiments dealing with the effect of ara-C on the mitotic rate in the corneal epithelium stems from a series of earlier unpublished experiments (Scheving et al., 1971). In those studies it was found that when ara-C was

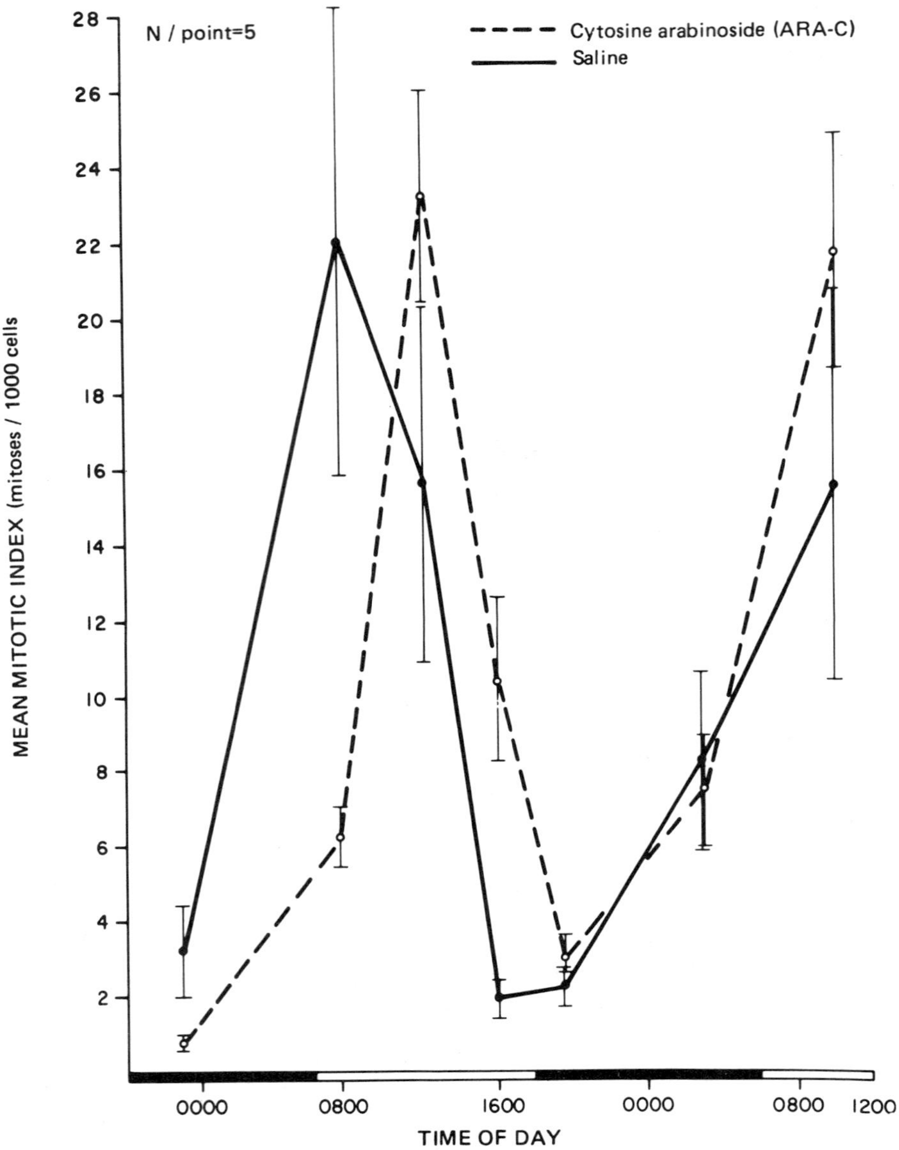


**FIGURE 18** Phase shift in the circadian rhythm of mitosis in mouse corneal epithelium induced by ara-C. For exaplanation see text, as well as Scheving and Pauly (1973).

injected during the circadian stage at which the rate of corneal epidermal mitosis was usually highest, no statistically significant decrease in the mitotic index could be detected 12 or 24 hr later; the peak mitotic index occurred 24 hr later at the expected time. When the same dose of ara-C was injected at the circadian stage when the corneal mitotic rate was usually low, there usually was a significant decrease in the mitotic rate 12 hr later, when one would expect the next daily high in the mitotic index to occur. The question asked was whether this apparent decrease was a result of a phase shift in the rhythm, or whether it actually represented a depression in the mitotic rate. As can be seen in Fig. 18, the apparent decrease is simply due to a phase delay of 4 hr brought about by the ara-C. Single time point sampling could be quite misleading in such a situation because it might suggest that a depression of mitosis has taken place when actually there is a phase delay in the rhythm. We recognize, of course, that ara-C will depress mitosis within the corneal epithelium if we give it in sufficiently high doses, for longer durations of time, or perhaps even at a different stage of the circadian system. The data in Fig. 18 do point out clearly the complexity of evaluating the effects of a drug on mitotic activity (Scheving and Pauly, 1973).

It has been reported, based primarily on data extrapolated from in vitro studies to the in vivo situation, that the duration of synthesis of DNA in the epithelium of the cornea is about 9-10 hr and that the peak of synthesis occurs about 10 hr prior to the mitotic index peak. Similar data on many other tissues claiming equally long intervals can be found in the literature, but we will use only the one example simply to illustrate a point. The point to be made is that our data and those of others simply do not support this concept of a 10 hr span between the two stages of the cell cycle. The phasing of DNA synthesis (S) rhythms and the mitotic index (M) rhythms are separated at most by a few hours in animals standardized to an LD: 12,12 schedule. Generally, the peak of the DNA rhythm occurs somewhere between the middle and the end of the dark span, whereas the mitotic index rhythm peaks around the time of transition from dark to light (Scheving and Pauly, 1967; Gasser et al., 1972a). This observation has significance because in the intact rodent there are spans of time (usually very narrow) during the 24 hr day when the peak activity of both the S and the M stages of the cell cycle is occurring, and possibly of even more importance is the fact that there are other time spans when the troughs of both phases of the cell cycle (or even other phases of the rhythms) are occurring and can be predicted (Scheving and Pauly, 1967). This phasing pattern has been found for corneal epithelium (Chumak, 1973), for the duodenum (Sigdestad et al., 1969; Scheving et al., 1972), for the liver (Scheving et al., 1975), and for the epidermis of hairless mice (Tvermyr, 1969). Gasser et al. (1972a) evaluated the rhythm in the mitotic index and in DNA synthesis in the epithelium of the tongue of the rat by cosinor analysis; they found that the acrophase of the S phase occurred about 4 hr prior to the acrophase of the mitotic index. When the phasing of the rhythm in DNA synthesis was compared with that of the rhythm in the mitotic index in the liver of mice bearing a tumor, only a short gap between the S and M phases was found (Scheving et al., 1975). These results from the liver might at first seem to differ from those reported by Barnum et al. (1957) in immature mouse liver, where the peaks of DNA synthesis and the mitotic index were separated by many hours. Of course, the two models used were quite different. Our results also were different from those described earlier by Nash and Echave Llanos in mouse hepatoma (1971) in that they describe a 12 hr gap in the peaks of the S and M stages. These findings are more difficult to reconcile with our own data, but it should be noted that the two phases of the cell cycle reported by Nash and Echave Llanos were from separate experiments done on different days and on different mice. In a more recent article, Rubin (1981), while using flow cytometry (FCM) to determine the rhythms in the S and $G_2$ phases of the hamster cheek pouch epithelium, commented that there was a perplexing coincidence of phasing of the rhythm of the S and $G_2$ phases and this finding would not have been anticipated from in vitro cell-kinetic studies. Obviously she was observing the same phenomenon. The FCM method used did not permit her to determine the M phase.

A possible explanation for this rather unexpected short duration of the S, $G_2$, and M stages of cell cycles, in some tissues at least, is that two different populations of cells

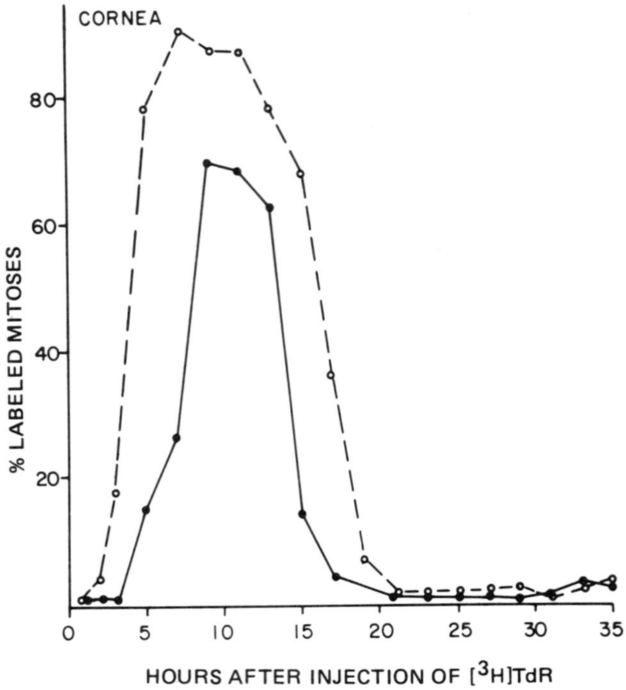


**FIGURE 19** The frequency-of-labeled mitosis method (FLM) was used to determine the duration of the S phase of the cell cycle in the epithelium of mouse cornea. The dashed line represents the data obtained from mice that were injected with [$^3$H]TdR at 0900 and killed at frequent intervals thereafter. The solid line represents data obtained from mice that were injected with [$^3$H]TdR at 2100 and killed at frequent intervals therafter; for complete details, see Burns and Scheving (1975) (see text also for further explanation).

may be present. A wave of synthesis occurring on one day may reflect the wave of mitoses on the following day. It also is possible, and not unlikely, that the $G_2$ phase in vivo may be of shorter duration than that customarily found in vitro, thus permitting rapid conversion from S to M in vivo. If [$^3$H] TdR is administered when the mitotic index is lowest in the corneal epithelium, about 2100, and the cell division in this tissue is analyzed by the once conventional, frequency-of-labeled mitosis (FLM) method, then $G_2 + 1/2M = 8$ hr, and $T_S = 5.4$ hr. If, however, [$^3$H] TdR is injected at the time of highest mitotic activity, at 0900, then $G_2 + 1/2M = 4$ hr and $T_S = 12.2$ hr (Fig. 19). (Burns and Scheving, 1975; Scheving and Burns, 1977). Tvermyr (1972) and Møller et al. (1974) have reported similar findings for the epidermis of the hairless mouse and the hamster cheek pouch, respectively. Unfortunately, the FLM method is still being used; one need only read the 1982 and 1983 issues of certain journals specializing in this type of data to confirm this. Some workers using the FLM technique claim to sample at the same time of day; apparently they do this thinking that they are avoiding the rhythmic problem! We believe that one simply cannot accept data found in tables throughout the literature giving precise durations of time for each phase of the classic cell cycle in various tissues; such data, especially for the in vivo situation, are of little value, especially if one is using them to plan treatment strategy.

Any technique used in the study of in vivo cell kinetics that cannot resolve the inherent rhythmicity in cell proliferation is unreliable. It should be emphasized that although cell cycle analysis by flow cytometry can complement older, established cytokinetic methods, such as [$^3$H] TdR incorporation, labeling indices, and the determination of mitotic indices, there are limitations. One serious shortcoming is that the FCM requires a homogeneous cell suspension. To date only a few tissues, other than the hemopoietic, can easily be so prepared. Among those that consistently have been adequately prepared are the skin of hairless mice, bone marrow, hamster cheek pouch epithelium, epithelium of the tongue, and esophagus in mice and Ehrlich ascites tumor. Among the studies using this technique that have been published are those of Møller and Larsen (1978), Clausen et al. (1979), Lareum and Aardal (1981), Rubin (1981), and Rubin et al. (1983a). Some of those workers are contributing further information in this volume on this potentially valuable technique (Chapters 7 and 8); thus I shall not discuss their findings.

Because at present it is impossible to obtain pure samples of proliferating cells from many tissues, we must depend on the older, more conventional methods of determining rates of cell proliferation in some tissues. That a rhythm is not found in a tissue with the FCM method, as is the case in the glandular stomach of mice, does not mean that random cell proliferation is occurring. Rubin et al. (1983b) suggests than when a single-cell suspension contains not only the epithelial cells of an organ, but a higher proportion of normal, nonproliferating cells, such as fibroblasts and muscle cells, both circadian variation and the true proliferative status of the tissue may be camouflaged.

### Chronochemotherapy

The question arises as to whether the circadian variation underlying cell proliferation can be utilized in chemotherapy or radiotherapy of cancer, since many of the agents used have as their target the cells in some stage of the cell cycle. In the past decade there has been a series of studies attempting to apply, on an experimental basis, chronochemotherapy to the treatment of different tumors. We shall describe some of these findings.

At least two approaches have been taken when exploring chronotherapy or radiotherapy. One has been to concentrate on rhythms (or lack of rhythms) found in the tumors themselves. This was the approach Badran and Echave Llanos (1965) took in treating mammary carcinoma of mice with cyclophosphamide in a chronobiological manner. They were the first, to the best of our knowledge, to claim benefits from chronochemotherapy of experimental cancer. The second strategy has been to minimize damage to normal cells by administering either or both modes of treatment when the normal cells of such target organs as the intestinal tract and bone marrow are least vulnerable to the treatment. The fact that the cells divide rhythmically permits such an approach.

The question as to whether circadian rhythms in DNA synthesis or mitotic activity occur in tumors has been investigated and reviewed (Scheving et al., 1983a). It is our opinion from examining our own data, as well as other studies, that the tumor models investigated are more likely to exhibit what might be described as an ultradian rather than a circadian frequency. However, occasionally a typical circadian pattern can be seen when data from one tumor study are collected and plotted over a 24 hr span. The waveform, however, of any pattern seen on one day in our study, is not likely to be repeatable in another study. In essence the fluctuation in cell proliferation in the tumor is behaving differently from that seen in normal tissues. The general tendency seems to be for the tumor to escape from circadian control, at least when it is in the exponential growth phase. However, without going into the details, it is not unreasonable to assume that those

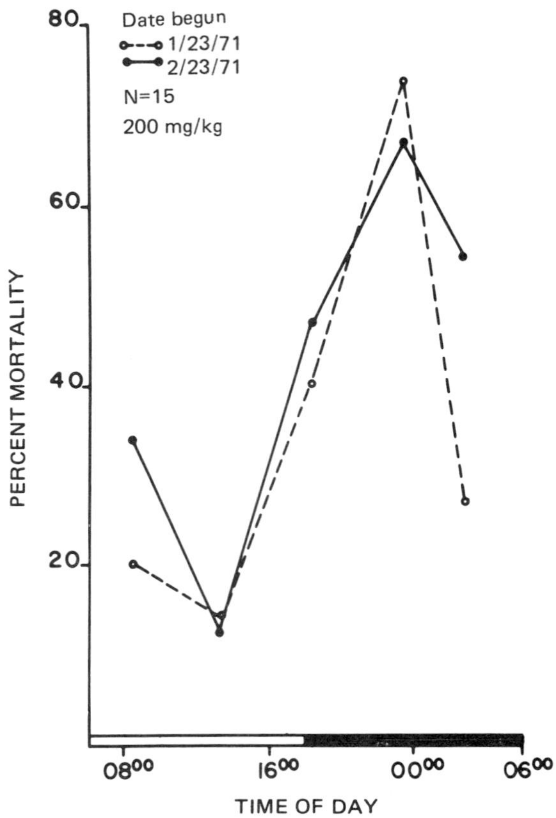


**FIGURE 20** Circadian susceptibility rhythm to $BD2F_1$ mice to ara-C given on 6 consecutive days at single, defined circadian system stages. (From Scheving et al., 1974d.)

tumors under strong hormonal control, such as mammary cancer, are more apt to show a more typical circadian variation, especially if they are slow-growing types (Echave Llanos and Nash, 1970).

Because of the unpredictable behavior of cell proliferation in tumors, we and others have followed the second strategy, i.e., to protect in time the normal tissues, such as the bone marrow and the intestinal tract, against the potent toxic effects of treatment while still attacking the tumor (a "shielding in time" of normal tissues) (Scheving et al., 1974a, d). This has been done experimentally by first determining the host toxicity rhythm to a particular chemotherapeutic agent and then applying that information to devising optimal schedules for the administration of those agents in tumor-bearing animals. A number of these studies have been a collaborative effort with Franz Halberg and Erhard Haus of the University of Minnesota and Ramsey Hospital, St. Paul, Minnesota, respectively.

*Host Toxicity Rhythms*

The host resistance to ara-C (Fig. 20) illustrates what is meant by circadian variation in host toxicity. A single dose of ara-C was administered once daily for 6 consecutive days to different groups of mice, but each group received the daily treatment at different circadian stages. It was found that as many as 78% of the animals injected at one time point died, compared with as few as 15% of those injected at another time point (Cardoso et al., 1970; Scheving et al., 1974d). Subsequent work by Haus et al. (1972) applied this finding to the treatment of leukemic mice. In an initial study, a potentially lethal dose of 240 mg/kg of ara-C was administered intraperitoneally to mice that had been inoculated with either $1 \times 10^6$ or $1 \times 10^7$ leukemic cells. Treatment was initiated 2 days (44 hr) after tumor cell inoculation and was repeated on days 6, 10, and 14. On each treatment day, a total dosage of the drug was divided into eight doses, which were given at 3 hr intervals over a 24 hr span; the doses given at the various injection times, however, varied in amount according to a sinusoidal pattern. The amounts ranged from 7.4 mg/kg given at the time of the animals' lowest resistance to the drug (as predicted from the earlier studies) to 67.5 mg/kg, given at the predicted time of highest resistance.

Such experimental chronotherapy succeeded in approximately doubling the survival time of leukemic mice compared with the controls that were treated without chronobiological consideration. A common treatment strategy at the time these studies were done was to administer eight equal doses at 3 hr intervals over the 24 hr span. Such a treatment schedule had been described by Skipper et al. (1967) as the current (at that time) best treatment for the mouse L1210 leukemia model.

Although the advantage of treating leukemic mice according to a sinusoidal schedule was clear-cut from the work reported by Haus et al. (1972), it seemed of interest to use a larger sample limited to tolerance per se in normal mice in order to attempt unequivocally to demonstrate rhythmic variation in host susceptibility to ara-C. Thus a study was done, using the same experimental design, on the same day in both Little Rock, Arkansas, and Minneapolis, Minnesota. In each laboratory, nine groups of 20 normal animals received four courses of ara-C treatment; in each course a total dose of 240 mg/kg of ara-C was divided among eight separate injections given at 3 hr intervals. One group of mice received equal doses of ara-C at each time point (the reference schedule). The eight other groups received the same total dose per course, but on one of several different "sinusoidal" schedules, as illustrated in Fig. 21. As predicted from the earlier work, survival time after treatment with ara-C on the different sinusoidal schedules differed significantly. Moreover, the sinusoidal schedules that yielded the longest as well as the shortest survival times were remarkably similar in the two locations (Scheving et al., 1976a).

### *Chronotherapy of L1210 Leukemia Using the Single Drug ara-C*

Skipper et al. (1967) reported that when mice bearing an inoculum of $1 \times 10^5$ L1210 tumor cells were treated on days 2,4, 10, and 14 after tumor inoculation with 120 mg/kg of ara-C (15 mg/kg every 3 hr) for 24 hr, generally most animals treated were cured. They also reported, using only eight animals, that 62% of the mice bearing $1 \times 10^6$ leukemia cells were cured by using the above treatment protocol. In a comparable repeat study using 10 mice, they reported that 50% were cured; however, in two subsequent studies (using 10 animals each), no cures were obtained from the same disease stage. While commenting on the merits of a chronobiological approach to chemotherapy, Rose et al. (1978) noted that Skipper and his coworkers could, without a chronobiological approach, cure 62% of animals bearing the abovementioned tumor load. However, none of our studies (using the same strain) or those of our collaborators, using hundreds of mice bearing the same tumor load and treated with this same drug, have resulted in such high cure rates as were only sometimes found, and with small members of animals, by Skipper and his coworkers. The claim of the 62% cure rate mentioned above should be evaluated in the light of all the data available from the laboratory cited (Rose et al., 1978).

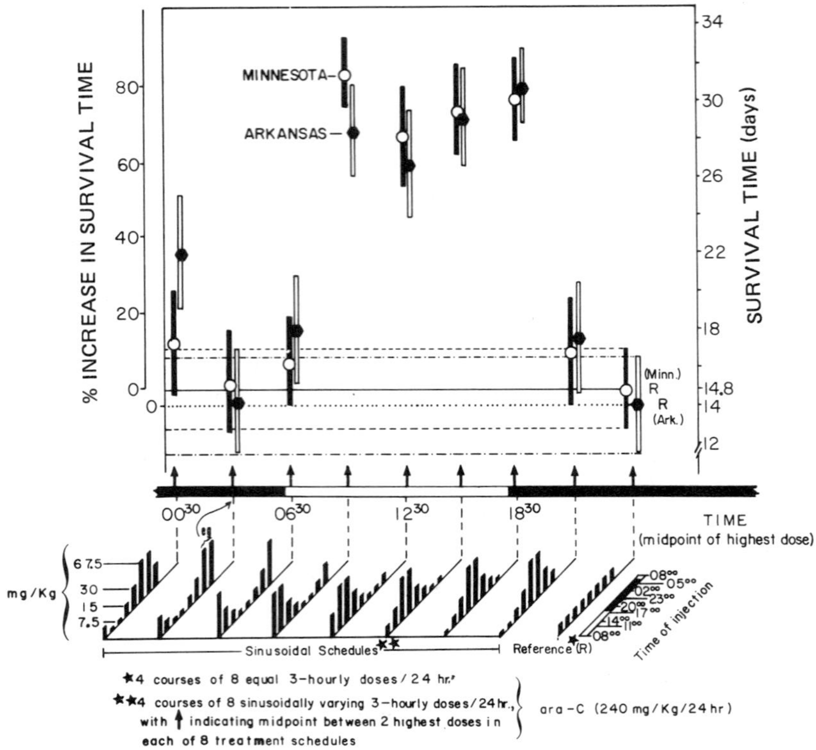


**FIGURE 21** Survival times of $CD2F_1$ mice on different drug administration schedules. All treatment schedules comprise four courses, each consisting of a total of 240 mg/kg per 24 hr. When the same total dose of ara-C is given, certain sinusoidal drug-adminstration schedules are definitely better tolerated by mice than are other sinusoidal schedules or a reference schedule (at the time the study was done the reference treatment schedule of eight equal doses over a 24 hr span was considered the best experimental treatment schedule). Also note the unequivocal reproducibility of chronotoxicity of the ara-C from experiments done on the same day in different laboratories in different geographic locations. (From Scheving et al., 1976a.)

The best cure rate that we or our collaborators consistently achieved with such an advanced disease state ($1 \times 10^6$ leukemia cells), when using the single drug ara-C (120 mg/kg per course for four courses of treatment given on days 2, 6, 10, and 14) is between 20 and 28%. Increasing or decreasing the dosage does not improve the cure rate; if the dosage is increased, overwhelming toxicity prevails; if it is decreased, the leukemia prevails. We believe that this is the realistic cure rate to be expected for this stage of the disease, using $CD2F_1$ or $BD2F_1$ mice, and even this cure rate cannot be obtained consistently using the nonchronobiological approach. For example, in one study using 282 leukemic mice (141 on each treatment schedule) the cure rate (60 day survivors) was 23% for those treated on a chronobiological schedule but only 11% for those treated on an nonchronobiological schedule (Kühl et al., 1974). Such findings have been consistent between Arkansas and Minnesota. Rose et al. (1978), in an attempt to confirm the findings of Haus et al. (1972), carried out studies from which they reported a variation in host toxicity using cyclophosphamide and confirmed what had been reported earlier.

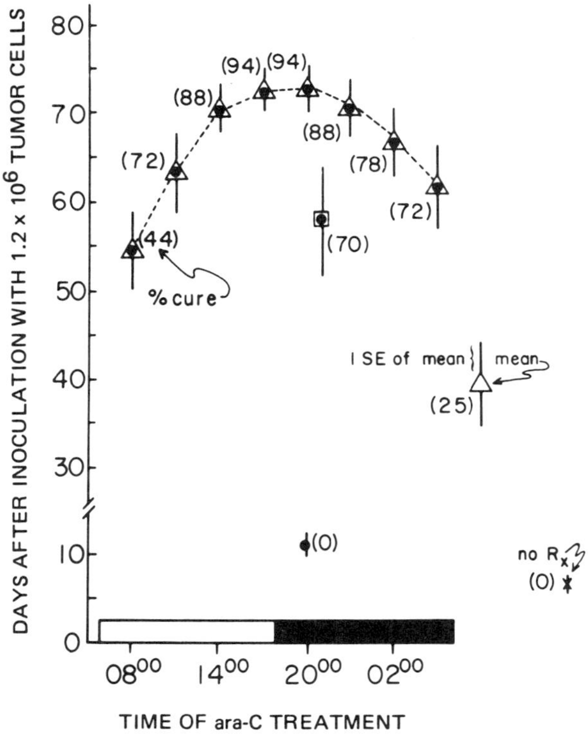


**FIGURE 22** Survival time and percentage of cures of L1210 leukemic mice treated with ara-C plus cyclophosphamide (CTX). The Δ implies that the best sinusoidal schedule was given. The □ implies that the reference treatment schedule was given. The ▲ or ▣ implies that CTX (•) was administered, in combination with ara-C once per course to each mouse; however, different groups received it at different circadian stages. The percentage cures (mice alive 75 days after tumor innoculation) are shown in parentheses. The time of CTX administration is indicated in the horizontal scale. The groups that did not receive CTX are shown just right to the time scale. N = 20 for each group. (From Scheving et al., 1977b.)

Rose and his coworkers, however, reported that they did not find the therapeutic advantage claimed by Haus et al. (1972) using ara-C on chronobiologically designed treatment schedules. It is important to recognize, however, that they did not utilize the same experimental design, administering only one or two courses of treatment, whereas Haus et al. had administered four courses. The fewer courses of treatment represented a significant alteration in design and would not bring about the full benefit of chronotherapy. For further comments relative to the studies of Rose et al. (1978), see Halberg et al. (1979), Burns and Scheving (1980), and Scheving et al. (1980a).

### *The Principle of Equal Toxicity*

A comment at this time is appropriate regarding the argument that one cannot compare two drug-treatment schedules even though the total dosage given is the same, if the resulting acute drug toxicity in the host turns out to be unequal. In light of chronobiological variation, the classic concept of equal toxicity has to be questioned. A pertinent situation may be the case of an extremely toxic dose of a drug one might elect not to use in a

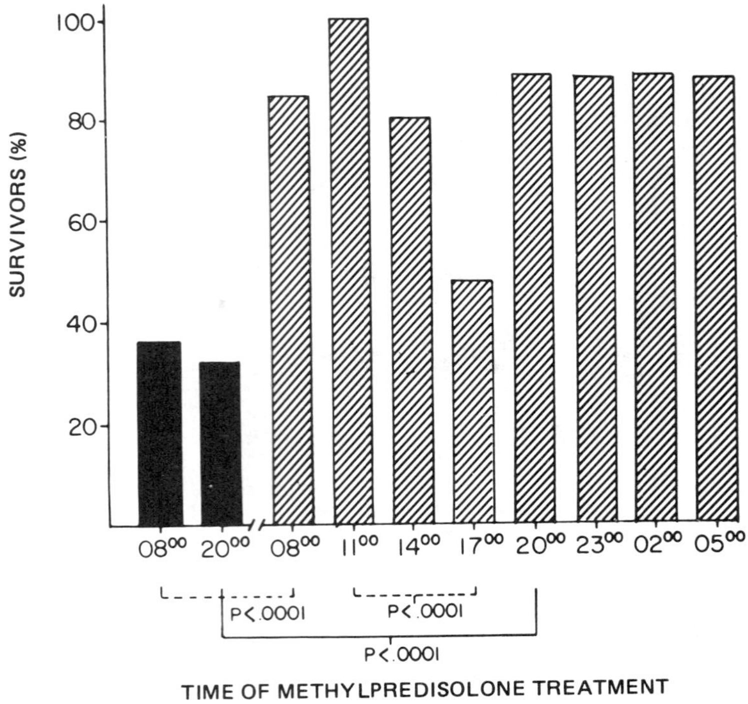


**FIGURE 23** Between 1130 and 1230 all ($CD2F_1$) mice were weighed and injected intraperitoneally with 5.0 × $10^6$ L1210 leukemia cells; the control group that received no subsequent treatment were injected with leukemia cells last. Treatment was initiated 45 hr later, beginning almost at 0800. Eight groups of animals received 135 mg/kg of ara-C over the 24 hr span administered on the sinusoidal schedule as follows: 15.88 mg/kg at 0800, then at 3 hr intervals, 38.25, 38.25, 15.88, 8.44, 4.22, 4.22, and 8.44 mg/kg. All these mice also received 39.6 mg/kg of cyclophosphamide (C) at 1400; 0.135 mg/kg of vincristine (V) at 0500, and 0.5 mg/kg of cis-platinum at 2000. Each of these eight groups was given 0.9 mg/kg of methylprednisolone (AA) once during the 24 hr span, one group received it at 0800, the second group at 1100, and so on, at 3 hr intervals. The ninth and tenth group (shown as black bars) received ara-C on a homeostatic (equal dose) treatment schedule of 16.88 mg/kg every 3 hr with 39.6 mg/kg of C at 1400, 0.135 mg/kg of vincristine at 0500, and cis-platinum at 1200. One of these groups (eight) received 0.90 mg/kg of 0800 and the other at 2000. Four courses of treatment were administered to all groups with 3 day intervals between courses; all animals were weighed at the beginning (0800) of each course. (From Scheving et al., 1980a.)

nonchronobiological treatment schedule, whereas its use in a time treatment approach would be feasible. With such a drug it might well be that the doses in a conventional regimen found to be equitoxic to the doses in a chemotherapeutically optimized regime would be so low they would be ineffective. A comparison of equitoxicity that disregards time may have little meaning (Haus et al., 1979).

### *Chronotherapy of L1210 Leukemia Using a Two-drug Combination: ara-C and Cyclophosphamide*

To obtain consistently better than about a 28% cure rate using ara-C in any manner against 1 × $10^6$ L1210 leukemic cells with treatment initiated 44 hr later, it was obvious that a second drug must be used in combination with ara-C. The second drug chosen was cyclophosphamide (CTX), and the results are illustrated in Fig. 22. In this study, mice were inoculated with $1.2 \times 10^6$ leukemia cells, and treatment was initiated 44 hr later. Four courses (at intervals of 4 days) of ara-C were given (120 mg/kg per course) to all mice in one of the "best sinusoidal" schedules previously described. Each mouse also received 15 mg/kg of CTX once per course, but each group of 20 received it at a different circadian stage. Only 2 mice out of 140 died from what might be attributed to acute toxicity; all others that died did so from leukemia, and herein lies an important observation. As is evident in Fig. 22, the cure rate ranged from 44 to 94% along the circadian time scale. When comparable animals bearing an identical tumor load were treated with both agents at the same dosage but given without chronobiological consideration, 30% died from acute drug toxicity. However, those that survived the acute drug toxicity were cured (Scheving et al., 1977b). Thus by giving drugs in a chronobiological manner it was possible to eliminate the death caused by toxicity from the drug, while still using an effective drug dose. There seems to be little merit in trying to treat with the conventional approach after adjusting the dosage so that its toxicity would be equal to that of the dosage used in the chronobiological approach; previous studies have shown that if we reduce the dosage of either drug as little as 10%, the leukemia will prevail.

*Chronotherapy of L1210 Leukemia Using More than Two Drugs*

Subsequent studies using more than two drugs against a larger tumor load than that described above have shown similarly that the chronobiological approach to treatment of L1210 leukemia was superior to the homeostatic approach (Burns and Scheving, 1980; Scheving et al., 1980a). For example, we found when $5 \times 10^6$ L1210 leukemic cells were injected and the mice were treated beginning 44 hr later with a combination of five drugs (those commonly used in the clinics) using the chronobiological approach, it was possible to obtain a cure rate as high as 88% (range 44–88%) (Scheving et al., 1980a). In this study there were no deaths due to drug toxicity; all animals dying did so from leukemia (Fig. 23). This cure rate was not possible using the same five drugs with a homeostatic treatment schedule; at the two times the treatment was given, the mortality was 54 and 64%, respectively, from acute drug toxicity (Fig. 24). Mean weight loss from the four courses of treatment given at the best circadian time (time highest cure rate) was 2.6 g, whereas in the homeostatically treated animals it was 6.8 g. Thus there was also evidence that the quality of life of the treated mouse also could be improved by chronotherapy.

*Circadian Variation and Sequencing of Two Drugs Used in Combination*

In still more recent studies using only two drugs and one course of treatment, it was shown that a circadian variation in response could be obtained. The treatment consisted of 100 mg/kg of CTX and 5 mg/kg of adriamycin (ADR) used against a tumor load resulting from an inoculum of $1 \times 10^5$ L1210 leukemia cells given 4 days prior to treatment. The variation in cure rate (mice alive and apparently free of disease 75 days post tumor inoculation) ranged from 8% to 68% in male animals standardized to 12 hr of light alternating with 12 hours of darkness (LD: 12,12) (Fig. 25). Similarly, in females standardized to LD: 8,16, the cure rate ranged from 0 to 56%, depending on the stage of the mouse circadian system at which drugs were administered. No cures were obtained with either drug alone. The maximum cure rate was recorded when the two drugs were administered during the early part of the dark span of the cycle (whether LD: 12,12 or 8,16), whereas maximum mortality occurred following treatment given early in the light span. Similar variation was seen in the length of survival time of those mice that died; thus both survival time and cure rate are circadian-stage dependent (Scheving et al., 1980b).

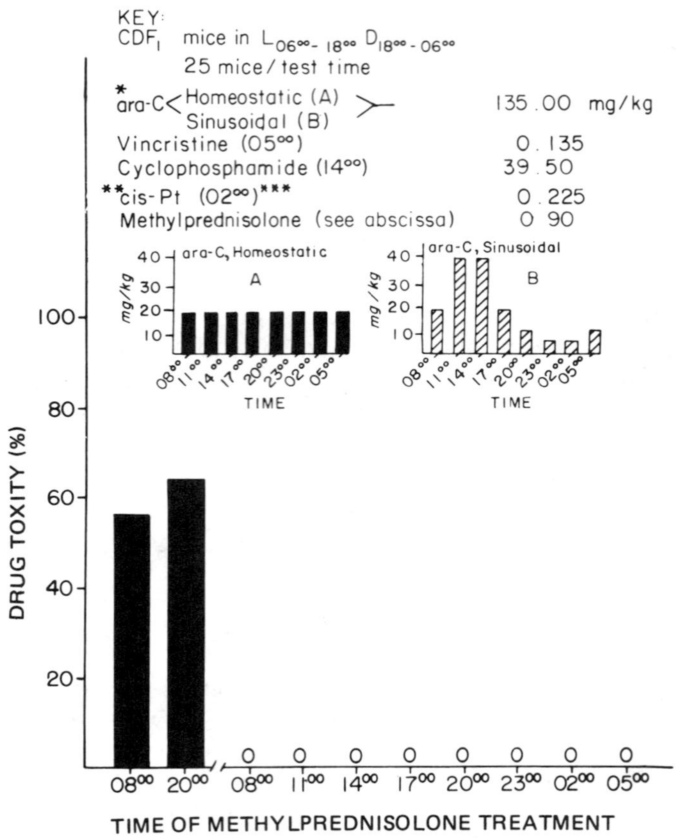


**FIGURE 24** Percentage of deaths from acute drug toxicity found in the same study described in Fig. 23. An important finding was that none of the chronobiologically treated animals died from acute drug toxicity, whereas 54 and 64% of those treated on the hemostatic schedule died. (From Scheving et al., 1980a.)

The data from the above study also documented that the maximum therapeutic advantage was obtained when the two drugs were separated by 2 or 3 hr intervals, and this effect of drug sequencing was strongly circadian-stage dependent (Fig. 26).

### *Additional Data Relative to Chronochemotherapy*

It had been questioned, when this work with ara-C began, if one should expect such variation in host resistance when using a long-acting drug such as adriamycin; some critics did not believe this would happen. To answer this, Kühl et al. (1973) gave a potentially lethal dose of ADR to both inbred Swiss albino mice and Fischer rats; they injected different subgroups from each species with 17.9 mg/kg at either 0800 or 2000 and found that 10% of those mice receiving the drug at 0800 died, whereas 55% of the mice receiving the drug at 2000 died. This finding was further investigated in great detail in the same laboratory (Sothern et al., 1977); it involved five different studies using a total of 858 inbred male mice and frequent intervals of sampling along the 24 hr time scale. The results of these studies demonstrated unequivocally that maximum mortality to adriamycin occurred during the middark and that the host was more resistant to the drug during mid-

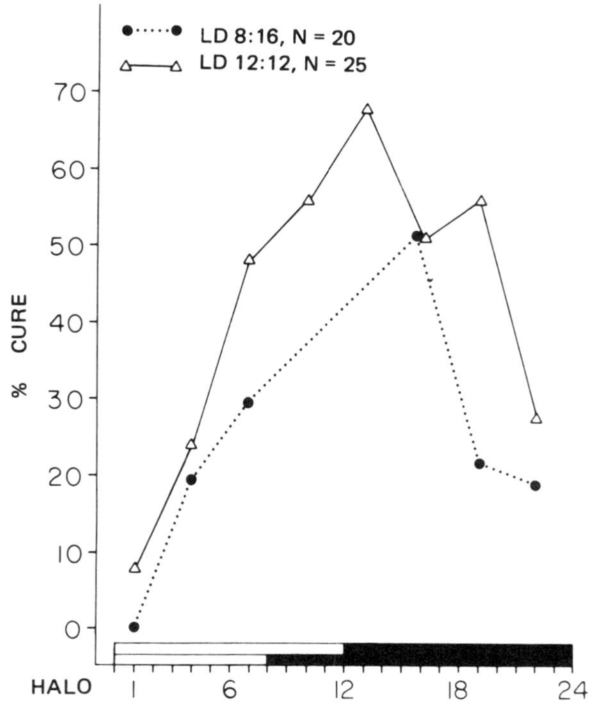


**FIGURE 25** Variation in cure rate of leukemic mice was dependent on timing of combined treatment with cyclophosphamide (CTX) and adriamycin (ADR). Study 1 (solid line) on male mice and LD 12:12; study 2 (dotted line) on female mice in LD: 8,16. HALO = hours after lights on. (From Scheving et al., 1980b.)

light. It was further found that this rhythm in susceptibility could be modified by restricting the intake of food to the first 4 hr of the daily light span.

Nelson et al. (1974) reported a therapeutic advantage after treating immunocytoma-bearing LOU rats with adriamycin at different stages of the circadian system. In ad libitum fed animals, the longest survival time was found in those animals treated at the beginning of the dark span; the shortest survival times were associated with those animals treated in the light span. Philippens and Scheving (1979) reported that the differences in survival rate between 10 day treated and nontreated Ehrlich ascites tumor-bearing NMRI mice were dependent on the stage of the mouse circadian system at which the treatment with adriamycin was initiated. For example, in a group of mice treated with 1.75 mg/kg per day of adriamycin for 10 days at 1900 (1 hr into the dark span), the mean survival time (calculated as the time when 50% of the animals have died) was 48 days longer than the mean survival time of the untreated mice. Those mice, however, treated 1 hr into the light phase (0700) of the daily light-dark cycle (LD: 12,12), died even before the control animals.

High-amplitude host-toxicity rhythms have also been documented for dauomycin and vincristine. Halberg (1980) and Simpson and Stoney (1977), while working with melphalan (1-phenylalanine mustard, L-pam), found that by giving the agent to different subgroups of mice at different circadian stages, a rhythm in the number of nucleated

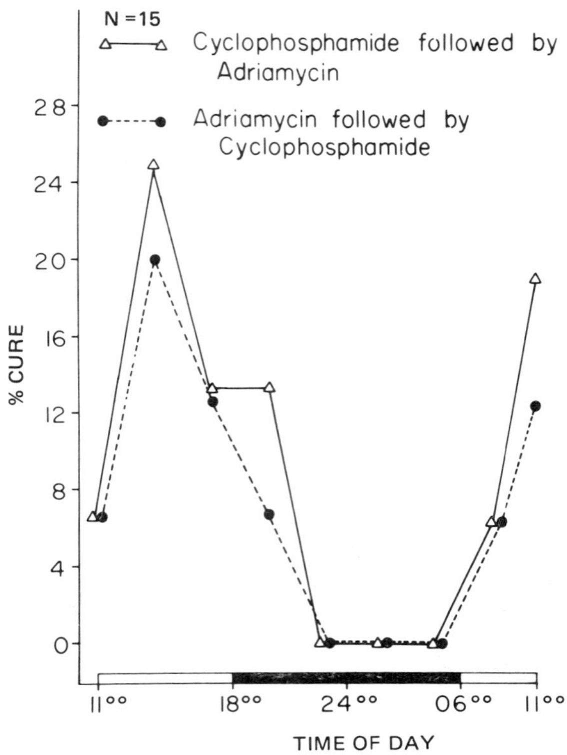


**FIGURE 26** Effect of drug sequence interval (probably confounded by circadian variation) in response of leukemic mice to cyclophosphamide and adriamycin. First drug given at 1100 (about midnight), second drug given either concomitantly or 3, 6, 9, 12, 15, 18, 21, or 24 hr later. We conclude from these data that the best way of administering these two drugs in combination is to separate them by 2 or 3 hr. Circadian variation response should be considered in any study of sequencing of drugs. (From Scheving et al., 1980b.)

femur marrow cells was evident 4 days after the treatment, thus demonstrating that the damage to normal bone marrow cells was circadian-stage dependent.

*Conclusions*

Clearly, the response to both short-acting and long-acting drugs is circadian-stage dependent. These differences in response to different drugs may be interpreted as reflecting the rhythmic changes in the biochemistry of a large number of variables within the organism over 24 hr span (Scheving, 1976, 1980a-d). Certainly the variables in the principal target organs of the drugs, such as DNA synthesis in the cells of the bone marrow and gut, play a prominent part in generating the rhythm in susceptibility to a drug seen in the host. At the present time, the mass of experimental data leads to the obvious conclusion that attempts must be made to bring the concept of chronochemotherapy into the clinics (Scheving, 1977a).

On the preventive side, there may be an advantage in monitoring a rhythmic variable, such as breast temperature; the rationale being that a tumor site will manifest a higher metabolic rate, thus a higher temperature, and even an altered tumor rhythm when compared with normal tissue (Simpson, 1976; Simpson and Mutch, 1982; Halberg et al.,

1979). Thus, one might be able to detect a tumor by monitoring temperature over it even before it becomes palpable. A great deal of interest has arisen regarding this concept of prevention; it has led to the development of the "chronobra," which is so constructed that continuous recording can be obtained simultaneously from many different sites of a mammary gland (Simpson and Mutch, 1982; Halberg et al., 1979).

Unfortunately, it took almost 50 years from the time of the first observation of circadian variation in cell proliferation to its application on an experimental basis to cancer chemotherapy or prevention. We have only now reached the stage when most scientists accept without question the existence of and the potential importance of such rhythms, even though they may still ignore them in experimental design. Based on the facts, we must abandon the erroneous concept that somehow sampling at the same time of day takes care of the rhythm problem (Scheving, 1976).

We are confident that there will not be a similar span of 50 years before the importance of rhythmicity to cancer biology and treatment is adequately explored. One pioneering study recently reported has shown in the clinics that kidney toxicity of cis-platinum can be dramatically reduced in cancer patients by timing the drug to body rhythms (Hrushesky et al., 1980, 1982; Levi et al., 1980). Halberg in collaboration with Gupta of India (Gupta and Deka, 1977; Halberg et al., 1973; Halberg, 1980) has reported that the timing of radiotherapy in relation to a marker rhythm (tumor temperature) can determine the *rate* of regression of perioral tumors. Moreover, it has been reported in another report (Hrushesky et al., 1981) that human bone marrow toxicity to adriamycin is statistically significantly reduced by circadian timing.

Even more important, Hrushesky et al. (1981) has reported a response rate of 100% in a preliminary study of 12 patients with advanced, stages 3 and 4, ovarian carcinoma treated with a chronobiologically designed two-drug chronotherapy regime. Over 80% of these responses were complete (total disappearance of all measurable disease). This 100% response rate compared favorably with a 42% response rate for 19 comparable ovarian cancer patients (controls) who received the same doses but at random circadian stages. The duration of response was also highly statistically different between the groups ($P < 0.0001$) (Hrushesky et al., 1981) in favor of those 12 cancer patients who received chronotherapy.

Circadian cytokinetics, pharmacokinetics, endocrinology, immunology, and toxicology can no longer be ignored but must be used to treat human cancer more effectively and eventually to prevent some classic varieties of malignancy. The major shortcoming of classic noncircadian cytokinetics has been the inability to translate in vitro advances directly from the Petri dish to the cancer patient. This was due to the fact that it was thought that the in vitro synchronized cell populations behaved differently from in vivo populations, which were classically assumed to be nonrhythmic with regard to cytokinetic parameters (Scheving et al., 1983b). The realization that there is relative synchrony along the circadian time scale within all in vivo cell populations is the only solution for cytokinetics as a practical, clinically relevant tool for treating cancer patients. If the basic rhythmicity inherent in the organism is not ignored, we believe the cancer patient will greatly benefit.

## ACKNOWLEDGMENTS

This work was supported in part by grant OH-00952 from the National Institute of Occupational Safety and Health and by the National Cancer Institute (CA 14388).

## REFERENCES


Andrew, W., and Andrew, N. V. (1949). Lymphocytes in the normal epidermis of the rat and man. *Anat. Rec.* 104:217-241.

Badran, A. F., and Echave Llanos, J. M. (1965). Persistence of mitotic circadian rhythm of the transplantable mammary carcinoma after 35 generations: Its bearing on the success of treatment within endoxan. *JCNI* 35:285-290.

Barnum, C. P., Jardetzky, C. D., and Halberg, F. (1957). Nucleic acid synthesis and regenerating liver. *Tex. Rep. Biol. Med.* 15:134-147.

Bertalanffy, F. D. (1960). Mitotic rates and renewal times of the digestive tract epithelia in the rat. *Acta Anat.* 40:130-148.

Bertalanffy, F. D., and Lau, C. (1962). Mitotic rate and renewal time of the corneal epithelium of the rat. *Arch. Opthalmol.* 68:546-550.

Bertalanffy, F. D., and Lau, C. (1963). Mitotic rates, renewal times and cytodynamics of the female genital tract epithelia in the rat. *Acta Anat.* 54:39-81.

Blumenfeld, C. M. (1939). Periodic mitotic activity in the epidermis of albino rat. *Science* 90:446-447.

Blumenfeld, C. M. (1942). Normal and abnormal mitotic activity; comparison of periodic mitotic activity in epidermis, renal cortex and submandibular salivary gland in albino rat. *Arch. Pathol.* 33:770-776.

Bostroem, E. (1928). *Der krebs des Menschen.* Georg Thieme Verlag, Leipzig.

Broders, A. C., and Dublin, W. B. (1939). Rhythmicity of mitosis in the epidermis of human beings. *Proc. Staff Meet. Mayo Clin.* 14:423-425.

Bucher, N. L. R., Patel, U., Cohen, S. (1978). Hormonal factors concerned with liver regeneration. In *Hepatrophic Factors.* R. Porter and J. Whelan (eds.), a CIBA Symposium, London, p. 95.

Burns, E. R., and Scheving, L. E. (1973). Isoproternol-induced circadian rhythm mitosis in murine corneal epithelium. *J. Cell Biol.* 56:605-608.

Burns, E. R., and Scheving, L. E. (1975). Circadian influence of the waveform of the frequency of labeled mitoses and mouse corneal epithelium. *Cell Tissue Kinet.* 8:61-66.

Burns, E. R., and Scheving, L. E. (1980). Circadian optimization of the treatment of L1210 leukemia with 1-β-D-arabinosylcytosine, cyclophosphamide vincristine and methylprednisolone. *Chronobiologia* 7:41-52.

Cameron, J. H. (1936). The origin of new epidermal cells in the skin of normal and x-rayed frogs. *J. Morphol.* 59:327-350.

Cardoso, S. S., Scheving, L. E., and Halberg, F. (1970). Mortality of mice is influenced by the hour of day of drug (ara-C) administration (abstract). *Pharmacologist* 12:301.

Carleton, A. (1934). A rhythmical periodicity in the mitotic division of animal cells. *J. Anat.* 68:251-263.

Chaudhry, A. P., and Halberg, F. (1960). Rhythms in blood eosinophils and mitoses of hamsters pinna and pouch; phase alterations by carcinogen. *J. Dent. Res.* 39:704.

Cherington, P. III, Smith, B. L., Pardee, A. B. (1979). Loss of epidermal growth factor requirement and malignant transformation (abstract). *Proc. Natl. Acad. Sci. USA* 70:3937.

Chumak, N. G. (1963). Mitotic cycle of epithelial cells of cornea in mice. Study by means of thymidine tagged with tritium. *Dokl. Acad. Nauk SSR* 1949:960-962.

Clausen, O. B. F., Thorud, E., Bjerknes, R., and Elgjo, K. (1979). Circadian rhythms in mouse epidermal basal cell proliferation. Variations in compartment size, flux and phase duration. *Cell Tissue Kinet.* 12:319-337.

Cooper, Z. K., and Franklin, H. C. (1940). Mitotic rhythms in the epidermis of the mouse. *Anat. Rec.* 78:1-8.

Cooper, Z. K., and Schiff, A. (1938). Mitotic rhythms in human epidermis. *Proc. Soc. Exp. Biol. Med.* 39:323-324.

Echave Llanos, J. M., and Nash, R. E. (1970). Mitotic circadian rhythm in a fast growing and a slow growing hepatoma. Mitotic rhythm in hepatomas. *JCNI* 44:581-586.

Edmunds, L. N., Jr. (1971). Persistent circadian rhythm of cell division in *Euglena*: Some theoretical considerations and the problem of intracellular communication. In *Biochemistry,* N. M. Menaker (ed.). National Academy of Sciences, Washington, D.C., pp. 594-611.

Fisher, L. B. (1968). The diurnal mitotic rhythm in the human epidermis. *Br. J. Dermatol.* 80:75-80.

Fortuyn-van Leyden, C. E. D. (1917). Some observations of periodic nuclear division in the rat. *Proc. Kon. Acad. Van Wetenschappen te (Amsterdam)* 19:38-44.

Fortuyn-van Leyden, C. E. D. (1926). Day and night period in nuclear division. *Proc. Kon. Acad. Van Wetenschappen te (Amsterdam)* 29:979-988.

Frieboes, W. (1920). Beitrage zur Anatomie und Biologie der Haut. *Dermatol. Z.* 31: 57-83.

Gasser, R. F., Scheving, L. E., and Pauly, J. E. (1972a). Circadian rhythm in the mitotic index in the uptake of [$^3$H] thymidine by the tongue of the rat. *J. Cell. Physiol.* 80: 437-442.

Gasser, R. F., Scheving, L. E., and Pauly, J. E. (1972b). Circadian rhythm in the cell division rate of the inner enamel epithelium and in the uptake of $^3$H-thymidine by the root tip of the rat incisors. *J. Dent. Res.* 51:740-746.

Gelfant, S., Ozawa, A., Chalker, D. K., and Smith, D. G., Jr. (1982). Circadian rhythm and differences in epidermal and in dermal cell proliferation in uninvolved and involved psosiatic skin. *J. Invest. Dermatol.* 78:58-62.

Goodrun, P. J., Sowall, J. G., and Cardoso, S. S. (1974). Characterization of the circadian rhythm of mitosis in the corneal epithelium of the immature rat. In *Chronobiology,* L. E. Scheving, F. Halberg, and J. E. Pauly (eds.). Igaku Shoin, Tokyo, pp. 29-32.

Gupta, B. D., and Dekka, A. C. (1975). Applications of chronobiology to radiotherapy of tumor of the oral cavity (abstract). *Chronobiologia (Suppl.)* 2:1, 125.

Halberg, E., Halberg, F., Cornelissen, G., Garcia-Sainz, M., Simpson, H. W., Tagatt-Anderson, M. A., and Haus, E. (1979). Toward a chronopsy. Part II. A thermopsy revealing asymmetrical circadian variation on surface temperature of human female breasts and related studies. *Chronobiologia* 6:231-257.

Halberg, F. (1964). Grundelagenforschung zur Aetiologie des Karvinoms. Mkurse arztl. *Forbild.* 14:67-77.

Halberg, F. (1980). Closing remarks: Chronobiology and health. In *Chronobiology: Principles and Applications to Shifts in Schedules,* L. E. Scheving and F. Halberg (eds.). NATO Advanced Study Institute Series, Sijthoff and Noordhoff 1980, Alphen aan den Rijn, The Netherlands, pp. 541-562.

Halberg, F., Gupta, B. D., Haus, E., Halberg, E., Deka, A. C., Nelson, W., Sothern, R. B., Cornelissen, G., Lee, J. K., Lakatua, D. J., Scheving, L. E., Burns, E. R. (1977). Steps toward a cancer chronopolytherapy. In *Proceedings of the Fourteenth International Congress of Therapeutics.* Montpelier, France. L'Expansion Scientifique Francaise, Paris, pp. 151-196.

Halberg, F., Haus, E., Cardoso, S. S., Scheving, L. E., Kuhl, J. F. W., Shiotsuka, R., Rosene, G., Pauly, J. E., Runge, W., Spalding, J. F., Lee, J. K., and Good, R. A. (1973). Toward a chronotherapy of neoplasia: Tolerance of treatment depneds on host rhythms. *Experientia* 29:909-1044.

Halberg, F., Johnson, E. A., Nelson, W., Runge, W., and Sothern, R. (1972). Autorhythmometry procedures for physiological self-measurements and their analysis. *Physiol. Teacher* 1:1-11.

Halberg, F., Nelson, W., Cornelissen, G., Haus, E., Scheving, L. E., and Good, R. A. (1979). On methods for testing and achieving cancer chronotherapy. *Cancer Treat. Rep.* 63:1428-1430.

Haus, E., Halberg, F., Scheving, L. E., Cardoso, S. S., Kuhl, J., Sothern, R., Shiotsuka, R., Hwang, D. S., and Pauly, J. E. (1972). Increased tolerance of leukemic mice to arabinosyl cytosine with scheduled adjusted to circadian system. *Science* 77:80-82.

Haus, E., Halberg, F., Scheving, L. E., and Simpson, H. (1979). Chronotherapy of cancer —a critical evaluation. *Int. J. Chronobiol.* 6:67-107.

Hrushesky, W. J. M., Borch, R., and Levi, F. (1982). A circadian time dependency of cis platin urinary pharmacokinetics. *Clin. Pharmacol. Ther.* 32:330-339.

Hrushesky, W. J. M., Levi, F., Halberg, F., Haus, E., Scheving, L. E., Sanchez, S., Medina, E., Brown, H., and Kennedy, B. J. (1980). Clinical chrono-oncology. In *Chronobiology: Principles and Applications to Shifts in Schedules,* L. E. Scheving and F. Halberg (eds.). NATO Advanced Study Institute Series, Sijthoff and Noordhoff, 1980, Alphen aan den Rijn, the Netherlands, pp. 513-533.

Hrushesky, W. J. M., Levi, F., Kennedy, B. J., Theoloides, A. A., and Frenning, D. (1981) *Proceedings of the 72nd Annual American Society of Clinical Oncology-Cancer Research,* vol. 22, Abstract 546, p. 472.

Karsten, G. (1918). Uber tagesteroide der kern-und zell teilung. *Z. Bot.* 10:1-20.

Killicott, W. E. (1904). The daily periodicity of cell division and the elongation in the root of *Allium. Bull. Tory Bot. Club* 31:529.

Killmann, S.-A., Cronkite, E. P., Fliedner, T. M., and Bond, B. T. (1962). Mitotic indices of human bone marrow cells. I. Number and physiological distribution of mitoses. *Blood* 19:743-750.

Klein, R. M. (1979). The third effective chemical sympathectomy on circadian variation of mitotic activity. *Cell Tissue Kinet.* 12:649-657.

Krieger, D. T., Hauser, H., Liotta, A., and Zelenetz, A. (1976). Circadian periodicity of epidermal growth factor and its abolition by superior cervical ganglionectomy. *Endocrinology* 99:1589-1596.

Kruger, W. A., Bo, W. J., and Hoops, P. C. (1975). Circadian rhythm of mitotic activity in the uterine luminal epithelium of the rat: Effect of estrogen. *Anat. Rec.* 183: 563-566.

Kühl, J. F. K., Grage, T. B., Halberg, F., Rosene, G., Scheving, L. E., and Haus, E. (1973). Ellen-effect: Tolerance of adriamycin of Bragg albino mice and Fischer rats depends on circadian timing of injection (abstract). *Int. J. Chronobiology* 1:333.

Kühl, J. F. K., Haus, E., Halberg, F., Scheving, L. E., Pauly, J. E., Cardoso, S. S., and Rosene, G. (1974). Experimental chronotherapy with Ara-C; comparison of marine Ara-C tolerance on differently timed treatment schedules. *Chronobiologia* 1:316-317.

Laerum, O. D., and Aardal, N. P. (1981). Chronobiological aspects of bone marrow and blood cells. In *Biological Rhythms in Structure and Function,* H. von Mayersbach, L. E. Scheving, and J. E. Pauly (eds.). Alan R. Liss, New York, pp. 87-98.

Levander, J. (1950). On the epithelium-regeneration in healing wounds. *Acta Chir. Scand.* 100:637-649.

Levi, F., Hrushesky, W., Haus, E., Halberg, F., Scheving, L. E., and Kennedy, B. J. (1980). Experimental chrono-oncology. In *Chronobiology: Principles and Applications in Shifts and Schedules,* L. E. Scheving, and F. Halberg (eds.). NATO Advanced Study Institute Series, Sijthoff and Noordhoff, 1980, Alphen aan den Rijn, The Netherlands, pp. 481-511.

Li, A. K., Schattenkerk, N. E., DeBries, J. E., Ford, W. E. A., and Malt, R. A. (1980). Submandibular sialoadenectomy retards dimethylhydrazine induced colonic carcinogenesis (abstract). *Gastroenterology* 78:1207.

Mauer, A. M. (1965). Diurnal variation of proliferative activity in the human bone marrow. *Blood* 26:1-6.

Møller, U., and Larsen, J. K. (1978). The circadian variations in the epithelial growth of the hamster cheek pouch. Quantitative Analysis of DNA distributions. *Cell Tissue Kinet.* 11:405-413.

Møller, U., Larsen, J. K., and Faber, M. (1974). The influence of injected tritiated thymidine on the mitotic circadian rhythm in the epithelium of the hamster cheek pouch. *Cell Tissue Kinet.* 7:231-239.

Nash, R. E., and Echave Llanos, J. M. (1971). Twenty-four hour variations in DNA synthesis of a fast-growing and a slow-growing hepatoma: DNA synthesis rhythm in hepatoma. *JNCI* 47:1007-1012.

Nelson, W., Zinneman, H., Selden, J. A., Schaber, K., Halberg, F., and Bazin, H. (1974). Circadian rhythm and Bence-Jones protein excretion by LOU rats bearing a transplantable immunocytoma, responsive to adriamycin treatment. *Int. J. Chronobiol.* 2:359-366.

Ortiz-Picón, J. M. (1934). Uber zellteilungsfrequenz und zellteillungsrhythms in der epidermis der Maus. *Z. Zellforsch. Mikrob. Anat.* 19:488-509.

Pauly, J. E., Burns, E. R., Halberg, F., Tsai, S., Betterton, H. O., and Scheving, L. E. (1975). Meal-timing dominates lighting regime as a synchronizer of the eosinophil rhythm in mice. *Acta Anat.* 93:60-68.

Phillipens, K. M. H., von Mayersbach, H., and Scheving, L. E. (1977). Effects of the scheduling of meal-feeding at different phases of the circadian system in rats. *J. Nutr.* 107:176-193.

Phillipens, K. M. H., and Scheving, L. E. (1979). Chronotoxic and therapeutic effects of adriamycin and Ehrlichs ascites tumor-bearing NMRI-mice (abstract). *Chronobiologia* 6:1943.

Powell, E. W., Pasley, J. N., Scheving, L. E., and Halberg, F. (1980). Amplitude-reduction and acrophase-advance of circadian mitotic rhythm in corneal epithelium of mice with bilaterally lesioned suprachiasmatic nuclei. *Anat. Rec.* 197:227-281.

Roberts, M. L., Friston, J. A., and Reade, G. C. (1976). Suppression of immune responsiveness by submandibular salivary gland factor. *Immunology* 30:811.

Rose, W. C., Trader, N. W., Laster, R. W., Jr., and Schabel, S. N., Jr. (1978). Chronotherapy of L1210 leukemic mice with cytosine arabinoside of cyclophosphamide. *Cancer Treat. Rep.* 62:1337-1349.

Rubin, N.H. (1980). Influence of the circadian rhythm in cell division on radiation-induced mitotic delay in vitro. Ph.D. dissertation, University of Arkansas for Medical Sciences, Little Rock.

Rubin, N. H. (1981). Circadian variation in the proportion of cells in cell-cycle phases in the hamster tongue epithelium as measured by flow microfluorometric technique. In *Biological Rhythms in Structure and Function,* H. von Mayersbach, L. E. Scheving, and J. E. Pauly (eds.). Alan R. Liss, New York, pp. 82-86.

Rubin, N. H. (1982). Influence of circadian rhythm in cell division on radiation-induced mitotic delay in vivo. *Radiat. Res.* 89:65-76.

Rubin, N. H., Hokanson, J. A., and Bodgon, G. (1983a). Circadian rhythms in phases of the cell cycle as demonstrated by flow cytometry. *Cell Tiss. Kinet.* 16:115-123.

Rubin, N. H., Hokanson, J. A., Mayschak, J. W., Tsai, T. H., Barranco, S. C., and Scheving, L. E. (1983b). Several cytokinetic methods for showing circadian variation in normal murine tissue and in a tumor. *Am. J. Anat.* 168:15-26.

Rubin, N. H., and Scheving, L. E. (2983). Letter to the editor. *J. Invest. Dermatol.* 80:79.

Scheving, L. A., Scheving, L. E., Tsai, T. H., and Pauly, J. E. (1982). Circadian stage-dependent effects of insulin and glucagon on incorporation of [$^3$H]thymidine into deoxyribonucleic acid in the esophagus, stomach, duodenum, jejunum, ileum, caecum, colon, rectum and spleen of the adult female mouse. *Endocrinology* 111:308-315.

Scheving, L. A., Yeh, Y. C., Tsai, T. H., Scheving, L. E. (1979). Circadian-phase dependent stimulatory effects of epidermal growth factor on DNA synthesis in the tongue, esophagus and stomach of the adult male mouse. *Endorcinology* 105:1475-1480.

Scheving, L. A., Yeh, Y. C., Tsai, T. H., and Scheving, L. E. (1980). Circadian-phase dependent stimulatory effects of epidermal growth factor on DNA synthesis of the duodenum, jejunum, ileum, caecum, colon, and rectum of the adult male mouse. *Endocrinology* 106:1498-1503.

Scheving, L. E. (1959). Mitotic activity in the human epidermis. *Anat. Rec.* 135:7-20.

Scheving, L. E. (1976). The dimension of time and biology in medicine. Chronobiology. *Endeavor* 35:(125):66-76.

Scheving, L. E. (1980a). Chronobiology, a new perspective for chronobiology in medicine. In *Proceedings of the Eleventh Collegium International Neuro-Psychopharmacologicum (CINP) Congress,* Vienna, Austria, 9-14 July, 1978, E. Saletu (ed.). Pergamon, Oxford, pp. 629-646.

Scheving, L. E. (1980b). Chronobiology—a new perspective in pharmacology and chemotherapy. *Trends Pharmacol. Sci.* 1(July):303-307.

Scheving, L. E. (1980c). Chronotoxicology in general and experimental chronotherapeutics of cancer. In *Chronobiology: Principles and Applications to Shifts of Schedules,* L. E. Scheving, and F. Halberg (eds.). NATO Advanced Institute Series, Sijthoff and Noordhoff, 1980, Aalphen aan den Rijn, The Netherlands, pp. 455-479.

Scheving, L. E. (1980d). Chronobiology and how it might be used to better understand and improve chemotherapy, H. B. Kaiser (ed.). Williams & Wilkins, Baltimore, pp. 317-332.

Scheving, L. E. (1981). Circadian rhythms of cell proliferation: Their importance when investigating the basic mechanism of normal vs. abnormal growth. In *Biological Rhythms in Structure and Function,* H. von Mayersbach, L. E. Scheving, and J. E. Pauly (eds.). Alan R. Liss, New York, pp. 39-79.

Scheving, L. E., and Burns, E. R. (1977). Some evidence that a consideration of chronobiology of the cell-cycle may improve chemotherapy. In *Die zeit und das leben,* J. H. Scharf and H. von Mayersbach (eds.). *Nova Acta Leopoldina* 46(225):227-283.

Scheving, L. E., Burns, E. R., Halberg, F., and Pauly, J. E. (1980a). Combined chronochemotherapy of L1210 leukemic mice using 1-β-arabinofuranosylcytosine, cyclophosphamide, vincristine, methylprednisolone and cis-diamminedichloroplatinum. *Chronobiologia* 8:33-40.

Scheving, L. E., Burns, E. R., and Pauly, J. E. (1972). Circadian rhythms and mitotic activity and $^3$H-thymidine uptake in the duodenum; effect of isoproterenol of the mitotic rhythm. *Am. J. Anat.* 135:311-317.

Scheving, L. E., Burns, E. R., and Pauly, J. E. (1975). Coincidence in timing between the synthetic and mitotic stages of the cell cycle in liver parenchymal cells and mice bearing 8-day Ehrlich ascites tumor (EAT). In *Twelfth International Conference Proceedings (International Society of Chronobiology,* Il Ponte, Milano, pp. 427-431.

Scheving, L. E., Burns, E. R., and Pauly, J. E. (1977a). Can chronobiology be ignored when considering the cancer problem? In *Prevention and Detection of Cancer, Part 1. Prevention. Vol. 1. Etiology,* H. E. Niebergs (ed.). Marcel Dekker, New York, pp. 1063-1079.

Scheving, L. E., Burns, E. R., Pauly, J. E., and Halberg, F. (1980b). Circadian bioperiodic response of mice bearing a L1210 leukemia to combination therapy with adriamycin and cyclophophamide. *Cancer Res.* 40:1511-1515.

Scheving, L. E., Burns, E. R., Pauly, J. E., Halberg, F., and Haus, E. (1977b). Survival and cure of leukemic mice after optimization of cancer treatment with cyclophosphamide and 1-β-D-arabinofuranosylcytosine. *Cancer Res.* 37:3648-3655.

Scheving, L. E., Burns, E. R., Pauly, J. E., and Tsai, T. H. (1978). Circadian variation and cell division of the mouse alimentary tract, bone marrow and corneal epithelium. *Anat. Rec.* 191:479-485.

Scheving, L. E., Burns, E. R., Pauly, J. E., Tsai, T. H., Betterton, H. O., and Halberg, F. (1976b). Meal-scheduling, cellular rhythms in the chronotherapy of cancer. In *Nutrition,* H. Koishi (ed.). Victory-sha Press, Kyoto, pp. 141-142.

Scheving, L. E., Cardoso, S.S., Pauly, J. E., Halberg, F., and Haus, E. (1974d). Variation and susceptibility of mice to the carcinostatic agent arabinosylcytosine. In *Chronobiology,* L. E. Scheving, F. Halberg, and J. E. Pauly (eds.). Igaku Shoin, Tokyo, pp. 213-217.

Scheving, L. E., and Gatz, A. J. (1955). Mitotic activity in the human epidermis (abstract). *Anat. Rec.* 121:363.

Scheving, L. E., Haus, E., Kuhl, J. F. W., Pauly, J. E., Halberg, F., and Cardoso, S. S. (1976a). Close reproduction by different laboratories of characteristics of circadian rhythms in 1-β-arabinofuranosylcytosine tolerance by mice. *Cancer Res.* 36:1133-1137.

Scheving, L. E., Mayersbach, H. v., and Pauly, J. E. (1974a). An overview of chronopharmacology (a general review). *J. Eur. Toxicol.* 7:203-227.

Scheving, L. E., and Pauly, J. E. (1967). Circadian phase relationship of thymidine-$^3$H uptake, labeled nuclei, grain counts and cell division rate and rat corneal epithelium. *J. Cell Biol.* 32:677-683.

Scheving, L. E., and Pauly, J. E. (1971). Pitfalls awaiting those who ignore circadian time structure in experimental design: A conclusion based on experiments to test the effect of cytosine arabinoside on mitosis. *Anat. Rec.* 169:419 (abst).

Scheving, L. E., and Pauly, J. E. (1973). Cellular mechanisms involving biorhythms with emphasis on those rhythms associated with the S and M stages of the cell cycle. *Int. J. Chronobiol.* 1:269-283.

Scheving, L. E., and Pauly, J. E. (1977). Several problems associated with the conduct of chronobiological research. In *Die Zeit und das Leben,* J. H. Scharf, and H. von Mayersbach (eds.). *Nova Acta Leopoldina,* 46(225):237-258.

Scheving, L. E., Pauly, J. E., Burns, E. R., Halberg, F., Tsai, T. H., and Betterton, H. O. (1974b). Lighting regime dominates enteracting meal schedules and synchronized mitotic rhythms in the mouse corneal epithelium. *Anat. Rec.* 180:447-452.

Scheving, L. E., Pauly, J. E., von Mayersbach, H., and Dunn, J. D. (1974c). The effect of continuous light or darkness on the rhythms of the mitotic index in the corneal epithelium of the rat. *Acta Anat.* 88:411-423.

Scheving, L. E., Pauly, J. E., and Scheving, L. A. (1983a). Chronobiology of cellular proliferation: Implication for cancer chemotherapy. In *Topics in Environmental Phsyiology and Medicine,* A. Reinberg and M. Smolensky (eds.). Springer Verlag, New York.

Scheving, L. E., Sohal, G., Enna, C. D., and Pauly, J. E. (1973). The persistence of the circadian rhythm in histamine response of guinea pigs maintained under continuous illumination. *Anat. Rec.* 175:1-6.

Scheving, L. E., Tsai, T. H., Pauly, J. E., and Halberg, F. (1983c). Circadian effect of ACTH 1-17 on mitotic index of the corneal epithelium of Balb/C mice. *Peptides* 4:183-190.

Scheving, L. E., Tsai, T. H., Pauly, J. E., and Halberg, F. (1984a). Circadian ACTH 1-17 effects on murine triatiated thymidine incorporation into DNA of thymus, bone marrow and spleen: also on total splenic RNA and DNA. *Chronobiol.* in press.

Scheving, L. E., Tsai, T. H., Pauly, J. E., and Halberg, F. (1984b). Circadian effect of ACTH 1-17 on DNA synthesis in the alimentary tract of Balb/C mice. *Peptides*: in press.

Scheving, L. E., Tsai, T. H., Powell, E. W., Pasley, J. N., and Halberg, F. (1983b). Bilateral lesions of suprachiasmatic nuclei affect circadian rhythms in [$^3$H]-thymidine incorporation into deoxyribonucleic acid in mouse intestine tract, in mitotic index of corneal epithelium, and in serum corticosterone. *Anat. Rec.* 205:239-249.

Sigdestad, C. T., Bouman, J., Lesher, S. W. (1969). Diurnal fluctuations in the number of cells and mitosis in DNA synthesis in the jejunum of the mouse. *Exp. Cell Res.* 58: 159-162.

Simpson, H. W. (1976). A new perspective: Chronobiochemistry. *Essays Med. Biochem.* 2:115-187.

Simpson, H. W., and Mutch, F. (1982). The fresh approach to human breast cancer. In *Commentaries on Research of Breast Disease,* R. D. Bulbork and J. D. Taylor (eds.). Alan R. Liss, New York, (in press).

Simpson, H. W., and Stoney, P. J. (1977). The circadian variation in melphalan (L-phenylalanine nitrogen mustard) toxicity to murine bone marrow: Relevance to cancer treatment protocols. *Br. J. Haematol.* 35:459-464.

Skipper, H. E., Schabel, F. M., Jr., and Walcox, W. S. (1967). Experimental evaluation of potential anticancer agents. XXI. Scheduling of arabinosylcytosine to take advantage of its S phase specificity against leukemic cells. *Cancer Chemother. Res.* 51: 125-165.

Sothern, R. B., Nelson, W. L., and Halberg, F. (1977). The circadian rhythm and susceptibility of mice to the anti-tumor, adriamycin. In *Proceedings Conference International Society for Chronobiology,* 1977, Washington, D.C. Il Ponte, Milano, pp. 433-438.

Stalfelt, M. G. (1921). Stud. uber die periodizitat d Zellteilung u sick daran anschliessende Erscheinungen kunst. *Svenska Vet. Hand.* 62:1-114.

Tvermyr, E. M. (1969). Circadian rhythms and epidermal mitotic activity. Diurnal variations of the mitotic index, mitotic rate and mitotic duration. *Virchows Arch. [Cell Pathol.]* 2:319-325.

Yeh, Y. C., Scheving, L. A., Tsai, T. H., and Scheving, L. E. (1981). Circadian phase dependent effects of epidermal growth factor desoxyribonucleic acid synthesis in 10 different organs of the adult male mouse. *Endocrinology* 109:644-651.

Zugula-Mally, Z. W., Cardoso, S. S., Williams, D., Simpson, H., and Reinberg, A. (1979). Time point differences in skin mitotic activity of actianic keratoses and skin cancer. Circadian reference: Plasma cortisol. In *Chronopharmacology,* A. Reinberg and F. Halberg (eds.), Pergamon, New York, pp. 399-402.

24

# Diurnal Rhythmicity Used in Experimental Cancer Research

**Ulla Møller** / The Finsen Institute, Copenhagen, Denmark

## INTRODUCTION

The topic of this paper is diurnal variations of temperature, $^{32}P$ uptake, and cell proliferation, mainly referring to studies performed at the Finsen Laboratory, Copenhagen.

During the past 50 years the interest in biological rhythms, especially diurnal rhythms, has increased. Rhythmicity is one of the basic manifestations of living systems. For humans a great many characteristics have been shown to change their values rhythmically, achieving a maximum and minimum value within a 24 hr period; however, the various biorhythms are different in their amplitude and the temporal position of their peaks and troughs (Wever, 1979; Scheving, 1981).

It is known that biorhythms have both an internal and an external regulation, but the biochemical and biophysical nature of the internal rhythm generator is still not understood. Comprehensive studies have been carried out to evaluate the effect of hormones on induction and control of biological rhythms, but it has not been shown that a general systemic release of hormones by itself could account for the rhythmic variations observed in normal tissue (Enright, 1980; Scheving, 1981).

Circadian variations are influenced by external conditions. The activity rhythm in all eukaryotic organisms can be entrained by a simple light-dark cycle, although social contacts play an important role in the circadian rhythm of humans (Czeisler et al., 1981). As the light-dark cycle seems to give an activity rhythm to which many biorhythms are coupled, biorhythms of the organism can be synchronized in a relatively simple way (Wever, 1979).

Studies of biorhythms for normal tissue seem to have reached the stage where most scientists accept the existence of such rhythms and the potential importance of them in experimental design (Møller et al., 1974; 1982; Halberg et al., 1978; Scheving, 1981).

Rhythmicity in cancer biology has not been explored adequately enough to determine to what extent circadian variations in cell proliferation are found in tumors (Tähti, 1956; Garcia-Sainz and Halberg, 1966; Nash and Echave Llanos, 1971; Gautherie and Gros, 1977; Møller and Bojsen, 1980). Expected inhomogeneities of tumors as well as problems concerning the kinetic effect of repeated taking of biopsies have made such studies less attractive. Recently, DNA flow cytometry applied to skin tumors has made it possible to obtain reproducible results, and this technique may lead to new knowledge of tumor rhythmicity (Frentz and Møller, 1981; Søndergaard et al., 1982).

Among the first biorhythms observed in tumors were circadian variations in $^{32}P$ uptake, and in tissue temperature of tumors, both interpreted as expressions of metabolic processes. Continuous recordings of $^{32}P$ uptake in tumors using implanted Geiger counters were first described by Hale (1961), followed by investigations of Bullen et al. (1963), Calcutt et al. (1967), Taylor et al. (1968), and Wolley-Hart et al. (1968). The clinical observations indicated that diurnal variations in $^{32}P$ uptake were chiefly restricted to hormone-dependent tumors; however, technical problems gave difficulties in the demonstration of the $^{32}P$ uptake rhythm. Stoll and Burch (1968) found a circadian variation in 9 of 19 mammary carcinomas. The continuous measurements of $^{32}P$ uptake for tumors gave evidence that this type of measurement might be used to optimize treatment with hormones and radiation using changes in tumor rhythms as an aid to the choice of the optimum time for therapy.

Studies on optimizing the hormone therapy in advanced breast cancer using biorhythms of tumors have been continued using temperature measurements, which involve fewer problems. Temperature curves of normal and abnormal breasts were recorded continuously by thermistors, either taped to the skin overlying the tumor or implanted in the tumor itself (Stoll and Burch, 1968; Stoll, 1971; Mansfield et al., 1970; Gillespie et al., 1971). Diurnal temperature rhythms were demonstrated in both normal and abnormal breasts with some differences, although results obtained by various research groups were not exactly equivalent. Administration of hormone therapy induced temperature changes which to some extent seemed to be correlated with response to therapy and with prognosis.

Thermorhythmicity has not been limited to studies on breast cancer. Discontinuous temperature measurements with clinical thermometers have been used as a marker rhythm for timing of radiotherapy of, for instance, perioral tumors (Deka et al., 1981). Diurnal variations in body temperature have also been used as a marker rhythm in cancer chemotherapy but have not always been sufficiently specific (Kuzel et al., 1978; Hrushesky et al., 1982).

Parallel to measurements of circadian metabolic rhythms attention has been drawn to studies on the application of proliferative tumor rhythmicity, e.g., cell cycle synchronization in cancer chemotherapy, the intention being to find an optimum time for therapy. Evidence was provided for an interrelationship between cell kinetic parameters and drug sensitivity for a number of cytotoxic drugs (Putten, 1974; Hill and Baserga, 1975). The drugs could be classified into two types; the cell cycle stage-specific drugs, which kill cells only if they are exposed in a sensitive phase of the cycle, and the cell cycle stage-nonspecific drugs, which kill cells to an almost similar extent in all phases of the cycle. A variation in the therapeutic efficacy of the drugs used according to their known cell kinetic effect was shown. Consequently, efforts were made to bring as many cells as possible in a tumor to a stage where one would expect those cells to be killed by the applied cytostatic drug.

Data concerning drug-induced partial synchrony and therapeutic success of chemotherapy after synchronization are conflicting (Klein and Lennartz, 1974; Focan et al., 1975; Nicolini, 1976; Putten et al., 1976; Camplejohn, 1980, Klein, 1980), and clinical protocols continue to be developed empirically. Data showing possible existing spontaneous rhythmicity in tumors are also scanty (Focan, 1976).

More information is needed on differential responses and differential kinetic characteristics of normal and malignant tissues. Successful experimental schedules have been extrapolated to the clinic with little regard for the often extensive kinetic differences between the model and the human situation. Rhythmicity in normal tissue could, however, perhaps be used as a marker rhythm for adjustment from the laboratory to the clinic.

The importance of rhythmicity or synchronization for cancer biology should be considered not only because it can permit more tumor cells to be eradicated at certain time points. In experimental cancer research compelling evidence has been given that, by appropriately timing the administration of a chemotherapeutic agent to the host's rhythms in toxicity, one can protect normal cells and significantly decrease toxicity, resulting in an increase in both survival time of the host and cure rate of the tumor (Haus et al., 1972; Halberg et al., 1973; Simpson and Stoney, 1977; Levi et al., 1982). Increased tolerance of the host can be obtained by schedules adjusted to the diurnal rhythm of susceptible normal tissue, and this illustrates that recognizing diurnal variations in normal tissue can have important implications for cancer treatment.

## BIORHYTHMS BASED ON MEASUREMENTS OF METABOLIC ACTIVITY

### Continuous Temperature Measurements by Implanted Equipment of Tumor and Normal Tissue

Continuous recordings on patients of $^{32}P$ uptake and temperature have demonstrated that circadian variations exist in tumors (see Introduction). The measurements were performed with detectors either located on the surface or implanted. Both positions, however, gave technical problems. A main problem seemed to be that because of the monitoring technique patients were immobilized for a period of time.

In order to make possible a more systematic study based on experimental tumors, our approach to the study of rhythmicity in metabolic processes in tumors has been that the heavy and stationary equipment, which required immobilization of the subject, should be avoided. An implantable biotelemetrical equipment was developed by which the circadian temperature rhythm in tumors could be measured on unrestrained animals without interfering with normal activity, with the result that measurements could continue for days and weeks (Bojsen et al., 1971).

Two experimental tumors were used. An adenocarcinoma induced by 7, 12 dimethylbenz(α)anthracene (DMBA), which is a hormone-dependent, rather benign mammary tumor, and the Walker 256 carcinosarcoma, which is a fast proliferating, malignant, hormone-independent carcinosarcoma (Møller and Bojsen, 1975; 1980).

The tumors were induced on female Sprague-Dawley rats. The rats were housed singly and had food and water freely available under a standardized lighting regimen (Fig. 1).

The tissue temperatures of tumor and surrounding subcutaneous tissue were monitored telemetrically by radiotransmitters giving a receiver signal proportional to the temperature of the thermistor location. The temperature measurements were performed continuously for a period of up to 1 month. Usually a transmitter with two thermistors was

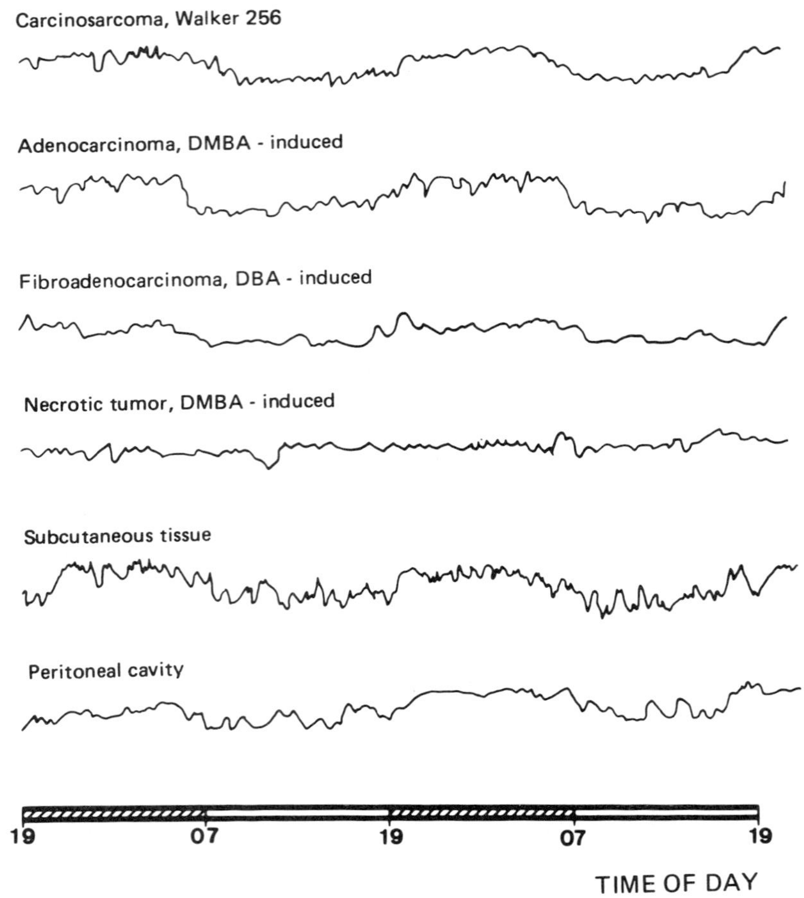


**FIGURE 1** Continuous temperature measurements performed by implanted biotelemetrical equipment on unrestrained animals. The lowest temperatures were found in the light period when the rats were at rest (Table 1). (From Møller and Bojsen, 1975, 1980).

used that permitted temperature measurements both from the tumor and from the surrounding subcutis on the same rat; a magnetic microswitch inside the electronic circuit could be switched from one thermistor to the other from outside the animal.

The temperature measurements showed that the same phase of temperature rhythm with a periodicity of 24 hr was found in tumor, subcutaneous tissue, and peritoneal cavity of normal rats (Fig. 1). The lowest temperatures were found in the light period when the rats were resting (Table 1). The temperature curves had a square wave shape with broad maximum and minimum levels. The increases and decreases were associated in time with the shift between dark and light. Curves for large fibrous tumors were smooth without the short-term temperature variations characteristic for the other tumors.

The tumors had the higher temperature level (Table 1). The periodicity of the temperature rhythm in itself did not indicate any difference between the hormone-dependent DMBA-induced mammary tumor and the hormone-independent Walker tumor. However, the temperature difference between rest and activity periods for the Walker tumor was

**TABLE 1** Temperatures (°C) Measured Telemetrically by Implanted Transmitters[a]

| Experimental group | Dark period | Light period | Difference dark-light |
|---|---|---|---|
| Normal rats | | | |
| Subcutaneous tissue | 36.2 ± 0.5 | 35.3 ± 0.5 | 0.9 ± 0.1 |
| Peritoneal cavity | 37.2 ± 0.5 | 36.3 ± 0.4 | 0.9 ± 0.1 |
| DMBA-induced adenocarcinoma | | | |
| Tumor | 38.4 ± 0.3 | 37.1 ± 0.3 | 1.3 ± 0.2 |
| Subcutaneous tissue | 36.3 ± 0.4 | 35.2 ± 0.3 | 1.1 ± 0.2 |
| Difference tumor/s.c. tissue | 2.1 ± 0.4 | 1.9 ± 0.5 | |
| Necrotic tumor | | | |
| Before necrosis | 38.2 | 36.8 | 1.4 |
| Early necrosis | 36.9 | 36.2 | 0.7 |
| Developed necrosis | 34.4 | 34.3 | 0.1 |
| Walker-256 carcinosarcoma | | | |
| Tumor | 38.3 ± 0.2 | 37.6 ± 0.2 | 0.7 ± 0.1 |
| Subcutaneous tissue | 36.3 ± 0.2 | 35.8 ±0.3 | 0.5 ± 0.1 |
| Difference tumor/s.c. tissue | 2.0 ± 0.02 | 1.8 ± 0.1 | |

[a]The results are given as mean values ± standard error. Student's t-test used on the light-dark differences gave values of $p < 0.001$ for normal rats, and $0.002 > p > 0.001$ for DMBA-induced tumors and $0.01 > p > 0.002$ for Walker tumors. For the difference between tumor and subcutaneous tissue the value was $0.01 > p > 0.002$. The difference of the temperature amplitude of the adenocarcinoma is about twice as high as for the carcinosarcoma; 1.3 compared to 0.7 ($0.02 > p > 0.01$, Student's t-test).

*Source*: Møller and Bjosen (1975, 1980).

only about half that found in the DMBA-induced tumor (Table 1). The same tendency was found in the subcutaneous tissue surrounding the tumors. Hence, other parameters, such as temperature level and amplitude, but not so much the phase of the temperature rhythms, seemed to be useful for characterization. A circadian rhythm was absent only in necrotic tissue.

The light-dark cycle to which the animal was exposed seemed to be the external regulator of the activity rhythm, which was indirectly measured by the temperature rhythm of the peritoneal cavity (Fig. 1; Table 1). The temperature rhythms of the tumors were coupled to this rhythm and were in phase. The higher amplitude of the temperature rhythm found in DMBA-induced tumors and the disappearance of the temperature rhythm and temperature decrease during necrosis were taken as an indication that the tumor had a temperature rhythm of its own. Blood flow and external temperature measurements indicated that the heat production of the tumor in some way must be returned to the body. The temperature of the skin covering the tumor was registrated as being 1-2°C lower than the neighboring skin areas; this could mean that the tumor represented an insulating cushion between the body and the skin (Møller and Bojsen, 1975). Similar results have in some cases been obtained in human breast cancer (Gillespie et al., 1971; Stoll, 1971).

### $^{32}P$ Uptake Measurements of Tumor and Normal Tissue

In addition to the temperature measurements we were interested in measuring $^{32}P$ uptake in experimental tumors. Stoll and Burch (1968) found in patients that in some of

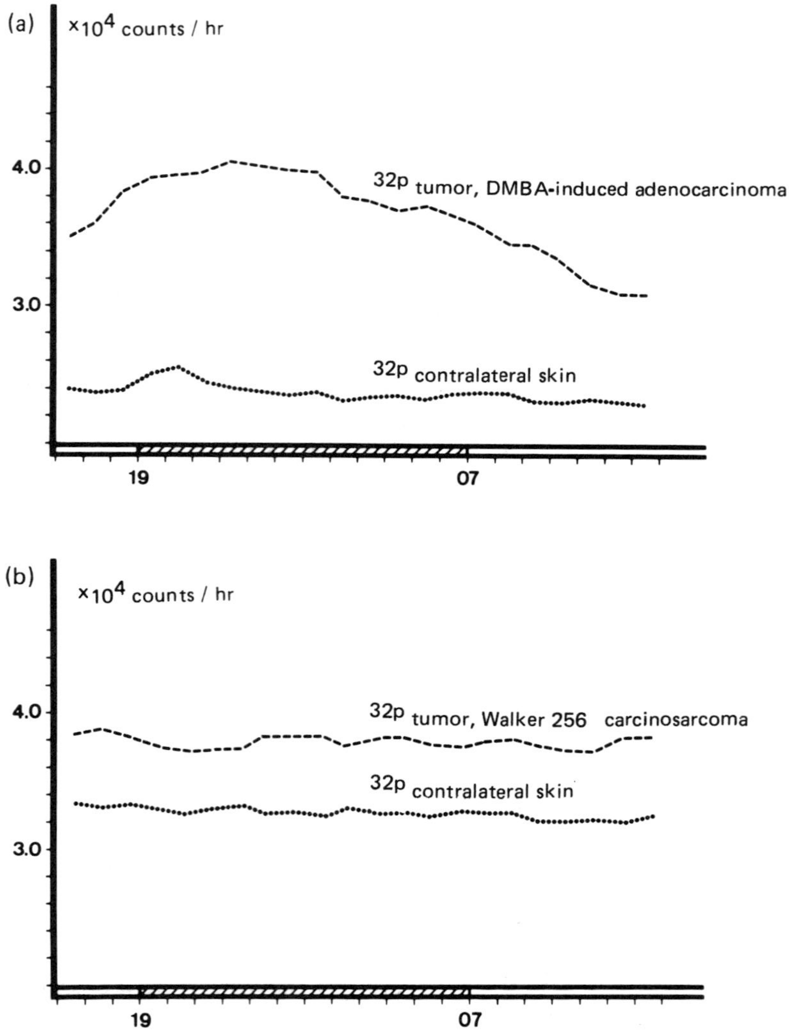


**FIGURE 2** $^{32}$P uptake in tumor and contralateral skin; 3-12 days before the start of the measurements the rats were injected with $^{32}$P (0.02 $\mu$Ci/g body weight, i.p.) in order to permit an equilibrium of $^{32}$P to be established. The mean uptake of $^{32}$P in both tumor types was 1.3 times higher than that of the skin. The uptake in the Walker carcinosarcoma (b) was, however, only significantly higher than in skin during the light periods. (From Møller and Bojsen, 1976.)

the tumorous breasts and in all normal breasts there was no $^{32}$P uptake periodicity, despite skin temperature periodicity.

Our intention was to develop implantable biotelemetrical equipment, but it was not possible to reduce the electronic detector unit sufficiently to permit its use in small laboratory animals (Bojsen et al., 1972). Nor was it possible to use storage telemetry by implantable thermoluminescent dosimeters in the relatively small experimental tumors, because they induced necrosis (Bojsen et al., 1977).

Thus the $^{32}P$ uptake in tumors and normal skin had to be measured as a surface detection by two Geiger tubes (Møller and Bojsen, 1976). The circadian temperature rhythm was monitored continuously (implanted thermistor) as a marker for adaptation of the animals to their restraints.

The mean utake of $^{32}P$ in both tumor types was 1.3 times higher than in skin. A difference between the two experimental tumors was also found in the $^{32}P$ uptake measurements: a significant difference between the $^{32}P$ uptake in the rest and the activity period was found only in the hormone-dependent DMBA-induced tumor (Fig. 2). The highest levels were measured during the activity period. The hormone-independent malignant tumor did not show such diurnal variations in $^{32}P$ uptake, nor did the skin of normal rats or skin of the contralateral flank of tumor bearing rats. The development of necrosis in the tumor resulted in a decrease in $^{32}P$ uptake and a cessation of rhythmicity.

### Conclusion

Our studies on experimental tumors strongly support the indications from clinical observations.

The hormone-dependent tumor was characterized both by a greater amplitude in temperature rhythm and by rhythmicity in $^{32}P$ uptake. Both tumors had higher temperatures and $^{32}P$ uptake levels than normal skin, and periodicity in $^{32}P$ uptake was not found in the hormone-independent tumor or in normal skin. These conclusions are in agreement with the findings of Stoll and Burch (1968), indicating that $^{32}P$ uptake might be specific for a controlling mechanism.

The phase of the temperature curves of the two experimental tumors was the same as found by Gillespie et al. (1971) in human breast cancer (implanted thermistors), but Mansfield et al. (1970) and Stoll and Burch (1968) recorded an inverse phase, with the highest temperature of human cancer appearing during the resting period (thermistors taped to skin). In a later study of Mansfield et al. (1973), the tumorous breasts had a temperature peak in the afternoon (thermistors taped to skin). For relatively slowly growing cancerous breast Gautherie and Gross (1977) found a fairly similar result.

The difference in the amplitude of the temperature rhythm between that of the rapidly growing Walker carcinosarcoma and of the slowly growing DMBA-induced adenosarcoma can be related to findings in human breast cancer, where temperature level (the diurnal low point being relatively high), amplitude, and presence of diurnal variation have been related to the degree of malignancy and growth rate of tumors (Stoll, 1971; Gauthrie and Gros, 1977). More studies are, however, needed to see how close the relation is.

The main technical problems in the temperature measurements of the experimental tumors were related to the broad distribution of the lifetimes of small mercury batteries. These problems are now solved by using either reliable miniature lithium batteries or external induction-powering systems for the implanted electronics (Singh et al., 1981). For investigation of patients, external equipment for continuous radionuclide tracer studies has been developed (Bojsen et al., 1981) and is now commercially available, making further investigations feasible.

The heterogeneity of the DMBA-induced tumor seen in histological studies did not appear to be reflected either in the temperature or in $^{32}P$ uptake measurements (Møller and Bojsen, 1975); this could help these studies, which have none of the problems inherent in making biopsies, to gain acceptance now that some of the technical difficulties have been overcome.

## BIORHYTHMS BASED ON MEASUREMENTS OF CELL PROLIFERATION

### Cell Kinetics of a Spontaneous Partially Synchronized Keratinized Epithelium

As evidence was provided for an interrelationship between cell kinetics and drug sensitivity, wherein anticancer drugs were classified as cell cycle stage–specific (CCSS) or cell cycle stage–nonspecific, interest in methodologies for synchronizing cells arose (see Introduction). The intention was to induce partial synchronization of tumor cell proliferation and thereby make the tumor more sensitive to subsequent administration of CCSS agents. Since the selective killing of tumor cells by cell cycle-specific drugs only is based on the feature that tumor cells proliferate more rapidly, the possibility of simultaneously inducing synchronization of normal proliferative cells could be a problem.

For drug-induced partial synchronization vincristine has been used widely both in experimental and clinical studies, but the degree and effect of the achieved synchronization is still being discussed (Camplejohn, 1980; Klein, 1980). In order to obtain further insight into synchronized cell growth with implication for future research on CCSS drugs, we studied the spontaneous partial synchrony in the hamster cheek pouch epithelium (Brown and Berry, 1968).

The cheek pouch epithelium of male Syrian hamsters was used. The animals were maintained in a standardized lighting regimen (Figs. 3 and 4) and had food and water freely available. To obtain as homogeneous a material as possible the animals were caged individually. The biological variation from cage to cage could thus be decreased (Nome, 1974); however, by caging the hamsters singly the circadian temperature rhythm of the peritoneal cavity disappeared (Møller and Bojsen, 1974), and thereby the possibility of using the body temperature as a host reference rhythm was lost. Judging from the disappearance of the temperature rhythm, other synchronized variables of the animals might also have been removed. As additional synchronizers, white noise in the light period and increased light intensity were therefore used (Figs. 3 and 4) (Møller, 1978; Møller et al., 1979).

The cell kinetic parameters were derived from countings on sectioned epithelium, and from flow cytometrical (FCM) measurements. The histological technique was combined with autoradiographic studies using [$^3$H]TdR as tracer (Møller et al., 1974). A double-labeling experiment using two injections of [$^3$H]TdR was used to obtain measurements of rate parameters of the S phase (Møller and Keiding, 1982). The stathmokinetic agent Colcemid was used when the rate parameters of the mitotic phase were measured (Møller et al., 1979). For the FCM measurements a method for disaggregation of stratified squamous epithelium into a suspension of single nuclei was developed (Møller and Larsen, 1978; 1979; Christensen et al., 1980; Frentz et al., 1980).

The cell kinetic parameters estimated were mitotic index (MI) (Møller et al., 1974); labeling index (LI) (Møller et al., 1982); mitotic influx, efflux, and mean transit time (Møller et al., 1979); and S-phase influx, efflux, and mean transit time (Møller and Keiding, 1982) (see also Chap. 8). Furthermore, estimates of the number of cells in $G_1$, S, and $G_2M$ fractions were obtained (Møller and Larsen, 1978; 1979). For all the cell kinetic parameters studied over a 24 hr period a diurnal variation was found, and for cells in S phase a diurnally varying age distribution could furthermore be demonstrated (Møller et al., 1982).

In Figs. 3, 4, 5, and 6 data obtained for cells in S phase are given, since they are regarded as being of the main interest for many studies of CCSS agents. As can be seen, the diurnal variation in number of cells in S phase was found to be rather stable for experi-

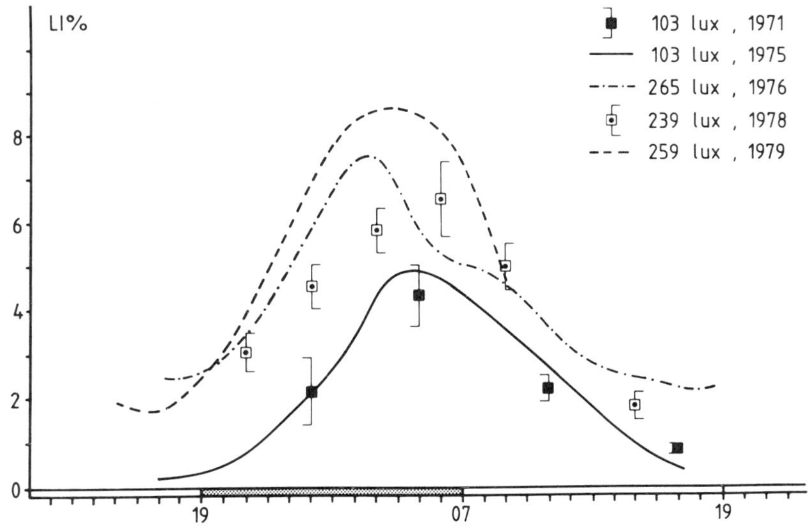


**FIGURE 3** Results obtained by autoradiography on sectioned epithelium over the years. LI, the percentage of epithelial cells incorporating the DNA precursor. The animals were injected i.p. with [$^3$H] TdR (1 $\mu$Ci/g body weight). The biopsy was taken 1 hr later. (From Møller and Larsen, 1978; Møller et al., 1982; Møller and Keiding, 1982.)

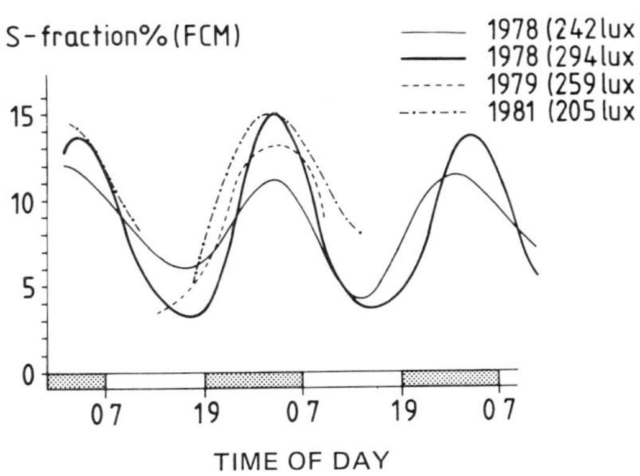


**FIGURE 4** Results obtained by flow cytometric DNA measurements of suspensions of epithelial nuclei. S fraction, the percentage of epithelial cells having DNA contents between 2 and 4c. (From Møller and Larsen, 1978; Møller et al., 1984b; and Møller, unpublished results, 1981.)

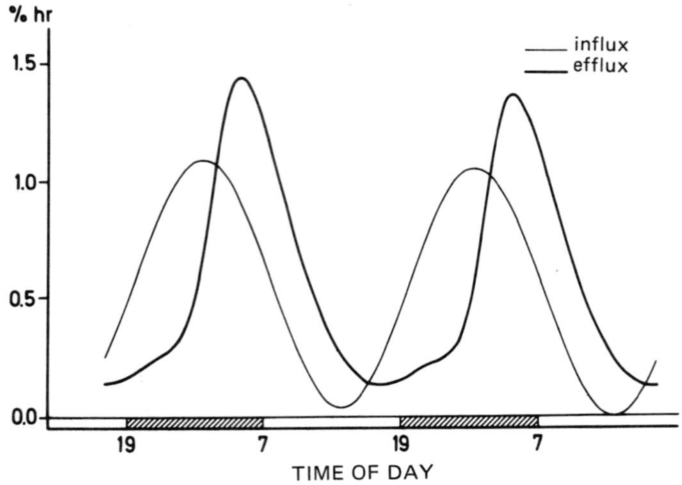


**FIGURE 5** Influx and efflux of the S phase. Results of a double-labeling experiment using two injections of [$^3$H]TdR. Diurnal variations in S-phase influx and efflux; the diurnal mean of both was estimated as 0.5%/hr, when based on measurements of all nucleated epithelial cells. Variations in S-phase influx seem mainly responsible for the diurnal variation in cell proliferation. (From Møller and Keiding, 1982; Keiding et al., 1983.)

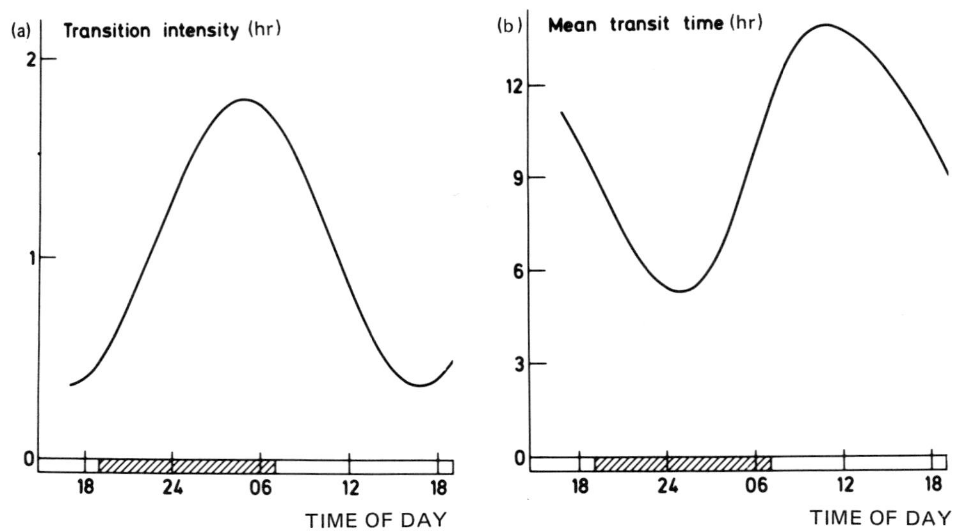


**FIGURE 6** Results of a mathematical cell kinetic modeling for cells in S phase based on a double-labeling experiment. The transition intensity (a) is proportional to DNA synthesis rate, which increases around the start of the dark period and reaches its maximum shortly before the end of this period. The mean transit time (b) for cells in S phase also showed a diurnal variation. The mean transit time is highest for cells entering S phase around noon. (From Keiding et al., 1983.)

ments performed under similar lighting intensities. The peak value for the number of cells in S phase occurred during the end of the dark period (0400-0500 hr), when the animals are active. The circadian variation in DNA synthesis rate and mean transit time of cells in S phase are given in Fig. 6a and b. In this naturally synchronized cell system the increase in S-phase influx, in the number of cells in S phase, and in the DNA synthesis rate are correlated to the beginning of the dark period (Hopper and Brockwell, 1978; Møller and Keiding, 1982 (see also Chap. 11).

By changing the lighting regimen the daily high and low points for cells in S phase changed correspondingly, which was taken as an indication that the light-dark cycle could be the external regulator.

The rhythm and level of MI were not found to be quite as reproducible as were the data for cells in S phase. Mitoses seemed to be more influenced by external conditions (Møller et al., 1974; Møller, 1978; Møller et al., 1979). The maximal level of MI varied from 0.8 to 2.0%, usually with a tendency for two peaks to be found during the beginning of the light period (about 1100-1500 hr) (Brown and Berry, 1968). For the mitotic phase no direct relationship between mitotic index, influx, and mean transit time was found (Møller et al., 1979).

The main proportion of cells (approximately 90%) are found in the $G_1$ phase as for other slowly growing epithelia in both experimental animals and in the human epidermis (Clausen et al., 1976; Frentz et al., 1980). Long mean transit times (days) for $G_1$ cells are responsible for this proportion (Iversen et al., 1968; Hopper and Brockwell, 1978).

The origin of the circadian wave seems to be in $G_1$, as flux from $G_1$ to S is mainly responsible for the partial synchrony of cells within the cell cycle (Møller and Keiding, 1982) (see also Chap. 11). As a consequence of S-phase influx and DNA synthesis rate being nearly in phase, S efflux varies more than S influx, increasing the synchrony of cells leaving S phase. A new cohort of cells is each 24 hr released from $G_1$ and passes through S, $G_2$, and M (Iversen et al., 1968; Møller and Larsen, 1978) (see also Chap. 11).

### Increased Cytokinetic Effect of Methotrexate Combined with Decreased Host Toxicity Using Chronotherapy

Since the cell cycle transverse in the spontaneous partially synchronized epithelium of the hamster cheek pouch seemed fairly well charted, it seemed reasonable to test one of the CCSS agents in an in vivo system without the problems associated with inducing synchrony.

Methotrexate (MTX) is a cell cycle phase-specific agent widely used in the treatment of a number of human malignancies. The toxicity of MTX is due to an inhibition of the enzyme dihydrofolate reductase. Inhibition of this enzyme leads to decreased tetrahydrofolate formation and subsequent inhibition of thymidylate (TMP) and purine biosynthesis (Fig. 7). Cells undergoing DNA synthesis are therefore susceptible to MTX. In clinical medicine MTX is normally regarded as an S phase-specific drug (Ernst and Killmann, 1971; Lampkin et al., 1971), although possibilities for MTX affecting also $G_1$ cells have been discussed (Wantzin et al., 1976). In studies in vitro using high doses of MTX, an inhibitory effect on $G_1$ purine synthesis has been demonstrated in addition to the S-phase inhibitory effect (Borsa and Whitmore, 1969; Taylor and Tattersall, 1981).

To supplement our cell kinetic studies of MTX toxicity, measurements of MTX clearance in plasma and tissue were performed together with survival experiments. In all the experiments bolus injections of MTX, both in lethal (10 g/m$^2$) and nonlethal

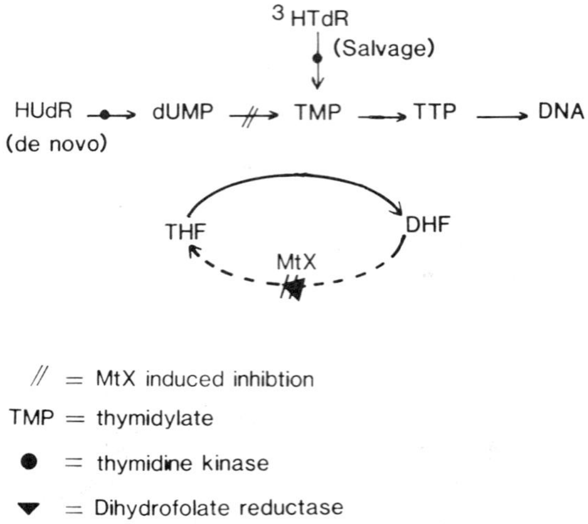


**FIGURE 7** Methotrexate (MTX) disturbs de novo biosynthesis of both thymidylate and purines and thereby interferes with DNA, RNA, and protein synthesis. MTX acts by blocking dehydrofolate, preventing the formation of tetrahydrofolate and thereby limiting the conversion of dUMP to TMP. Thymidine kinase is responsible for the phosphorylation of both deoxyuridine and thymidine.

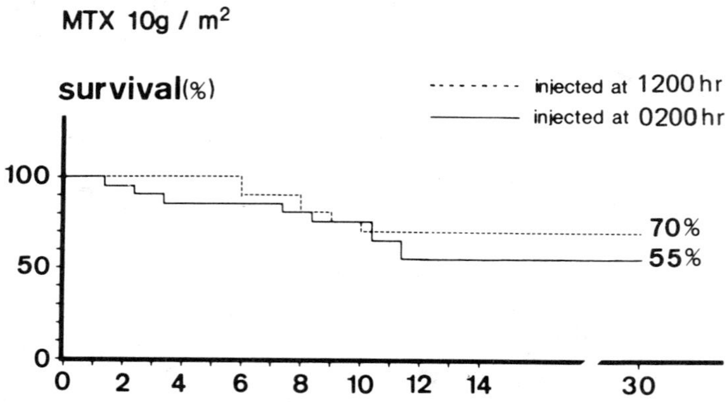


**FIGURE 8** In the survival experiments MTX (10 $g/m^2$) was injected either in the light period (1200 hr) or in the dark period (0200 hr) when the animals are motor active. The bolus injection at 0200 hr tended to be the more toxic, but only when external conditions were interacting did this tendency become significant. (From Møller and Keiding, 1983.)

**TABLE 2** MTX Plasma Elimination Data[a]

| Group | Mtx bolus injection (hr) | Dose ($g/m^2$) | Second half-life (hr) | Terminal half-life (hr) | $C_0$ ($10^{-3}$ M) | $C_{24}$ ($10^{-6}$ M) |
|---|---|---|---|---|---|---|
| Hamster | 02.00 | 10 | 1.3 ± 0.1 | 12.6 ± 3.4 | 9.7 ± 2.8 | 1.3 ± 0.3 |
| | 12.00 | | 1.8 ± 0.2 | 17.3 ± 13.9 | 9.6 ± 4.3 | 4.9 ± 1.9 |
| Human data | | | | | | |
| Isacoff et al. | | 5.4 | 2.0 ± 0.1 | 10.0 ± 1.9 | 2.6 ± 0.6 | 1.3 |
| (1977) | | 7.2 | 20.0 ± 0.2 | 7.3 ± 0.8 | 2.8 ± 0.7 | 2.5 |
| Pratt et al. | | 10 | | | | 2.6 ± 1.6 |
| (1980) | | 15 | | | | 2.5 ± 0.9 |

[a]Results of MTX plasma elimination obtained for hamsters after i.p. injection were comparable to what is reported for humans. $C_0$, extrapolated initial concentration; $C_{24}$, estimated concentration 24 hr after injection. The MTX elimination for hamsters tended to be faster when the bolus injection was given at 0200 hr.
*Source*: Møller and Keiding (1983).

($2\ g/m^2$) dosage were used as it was intended that the problems of high-dose MTX treatment as used in therapy should be mimicked as closely as possible. The maximal tolerated dose of MTX in humans varies from 0.08 to 0.9 $g/m^2$ surface area, when rescue therapy is not administered (Hansen et al., 1971), and seems in fair accordance with the tolerated dose of 2 $g/m^2$ for the hamster, where the risk for renal failure is decreased because of high urinary pH (8.5) (Møller and Keiding, 1983). When the human MTX dosage is 8-15 $g/m^2$, rescue therapy must be given (Vendrik et al., 1981).

In the survival experiments when we studied a possible diurnal effect of MTX, the drug was either injected in the light (1200 hr) or in the dark period (0200 hr). As seen in Fig. 8, about 60% survived 12 days after a bolus injection of 10 $g/m^2$. The bolus injection at 0200 hr tended to be the more toxic. The tendency to decreased survival when MTX was given at 0200 hr was supported also by other experiments, but only when external conditions were interacting was this tendency significant (Møller and Keiding, 1983).

The rate of MTX elimination after intraperitoneal injection in hamsters was comparable with that reported for humans (Table 2; Fig. 10) (Møller and Keiding, 1983). The MTX elimination for hamsters tended to be faster when the bolus injection was given at 0200 hr, in agreement with an assumed higher glomerular filtration rate accompanying high motor activity during the dark period (Liegler et al., 1969).

In the cell kinetic studies of the MTX-induced alterations in the partially synchronized epithelium of the cheek pouch, the bolus injections were given at time periods with either minimal (1200 hr) or maximal number of cells in S phase (0200 hr) (Figs. 3 and 4). These studies were mainly assessed by FCM; however, in order to strenghten the interpretation of the flow data some autoradiographic studies on histological specimens were also performed. For the autoradiographic studies two tracers were used: [$^3$H] TdR was used as precursor for salvage thymidylate synthesis, and [$^3$H] UdR as tracer for de novo thymidylate synthesis (Fig. 7).

In the cell kinetic experiments the bolus injections of MTX were found to decrease cell cycle progression in the cheek pouch epithelium for at least 48 hr, mainly by impairing the S-phase influx. Both the FCM and the autoradiographic analysis indicated that the

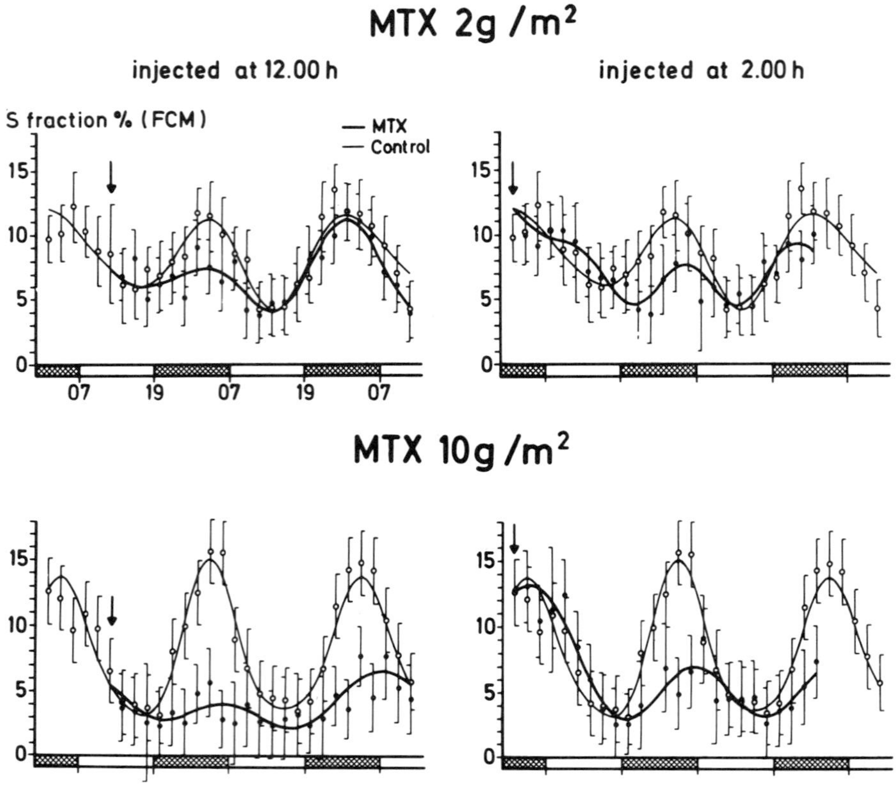


**FIGURE 9** Chronobiological studies on the hamster cheek pouch epithelium. Results of the effect of a bolus injection of MTX on cells in S phase, assessed by flowcytometry. Two injection times (1200 hr and 0800 hr) and two doses (2 g/m$^2$ and 10 g/m$^2$) were used. The circadian-dependent effect of the MTX injection concerned both differences in the time passing before the decrease in S fraction could be detected (10 hr versus 20 hr) and the degree of inhibitory effect of MTX (Table 3). The decrease in S fraction was most pronounced by MTX injection at the time with minimal number of cells in S phase (1200 hr). (From Møller et al., 1984b.)

MTX injection led to a decrease in the number of cells in S phase; the decrease was dependent both on dose and time of injection (Figs. 9 and 10) (Møller et al., 1984a,b). In Table 3 the cytokinetic effect of MTX is expressed as the cumulative difference between the size of the S fractions for controls and treated animals—equal to the area between curves depicting the values of controls and treated animals. By increasing the dose of MTX by a factor of 5, the inhibitory effect, as expressed by CHD over a 48 hr period, increased by an insignificant factor of 1.5 when the injection was given at 0200 hr, but by a highly significant factor of 3.7 when the injection was given at 1200 hr ($p < 0.0001$).

The diurnally dependent effect of the MTX injection concerns both differences in the time before a decrease in S fraction could be detected (10 versus 20 hr) and the degree of inhibitory effect of MTX (Table 3). The decrease in S fraction after MTX injection was most pronounced at the time with minimal number of cells in S phase (1200 hr). Injection of 10 g/m$^2$ of MTX at 1200 hr yielded an inhibitory effect that was larger by a factor of 2.4 ($p < 0.0001$) than that obtained after injection at 0200 hr. At the lower dose the CHD for the two injection times were equal.

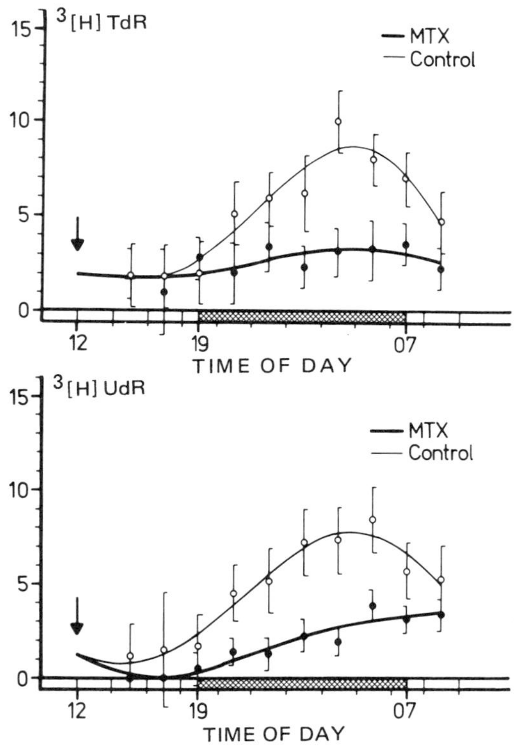


**FIGURE 10** Results from autoradiographic studies on the hamster cheek pouch epithelium using tracers for both salvage ([$^3$H] TdR) and de novo ([$^3$H] UdR) thymidylate synthesis. The decrease in [$^3$H] TdR-LI after MTX injection was similar to the decrease in S fraction found by flow cytometry (Fig. 9), and was not due to a decrease in the thymidylate synthesis rate via salvage pathway (Table 4), but to a decreased S-phase influx. The de novo pathway ([$^3$H] UdR), however, was completely blocked during the first 10 hr after MTX injection. (From Møller et al., 1984a.)

**TABLE 3** S Fraction: Cumulative Hourly Decrease Induced by MTX[a]

| | Methotrexate | |
|---|---|---|
| Injected (hr) | 2 g/m$^2$ | 10 g/m$^2$ |
| 12.00 | 58 ± 16 | 214 ± 18 |
| 0200 | 60 ± 16 | 91 ± 18 |

[a]The cytokinetic effect of MTX is expressed as the cumulative difference between the S fractions for control and treated animals. Since measurements were taken every second hour, multiplication of the cumulative difference by 2 yields the cumulative hourly decrease (CHD), equal to the area between curves spanned by the values of control and treated animals.

*Source*: Møller et al. (1984b).

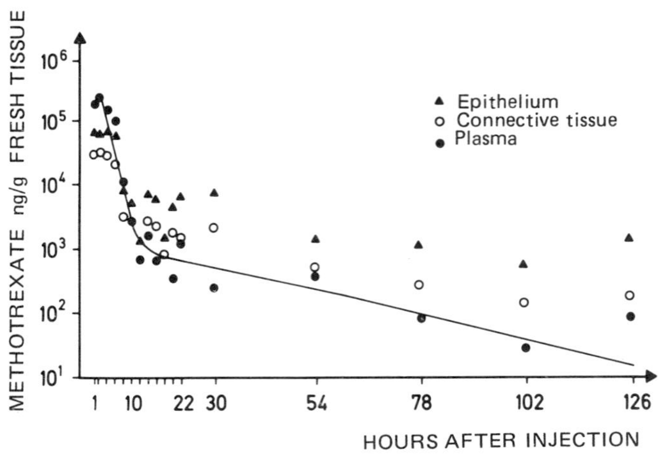


**FIGURE 11** Clearance of MTX in plasma and in the hamster cheek pouch epithelium and corresponding subcutaneous tissue after an i.p. injection of MTX (2 g/m$^2$). Ten hours after MTX injection the highest concentration is found in the epithelium of the cheek pouch. A diphasic elimination curve was fitted to the plasma data. (From Møller et al., 1980; Møller and Keiding, 1983.)

Independent of the time of MTX injection, and in both 24 hr spans, the MTX-induced decrease in the number of cells in S phase was detected during an interval with—for unperturbed conditions—increasing diurnal flux from $G_1$ to S phase (Fig. 5).

Preliminary experiments using lower MTX doses (factor of 10) showed no effect on the DNA histograms (Møller et al., 1980). A high initial plasma concentration of MTX seems to be necessary to provoke a cytokinetic effect. The MTX plasma concentration found, for instance, 10 hr after MTX injection (Fig. 11; Table 2) (Møller and Keiding, 1983), is not assumed to be high enough to provoke a kinetic reaction measurable by FCM. Damage to the cells that at the time of MTX injection were in $G_1$ seems to be mainly responsible for the impaired S-phase influx. Because the $G_1$ mean transit time is long, the decrease in S fraction in the second 24 hr span is not influenced by cells re-entering the S phase. From the results of the histological study it was concluded that neither cell death nor reduced transit time in S phase could be a major cause of the MTX-induced decrease in S fraction.

From the histological study (MTX, 2 g/m$^2$) it was also concluded that for cells in S phase at the time of MTX injection, and for the reduced number of cells entering S phase after MTX injection, the salvage pathway for thymidylate synthesis was unaffected (grains per nucleus, [$^3$H] TdR) (Figs. 7 and 10; Table 4) (Møller et al., 1984a,b). The de novo pathway, however, was completely blocked during the first 10 hr after MTX injection ([$^3$H] UdR) (Fig. 10). During the rest of the experiment the de novo synthesis was partially blocked, i.e., a reduced number of cells were synthesized at a reduced rate (Table 4).

For the MTX injection at the time with minimal number of cells in S phase the flow cytometrically measured decrease in S fractions seemed sufficiently explained by a decreased S-phase influx, assuming that the thymidylate synthesis by the salvage pathway was able to compensate for the blocked de novo synthesis in the relative few S-phase cells synthesizing DNA at a lower rate (Fig. 6a).

**TABLE 4** Grains per Labeled Cell After MTX (2 g/m²) Injected at 1200 hr[a]

| Tracer injected (hr) | Tracer [³H]TdR Average grain count | | Tracer [³H]UdR Average grain count | |
|---|---|---|---|---|
| | Control | MTX | Control | MTX |
| 19 | 17.7 | 23.0 | 8.4 | 6.3 |
| 21 | 22.4 | 25.0 | 12.4 | 6.5 |
| 23 | 32.0 | 28.0 | 15.4 | 7.0 |
| 03 | 27.0 | 28.7 | 17.7 | 7.8 |
| 07 | 37.2 | 36.3 | 18.3 | 8.7 |
| 09 | 21.5 | 27.4 | 19.8 | 10.0 |

[a]Results from an autoradiographic study on the hamster cheek pouch epithelium using tracers for both salvage ([³H]TdR) and de novo ([³H]UdR) thymidylate synthesis. From the average grain counts it can be seen that in general no significant difference in incorporation of [³H]TdR was found between controls and cells exposed to MTX. When the partial recovery of [³H]UdR incorporation into DNA was appearant (Fig. 10), significantly lower average grain counts were found in MTX-exposed cells, in general about half the values of those found for the controls. The less effective phosphorylation of deoxyuridine by thymidine kinase was not significantly reflected in the LI curve, as the control curves obtained by [³H]TdR and [³H]UdR were practically identical (Fig. 9). However, the number of grains per nucleus was reduced when [³H]UdR was used as tracer.

However, when the bolus injection was given at an interval with maximal number of cells in S phase—and high DNA synthesis rate and short S-phase mean transit times (Fig. 6a and b)—two MTX-induced effects were registered: an initial effect observed on cells in S phase immediately after MTX injection, and an effect observed 20 hr later and traced to cells that were in $G_1$ at the time of MTX injection. Regarding the first effect, an increase in S fraction during the interval 8-12 hr after injection was seen for both MTX doses (Fig. 9). The most likely explanation for this increase is a lengthening of the mean transit time for cells in S phase at the time of MTX injection, giving a decreased S-phase efflux (Torres et al., 1980). This decreased S-phase efflux amplifies the circadian-dependent decrease in S-phase efflux (Fig. 5). Regarding the second effect, a decrease in S fraction was observed 20 hr after MTX injection due to a reduced S-phase influx, the reduced influx is regarded as the main effect.

The partial block at the $G_1$/S border could in the FCM measurements be registered also as an increased number of cells in $G_1$ at intervals corresponding to the decrease in S fraction. Whether, in what way, or at what speed, the blocked $G_1$ cells continued in the cell cycle cannot be explained by the present experiments. No signs of cell death were registered during the first 48 hr after injection, nor was the proportion between basal and differentiating cells altered.

The MTX-induced effects on cells in $G_2$M and M were in principle assumed to be indirect effects of the reduced S-phase influx and prolonged S transit.

## CONCLUSION

Our cell kinetic and toxicological studies on methotrexate effects raise questions concerning both the complex action of MTX on cellular proliferation and the circadian optimization of both cytokinetic and toxic effects (Møller and Keiding, 1983; Møller et al., 1984a, b).

The phase specificity of MTX was studied in a spontaneously synchronized in vivo system, thus avoiding problems of induced synchrony. The study was done on nontumor cells, as it was regarded essential to have a reproducible cell system with well characterized cell kinetic parameters. However, as all cancer cells are sufficiently similar to some normal cells, the basic information obtained by these studies should also be relevant for cancer therapy (Pardee, 1982). Most effective antineoplastic drugs take advantage of one of the neoplastic properties—the larger growth fraction of cells in tumors.

The question whether there is any major difference between normal and malignant cells in the proportion of cellular ribonucleotides contributed by the de novo and the salvage pathways has still not been answered, although these pathways are target sites for a number of antimetabolites used in the treatment of cancer—and is of great interest for the understanding of the effectiveness of antitumor drugs (Jackson and Harkrader, 1981).

The main effect of MTX is regarded to be on S-phase cells, since a block of the de novo thymidylate synthesis is the first result from the inhibition of the enzyme dihydrofolate reductase, but in vitro experiments have demonstrated an inhibitory effect also on $G_1$ cells due to disturbances of the purine synthesis (Ernst and Killmann, 1971; Bleyer, 1978; Bertino, 1981; Taylor and Tattersall, 1981; Chamber and Myers, 1982). The cytotoxic effect of MTX seems to be associated with disturbances of both thymidylate and purine biosynthesis (Taylor and Tattersall, 1981; Weizsaecker et al., 1981).

The inhibitory effect on the de novo thymidylate synthesis is usually measured by the incorporation of [$^3$H] UdR into DNA (Ernst and Killmann, 1971; Weinstein and Velasco, 1972; Tew and Taylor, 1977; Sirotnak and Moccio, 1980). The long suppression of the de novo thymidylate synthesis in the tissue studied in our in vivo system is comparable to previously published results, and on this basis the cheek pouch epithelium is very sensitive to MTX.

In the cheek pouch epithelium the main effect of MTX is on $G_1$ cells. When in these in vivo studies the $G_1$-inhibitory effect seems predominant, it must be due to the capability of the epithelial cells to bypass the MTX-induced block of the de novo thymidylate synthesis by using the salvage pathway.

A bypass of the MTX-induced reduction in the de novo thymidylate synthesis by utilizing a salvage supply of thymidine has been demonstrated for skin (DeLapp and Karasek, 1976) and is used in rescue therapy. Some tumor cells (in vivo) also have this capability (Tew and Taylor, 1977).

What seems particularly interesting in our in vivo system is the $G_1$ effect, which is repeated in the second 24 hr period; this effect could also be considered in the context of a need for a drug affecting resting cells in slowly proliferating tumors.

For perturbed cheek pouch epithelium the balance between the de novo and the salvage pathways is not known; the control curves using [$^3$H] TdR or [$^3$H] UdR demonstrate only that the same number of cells can be labeled, independently of the tracer used. For rabbit skin it has been suggested that DNA synthesis is primarily dependent on salvage mechanisms (DeLapp and Karasek, 1976). For the skin disease psoriasis it has recently been suggested that the high sensitivity of psoriatic cells to MTX could be due to a limited capability for utilizing the salvage pathway (McCullough and Weinstein, 1983).

Considerable evidence has, however, also been accumulated to suggest that selective uptake and retention of MTX—perhaps as polyglutamate (Bertino, 1981; Poser et al., 1981)—is decisive for the inhibition of DNA synthesis as measured by incorporation of [$^3$H] UdR (Hill and Montgomery, 1980; Sirotnak and Moccio, 1980). For the cheek pouch epithelium, our MTX clearance experiments indicate a high retention of MTX in epithelial cells (Fig. 11).

The circadian-dependent cytokinetic effect of MTX concerns both differences in the time before a decrease in S fraction could be detected (10 versus 20 hr) and the degree of inhibitory effect of MTX. By carefully selecting the time of injection, the inhibitory effect of MTX could be increased by a highly significant factor of 2.4. Maximal cytokinetic effect was registered when the bolus injection was given during the interval having the maximal number of cells in $G_1$ and minimal number of cells in S phase.

The protracted effect on $G_1$ cells, repeating the decrease in S fraction in the first part of the dark period, did not seem to be influenced by the differences in plasma concentrations 10 and 20 hr after MTX injection. The $G_1$ cells are assumed to have acquired a high intracellular concentration of MTX during the initially high plasma concentration—in agreement with the rationale for administration of high-dose MTX therapy (Bleyer, 1978). Cells dependent on the de novo thymidylate synthesis might be more sensitive to lower MTX plasma concentrations.

The toxic side effects of MTX were evaluated by their influence on survival. The MTX injection during the interval with motor activity (the dark period) tended to be more toxic. Interferences with diurnal variations in the hematopoietic tissue (Lærum and Aardal, 1981), gastrointestinal system (Scheving et al., 1978), and renal function are probably responsible for the increased host toxicity of the bolus injection at 0200 hr (Wesson, 1964).

The cell kinetic and toxicological experiments on hamsters demonstrated the possibility of increasing the cytokinetic effect of methotrexate and of decreasing host toxicity using chronotherapy. Thus, wanted and unwanted effects of the antitumor drug could be partly separated by taking advantage of phase differences between different biorhythms in tumor and normal tissue (Scheving, 1981).

The results of the presented studies on diurnal variations in tumor temperature and $^{32}$P uptake and on the cytokinetic and toxicologic effects of MTX strongly support the idea of chronotherapy. One might hope that the time will soon come when chronotherapy could be seriously employed in the clinic.

## ACKNOWLEDGMENTS

This work was supported by grants from the Danish Medical Research Council, The Danish Cancer Society, Director Jacob Madsens og Hustru Olga Madsens Fond, and Fru Astrid Thaysens Legat for Laegevidenskabelig Grundforskning.

## REFERENCES

Bertino, J. R. (1981). Methotrexate: Molecular pharmacology (Chap. 18); Methotrexate: Clinical pharmacology and therapeutic application (Chap. 22). In *Cancer and Chemotherapy,* vol. III, S. I. Crooke, A. W. Prestayko (eds.). Academic Press, New York, pp. 311-322; 359-375.

Bleyer, W. A. (1978). The clinical pharmacology of methotrexate. New applications of an old drug. *Cancer* 41:36-51.

Bojsen, J., Groth, S., and Rossing, N. (1981). Portable cadmium telluride detectors and their applicability for external measurement of $^{51}$Cr-Edta clearance. *Int. J. Appl. Radiat. Isot.* 32:719-727.

Bojsen, J., Møller, U., Christensen, P., and Lippert, J. (1977). Storage telemetry of radionuclide tracers by implantable thermoluminescent dosimeters. *Am. J. Physiol.* 233: E479-E482.

Bojsen, J., Møller, U., and Faber, M. (1971). Radiotelemetrical equipment for continuous subcutaneous measurements of the circadian temperature rhythm in rats. *Pflügers Arch.* 328:176-184.

Bojsen, J., Wallevik, K., and Møller, U. (1972). An implantable radiotelemetrical measuring device for simultaneous long term measurements of body temperature and turnover of rabbit serum albumin-125-J in unrestrained rabbits. In *Biotelemetry, International Symposium,* H. P. Kimmich and J. Vos (eds.). Meander, Leiden, pp. 337-344.

Borsa, J., and Whitmore, G. F. (1969). Cell killing studies on the mode of action of methotrexate on α-cells in vitro. *Cancer Res.* 29:737-744.

Brown, J. M., and Berry, R. J. (1968). The relationship between diurnal variation of the number of cells in mitosis and of the number of cells synthesizing DNA in the epithelium of the hamster cheek pouch. *Cell Tissue Kinet.* 1:23-29.

Bullen, M. A., Freundlich, H. F., Hale, B. T., Marshall, D. H., and Tudway, R. C. (1963). The activity of malignant tumours and response to therapeutic agents, studied by continuous records of radioactive phosphorus uptake. *Postgrad. Med. J.* 39:265-277.

Calcutt, G., Bullen, M. A., Marshall, D. H., and Godden, T. J. (1967). The continuous countings of phosphorus-32 from transplanted rat tumours and the effects of radiosensitisers and radioprotective agents. *Br. J. Cancer* 21:438-447.

Camplejohn, R. S. (1980). A critical review of the use of vincristine (VRC) as a tumour cell synchronizing agent in cancer therapy. *Cell Tissue Kinet.* 13:327-335.

Chamber, B. A., and Myers, C. E. (1982). Clinical pharmacology of cancer chemotherapy. In *Cancer–Principles and Practice of Oncology.* V. T. DeVita, S. Hellman, and S. A. Rosenberg (eds.) Lippincott, Philadelphia, pp. 156-197.

Christensen, I. J., Møller, U., and Larsen, J. K. (1980). The choice of model for deconvolution of DNA distributions measured by flow cytometry. *Flow Cytometry* IV: 147-151.

Clausen, O. P. F., Lindmo, T., Sandnæs, K., and Thorud, E. (1976). Separation of mouse epidermal basal and differentiating cells for micro-flow fluorometric measurements. *Virchows Arch. B. [Cell Pathol.]* 20:261-271.

Czeisler, C. A., Richardson, G. S., Zimmerman, J. C., Moore-Ede, M. C., and Weitman, E. D. (1981). Entrainment of human circadian rhythms by light-dark cycles: A reassessment. *Photochem. Photobiol.* 34:239-247.

Deka, B., Deka, A. C., Patil, R. B., and Halberg, F. (1981). In-phase synchronization of circadian temperature rhythm of cancerous and healthy mouth tissue. *Int. J. Chronobiol.* 7:226.

DeLapp, N. W., and Karasek, M. A. (1976). Importance of pyrimidine nucleotide salvage pathways for DNA synthesis in skin. *J. Invest. Dermatol.* 66:306-312.

Enright, J. T. (1980). *The Timing of Sleep and Wakefulness.* Springer-Verlag, Berlin.

Ernst, P., and Killmann, S.-A. (1971). Perturbation of generation cycle of human leukemic myeloblasts in vivo by methotrexate. *Blood* 38:689-705.

Focan, C. (1976). Circadian rhythm and chemotherapy for cancer. *Lancet* 2:638.

Focan, C., Barbason, H., and Betz, E. H. (1975). Use of synchronization induced by cyclophosphamide in a methylcholanthrene sarcoma with circadian proliferation to rational sequential chemotherapy. *Biomedicine* 23:230-235.

Frentz, G., and Møller, U. (1981). DNA measurements by single nuclei flow cytometry in human actinic skin lesions. *J. Invest. Dermatol.* 77:431-433.

Frentz, G., Møller, U., and Christensen, I. (1980). DNA flow cytometry for human epidermis. I. Methodological studies on normal skin. *J. Invest. Dermatol.* 74:119-121.

Garcia-Sainz, M., and Halberg, F. (1966). Mitotic rhythms in human cancer, reevaluated by electronic computer programs—evidence for chronopathology. *JNCI* 37:279-292.

Gautherie, M., and Gros, C. (1977). Circadian rhythm alteration of skin temperature in breast cancer. *Chronobiologia* 4:1-17.

Gillespie, P. J., Burrows, B. D., and Edelstyn, G. A. (1971). Possible clinical implications of therapeutically induced temperature changes in continuously monitored tumour mass. *Br. J. Cancer* 25:85-92.

Halberg, F., Haus, E., Cardoso, S. S., Scheving, L. E., Kühl, J. F. W., Shiotsuka, R., Rosene, G., Pauly, J. E., Runge, W., Spalding, J. F., Lee, J. K., and Good, R. A. (1973). Toward a chronotherapy of neoplasia: Tolerance of treatment depends upon host rhythms. *Experientia* 29:909-1044.

Halberg, F., Haus, E., and Scheving, L. E. (1978). Sampling of biologic rhythms, chronocytokinetics and experimental oncology. In *Biomathematics and Cell Kinetics,* A.-J. Valleron and P. D. M. Macdonald (eds.). Elsevier/North-Holland Biomedical Press, Amsterdam, pp. 175-190.

Hale, B. T. (1961). A technique for studying human tumour growth in vivo. *Lancet* 2: 345-347.

Hansen, H. H., Selawry, O. S., Holland, J. F., and McCall, C. B. (1971). The variability of individual tolerance to methotrexate in cancer patients. *Br. J. Cancer* 75:298-305.

Haus, E., Halberg, F., Scheving, L. E., Pauly, J. E., Cardoso, S., Kühl, J. W. F., Sothern, R. B., Shiotsuka, R. N., and Hwang, D. S. (1972). Increased tolerance of leukemic mice to arabinosyl cytosine with schedule adjusted to circadian system. *Science* 177:80-82.

Hill, B. T., and Baserga, R. (1975). The cell cycle and its significance for cancer treatment. *Cancer Treat. Rev.* 2:159-175.

Hill, D. L., and Montgomery, J. A. (1980). Selective uptake and retention of anticancer agents by sensitive cells. *Cancer Chemother. Pharmacol.* 4:221-225.

Hopper, J. L., and Brockwell, P. J. (1978). A stochastic model for cell populations with circadian rhythms. *Cell Tissue Kinet.* 11:205-225.

Hrushesky, W. J. M., Levi, F. A., Halberg, F., and Kennedy, B. J. (1982). Circadian stage dependence of cis-diamminedichloroplatinum lethal toxicity in rats. *Cancer Res.* 42:945-949.

Iversen, O. H., Bjerknes, R., and Devik, F. (1968). Kinetics of cell renewal, cell migration and cell loss in the hairless mouse dorsal epidermis. *Cell Tissue Kinet.* 1:351-367.

Jackson, R. C., and Harkrader, R. J. (1981). The contributions of de novo and salvage pathways of nucleotide biosynthesis in normal and malignant cells. In *Nucleosides and Cancer Treatment. Rational Approaches to Antimetabolite Selectivity and Modulation.* M. H. N. Tattersall and R. M. Fox (eds.). Academic Press, New York, pp. 18-31.

Klein, H. O. (1980). Cell kinetic alterations in normal and neoplastic cell populations in vitro and in vivo following vincristine. *Cell Tissue Kinet.* 13:425-434.

Klein, H. O., and Lennartz, K. J. (1974). Chemotherapy after synchronization of tumor cells. *Semin. Hematol.* 11:203-227.

Kuzel, M., Zinneman, H., Nelson, W., Halberg, H., Rosen, G., Scheving, L. E., and Haus, E. (1978). Competing circadian effects of methylprednisolone and rat weight, light chains, immunocytoma size and survival. *Chronobiologia* 5:295-311.

Lærum, O. D., and Aardal, N. P. (1981). Chronobiological aspects of bone marrow and blood cells. In *Biological Rhythms in Structure and Function.* H. von Mayersbach, L. E. Scheving, and J. E. Pauly (eds.). Alan R. Liss, New York, pp. 87-97.

Lampkin, C. B., Takeshi, N., and Mauer, A. M. (1971). Synchronization and recruitment in acute leukemia. *J. Clin. Invest.* 50:2204-2214.
Levi, F. A., Hrushesky, W. J. M., Blomquist, C. H., Lakatua, D. J., Haus, E., Halberg, F., and Kennedy, B. J. (1982). Reduction of cis-diamminedichloroplatinum nephrotoxicity in rats by optimal circadian drug timing. *Cancer Res.* 42:950-955.
Liegler, D. G., Henderson, E. S., Holm, M. A., and Oliverio, V. T. (1969). The effect of organic acid on renal clearance of methotrexate in man. *Clin. Pharmacol. Ther.* 10: 849-853.
Mansfield, C. M., Carabasi, R. A., Wells, W., and Borman, K. (1973). Circadian rhythm in the skin temperature of normal and cancerous breasts. *Int. J. Chronobiol.* 1:235-243.
Mansfield, C. M., Wallace, J. D., Curley, R. F., Kramer, S., Southard, M. E., and Driscoll, D. (1970). A comparison of the temperature curves recorded over normal and abnormal breasts. *Radiology* 94:697-698.
McCullough, J. L., and Weinstein, G. D. (1983). The action of cytotoxic drugs on cell proliferation in psoriasis. In *Psoriasis: Cell Proliferation.* N. A. Wright and R. S. Camplejohn (eds.). Churchill Livingstone, Edinburgh, pp. 347-354.
Møller, U. (1978). Interaction of external agents with the circadian mitotic rhythm in the epithelium of the hamster cheek pouch. *J. Interdiscipl. Cycle Res.* 9:105-114.
Møller, U., and Bojsen, J. (1974). The circadian temperature rhythm in Syrian hamsters as a function of the number of animals per cage. *J. Interdiscipl. Cycle Res.* 5:61-69.
Møller, U., and Bojsen, J. (1975). Temperature and blood flow measurements in and around 7,12-dimethylbenz(α)anthracene-induced tumors and Walker 256 carcinosarcomas in rats. *Cancer Res.* 35:3116-3121.
Møller, U., and Bojsen, J. (1976). Circadian variations in $^{32}P$ uptake of DMBA-induced mammary tumors and Walker 256 carcinomas in rats. *Br. J. Cancer* 33:646-654.
Møller, U., and Bojsen, J. (1980). Heat transfer and blood flow in experimental tumors in rats compared with the circadian temperature rhythm of the body. *Ann. N.Y. Acad. Sci.* 335:22-34.
Møller, U., Hartmann, N. R., and Faber, M. (1979). Mitotic index, influx and mean transit time in the hamster cheek pouch epithelium, a partially synchronized cell system. *Cell Tissue Kinet.* 12:581-596.
Møller, U., and Keiding, N. (1982). Circadian variations in influx and efflux of the S phase in a partially synchronized cell system. *Cell Tissue Kinet.* 15:341-350.
Møller, U., and Keiding, N. (1983). Circadian dependent toxicity of methotrexate. In *Psoriasis: Cell Proliferation.* N. A. Wright and R. S. Camplejohn (eds.). Churchill Livingstone, Edinburgh, pp. 385-392.
Møller, U., Keiding, N., and Engel, F. (1982). Subclassification of cells in S phase in a partially synchronized cell system. *Cell Tissue Kinet.* 15:157-168.
Møller, U., Keiding, N., and Larsen, J. K. (1984a). The effect of methotrexate upon incorporation of de novo and salvage precursors into DNA of a partially synchronized keratinized epithelium. (Submitted).
Møller, U., and Larsen, J. K. (1978). The circadian variations in the epithelial growth of the hamster cheek pouch: Quantitative analysis of DNA distributions. *Cell Tissue Kinet.* 11:405-413.
Møller, U., and Larsen, J. K. (1979). DNA-flowcytometry of isolated keratinized epithelia: A methodological study based on ultrasonic tissue disaggregation. *Cell Tissue Kinet.* 12:203-211.
Møller, U., Larsen, J. K., and Faber, M. (1974). The influence of injected tritiated thymidine on the mitotic circadian rhythm in the epithelium of the hamster cheek pouch. *Cell Tissue Kinet.* 7:231-239.

Møller, U., Larsen, J. K., Hendel, J., and Christensen, I. J. (1980). The cell kinetic effect of methotrexate on the epithelium of the hamster cheek pouch, a circadian dependent effect. *Flow Cytometry* IV:372-377.

Møller, U., Larsen, J. K., Keiding, N., and Christensen, I. J. (1984b). Chronotherapeutically increased cytokinetic effect of methotrexate in a keratinized epithelium assessed by flow cytometry. (Submitted).

Nash, R. E., and Echave Llanos, M. (1971). Twenty-four-hour variations in DNA synthesis of a fast-growing and a slow-growing hepatoma: DNA synthesis rhythm in hepatoma. *JNCI* 47:1007-1012.

Nicolini, C. (1976). The principles and methods of cell synchronization in cancer chemotherapy. *Biochim. Biophys. Acta* 458:243-282.

Nome, O. (1974). Tissue specificity of the epidermal chalones. M.D. Thesis. Oslo University, Norway.

Pardee, A. B. (1982). Principles of cancer biology: Cell biology and biochemistry of cancer. In *Cancer–Principles and Practice of Oncology.* V. T. DeVita, S. Hellman, and S. A. Rosenberg (eds.). Lippincott, Philadelphia, pp. 59-69.

Poser, R. G., Sirotnak, F. M., and Chello, P. L. (1981). Differential synthesis of methotrexate polyglutamates in normal proliferative and neoplastic mouse tissues in vivo. *Cancer Res.* 41:4441-4446.

van Putten, L. M. (1974). Are cell kinetic data relevant for the design of tumor chemotherapy schedules? *Cell Tissue Kinet.* 7:493-504.

van Putten, L. M., Keizer, H. J., and Mulder, J. H. (1976). Perspectives in cancer research. Synchronization in tumour chemotherapy. *Eur. J. Cancer* 12:79-85.

Rütger, A., and Wever, R. A. (1980). Phase shifts of human circadian rhythms due to shifts of artificial zeitgebers. *Chronobiologia* 7:303-317.

Scheving, L. E. (1981). Circadian rhythms in cell proliferation: Their importance when investigating the basic mechanism of normal versus abnormal growth. In *Eleventh International Congress of Anatomy: Biological Rhythms in Structure and Function.* H. von Mayersbach, L. E. Scheving and J. E. Pauly (eds.). Alan R. Liss, New York, pp. 39-79.

Scheving, L. E., Burns, E. R., Pauly, J. E., and Tsai, T. H. (1978). Circadian variation in cell division of the mouse alimentary tract, bone marrow and corneal epithelium. *Anat. Rec.* 191:479-486.

Simpson, H. W., and Stoney, P. J. (1977). A circadian variation of melphalan (L-phenylalanine nitrogen mustard) toxicity to murine bone marrow: Relevance to cancer treatment protocols. *Br. J. Haematol.* 35:459-464.

Singh, B., Vadstrup, S., Pearce, G., Bunce, R. W., and Bojsen, J. (1981). A mercuric iodide detector unit implantable and externally powered for use in radionuclide tracer studies in small animals. *Biotelem. Patient Monit.* 8:204-212.

Sirotnak, F. M., and Moccio, D. M. (1980). Pharmacokinetic basis for differences in methotrexate sensitivity of normal proliferative tissues in the mouse. *Cancer Res.* 40:1230-1234.

Stoll, B. A. (1971). The thermoprofile as an early indicator of breast cancer response to hormonal therapy. *Cancer* 27:1379-1383.

Stoll, B. A., and Burch, W. M. (1968). Surface detection of circadian rhythm in $^{32}P$ content of cancer of the breast. *Cancer* 21:193-196.

Søndergaard, K., Larsen, J. K., Møller, U., Christensen, I. J., and Hou-Jensen, K. (1983). DNA ploidy-characteristics of human malignant melanoma analysed by flow cytometry and compared with histology and clinical course. *Virchows Arch. [Cell Pathol.]* 42:43-52.

Tähti, E. (1956). Studies of the effect of X-radiation on 24-hour variations in the mitotic activity in human malignant tumours. *Acta Pathol. Microbiol. Scand. (Suppl.)* 117: 1-61.

Taylor, D. M., Parker, R. P., Field, E. O., and Greatorex, C. A. (1968). An interpretation of the results of measurements of the uptake of $^{32}P$ in human tumours. *Br. J. Radiol.* 41:432-439.

Taylor, I. W., and Tattersall, M. N. H. (1981). Methotrexate cytotoxicity in cultured human leukemic cells studied by flow cytometry. *Cancer Res.* 41:1549-1558.

Tew, K. D., and Taylor, D. M. (1977). The effect of methotrexate on the uptake of de novo and salvage precursors into the DNA of rat tumours and normal tissues. *Eur. J. Cancer* 13:279-289.

Torres, J., Klein, R. M., Tung, H.-N., and Chapman, A. L. (1980). Cell cycle kinetics of uninfected and feline leukemia virus-induced canine lymphoma cell lines: Effects of methotrexate treatment. *Virchows Arch. [Cell Pathol.]* 33:139-152.

Vendrik, C. P. J., Pinedo, H. H., Jasmin, C., Sylvester, R., and Staquet, H. (1981). Phase II study of low dose methotrexate in advanced osteosarcoma, followed by escalation if progression occurs. EORTC Protocol number 62782.

Wantzin, G. L., Karle, H., and Killmann, S.-A. (1976). Nuclear DNA-polymerase estimation in human leukaemic myeloblasts. *Br. J. Haematol.* 33:329-334.

Weinstein, G. D., and Velasco, J. (1972). Selective action of methotrexate on psoriatic epidermal cells. *J. Invest. Dermatol.* 59:121-127.

Weizsaecker, H., Hoshino, T., Dennis, F. D., and Kobayashi, S. (1981). Cytokinetic and cytotoxic effects of antimetabolites on 9 L rat brain tumor cells in vitro. *Cancer Res.* 41:3835-3839.

Wesson, L. G. (1964). Electrolyte excretion in relation to diurnal cycles of renal function. Plasma electrolyte concentrations and aldosterone secretion before and during salt and water balance changes in normotensive subjects. *Medicine (Baltimore)* 43:547-588.

Wever, R. A. (1979). The circadian system of man. Results of experiments under temporal isolation. Springer Verlag, New York.

Woolley-Hart, A., Twentyman, P., Corfield, J., Joslin, C., Morrison, R., and Fowler, J. F. (1968). Changes in $^{32}P$ counting rate in human and animal tumours. *Br. J. Radiol.* 41:440-447.

# PART VI
# CELL CYCLE CLOCKS IN DEVELOPMENT AND AGING

It seems somehow most appropriate that in the last part of this book we deal briefly with the role of cell cycle clocks in development and in aging—the $\alpha$ and the $\Omega$ of biological existence. In his state-of-the-art overview, J. M. Mitchison (Chap. 9) noted that one line of evidence supporting the existence of "independent-timer" periodic controls, which continue to function even if growth or division is blocked or perturbed, is afforded by the development of eggs and early embryos. As he states, these systems are particularly interesting in that "they break many of the normal rules of the cell cycle." Noriyuki Satoh continues this theme in Chap. 25, reviewing the evidence that cell division cycles (CDS) serve as the basis for timing mechanisms in early embryonic development of animals. He concludes that the developmental clock that determines the time of cellular differentiation is closely associated with the DNA replication cycle and appears to be independent of the cytoplasmic clock, which times early morphogenetic events. Similarly, B. Wolfanger Belisle et al. (Chap. 26) give experimental evidence implicating a central timer mechanism in the regulation of the CDC during the development of sea urchin embryos. These authors propose a "sequential interactive network" (SIN) model, which couples (by gene action) a semiautonomous chromosome cycle with an independent metabolic cycle that would regulate the former.

In the final contribution to this volume, Glenn A. Zorn and Bonnie Smith (Chap. 27) consider cell cycle clocks and aging: Is there a life cycle clock? The viewpoint that aging can be viewed as the loss of temporal organization has been treated comprehensively by Samis and Capobianco (1978); but what does this mean at the level of the cell? Theories of cellular aging can be divided into two general categories: the genetic "program" theories, in which aging would be an event caused by the active expression of specific "aging genes" or "cell clock longevity genes," or by the passive exhaustion of vital genetic information; and cumulative error, "wear-and-tear" theories, in which senescence would result from the progressive damage to organelles or to the accumulating error in cytoplasmic, information-carrying molecules (Finch and Hayflick, 1977). Ever since the report of Hayflick and Moorehead (1961) that normal human embryonic fibroblasts cultured in vitro were limited in their capacity to replicate (50 ± 10 population doublings), there has been considerable controversy as to whether the so-called phase III phenomenon (the period of greatly reduced replicative capacity) reflects cellular aging in vivo, and therefore whether the study of in vitro cellular senescence has any relevance to the normal mammalian life span. Many investigators feel, nevertheless, that the biological changes preceding phase III in vitro and mortality in vivo may be at least indirectly responsible for both phenomena.

Zorn and Smith are among this group and report recent experiments with reconstructed cells (in which "old" fibroblast nuclei, or karyoplasts, were fused with "young" cytoplasms, or cytoplasts, and vice versa) designed to determine the relative contributions of the nucleus and cytoplasm to the phase III phenomenon. Their results, supporting a predominantly nuclear role in cellular senescence, provide evidence that the expression of the phase III "clock" is encoded into the genetic material, presumably on one or more chromosomes. If so, the authors note that it should be feasible to transfer this "ticking" clock from an old cell to another younger one that has gone through fewer population doublings (as they did with whole-cell hybrids) and then identify by karyotype analysis those chromosomes responsible for phase III—i.e., the senescent clock could be localized. Finally, they note the intriguing possibility that cellular aging and cellular transformation might map on the same chromosome(s) and be related by differential expression of the same genes.

## REFERENCES

Finch, C. E., and Hayflick, L. (eds.) (1977). *Handbook of the Biology of Aging.* Van Nostrand-Reinhold, Princeton, New Jersey.

Hayflick, L., and Moorhead, P. S. (1961). The serial cultivation of human diploid cell strains. *Exp. Cell Res.* 25:585-621.

Samis, H. V., Jr., and Capobianco, S. (eds.) (1978). *Aging and Biological Rhythms.* Plenum, New York.

# 25

# Cell Division Cycles as the Basis for Timing Mechanisms in Early Embryonic Development of Animals

**Noriyuki Satoh** / Kyoto University, Kyoto, Japan

## INTRODUCTION

Since embryological development is a dynamic process that takes place in four dimensions, time is an indispensable element in any interpretation of development. Not only the precision of the absolute timing of events during development but also the synchrony observed between embryos growing under specific environmental conditions gives the impression that an embryo is intrinsically provided with time-measuring systems within it. Because an alteration in the timing of elementary processes of development brings about various anomalies in the following differentiation and morphogenesis (Freeman, 1976; Niki and Okada, 1981), attention has been drawn recently to the timing mechanisms (Wilde, 1974; Soll, 1979; Snow and Tam, 1980; Satoh, 1982). Nevertheless, we still have little understanding of their mechanisms. One of the reasons it is difficult to study timing mechanisms is that a decision regarding the timing of a certain event usually takes place several hours (or days) before the event actually occurs. However, evidence is accumulating to suggest that events of cell division cycles are closely associated with the timing of emergence of many developmental processes. In this chapter I will review the available information in this field, along the line of my recent proposal that the clock mechanism that determines the time of initiation of cellular differentiation may be independent of that for the timing of morphogenesis.

A "developmental clock" is a designation of a possible system within the embryo, which measures the time flow and determines the timing of initiation of developmental events. The developmental clock is not a clock of the type envisaged in the work on circadian rhythms, because the developmental clock has a dependency on temperature (Fig. 1). In developmental biological terms little is known about the driving force or the homeostatic mechanism, the components of clock system. Studies have been focused on the underlying, regulating oscillator of the clock and the overt, measurable rhythms driven by the oscillator.

Evidence has been accumulated to show that the timing of various events during early embryonic development is not governed by chronological time or elapsed time since fertilization but is more plausibly related to the number of cell division cycles, as shown later. Therefore, the oscillatory mechanisms are most likely to be ascribed to events in the cell division cycle, i.e., cytokinesis, nuclear division, or DNA replication. In addition, recent studies have revealed the presence of cytoplasmic clocks and their significance for the timing of morphogenetic events.

## DNA REPLICATION CYCLES AS A CLOCK MECHANISM FOR CELLULAR DIFFERENTIATION

It is generally accepted that the differentiated state of a given type of cells is associated with the activity of a particular set of genes and that the establishment of a differentiated state is characterized by the appearance of tissue-specific enzymes. In normal development cellular differentiation takes place coincidentally with morphogenesis. However, there is considerable evidence to show that the process of differentiation is not necessarily coupled with that of morphogenesis (Whittaker, 1973; Spooner et al., 1977), suggesting that the developmental clock for cellular differentiation may be independent of that for morphogenesis. The nucleus is generally regarded as the control center of cellular function and has been implicated in the timing of various developmental events. Results of our recent studies on a clock mechanism for histospecific enzyme development during ascidian embryogenesis strongly support the concept that the cycle of DNA replication may be the clock mechanism that determines the time when specific gene activations occur; and these results will be described below.

### Acetylcholinesterase Development in Ascidian Embryos

Acetylcholinesterase (AChE) is a tissue-specific enzyme of the muscle cells of the tail of the developing ascidian embryo; the enzyme activity is found histochemically only in muscle cells and not in the nervous system (Durante, 1956; Meedel and Whittaker, 1979), although Minganti and Falugi (1980) reported an additional occurrence of AChE activity in ectodermic epithelial cell membrane. Distinct AChE activity can first be detected histochemically and biochemically in the presumptive muscle cells of the neurula, and the enzyme activity increases progressively with time of development. Studies with puromycin and actinomycin D suggest that the histospecific protein is synthesized at the time it is first detected and that AChE development requires new RNA synthesis, which begins between the early and late gastrula stage (Whittaker, 1973; Meedel and Whittaker, 1979; Satoh, 1979).

AChE development is dependent on temperature and is associated more closely with developmental age than with chronological age or elapsed time since fertilization. As shown in Fig. 1, when embryos are raised slowly at a low temperature, the first appearance of AChE is delayed. When embryos are allowed to develop rapidly at a high temperature, the first AChE occurrence is hastened. But in either case AChE activity is first detected in the presumptive muscle cells of the neurula. The number of muscle cells in embryos developed at lower temperatures, however, is the same as that of embryos raised at higher temperature; the average numbers of muscle cells in the tailbud-stage embryo

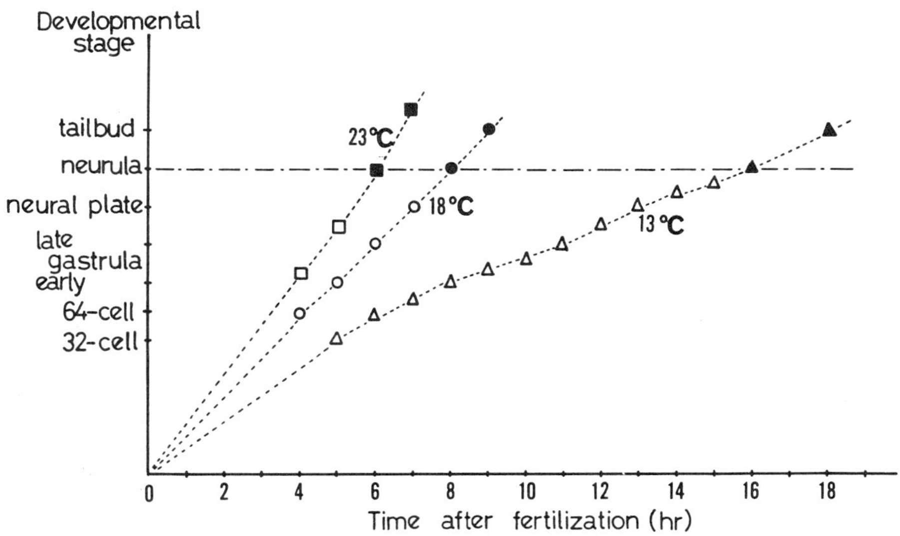


**FIGURE 1** The time of first AChE development in *Ciona intestinalis* embryos raised at 23°C (squares), 18°C (circles), and 13°C (triangles), respectively. Open symbols show embryos with no histochemically detectable AChE activity, and solid symbols those with AChE activity. Development of AChE is dependent on temperature, but AChE activity is first detected in the presumptive muscle cells of the neurula, irrespective of time elapsed after fertilization.

developed at 13, 18, and 23°C are 35.8 ± 1.4 (mean ± standard deviation, n = 10), 36.4 ± 1.8 (n = 10), and 35.6 ± 1.6 (n = 8), respectively.

As successfully discovered by Whittaker (1973), if cleavage-stage embryos are permanently prevented from undergoing further cytokinesis by treatment with cytochalasin B and histochemical reactions for AChE are carried out when normal embryos reach the middle tailbud stage, these cleavage-arrested embryos are found to be able to produce this histospecific enzyme. The time of first AChE occurrence in embryos that have been arrested in the 32-cell stage with cytochalasin is about the same as in normal embryos (Whittaker, 1973; Satoh, 1979). This result clearly indicates that the clock mechanism that counts the time when the enzyme first develops is not regulated by the events of cytokinesis. Cytochalasin, however, permits nuclear divisions; the nucleus in the cells of cytochalasin-arrested embryos divides in good synchrony with that of normal embryos (Satoh and Ikegami, 1981b).

Embryos that have been continuously arrested at the 32-cell stage with colchicine or with colcemid also develop AChE activity at nearly the same time as do normal control embryos (Whittaker, 1973; Satoh and Ikegami, 1981b). Colchicine blocks the formation of mitotic apparatus, and then normal nuclear divisions do not occur in colchicine-arrested embryos. Therefore, the clock does not seem to be controlled by the mitotic cycle of the nucleus, as originally pointed out by Whittaker (1973). The cells of colchicine-arrested embryos, however, show repeated cycles of nuclear envelope breakdown and reformation virtually in parallel with the mitotic cycle of normal embryos (Satoh and Ikegami, 1981b). In addition, the cells of colchicine-arrested embryos incorporate labeled thymidine, and its uptake increases with developmental time in a fashion similar to that of normal embryos (Satoh and Ikegami, 1981b).

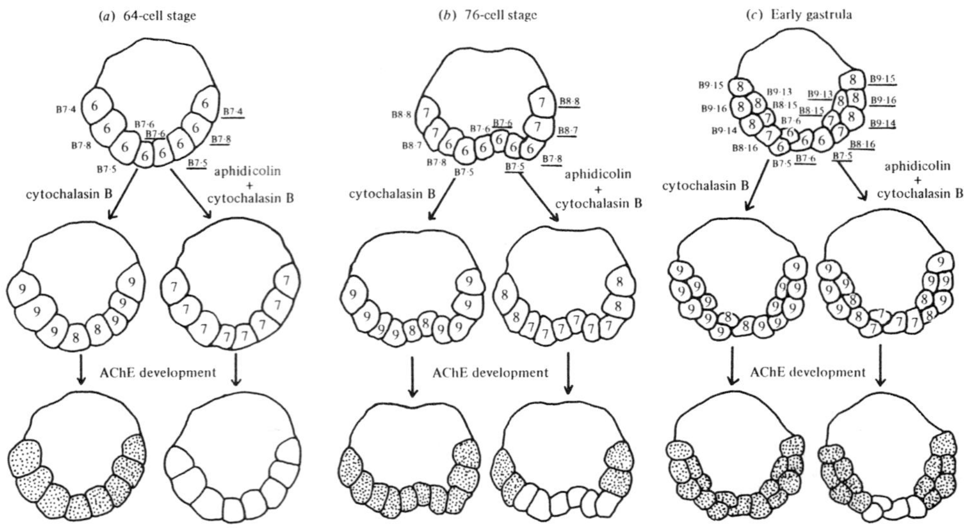


**FIGURE 2** Illustration demonstrating the relationship between the number of DNA replication cycles and AChE development in the presumptive muscle cells at three embryonic stages of *Halocynthia roretzi* embryos. Embryos were permanently cleavage arrested with aphidicolin and cytochalasin, and histochemical reactions for AChE were carried out at the time of middle tailbud stage of normal control embryos. Blastomeres differentiating AChE are dotted. The numerals within the blastomeres indicate the supposed number of DNA replications. Cytochalasin B prevents cytokinesis but not nuclear division or DNA synthesis. Aphidicolin is a specific inhibitor of DNA synthesis, which blocks ascidian development but allows one more division in eggs kept in aphidicolin and stops division thereafter, no matter how soon after a cleavage it is applied. See text for detail. (From Satoh and Ikegami, 1981a.)

Aphidicolin, a specific inhibitor of DNA polymerase α, blocks DNA synthesis but does not inhibit synthesis of RNA and protein in ascidian embryos (Satoh and Ikegami, 1981a). Figure 2 summarizes the results of experiments with aphidicolin on AChE development (Satoh and Ikegami, 1981a). The 64-cell stage embryo contains eight presumptive muscle cells, all of which are at the seventh generation, counting the unsegmented egg as the first generation. If the 64-cell stage embryos are arrested with cytochalasin B, the cleavage-arrested embryos can develop AChE in all eight presumptive muscle cells. However, embryos at the same stage, which have been prevented from undergoing further divisions by simultaneous treatment with aphidicolin and cytochalasin, do not produce the enzyme activity (Fig. 2a). From the 76-cell stage onward, divisions of the presumptive muscle cells become asynchronous; the 76-cell stage embryos has 10 presumptive muscle cells of two different (seventh and eighth) generations. Cytochalasin-arrested 76-cell stage embryos are able to synthesize AChE in the 10 presumptive muscle cells. Aphidicolin-cytochalasin-arrested 76-cell stage embryos can also synthesize AChE, but the maximum number of AChE-containing cells is four (Fig. 2b). The relative positions of the blastomeres in these cleavage-arrested embryos remain fixed. Judging from the relative positions of the blastomeres, the AChE-containing cells in aphidicolin-cytochalasin-arrested embryos are always at the eighth generation, but those with no enzyme activity are certainly at the seventh generation. The early gastrula has 16 presumptive muscle cells of three different (seventh, eighth, and ninth) generations. Early gastrulae that have been arrested with cytochalasin produce AChE in all 16 presumptive muscle cells, but the same stage embryos arrested with aphidicolin and cytochalasin produce AChE in as many as 12 cells (Fig. 2c). These AChE-containing cells in aphidicolin-cytochalasin-arrested gastrulae are always at the eighth or ninth generation; the cells of the seventh generation do not differentiate enzyme activity.

These results suggest a clock mechanism involved in the timing of cellular differentiation. Neither cytokinesis nor nuclear division is required for the differentiation, but DNA replication is inevitably necessary for the tissue-specific enzyme development. Therefore, cytokinesis and nuclear division are not directly associated with the clock mechanism. Second, the results clearly indicate that histospecific enzyme development is not governed by elapsed time since fertilization but is closely related to the number of aphidicolin-sensitive cell cycle events or DNA replications. Third, in this case, activation of the genes may take place during the eighth DNA replication cycle. Whittaker (1973) discovered that cells that showed AChE activity in cleavage-arrested embryos are always, at each cleavage stage, the presumptive muscle cells, strongly suggesting the presence of cytoplasmic information for the tissue-specific enzyme development, which segregates during cleavage into muscle lineage blastomeres. Therefore, the egg counts the time of initiation of differentiation by the cycle of DNA replication, in the manner that one round of DNA replication cycle is ticked as one o'clock, and the clock must be determining the time at which cytoplasmic information can interact with the genome.

### A Possible Molecular Mechanism

If the cycle of DNA replication is a clock mechanism for cellular differentiation or stage-specific protein synthesis, how does the DNA keep track of its number of replication? Related to this point, Holliday and Pugh (1975) have advanced theoretical models intended to show that DNA itself can enumerate its replications by the enzymatic modification of specific bases in repeated sequences of the DNA. One of the mechanisms they proposed involved DNA methylation. Evidence has been accumulating over the past few years to suggest that in most higher eukaryotic organisms DNA methylation is intimately related to gene function and differentiation; see Razin and Riggs (1980) for a review. Basic facts revealed are that (1) the occurrence of the methylated base, 5-methylcytosine, has been demonstrated in a wide variety of organisms, including all vertebrates and plants that have been studied; (2) the majority of 5-methylcytosine residues in eukaryotic DNA are found in the dinucleotide sequence CpG (C, cytosine; G, guanine); (3) specific methylation patterns exist; (4) methylation patterns are tissue specific; (5) methylation patterns are clonally heritable; and (6) the DNA in mammalian sperm and early embryos is highly methylated, and undermethylation is associated with gene activity. Based on these facts, Razin and Riggs (1980) have proposed a demethylation model for differentiation.

Coupling these two proposals, one can envisage a rather simple model for counting cell divisions and for establishing the differentiated state, which is shown in Fig. 3. In the fertilized egg the DNA in certain repeated sequences is postulated to be fully or uniformly methylated. During cleavage and/or proliferative cell cycles sequence-specific proteins would inhibit methylation step by step at each DNA replication cycle. Thus, the DNA could count its replication number by the degree of methylation at specific sites. After a specific number of DNA replications, a quantal cell cycle occurs in which the specific demethylation of certain genes would bring about the interaction between the cytoplasmic information and the genome, leading the gene activation. After the establishment

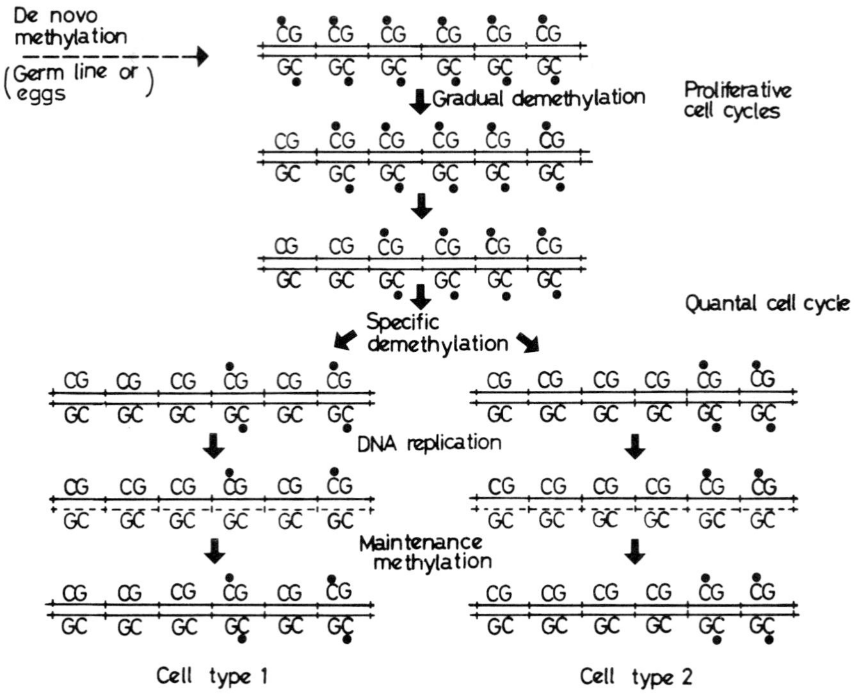


**FIGURE 3** A demethylation model of DNA for counting cell divisions and establishment and maintenance of a differentiated state, based on the proposals of Holliday and Pugh (1975) and Razin and Riggs (1980). Closed circles represent methyl group; C, cytosine; G, guanine. See text for details. (Modified from Razin and Riggs, 1980.)

of stable methylation patterns in various cell types, the maintenance methylation system ensures heritability of the methylation pattern. It is clearly too early to judge if this model has any validity, but it is worth mentioning that this model incorporates features that could allow the cell to count its number of DNA replication. Most of the details, however, remain to be solved in future.

## CYTOPLASMIC CLOCKS

It has been demonstrated that nonnucleate cytoplasmic fragments of animal eggs show periodic changes in their shape, and such oscillatory activities proper to the cytoplasm are designated as "cytoplasmic clocks." The start of the clock is triggered without the activity of the nucleus or the centriole and can be maintained for at least several cleavage cycles without nuclear regulation. Here, evidence for the presence of cytoplasmic clock is presented and a possibility that cytoplasmic clocks may be related to the timing of morphogenesis during early development is discussed.

### Evidence for Cytoplasmic Clocks

In 1904, Wilson described that the polar lobe that is isolated from trefoil-stage *Dentalium* (molluscan) eggs exhibits cyclic constriction and rounding-up in parallel with the

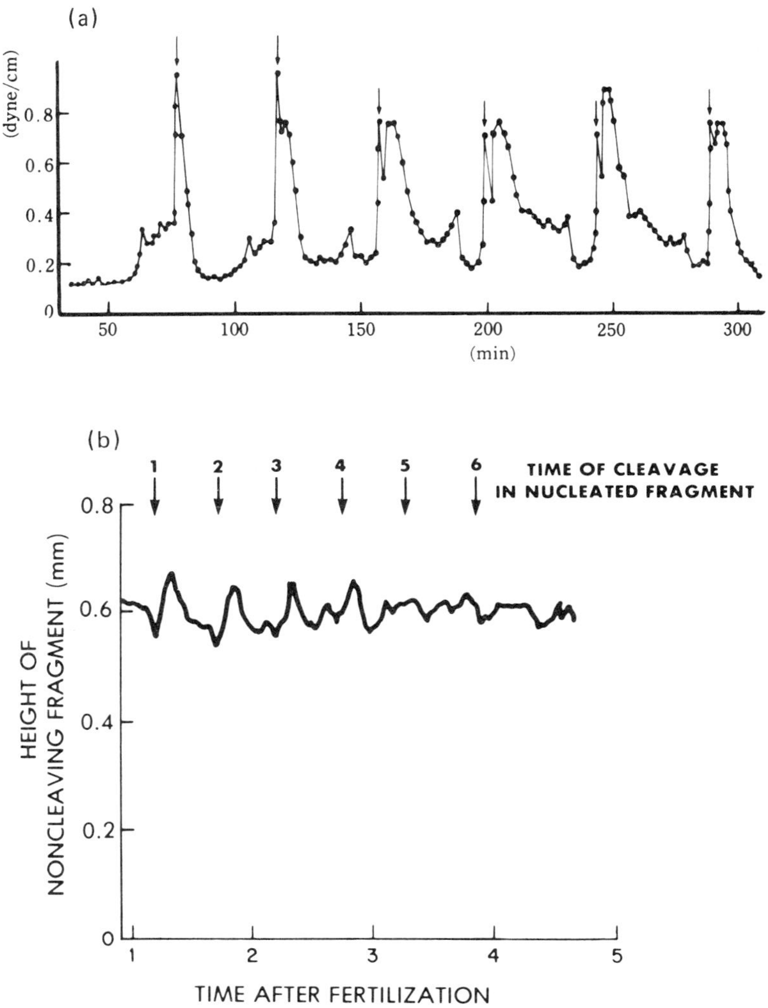


**FIGURE 4** Evidence for cytoplasmic clocks. (a) Changes in tension at the surface of activated nonnucleate fragment of a sea urchin egg. The egg of *Hemicentrotus pulcherrimus* was separated into two (with and without nucleus) fragments by centrifugal force. Subsequently, the nonnucleate fragment was activated by treatment with butyric acid and the surface tension was measured continuously by a compression method. Timings of sharp peaks in each cycle (arrows) almost coincide with divisions of normal eggs. (b) Periodic surface contraction waves in a nonnucleate fragment obtained by constriction of a fertilized *Xenopus* egg. The height of the egg fragment measured from the projected image is plotted versus the time after fertilization. The arrows 1-6 indicate the time of onset of cleavage in the cleaving (nucleate) fragment. [(a) From Yoneda et al., 1978; (b) from Kirschner et al., 1980.]

cleavage cycle of normal eggs. Since then, similar morphological changes in the cytoplasmic fragments have been reported in various kinds of invertebrate eggs, including *Ilyanassa* (molluscan) (Morgan, 1935), sea urchin (Kijima, 1960), ascidian (Bell, 1962), annelida (Shimizu, 1981), and starfish (Yamamoto and Yoneda, 1981). During cleavage of sea urchin egg, the tension at the egg surface reaches a maximum at the time of segmentation, after which it steadily decreases through the course of cleavage. Recently, Yoneda et al. (1978) separated an unfertilized egg into two fragments (with and without the nucleus) by centrifugal force. Subsequently they activated the nonnucleate fragment by treating it with butyric acid and measured the surface tension continuously. As shown in Fig. 4a, the activated nonnucleate fragment was found to exhibit cyclic changes in tension, with a temporal pattern very similar to that of the changes accompanying the division cycle of normally fertilized eggs.

Recently an autonomy of cyclic activity in the cytoplasm has been established in vertebrate eggs as well. Preceding the first cleavage in *Xenopus,* the egg undergoes a contraction of its cortex, called the *surface contraction wave,* which can be visualized by time-lapse cinematography (Hara, 1971). Amphibian eggs deprived of the vitelline membrane show periodic rounding-up and relaxation at each cleavage cycle. As shown in Fig. 4b, Hara et al. (1980) and Kirschner et al. (1980) have found that a nonnucleate fragment of *Xenopus* egg, produced by constricting a fertilized egg, shows periodic occurrence of the contraction waves and of rounding-up and relaxation, with approximately the same frequency as those in the cleaving (nucleate) fragment. Similar observations have been reported independently by Sakai and Kubota (1981), in which they pointed out that the rounding-up intervals of the nonnucleate fragment are slightly but definitely longer than the cleavage intervals of the nucleate fragment and the whole egg. In addition, Sawai (1979) also observed a very similar set of periodic changes in the tension and height of the cortex in nonnucleate egg fragments of the Japanese newt.

The biochemical basis for cytoplasmic clocks has not been defined yet, so that the term rather vaguely refers to the changes in properties of the cytoplasm that take place periodically in correlation with the cleavage cycle. An important manifestation of the clock at the molecular level is the periodic changes in sea urchin eggs activated by treatment with ammonia (Mazia, 1974). In these experiments the activated eggs do not divide but show periodic changes in chromosome condensation and in enzyme activities. Recently, Nishioka and Mazia (1977) reported that the activation of thymidine kinase and thymidylate kinase may occur in the absence of a nucleus and that the schedule of turn-on and turn-off of the phosphorylation steps in enucleated eggs is the same as that of S phase in nucleated cells and independent of DNA replication, mitosis, or cell division. Another significant part of this timing mechanism may be the periodic fluctuation of the sulfhydryl (SH) contents of egg proteins in sea urchin embryos (Sakai and Dan, 1959). This fluctuation persists even when the nuclear division is suppressed by ultraviolet irradiation, 2,4-dinitrophenol (DNP), or 4-nitroquinolinoxide (4NQO). Furthermore, as will be described later, it has been shown experimentally that the formation of micromeres in sea urchin eggs is coupled with a fourth SH cycle and is not related to the number of nuclear divisions. An intriguing report of Mano (1970) shows that there is a cycling of protein synthesis in the 12,000 g supernatant of sea urchin egg homogenates. He postulated a factor leading the synthetic rhythm in the oscillations of SH groups of the KCl-soluble fraction of the supernatant. Anyway, the relationships between cytoplasmic clocks and biochemical oscillators within cells deserve more attention.

### The SH Clock for Micromere Formation in Sea Urchin Embryos

Good evidence has been presented to indicate a causal relationship between a cytoplasmic clock and the timing of morphogenesis in sea urchin eggs. The first three divisions of the egg are equal, but the fourth cleavage of the vegetal blastomeres takes place unequally,

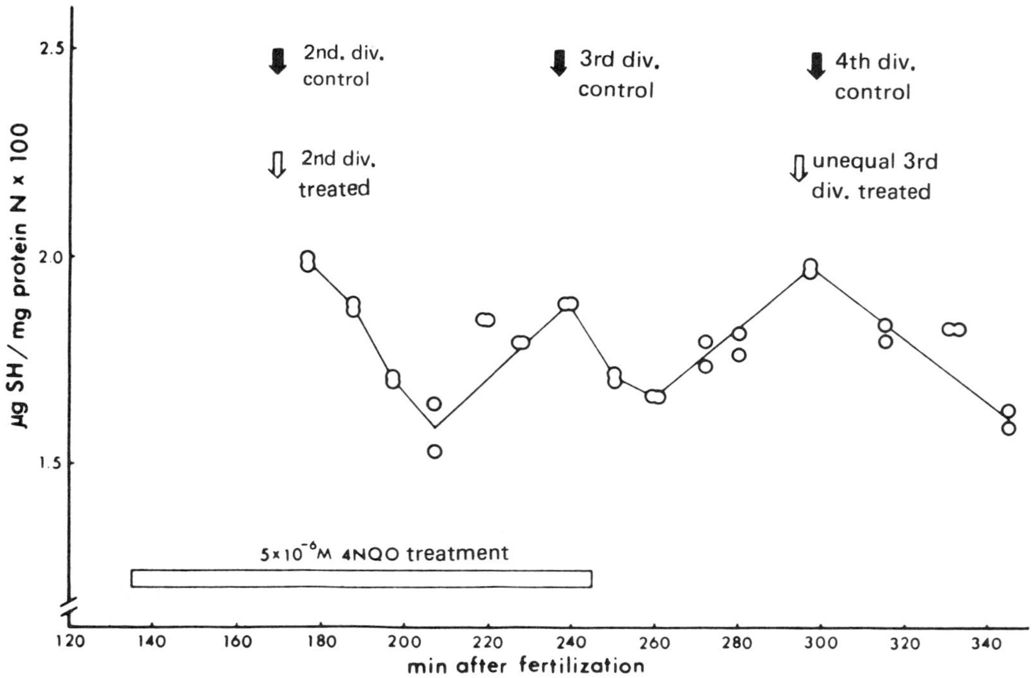


**FIGURE 5** Relationship between periodic fluctuation of sulfhydryl content in the KCl-soluble fraction of egg proteins and timing of micromere formation in sea urchin eggs. Eggs are treated with 4NQO from 135 to 245 min after fertilization and later removed from the medium. A maximum of SH content takes place in spite of the absence of cleavage in 4NQO at the time corresponding to the third cleavage of the control embryos. The treated egg forms precocious micromeres at the third cleavage, the time of which almost coincides with the fourth cleavage of normal eggs. (From Dan and Ikeda, 1971.)

resulting in four macromeres and four micromeres. The descendants of the micromeres form the larval skeleton. It has been reported that if an early cleavage is suppressed or retarded by various means and then the eggs are returned to normal seawater, when the treated eggs reach the delayed eight-cell stage often the four cells on the animal side become larger and those on the vegetal side are smaller as if they simulate the micromere formation of the normal 16-cell stage (Painter, 1915; Rustad, 1960), which is designated as "precocious" micromere formation. As described earlier, rhythmic fluctuations of the SH contents of egg proteins accompanying the cleavage cycles have been verified. Ikeda (1965) found that when a cleavage of the eggs of *Clypeaster japonicus* is suppressed by ultraviolet irradiation, the cytoplasmic SH fluctuation continues in spite of the fact that the eggs show no visible changes. A hypothesis advanced by Dan and Ikeda (1971) is that the cyclic fluctuation of the SH contents is the clock system and micromere formation is connected with the fourth SH cycle after fertilization. This hypothesis is based on the following experiments: (1) If eggs are treated under the conditions that stop the nuclear division but preserve the SH cycle (for example, treatment with DNP or 4NQO or by ultraviolet irradiation), followed by a release from these conditions, precocious micromeres are formed (Fig. 5). (2) Conditions that freeze both nuclear and cytoplasmic rhythms (for example, treatment with $NaN_3$ or with puromycin), will not induce precocious micromere formation. (3) Conditions that leave nuclear activities intact but prevent cytoplasmic rhythms (i.e., treatment with ether seawater), followed by normal culturing do not lead to micromere formation at the 16-cell stage. Although several details remain to be resolved in the future, these experiments clearly indicate that the clock for the timing of micromere formation is intimately related to the cyclic fluctuation of the SH contents in the cytoplasm but not to cytokinesis, nuclear division, and the cycle of DNA replciation.

## PROBLEM OF CRITICAL TIME

Regardless of the exact mechanism by which the clock operates, we can justifiably discuss what properties of the organism determine the timing of a certain event during early embryonic development. For example, it has been argued that the nucleocytoplasmic ratio may be a possible critical factor for timing (Smith and McLaren, 1977), and recent studies have revealed that the nucleocytoplasmic ratio does affect a duration of synchronous cleavages in vertebrate eggs. In sturgeon embryos injected with cytoplasm or the supernatant of the homogenate of the mature unfertilized eggs, Chulitskaia (1970) observed that desynchronization started later as compared with normal untreated embryos but that the injection had no effect on the timing of the onset of gastrulation. Kobayakawa and Kubota (1981), on the other hand, obtained newt egg fragments with a half or a quarter of the volume of the cytoplasm by cutting a fertilized unsegmented egg into two or four equal-sized parts. They found that desynchronization of cleavage started one or two divisions earlier in half or quarter embryos compared with the whole embryos. They also found that the time from the first cleavage to the onset of gastrulation, however, did not widely vary among quarter, half, and whole embryos, but the numbers of blastomeres that constitute embryos at the initial gastrula decreased in proportion to the egg volume. Further support for a close relationship between nucleocytoplasmic ratio and duration of synchronous cleavages in amphibian eggs comes from studies of the developmental rate of haploid embryos compared with normal diploid embryos: the decrease in the high division frequency in haploid embryos occurs one cell generation later than in diploid embryos and thus haploid cells achieve the same nucleocytoplasmic ratio as the control diploid cells (Hronowski et al., 1979).

It is uncertain, however, that the nucleocytoplasmic ratio is related to the timing of cellular differentiation and of morphogenesis. Mechanisms of the clock operation that depend on counting seem the most amenable to study and experimentation. The mechanisms that depend on a "critical time" rather than on counting, however, might be less accessible experimentally.

## CONCLUSIONS

Embryological development takes place in four dimensions and requires the existence of time-measuring processes within the embryo. Evidence is accumulating that suggests that the emergence of events during the early embryonic development is controlled by timing mechanisms or developmental clocks, which might be attributed to cell division cycles. It is an attractive idea that the development of an embryo is timed by a single clock set in motion at fertilization, but this seems unfeasible. The clock mechanism that determines the time of initiation of cellular differentiation may be independent of that for the timing of morphogenesis. The clock mechanism for cellular differentiation may be closely associated with the cycles of DNA replication, but the clock that counts the time to onset of early morphogenetic events is found in the cytoplasm.

## ACKNOWLEDGMENT

This work was supported in part by a Grant-in-Aid from the Ministry of Education, Science, and Culture of Japan.

## REFERENCES

Bell, L. G. E. (1962). Some mechanisms involved in cell division. *Nature* 193:190-191.

Chulitskaia, E. V. (1970). Desynchronization of cell divisions in the course of egg cleavage and an attempt at experimental shift of its onset. *J. Embryol. Exp. Morphol.* 23:359-374.

Dan, K., and Ikeda, M. (1971). On the system controlling the time of micromere formation in sea urchin embryos. *Dev. Growth Differ.* 13:285-301.

Durante, M. (1956). Cholinesterase in the development of *Ciona intestinalis* (Ascidia). *Experientia* 12:307-308.

Freeman, G. (1976). The effects of altering the position of cleavage planes on the process of localization of developmental potential in *Ctenophore. Dev. Biol.* 51:332-337.

Hara, K. (1971). Cinematographic observation of "surface contraction waves" (SCW) during the early cleavage of Axolotl eggs. *Wilhelm Rouxs Arch.* 167:183-186.

Hara, K., Tydeman, P., and Kirschner, M. (1980). A cytoplasmic clock with the same period as the division cycle in *Xenopus* eggs. *Proc. Natl. Acad. Sci. USA* 77:462-466.

Holliday, R., and Pugh, J. E. (1975). DNA modification mechanisms and gene activity during development. *Science* 187:226-232.

Hronowski, L., Gillespie, L. L., and Armstrong, J. B. (1979). Development and survival of haploids of the mexican axolotl, *Ambystoma mexicanum. J. Exp. Zool.* 209: 41-48.

Ikeda, M. (1965). Behaviour of sulfhydryl groups of sea urchin eggs under the blockage of cell division by UV and heat shock. *Exp. Cell Res.* 40:282-291.

Kirschner, M., Gerhart, L. C., Hara, K., and Ubbels, G. A. (1980). Initiation of the cell cycle and establishment of bilateral symmetry in *Xenopus* eggs. In *The Cell Surface: Mediator of Developmental Processes,* S. Subtelny and N. K. Wessells (eds.). Academic Press, New York, pp. 187-215.

Kobayakawa, Y., and Kubota, H. Y. (1981). Temporal pattern of cleavage and the onset of gastrulation in amphibian embryos developed from eggs with the reduced cytoplasm. *J. Embryol. Exp. Morphol.* 62:83-94.

Kojima, M. K. (1960). Cyclic changes of the cortex and the cytoplasm of the fertilized and the activated sea urchin eggs. I. Changes in the thickness of the hyaline layer. *Embryologia* 5:1-7.

Mano, Y. (1970). Cytoplasmic regulation and cyclic variation in protein synthesis in the early cleavage stage of the sea urchin embryo. *Dev. Biol.* 22:433-460.

Mazia, D. (1974). Chromosome cycle turned on in unfertilized sea urchin eggs exposed to $NH_4OH$. *Proc. Natl. Acad. Sci. USA* 71:690-693.

Meedel, T. H., and Whittaker, J. R. (1979). Development of acetylcholinesterase during embryogenesis of the ascidian *Ciona intestinalis. J. Exp. Zool.* 210:1-10.

Minganti, A., and Falugi, C. (1980). An epithelial localization of acetylcholinesterase in the ascidian *Ciona intestinalis* embryos and larvae. *Acta Embryol. Morphol. Exp.* 1: 143-155.

Morgan, T. H. (1935). The rhythmic changes in form of the isolated antipolar lobe of *Ilyanassa. Biol. Bull.* 68:296-299.

Niki, Y., and Okada, M. (1981). Isolation and characterization of Grandchildless-like mutants in *Drosophila melanogaster. Wilhelm Rouxs Arch.* 190:1-10.

Nishioka, D., and Mazia, D. (1977). The phosphorylation of thymidine and the synthesis of histones in ammonia-treated eggs and egg fragments of the sea urchin. *Cell Biol. Int. Rep.* 1:23-30.

Painter, T. S. (1915). An experimental study in cleavage. *J. Exp. Zool.* 18:299-323.

Razin, A., and Riggs, A. D. (1980). DNA methylation and gene function. *Science* 210: 604-610.

Rustad, R. C. (1960). Dissociation of the mitotic time-schedule from the micromere "clock" with X-rays. *Acta Embryol. Morph. Exp.* 3:155-158.

Sakai, H., and Dan, K. (1959). Studies on sulfhydryl groups during cell division of sea urchin egg. *Exp. Cell Res.* 16:24-41.

Sakai, M., and Kubota, H. Y. (1981). Cyclic surface changes in the non-nucleate egg fragment of *Xenopus laevis. Dev. Growth Differ.* 23:41-49.

Satoh, N. (1979). On the "clock" mechanism determining the time of tissue-specific enzyme development during ascidian embryogenesis. I. Acetylcholinesterase development in cleavage-arrested embryos. *J. Embryol. Exp. Morphol.* 54:131-139.

Satoh, N. (1982). Timing mechanisms in early embryonic development. *Differentiation* 22:156-163.

Satoh, N., and Ikegami, S. (1981a). A definite number of aphidicolin-sensitive cell-cyclic events are required for acetylcholinesterase development in the presumptive muscle cells of ascidian embryos. *J. Embryol. Exp. Morphol.* 61:1-13.

Satoh, N., and Ikegami, S. (1981b). On the "clock" mechanism determining the time of tissue-specific enzyme development during ascidian embryogenesis. II. Evidence for association of the clock with the cycle of DNA replication. *J. Embryol. Exp. Morphol.* 64:61-71.

Sawai, T. (1979). Cyclic changes in the cortical layer of non-nucleated fragments of the newt's egg. *J. Embryol. Exp. Morphol.* 51:183-193.

Shimizu, T. (1981). Cyclic changes in shape of a non-nucleate egg fragment of *Tubifex* (Annelida, Oligochaeta). *Dev. Growth Differ.* 23:101-109.

Smith, R., and McLaren, A. (1977). Factors affecting the time of formation of the mouse blastocoele. *J. Embryol. Exp. Morphol.* 41:79-92.

Snow, M. H. L., and Tam, P. P. L. (1980). Timing in embryological development. *Nature* 286:107.

Soll, D. R. (1979). Timers in developing systems. *Science* 203:841-849.

Spooner, B. S., Cohen, H. I., and Faubion, J. (1977). Development of the embryonic mammalian pancreas: The relationship between morphogenesis and cytodifferentiation. *Dev. Biol.* 61:119-130.

Whittaker, J. R. (1973). Segregation during ascidian embryogenesis of egg cytoplasmic information for tissue-specific enzyme development. *Proc. Natl. Acad. Sci. USA* 70: 2096-2100.

Wilde, C. E., Jr. (1974). Time flow in differentiation and morphogenesis. In *Concepts of Development,* J. Lash and J. R. Whittaker (eds.). Sinauer Associates, Inc., Stamford, Conn., pp. 241-260.

Wilson, E. B. (1904). Experimental studies on germinal localization. *J. Exp. Zool.* 1:1-72.

Yamamoto, K., and Yoneda, M. (1981). Cytoplasmic cycle in meiotic division of starfish oocytes. *Dev. Growth Differ.* 23:454.

Yoneda, M., Ikeda, M., and Washitani, S. (1978). Periodic changes in the tension at the surface of activated non-nucleate fragments of sea-urchin eggs. *Dev. Growth Differ.* 20:329-336.

# 26

# Central Timer Versus Dependent Pathway Regulation of Cell Division in Sea Urchin Cell Cycles

**B. Wolfanger Belisle*** / Georgetown University, Washington, D.C.

**John J. Wille, Jr.** / Mayo Clinic/Foundation, Rochester, Minnesota

**William Byrd** / The University of Texas Medical School, Houston, Texas

## INTRODUCTION

Early embryonic cell cycles are generally characterized by rapid rates (Bragg, 1938; Callan, 1949; Sonnenblick, 1950; Graham and Morgan, 1966), a short or absent $G_1$ (Graham and Morgan, 1966; Gaulden, 1956; Fujita, 1962; Hinegardner et al., 1964; Baserga, 1965; Wimber and Lameston, 1965; Kauffman, 1966; Dan et al., 1980), and relative growth independence (Graham and Morgan, 1966). Despite cell cycles that are atypical relative to normal adult cycles, temporal patterns of gene expression emerge as one studies development. Temporal regulatory patterns are apparent during differentiation in *Drosophila* (Cherbas and Kafatos, 1975; Lewis et al., 1975), during slime mold development (Newell and Sussman, 1970; Mitchison et al., 1975), and in several other developmental systems (Palmiter, 1975; Paul et al., 1972; Paul and Kafatos, 1975). When the puffing patterns of the giant salivary gland chromosomes found in *Drosophila* are studied with combinations of positive and negative effectors, the observed patterns of posteffector lag times and response-peak displacements suggest an endogenous capacity for "timekeeping" in addition to any exogenous regulatory mechanisms; reviewed in Mitchison et al. (1975). In sea urchin development, one notable example of temporal gene expression is the pattern of histone switching seen throughout early embryogenesis, where early cleavage-stage histone subtypes are differentially replaced by newly synthesized variants at specific developmental stages (Thaler et al., 1970; Newrock et al., 1978; Kedes, 1979; Wu et al., 1982). In addition, the temporal progression of cell cycle events during egg activation and early

*Current affiliation: Uniformed Services University of the Health Sciences, Bethesda, Maryland.

cleavage stages suggests not only that timekeeping plays a role in the earliest embryonic cell cycles but also that it is critical to progression of normal development. The aim of this chapter is to review briefly current facts regarding sea urchin cell cycles during early development, and to offer a working model for its regulation.

## EGG ACTIVATION AND EARLY CLEAVAGE IN SEA URCHINS

It is beyond the scope of this chapter to cover in detail all the many and varied aspects of fertilization and egg activation in sea urchins. For this readers are referred to several useful review articles that offer in-depth analyses of these topics (Harvey, 1956; Epel, 1975; 1978; Schatten, 1982; Vacquier, 1981; Nishioka, 1982; Giudici, 1973).

### Fertilization and Egg Activation

The ripe sea urchin egg, having completed oogenesis and meiotic reduction divisions becomes metabolically repressed (Giudici, 1973). It remains in a relatively quiescent state (M or $G_0$) until fertilization occurs, at which time the egg becomes metabolically reactivated, and the first embryonic cell cycle is initiated. The pluripotent zygote subsequently begins a series of rapid, synchronous divisions, resulting in the production of anywhere from 200 to 1000 daughter cells in approximately 8-12 hr (i.e., 8-10 cell cycles, depending on the species and temperature) (Stearns, 1974; Davidson et al., 1982). This transition from a resting to a cycling state, combined with the great division potential and the high degree of natural synchrony found within a population of eggs, makes the sea urchin egg an attractive model system for use in studies involving both the initiation and timing of cell division.

The initiation of the first cell cycle as a fertilization event has been well characterized. The process strongly suggests the involvement of a series of interacting ionic mechanisms, which operate to initiate and subsequently regulate egg activation. Following sperm-egg interaction, the egg becomes metabolically and biochemically activated. A chronological list of developmental events (Epel, 1980a) is shown in Table 1. Briefly, a series of early fertilization events under the control of $Ca^{2+}$ leads to cortical granule exocytosis, followed by elevation and hardening of the fertilization coat. Concomitant with calcium ion release, $Na^+$ influx and $H^+$ efflux leads to decreased intracellular levels of $H^+$, which is followed by increased rates of protein synthesis, enzyme activation, induction of DNA synthesis, and eventual chromosome condensation. Approximately 90-120 min later, the zygote cleaves equally into two daughter cells.

### The Cell Cycle

The cell cycle phases have been analyzed in two species, *Strongylocentrotus purpuratus* (Hinegardner et al., 1964) and *Hemicentrotus pulcherrimus* (Dan et al., 1980). The first S phase begins approximately 30 min after fertilization, about the time of pronuclear fusion. Subsequent DNA phases after the first S begin either in anaphase (Dan et al., 1980) or in telophase (Hinegardner et al., 1964) and end as "interphase" begins. This overlap of S with M indicates an apparent lack of $G_1$ and is characteristic of developmental cell cycles (Graham and Morgan, 1966; Gaulden, 1956; Fujita, 1962; Hinegardner et al., 1964; Baserga, 1965; Wimber and Lameston, 1965; Kauffman, 1966; Dan et al., 1980) and some other cell types (Baserga, 1965; Baserga, 1976). Interphase ($G_2$) is difficult to

**TABLE 1** Temporal Sequence of Developmental Cell Cycle Events

| Developmental event | Time after activation (sec) | Type of egg activation | | | |
|---|---|---|---|---|---|
| | | Sperm | $NH_4$ | Procaine | Ionophore A23187 |
| Sperm binding to egg surface | 0 | + | – | – | – |
| Minor influx of $Na^+$ ions; membrane potential change | 1 | + | – | ? | + |
| $Ca^{2+}$ liberation (from intracellular depots) | 8 | + | – | ? | + |
| Cortical reaction; release of $H^+$; major $Na^+$ influx | 20 | + | – | – | + |
| NAD → NADP | 30 | + | + | ? | ? |
| Increased $O_2$ consumption | 40 | + | – | ? | ? |
| Completion of fertilization membrane | 60 | + | – | – | + |
| Increased intracellular pH | 120 | + | + | + | + |
| Protein synthesis activation | 350 | + | + | + | + |
| Activation of transport systems | 400 | + | + | + | + |
| Pronuclear fusion | 1200 | + | – | – | – |
| Initiation of DNA synthesis | 1500 | + | + | + | + |
| First cell division in activated eggs | 3000 | – | + | + | + |
| First cell division | 5500 | + | – | – | – |

*Source*: Modified from Epel (1980a).

define, being of very short duration. It is described in sea urchins (Dan et al., 1980) as the period occurring between the end of telophase and the beginning of the next prophase, which ranges from 8 to 21 min in length in *Hemicentrotus* (Dan et al., 1980) and is approximately 20 min in duration in *Strongylocentrotus* at 15°C (Hinegardner et al., 1964).

### Early Cleavage Divisions

The first three divisions in sea urchin development are equal, resulting in production of uniformly sized blastomeres. During this portion of embryonic development, a high degree of division synchrony among blastomeres occurs. At the fourth cleavage, cell division becomes unequal, and 16-cell embryos containing three cell types separable on the basis of size (micromeres, mesomeres, and macromeres) are formed (Harvey, 1956). At this time in development, division synchrony is lost, and cells exhibit overall increases in cycle duration and increased asynchrony in cycle times among the three cell types. Synchrony within a particular cell type, however, appears to remain high (Dan et al., 1980; Stearns, 1974). Asynchronies among the blastomeres are thought to reflect differences in the length of S, and in the $G_1$ portion of the cycle, which becomes incorporated into the cell cycle during these later divisions (Dan et al., 1980).

## CELL CYCLE REGULATION DURING EARLY CLEAVAGE STATES

### Parthenogenetic Activation

The sea urchin egg lends itself to exploitation from an experimental point of view. Not only can naturally fertilized eggs and embryos be manipulated by a variety of chemical and mechanical treatments (i.e., treatment with radiation, microtubule inhibitors, metabolic poisons, and exposure to high pressure and high centrifugal forces), but unfertilized eggs can also be differentially activated with certain chemicals (particularly those of the amine group). Eggs stimulated artificially are capable of initiating several of the metabolic events normally seen in naturally fertilized eggs (Harvey, 1956; Loeb, 1913; Steinhardt and Epel, 1974; Epel et al., 1974; Mazia, 1974a; Mazia and Ruby, 1974). In many instances, activated eggs can still be fertilized, and may even develop into normal larval forms (Brandriff et al., 1975). This is particularly useful in studies concerning cell division regulation, as eggs can be fertilized at various points within the cell cycle and the effects on cycle-related events determined (i.e., for example, the effects on paternal chromosome condensation and cell cycle length).

The most commonly used activators for sea urchin eggs are ammonia (either $NH_4OH$ at pH 9.0 or $NH_4Cl$ at pH 8.0), procaine hydrochloride (pH 8.0), and the calcium ionophore A23187. Ammonia and procaine induce only the late fertilization events, bypassing the early calcium events. Ionophore A23187, on the other hand, induces both early and late fertilization events. In the case of $NH_4Cl$, egg activation is induced by increased intracellular pH resulting from a shift in the uncharged species ($NH_3$) across the cell membrane as a new transmembrane equilibrium is established (Shen and Steinhardt, 1978; Winkler and Grainger, 1978; Grainger et al., 1979; Johnson et al., 1976). Neither $Na^+/H^+$ exchange, nor the early $Ca^{2+}$ events occur in these eggs. No $Ca^{2+}$ release from intracellular stores occurs, and $Ca^{2+}$ influx from an extracellular source is required (Zucker et al., 1978). The late-phase fertilization events, including increased protein synthesis, and stimulation of S and chromosome condensation, are stimulated, suggesting a role for pH in the initiation of these events (Grainger et al., 1979).

### Cell Cycle Regulation

Although the sequence of morphological events occurring during cell division of early cleavage stage embryos has been described extensively, the division cycle has historically been less well studied in terms of its regulation. Most studies of this nature have approached the problem by looking at regulatory mechanisms of specific cycle events.

#### *The Control of Chromosome Condensation*

Chromosome condensation in fused cells has been studied by both sperm-egg interaction (Mazia, 1974; Krystal and Poccia, 1979) and by egg-egg hybridization techniques (Bennett and Mazia, 1981a, b). In both approaches the results are similar to those reported in cell fusion studies involving cultured cells (Rao and Johnson, 1970). Fusion of a cell in interphase with a cell in mitosis results in stimulation of mitotic events, and particularly chromosome condensation, in the interphase cells.

Unfertilized eggs treated with ammonia for extended periods of time display cycles of [$^3$H] thymidine incorporation representative of repetitive S phases of the cell cycle (Mazia and Ruby, 1974). Successive chromosome cycles also occur, although there is no spindle formation, and normal anaphase movements do not occur (Paweltz and Mazia, 1978). When sperm are added to a suspension of activated eggs, the eggs become polyspermically fertilized, and the paternal chromatin undergoes premature condensation (PCC) (Mazia, 1974; Krystal and Poccia, 1979; Poccia et al., 1978). When whole, activated eggs have entered prophase, the paternal chromatin condensation conforms to that of the maternal pronuclear cycle. If unfertilized eggs are broken into nucleated and nonnucleated halves (a technique performed by centrifugation of the eggs through sucrose gradients) (Harvey, 1956; Nishioka and Mazia, 1977) and then treated with ammonia, both halves activate (Epel, 1975), and chromosome cycles are visible in the nucleated halves. If the egg halves are fertilized when the nucleated halves are in prophase, both halves show PCC at the same time, but nonnucleated halves do not give as strong a response and the lifetime of PCC-promoting activity is less than in nucleated halves (Krystal and Poccia, 1979; Poccia et al., 1978). It seems from these results that cytoplasmic components affect PCC-promoting activity, but these effects are modulated by the maternal pronucleus (Poccia et al., 1978).

The tendency of unfertilized eggs to become polyspermically fertilized after chemical treatments has been utilized to study the biochemical transformation of the male pronucleus. During the first cell cycle, the male pronucleus must also undergo a transition from a highly compacted, molecularly inactive state. This appears to be accomplished in part by the selective and sequential replacement of sperm-specific nuclear proteins with egg-specific nuclear proteins (Poccia, 1982; Imschenetzky et al., 1980). This replacement of the histones found on sperm chromatin with maternal cleavage-stage variants appears to be independent of protein synthesis (Poccia et al., 1981; Salik et al., 1981) and DNA synthesis (Poccia, 1982) and follows a temporal pattern with respect to histone type (Poccia et al., 1981).

The effects of the activator, procaine hydrochloride, on maternal chromosome condensation and on S have also been studied (Vacquier and Brandriff, 1975). Procaine, like $NH_4Cl$, stimulates late-phase fertilization events, and also induces cytaster formation (Vacquier and Brandriff, 1975), the assembly and disassembly of which follows the procaine-induced chromosome cycle (Moy et al., 1977). Vacquier and Brandriff (1975) showed that DNA synthesis can be turned on and off by addition and removal of procaine. When procaine is removed by washing the treated eggs with fresh seawater after chromosome condensation has started, chromosomes continue to condense, but do not decondense. Since the following S does not occur, they suggest chromosome decondensation is controlling entrance into S.

Pulse treatment of eggs with $NH_4OH$ (pH 9.0) seems to result in irreversible activation of the egg (Mazia, 1974), and chromosomes are reported to continue to cycle through condensation and decondensation even after $NH_4OH$ removal. We have found that $NH_4Cl$ (pH 8.0) treatment is similar to procaine activation (Vacquier and Brandriff, 1975) in that when $NH_4Cl$ is removed after condensation has started, condensation will proceed, but decondensation does not occur (Belisle, 1981; Belisle et al., 1981). However, if decondensation has started when $NH_4Cl$ is removed, chromosomes continue to decondense and usually even undergo a second cycle of condensation (Belisle, 1981). We also find that protein synthesis is turned off after $NH_4Cl$ removal, but always at the same time of postactivation, regardless of when $NH_4Cl$ is removed (Belisle, 1981; Belisle et al., 1981). Others (Whitaker and Steinhardt, 1981) also report that activator removal stops [$^3H$] thymidine incorporation.

Other recent experiments suggest that chromosome cycles can be initiated by activating eggs with $NH_4Cl$ in $Ca^{2+}$-free seawater (Zucker et al., 1978; Belisle, 1981; Belisle et al., 1981), ionic conditions that are supposed to prevent any intracellular increases in $Ca^{2+}$ concentration during $NH_4Cl$ activation (Zucker et al., 1978).

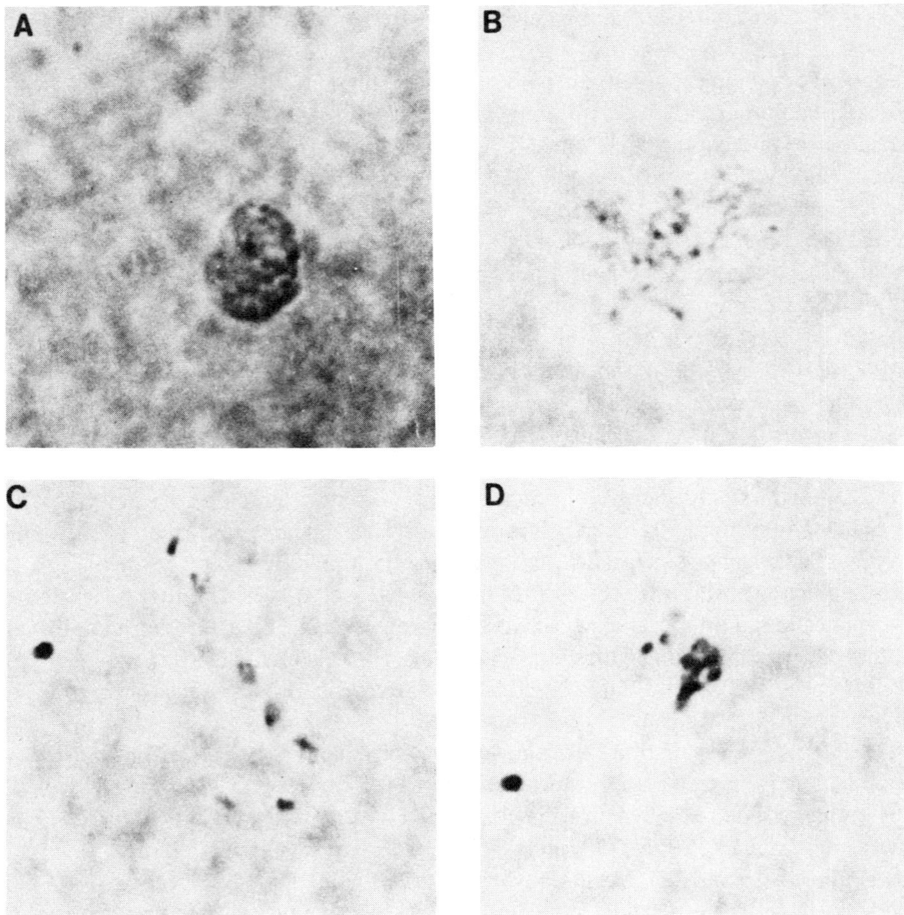


FIGURE 1

*Effects of Protein Synthesis Inhibitors*

When fertilized eggs are treated with emetine at concentrations shown to inhibit 95-99% of protein synthesis levels seen in control, fertilized eggs (Hogan and Gross, 1971), only low levels of chromosome condensation occur (Wagenaar and Mazia, 1978). In inhibited eggs, a specific temporal sequence of protein synthesis appears to be required for complete and normal progression of the chromosome cycle. Proteins produced at different times in the cell cycle are required for chromosome condensation, nuclear envelope breakdown, and formation of a normal nuclear mitotic apparatus (Wagenaar and Mazia, 1978).

When unfertilized eggs are preloaded with similar concentrations of cycloheximide or emetine, and then activated with $NH_4Cl$ while in the continuous presence of the inhibitors, chromosome condensation occurs, but no decondensation is observed (Fig. 1) (Belisle, 1981; Belisle et al., manuscript submitted). These results must be interpreted with care, as concentrations of inhibitor that reduce protein synthesis in fertilized eggs by greater than 95%, reduce protein synthesis in $NH_4Cl$ activated eggs by only about

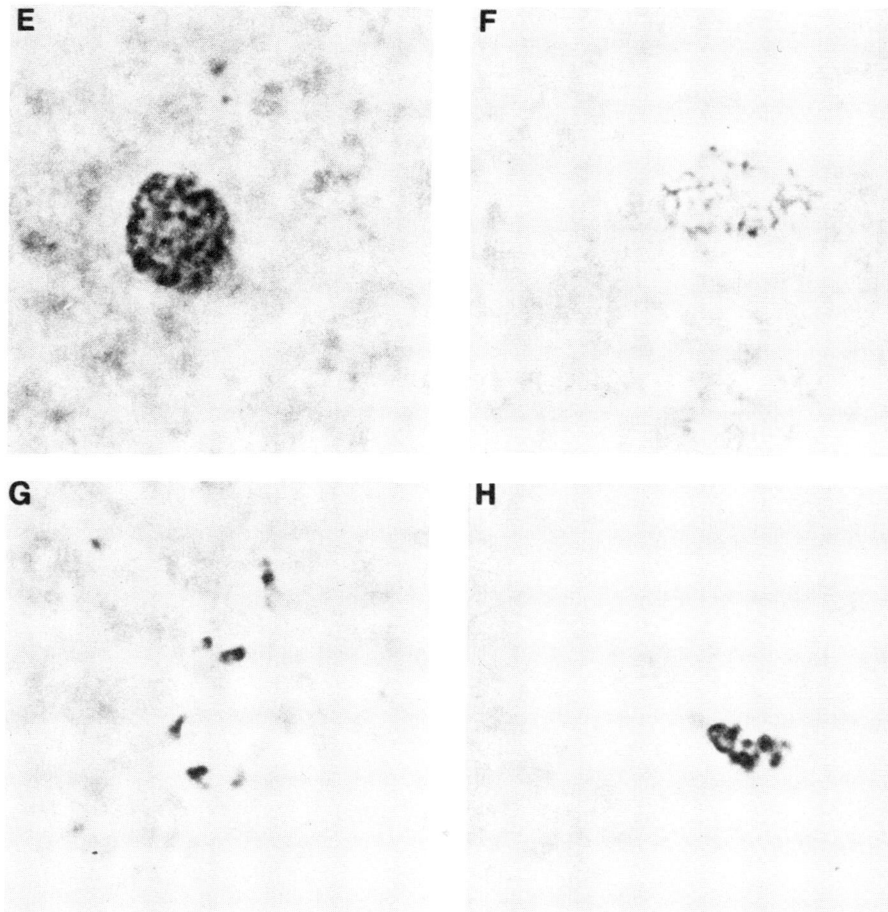

**FIGURE 1 (continued)** Chromosome cycles in eggs from *L. pictus* treated with 1 mM $NH_4Cl$ (a-d) or with 5 mM cycloheximide and 10 mM $NH_4Cl$ (e-h) for 120 min. Egg samples were collected at 90, 100, 110, and 120 min.

60-90% (unpublished results), depending on the species and inhibitor, and higher concentrations result in increased cell death.

*Summary*

Results from activator removal experiments (which turn on and off metabolic events), and studies involving use of metabolic inhibitors, suggest that chromosome cycles in sea urchins may be multiphasic, with each phase independently regulated by separate metabolic events. The evidence of PCC in nonnucleated egg halves, coupled with data obtained from other types of cell fusion studies, suggests cytoplasmic factors are involved, and protein synthesis inhibitor experiments suggest cytoplasmic factors are, in fact, proteins.

## THE CONTROL OF CYCLE TIMING

The study of cell cycle timing in sea urchin embryos has been approached from several independent directions.

## The Concept of a Central Timer

### *pH and the Control of Sea Urchin Cell Division*

pH has been demonstrated to be involved in the initiation of the first cell cycle in sea urchin eggs (Shen and Steinhardt, 1978; Winkler and Grainger, 1978; Grainger et al., 1979; Johnson and Epel, 1981). However, there are conflicting data regarding the role of pH in regulating cell cycle events subsequent to the initial transition from a resting to a metabolically active state. There is limited evidence that pH fluctuates with respect to the cell cycle. When pH is measured by the egg homogenization method (Belisle, 1981) or by direct measurement with microelectrodes (Grainger et al., 1979) fluctuations are seen that appear to correlate with the changing phases of the cell cycle.

Recently, increases in the intracellular pH similar to egg activation pH changes have been observed in murine lymphocytes after mitogen stimulation (Gerson and Kiefer, 1982). Since pH is known to fluctuate with the cell cycle in *Physarum* (Gerson, 1978; Gerson and Burton, 1977), in mammalian cells (Gerson, 1978), and in *Tetrahymena* (Gillies and Deamer, 1979), fluctuations in intracellular pH seem to be a likely candidate for cell cycle regulation. However, when egg pH measurements are made using the dimethadione (dimethyloxazolidinedione) method, no correlation between pH and the cell division cycle is observed (Johnson and Epel, 1981).

### *Intracellular Calcium as a Mitotic Trigger*

It has been suggested that mitosis is controlled by the temporal and spatial effects of oscillating intracellular calcium concentrations on microtubule assembly and disassembly (Harris, 1978). This hypothesis suggests that unknown factors (i.e., pH, critical tubulin levels, cAMP, and others) stimulate a self-propagated trigger wave of calcium release that initiates at the aster center and radiates outward, causing in its path a change in the state of tubulin polymerization. It is generally accepted that microtubules assemble and disassemble and that the pattern of assembly can be correlated with the cell cycle (Lloyd, 1979). Microtubules are universally associated with chromosomes during eumitosis. The sea urchin aster is primarily composed of microtubules and smooth endoplasmic reticulum. Although microtubules themselves comprise only a small portion of the sea urchin apparatus (the mitotic gel is actually made up of membrane-bound vesicles within which the microtubules are embedded), a close relationship is maintained between these two structures throughout the mitotic cycle (Harris, 1975). Microtubule assembly is influenced by calcium, and vesicles with calcium-ion sequestering activity have been identified in the isolated sea urchin mitotic apparatus (Silver et al., 1980). A calcium-activated ATPase has also been implicated in the mitotic cycle. Mazia et al. (1972) identified this enzyme in the mitotic apparatus isolated from echinoderm eggs, and Petzelt (1972) showed initiation of a calcium ATPase activity cycle in fertilized sea urchin eggs correlates with the cell cycle. Enzyme activity peaks near the start of chromosome condensation and decreases when nuclear membranes are reformed (Petzelt, 1972).

Although calcium is implicated in the above-mentioned processes, there is no clear evidence that calcium is capable of directing the initiation of protein synthesis, DNA synthesis, or chromosome condensation. In fact, under ionic conditions where a calcium influx has occurred, but corresponding changes in intracellular pH are prevented (as with activation by ionophore A23187 in Na-free seawater), no late-phase metabolic events are stimulated (Winkler et al., 1980). In addition, several mitotic events can be stimulated under ionic conditions where intracellular calcium release is prevented. For example,

$NH_4Cl$ activation in $Na^+$, $Ca^{2+}$ free EGTA-chelated sea water stimulates both chromosome condensation (Belisle, 1981) and protein synthesis (Winkler et al., 1980). Calcium does appear to modulate the pH effects on protein synthesis (Winkler et al., 1980). Also, during normal fertilization, activation of a calcium-dependent NAD-kinase appears to be required for DNA synthesis to occur (Whitaker and Steinhardt, 1981; Epel, 1980b).

### *Summary*

Both of the above suggest that a central timer mechanism involving oscillations in the intracellular concentrations of specific ions is regulating cell division cycles in sea urchin embryos. Yet calcium alone appears to be inadequate to stimulate the initiation of egg activation. Eggs treated with the calcium ionophore A23187 in sodium-free seawater show cortical granule exocytosis, elevation of the fertilization membrane, and increased levels of NADP (Epel et al., 1981) but do not show activation of late-phase events. If sodium is readded to the external medium, egg pH increases and the eggs become metabolically activated (Johnson et al., 1976; Nishioka and Magagna, 1981; Nishioka and McGwin, 1980). On the other hand, $NH_4Cl$, which stimulates late-phase events but no calcium-related events, does not appear to stimulate aster formation (Paweltz and Mazia, 1978). It appears from these results that both ions are responsible for full activation of the sea urchin egg, and neither can alone account for the many and varied responses of the egg when it undergoes normal fertilization. There is little evidence that either is causally involved in maintenance of cell cycle timing.

## Cycles Within Cycles: Evidence Favoring Dependent Pathway Regulation

### *The Microtubule Cycle and Cycle Timing*

The application of drugs causing depolymerization of microtubules to cultured cells has been reported to stimulate DNA synthesis in a variety of cell types (Vasiliev et al., 1971; McClain and Edelman, 1980; Crossin and Carney, 1981), suggesting microtubule depolymerization may be involved in initiation of events leading to cell division. In sea urchins there is evidence that microtubules play not only a mechanical role in cell division, but are also involved in the timing of execution of certain cell cycle events. For example, it has been shown that fertilized sea urchin eggs treated with Colcemid show increases in the duration of the cell cycle that appear to result from a prolongation of prometaphase (Sluder, 1979). Chromosome cycles continue in Colcemid-treated eggs, but no spindle assembly or cell division occurs. When the Colcemid is inactivated by irradiation, spindle assembly occurs, but is delayed relative to nuclear envelope breakdown. Delays in spindle assembly are reflected as delays in the onset of anaphase and time to cell division. In the Colcemid-inactivated eggs, other cell cycle events (e.g., spindle formation, chromosome condensation) show a normal temporal coordination (Sluder, 1979),.

Other workers have also demonstrated a continuation of cell cycle events in the absence of microtubule assembly-disassembly cycles, indicating the relative independence of several cycle events (Schatten et al., 1982). Eggs treated with griseofulvin (a reversible inhibitor of microtubule assembly) show cycles of DNA synthesis that are delayed relative to control eggs, suggesting microtubules may have some role in timing. However, in these experiments eggs treated with subthreshold levels of inhibitor (which prevents spindle formation but allows formation of a miniature mitotic apparatus) or those eggs undergoing recovery after griseofulvin treatment, attempted cleavage at the normal time (Schatten et al., 1982).

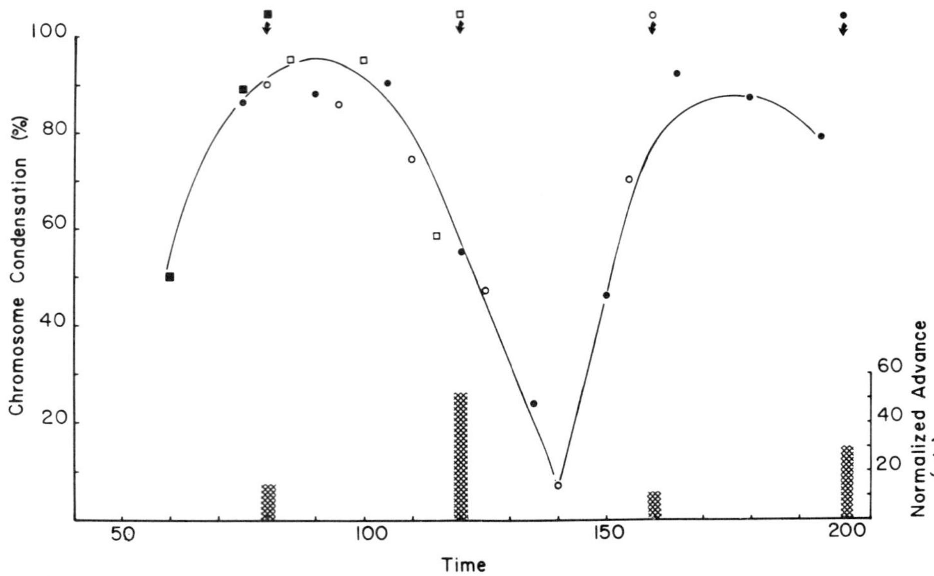


**FIGURE 2** Eggs from *L. pictus* treated with 10 mM $NH_4Cl$ for a time sufficient to produce two consecutive chromosome cycles. Eggs were treated for 80 (■), 120 (□) and 160 (o), and 200 (●) min and fertilized. Cleavage advance is expressed as the difference (in minutes) between the time to cleavage of 50% of control versus treated eggs, normalized to the total number of eggs cleaved. The percentage of chromosome condensation is the percentage of eggs scored containing condensed chromosomes in a random group of 100 or more eggs.

### *Radiation-Induced Mitotic Delay*

In sea urchin eggs, mitotic delay can also be induced by irradiating eggs or sperm. The effects are complex and depend to some degree on the type and quantity of radiation used, and on the timing of administration (reviewed in Rustad, 1971). In general, irradiation-induced mitotic delay seems to reflect a prolongation of "streak" stages (Henshaw, 1940; Yamashita et al., 1940) and delays in chromosome condensation (Rao, 1963). There is no effect on pronuclear fusion (Zeitz et al., 1968). Eggs show varying response patterns to irradiation, initially showing increasing sensitivity, followed by a sensitivity plateau, and eventually becoming insensitive (Rustad, 1971).

A large body of literature suggests that irradiation-induced mitotic delay may be caused by the effects of radiation on synthesis of proteins required for completion or critical cell cycle events (Rustad, 1978; Rustad et al., 1975; Oleinick and Rustad, 1976). For example, puromycin prevents x-irradiation recovery (Rustad and Burchill, 1966; Failla and Rustad, 1970) and also extends the period of sensitivity to irradiation (Rustad and Burchill, 1966). Actinomycin D has no effect on irradiation-induced delay (Rustad and Burchill, 1966), indicating protein synthesis from preexisting maternal mRNA is involved in repair of radiation-induced damage.

### *Cell Division Advance in Fertilized Activated Eggs*

Sea urchin eggs activated with monogen or Duponol, and then fertilized, cleave earlier than untreated, fertilized controls. This decrease in the time to first cleavage is dependent

**TABLE 2** Cleavage Advance in $NH_4Cl$-Treated Eggs and Egg Halves[a]

| Treatment | 10 mM $NH_4Cl$ | 50% Cleavage (min) |
|---|---|---|
| Whole eggs | – | 97 |
| Whole eggs | + | 72 |
| Whole eggs (+5 mM cycloheximide) | + | 93 |
| Nucleated halves | – | 95 |
| Nucleated halves | + | 75 |
| Anucleated halves | + | 95 |

[a]Egg or egg halves from *S. purpuratus* treated for 100 min, washed into normal seawater, and fertilized at 120 min postactivation.

on the length of activator treatment (Kojima, 1967, 1969). Sea urchin eggs activated with $NH_4Cl$ or procaine and then fertilized show this same time dependency (cleavage advance) relative to control eggs (Belisle, 1981; Belisle et al., 1981). In these eggs, cleavage advance seems to correlate on a temporal basis with the parthogenetically induced chromosome cycle. If eggs are fertilized before pronuclear prophase has begun, there is no cleavage advance. As eggs begin to undergo chromosome condensation, a slight advance occurs, and increases to a maximum at a time when 100% of the eggs contain condensed chromosomes. The length of cleavage advance remains constant as chromatin begins to decondense, but decreases again at interphase and early prophase stages of the second cell cycle (Fig. 2). The pattern of cleavage advance is repeated within successive chromosome cycles, and suggests two things. First, cleavage advance is dependent on factors involved in chromosome condensation, but not on decondensation. Second, the capacity to cleave in advance is reset with each successive chromosome cycle (Belisle et al., manuscript in preparation).

If chromosome decondensation is prevented by activator removal, cleavage advance is not affected. As long as the chromosomes remain condensed, maximal advance occurs (Belisle, 1981; Belisle et al., 1981), suggesting some permissive signal to cleavage is "turned-on" indefinitely in these eggs (Belisle et al., manuscript submitted).

Eggs activated with $NH_4Cl$ while in the presence of cycloheximide or emetine show chromosome condensation but no decondensation. When eggs treated with cycloheximide and $NH_4Cl$ are washed into normal seawater to remove the inhibitor, and then fertilized, no cleavage advance occurs (Table 2). This suggests timing of cell division in these eggs is either only indirectly linked to the chromosome cycle through other metabolic events (for example, synthesis of mitotic proteins) or, alternatively, is merely paralleling the chromosome cycle on a temporal basis. However, in preliminary experiments using activated egg halves, nucleated egg halves show cleavage advance similar to that obtained with whole eggs, but nonnucleated egg halves cleave at the same time as whole control eggs (Table 2; unpublished results), thus lending validity to the concept that the capacity of the egg to cleave in advance is somehow linked to the chromosome cycle. Inhibitor experiments suggest that critical factors involved in cycle timing are proteins produced early in the cell cycle that are later involved in chromosome decondensation.

### *Summary—A Sequential Interactive Network Model for Cycle Regulation*

A survey of results obtained from studies involving the manipulation of the cell cycle suggests a model analogous to a genus of the dependent pathway model described by Hartwell for the yeast cell cycle (Hartwell et al., 1974) may also be an appropriate model

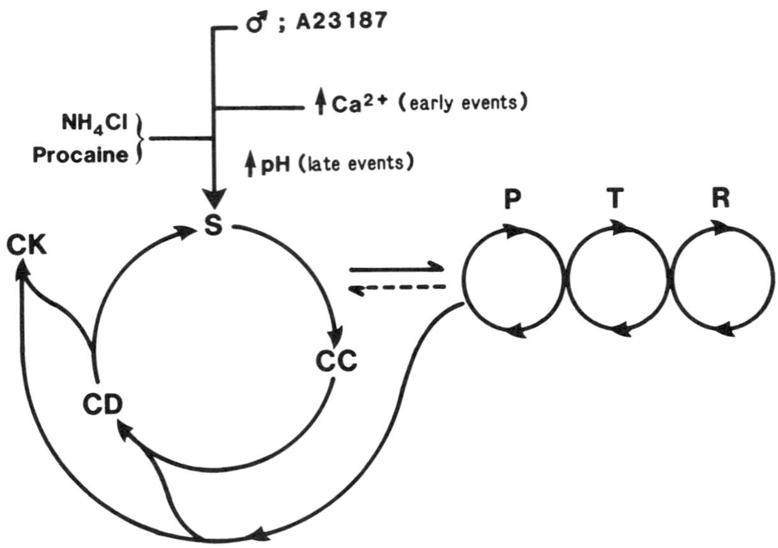


**FIGURE 3** Sequential interactive network (SIN) model for regulation of early division cycles in sea urchin embryos. Fertilization or egg activation results in an alteration of intracellular ionic composition and reactivation of a start event (S). The chromosome condensation-decondensation loop cycles semiautonomously, but is influenced by $G_2$-phase events occurring on an independent metabolic loop (outer loop). Coupling between the chromosome loop and the metabolic loop is mediated at two sites: (1) a pre-CC event (→) that "switches on" protein synthesis, and that in turn supplies gene products necessary for switching on the chromosome loop (- - -); and (2) a $G_2$ event associated with gene products that supply the chromosome loop with proteins needed for chromosome decondensation. CC, chromosome condensation; CD, chromosome decondensation; CK, cytokinesis; P, protein synthesis; R, replication events; T, transcription events.

for early embryonic sea urchin cell cycles (Fig. 3). This sequential interactive network (SIN) model involves the interaction of a semiautonomous chromosome cycle through gene action with an independent metabolic cycle that regulates the former through, for example, feedback interaction, enzyme activation, and production of critical proteins. The two loops close at a late $G_2$ event to cause cytokinesis (CK). In this scheme (Fig. 3), fertilization results in an intracellular shift in ionic composition that creates an environment favorable for reinitiation of some start event. Activated at this time are translation of proteins from preexisting stored mRNA (P loop) and from newly transcribed mRNA (T loop). Replication events of the first cell cycle are initiated, and all three loops interact (feedback) to regulate cycles subsequent to the first. The chromosome loop cycles independently of $G_2$-phase events, but can be regulated by coupling to other events (see Fig. 3, legend) occurring on other pathways. Alterations in timing of events on any loop can result in increases or decreases in overall length of the cell cycle, and could explain the many varied results obtained in cell cycle studies to date.

## REFERENCES

Baserga, R. (1965). The relationship of the cell cycle to tumor growth and control of cell division: A review. *Cancer Res.* 25:581-595.

Baserga, R. (ed.) (1976). *Multiplication and Division in Mammalian Cells.* Marcel Dekker, New York.

Belisle, B. W. (1981). Mitotic triggers in the cell cycle: Phase dependency of cleavage advance on the chromosome cycle in parthenogenically activated sea urchin eggs. Ph.D. Dissertation, Louisiana State University, Baton Rouge, Louisiana.

Belisle, B. W., Byrd, W., and Wille, J. (1981). Cleavage advance and chromosome cycles in parthenogenically activated sea urchin eggs. *J. Cell Biol.* 91:11a.

Belisle, B. W., Byrd, W., and Wille, J. Manuscript submitted.

Bennett, J., and Mazia, D. (1981a). Interspecific fusion of sea urchin eggs. Surface events and cytoplasmic mixing. *Exp. Cell Res.* 131:197-207.

Bennett, J., and Mazia, D. (1981b). Fusion of fertilized and unfertilized sea urchin eggs. Maintenance of cell surface integrity. *Exp. Cell Res.* 134:494-498.

Bragg, A. N. (1938). The organization of the early embryo of *Bufo cognatus* as revealed especially by the mitotic index. *Z. Zellforsch. Microskop. Anat.* 28:154-178.

Brandriff, B., Hinegardner, R. T., and Steinhardt, R. (1975). Development and life cycle of the parthenogenetically activated sea urchin embryo. *J. Exp. Zool.* 192:13-24.

Callan, H. G. (1949). Cleavage rate, oxygen consumption and ribose nucleic acid content of sea urchin eggs. *Biochim. Biophys. Acta* 3:92-102.

Cherbas, P. T., and Kafatos, F. C. (1975). Temporal programs of gene expression during cell differentiation. In *The Molecular Basis of Circadian Rhythms.* J. W. Hastings and H. Schweiger (eds.). Dahlem Workshop, Berlin, pp. 373-394.

Crossin, K. L., and Carney, D. H. (1981). Evidence that microtubule depolymerization in the cell cycle is sufficient to initiate DNA synthesis. *Cell* 23:61-71.

Dan, K., Tanaka, S., Yamazaki, K., and Kato, Y. (1980). Cell cycle study up to the time of hatching in the embryos of the sea urchin, *Hemicentrotus pulcherrimus. Dev. Growth Differ.* 22:589-598.

Davidson, E. H., Hough-Evans, B. R., and Britten, R. J. (1982). The molecular biology of the sea urchin embryo. *Science* 217:17-26.

Epel, D. (1975). The program and mechanisms of fertilization in the echinoderm egg. *Am. Zool.* 15:507-522.

Epel, D. (1978). Mechanisms of activation of sperm and egg during fertilization of sea urchin gametes. In *Current Topics in Developmental Biology,* vol. 12, A. A. Moscona and A. Monroy (eds.). Academic Press, New York, pp. 185-246.

Epel, D. (1980a). Ionic triggers in the fertilization of sea urchin eggs. *Ann. N.Y. Acad. Sci.* 339:74-85.

Epel, D. (1980b). In *The Cell Surface: Mediator of Developmental Processes,* S. Sobtilny and N. K. Wessels (eds.). Academic Press, New York, p. 169.

Epel, D., Patton, C., Wallace, R. W., and Cheung, W. Y. (1981). Calmodulin activates NAD kinase of sea urchin eggs: An early event of fertilization. *Cell* 23:543-549.

Epel, D., Steinhardt, R. A., Humphreys, T., and Mazia, D. (1974). An analysis of the partial metabolic derepression of sea urchin eggs by ammonia; The existence of independent pathways in the program of activation at fertilization. *Dev. Biol.* 40:245-255.

Failla, P. M., and Rustad, R. C. (1970). Protein synthesis and recovery from $\gamma$-ray-induced mitotic delay in fertilized sea urchin eggs. *Int. J. Radiat. Biol.* 17:385-388.

Fujita, S. (1962). Kinetics of cell proliferation. *Exp. Cell Res.* 28:52-60.

Gaulden, M. E. (1956). DNA synthesis and x-ray effects at different mitotic stages in grasshopper neuroblasts. *Genetics* 41:645.

Gerson, D. F. (1978). Intracellular pH and the mitotic cycle in *Physarum* and mammalian cells. In *Cell Cycle Regulation,* J. R. Jeter, I. L. Cameron, G. M. Padilla, and A. M. Zimmerman (eds.). Academic Press, New York, pp. 105-131.

Gerson, D. F., and Burton, A. C. (1977). The relation of cycling of intracellular pH to mitosis in the cellular slime mold *Physarum polycephalum. J. Cell Physiol.* 91:297-304.

Gerson, D. F., and Kiefer, H. (1982). High intracellular pH accompanies mitotic activity in murine lymphocytes. *J. Cell. Physiol.* 112:1-4.

Gillies, R. J., and Deamer, D. W. (1979). Intracellular pH: Methods and applications. *Curr. Top. Bioenerg.* 9:63-89.

Giudici, G. (1973). *Developmental Biology of the Sea Urchin Embryo.* Academic Press, New York.

Graham, C. F., and Morgan, R. W. (1966). Changes in the cell cycle during early amphibian development. *Dev. Biol.* 14:439-460.

Grainger, J. L., Winkler, M. M., Shen, S. S., and Steinhardt, R. A. (1979). Intracellular pH controls protein synthesis rate in the sea urchin egg and early embryo. *Dev. Biol.* 68:396-406.

Harris, P. (1975). The role of membranes in the organization of the mitotic apparatus. *Exp. Cell Res.* 94:409-425.

Harris, P. (1978). Triggers, trigger waves, and mitosis: A new model. In *Cell Cycle Regulation,* J. R. Jeter, I. L. Cameron, G. M. Padilla, and A. M. Zimmerman (eds.). Academic Press, New York, pp. 76-104.

Hartwell, L. H., Culotti, J., Pringle, J. R., and Reid, B. J. (1974). Genetic control of the cell division cycle in yeast. *Science* 183:46-51.

Harvey, E. B. (1956). *The American Arbacia and Other Sea Urchins.* Princeton University Press, Princeton, New Jersey.

Henshaw, P. S. (1940). Further studies on the action of Roentgen rays on the gametes of *Arbacia punctulata.* II. Modification of the mitotic time schedule in the eggs by exposure of the gametes to Roentgen rays. *J. Roentgenol. Radium Ther.* 43:907-912.

Hinegardner, R. T., Rao, B., and Feldman, D. E. (1964). The DNA synthetic period during early development of the sea urchin egg. *Exp. Cell Res.* 36:53-61.

Hogan, B., and Gross, P. R. (1971). The effect of protein synthesis inhibition on the entry of messenger RNA into the cytoplasm of sea urchin embroys. *J. Cell Biol.* 49:692.

Imschenetzky, M., Puchi, M., and Massone, R. (1980). Histone analysis during the first cell cycle of development of the sea urchin *Tetrapygus niger. Differentiation* 17:111-115.

Johnson, C. H., and Epel, D. (1981). Intracellular pH of sea urchin eggs measured by the dimethyloxazolidinedione (DMO) method. *J. Cell Biol.* 89:284-291.

Johnson, J. D., Epel, D., and Paul, M. (1976). Intracellular pH and activation of sea urchin eggs after fertilization. *Nature* 262:661-664.

Kauffman, S. L. (1966). An autoradiographic study of the generation cycle in ten day mouse embryo neural tube. *Exp. Cell Res.* 42:67-73.

Kedes, L. H. (1979). Histone genes and histone messengers. *Annu. Rev. Biochem.* 48:837-870.

Kojima, M. K. (1967). Acceleration of the cleavage in sea urchin eggs by insufficient treatments with activating reagents before fertilization. *Embryologia* 10:75-82.

Kojima, M. K. (1969). Acceleration of the cleavage in sea urchin eggs by repeated insufficient stimulations with various activating reagents and spermatozoa. *Embryologia* 10:323-333.

Krystal, W., and Poccia, D. (1979). Control of chromosome condensation in the sea urchin egg. *Exp. Cell Res.* 123:207-219.

Lewis, M., Helmsing, P. J., and Ashburner, M. (1975). Parallel changes in puffing activity and patterns of protein synthesis in salivary glands of *Drosophila. Proc. Natl. Acad. Sci. USA* 72:3604-4608.

Lloyd, C. (1979). Primitive model for cell cycle control. *Nature* 280:631-632.
Loeb, J. (1913). The comparative efficiency of weak and strong bases in artificial parthenogenesis. *J. Exp. Zool.* 13:577-590.
Mazia, D. (1974a). Chromosome cycles turned on in unfertized sea urchin eggs after exposure to $NH_4OH$. *Proc. Natl. Acad. Sci. USA* 71:690-693.
Mazia, D. (1974b). The cell cycle. *Sci. Am.* 230:54-64.
Mazia, D., Petzelt, C., Williams, R. O., and Meza, I. (1972). A $Ca^{+2}$ activated ATPase in the mitotic apparatus of the sea urchin (isolated by a new method). *Exp. Cell Res.* 70:325-332.
Mazia, D., and Ruby, A. (1974). DNA synthesis turned on in unfertilized sea urchin eggs by treatment with $NH_4OH$. *Exp. Cell Res.* 85:167-172.
McClain, D. A., and Edelman, G. M. (1980). Density-dependent stimulation and inhibition by agents that disrupt microtubules. *Proc. Natl. Acad. Sci. USA* 77:2748-2753.
Mitchison, J. M., Gerisch, G., Kafatos, F. C., Kauffman, S. A., Matthews, H. R., Sachsenmaier, W., and Zeuthen, E. (1975). Other types of cellular periodic systems. Group Report. In *The Molecular Basis of Circadian Rhythms,* J. W. Hastings and H. Schweiger (eds.). Dahlem Workshop, Berlin, pp. 361-372.
Moy, G. W., Brandriff, B., and Vacquier, V. D. (1977). Cytasters from sea urchin eggs parthenogenetically activated by procaine. *J. Cell Biol.* 73:788-793.
Newell, P. C., and Sussman, M. (1970). Regulation of enzyme synthesis by slime mold cell assemblies embarked upon alternative developmental programs. *J. Mol. Biol.* 49:627-637.
Newrock, K. M., Alfagame, R., Nardi, R. V., and Cohen, L. H. (1978). Histone changes during chromatin remodeling in embryogenesis. *Cold Spring Harbor Symp. Quant. Biol.* 42:421-431.
Nishioka, D. (1982). The ionic basis of fertilization. *J. Wash. Acad. Sci. 72:1-11.*
Nishioka, D., and Magagna, L. S. (1981). Increased uptake of thymidine in the activation of sea urchin eggs: Specificity of uptake on internal pH, the cortical reaction, and external sodium. *Exp. Cell Res.* 133:363-372.
Nishioka, D., and Mazia, D. (1977). The phosphorylation of thymidine and the synthesis of histones in ammonia-treated eggs and egg fragments of the sea urchin. *Cell Biol. Int. Rep.* 1:23-30.
Nishioka, D., and McGwin, N. F. (1980). Relationships between the release of acid, the cortical reaction, and the increase of protein synthesis in sea urchin eggs. *J. Exp. Zool.* 212:215-223.
Oleinick, N. L., and Rustad, R. C. (1976). Interrelationships between ionizing radiation, protein synthesis, and the physiological expression of radiation damage. *Adv. Radiat. Biol.* 6:107-160.
Palmiter, R. D. (1975). Quantitation of parameters that determine the rate of ovalbumin synthesis. *Cell* 4:189-197.
Paul, M., Goldsmith, M. R., Hunsley, J. R., and Kafatos, F. C. (1972). Cellular differentiation and specific protein synthesis: Production of egg shell proteins by silkmoth follicular cells. *J. Cell Biol.* 55:653-680.
Paul, M., and Kafatos, F. C. (1975). Specific protein synthesis in cellular differentiation. II. The program of protein synthetic changes during chorion formation by silkmoth follicles, and its implementation in organ culture. *Dev. Biol.* 42:141-159.
Paweltz, N., and Mazia, D. (1978). The "nuclear mitotic apparatus" in sea urchin eggs. In *Cell Reproduction,* E. R. Dirkson, D. M. Prescott, and C. F. Fox (eds.). Academic Press, New York, pp. 495-504.
Petzelt, C. H. (1972). $Ca^{+2}$ activated ATPase during the cell cycle of the sea urchin *Strongylocentrotus purpuratus. Exp. Cell Res.* 70:332-339.
Poccia, D. (1982). Biochemical aspects of sperm nucleus activation by egg cytoplasm. *J. Wash. Acad. Sci.* 72:24-33.

Poccia, D., Krystal, G., Nishioka, D., and Salik, J. (1978). Control of sperm chromatin structure by egg cytoplasm in the sea urchin. In *Cell Reproduction,* E. R. Dirkson, D. M. Prescott, and C. F. Fox (eds.). Academic Press, New York, pp. 197-206.

Poccia, D., Salik, J., and Krystal, G. (1981). Transitions in histone variants of the male pronucleus following fertilization and evidence for a maternal store of cleavage stage histones in the sea urchin egg. *Dev. Biol.* 82:287-296.

Rao, B. (1963). Analysis of x-ray induced mitotic delay in sea urchin eggs. Ph.D. Thesis, University of California, Berkeley, California.

Rao, P. N., and Johnson, R. T. (1970). Mammalian cell fusion: Studies on the regulation of DNA synthesis and mitosis. *Nature* 225:159-164.

Rustad, R. C. (1971). Radiation responses during the mitotic cycle of the sea urchin egg. In *Developmental Aspects of the Cell Cycle,* I. L. Cameron, G. M. Padilla, and A. M. Zimmerman (eds.). Academic Press, New York, pp. 128-159.

Rustad, R. C. (1978). Protein synthesis and radiation induced mitotic delay. In *Cell Reproduction,* E. R. Dirkson, D. M. Prescott, and C. F. Fox (eds.). Academic Press, New York, pp. 591-597.

Rustad, R. C., and Burchill, B. R. (1966). Radiation-induced mitotic delay in sea urchin eggs treated with puromycin and actinomycin D. *Radiat. Res.* 29:203-210.

Rustad, R. C., Oleinick, N. L., and Brewer, E. (1975). A new mitotic cycle marker. *Exp. Cell Res.* 93:477-479.

Salik, J., Herlands, L., Hoffman, H. P., and Poccia, D. (1981). Electrophoretic analysis of stored histone pool in unfertilized sea urchin eggs: Quantification and identification by antibody binding. *J. Cell Biol.* 90:385-395.

Schatten, G. (1982). The movements during fertilization. *J. Wash. Acad. Sci.* 72:12-23.

Schatten, H., Schatten, G., Petzelt, C., and Mazia, D. (1982). Effects of griseofulvin on fertilization and early development in sea urchins. Independence of DNA synthesis, chromosome condensation and cytokinesis from microtubule mediated events. *Eur. J. Cell Biol.* 27:74-83.

Shen, S. S., and Steinhardt, R. A. (1978). Direct measurement of intracellular pH during metabolic derepression at fertilization and ammonia activation of the sea urchin egg. *Nature* 272:253-254.

Silver, R. B., Cole, R. D., and Cande, W. Z. (1980). Isolation of mitotic apparatus containing vesicles with calcium sequestration activity. *Cell* 19:505-516.

Sluder, G. (1979). Role of spindle microtubules in the control of cell cycle timing. *J. Cell Biol.* 80:674-691.

Sonnenblick, B. P. (1950). The early embryology of *Drosophila melanogaster.* In *The Biology of Drosophila,* M. Demerec (ed.). Wiley, New York, pp. 62-165.

Stearns, L. W. (1974). *Sea Urchin Development: Cellular and Molecular Aspects.* Dowden, Hutchison & Ross, Penn, p. 339.

Steinhardt, R., and Epel, D. (1974). Activation of sea urchin eggs by calcium ionophore. *Proc. Natl. Acad. Sci. USA* 71:1915-1919.

Thaler, M. M., Cox, M. C. L., and Villee, C. A. (1970). Histones in early embryogenesis. *J. Biol. Chem.* 245:1479-1483.

Vacquier, V. D. (1981). Dynamic changes of the egg cortex. *Dev. Biol.* 84:1-26.

Vacquier, V. D., and Brandriff, D. (1975). DNA synthesis in unfertilized sea urchin eggs can be turned on and turned off by the addition and removal of procaine hydrochloride. *Dev. Biol.* 47:12-31.

Vasiliev, J. M., Gelfand, I. M., and Guelstein, V. I. (1971). Initiation of DNA synthesis in cell cultures by colcemid. *Proc. Natl. Acad. Sci. USA* 68:977-979.

Wagenaar, E. B., and Mazia, D. (1978). The effect of emetine on first cleavage division in the sea urchin, *Strongylocentrotus purpuratus.* In *Cell Reproduction,* E. R. Dirkson, D. M. Prescott, and C. F. Fox (eds.). Academic Press, New York, pp. 539-545.

Whitaker, M. J., and Steinhardt, R. A. (1981). The relation between the increase in reduced nicotinamide nucleotides and the initiation and maintenance of DNA synthesis in the egg of the sea urchin *Lytechinus pictus*. *Cell* 25:95-103.
Wimber, D. E., and Lameston, L. F. (1965). Cell cycle of mouse embryonic tissue under continuous gamma irradiation. *Nature* 207:432-433.
Winkler, M. M., and Grainger, J. L. (1978). Mechanism of action of $NH_4Cl$ and other weak bases in the activation of sea urchin eggs. *Nature* 273:536-538.
Winkler, M., Steinhardt, R. A., Grainger, L., and Minning, L. (1980). Dual controls for the activation of protein synthesis at fertilization. *Nature* 287:558-560.
Wu, R. S., Nishioka, D., and Bonner, W. M. (1982). Differential conservation of histone 2A variants between mammals and sea urchins. *J. Cell Biol.* 93:426-431.
Yamashita, H., Mori, K., and Miwa, M. (1940). The action of ionizing rays on sea urchin. V. The mitotic observations on the effects of Roentgen rays upon the unfertilized eggs and sperm. *Gann* 34:239-245.
Zeitz, L., Ferguson, R., and Gardinkel, E. (1968). Radiation-induced effects on DNA synthesis in developing sea urchin eggs. *Radiat. Res.* 34:200-208.
Zucker, R. S., Steinhardt, R. A., and Winkler, M. M. (1978). Intracellular calcium and the mechanisms of parthenogenetic activation of sea urchin eggs. *Dev. Biol.* 65:285-295.

# 27
# Cell Clocks and Cellular Aging

**Glenn A. Zorn and Bonnie Smith** / John Muir Memorial Hospital, Walnut Creek, California

## INTRODUCTION

Early studies conducted with purportedly normal vertebrate fibroblast cells demonstrated that in sophisticated cell culture systems in vitro, the cells could replicate indefinitely (Ebeling, 1913; Carrel and Ebeling, 1921; Parker, 1961). The conclusion that was reached from these investigations was obvious: aging did not occur at the cellular level, and individual cells were immortal.

Swim and Parker (1957) were the first to observe that the capacity to divide indefinitely was not consistent among all species. Indeed, with many cells in culture the ability to replicate in vitro slowed dramatically with time and was eventually lost. It remained for Hayflick and Moorhead (1961) to show that for normal human embryonic fibroblasts, cultured in vitro, the capacity to replicate was limited to 50 ± 10 population doublings (PD). The life cycle of cells in culture was divided into three distinct periods. The initial period of isolation and plating of the cells in culture vessels was known as phase I. The period immediately following, characterized by rapid cell proliferation, was known as phase II. The final period of greatly reduced replicative ability was called phase III. Cultures of phase III cells can survive for many months without ever becoming confluent. The replicative potential could be determined for any culture if the population doubling level (PDL) were known. For example, if a particular culture had been divided 30 times, approximately 20 PD remained (50 - 30) before the culture entered phase III. We should note that the PDL does not indicate the number of times individual cells in a particular culture have replicated. Some cells at any given PDL do not have the ability to replicate (Smith and Hayflick, 1974). Even if the cells are meticulously divided at the time of passage, the size of the culture vessels, the dilution ratio, and the number of cells destroyed by the trauma of trypsinization are all variables that contribute to inaccuracies in relating PDL to individual cell division. In recent years, many laboratories have overinterpreted the Hayflick and Moorhead findings and tried to correlate PDL to the number of cell divisions. Often the PDL is reported in fractional numbers (PDL = 25.7). Such attempts at increased accuracy are purely speculative and do not express the views of the original authors. PDL is a crude indication of the division potential of the entire culture, which is composed of cells with different division potentials (Smith and Hayflick, 1974).

Cultures of cells that demonstrate the phase III phenomenon are referred to as cell strains. Cultures of cells that have the potential to replicate indefinitely are called cell lines. Cell lines are usually morphologically and always cytogenetically distinguishable from cell strains. Cell lines, whether derived from normal or cancerous tissue, are always abnormal in one or more biological properties, cell strains retaining many of the properties associated with normal cells in vivo. Unfortunately, few authors adhere to the definitions of cell strain and cell line. Consequently, both terms have become interchangeable, obscuring their meaning. The original definitions of these terms will be followed in this report.

Since the initial findings of Hayflick and Moorhead (1961) it has been observed that most (if not all) normal cells in culture have a finite life span. Murine cells, however, have a high capacity to change spontaneously from a cell strain to a cell line (Todaro and Green, 1963) and concomitantly are able to divide indefinitely. This alteration is usually associated with identifiable cytogenetic alterations as well as morphological changes. The fact that murine cells are easy to cultivate has resulted in their widespread use. Thus, the behavior of unlimited division potential of "normal" murine cells is accepted as fact by many investigators.

In contrast, even clones derived from human cell lines may not have the ability to replicate indefinitely (Martinez et al., 1978).

There is considerable debate whether the phase III phenomenon is related to aging in vivo. Cell differentiation is offered as an alternate explanation of loss of division potential (Martin, 1977a). We doubt that such arguments are meaningful, since aging and cell differentiation are also viewed as related phenomenon. Hayflick (1980b) has compiled several lists of the functional changes that occur in cells as they approach phase III in vitro. Many biological functions have been measured and observed to either increase, decrease, or remain unchanged with increasing PDL. Not surprisingly, these changes mimic the alterations in functional capacity observed in vivo with increasing age.

Because there is no evidence that the phase III phenomenon observed in vitro is related to mortality in vivo, skeptism exists concerning the study of cellular senescence and its relation to the normal human life span. Macieira-Coelho (1981) recently constructed an analogy between aging and cancer research. He stated,

> Although one should not extrapolate completely to conditions in vivo from the studies of cells in vitro, an immense amount of knowledge has accumulated from the studies . . . on cells in culture; this knowledge has been extremely useful in understanding the basic mechanisms involved in oncogenicity and would have been impossible to gather by other methods . . . we cannot see why the field of aging should not benefit from tissue culture. . . .

Support for cellular aging research comes from investigators who believe that the biological changes preceding phase III in vitro and mortality in vivo may be indirectly responsible for both phenomena. Thus, phase III and mortality may be observable and predictable consequences of a series of functional alterations that appear in vitro and in vivo, respectively. Whether or not phase III is related to aging in vivo, the phenomenon is worthy of continued scientific investigation.

Cellular senescence is not limited to human cells in culture. Aging in culture has been observed in fungi (Marcou, 1961; Holliday, 1969; Smith and Rubenstein, 1973), slime molds (Poulter, 1969), protozoa (Sonneborn, 1954; Nanney, 1974), algae (Berger and Schweiger, 1975), *Drosophila* (Miguel et al., 1979), and vertebrate cells (Comfort, 1979). Our purpose is not to provide an exhaustive review of cellular aging (cytogerontology) but to discuss some of the recent experimental data collected in our own and other laboratories with normal human cells. In addition, we recommend that other reviews be consulted (Macieira-Coelho, 1973; Martin, 1977a,b; Rosen, 1978; Muggleton-Harris, 1979; Comfort, 1979; Hayflick, 1980a,b; Macieira-Coelho, 1981).

## THEORIES OF AGING AND THE ANALYSIS OF CELL HYBRIDS

Theories of cellular aging can be divided into two general categories. The first category considers aging as a programmed genetic event caused by (1) the active expression of specific "aging genes," (2) the active expression of "cell clock longevity genes," or (3) the passive exhaustion of accurate genetic information. Within this group of theories the fundamental cause of cellular senescence would involve mostly nuclear rather than cytoplasmic events. The second category of theories considers senescence to be the result of progressive and cumulative damage to organelles or errors in cytoplamsic information-containing molecules. Thus, decrements that occur in cytoplasmic factors are encompassed by these theories; for review, see Hayflick (1980a,b). In nonvertebrate cells aging appears to be a combination of nuclear and cytoplasmic factors; for review, see Muggleton-Harris (1979).

Littlefield (1973) was the first to probe cellular aging in normal human diploid fibroblasts by the use of cell hybridization techniques. Early to midphase II cultures are referred to as "young" cultures, composed of a majority of young cells. Cultures about to enter phase III or already in phase III are referred to as "old" cultures, composed entirely of old cells.

He fused pairs of young and old fibroblasts, and pairs of old fibroblasts with ultraviolet (UV)-inactivated Sendai virus hoping to observe complementation and prolongation of the in vitro life span of the hybrid cells. The cells were fused and passaged until phase III was reached and the cultures no longer replicated. He did not observe complementation in any of the cell hybrids he examined. He concluded that the senescent phenotype behaved as if it were dominant in all combinations of hybrid colonies between young and old cells.

Norwood et al. (1974) and Rabinovitch and Norwood (1980) reached essentially the same conclusion using heterokaryons of senescent and young human diploid fibroblasts. One of the four cell strains that Norwood et al. (1974) examined was karyotypically 47, XYY, and therefore not normal diploid. Rather than measuring in vitro life span, they examined [$^3$H] thymidine incorporation in di- and polykaryons. Again, the senescent phenotype was observed to be dominant. Hybrid cells consisting of young and old cells behaved as if they were senescent and failed to incorporate [$^3$H] thymidine. Later, Hoehn et al. (1978) fused young human fibroblasts from different sources with different total population doubling potentials. They concluded that senescence behaved codominantly. Hybrids displayed growth potentials intermediate between the parental cells. The results of all of these studies were consistent with either group of theories of aging just described.

Wright and Hayflick (1975a,b) and Muggleton-Harris and Hayflick (1976) were the first to provide a tentative answer to the question of the cytoplasmic or nuclear location of the cause of the Phase III phenomenon. Proof of the location of the initial events would serve to distinguish between the two main categories of theories of cellular aging. They found that fusion of cytochalasin B induced anucleate cells (cytoplasts) obtained from young or old human diploid fibroblasts with young or old (cytoplasmically poisoned) whole cells failed to alter the time of expected occurrence of phase III when fused to whole cells (Wright and Hayflick, 1975a, b). The results suggested that nuclear, rather than cytoplasmic factors, were involved in control of in vitro cellular senescence (Wright and Hayflick, 1975a, b). The experiments were criticized because the cytoplasts and the whole cells were subjected to extremely toxic treatment to assure survival of only the appropriate fusion products.

The second study (Muggleton-Harris and Hayflick, 1976) employed reconstructed cells composed of cytoplasts and karyoplasts (nuclear-containing bodies) from young and old cells fused by UV-inactivated Sendai virus. The reconstructed cells were followed individually by microscopic examination until the division potential of each cell was exhausted. This study confirmed the dominance of the senescent nucleus. However, cytoplasts derived from senescent cells contained factors that reduced the doubling potential of young nuclei. In vitro cellular aging of human diploid fibroblasts may, therefore, have involved a combination of both nuclear and cytoplasmic factors. A criticism can be made of their results because of the limited number of cells examined and the fact that at any PDL, a normal population of human cells is composed of cells with a wide range of division potential (Smith and Hayflick, 1974).

## RECENT EXPERIMENTS WITH RECONSTRUCTED CELLS AND CELLULAR AGING

We were obligated to reexamine the behavior of cell hybrids composed of young and old cells before embarking on cell reconstruction experiments with young and old cell components (cytoplasts and karyoplasts). A preliminary report of our results has been published (Zorn et al., 1982).

We used two different strains of human diploid fibroblasts. The cell strain WI-38, a human diploid lung fibroblast, was derived by Hayflick (1965) from a female fetus and kindly supplied for this study. GM1362, a human fibroblast cell strain, was obtained from the Institute for Medical Research, Camden, New Jersey. GM1362 were derived from a 10-year-old male with Lesch-Nyhan syndrome. The cells are deficient for the enzyme hypoxanthine-guanine phosphoribosyltransferase (HGPRT) and will grow in media containing 8-azaguanine, but not in media containing hypoxanthine, methotrexate (amethopterin), and thymidine (HAT) (Littlefield, 1964). Figure 1 illustrates the morphological changes that occur with aging in vitro in both WI-38 and GM1362 cells. The most noticeable differences as the cells approach phase III include, (1) increased nuclear diameter, (2) increased volume and surface area of the cytoplasm, and (3) decreased cell density.

Hybrids were produced by fusion with polyethylene glycol (PEG), 1000 molecular weight (Pontecorvo, 1975). After hybridization the cells were allowed to recover in normal growth media for several days and then placed in media containing HAT. WI-38 cells (which contain the enzyme HGPRT) will survive and continue to proliferate as determined by experiments conducted in our laboratory (data not shown). GM1362 cells will not survive in HAT media and are not capable of extensive proliferation. Consequently, the single biochemical marker (HGPRT) is sufficient to select for hybrid cells. As shown in Table 1, hybrids between WI-38 cells near the end of their life span and relatively young GM1362 cells produced hybrids that rapidly entered phase III. In the reverse hybridization, between young WI-38 cells and old GM1362 cells, we were unable to select for the hybrid cells without allowing for the continued propagation of the parental WI-38 cells. Despite this limitation, it would appear that in hybrids between young and old cells, the senescent phenotype is dominant.

Once we established the dominant behavior of the senescent phenotype in whole-cell hybrids, our goal was to determine the relative contribution of the nucleus and cytoplasm to cellular aging in human cells. Reports from the same laboratory (Wright and Hayflick, 1975b; Muggleton-Harris and Hayflick, 1976) disputed the relative contribution of cytoplasmic factors to cellular senescence. In each report, the experimental protocols set limitations on interpretability of their results. We were determined not to repeat these errors by (1) not subjecting the reconstructed cells to potentially toxic substances, and (2) establishing a *mass* population of reconstructed cells composed of cell components and from old and young cells.

Cytochalasin B (CB) causes the cell to extrude its nucleus (Carter, 1967), and when CB-treated cells are exposed to centrifugal forces, mass enucleation of mammalian cells can be accomplished; for reviews see Wright and Hayflick (1972, 1975a,b), and Wright (1973), Prescott and Kirkpatrick (1973), and Veomett et al. (1973). The karyoplasts (nucleus-containing bodies) and the cytoplasts (anucleate cells) can be fused to their opposite counterparts derived from different cell types (cell reconstruction) or to whole cells (hybridization) to study a variety of biological phenomena (Poste and Reeve, 1971; Wright and Hayflick, 1972, 1975a,b; Shay et al., 1974; Lucas and Kates, 1976; Lucas et al., 1977; Ringertz et al., 1978; Nette et al., 1979; McBurney and Strutt, 1979; Lipsich et al., 1979; Bunn and Tarrant, 1980; Zorn and Anderson, 1981).

Many procedures have been developed to enucleate cells using cytochalasin B and centrifugation. These are based on the original method of Prescott and Kirkpatrick (1973). We elected to use procedures as modified by Lucas and Kates (1976), in which cells were enucleated when grown on plastic sheets (Zorn et al., 1979; Zorn and Anderson, 1981).

Immediately after enucleation, we determined that the efficiency of enucleation was greater than 95% (the number of whole cells in the cytoplast monolayer was less than 5% for either WI-38 or GM1362 cells). This highly pure cytoplast monolayer could be further purified after fusion by plating in appropriate selective media (either HAT or 8-azaguanine-containing media).

The karyoplast populations, however, presented a severe problem. We previously reported the karyoplast pellet consisted of (1) some few contaminating whole cells, (2) dead karyoplasts, and, (3) cytoplasmic fragments. We had devised procedures for eliminating each of these contaminants (Zorn et al., 1979). Briefly, the purification procedure consisted of (1) a 2-4 hr plating period to allow whole cells to attach (karyoplasts have lost this ability immediately after enucleation), (2) gravity sedimentation through 1-6% Ficoll to separate cytoplasmic fragments from karyoplasts, and (3) a mild centrifugation through Ficoll-Paque (Pharmacia) to separate living and dead karyoplasts. We had not used this purification procedure with normal human diploid fibroblasts (either

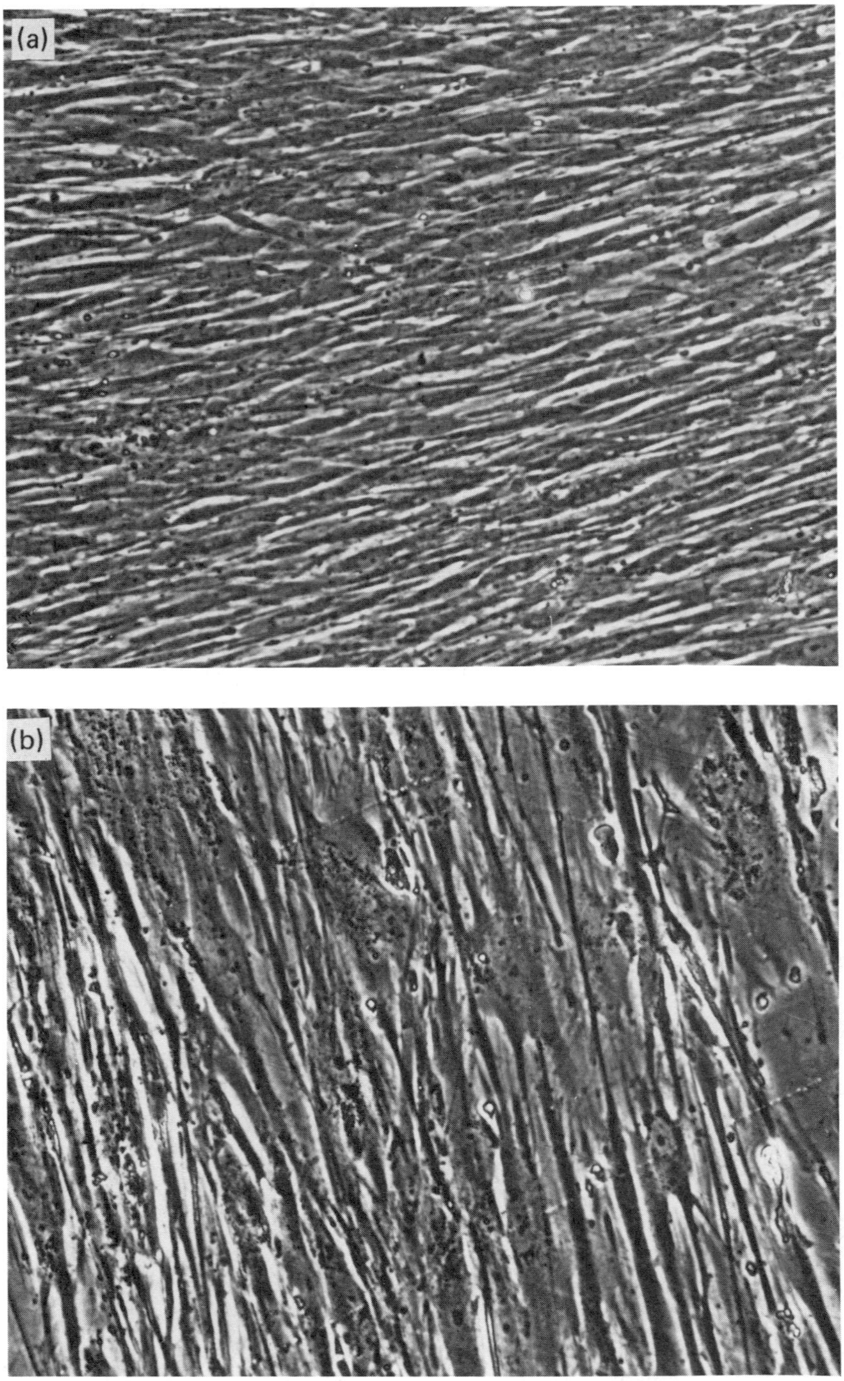


**FIGURE 1** Phase II and phase III cells. (a) young WI-38 cells (PDL = 25); (b) old WI-38 cells (PDL = 54).

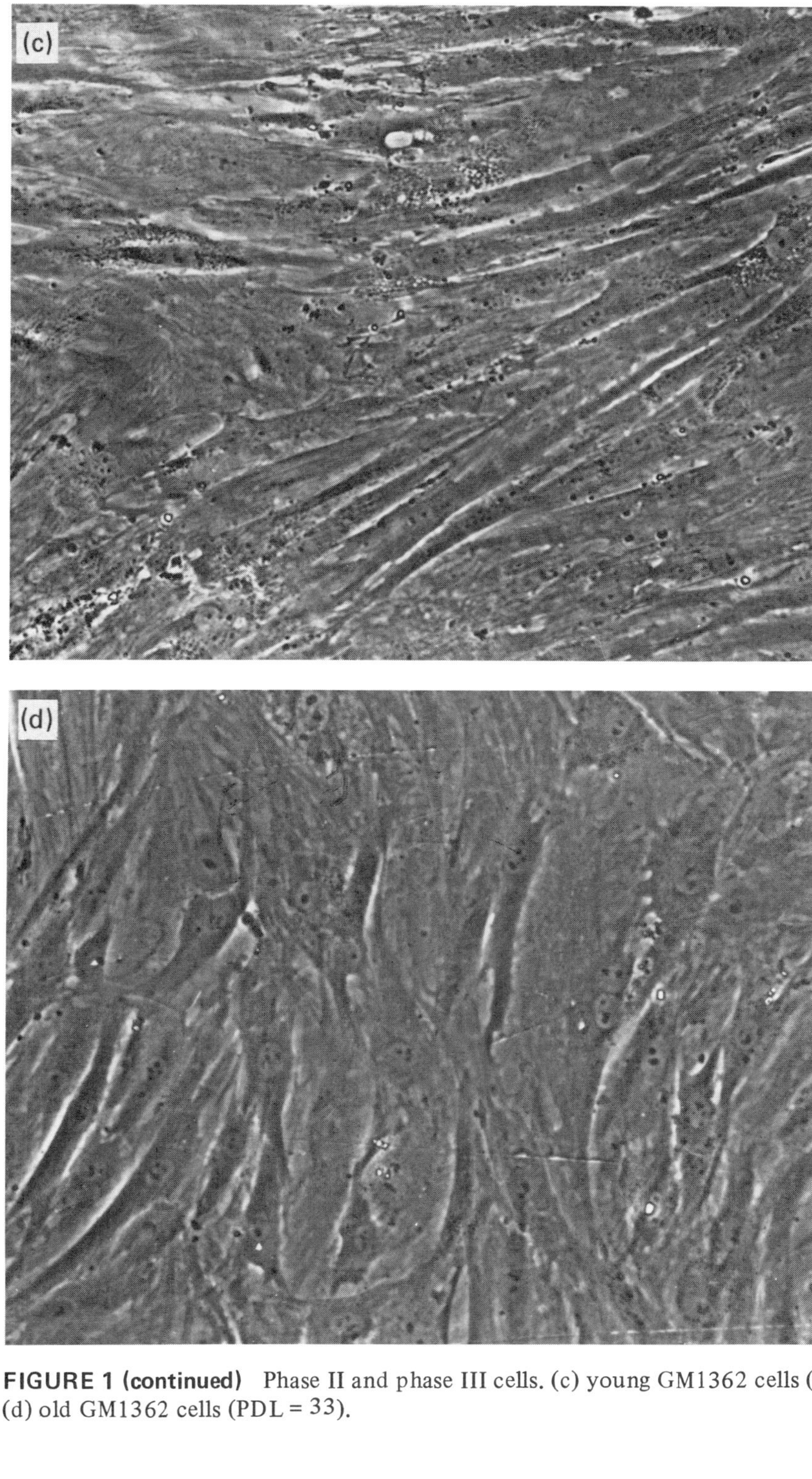


**FIGURE 1 (continued)** Phase II and phase III cells. (c) young GM1362 cells (PDL = 12); (d) old GM1362 cells (PDL = 33).

**TABLE 1** Whole-Cell Hybrids–Old WI-38 X Young GM1362

| Experiment number | PDL of WI-38 cells at time of fusion[a] | PDL of GM1362 cells at time of fusion[a] | Maximum PDL of WI-38 control cultures[b] | Maximum PDL of GM1362 control cultures[b] | PDL of Hybrid WI-38 X GM1362 cultures[c] |
|---|---|---|---|---|---|
| 1 | 54 | 12 | 55 | 32 | 0 |
| 2 | 45 | 12 | 51 | 33 | 2 |

[a]The cells were fused in 40% PEG 1000. Fusion was approximately 30% efficient.

[b]Maximum PDL in vitro is inversely related to donor age; see Hayflick (1980a) for review. GM1362 cells were obtained from a 10 year old and are capable of about 30-35 PD. WI-38 cells were obtained from a fetus and undergo 50 ± 10 PD.

[c]Hybrid cells were plated in HAT media 72-96 hr after fusion. Any unfused GM1362 cells would be killed by this treatment; unfused WI-38 cells would survive and continue to replicate until phase III. The hybrid culture then consisted of a mixture of WI-38 cells and WI-38 X GM1362 cells.

WI-38 or GM1362 cells). In preliminary experiments we determined that in the crude WI-38 karyoplast pellets and in some of the crude GM1362 karyoplast pellets, a significant proportion of the presumed karyoplasts attached to tissue culture dishes after an initial 4 hr "preplating period" designed to eliminate whole cells. They attached slowly over a period of 12-20 hr, well in advance of expectations if the karyoplasts were regenerating their lost cytoplasm (Zorn et al., 1979). We believe that these were either whole cells temporarily "shocked" by the cytochalasin B treatment or extraordinarily large karyoplasts (containing significantly more than 10% of the cytoplasm of whole cells) that were capable of attachment earlier than normal. Most karyoplasts, for an extended period of up to 72 hr, have lost the ability to attach to tissue culture surfaces (Lucas et al., 1976; Lucas et al., 1977; Zorn et al., 1979). To overcome this problem, we devised a gravity filtration procedure through 8-10 $\mu$m filters (2-3X) which resulted in WI-38 and GM1362 karyoplast populations that were incapable of regeneration into whole cells and did not contain the "shocked" whole cells or extraordinarily large karyoplasts. With this addition to our standard purification procedure (Zorn et al., 1979), no bodies were observed to attach to tissue culture dishes from populations of WI-38 or GM1362 karyoplasts. No regeneration of karyoplasts was observed when cultures were examined after 72-96 hr or later.

Our next step was to determine whether an old nucleus (karyoplast) could be rescued from entrance into phase III by fusion to a young cytoplast. Old GM1362 cells and young WI-38 cells were enucleated by modification of our procedures as reported in detail elsewhere (Zorn and Smith, manuscript in preparation). The cytoplasts and karyoplasts were purified as described. The appropriate cell components were mixed in monolayer culture using phytohemagglutinin P (PHP) to promote adhesion of karyoplasts (in suspension) to the cytoplasts that were attached to the vessel surface. The cell components were rapidly fused with PEG 1000 and allowed to recover for at least several hours before examination. Reconstructed cells consistently represented about 10% of the entire population, and were selected in either HAT- or 8-azaguanine-containing media, depending on the cell components fused.

Table 2 is a summary of experiments in which old GM1362 karyoplasts were fused to young WI-38 cytoplasts. The components were plated in media containing 8-azaguanine 72-96 hr after fusion. Any young WI-38 cells in the cytoplast monolayer would have been killed in the selection process. The old parental GM1362 cells had almost reached phase III and were not capable of many additional doublings. Consequently, any growth in the hybrid cultures would have represented a rescue of the old GM1362 karyoplasts by a young WI-38 cytoplast. We did not observe cell replication. Figure 2 illustrates that the reconstructed cells have a morphology consistent with phase III cells. Large cells with large nuclei composed the entire culture.

Although our whole-cell hybridization (GM1362 and WI-38) indicated that whole hybrid cells were viable, we could not ascertain for certain whether the young WI-38 cytoplasm (cytoplasts) was deficient in important "ingredients" necessary for cell replication of the old GM1362 karyoplast due to differences in distribution in essential organelles between young and old cells at the time of enucleation. To be certain our findings were not an artefact of our procedures, we would have to await analysis of young GM-1362 karyoplasts fused to WI-38 cytoplasm.

Table 3 summarizes a series of experiments conducted with the remaining cell components, and represent the fusion of young WI-38 karyoplasts × old GM1362 cytoplasts.

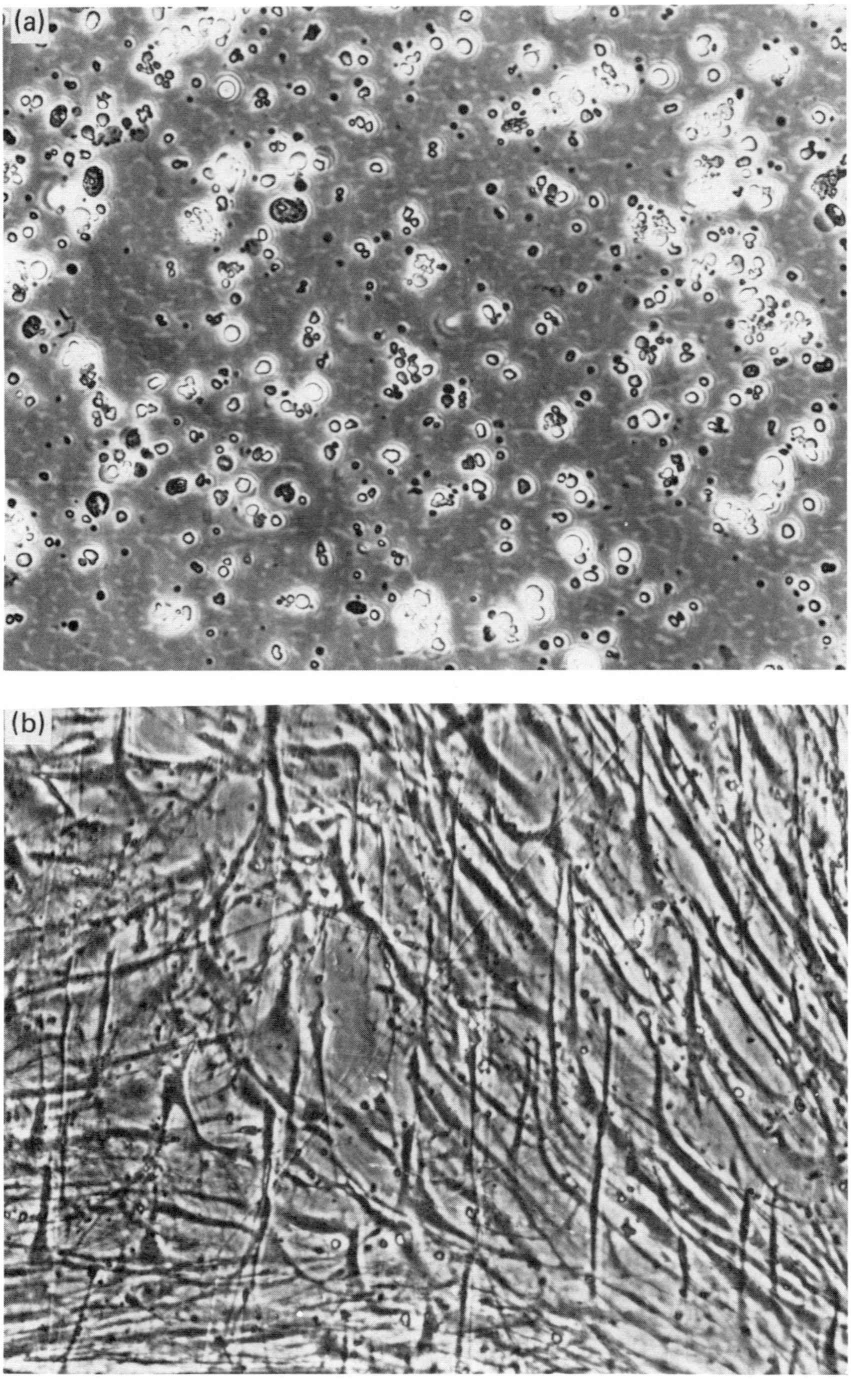


FIGURE 2

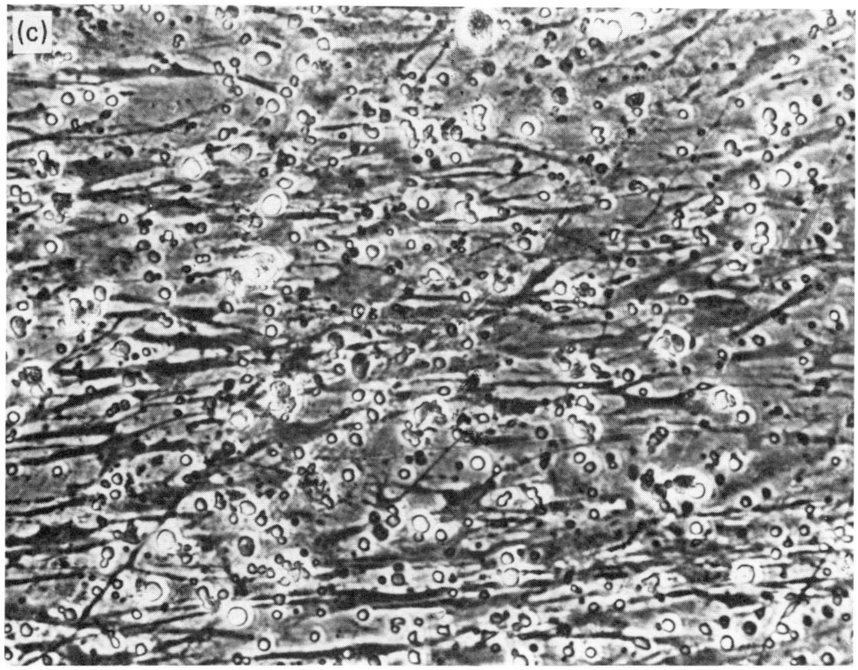

**FIGURE 2 (continued)** Old GM1362 karyoplast X young WI-38 cytoplasts. (a) GM1362 karyoplasts during the plating period to allow attachment of any contaminating whole cells. The karyoplast pellet is plated at low densities to prevent the various components from blocking the attachment of whole cells. The number of cytoplasmic fragments greatly exceeds the number of karyoplasts in the photomicrograph. (b) WI-38 cytoplasts 4 hr after enucleation and recovery in normal growth media. The cytoplasts of WI-38 (and GM1362 cells) remained "extended" in the direction of the centrifugal force at the time of enucleation. (c) GM1362 karyoplasts X WI-38 cytoplasts immediately after fusion with PEG 1000. Before fusion the karyoplasts were attached to the cytoplasts by incubation with phytohemagglutinin P. The cell components were fused with 40% PEG 1000 in growth media without serum.

The reconstructed cells were selected in HAT media. As shown in all five experiments, the reconstructed cells were viable and capable of extensive replication (see Fig. 3). The division potential of the young WI-38 karyoplast X old GM1362 cytoplast cultures paralleled, but did not equal, the replicative ability of young control WI-38 cultures. We believe that the decrease in the number of PDL observed in young WI-38 karyoplasts fused to old GM-1362 cytoplasts as opposed to control WI-38 cultures is merely an indication of the efficiency of fusion (approximately 1-10%) and the decreased survivability of cells reconstructed with cell components in different stages of the cell cycle. Such decreases have been reported for whole-cell hybrids (Rao and Johnson, 1972). Figure 3d is a light micrograph of the young WI-38 karyoplast X old GM1362 cytoplast culture approximately 4 weeks after fusion. The culture morphology resembles cultures of young WI-38 cells of GM1362 cells rather than senescent cultures of either cell type (see Fig. 1). We cannot rigorously ex-

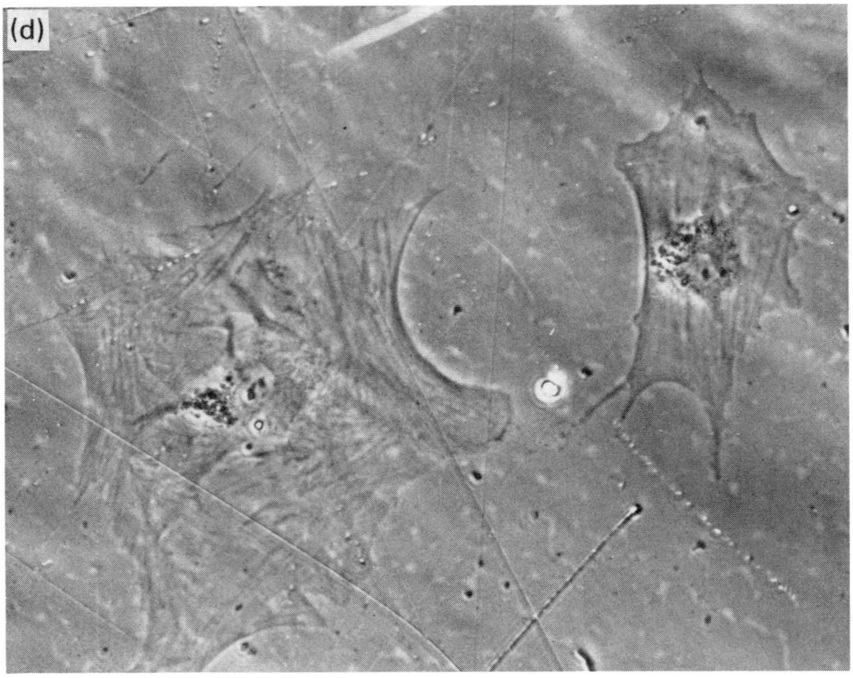


**FIGURE 2 (continued)** (d) GM1362 karyoplast X WI-38 cytoplast cultures 4 weeks after fusion in media containing 8-azaguanine. Little or no growth was observed during this time interval. The cultures never became confluent. The morphology of the reconstructed cells resembled that of senescent cells. Depending upon the experiment, control dishes of unfused karyoplasts and/or cytoplasts were established for each reconstruction experiment in selective media. Magnification (in all photomicrographs) is 200X.

clude the possibility that the old cytoplasm does have an adverse effect on the young nucleus. Our data do demonstrate that the overriding component of cellular senescence lies within the cell nucleus (karyoplast).

We have presented data from experiments with young WI-38 cells and old GM1362 cells. Ideally, experiments should be conducted with old WI-38 cell components and young GM1362 cell components. These experiments are now in progress in our laboratory. Table 4 summarizes our results with the reconstructed cells obtain by fusion of old WI-38 karyoplasts with young GM1362 cytoplasts. The reconstructed cells were selected in HAT media. The young cytoplasm of GM1362 cells failed to rejuvenate the old nucleus of the WI-38 cells. As discussed in Table 2, the young cytoplasm of WI-38 cells failed to rescue the old nuclei in the karyoplasts of old GM1362 cells. Figure 4d is a light micrograph of old WI-38 karyoplasts X young GM1362 cytoplasts several months after fusion. The culture never became confluent or was passaged.

It is interesting to note that the young cytoplasm of GM1362 cells failed to rescue an old WI-38 karyoplast from phase III (Table 4). As we mentioned for old GM1362 kayoplasts X young WI-38 cytoplasts (Table 2), the failure of the reconstructed cells to replicate could be that the cytoplasts and karyoplasts of the two cell types are incompatible. The reconstructed cells then failed to replicate for reasons not related to the age of the nucleus. However, an old GM1362 cytoplast was able to support the biological

**TABLE 2** Old GM1362 Karyoplasts × Young WI-38 Cytoplasts

| Experiment number | PDL of GM1362 cells at time of enucleation | PDL of WI-38 cells at time of enucleation | Maximum PDL of GM1362 control cultures | Maximum PDL of WI-38 control cultures | Maximum PDL of reconstructed cell cultures[a] |
|---|---|---|---|---|---|
| 1 | 27 | 24 | 32 | 51 | 0 |
| 3 | 27 | 25 | 32 | 51 | 0 |
| 4 | 28 | 18 | 32 | 40 | 0 |
| 5 | 29 | 21 | 32 | 48 | 0 |
| 6 | 30 | 23 | 32 | 48 | 0 |
| 7 | 30 | 23 | 32 | 48 | 0 |
| 8 | 30 | 23 | 32 | 48 | 0 |

[a]The reconstructed cell cultures never became confluent.

**TABLE 3** Young WI-38 Karyoplasts × Old GM1362 Cytoplasts

| Experiment number | PDL of WI-38 cells at time of fusion | PDL of GM1362 cells at time of fusion | Maximum PDL of WI-38 control cultures | Maximum PDL of GM1362 control cultures | Maximum PDL of reconstructed cell cultures | Maximum PDL of control WI-38 cultures after fusion[a] |
|---|---|---|---|---|---|---|
| 3 | 25 | 27 | 51 | 32 | 16 | 26 |
| 6 | 21 | 30 | 48 | 32 | 18 | 27 |
| 7 | 23 | 30 | 48 | 32 | 20 | 25 |
| 8 | 23 | 30 | 48 | 32 | 14 | 25 |
| 9 | 23 | 30 | 48 | 32 | 16 | 25 |

[a]Determined by simple subtraction of PDL of WI-38 cells at time of fusion from the maximum PDL of WI-38 control cultures. This value should be compared to maximum PDL of reconstructed cell cultures. Differences in PDL between these cultures is related to the efficiency of fusion rather than to premature senescence. See text for explanation.

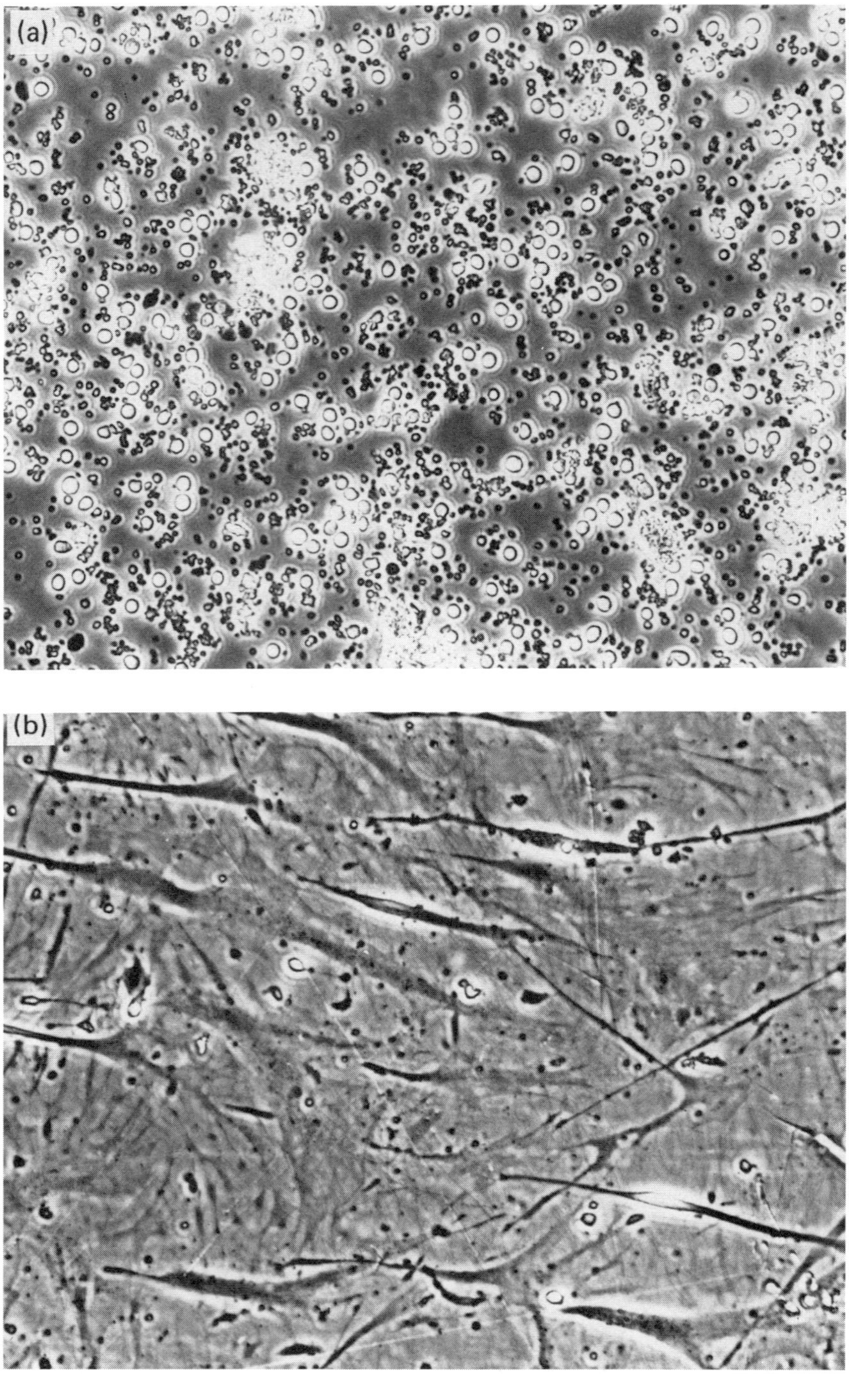


**FIGURE 3** Young WI-38 karyoplasts X old GM1362 cytoplasts. (a) WI-38 karyoplasts during the plating period. See Fig. 2a for details. (b) GM1362 cytoplasts 4 hr after enucleation and recovery in normal growth media. Notice that the cytoplasts are less dense and larger than cytoplasts from young WI-38 cells (see Fig. 2b).

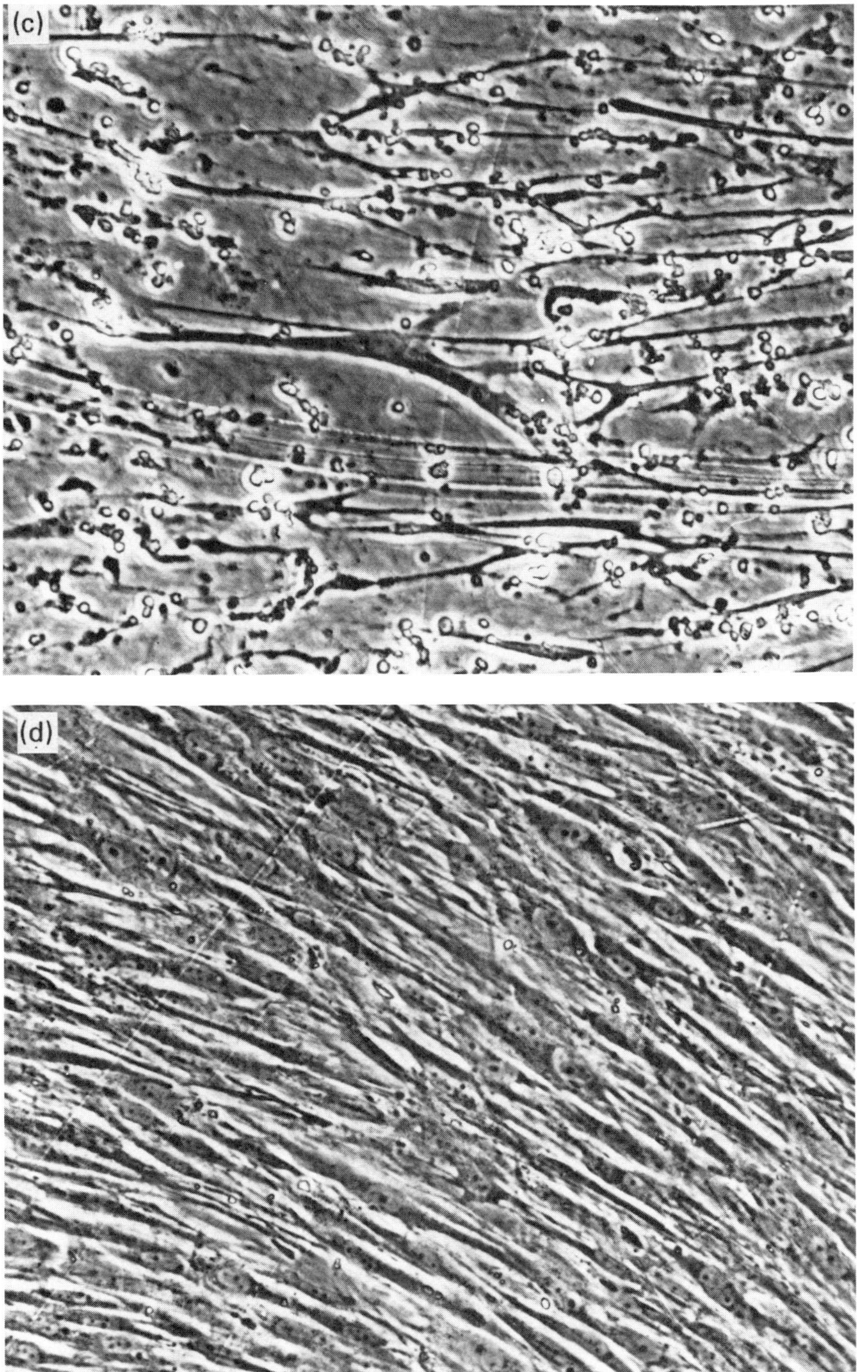


**FIGURE 3 (continued)** (c) WI-38 karyoplasts X GM1362 cytoplasts immediately after fusion. (d) WI-38 karyoplasts X GM1362 cytoplasts 4 weeks after fusion in HAT media. The culture had already been passaged several times. By this time the control WI-38 cells and the reconstructed cells were approaching phase III (see Table 2).

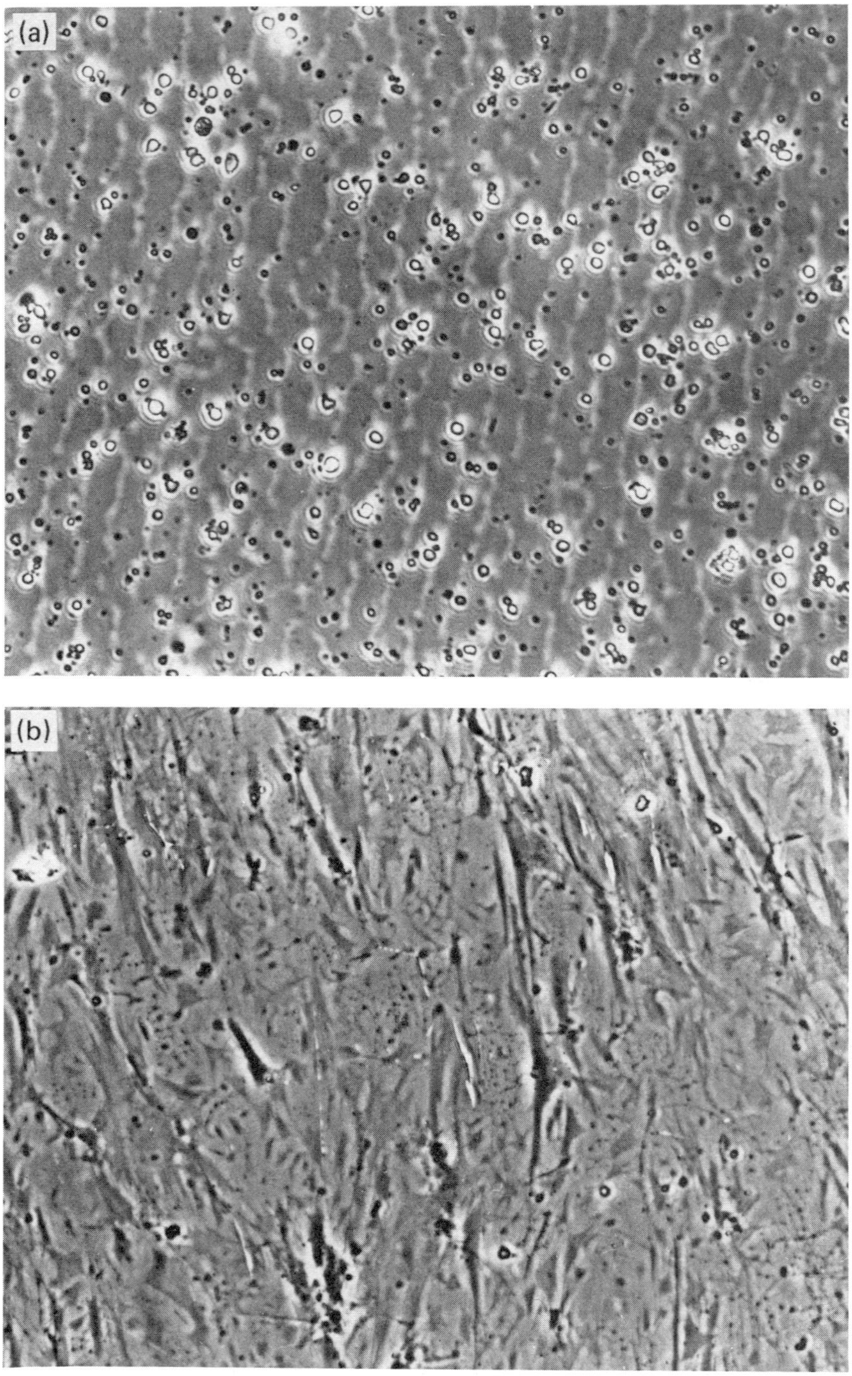


**FIGURE 4** Old WI-38 karyoplasts X young GM1362 cytoplasts. (a) WI-38 karyoplasts plated as in Fig. 2a. (b) GM1362 cytoplasts 4 hr after enucleation and recovery in growth media.

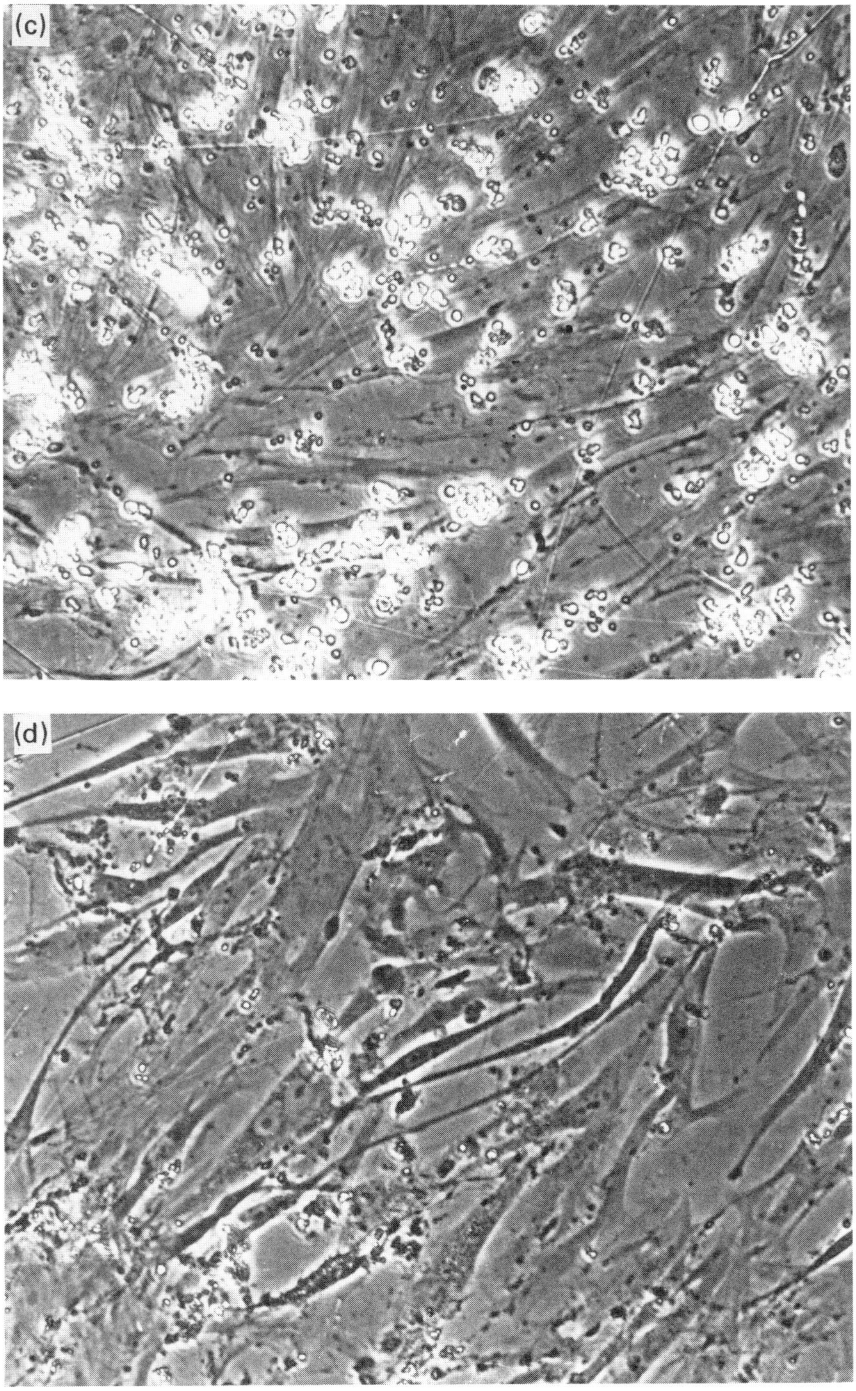

**FIGURE 4 (continued)** (c) WI-38 karyoplasts X GM1362 cytoplasts immediately after fusion (see Fig. 2c for details). (d) WI-38 karyoplasts X GM1362 cytoplasts 4 weeks after fusion in HAT media. No increase in cell number was observed during this time interval. The cells never became confluent and resembled cultures of senescent cells (see Fig. 1). Controls were established at the time of fusion, as explained in Fig. 2d.

**TABLE 4** Old WI-38 Karyoplasts X Young GM1362 Cytoplasts

| Experiment number | PDL of WI-38 cells at time of fusion | PDL of GM1362 cells at time of fusion | Maximum PDL of WI-38 control cultures | Maximum PDL of GM1362 control cultures | Maximum PDL of reconstituted cell cultures[a] |
|---|---|---|---|---|---|
| 11 | 49 | 12 | 49 | 33 | 0 |
| 12 | 48 | 13 | 49 | 33 | 0 |
| 16 | 46 | 13 | 49 | 33 | 0 |
| 17 | 45 | 13 | 49 | 33 | 0 |
| 18 | 45 | 14 | 49 | 33 | 0 |
| 20 | 45 | 14 | 49 | 33 | 0 |

[a]The reconstructed cells never became confluent.

functions of a young WI-38 karyoplast. The cells divided, and the nucleus and cytoplasm were compatible. The lack of growth observed in old karyoplasts fused to young cytoplasm appears to be an age-related phenomenon. The young cytoplasm is not capable of rejuvenating an old nucleus. We cannot eliminate the possibility that old and young cells enucleate differently, and young cytoplasm lack factors essential for cell replication. The fusion of young karyoplasts to young cytoplasts would not provide a clear answer to this question.

Young GM1362 karyoplasts × old WI-38 cytoplasts reconstructed cell cultures are still undergoing analysis. Our preliminary results indicate that these hybrids will behave similarly to young WI-38 karyoplasts × old GM1362 cytoplasts. Our results will be the subject of a future report (Smith and Zorn, manuscript in preparation).

## CONCLUSIONS AND A GLANCE AT THE FUTURE

Macieira-Coelho and Berumen (1973) demonstrated that as cells approached phase III in culture, the $G_2$ period of the cell cycle was progressively lengthened. A similar finding had been reported, in vivo, in tissues of older (as compared to younger) animals (Gelfant and Smith, 1978). The goal of our experiments was not to further define either the site of blockage in the $G_2$ period or to determine the factors directly responsible for the arrest in vitro. Rather, we sought a modest goal of trying to determine the relative contribution of the nucleus and cytoplasm to the phase III phenomenon. In our experiments, an old fibroblast cell nucleus (karyoplast) could not be rescued from entering phase III by fusion to a young cytoplasm (cytoplast). Neither old WI-38 or old GM1362 karyoplasts responded by increased cell replication when fused to young GM1362 or WI-38 cytoplasts respectively. The young cell cytoplasm could not rescue a senescent nucleus. An alternate explanation is that the fusion products were nonviable, lacking an essential factor in the karyoplasts or cytoplasts necessary for cell replication independent of the fact that the karyoplasts were obtained from senescent cells. Although this is a possible explanation for the inability of these cells to divide, then we would have anticipated that reconstructed cells composed of young karyoplasts and old cytoplasts should not be capable of division. However, young WI-38 karyoplasts fused to old GM1362 cytoplasts divided in parallel with cultures of control WI-38 cytoplasts. The cytoplasm (cytoplast) of an old cell could support the functions of a young nucleus. Further, the young nucleus was not severely inhibited by the cytoplasm of the old cell. The reconstructed cells did divide somewhat less than WI-38 control cultures. However, the efficiency of fusion was approximately 10%. Survival and continued replication of the reconstructed cells probably occur at a much lower frequency due to cell damage or cell cycle incompatibility of the karyoplast and cytoplast. Therefore, the experimental cultures have many fewer cells than the control WI-38 cultures. The subsequent loss of several PDL would be expected (see Table 2). The cytoplasm of an old cell may have a slight inhibitory effect on the replicative capability of a young nucleus, as was observed by Muggleton-Harris and Hayflick (1976). Our results are more consistent with the predominant role of a nuclear function being responsible for cellular senescence as first reported by Wright and Hayflick (1975b).

Using whole human cell hybrids, we have confirmed the experiments of Littlefield (1973), which suggested that the senescent phenotype was dominant in hybrids between young and old cells. We also know that the nucleus is predominantly responsible for expression of the senescent phenotype (from the reconstruction experiments). But, we do not know if the entire nucleus is necessary for the expression of phase III. It is logical to assume that the expression of a phase III "clock" is encoded into the genetic material. Presumably, one or more chromosomes, or a group of chromosomes, carry this genetic information. It should then be possible to transfer the "ticking" clock from an old cell to another cell that has gone through fewer PDL as we did with the whole-cell hybrids. If we added less than the entire diploid genome, we could then karyotype the population of surviving hybrids and thereby identify those chromosomes or group of chromosomes not responsible for phase III. By the process of elimination, the chromsome(s) that carry the senescent clock could be identified. In a more direct manner it should also be possible to add specific chromosomes from transformed cell lines into normal cell strains, karyotype the cells that transform and escape phase III, and identify the chromosome or group of chromosomes responsible for transformation. The intriguing possibility exists that cellular aging and cellular transformation would map on the identical chromosome(s) and are related by differential expression of the same genes. This, of course, remains to be determined. However, several techniques are available for adding less than the full diploid complement of chromosomes into other cell types. Pontecorvo (1971) developed a technique based on BUdR incorporation to selectively eliminate chromosomes in hybrid cells. Microcells (Ege and Ringertz, 1974; Fournier and Ruddle, 1977; McNeil and Brown, 1980) selectively add chromosomes of one cell type to another with subsequent expression of the genetic information contained on the transferred chromosomes; for review see Shows and Sakaguchi (1980).

We have already developed procedures for purifying different size classes of microcells (and presumable microcells containing different chromosome numbers) based on gravity filtration (Zorn, unpublished data) from both normal and transformed cells. We are now in the process of fusing these microcells to normal human fibroblasts.

## ACKNOWLEDGMENTS

The authors are grateful to Dr. L. Hayflick for helpful discussions and providing laboratory space to conduct these experiments. We would also like to thank Marion Douglass for typing this manuscript. This research was supported by grant from the National Institute on Aging No. AG00850 awarded to Dr. Hayflick.

## REFERENCES

Berger, S., and Schweiger, H. C. (1975). Cytoplasmic induction of changes in the ultrastructure of the *acetabularia* nucleus and perinuclear cytoplasm. *J. Cell Sci.* 17: 517-529.

Bunn, C. L., and Tarrant, G. M. (1980). Limited lifespan in somatic cell hybrids and cybrids. *Exp. Cell Res.* 127:385-396.

Carrel, A., and Ebeling, A. H. (1921). Age and multiplication of fibroblasts. *J. Exp. Med.* 34:599-623.

Carter, S. B. (1967). Effects of cytochalasins on mammalian cells. *Nature* 213:261-264.

Comfort, A. (1979). *The Biology of Senescence,* 3rd ed. Elsevier-North Holland, Amsterdam.

Ebeling, A. H. (1913). The permanent life of connective tissue outside the organism. *J. Exp. Med.* 17:273-285.

Ege, T., and Ringertz, N. R. (1974). Preparation of microcells by enucleation of micronucleated cells. *Exp. Cell Res.* 87:378-382.

Fournier, E. K., and Ruddle, F. H. (1977). Microcell-mediated transfer of murine chromosomes into mouse, chinese hamster, and human somatic cells. *Proc. Natl. Acad. Sci. USA* 74:319-323.

Gelfant, S., and Smith, J. C., Jr. (1978). Aging noncycling cells, an explanation. *Science* 178:357-361.

Hayflick, L. (1965). The limited in vitro lifetime of human diploid cell strains. *Exp. Cell Res.* 37:614-636.

Hayflick, L. (1980a). The cell biology of human aging. *Sci. Am.* 242:58-65.

Hayflick, L. (1980b). Cell aging. In *Annual Review of Gerontology and Geriatrics,* Chap. 2. Springer, New York, pp. 26-67.

Hayflick, L., and Moorhead, P. S. (1961). The serial cultivation of human diploid cell strains. *Exp. Cell Res.* 25:585-620.

Hoehn, H., Bryant, E. M., and Martin, G. M. (1978). The replicative life spans of euploid hybrids derived from short-lived and long-lived human skin fibroblast cultures. *Cytogenet. Cell Genet.* 21:282-295.

Holliday, R. (1969). Errors in protein synthesis and clonal senescence in fungi. *Nature* 221:1224-1228.

Lipsich, L. A., Kates, J. R., and Lucas, J. J. (1979). Expression of a liver-specific function by mouse fibroblasts nuclei transplanted into rat hepatoma cytoplasts. *Nature* 281:74-76.

Littlefield, J. W. (1964). Selection of hybrids from matings of fibroblasts in vitro and their presumed recombinants. *Science* 145:709-710.

Littlefield, J. W. (1973). Attempted hybridizations with senescent human fibroblasts. *J. Cell. Physiol.* 82:129-132.

Lucas, J. J., and Kates, J. R. (1976). The construction of viable nuclear-cytoplasmic hybrid cells by nuclear transplantation. *Cell* 7:397-405.

Lucas, J., and Kates, J. (1977). Nuclear transplantation with mammalian cells. In *Methods in Cell Biology,* vol. 15, D. M. Prescott, (ed). Academic Press, New York, pp. 359-363.

Lucas, J. J., Zorn, G. A., Brings, A., Szekely, E., and Kates, J. R. (1977). Recent developments with karyoplast regeneration and nuclear transplantation. *Nat. Cancer Inst. Monogr.* 48:37-44.

Macieira-Coelho, A. (1973). Aging and cell division. *Front. Matrix Biol.* 1:46-77.

Macieira-Coelho, A. (1981). Tissue culture in aging research: Present status and prospects. *Experientia* 37:1050-1053.

Macieira-Coelho, A., and Berumen, L. (1973). The cell cycle during growth inhibition of human embryonic fibroblasts in vitro. *Proc. Soc. Exp. Biol. Med.* 144:43-47.

Marcou, D. (1961). Notion de longévité et nature cytoplasmique du determinant de la sénescence chez quelques champignons. *Ann. Sci. Nat. Bot. Bol. Veg., Ser. 12* 2: 653-763.

Martin, G. M. (1977a). Cellular aging-clonal senescence. *Am. J. Pathol.* 89:484-511.

Martin, G. M. (1977b). Cellular aging-postreplicative cells. *Am. J. Pathol.* 89:513-550.

Martinez, A. O., Norwood, T. H., Prothero, J. W., and Martin, G. M. (1978). Evidence for clonal attenuation of growth potential in HeLa cells. *In Vitro* 14:996-1002.

McBurney, M. W., and Strutt, B. (1979). Fusion of embryonal carcinoma cells to fibroblast cells, cytoplasts, and karyoplasts. Developmental properties of viable fusion products. *Exp. Cell Res.* 124:171-180.

McNeil, C. A., and Brown, R. L. (1980). Genetic manipulation by means of microcell-mediated transfer of hormonal human chromosomes into recipient mouse cells. *Proc. Natl. Acad. Sci. USA* 77:5394-5398.

Miguel, J., Economos, A. C., Bensch, K. G., Altan, H., and Johnson, J. E., Jr. (1979). Review of cell aging in drosophila and mouse. *Age* 2:78-88.

Muggleton-Harris, A. (1979). Reassembly of cellular components for the study of aging and finite lifespan. In *International Review of Cytology,* Supplement 9, G. H. Bourne and J. F. Danielli (eds.) Academic Press, New York, pp. 279-301.

Muggleton-Harris, A. L., and Hayflick, L. (1976). Cellular aging studied by the reconstruction of replicating cells from nuclei and cytoplasms isolated from normal human diploid cells. *Exp. Cell Res.* 103:321-330.

Nanney, D. L. (1974). Aging and longterm temporal regulation in ciliated protozoa. A critical review. *Mech. Aging Dev.* 3:81-105.

Nette, E. G., Sit, H. L., Clavey, W., and King, D. W. (1979). Isolation of viable reconstituted cells from human karyoplasts fused to mouse cytoplasts. *Exp. Cell Res.* 121: 143-151.

Norwood, T. H., Pendergrass, W. R., Sprague, C. A., and Martin, G. M. (1974). Dominance of the senescent phenotype in heterokaryons between replicative and postreplicative human fibroblast-like cells. *Proc. Natl. Acad. Sci. USA* 71:2231-2235.

Parker, R. C. (1961). *Methods of Tissue Culture.* Harper and Row, New York.

Pontecorvo, G. (1971). Induction of directional chromosome elimination in somatic cell hybrids. *Nature* 230:467-369.

Pontecorvo, G. (1975). Production of mammalian somatic cell hybrids by means of polyethylene glycol treatment. *Somatic Cell Genet.* 1:397-400.

Poste, G., and Reeve, P. (1971). Formation of hybrid cells and heterokaryons by fusion of anucleated and nucleated cells. *Nature [New Biol.]* 229:123-125.

Poulter, R. T. M. (1969). Ph.D. Thesis. University of Leicester, England.

Prescott, D. M., and Kirkpatrick, J. B. (1973). Mass enucleation of cultured animal cells. In *Methods in Cell Biology,* vol. 7, D. M. Prescott, (ed.). Academic Press, New York, pp. 189-202.

Rabinovitch, P. S., and Norwood, T. H. (1980). Comparative heterokaryon study of cellular senescence and the serum deprived state. *Exp. Cell Res.* 130:101-109.

Rao, P. N., and Johnson, R. T. (1972). Cell fusion and its application to studies on the regulation of the cell cycle. In *Methods in Cell Physiology,* vol. 5, D. M. Prescott (ed.). Academic Press, New York, pp. 75-126.

Ringertz, N. R., Krondahl, V., and Coleman, J. R. (1978). Reconstitution of cells by fusion of cell fragments. *Exp. Cell Res.* 113:233-246.

Rosen, R. (1978). Cellular senescence. In *International Review of Cytology,* vol. 54, G. H. Bourne and J. F. Danielli, (eds.). Academic Press, New York, pp. 162-191.

Shay, J., Porter, K. R., and Prescott, D. M. (1974). The surface morphology and fine structure of CHO (Chinese hamster ovary) cells following enucleation. *Proc. Natl. Acad. Sci. USA* 71:3059-3063.

Shows, T. B., and Sakaguchi, A. Y. (1980). Gene transfer and gene mapping in mammalian cells in culture. *In Vitro* 16:55-76.

Smith, J. R., and Rubenstein, I. (1973). The development of "senescence" in *Podospora anserina. J. Gen. Microbiol.* 76:283-296.

Smith, J. R., and Hayflick, L. (1974). Variation in the lifespan of clones derived from human diploid cell strains. *J. Cell Biol.* 62:48-53.

Sonneborn, T. M. (1954). The relation of autogamy to senescence and rejuvenescence in *Paramecium aurelia. J. Protozool.* 1:38-53.

Swim, H. E., and Parker, R. F. (1957). Culture characteristics of human fibroblasts propagated serially. *Am. J. Hyg.* 66:235-243.

Todaro, G. J., and Green, H. (1963). Quantitative studies of the growth of mouse embryo cells in culture and their development into established lines. *J. Cell Biol.* 17:299-313.

Veomett, G., Shay, J., Hough, P. V. C., and Prescott, D. M. (1973). Large-scale enucleation of mammalian cells. In *Methods in Cell Biology,* vol. 7, D. M. Prescott (ed.). Academic Press, New York, pp. 1-6.

Wright, W. E. (1973). The production of mass populations of anucleate cytoplasms. In *Methods in Cell Biology,* vol. 7, D. M. Prescott (ed.). Academic Press, New York, pp. 203-210.

Wright, W. E., and Hayflick, L. (1972). Formation of anucleate and multinucleate cells in normal and SV40 transformed WI-38 by cytochalasin B. *Exp. Cell Res.* 74:187-194.

Wright, W. E., and Hayflick, L. (1975a). The use of biochemical lesions for the selection of human cells with hybrid cytoplasms. *Proc. Natl. Acad. Sci. USA* 72:1812-1816.

Wright, W. E., and Hayflick, L. (1975b). Contributions of cytoplasmic factors to in vitro cellular senescence. *Fed. Proc.* 34:76-79.

Zorn, G. A., and Anderson, C. W. (1981). Adenovirus type 2 expresses fiber in monkey human hybrids and reconstructed cells. *J. Virol.* 37:759-769.

Zorn, G. A., Lucas, J. J., and Kates, J. R. (1979). Purification and characterization of regenerating mouse L929 karyoplasts. *Cell* 18:659-672.

Zorn, G. A., Smith, B., and Hayflick, L. (1982). Nuclear control of cellular senescence in reconstructed human diploid fibroblasts. *J. Cell Biol.* 95:449a.

# Author Index

*Numbers in italics give the page on which the complete reference is listed.*

# Subject Index